McGraw-Hill Ryerson

# Physics 11

## Authors

**Greg Dick**
Galt Collegiate Institute and Vocational School
Cambridge, Ontario

**Arthur N. Geddis**
Formerly of University of Western Ontario
London, Ontario

**Ed James**
Waterloo District Board of Education
Waterloo, Ontario

**Tom McCaul**
Bayview Secondary School
Richmond Hill, Ontario

**Barry McGuire**
Formerly of Western Canada High School
Calgary, Alberta

**Richard Poole**
Formerly of Parry Sound High School
Parry Sound, Ontario

**Bob Holzer**
Formerly of Edmonton School District #7
Edmonton, Alberta

## Contributing Author

**Rob Smythe**
Iroquois Ridge High School
Oakville, Ontario

## Consultants

**John Caranci**
Earl Haig Secondary School
Toronto, Ontario

**Terry Price**
Huron Heights Secondary School
Newmarket, Ontario

**Douglas Hayhoe**
Toronto District School Board
Toronto, Ontario

## Probeware Specialist

**John Braun**
Formerly of Kingsville High School
Kingsville, Ontario

## Technology Consultant

**Khoi Trinh**
Sir Oliver Mowat Collegiate Institute
West Hill, Ontario

**McGraw-Hill
Ryerson**

Toronto   Montréal   New York   Burr Ridge   Bangkok   Beijing   Bogotá   Caracas   Dubuque
Kuala Lumpur   Lisbon   London   Madison   Madrid   Mexico City   Milan   New Delhi
San Francisco   Santiago   St. Louis   Seoul   Singapore   Sydney   Taipei

# McGraw-Hill Ryerson Limited

A Subsidiary of The **McGraw·Hill** Companies

COPIES OF THIS BOOK
MAY BE OBTAINED BY
CONTACTING:
McGraw-Hill Ryerson Ltd.

E-MAIL
orders@mcgrawhill.ca

TOLL FREE FAX:
1-800-463-5885

TOLL FREE CALL:
1-800-565-5758

OR BY MAILING YOUR
ORDER TO:
McGraw-Hill Ryerson
Order Department
300 Water Street
Whitby ON, L1N 9B6

Please quote the ISBN
and title when placing
your order.

## McGraw-Hill Ryerson Physics 11

The information and activities in this textbook have been carefully developed and reviewed by professionals to ensure safety and accuracy. However, the publishers shall not be liable for any damages resulting, in whole or in part, from the reader's use of the material. Although appropriate safety procedures are discussed in detail and highlighted throughout the textbook, safety of students remains the responsibility of the classroom teacher, the principal, and the school board/district.

0-07-088691-1

*http://www.mcgrawhill.ca*

1 2 3 4 5 6 7 8 9 0 TRI 0 9 8 7 6 5 4 3 2 1

Printed and bound in Canada

Care has been taken to trace ownership of copyright material contained in this text. The publisher will gladly take any information that will enable it to rectify any reference or credit in subsequent printings. Please note that products shown in photographs in this textbook do not reflect an endorsement by the publisher of those specific brand names.

### National Library of Canada Cataloguing in Publication Data

Main entry under title:
McGraw-Hill Ryerson physics 11

Includes index.
ISBN 0-07-088691-1

1.Physics. I. Dick, Greg. II. Title: Physics 11. III. Title: McGraw-Hill Ryerson physics eleven.

### The Physics 11 Team

SCIENCE PUBLISHERS: Trudy Rising, Jane McNulty
PROJECT MANAGERS: Lois Edwards, Greg Dick
SENIOR DEVELOPMENTAL EDITOR: Lois Edwards
DEVELOPMENTAL EDITORS: June Trusty, Greg Dick, Tom Gamblin, Barry McGuire,
    Michael Stein, Daniel Hudon
SENIOR SUPERVISING EDITOR: Linda Allison
PROJECT CO-ORDINATORS: Valerie Janicki, Shannon Leahy
COPY EDITORS: Dianne Brassolotto, Trish Brown
PERMISSIONS EDITORS: Pronk&Associates Inc.
SPECIAL FEATURES CO-ORDINATORS: Jill Bryant
PRODUCTION CO-ORDINATOR: Jennifer Vassiliou
COVER DESIGN, INTERIOR DESIGN, AND ART DIRECTION: Pronk&Associates Inc.
ELECTRONIC PAGE MAKE-UP: Pronk&Associates Inc.
TECHNICAL ILLUSTRATION: Pronk&Associates Inc., Jun Park, Imagineering Scientific
    and Technical Artworks Inc.
ILLUSTRATORS: Steve Attoe, Brett Clayton, Deborah Crowle, Frank Zsigo
SET-UP PHOTOGRAPHY: Ian Crysler
SET-UP PHOTOGRAPHY CO-ORDINATOR: Shannon O'Rourke
COVER IMAGE: Artbase Inc.

# Acknowledgements

Producing a textbook of high quality is a true team effort, requiring the input and expertise of a very large number of people. The authors, consultants, and publishers of this book would like to convey our sincere thanks, first and foremost, to the reviewers listed below who provided critical analyses of our draft manuscript, and often provided reviews of designed pages, as well. Their assistance was invaluable in helping us to develop a text that we hope you will find completely appropriate for your teaching and your students' learning. We also thank the following writers who authored the Special Features in *Physics 11*: Meaghan Craven, Ann Douglas, Jill Lazenby, Natasha Marko, Celeste Peters, Andrea Rutty, Elma Schemenauer, Rob Smythe, and Erik Spigel. Special thanks go to the following people who wrote additional questions and problems for our use — Chyrl Bourassa, Elio Covello, Doug Hayhoe, Bruce Hickey, Tony Kwan, and Ingrid Macey — and to Dr. Daniel Hudon and his team of teachers and graduate students, Suresh Pereira, Mike Allen, Mark Brodwin, Dudley Brown, Craig Burrell, Ingrid Macey, Isamu Matsuyama, and Rob Reid, who together ensured the answers to all of the problems in this book were double-checked. Dr. Douglas Roberts, formerly of the University of Calgary, is sincerely thanked for his ongoing support throughout the project, but in particular with the development of Chapter 1.

Finally, we thank the wonderfully co-operative design studio, Pronk&Associates, and its talented staff, who worked with us closely under very difficult timelines. This close teamwork was critical throughout the book's final stages of development.

## Pedagogical and Academic Reviewers

**Chyrl Bourassa**
Twin Lakes Secondary School
Orillia, Ontario

**Robert Callcott**
Formerly with Sutton District
High School
Sutton, Ontario

**Ron Chisholm**
Holy Heart of Mary High School
St. John's, Newfoundland

**Elio Covello**
Huron Heights Secondary
School
Newmarket, Ontario

**Cliff Coveyduc**
Atlantic Provinces Education
Foundation
Dartmouth, Nova Scotia

**Bruce Hickey**
Holy Spirit High School
Conception Bay South
Newfoundland

**Stephen Houlden**
Formerly with Toronto District
School Board
Toronto, Ontario

**David Knox**
Canterbury High School
Ottawa, Ontario

**Elizabeth Kozoriz**
Daniel McIntyre Collegiate
Winnipeg, Manitoba

**Tony Kwan**
Earl Haig Secondary School
Toronto, Ontario

**Ping Lai**
University of Toronto Schools
Toronto, Ontario

**Kathleen Lostracco**
Notre Dame Secondary School
Brampton, Ontario

**Alden McEachern**
M.M Robinson High School
Burlington, ON

**Robert Miller**
Mother Teresa High School
Ottawa, Ontario

**Geoff Orton**
Milton District High School
Milton, Ontario

**Otto Pike**
Formerly with College of the
North Atlantic
St. John's, Newfoundland

**Douglas A. Roberts**
Formerly of the
University of Calgary
Calgary, Alberta

**Jim Scarth**
Anderson College of
the North Atlantic
Whitby, Ontario

**Richard Wahrer**
Newtonbrook Secondary
School
Willowdale, Ontario

**Sarah Zeegen**
West Humber Collegiate
Institute
Etobicoke, Ontario

## Safety Reviewer

**Margaret Redway**
Fraser Scientific & Business
Services
Delta, British Columbia

# Contents

## UNIT **2**    Energy, Work, and Power    192

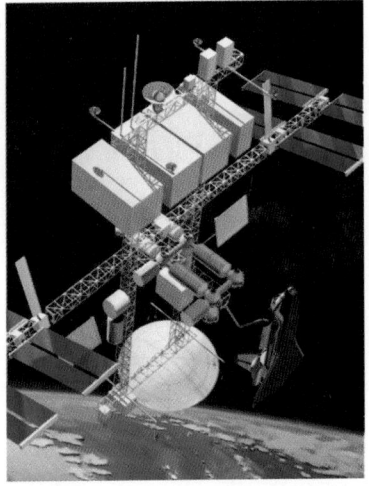

## UNIT **4**    Light and Geometric Optics    458

Here is a quick glimpse of the learning that you will encounter in this course. Expand your knowledge, and build on skills learned in previous courses as you experience physics in action.

In Unit 1, you will consider *Force and Motion*, and how our knowledge of these phenomena can be applied. What happens to structures during an automobile collision? How are forces involved in enabling us to walk? How can a space shuttle be launched? Answers to all of these questions rely on knowledge of forces and motion.

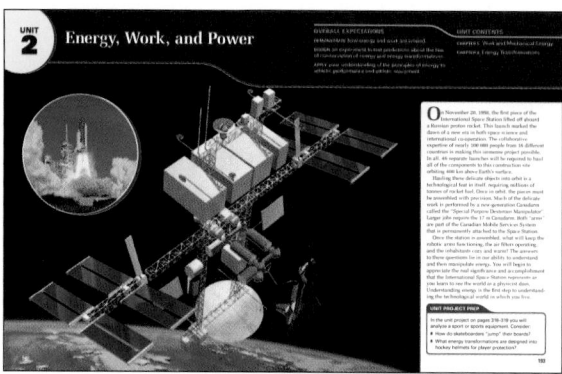

Everything you do requires energy. In Unit 2 (*Energy, Work, and Power*), you will study energy — the essence of our universe — and how its transformations power our world.

Lightning flashes, then a pause, and then the thunder crackles — and it's all due to the nature of waves. Unit 3 (*Waves and Sound*) offers insights and explanation about the nature of speech, melodic sounds, and the sounds you hear in nature.

A hot shimmering pavement and quiet twinkling stars reveal their secrets in Unit 4 (*Light and Geometric Optics*). You will also explore a wide span of technologies that includes corrective eyeglasses and space-borne telescopes.

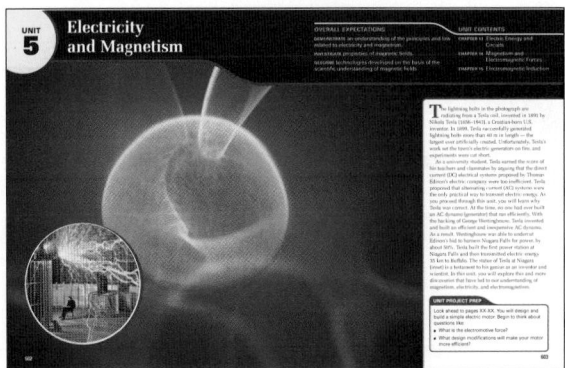

Lodestone, a naturally magnetic rock, is able to make objects move without ever touching them. This force at a distance seems almost magical, yet it is a vital part of your life. The study of electricity and magnetism will show you how the electric current that operates so much of our society is created — by technologies that use magnetic fields and electrons moving through wires.

After the fifth unit in the textbook, you will find a Physics Course Challenge. You will apply your skills of inquiry, communicate ideas, and analyze connections among science, technology, society, and the environment. Participating in the Physics Course Challenge allows you and your teacher to complete a rewarding end-of-course performance task that can be the culminating achievement of your course.

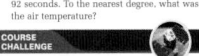

Watch for this feature in text margins throughout the textbook and in the Unit Reviews; it will help you begin planning for your Challenge early in the course. The cues are designed to trigger your thought processes, to point you to a line of research, and/or point you to a short activity that will help you plan and design a successful Challenge for assessment.

You will probably be designing rubrics to assess your Course Challenge. When you design or choose the rubrics that your class will use to assess the Challenges, remember to include criteria that will address all of the achievement categories, as shown in Chapter 1, page 18. As you work on the Challenge, and on activities and investigations earlier in the year, remember to refer to the rubric that will be used to assess your progress throughout the process.

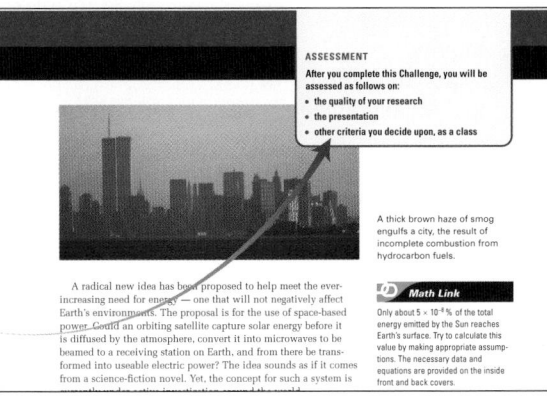

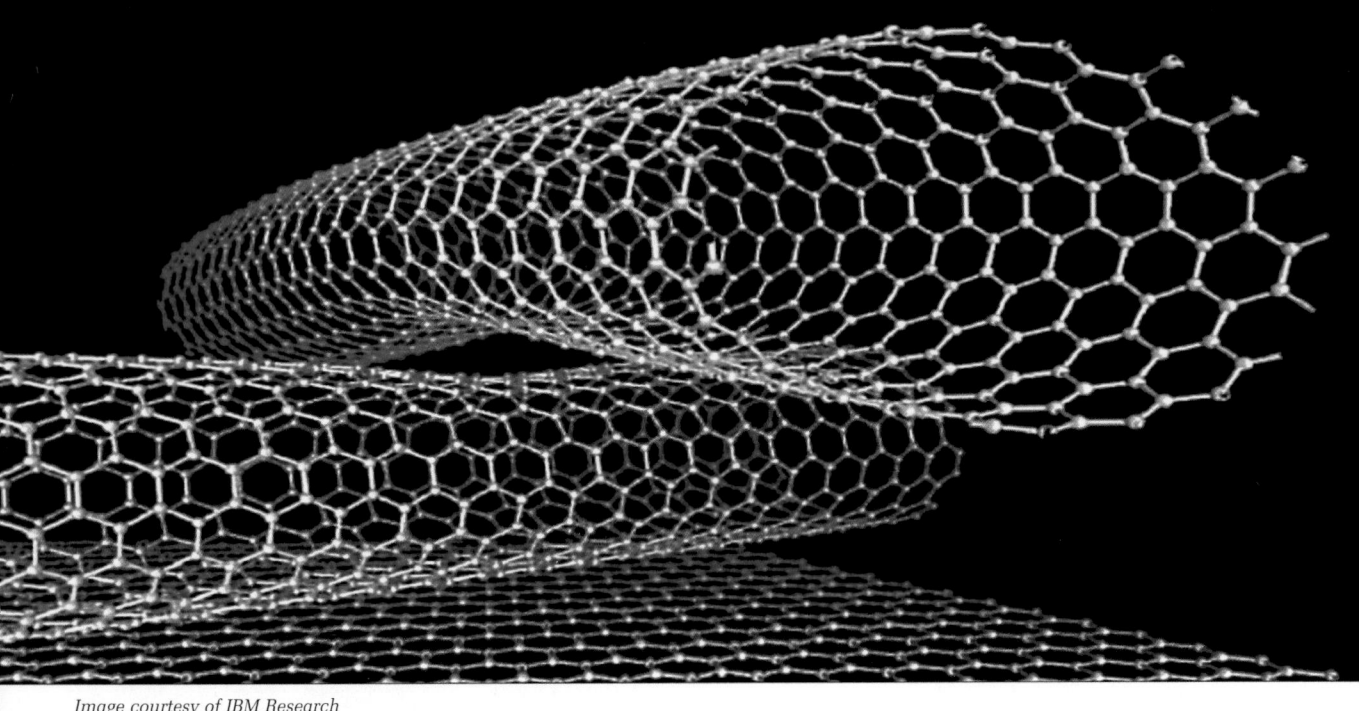

*Image courtesy of IBM Research*

You are looking at two different views of a computer-generated model of a carbon nanotube — a straw on an atomic scale. Built one carbon atom at a time, this nanotube is a pioneering example of a new class of machines, so tiny they cannot be seen by the unaided eye, or even through most microscopes. Extraordinarily strong, yet only a few atoms in diameter, minuscule devices like this one may dramatically alter our lives in the years to come. In fact, some leading researchers believe the "nano age" has already begun. The inset "molecule man" made of 28 carbon monoxide molecules, and the "guitar" shown on page 10 are the results of researchers having fun with nanotechnology.

*Nanotechnology*, the emerging science and technology of building mechanical devices from single atoms, seeks to control energy and movement at an atomic level. Once perfected, nanotechnology would permit microscopic machines to perform complex tasks atom-by-atom, molecule-by-molecule. Imagine a tiny robotic device that could be programmed to produce specific products, like paper or steel, simply by extracting the required atoms from the atmosphere, in much the same way a potato plant absorbs nutrients from the soil, water, and air, and reorganizes them to create more potatoes.

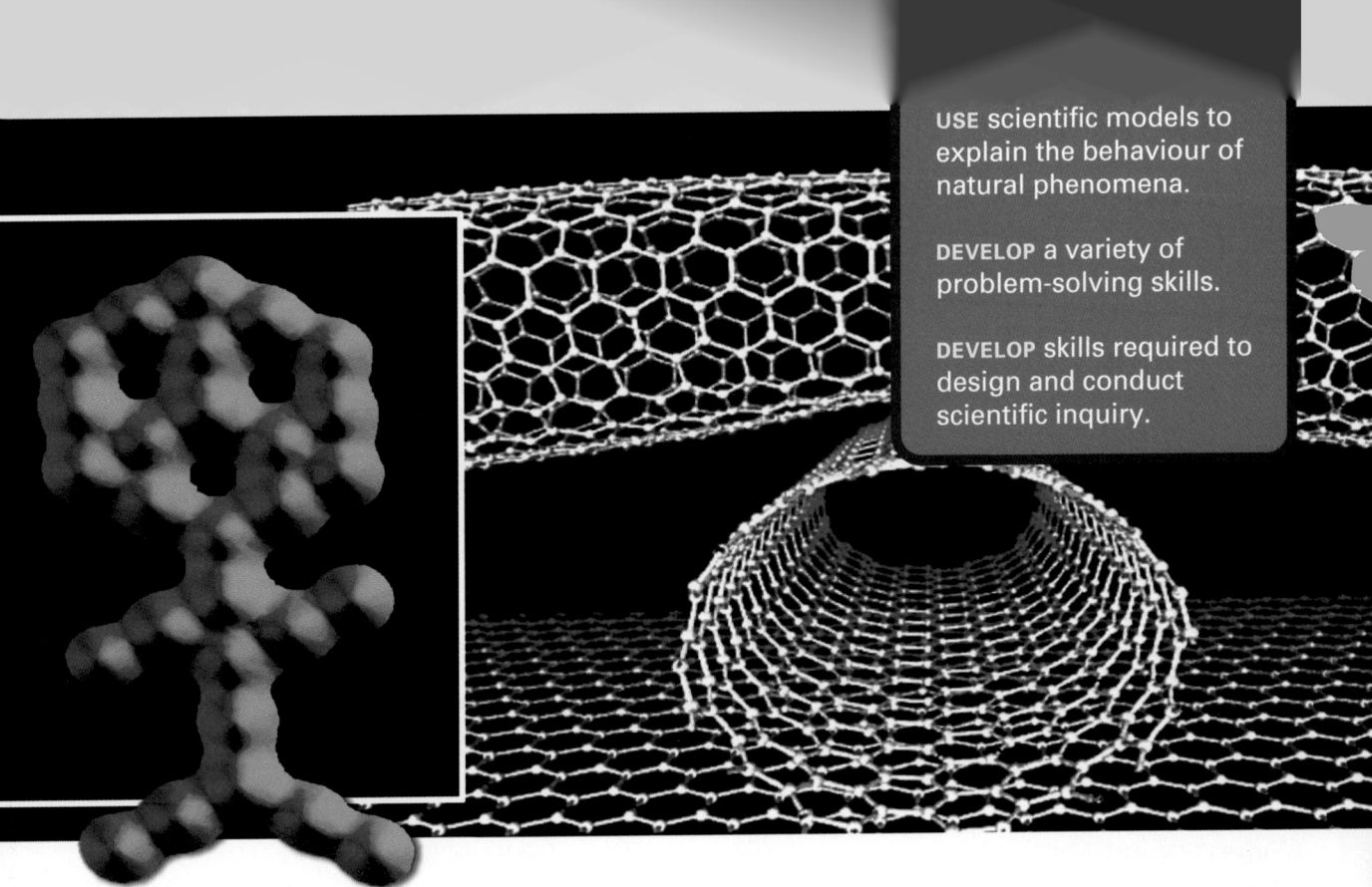

**USE** scientific models to explain the behaviour of natural phenomena.

**DEVELOP** a variety of problem-solving skills.

**DEVELOP** skills required to design and conduct scientific inquiry.

Imagine if a machine could produce diamonds by rearranging atoms of coal or produce fresh water by coupling atoms of hydrogen and oxygen. What if such a machine could be programmed to clean the air by rearranging atoms in common pollutants, or heal the sick by repairing damaged cells? It is difficult even to begin to understand the impact such technology could have on our everyday lives, and on the countless chemical, biological, and physical relationships and processes that govern our world. However, one thing is certain: nanotechnology represents a new way of harnessing and transforming matter and energy, making it an important application of the science we call physics.

Throughout this course you will be involved in the processes of doing physics. You will be asking questions, forming hypotheses, designing and carrying out investigations, creating models and using theories to explain your findings, and solving problems related to physics. In short, you will be learning to think like a physicist. The activities in this course will be carried out at many levels of sophistication. In science, as well as in other disciplines, the simplest questions and investigations often reveal the most interesting and important answers.

## Web Link

www.school.mcgrawhill.ca/resources/
To learn more about nanotechnology and view pictures of nanomachines, go to the above site. Click on **Science Resources** and **Physics 11** to find out where to go next.

An important part of physics is creating models that allow us to develop explanations for phenomena. Models are helpful in making predictions based on observations. Try the following labs, creating your own models and making your own predictions based on what you already know. Keep these definitions in mind as you proceed.

## Black Box

Pull the strings on the black box and observe what happens. Try several combinations, noting the motion and tension of the strings, any noises you hear, and anything else that strikes you. Record your observations.

1. Based on your observations, draw a model showing how you think the strings are connected inside the black box.

2. Test the accuracy of your prediction by once again pulling the strings on the black box.

3. How can this experiment be used to explain the process of scientific inquiry?

## Beach Ball

With a partner, observe what happens to a beach ball when you throw it back and forth while applying various spins. Record your observations.

1. Describe the effects of each spin.

2. Draw a model representing what you observed.

## Van de Graaff Generator

Place scraps of paper from a 3-hole punch onto the Van de Graaff generator as shown. Switch on the generator and observe what happens. Record your observations.

1. Based on your observations, draw a model showing what happened to the paper.

## Super Ball

Drop a super ball from a specific height. Conduct several trials, changing variables like the initial velocity of the ball and its rate of spin. Record your observations. Then, develop rules that will allow you to predict whether the ball, based on its initial velocity and rate of spin, will bounce to a height above its starting point.

1. Test your predictions.

2. Describe the motion of the super ball using a model about the conservation of energy.

## Radiometer

Shine a light on the radiometer and observe what happens. Repeat the process using a hair dryer on cool and hot settings. Record your observations.

1. What causes the vanes to spin? Formulate a hypothesis.

2. How was the energy transferred?

3. What similarities exist between heat and light?

4. Test your hypothesis.

## Multiple Images with Two Plane Mirrors

Use a protractor to create a template similar to the one shown. Set up the mirrors and coin as shown. Then, create a table like this one. Count the number of images you see when the mirrors are set to specific angles. Record your observations.

| Number of objects | Angle between mirrors | Number of images |
|:---:|:---:|:---:|
| 1 | 180° | |
| 1 | 120° | |
| 1 | 90° | |
| 1 | 60° | |

1. Develop a mathematical equation that predicts the number of images that will appear when the angle between the plane mirrors is known. Hint: there are 360° in a circle.

 **Web Link**

www.school.mcgrawhill.ca/resources/

Go to the above web site for other Quick Labs to help you get started. Click on **Science Resources** and then **Physics 11**.

- Use appropriate scientific models to explain and predict the behaviour of natural phenomena.

- Identify and describe science- and technology-based careers related to physics.

**KEY**
**TERMS**

- physics
- scientific inquiry
- observation
- qualitative
- quantitative
- theory
- model

**MISCONCEPTION**

**From X-rays to Nerve Impulses**

Many people think that physics is very difficult and highly mathematical. While mathematics is very much a part of physics, the basics of physics need not be difficult to understand. No matter what field of study is most interesting to you, it is likely that physics concepts will help you better understand some facet of it. You may be especially interested in another science, such as biology or chemistry. As your study of science progresses, you will discover that each science depends on the others. For example, chemists use X-rays to study the structure of large molecules. Biologists use the theory of electricity to study the transmission of nerve impulses.

What makes physics so exciting is that you will be involved in thinking about how the universe works and why the universe behaves as it does. When asked to define science, Albert Einstein once replied, "science is nothing more than refinement of everyday thinking." If you substitute "physics" for "science" in Einstein's definition, just what is the refinement he is referring to? Using the language of mathematics to construct models and theories, **physics** attempts to explain and predict interactions between matter and energy. In physics, the search for the nature of these relationships takes us from the submicroscopic structure of the atom to the supermacroscopic structure of the universe. All endeavours in physics, however, have one thing in common; they all aim to formulate fundamental truths about the nature of the universe.

Your challenge in this course will be to develop a decision-making process for yourself that allows you to move from Einstein's "everyday thinking" to his "refinement of everyday thinking." This refinement, the systematic process of gathering data through observation, experimentation, organizing the data, and drawing conclusions, is often called **scientific inquiry**. The approach begins with the process of hypothesizing. A good scientist tries to find evidence that is *not* supported by a model. If contradictory evidence is found, the model was inadequate.

Throughout the textbook, you will find scientific misconceptions highlighted in the margins. See if your current thinking involves some of these misconceptions. Then, by exploring physics through experimentation throughout the course, develop your own understanding.

How did our present understanding of the universe begin? What was the progress over the centuries before present time? The thinking that we know about started with Aristotle.

## Two Models from Aristotle

Over 2300 years ago, two related models were used as the basis for explaining why objects fall and move as they do. Aristotle (384–328 B.C.E.) used one model to account for the movement of objects on Earth, and a second model (see the diagram opposite) for the movement of stars and planets in the sky. We do not accept these models today as the best interpretation of movement of objects on Earth and in space. However, at the time they were very intelligent ways to explain these phenomena as Aristotle observed them.

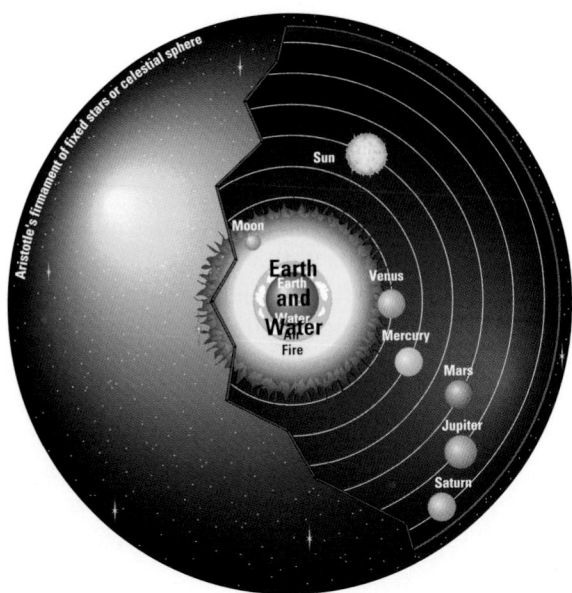

**Figure 1.1** In Aristotle's cosmology, Earth is at the centre of the universe.

## Aristotle and Motion

The model for explaining movement on Earth was based on a view advanced by the Greeks, following Aristotle's thinking. Aristotle accepted the view of Empedocles (492–435 B.C.E.) that everything is made of only four elements or essences — earth, water, air, and fire. All objects were assumed to obey the same basic rules depending on the essences of which they were composed. Each essence had a natural place in the cosmic order. Earth's position is at the bottom, above that is water, then air and fire. According to this model, every object in the cosmos is composed of varous amounts of these four elements. A stone is obviously earth. When it is dropped, a stone falls in an attempt to return to its rightful place in the order of things. Fire is the uppermost of the essences. When a log burns, the fire it trapped from the sun while it was growing is released and rises back to its proper place. Everything floats, falls, or rises in order to return to its proper place in the world, according to Aristotle. These actions were classified as natural motions. When an object experiences a force, it can move in directions other than the natural motions that return them to their natural position. A stone can be made to move horizontally or upward by exerting a force in the desired direction. When the force stops so does the motion.

The model for explaining movement in the sky was somewhat different. Greek astronomers knew that there were two types of "stars," the fixed stars and the planets (or wanderers), as well as the Sun and the Moon. These objects seemed not to be bound by the same rules as objects formed of the other essences. They

**PHYSICS FILE**

Richard Feynman (1918–1988), a Nobel Prize winner and the father of nanotechnology, was one of the most renowned physicists of the twentieth century. In 1959, while presenting a paper entitled "There's Plenty of Room at the Bottom" on the then little-known characteristics of the submicroscopic world, Feynman remarked: "There is nothing besides our clumsy size that keeps us from using [that] space." When he spoke those words, nanotechnology was still a distant dream. That dream now appears to be verging on reality. Indeed, twenty-first century medicine and computer science could well see the first applications of nanotechnology, as both disciplines race to develop tools that will one day allow them to manipulate individual atoms.

**TRY THIS...**

**Physics in the News**

Using print and electronic resources, research a current or historical article that discusses some aspect of physics. Summarize the article in two or three paragraphs, highlighting why you think the topic is significant. Provide as much information about the source of the article as possible.

moved horizontally across the sky without forces acting on them. The Greeks placed them in a fifth essence of their own. All objects in this fifth essence were considered to be perfect. The Moon, for example, was assumed to be a perfect sphere. Aristotle's model assumes that perfect crystal, invisible spheres existed, supporting the celestial bodies.

Later, when Ptolemy (87–150 C.E.) developed his Earth-centred universe model, he used this idea as a base and expanded upon it to include wheels within wheels in order to explain why planets often underwent retrograde (backward) motion. A single spherical motion could explain only the motions of the Sun and the Moon.

To European cultures, Aristotle's two models were so successful that for almost 2000 years people accepted them without question. They remained acceptable until challenged by the revolutionary model of Copernicus (1473–1543) and the discoveries of Galileo Galilei (1564–1642).

## Galileo and Scientific Inquiry

In 1609, using a primitive telescope (Figure 1.2), Galileo observed that the Moon's surface was dotted with mountains, craters, and valleys; that Jupiter had four moons of its own; that Saturn had rings; that our galaxy (the Milky Way) comprised many more stars than anyone had previously imagined; and that Venus, like the Moon, had phases. Based on his observations, Galileo felt he was able to validate a revolutionary hypothesis — one advanced previously by Polish astronomer Nicolaus Copernicus — which held that Earth, along with the other planets in the Solar System, actually orbited the Sun.

What the Greeks had failed to do was test the explanations based on their models. When Galileo observed falling bodies he noted that they didn't seem to fall at significantly different rates. Galileo built an apparatus to measure the rate at which objects fell, did the experiments, and analyzed the results. What he found was that all objects fell essentially at the same rate. Why had the Greeks not found this? Quite simply, the concept of testing their models by experim entation was not an idea they found valuable, or perhaps it did not occur to them.

Since Galileo's time, scientists the world over have studied problems in an organized way, through observation, systematic experimentation, and careful analysis of results. From these analyses, scientists draw conclusions, which they then subject to additional scrutiny in order to ensure their validity.

As you progress through this course, keep the following ideas about theories, models, and observations in mind. Use them to stimulate your own thinking, and questioning about current ideas.

**Figure 1.2** The telescope through which Galileo first observed Jupiter's moons and other celestial bodies in our solar system.

## • *Think It Through*

- A log floats partially submerged on the surface of a lake. The log is obviously wood, a material which clearly grows out of the essence "earth" and is a fairly dense solid like other earth objects. If you were an ancient Greek who believed in the Aristotelian Cosmology, how could you explain why the log floats rather than sinking like rocks or other earth materials?

## Thinking about Science, Technology, Society and the Environment

In the middle of the twentieth century, scientific progress seemed to go forward in leaps and bounds. The presence of figures like Albert Einstein gave science in general, and physics in particular, an almost mystical aura. Too often physics was seen as a pure study isolated from the "real" world. Contrary to that image, science is now viewed as part of the world and has the same responsibilities, perhaps even greater, to the world as any other form of endeavour. Everything science does has a lasting impact on the world. Part of this course is to explore the symbiotic relationship that exists between science, technology, society and the environment (STSE).

To many people, science and technology are almost one and the same thing. There is no doubt that they are very closely related. New discoveries in science are very quickly picked up by technology and vice versa. For example, once thought of as a neat but rather impractical discovery of physics, the laser is a classic example of how science, technology, society, and the environment are inseparable. The laser's involvement in our lives is almost a daily occurrence. Technology has very quickly refined and improved its operation. Today, laser use is widespread. Supermarket scanners, surveying, communications, holography, metal cutters, surgery, and the simple laser pointer are just a few examples of the innovations that technology has found for the laser. Clearly it would be impossible to separate the importance of science and technology to society. Figure 1.3 on the following page shows just a sfew of the many applications of physics in today's world.

Often the same developments have both positive and negative impacts. Our society's ever increasing demand for energy has strained our environment to its limits. Society, while demanding more and more energy, has also demanded that science and technology find alternate sources of energy. This has led to the technological development of nuclear, solar, wind, hydro, geothermal, and fossil fuel as energy sources. Society's and the environment's relationship with science and technology seems to be a two-edged sword.

**TRY THIS...**

**Was Aristotle Right?**
Do heavy objects fall faster than lighter ones? Drop an eraser and a sheet of paper simultaneously from about eye level to the floor. Which gets there first? Is there anything about the motion of the paper that makes you think that this was not a good test? Now crumple the paper up into a small ball and repeat the experiment. Is there a significant difference in the time they take to reach the floor? Describe the variables that you attempted to test.

**PHYSICS FILE**

Aristotle's models had been used to explain the nature of falling for centuries. According to Aristotle, since a large rock has more of the essence "earth" in it than a small one it has a greater tendency to return to the ground. This causes the big rock to weigh more and thus it must fall faster than a small rock. This is a classic application of a model to explain a phenomenon. However, it should not surprise you to find that since the model is in error so is the explanation based on the model.

**Web Link**

***www.school.mcgrawhill.ca/ resources/***
To learn more about careers in physics, go to this web site. Click on **Science Resources** and **Physics 11** to find out where to go next.

**Figure 1.3** Some applications of physics discoveries

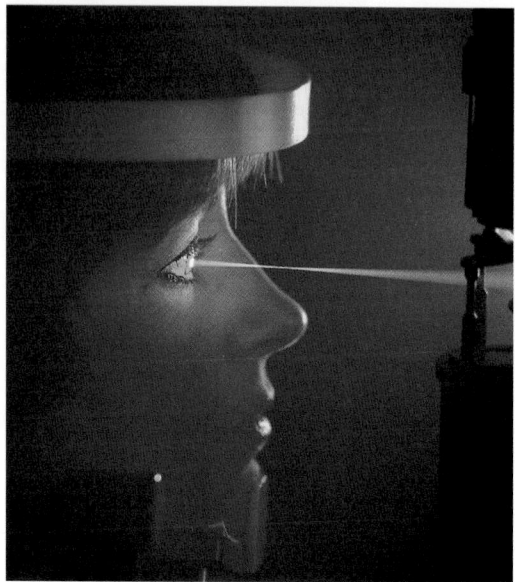

Laser eye surgery is one of many applications that technology has found for lasers.

This tiny "guitar" (about the size of a red blood cell) was built using nanotechnology. This technology will help scientists explore the processes by which atoms and molecules can be used individually as sub-microscopic building blocks.

Hybrid autos that run on both electricity and gasoline can greatly reduce pollution. Cars built of carbon composite materials are lighter and stronger that cars made of traditional materials. Computer controlled ignition and fuel systems increase motor efficiency. All these factors can assist in protecting the environment.

Physics research into thermal properties of materials and technological advances in structural design have combined to produce energy efficient houses that greatly reduce our demand for heating fuels.

Innovations in technology have resulted in the ability to put more and more powerful computers into smaller and smaller spaces.

Technology reaches into the most mundane aspects of our lives. Micro-layers of Teflon on razor blades make them slide more smoothly over the skin.

# Thinking Scientifically

Knowledge begins with observations and curiosity. Scientists organize their thinking by using observations, models, and theories, as summarized below.

## Theory

A **theory** is a collection of ideas, validated by many scientists, that fit together to explain and predict a particular natural phenomenon. New theories often grow out of old ones, providing fresh, sometimes radical ways of looking at the universe. One such example, still in the process of development, is the GUT, or Grand Unified Theory, being sought by researchers across the different fields of physics. Through the GUT, physicists hope one day to be able describe all physical phenomena in the universe by using the same set of laws.

## Model

A **model** is a representation of phenomena and can come in a variety of forms, including a list of rules, pencil lines on a piece of paper, an object that can be manipulated, or a mathematical formula. An observation may be explained using more than one model; however, in most cases, one model type is more effective than others.

## Observations

An **observation** is information gathered by using one or more of the five senses. Observations may yield a variety of explanations, as participants in the same event often report different things. It takes hundreds of observations of a single phenomenon to develop a theory. There are two kinds of observations that can be made. The first are **qualitative**, which describe something using words: "A feather is falling slowly to the ground." The second are **quantitative**, which describe something using numbers and units: "The rock fell at 2 m/s."

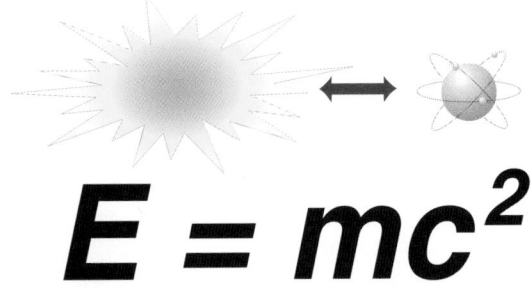

$$E = mc^2$$

**Figure 1.4** You have undoubtedly heard of Einstein's theory of special relativity. One part of the theory states that the speed of light, $c$, is the only thing in the universe that is constant. All other measurements are relative, depending on the observer's frame of reference. The famous formula (model) associated with the theory is $E = mc^2$.

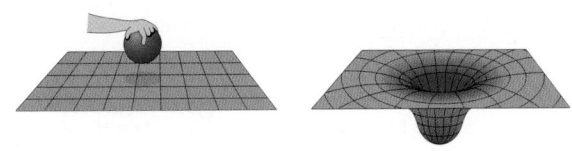

**Figure 1.5** This "rubber sheet model" is often used to simulate Albert Einstein's idea of curved space. The model shows that a central mass can cause the space around the mass to curve.

**Figure 1.6** Observations can be quantitative or qualitative. The cyclist can determine her speed by applying the mathematical model, $v = \Delta d/\Delta t$, to her observable data of distance and time.

## CAREERS IN PHYSICS

As you have read in this introductory section to the chapter, your world, from the natural cycles of weather to the high-tech gadgets of communication, relies on basic principles of physics. The wide scope of what physics is translates into a very long list of careers that involve the study of physics. For example, are you interested in theatre? Knowledge of how light acts is crucial to the intricate lighting techniques used in theatres today. Are you a musician? You will be able to achieve better musical effects by understanding more about the nature of sound. Study the diagram shown here to note career opportunities in physics or that use much of the knowledge and skills you will gain in this physics course. Consider one or more that might be especially appealing to you, and begin research on educational requirements to attain it. People succeed and are happiest when in a career that really interests them, not just one they are good at, so keep that in mind as you explore opportunities.

---

## 1.1    Section Review

1. **MC** What is nanotechnology? Cite specific examples of how this technology could affect our lives.

2. **C** How would you define physics?

3. **K/U** Why do scientists employ scientific inquiry to investigate problems?

4. **K/U** What is the difference between a theory, a model, and an observation? What is the significance of each?

5. **C** Describe the difference between qualitative and quantitative observations, and provide an example of each.

Problem-solving skills are important in everyday life, in school, and in the workplace. Some problems, like deciding whether to walk or ride your bike, are easier to solve than others. In each case, however, you develop a process to help you make up your mind. In physics, understanding a concept is more important than simply doing the math; hence, the need for creativity and adaptability. As you apply the problem-solving strategies contained in this textbook, remember that your answer to any one question is less important than the reasoning you use.

**SECTION EXPECTATIONS**

• Select and use appropriate numeric, symbolic, graphical, and linguistic modes of representation to communicate science.

• Analyze and synthesize information in the process of developing problem-solving skills.

**KEY TERMS**

• framing a problem

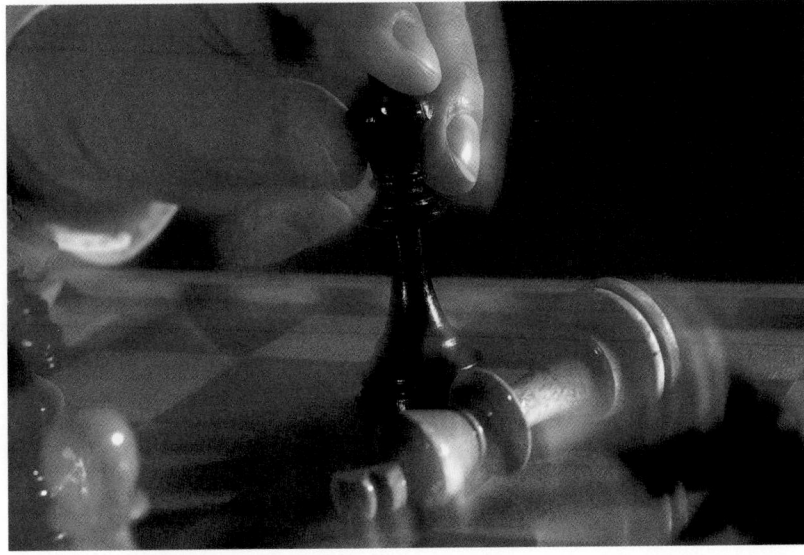

**Figure 1.7** Chess is a game of intricate strategy. Victory belongs to the player who can visualize how the game will progress several moves into the future.

## Framing A Problem

**Framing a problem** is a way to set parameters (important boundaries) and organize them in a way best suited to a particular problem. There is rarely only one way to frame a problem, and how you do so depends on each situation; you must determine which methods work best for you, and for each problem. Often, simply framing a problem will help the solution to become apparent to you.

Framing a problem, whether it is a physics question or a typical household problem, is a creative and systematic process designed to clarify what is known, what restrictions exist, and what the ultimate goal is. Most people have a preferred method of organizing information. Often the method used to organize information is topic specific rather than personal preference.

**Figure 1.9** Framing a problem and developing solution strategies is applicable to all types of problems.

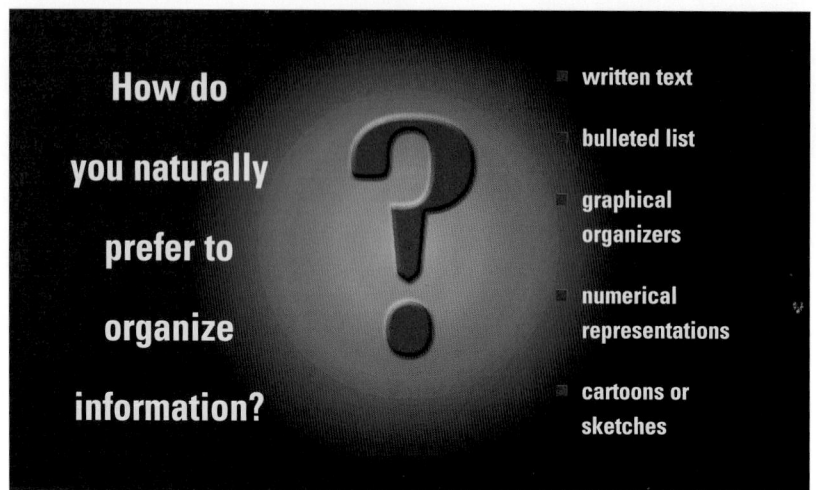

How do you naturally prefer to organize information?

- written text
- bulleted list
- graphical organizers
- numerical representations
- cartoons or sketches

**Figure 1.8** Recognizing the modes of organization that you prefer will help you develop your problem-solving strategies.

### Example 1: Organizing Data Using Text

You can represent your thinking process in the form of questions.

In this way, you have framed the problem by posing key questions about your available time. Your solution must fit within these parameters.

## A Typical Problem

A friend calls you on a Tuesday evening at 6:00 p.m. and asks if you want to join two other friends for two-and-a-half hours of "down time," to play an ongoing game you have all been enjoying. Your friends plan to begin at 7:30 p.m. You know that you have two homework assignments that must be completed before tomorrow. Before you are able to answer, you need to decide if you have enough time to complete your homework and to take time out to play the game. You also need to prioritize your feelings about the benefits of taking time out to play the game.

This scenario has been framed graphically using different strategies. As you examine them, consider their effectiveness. Develop your own strategies for framing problems, and for setting parameters that work best for you.

**(a)** Written Text

I feel like playing the game. It would be an enjoyable break, but I also have two homework assignments due in the morning. How long do my friends intend to play the game? Two-and-a-half hours. How much time do my assignments require? Physics: thirty minutes. Math: no homework tonight. English: thirty minutes. I should be home by 11:00.

**(b)** Bulleted List

- Ongoing game
- Fun, and provides a break
- Homework to do
- Thirty minutes of Physics homework
- Thirty minutes of English
- Two-and-a-half hours
- Home by 11:00

## Example 2: Organizing Data Using Diagrams

You have framed the problem by generating diagrams (a) and (b) which outline the parameters. Your solution must fit within these parameters.

**(a)** Tree Diagram

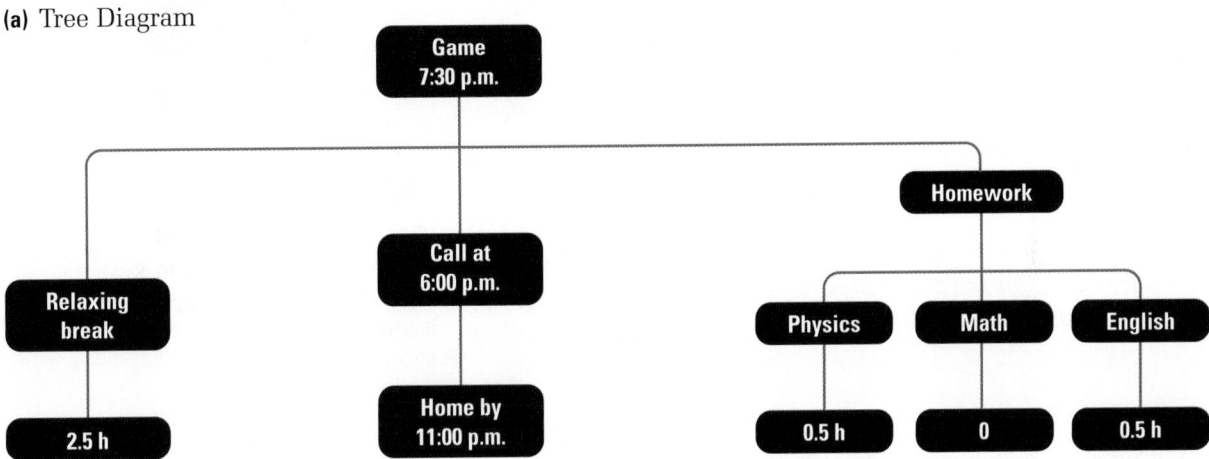

**(b)** Temporal Diagram

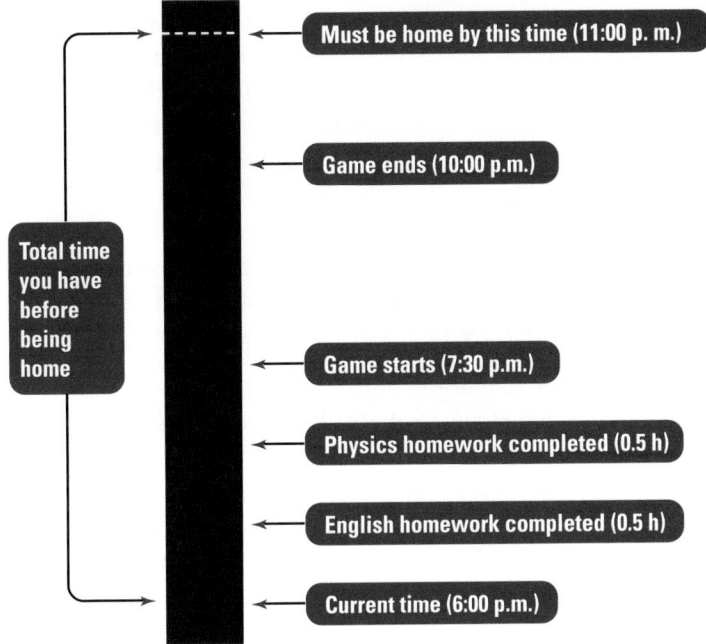

## Model Problems

Throughout this resource, you will find a feature called Model Problem. Each one presents a specific physics problem and its solution. Model Problems follow a step-by-step approach, identical to the one below. Become familiar with these steps, and integrate them into your own bank of problem-solving strategies. Throughout the book, the Model Problems are followed by Practice Problems to help you develop your skills. Answers to these are placed at the end of the Chapter Review.

### MODEL PROBLEM

*A problem is posed.*

### Frame the Problem

*This section describes the problem and defines the parameters of the solution. Consider statements made in this section very carefully.*

### PROBLEM TIP

Often you will find problem tips embedded in model problems. The problem tips are designed to highlight strategies to help you successfully navigate a specific type of problem.

### Identify the Goal

*Narrow your focus and determine the precise goal.*

### Variables

| Involved in the problem | Known | Unknown |
|---|---|---|
| *Lists each variable that was mentioned in Frame the Problem.* | *Lists variables about which information is known or implied.* | *Lists variables that are unknown and must be determined in the solution.* |

### Strategy

*A step-by-step description of the mathematical operations involved.*

### Calculations

*Use the data you have accumulated to complete the solution. Simplify the units required in your final answer.*

A concluding statement verifies that the goal has been accomplished. The number of significant digits in the solution statement must match those in the question statement.

### Validate

*This provides an opportunity to clarify the steps used in calculating the solution. Validating the solution helps catch numerical and conceptual errors.*

## Average Speed

**A student runs 15 km in 1.5 h. What was the student's average speed?**

### Frame the Problem

- The student may or may not have stopped for a rest, but the term average implies that only total time and total distance are to be considered.

- Speed has units of distance/time.

- Use the distance/time information to help build a formula for speed (or verify that the one you have memorized is correct).

- Total distance/total time will provide the average speed.

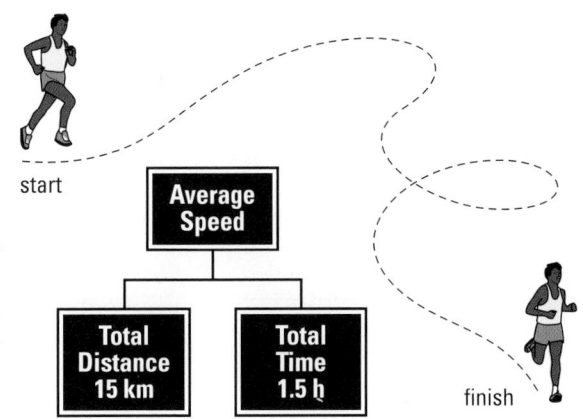

### Identify the Goal

The average speed, $v_{ave}$

### Variables and Constants

| Involved in the Problem | Known | Unknown |
|---|---|---|
| $\Delta d$ | $\Delta d = 15$ km | $v_{avg}$ |
| $v_{ave}$ | $\Delta t = 1.5$ h | |
| $\Delta t$ | | |

> **PROBLEM TIP**
>
> Be sure to identify the number of *significant figures* provided in the question as they will vary from one question to the next. Carry excess significant figures through during calculations, and then round your final answer to the correct number of significant figures. See Skill Set 2 at the back of this textbook for significant digits and rounding information.

### Strategy

Use the average speed formula

Substitute in the known values, and solve

### Calculations

$$v_{ave} = \frac{\Delta d_{Total}}{\Delta t_{Total}}$$

$$v_{ave} = \frac{(15 \text{ km})}{(1.5 \text{ h})}$$

$$= \frac{10 \text{ km}}{\text{h}}$$

Therefore, the student ran at an average speed of 10 km/h.

### Validate

The value for speed is given in distance (km) per time (h) which is correct.

## Achieving in Physics

The following Achievement Chart identifies the four categories of knowledge and skills in science that will be used in all science courses to assess and evaluate your achievement. The chart is provided to help you in assessing your own learning, and in planning strategies for improvement, with the help of your teacher.

You will find that all written text, problems, investigations, activities, and questions throughout this textbook have been developed to encompass the curriculum expectations of your course. The expectations are encompassed by these general categories: Knowledge/Understanding, Inquiry, Communication, and Making Connections. You will find, for example, that questions in the textbook have been designated under one of these categories to enable you to determine if you are able to achieve well in each category (some questions could easily fall under a different category; we have selected, for each question, one of the categories with which it best complies). Keep a copy of this chart in your notebook as a reminder of the expectations of you as you proceed through the course. (In addition, problems that involve calculation have been designated either Practice Problems or, in Chapter and Unit Reviews, Problems for Understanding.)

**Table 1.1** Achievement Chart

| Knowledge and Understanding | Inquiry | Communication | Making Connections |
|---|---|---|---|
| ■ Understanding of concepts, principles, laws, and theories<br><br>■ Knowledge of facts and terms<br><br>■ Transfer of concepts to new contexts<br><br>■ Understanding of relationships between concepts | ■ Application of the skills and strategies of scientific inquiry<br><br>■ Application of technical skills and procedures<br><br>■ Use of tools, equipment, and materials | ■ Communication of information and ideas<br><br>■ Use of scientific terminology, symbols, conventions, and standard (SI) units<br><br>■ Communication for different audiences and purposes<br><br>■ Use of various forms of communication<br><br>■ Use of information technology for scientific purposes | ■ Understanding of connections among science, technology, society, and the environment<br><br>■ Analysis of social and economic issues involving science and technology<br><br>■ Assessment of impacts of science and technology on the environment<br><br>■ Proposing courses of practical action in relation to science- and technology-based problems |

At the end of each unit, you will have the opportunity to tie together the concepts and skills you have learned through the completion of either an investigation, an issue, or a project. Throughout each unit, one of the logos below will remind you of the end-of-unit performance task for that unit. Ideas are provided under each logo to help you prepare and plan for the task. Assessment of your work for each of the end-of-unit tasks, like all assessment in the course, will be based on the Achievement Chart shown in Table 1.1.

**UNIT ISSUE PREP**

**UNIT PROJECT PREP**

**UNIT INVESTIGATION PREP**

The Physics Course Challenge will allow you to incorporate concepts and skills learned from every unit of this course. This culminating assessment task will be developed during the year, but completed at or near the end of the course. Course Challenge logos exist throughout the text, cueing you to relate specific concepts and skills to your end-of-course task. The units in this course may seem to be largely unrelated. By investigating Space-Based Power in the Course Challenge, however, you will find some intriguing interactions among many concepts. Again, use the Achievement Chart in Table 1.1 as your guide to how your work will be assessed.

**COURSE CHALLENGE**

## 1.2 Section Review

1. **C** Explain why problem solving is a creative process. State the importance of framing a problem.

2. **K/U** Reflect on the game scenario. Which framing method most closely matches the thought process you would use to solve the same problem?

3. **I** Develop a different framing technique for the game problem. Share your model with the class.

4. **I** You have been offered a part-time job at the mall on weekends. However, you are determined to pursue a post-secondary education and have been devoting extra time to your studies. Should you accept the job? Frame the problem to help you decide.

5. **I** A friend asks you if warm water freezes faster than cold water. Frame the problem.

6. **I** Another friend tells you that astronauts are weightless when they orbit Earth. You know this to be inaccurate. Frame the problem to help dispel the misconception.

- Select and use appropriate equipment to accurately collect scientific data.

- Design and conduct experiments that control major variables.

- Hypothesize, predict, and test phenomena based on scientific models.

KEY
TERMS

- period
- frequency
- percent difference
- percent deviation

Analyzing "real" world phenomena, as you will be doing throughout this course, requires the ability to take measurements — from very small to very large. It also requires that you be able to visualize the data in various ways, and to determine how accurately current models can predict actual events. In this section you will do two experiments that give you an opportunity to start having experience at measuring actual events, and analyzing the data generated in the experiments.

In the first investigation, you will design your own experiment to investigate the variables that determine the rate of the swing of a pendulum. In the second investigation, you will compare your experimental results from the first investigation to an existing model that predicts how the swing rate of a pendulum is controlled. You will then have the opportunity to practise using some of the mathematical tools of a physicist, comparing your data with the predictions of a mathematical model.

Before you conduct the investigations on the next two pages, think about the motion of a swing, like the one shown in Figure 1.10. See if you can apply the terms that follow the photograph to the child's motion.

**Figure 1.10** A swing is an excellent example of periodic motion.

The time required for one complete oscillation is called the **period**.
   Period = time interval / 1 cycle
   The SI unit for period, T, is seconds (s).

The number of oscillations in a specific time interval is called the **frequency**.
   $f$ = number of oscillations / time interval
   The SI unit for frequency, f, is 1/s or Hertz (Hz)

# INVESTIGATION 1-A

## Analyzing a Pendulum

TARGET SKILLS
- Hypothesizing
- Predicting
- Identifying variables
- Performing and recording
- Analyzing and interpreting
- Communicating results

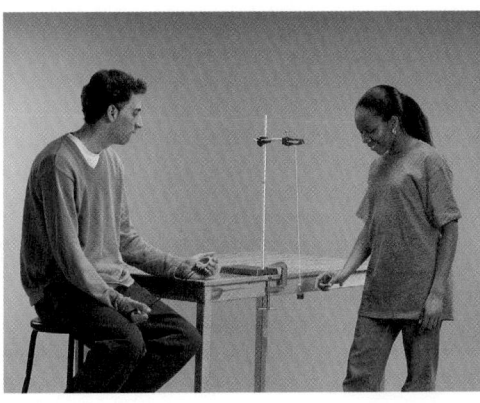

Grandfather clocks are not merely timepieces, they are also works of art. A key feature of a grandfather clock is the ornate pendulum that swings back and forth.

### Problem

**Part 1:** What factors affect the period of oscillation of a pendulum?

**Part 2:** Compare your results with your predictions.

### Hypothesis

Formulate a hypothesis listing variables that will affect the period of oscillation of a pendulum. Predict how each variable will affect the period of oscillation.

### Equipment

- various masses (50 g to 100 g)
- string (1 m)
- stopwatch
- retort stand

### Procedure

1. With a partner, design an experiment to determine variables that will affect the period of oscillation of a pendulum. Investigate a minimum of three variables.

2. Provide step-by-step procedures.

3. Predict and record the effect of each variable, and have your teacher initial each prediction.

4. Following your school's safety rules, carry out the experiment and record your observations.

### Analyze and Conclude

1. How many oscillations did you use to determine the period of the pendulum?

2. How many trials did you run before changing variables? Was this enough? Explain.

3. Did your hypothesis include length as a variable? If so, why? If not, why not? Explain your choice of variables.

4. Determine the uncertainty *within* your data by calculating the **percent difference** between your maximum and minimum values for the period of oscillation for each controlled variable. Refer to Skill Set 1 for an explanation of percent difference.

5. According to your results, what variables affect the period of oscillation of a pendulum? Explain, providing as much detail as possible.

# INVESTIGATION 1-B

## Analyzing Pendulum Data

### TARGET SKILLS
- **Hypothesizing**
- **Performing and recording**
- **Analyzing and interpreting**
- **Communicating results**

Physicists and clock designers have used results from experiments like the previous one to develop a relationship between the period of oscillation of a pendulum and its length. The mathematical model for this relationship is approximated by the following equation:

$$T = 2\pi\sqrt{\frac{l}{g}}$$

where: $T$ = period of oscillation
$l$ = pendulum length
$g$ = 9.8 m/s$^2$ (acceleration due to gravity near Earth's surface)

## Problem

How should experimental data be analyzed to test for (a) error within the data set and (b) accuracy when compared to a theoretical value?

## Hypothesis

Formulate a hypothesis predicting how closely your experimental results from Investigation 1A will match the mathematical model shown above.

## Procedure

1. Set up a table identical to the one shown.

2. Use the theoretical equation and the data you collected in the previous investigation to complete the table. Refer to Skill Set 1 for an explanation of **percent deviation**.

3. If length was not one of the variables that you and your partner tested, borrow data from tests carried out by your classmates.

## Analyze and Conclude

1. Generate the following graphs on one set of axes:
   (a) $T_{Experimental}$ vs. l
   (b) $T_{Theoretical}$ vs. l

2. Analyze the graph. Is it possible to qualitatively determine whether your experimental data were similar to the results predicted by the theory?

3. Do the percent deviation values allow you to quantitatively determine whether your experimental data were similar to the results predicted by the theory? Again, refer to Appendix B for an explanation of percent deviation.

4. Suggest a method of determining whether the experimental deviation of your data is within acceptable parameters.

5. Suggest techniques to reduce the experimental deviation between your data and the theoretical period values.

6. Explain the difference between percent deviation and percent difference. When should each one be used?

| Trial | Length (m) | Experimental results | | Theoretical results $\left(T = 2\pi\sqrt{\frac{l}{g}}\right)$ | Percent deviation |
|---|---|---|---|---|---|
| | | Time 5 cycles (s) | Time 1 cycle (s) | | |
| Sample 1 | 0.80 m | $1.0 \times 10^1$ s | 2.0 s | $T = 2\pi\sqrt{\frac{0.80 \text{ m}}{9.81 \text{ m/s}^2}} = 1.8$ s | 11 % |
| Sample 2 | | | | | |

# Physics: an Active Endeavour

Understanding physics concepts requires making good observations and analyses. Thus, this book provides numerous active investigations, less formal Quick Labs that require few materials to carry them out, and marginal Try This activities that are just that — actions that won't take much time to do, but will help make concepts clearer. Watch for the following designations throughout the text:

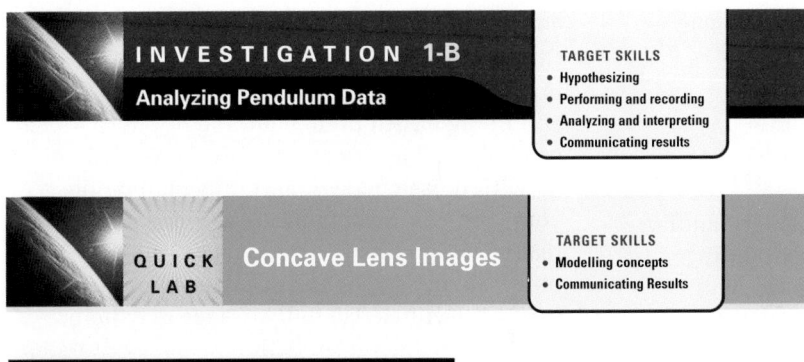

## 1.3 Section Review

1. **K/U** When should percent deviation be used to analyze experimental data?

2. **K/U** When should percent difference be used to analyze experimental data?

3. **I** A group of science students hypothesize that the ratio of red jellybeans to green jellybeans is the same in packages with the same brand name, regardless of size. Their results are provided in Table 3.

**Table 3** Jellybean data

| Package | 1 | 2 | 3 | 4 | 5 |
|---------|-----|-----|-----|-----|-----|
| Red | 23 | 18 | 50 | 62 | 19 |
| Green | 35 | 24 | 2 | 81 | 23 |

(a) Compute a red-to-green jellybean ratio for each package.

(b) Is there a general trend in the data?

(c) Is there a data set that, while properly recorded, should not be considered when looking for a trend? Explain.

## REFLECTING ON CHAPTER 1

- Physics is the study of the relationships between matter and energy. As a scientific process, physics helps us provide explanations for things we observe. Physicists investigate phenomena ranging from subatomic particles, to everyday occurrences, to astronomical events.

- Like all science, physics is:
  1. a search for understanding through inquiry;
  2. a process of crafting that understanding into laws applicable to a wide range of phenomena; and
  3. a vehicle for testing those laws through experimentation.

- Aspects of physics are found in a wide range of careers. Engineering and academic research positions may be the first to come to mind, but medical and technological professions, science journalism, and computer science, are other fields that may require a physics background.

- A theory is a collection of ideas that fits together to describe and predict a particular natural phenomenon. New theories often grow out of old ones, providing fresh, sometimes radical ways of looking at the universe. A theory's value is determined by its ability to accurately predict the widest range of phenomena.

- A model is the representation of a theory. Models may take different forms, including mathematical formulas, sketches, and physical or computer simulations.

- An observation is information gathered by using one or more of the five senses. Models and theories attempt to predict observations.

- Changes in science and technology can have huge impacts on our society and on the global environment. An understanding of physics can help you assess some of the risks associated with those changes, and thus help guide your decision-making process. Since most real-world problems involve economic, political, and social components, applying scientific knowledge to the issues may help you separate fact from fiction.

- A learned skill, problem solving is a thought process specific to each of us and to each problem. Several problem-solving techniques are modelled in this chapter, each illustrating the conceptual thinking involved in framing the parameters within which the solution must fit.

- Experimental design requires a clear understanding of the hypothesis that is to be tested. Whenever you are designing your own experiments, your challenge will be to ensure that only one variable at a time is being tested. The number of trials that you run depends on the results. Enough trials have been run when there is a clear trend in the data. If, during your analyses, a clear trend is not evident, more data must be collected. Refer to the Skill Sets at the back of this textbook to help you with data analysis.

### Knowledge and Understanding

1. Describe how nanotechnology is the product of both scientific inquiry and technology.
2. In general terms, describe the factors involved in the study of physics.
3. Describe how the Black Box activity can be used to explain the process of scientific inquiry.
4. State one definition of scientific inquiry.

5. Who first discussed the concept of nanotechnology?
6. What observation caused Aristotle to assume that the planets and the Moon were made of material different than Earth?
7. Why was Galileo able to observe the mountains and craters on the Moon, and four moons orbiting Jupiter?

## Inquiry

**8.** While stargazing with friends, you observe a strange light in the sky. The following list of observations details information collected by you and your friends.

- The light moved from that distant hilltop in the east to the TV tower over there to the west.
- As the light moved, it seemed to be hovering just above the ground.
- As it moved from east to west, it got really bright and then faded again.
- It took about 3.0 s for it to move from the hilltop to the TV tower.
- The hilltop is about 15 km from the TV tower.
- It moved at a constant speed from point to point, and then stopped instantaneously.

What was the source of this light? Frame the problem using two different methods; incorporate the data provided and include any other parameters that you feel are relevant. You do not need to reach a solution.

## Communication

**9.** Define scientific inquiry.

**10.** Generate two specific questions that you would like to have answered during this Physics course. Flip through the text to determine which unit(s) might contain the answers.

**11.** Briefly describe the purpose of a theory, a model, and an observation.

**12.** Describe how physics has evolved and continues to evolve.

**13.** Refer to Table 1.1. Provide one type of activity (for example, test, lab, presentation, debate) that would best allow you to demonstrate your strengths in each category (Knowledge and Understanding; Inquiry; Communication; and Making Connections).

## Making Connections

**14.** Are there any scientific theories or models that you believe will eventually be proven false? Explain.

**15.** Read the Course Culminating Challenge on page 756. Generate a list of topics that you believe would be suitable as an independent study for this activity.

## Problems for Understanding

**16.** A student conducts an experiment to determine the density of an unknown material. Use the data collected from both trials to calculate the percent difference in the density measurements.

**Trial 1**  19.6 g/mL
**Trial 2**  19.1 g/mL

**17.** A student decides to compare the theoretical acceleration due to gravity at her location ($g = 9.808$ m/s$^2$) to experimental data that she collects using very sensitive equipment. She runs 15 trials and then averages her results to find $g = 9.811$ m/s$^2$.

**(a)** Calculate the percent deviation in her calculation.

**(b)** Is the percent deviation reasonable? Explain.

**18.** The following data are collected during an experiment.

| Trial # | 1 | 2 | 3 | 4 | 5 | 6 | 7 | 8 | 9 | 10 | 11 |
|---|---|---|---|---|---|---|---|---|---|---|---|
| Frequency (Hz) | 12 | 11 | 13 | 9 | 12 | 11 | 11 | 14 | 13 | 11 | 10 |

Refer to Skill Set 4 for reference on the following calculations:

**(a)** Find the mean of the data.
**(b)** Find the median of the data.
**(c)** Find the mode of the data.

# Forces and Motion

## OVERALL EXPECTATIONS

**DEMONSTRATE** an understanding of the relationship between forces and acceleration of an object.

**INVESTIGATE** and analyze force and motion using free-body diagrams and vector diagrams.

**DESCRIBE** contributions made to our understanding of force and motion, and identify current safety issues.

A blink of an eye is a lifetime compared to the time that has elapsed between the first and last image in the inset photos of a bullet impacting on body armour. This sequence of motion was captured by a device composed of eight digital cameras, ingeniously wired together, that produces the fastest frame-by-frame images to date. Such technology is allowing scientists to examine an object's motion in the kind of minute detail that previously could only be hypothesized by using computer modelling.

Not all motion is quite this fast. Scientists have discovered, for example, that the continents of the world are adrift. What are now rugged mountain ranges were once buried deep beneath the sea. These very mountains might one day become rolling hills or farmland. The movements of the continents escaped the notice of scientists until recently, because the rate of the motion of continents is extremely slow.

This unit examines how physicists describe motion and how they provide explanations for the forces that cause it. The unit ends by giving you an opportunity to consider motion from the director's chair. Based on your expanded knowledge, you will be challenged to create your own realities by either speeding up imperceptibly slow motion through animation, or slowing down events that normally escape vision because they happen in the blink of an eye.

### UNIT PROJECT PREP

Refer to pages 190–191 before beginning this unit. In this unit project, you will create a cartoon, video, or special effects show.

■ How can you manipulate frames of reference to create the illusion of motion?

■ What ideas can you get from amusement parks to simulate natural forces?

*Photo insets courtesy of Professor Arun Shukla, University of Rhode Island*

The world of entertainment thrives on our passion for thrill and adventure. Many movies provide experiences that make you feel as though you are part of the action. Why do you get that sick feeling in your stomach when a car in a movie races up to the edge of a cliff, giving you a sudden panoramic view over the edge? How do live theatre productions such as the one in the photograph create the impression that an actor is travelling in relation to the other actors and audience? How are cartoons created to look like free-flowing action, when they are simply a series of individual pictures? How do cartoonists design a series of pictures so that they will simulate someone speeding up, slowing down, or travelling at a steady pace?

In this chapter, you will begin a detailed analysis of motion, which will lead you to the answers to some of the questions above. You will learn to apply models developed by physicists for understanding different types of motion.

# Winning the Race

**TARGET SKILLS**

- Identifying variables
- Performing and recording
- Analyzing and interpreting

## The Tortoise and the Hare

Assemble a 1 m long ramp that has a groove to guide a marble that will roll down the ramp. For example, you may use a curtain track or tape two metre sticks together in a "V." Stabilize the ramp so it is at an angle of 30° with the horizontal. Hold one marble (the hare) at the top of the ramp. Roll another marble (the tortoise) along the bench beside the ramp. Start the "tortoise" rolling from behind the ramp. Release the "hare" when the "tortoise" is rolling along beside the ramp. Change the angle of the ramp in an attempt to find an angle at which the "hare" will beat the "tortoise" and win the race.

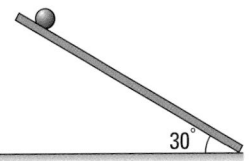

### Analyze and Conclude

1. How is the motion of the "hare" different from the motion of the "tortoise?"

2. Did the "hare" ever win the race?

3. Give a possible explanation for the outcome of all of the races.

## Caught in a Rut

Use the same ramp as above. Mark points on the ramp that are 60 cm and 90 cm from the base of the ramp as shown in the figure. Stabilize the ramp at an angle of 60° with the horizontal. Hold one marble at the 90 cm mark and another at the 60 cm mark. Observe the motion of the marbles under the following conditions.

**CAUTION** Do not drop the marbles from a greater height.

- use marbles of same mass; release at same time

- use marbles of same mass; release marble at 90 cm first; release 60 cm marble when first marble has rolled 10 cm

- use larger marble at 90 cm than at 60 cm; release both marbles at same time

- use larger marble at 90 cm than at 60 cm; release 90 cm marble first; release 60 cm marble when first marble has rolled 10 cm

Choose another angle for the ramp and repeat the procedure.

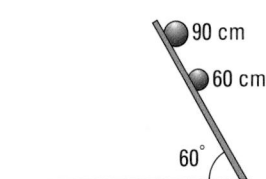

### Analyze and Conclude

1. Describe any effect that the starting position of the marbles had on the rate at which their speed increased.

2. Was there any case in which the 90 cm marble caught up with the 60 cm marble? Give a possible reason for these results.

3. Describe any way in which the mass of the marble affected the outcome of the race.

4. Describe any way in which the angle of the ramp influenced the outcome of the race.

5. Write a summary statement that describes the general motion of marbles rolling, from rest, down a ramp.

- Describe motion with reference to the importance of a frame of reference.

- Draw diagrams to show how the position of an object changes over a number of time intervals in a particular frame of reference.

- Analyze position and time data to determine the speed of an object.

KEY
TERMS

- frame of reference
- at rest

**COURSE CHALLENGE**

**Getting into Orbit**

Research the current types of orbits given to satellites. Investigate which type of orbit would best meet the needs of a satellite that was to be used for a space based power system. Should the satellite be in motion relative to Earth?

Begin your research at the **Science Resources** section of the following web site: *www.school.mcgrawhill.ca/resources/* and go to the *Physics 11 Course Challenge*.

In the introductory investigations, you observed marbles moving in different ways. Some were moving at a constant rate or speed. Some were starting at rest and speeding up, while others were slowing down. How easy or difficult did you find it to describe the relative motion of the marbles and the pattern of their motion?

Although you might think you can readily identify and describe motion based on everyday experiences, when you begin to carefully examine the physical world around you, motion can be deceiving. A few examples were discussed in the chapter introduction. To describe motion in a meaningful way, you must first answer the question, "When are objects considered to be moving?" To answer this seemingly obvious question, you need to establish a **frame of reference**.

## Frames of Reference

**Figure 2.1** For more than 50 years, movie producers have used cameras that more to give you, the viewer, the sense that you are moving around the set of the movie and interacting with the actors.

Movie producers use a variety of reference clues to create images that fool your senses into believing that you are experiencing different kinds of motion. In the early years of moving making, film crews such as those in Figure 2.1, could drive motorized carts carrying huge cameras around a stage. To create the sense that the actors were in a moving car, the crew would place a large screen behind a stationary car and project a moving street scene on the screen so the viewer would see it through the car windows. Today, the movie crew might ride on a moving dolly that is pulling the car down an active street. In this case, the crew and the viewer would be **at rest** relative to the car in which the actors are riding. The buildings and street would be moving relative to the stationary actors.

**Figure 2.2** Does this astronaut appear to be hurtling through space at 28 000 km/h?

In everyday life, Earth's surface seems to provide an adequate frame of reference from which to consider the motion of all objects. However, Earth's reference frame is limited when you consider present-day scientific endeavours such as the flight of aircraft and space shuttles. As you examine the meanings of terms such as "position," "velocity," and "acceleration," you will need to consider the frame of reference within which objects are considered to be moving.

### Think It Through

- For each picture shown here, describe a frame of reference in which the pencil

  **(a)** is moving

  **(b)** is not moving

## Illustrating Motion

Because you need to establish a frame of reference to study motion, diagrams and sketches are critical tools. Diagrams show how the object's position is changing in relation to a stationary frame, during a particular time interval or over several time intervals. When comparing the object's position in each of a series of pictures, you can determine whether the object is at rest, speeding up, slowing down, or travelling at a constant speed.

Your diagrams could be as elaborate as pictures taken by a camcorder (see the Physics File on page 31), as simple as stick-people, or even just dots. In any case, you would superimpose (place one on top of the other) each image, ensuring that something visible in the background is in the same place in each frame. This point provides your frame of reference. Knowing the time that passed between the recording of each image, you can analyze the composite picture or diagram and determine the details of the motion.

The four stick diagrams in Figure 2.3 illustrate four different kinds of motion. Each diagram shows the position of a sprinter after five equal time intervals. In diagram A, the sprinter has not changed position, and is therefore at rest. In diagram B, she changes her position by an equal amount during each time period, and therefore she is travelling at a constant speed. In diagram C, she is changing her position by an increasing amount in each time interval, and therefore she is speeding up. In diagram D, she is changing her position by a decreasing amount, and therefore is slowing down.

**Figure 2.3** Stick diagrams illustrating the position of a sprinter after five equal time intervals

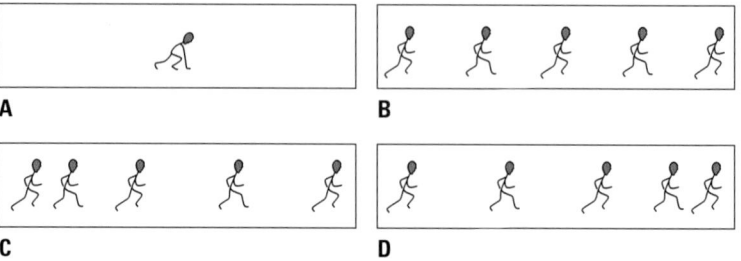

A digram of the composite picture of the sprinter in motion can be made even simpler by considering a single point on her waist. This point is approximately her centre of mass. In other words, this point moves as though the sprinter's entire mass was concentrated there. You can measure the distance between points, and the analysis of her motion then becomes straightforward. The diagrams in Figure 2.4 show how a picture can be drawn simply as a set of dots to show how the position of an object changes over a number of time intervals in a particular frame of reference.

**Figure 2.4** Dots can be used to show how the position of an object changes over a number of time intervals in a particular frame of reference.

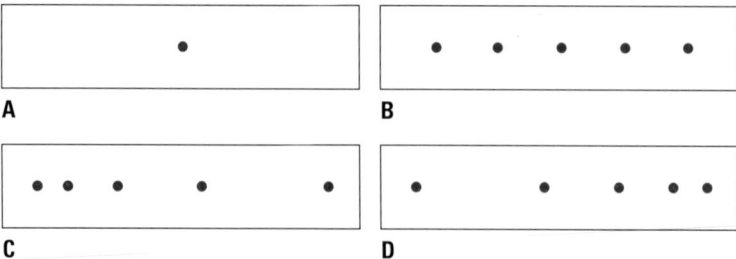

# The Importance of Relative Motion

Assume that you have selected a frame of reference in which you are at rest. For example, when you are dozing off in the back seat of a car that is traveling smoothly along a super highway, you may be unaware of your motion relative to the ground. The sensation is even more striking when you are in a large, commercial airliner. You are often entirely unaware of any motion at all relative to the ground. You become aware of it only when the motion of your frame of reference changes. If the car or airplane speeds up, slows down, or turns, you become very aware of the change in the motion of your reference frame. Physicists and engineers need to understand these relative motions and their effects on objects that were at rest in that reference frame. As you solve motion problems and move on into a study of forces, always keep the reference frame and its motion in mind.

---

## PHYSICS & SOCIETY

### The Physics of Car Safety

When a car stops suddenly, you keep going. This example of Newton's first law of motion has been responsible for many traffic injuries. Countless drivers and passengers have survived horrible crashes because they were wearing seat belts, and air bags have also played a major role.

To understand the physics behind the design of air bags, imagine that the car you are driving is suddenly involved in a head-on collision. At the instant of impact, the car begins to decelerate. Your head and shoulders jerk forward, and the air bag pops out of its compartment. The bag must inflate rapidly, before your head reaches the wheel, and then start to deflate as your head hits it. This causes your head to decelerate at a slower rate. In addition, the force of your impact with the air bag is exerted over a wider area of your body, instead of being concentrated at the impact site of your head with something small, such as the top of the steering wheel.

Physics is also involved in the design of car tires. The key consideration is the amount of tire area that stays in contact with the road during braking and turning; the more tire contact, the better your control of the car. Also important is having tires that resist "hydroplaning" on wet roads — at slow speeds, water skiers sink; at

high speeds, they glide over the surface of the water. That's just what you do not want your car tires to do in the rain. Engineers used various physics principles to design tires with a centre groove that pumps water away from the surface as the tires roll over wet pavement.

**TARGET SKILLS**
- Analyzing and interpreting
- Hypothesizing

### Analyze

1. Air bags have come under increased scrutiny. Research the reasons for this debate.

2. To keep more rubber in contact with the road, tires could be made wider. The ultimate would be a single tire as wide as the car. What would be the disadvantage of this type of tire? What might limit the maximum width of a tire?

1. **K/U** Draw dot diagrams, such as those illustrated in Figure 2.4 on page 32, of the motion described in the following situations.

   (a) A sprinter running at a constant speed.

   (b) A marble starting from rest and rolling down a ramp.

   (c) A car starting from rest, speeding up, and then travelling at a constant speed. Finally, the car slows down and stops.

2. **K/U** Alex is sitting at a bus stop facing north. Darcy walks by heading west. Jennifer jogs by going east. Draw dot diagrams of the motion of each person from:

   (a) Alex' frame of reference,

   (b) Darcy's frame of reference, and

   (c) Jennifer's frame of reference.

3. **I** Draw dot diagrams according to the following directions then write two scenarios for each diagram that would fit the motion.

   (a) Draw seven, evenly spaced dots going horizontally. Above the fourth dot, draw five vertical dots that are evenly spaced.

   (b) Draw a square. Make a diagonal line of dots starting at the upper left corner to the lower right corner. Make the dots closer together at the upper right and getting further apart as they progress to the lower right.

   (c) Draw a horizontal line of dots starting with wide spacings. The spacing becomes smaller, then, once again gets wider on the right end.

   (d) Start at the lower left with widely spaced dots. The dots start going upward to the right and get closer together. They then go horizontally and become evenly spaced.

4. **C** Sketch two frames of reference for the each of the following:

   (a) a ferry boat crossing a river.

   (b) a subway car moving through a station.

   (c) a roller coaster cart at Canada's Wonderland.

5. **C** Use single points (centre of mass) to sketch the motion in the following situations: (The distance between dots should represent equal time intervals.)

   (a) a person on a white water rafting trip jumps of a cliff.

   (b) a person hops across the length of a trampoline.

   (c) an Olympic diver jumps off a high dive board, hits the water and comes back to the surface.

6. **MC** Explain how the frequency of frames affects the quality of a Disney cartoon.

7. **I** A marble rolls down a 1.0 m ramp that is at an angle of 30° with a horizontal bench. The marble then rolls along the bench for 2.0 m. Finally, it rolls up a second 1.0 m ramp that is at an angle of 40° with the bench.

   (a) Draw a scale diagram of this situation and use dots to illustrate your predictions of the marble's motion. Use at least four position dots on each ramp.

   (b) Design and conduct a brief investigation to determine the accuracy of your predictions.

   (c) Describe your observations and explain any discrepancies with your predictions.

In the last section, you saw how diagrams allow you to describe motion qualitatively. It is not at all difficult to determine whether an object or person is at rest, speeding up, slowing down, or moving at a constant speed. Physicists, however, describe motion quantitatively by taking measurements.

From the diagrams you have analyzed, you can see that the two fundamental measurements involved in motion are distance and time. You can measure the distance from a reference point to the object in each frame. Since a known amount of time elapsed between each frame, you can determine the total time that passed, in relation to a reference time, when the object reached a certain location. From these fundamental data, you can calculate an object's position, speed, and rate of change of speed at any particular time during the motion.

## Vectors and Scalars

Most measurements that you use in everyday life are called scalar quantities, or **scalars**. These quantities have only a magnitude, or size. Mass, time, and energy are scalars. You can also describe motion in terms of scalar quantities. The distance an object travels and also the speed at which it travels are scalar quantities.

In physics, however, you will usually describe motion in terms of vector quantities, or **vectors**. In addition to magnitude, vectors have *direction*. Whereas distance and speed are scalars, the **position**, **displacement**, **velocity**, and **acceleration** of an object are vector quantities. Table 2.1 lists some examples of vector and scalar quantities. A vector quantity is represented by an arrow drawn in a frame of reference. The length of the arrow represents the magnitude of the quantity and the arrow points in the direction of the quantity within that reference frame.

**SECTION EXPECTATIONS**

- Differentiate between vector and scalar quantities.
- Describe and provide examples of how the position and displacement of an object are vector quantities.
- Analyze problems with variables of time, position, displacement, and velocity.

**KEY TERMS**

- scalar
- vector
- position
- displacement
- velocity
- acceleration
- time interval
- speed

**Table 2.1** Examples of Scalar and Vector Quantities

| Scalar quantities | | Vector quantities | |
|---|---|---|---|
| Quantity | Example | Quantity | Example |
| distance | 15 km | displacement | 15 km[N45°E] |
| speed | 30 m/s | velocity | 30 m/s [S] |
| | | acceleration | 9.81 m/s$^2$[down] |
| time interval | 10 s | | |
| mass | 6 kg | | |

**Note:** There is no scalar equivalent of acceleration.

## Position Vectors

A position vector locates an object within a frame of reference. You will notice in Figure 2.5 that an x-y coordinate system has been added to the diagram of the sprinting stick figures. The coordinate system allows you to designate the zero point for the variables under study and the direction in which the vectors are pointing. It establishes the *origin* from which the position of an object can be measured. The position arrow starts at the origin and ends at the location of the object at a particular instant in time. In this case, the sprinter is the object.

**Figure 2.5** Stick diagram with coordinate system and position vectors added

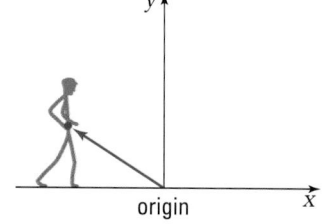

**Figure 2.6** As the sprinter walks toward the origin, the sprinter's position is negative in this coordinate system.

---

### POSITION VECTOR

A position vector, $\vec{d}$, points from the origin of a coordinate system to the location of an object at a particular instant in time.

---

As you can see in Figure 2.5, vectors locate the sprinter's position for two of the five different points in time. Time zero is selected as the instant at which the sprint started. However, as shown in Figure 2.6, you can show the sprinter several seconds before the race. Her position is to the left of the origin as she is walking up to the starting position. Thus, it is possible to have negative values for positions and times in a particular frame of reference.

## Displacement

Although you might think you know when an object is moving or has moved, you can be fooled! Pay close attention to the scientific definition of displacement and you will have a ready denial for the next time you are accused of lying around all day.

The **displacement** of an object, $\Delta\vec{d}$, is a vector that points from an initial position, $\vec{d}_1$, from which an object moves *to* a second

**Figure 2.7** Could this be an example of displacement?

position, $\vec{d_2}$, in a particular frame of reference. The vector's magnitude is equal to the straight-line distance between the two positions.

---

**DISPLACEMENT**

Displacement is the vector difference of the final position and the initial position of an object.

$$\Delta\vec{d} = \vec{d_2} - \vec{d_1}$$

| Quantity | Symbol | SI unit |
|---|---|---|
| displacement | $\Delta\vec{d}$ | m (metre) |
| final position | $\vec{d_2}$ | m (metre) |
| initial position | $\vec{d_1}$ | m (metre) |

---

Notice in the boxed definition that displacement depends only on the initial and final positions of the object or person. It is like taking snapshots of a person at various points during the day and not knowing or caring about anything that happened in between.

To see how the definition of displacement affects your perception of motion, follow a typical student, Freda, through a normal day. Figure 2.8 is a map of Townsville, where Freda lives. The map is framed by a coordinate system with its origin at Freda's home, position $\vec{d_0}$. In Table 2.2, you are given her position at six times during the day. What can you learn about her displacement from these data?

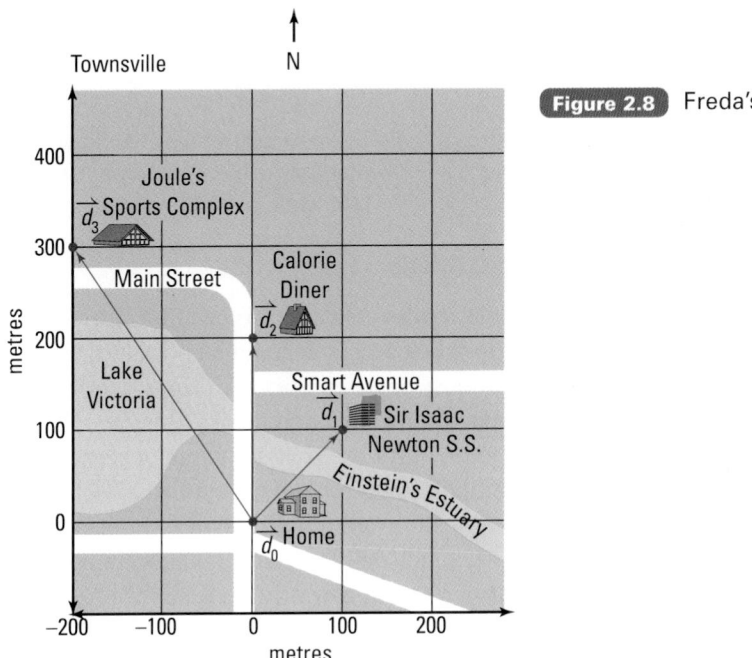

**Figure 2.8** Freda's town

**Table 2.2** Freda's Typical Daily Schedule

| Time | Location | Position | Activity |
|------|----------|----------|----------|
| 6:30 a.m. | home | $\vec{d_0}$ | sleeping |
| 9:00 a.m. | school | $\vec{d_1}$ | studying physics |
| 12:00 noon | diner | $\vec{d_2}$ | eating lunch |
| 2:00 p.m. | school | $\vec{d_1}$ | studying physics |
| 5:00 p.m. | sports complex | $\vec{d_3}$ | playing squash |
| 10:00 p.m. | home | $\vec{d_0}$ | sleeping |

**Figure 2.9** Freda's displacement from **(A)** home to school, **(B)** school to diner, and **(C)** school to sports complex

You can determine Freda's displacement for any pair of position vectors. To develop a qualitative understanding of displacement, consider the following examples.

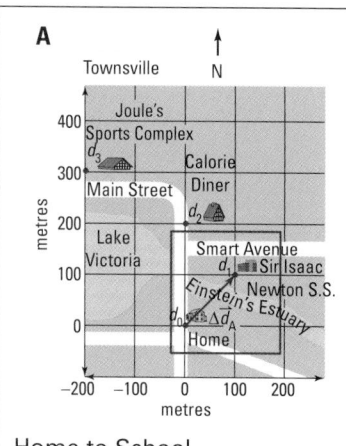

**Home to School**

6:30 a.m. to 9:00 a.m.

$\vec{d_0}$ to $\vec{d_1}$

$\Delta\vec{d_A} = \vec{d_1} - \vec{d_0}$

Since $\vec{d_0} = 0$, $\Delta\vec{d_A} = \vec{d_1}$

Scale measurement would show that $\Delta\vec{d_A} = 140$ m and points northeast.

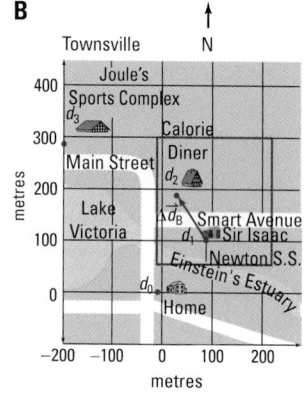

**School to Lunch**

9:00 a.m. to 12:00 noon

$\vec{d_1}$ to $\vec{d_2}$

$\Delta\vec{d_B} = \vec{d_2} - \vec{d_1}$

Scale measurement would show that $\Delta\vec{d_B} = 140$ m, but points northwest.

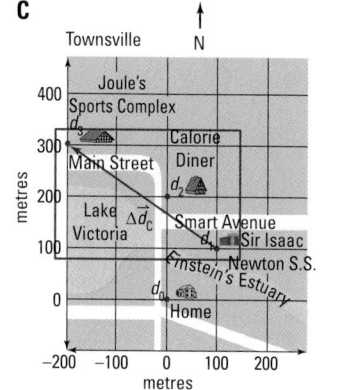

**School to Sports Complex**

9:00 a.m. to 5:00 p.m.

$\vec{d_1}$ to $\vec{d_3}$

$\Delta\vec{d_C} = \vec{d_3} - \vec{d_1}$

Scale measurement would show that $\Delta\vec{d_C} = 360$ m, and points a little west of northwest.

By now, you have probably discovered the important difference between measuring the *distance* a person travels and determining the person's *displacement* between two points in time. You know that Freda covered a much greater distance during the day than these displacements indicate. Suppose that someone observed Freda only at 6:30 a.m. and at 10:00 p.m. Her position at both of those times was the same — she was in bed. Despite the fact that she had a very energetic day, her displacement for this time interval is zero. Imagine what her reaction would be if she was accused of lying around all day.

- Use the scale map of Townsville in Figure 2.8 to estimate the minimum distance that Freda would walk while following her daily schedule. Compare this value to her displacement for the day.

- Determine Freda's displacement when she walks from the sports complex to her home.

- On a piece of graph paper, draw a scale map of your home and school area. Mark on it the major locations that you would visit on a typical school day. Frame the map with a coordinate system that places your home at the origin, the positive x-axis pointing east and the positive y-axis pointing north. Label your home position $\vec{d_0}$ and designate the other locations $\vec{d_1}$, $\vec{d_2}$, and so on. Determine your displacement and estimate the distance you travel

  **(a)** from home to school

  **(b)** from school to home

  **(c)** from school to a location that you visit after school

  **(d)** from a location that you visit after school to home

  **(e)** from the time you get out of bed to the time you get back into bed

- In the following situation, choose the correct answer and explain your choice. A basketball player runs down the court and shoots at the basket. After she arrives at the end of the court, her displacement is

  **(a)** either greater than or equal to the distance she travelled

  **(b)** always greater than the distance she travelled

  **(c)** always equal to the distance she travelled

  **(d)** either smaller than or equal to the distance she travelled

  **(e)** always smaller than the distance she travelled

  **(f)** either smaller or larger than, but not equal to, the distance she travelled

## Time and Time Intervals

The second fundamental measurement you will use to describe motion is *time*. In the example of Freda's schedule, you used clock time. However, in physics, clock time is very inconvenient, even if you use the 24 h clock. In physics, the time at which an event begins is usually designated as time zero. You might symbolize this as $t_0 = 0$ s. Other instants in time are measured in reference to $t_0$ and designated as $t_n$. The subscript "n" indicates the time at which a certain incident occurred during the event.

The elapsed time between two instants of time is called a **time interval**, $\Delta t$. Notice the difference between $t_n$ and $\Delta t$: $t_n$ is an instant of time and $\Delta t$ is the time that elapses between two incidents.

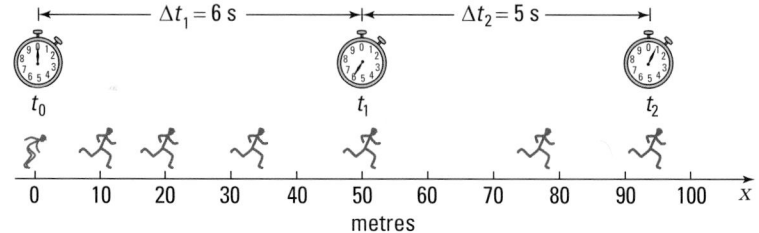

**Figure 2.10** A time interval is symbolized as $\Delta t$. The symbol $t$ with a subscript indicates an instant in time related to a specific event.

### Think It Through

- Write an equation to show the mathematical relationship between the time interval $\Delta t$ that elapsed while you were travelling to school this morning and the instants in time at which you left home and at which you arrived at school.

- Draw a sketch, similar to Figure 2.10, of a sprinter running a 100 m race and label it with the following information. (Remember, if you are not a good artist, you can use dots to show the sprinter at the specified positions.)

| Time (s) | Position (m) |
|----------|--------------|
| 0 | 0 |
| 3.6 | 10 |
| 5.7 | 25 |
| 10.0 | 50 |
| 12.8 | 80 |
| 14.0 | 100 |

(a) Determine the time interval that elapsed between the runner passing the following positions.
  - the beginning of the race and the 10 m point
  - the 10 m point and the 80 m point
  - the 80 m point and the 100 m point

(b) Compare the time interval taken for the first 50 m and the second 50 m of the sprint. Explain why they are different.

# Velocity

You have probably known the meaning of "speed" since you were very young. Speed is a scalar quantity that is simply defined as the distance travelled divided by the time spent travelling. For example, a car that travels 250 m in 10 s has an average speed of 25 m/s.

In physics, you will use the vector quantity *velocity* much more frequently than *speed*. Velocity not only describes how fast an object moves from one position to another, but also indicates the direction in which the object is moving. Physicists define velocity as the *rate of change of position.*

As you have discovered, when you determine the displacement (change of position) of an object (or person), you do not consider anything that has occurred between the initial and final positions. Consequently, you do not know whether the velocity has been changing during that time. Therefore, when you calculate velocity by dividing displacement by time, you are, in reality, finding the *average* velocity and ignoring any changes that might have occurred during the time interval.

---

### VELOCITY

Velocity is the quotient of displacement and the time interval.

$$\vec{v}_{ave} = \frac{\Delta \vec{d}}{\Delta t} \quad \text{or} \quad \vec{v}_{ave} = \frac{\vec{d_2} - \vec{d_1}}{t_2 - t_1}$$

| Quantity | Symbol | SI unit |
|---|---|---|
| average velocity | $\vec{v}_{ave}$ | $\dfrac{m}{s}$ (metres per second) |
| displacement | $\Delta \vec{d}$ | m (metres) |
| time interval | $\Delta t$ | s (seconds) |

---

## Think It Through

- Consider the speedometer of a car. Does it provide information about speed or velocity?

- A student runs around a 400 m oval track in 80 s. Would the average velocity and average speed be the same? Explain this result using both the definition of average velocity and a distinction between scalars and vectors.

- Consider the definition of average velocity. Describe the effect of reducing the time interval over which average velocity is calculated from a very large value such as several hours compared to a very short interval such as a fraction of a second.

## Calculating Average Velocity

1. A dragster in a race is timed at the 200.0 m and 400.0 m points.
   The times are shown on the stopwatches in the diagram.
   Calculate the average velocity for (a) the first 200.0 m,
   (b) the second 200.0 m, and (c) the entire race.

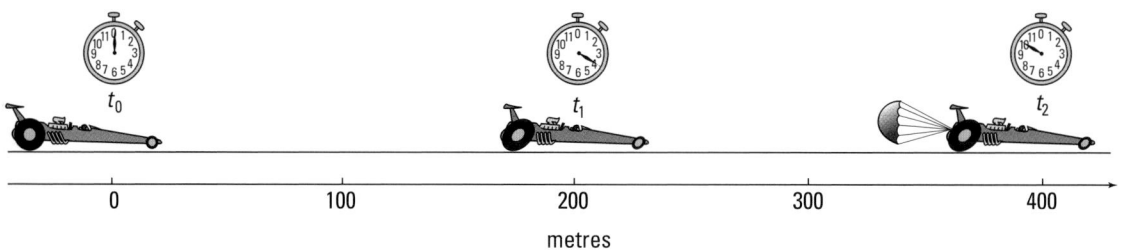

### Frame the Problem

- The dragster undergoes a *change in position*.

- The stopwatch shows a reading at *three instants in time*.

- Since you have data for only three instants, you can determine only the *average* velocity.

- The equation for average velocity applies to this problem.

### Identify the Goal

(a) The average velocity, $\vec{v}_{ave}$, for the displacement from 0.0 m to 200.0 m

(b) The average velocity, $\vec{v}_{ave}$, for the displacement from 200.0 m to 400.0 m

(c) The average velocity, $\vec{v}_{ave}$, for the displacement from 0.0 m to 400.0 m

### Variables and Constants

| Involved in the problem | | | Known | | Unknown | | |
|---|---|---|---|---|---|---|---|
| $\vec{d}_0$ | $t_0$ | $\vec{v}_{ave(0\to1)}$ | $\vec{d}_0 = 0.0$ m[E] | $t_0 = 0.0$ s | $\Delta\vec{d}_{0\to1}$ | $\Delta t_{0\to1}$ | $\vec{v}_{ave(0\to1)}$ |
| $\vec{d}_1$ | $t_1$ | $\vec{v}_{ave(1\to2)}$ | $\vec{d}_1 = 200.0$ m[E] | $t_1 = 4.3$ s | $\Delta\vec{d}_{1\to2}$ | $\Delta t_{1\to2}$ | $\vec{v}_{ave(1\to2)}$ |
| $\vec{d}_2$ | $t_2$ | $\vec{v}_{ave(0\to2)}$ | $\vec{d}_2 = 400.0$ m[E] | $t_2 = 11$ s | $\Delta\vec{d}_{0\to2}$ | $\Delta t_{0\to2}$ | $\vec{v}_{ave(0\to2)}$ |
| $\Delta\vec{d}_{0\to1}$ | $\Delta t_{0\to1}$ | | | | | | |
| $\Delta\vec{d}_{1\to2}$ | $\Delta t_{1\to2}$ | | | | | | |
| $\Delta\vec{d}_{0\to2}$ | $\Delta t_{0\to2}$ | | | | | | |

### Strategy

Find the displacement for the first 200.0 m, using the definition of displacement.

### Calculations

$$\Delta d_{0\to1} = \vec{d}_1 - \vec{d}_0$$

$$\Delta d_{0\to1} = 200.0 \text{ m[E]} - 0.0 \text{ m[E]}$$

$$\Delta d_{0\to1} = 200.0 \text{ m[E]}$$

Find the time interval for the first 200.0 m, using the definition of time interval.

$$\Delta t = t_1 - t_0$$
$$\Delta t = 4.3 \text{ s} - 0.0 \text{ s}$$
$$\Delta t = 4.3 \text{ s}$$

Find the average velocity for the first 200.0 m, using the definition of average velocity.

$$\vec{v}_{ave(0 \to 1)} = \frac{\Delta \vec{d}_{0 \to 1}}{\Delta t_{0 \to 1}}$$
$$\vec{v}_{ave(0 \to 1)} = \frac{200.0 \text{ m[E]}}{4.3 \text{ s}}$$
$$\vec{v}_{ave(0 \to 1)} = 46.51 \frac{\text{m}}{\text{s}} \text{[E]}$$

**(a)** The average velocity for the first 200.0 m was 47 m/s[E].

Find the displacement for the second 200.0 m, using the definition of displacement.

$$\Delta d_{1 \to 2} = \vec{d}_2 - \vec{d}_1$$
$$\Delta d_{1 \to 2} = 400.0 \text{ m[E]} - 200.0 \text{ m[E]}$$
$$\Delta d_{1 \to 2} = 200.0 \text{ m[E]}$$

Find the time interval for the second 200.0 m, using the definition of time interval.

$$\Delta t = t_2 - t_1$$
$$\Delta t = 11 \text{ s} - 4.3 \text{ s}$$
$$\Delta t = 6.7 \text{ s}$$

Find the average velocity for the second 200.0 m, using the definition of average velocity.

$$\vec{v}_{ave(1 \to 2)} = \frac{\Delta \vec{d}_{1 \to 2}}{\Delta t_{1 \to 2}}$$
$$\vec{v}_{ave(1 \to 2)} = \frac{200.0 \text{ m[E]}}{6.7 \text{ s}}$$
$$\vec{v}_{ave(1 \to 2)} = 29.85 \frac{\text{m}}{\text{s}} \text{[E]}$$

**(b)** The average velocity for the second 200.0 m was 30 m/s[E].

Find the displacement for the entire race.

$$\Delta d_{0 \to 2} = \vec{d}_2 - \vec{d}_0$$
$$\Delta d_{0 \to 2} = 400.0 \text{ m[E]} - 0.0 \text{ m[E]}$$
$$\Delta d_{0 \to 2} = 400.0 \text{ m[E]}$$

Find the time interval for the entire race.

$$\Delta t = t_2 - t_0$$
$$\Delta t = 11 \text{ s} - 0.0 \text{ s}$$
$$\Delta t = 11 \text{ s}$$

Find the average velocity for the entire race.

$$\vec{v}_{ave(0 \to 2)} = \frac{\Delta \vec{d}_{0 \to 2}}{\Delta t_{0 \to 2}}$$
$$\vec{v}_{ave(0 \to 2)} = \frac{400.0 \text{ m[E]}}{11 \text{ s}}$$
$$\vec{v}_{ave(0 \to 2)} = 36.36 \frac{\text{m}}{\text{s}} \text{[E]}$$

**(c)** The average velocity for the entire race was 36 m/s[E].

*continued* ▶

## Validate

Velocities with magnitudes between 30 m/s and 47 m/s are very large (108 km/h to 169 km/h), which you would expect for dragsters. As well, the units gave m/s, which is correct for velocity.

2. A basketball player gains the ball in the face-off at centre court. He then dribbles down to the opponents' basket and scores 6.0 s later. After scoring, he runs back to guard his own team's basket, taking 9.0 s to run down the court. Using centre court as his reference position, calculate his average velocity (a) while he is dribbling up to the opponents' net, and (b) while he is running down from the opponents' net to his own team's net. (A basketball court is $3.0 \times 10^1$ m long.)

## Frame the Problem

- The basketball player's *starting position* is at *centre court*.

- Two additional *positions* are identified, one up-court and one down-court from his starting position.

- Two *time intervals* are given in the description of his play.

- *Average velocity* is a calculation of his *displacement* for particular *time intervals*.

- Centre court is the *origin* of the coordinate system.

- Since *directions* are required in order to determine *velocities*, define the direction of the opponents' net as *positive* and the direction of the player's own net as *negative*.

## Identify the Goal

(a) The average velocity, $\vec{v}_{\text{ave}}$, for the first event

(b) The average velocity, $\vec{v}_{\text{ave}}$, for the second event

## Variables and Constants

| Involved in the problem | | | Known | Implied | Unknown |
|---|---|---|---|---|---|
| $\vec{d}_0$ | $t_0$ | $\vec{v}_{\text{ave}(0 \to 1)}$ | $\vec{d}_0 = 0.0$ m | $\vec{d}_1 = +15$ m | $\vec{v}_{\text{ave}(0 \to 1)}$ |
| $\vec{d}_1$ | $t_1$ | $\vec{v}_{\text{ave}(1 \to 2)}$ | $\Delta t_1 = 6.0$ s | $\vec{d}_2 = -15$ m | $\vec{v}_{\text{ave}(1 \to 2)}$ |
| $\vec{d}_2$ | $t_2$ | | $\Delta t_2 = 9.0$ s | | $\Delta \vec{d}_{0 \to 1}$ |
| $\Delta \vec{d}_{0 \to 1}$ | $\Delta t_{0 \to 1}$ | | | | $\Delta \vec{d}_{1 \to 2}$ |
| $\Delta \vec{d}_{1 \to 2}$ | $\Delta t_{1 \to 2}$ | | | | $t_0$ |
| | | | | | $t_1$ |
| | | | | | $t_2$ |

**Note:** Since the court is 30 m long and the position zero is defined as centre court, each basket must be half of 30 m, or 15 m, from position zero.

## Strategy

Find the displacement for the first event, using the definition of displacement.

The time interval is given, so calculate the average velocity of the first event by using the definition of average velocity.

## Calculations

$$\Delta \vec{d}_{0\to1} = \vec{d}_1 - \vec{d}_0$$
$$\Delta \vec{d}_{0\to1} = +15 \text{ m} - 0.0 \text{ m}$$
$$\Delta \vec{d}_{0\to1} = +15 \text{ m}$$

$$\vec{v}_{0\to1} = \frac{\Delta \vec{d}_{0\to1}}{\Delta t_{0\to1}}$$
$$\vec{v}_{0\to1} = \frac{+15 \text{ m}}{6.0 \text{ s}}$$
$$\vec{v}_{0\to1} = +2.5 \frac{\text{m}}{\text{s}}$$

(a) The average velocity for the first event was +2.5 m/s. The positive sign indicates that the direction of the player's velocity was toward the opponents' net.

Find the displacement for the second event by using the definition of displacement.

The time interval is given, so calculate the average velocity of the second event by using the definition of average velocity.

$$\Delta \vec{d}_{1\to2} = \vec{d}_2 - \vec{d}_1$$
$$\Delta \vec{d}_{1\to2} = -15 \text{ m} - (+15 \text{ m})$$
$$\Delta \vec{d}_{1\to2} = -30 \text{ m}$$

$$\vec{v}_{1\to2} = \frac{\Delta \vec{d}_{1\to2}}{\Delta t_{1\to2}}$$
$$\vec{v}_{1\to2} = \frac{-30 \text{ m}}{9.0 \text{ s}}$$
$$\vec{v}_{1\to2} = -3.33 \frac{\text{m}}{\text{s}}$$

(b) The average velocity for the second event was −3.3 m/s. The negative sign indicates that the direction of the player's velocity was toward the player's own net.

## Validate

The units in the answer were m/s, which is correct for velocity. The player's velocity was faster when he was going to guard his own net than when he was dribbling toward his opponents' net to make a shot. This is logical because, when planning a shot, a player would take more time. When guarding, it is critical to get to the net quickly.

### PRACTICE PROBLEMS

1. Calculate the basketball player's average velocity for the entire time period described in Model Problem 2.

2. Freda usually goes to the sports complex every night after school. The displacement for that walk is 360 m[N57°W]. What is her average velocity if the walk takes her 5.0 min?

*continued* ▶

3. Imagine that you are in the bleachers watching a swim meet in which your friend is competing in the freestyle event. At the instant the starting gun fires, the lights go out! When the lights come back on, the timer on the scoreboard reads 86 s. You observe that your friend is now about halfway along the length of the pool, swimming in a direction opposite to that in which he started. The pool is $5.0 \times 10^1$ m in length.

(a) Determine his average velocity during the time the lights were out.

(b) What are two possible distances that you might infer your friend swam while the lights were out?

(c) Given that the record for the 100 m freestyle race is approximately 50 s, which is the most likely distance that your friend swam while the lights were out? Explain your reasoning.

(d) Based on your conclusions in (c), calculate your friend's average speed while the lights were out.

## 2.2    Section Review

1. **K/U** List four scalar quantities and five vector quantities.

2. **K/U** Describe the similarities and difference between:

(a) time and time interval.

(b) position, displacement, and distance.

(c) speed and velocity.

3. **K/U** What is the displacement of Earth after a time interval of 365 $\frac{1}{4}$ days?

4. **MC** Create a scale diagram of your route to school. What is the displacement of your house from the school?

5. **C** Draw a displacement **and** a velocity scale diagram for the following:

(a) a farmer drives 3 km[N43°W] at 60 km/h.

(b) a swimmer crosses a still river heading [S56°W] at 3 m/s.

(c) an easterly wind blows a plastic bag at 6 km/h over a distance of 100 m.

6. **K/U** The following diagram represents a putting green at a 9 hole golf course.

(a) What is the displacement from the furthest hole to the ball?

(b) What is the displacement from the ball to furthest hole?

(c) What is the displacement from the closest hole to the ball?

(d) What is the displacement from the ball to the closest hole?

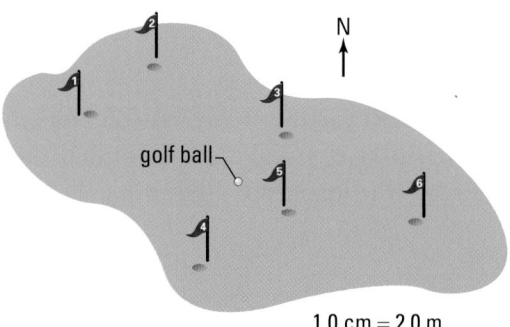

golf ball

N

1.0 cm = 2.0 m

### UNIT ISSUE PREP

Displacement, velocity, and the time interval are important features of any motion picture.

- How do film-makers manipulate these quantities to simulate motion?

- Have you ever been in a vehicle that was stopped but seemed like it was moving?

- What situations and effects can you create by lengthening or decreasing the time interval?

How is the motion of a tortoise similar to the motion of a jet aircraft? How does the motion of a jet cruising at 10 000 m differ from that of a space shuttle lifting off?

Physicists have classified different patterns of motion and developed sets of equations to describe these patterns. For example, **uniform motion** means that the velocity (or rate of change of position) of an object remains constant. A tortoise and a cruising jet might be travelling with extremely different velocities, yet they move for long periods without *changing* their velocities. Therefore, both tortoises and jet airplanes often travel with uniform motion.

**Non-uniform motion** means that the velocity *is* changing, either in magnitude or in direction. A cruising jet is travelling at a constant velocity, or with uniform motion, while a space shuttle changes velocity dramatically during lift-off. The shuttle travels with non-uniform motion.

In the last section, you studied motion by looking at "snapshots" in time. You had no way of knowing what was happening between the data points. The only way you could report velocity was as an *average* velocity. Clearly, you need more data points to infer that an object is moving with uniform motion, or at a constant velocity. Graphing your data points provides an excellent tool for analyzing patterns of motion and determining whether the motion is uniform or non-uniform.

## Constant Velocity

Is the motion of the skateboarder in Figure 2.11 uniform or non-uniform? To answer questions such as this, you should organize the data in a table (see Table 2.3) and then graph the data as shown in Figure 2.12. This plot is called a "position-time graph." When you plot the points for the skateboarder, you can immediately see that the plot is a straight line, with an upward slope.

**SECTION EXPECTATIONS**

- Describe motion with reference to the importance of a frame of reference.

- Draw diagrams to show how the position of an object changes over a number of time intervals in a particular frame of reference.

- Analyze position and time data to determine the speed of an object.

**KEY TERMS**

- uniform motion
- non-uniform motion
- instantaneous velocity
- tangent

**Figure 2.11** Is a skateboarder's motion uniform or non-uniform?

| Time (s) | Position (m[E]) |
|----------|-----------------|
| 0.0      | 0.0             |
| 1.3      | 12              |
| 2.6      | 24              |
| 3.9      | 36              |
| 5.2      | 48              |
| 6.5      | 60              |

**Table 2.3** Position versus Time for a Skateboarder

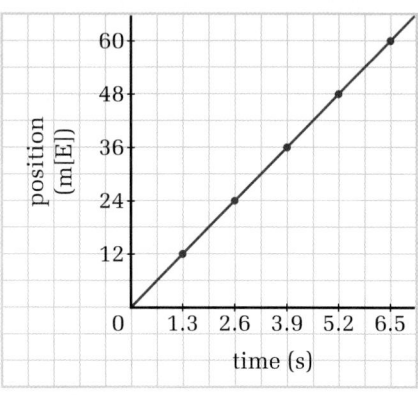

**Figure 2.12** Position-time graph for a skateboarder's motion

To determine the significance of the straight line, consider the meaning of the slope. Start with the mathematical definition of slope.

- Slope is the rise over the run.

$$\text{slope} = \frac{\text{rise}}{\text{run}}$$

- On a typical $x$-$y$ plot, the slope is written as

$$\text{slope} = \frac{\Delta y}{\Delta x}$$

- However, on a position-time plot, the slope is

$$\text{slope} = \frac{\Delta \vec{d}}{\Delta t}$$

- The definition of velocity is

$$\vec{v} = \frac{\Delta \vec{d}}{\Delta t}$$

- Since the slope and the velocity are equal to the same expression, the slope of the line on a position-time graph must be the velocity of the object.

$$\vec{v} = \text{slope}$$

Since you now know that the slope of a position-time graph is the velocity of the moving object, the straight line on the graph of the skateboarder's motion is very significant. The slope is the same everywhere on a straight line, so the skateboarder's velocity must be the same throughout the motion. Therefore, the skateboarder is moving with a constant velocity, or uniform motion. You could take any two points on the graph and calculate the velocity. For example, use the first and last points.

$$\text{slope} = \frac{\vec{\Delta d}}{\Delta t}$$

$$\text{slope} = \frac{60 \text{ m[E]}}{6.5 \text{ s}}$$

$$\vec{v} = 9.2 \frac{\text{m}}{\text{s}} \text{[E]}$$

$$\vec{v} = \text{slope}$$

$$\vec{v} = 9.2 \frac{\text{m}}{\text{s}} \text{[E]}$$

Mathematically, this is the same way that you calculated average velocity. What, then, is the difference between *average velocity* and *constant velocity*? Think back to the snapshot analogy and the case of Freda's typical school day. Over several hours, you had only two points to consider — the beginning and the end of the motion.

In the case of the skateboarder, you have several points between the beginning and end of the motion and they are all consistent, giving the same velocity. Nevertheless, you still cannot know exactly what happened between each measured point. Although it is reasonable to think that the motion was uniform throughout, you cannot be sure. Strictly speaking, you can calculate only an average velocity for each small interval. Without continuous data, you cannot be certain that an object's velocity is constant.

To emphasize this point, use your imagination to fill in what could be happening between your observation points for a master skateboarder and for a novice struggling to stay on the board. Examine the graphs in Figures 2.13 and 2.14. What is the average velocity for each time interval on each graph? Is the rate of change of position between time intervals constant on each graph? Based on the data, is there a difference between the average velocities of each skateboarder?

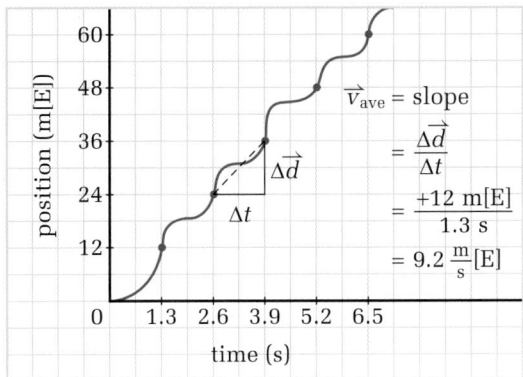

**Figure 2.13** The sharp curves in the graph indicate that the skate boarder's velocity was constantly changing. You would expect this jerky motion from a novice skateboarder.

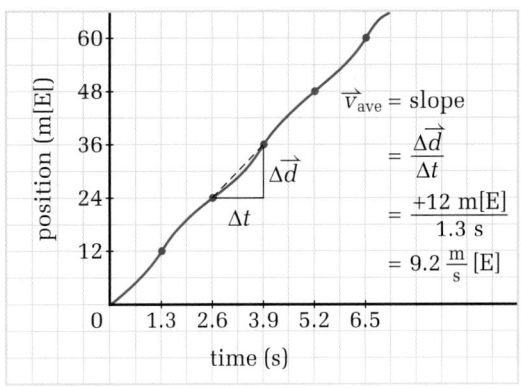

**Figure 2.14** Since the line is nearly straight the velocity is almost constant. This motion is what you would expect from a master skateboarder.

# Maintaining a Constant Pace

Generate a position-time graph using Method A or Method B.

## Method A: Motion Sensor

Use a motion sensor and computer interface to generate a position-time graph of your own motion as you walk toward the sensor, while trying to maintain a constant pace.

## Method B: Spark Timer

Pull a piece of recording tape, about 1 m long, through a spark timer, while trying to maintain a constant velocity. Examine the recording tape and locate a series of about 10 dots that seem to illustrate a period of constant velocity. On the recording tape, label the first dot in the series $\vec{d}_0$. Mark an arrow on your tape to show the direction of the motion. Make a data table to record the position and time of the 10 dots that immediately follow your labelled starting point, $\vec{d}_0$. Draw a position-time graph based on your data table.

## Analyze and Conclude

1. Explain, using the graph as evidence, whether you were successful in maintaining a constant velocity. If you were not successful for the entire timing, were you successful for at least parts of it?

2. Determine your average velocity for the entire timing period and your constant velocity for appropriate segments of your walk by calculating the relevant slopes of the graph.

## Apply and Extend

3. Explain for each of the following situations whether you can determine if the person or object is maintaining a constant velocity.

   (a) A student leaves home at 8:00 a.m. and arrives at school at 8:30 a.m.

   (b) A dog was observed running down the street. As he ran by the meat store, the butcher noted that it was 10:00 a.m. When he ran by the bakery, it was 10:03 a.m. A woman in the supermarket saw him at 10:05 a.m. Finally, he reached home at 10:10 a.m.

   (c) A swimmer competes in a 50 m back-stroke race. Three judges, each with a stopwatch, timed her swim. Their stop-watches read 28.65 s, 28.67 s, and 28.65 s.

4. A spark timer generated the recording tape, shown here, as a small cart rolled across a lab bench.

   (a) Set up a data chart to record the positions and times for the 8 dots after $\vec{d}_0$. The spark timer was set at a frequency of 60 Hz, thus making 60 dots per second.

   (b) Draw a position-time graph.

   (c) State whether the graph shows a constant velocity for the whole time period under observation or for only segments of it. Explain your reasoning.

   (d) Calculate the value of a segment of constant velocity.

   (e) Calculate the average velocity for the entire time period.

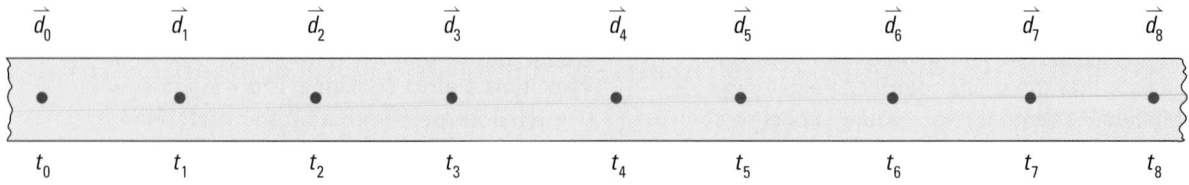

**Average Velocity**

**Constant Velocity**

**or**

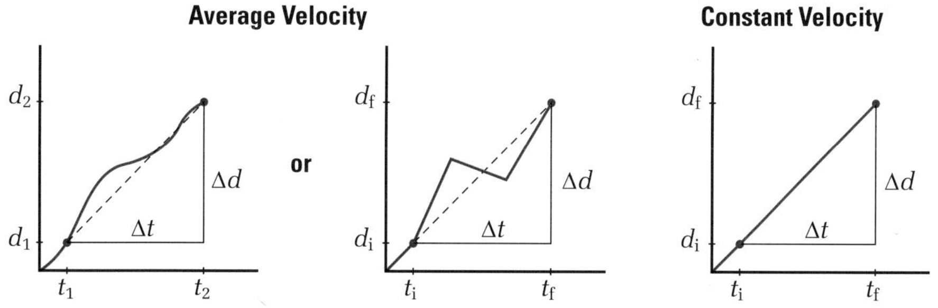

Notice that two different ways of writing subscripts are used for the position and time symbols. One graph uses $\vec{d}_1$, $\vec{d}_2$, $t_1$, and $t_2$, to designate consecutive positions and times. The other two graphs use $\vec{d}_i$, $\vec{d}_f$, $t_i$, and $t_f$, to designate initial and final positions and times. The use of subscripts to designate different positions and velocities varies in physics literature. The important point to remember is to use a system of subscripts that allows you to be clear about the meaning. For example, you might want to calculate the average velocity for several pairs of points, such as point 1 to point 2, and then from point 2 to point 3. Is point 2 the initial or final point? In the first case, point 2 is the final point, and in the second case, it is the initial point.

**Figure 2.15** The numerical value of the velocities represented in the graphs are all the same but the concepts are quite different.

## Average Velocity and Changing Directions

You have seen how a position-time graph helps you to determine whether motion is uniform or non-uniform. These graphs are even more helpful when doing a detailed analysis of non-uniform motion. Consider the situation illustrated in Figure 2.16. Adrienne drives her friend Jacques home from school. Jacques lives 800 m east of the school and Adrienne lives 675 m west. The diagram shows Adrienne's position at several specific times.

The data in Figure 2.16 are organized in Table 2.4 and graphed in Figure 2.17. Notice that Adrienne stops for 10 s to let Jacques out and then turns around and goes in the opposite direction.

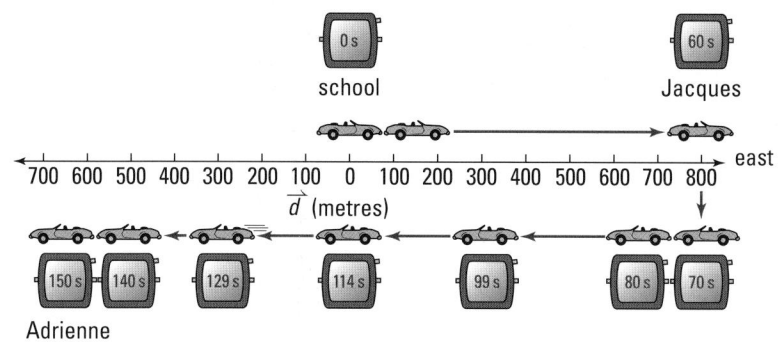

**Figure 2.16** Motion diagram of Adrienne's car trip

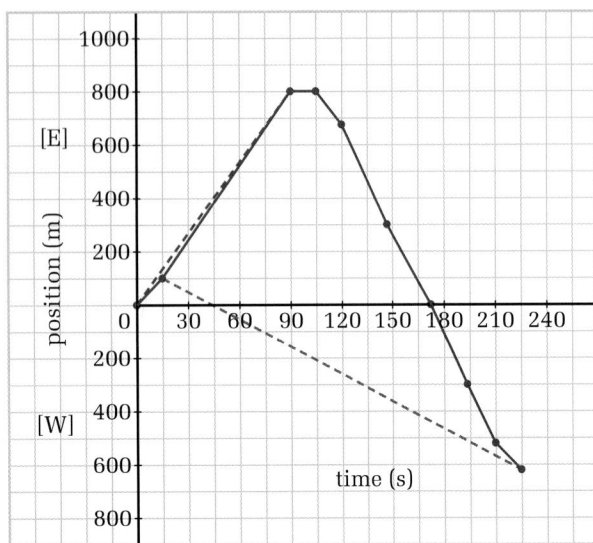

**Table 2.4** Data for Adrienne's Trip

| Time (s) | Position (m) |
|----------|--------------|
| 0.0 | 0.0 |
| 15 | 100 [E] |
| 90 | 800 [E] |
| 105 | 800 [E] |
| 120 | 675 [E] |
| 148 | 300 [E] |
| 171 | 0.0 |
| 194 | 300 [W] |
| 210 | 525 [W] |
| 225 | 625 [W] |

**Figure 2.17** Position-time graph of Adrienne's car trip

Clearly, the position-time graph of Adrienne's journey shows that her velocity is not constant for her entire trip home. You can see from the changing slopes on the graph that not only is she speeding up and slowing down, she is also changing direction.

**Note:** In the calculations to the right, you will see that the choice of time intervals for determining average velocity can lead to unreasonable results.

**Segment of Trip**

From school to Jacques' home | From the 15 s mark to Adrienne's home

**Initial and Final Times**

$t = 0$ s to $t = 90$ s
(See dashed line on graph)

$t = 15$ s to $t = 225$ s
(See dashed line on graph)

**Average Velocity**

$$\text{slope} = \frac{\text{rise}}{\text{run}}$$

$$\text{slope} = \frac{\Delta \vec{d}}{\Delta t}$$

$$\text{slope} = \frac{800 \text{ m[E]} - 0.0 \text{ m}}{90 \text{ s} - 0.0 \text{ s}}$$

$$\text{slope} = \frac{800 \text{ m[E]}}{90 \text{ s}}$$

$$\text{slope} = 8.88 \, \frac{\text{m}}{\text{s}} \text{[E]}$$

$$\vec{v} \approx 9 \, \frac{\text{m}}{\text{s}} \text{[E]}$$

$$\text{slope} = \frac{\text{rise}}{\text{run}}$$

$$\text{slope} = \frac{\Delta \vec{d}}{\Delta t}$$

$$\text{slope} = \frac{625 \text{ m[W]} - 100 \text{ m[E]}}{225 \text{ s} - 15 \text{ s}}$$

$$\text{slope} = \frac{(-625 \text{ m[E]}) - 100 \text{ m[E]}}{210 \text{ s}}$$

$$\text{slope} = \frac{-725 \text{ m[E]}}{210 \text{ s}}$$

$$\text{slope} = -3.45 \, \frac{\text{m}}{\text{s}} \text{[E]}$$

$$\vec{v} \approx 3 \, \frac{\text{m}}{\text{s}} \text{[W]}$$

Notice that, if you consider east to be positive, then west is equivalent to negative east.

You have probably concluded that the average velocity from the 15 s point to Adrienne's home does not seem reasonable. If you convert 3 m/s to km/h, the result is approximately 11 km/h. As well, the direction is west, but you know that Adrienne started the trip going east. The seemingly unreasonable average velocity is due to the *definition* of average velocity.

The diagram records the flight of a hawk soaring in the air looking for prey. Using videotape footage, the observer recorded the position of the hawk at 2.0 s intervals. He plotted the points and connected them with a smooth curve.

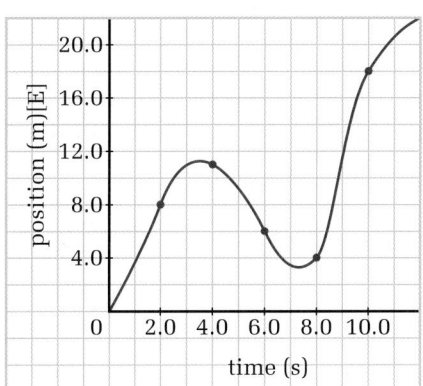

To estimate the hawk's velocity at precisely $t = 8.0$ s, determine its average velocity at several intervals that include the 8.0 s mark. Observe any changes in the calculated values for average velocity as the interval becomes smaller. Then, draw conclusions based on the following steps.

1. Reconstruct the graph on a piece of graph paper.

2. Draw straight lines on the curve connecting the following pairs of points.

   (a) 5.0 s and 11.0 s

   (b) 6.0 s and 10.0 s

   (c) 7.0 s and 9.0 s

   (d) 7.5 s and 8.5 s

3. Determine the hawk's average velocity for each of the pairs of points in step 2.

4. Draw one more straight line between points on opposite sides of the 8.0 s point and as close as possible to the 8.0 s mark. Extend the straight line as far as possible on the graph. Determine the slope of the straight line by choosing any two points on the line and calculating $\frac{\vec{d}_2 - \vec{d}_1}{t_2 - t_1}$, for those points.

**Analyze and Conclude**

1. Why were the calculated average velocities different for the different pairs of points?

2. What do you think is the meaning of the slope that you calculated in step 4?

3. Describe the relationship among the five velocities that you calculated.

4. Evaluate, in detail, the process you just performed. From your evaluation, propose a method for determining the velocity of an object at one specific time, rather than an average between two time points.

5. Using your method, determine the velocity of the hawk at exactly 3.0 s and 5.0 s.

- Describe circumstances in which the average velocity of a segment of a trip is very close to the reading you would see on the speedometer of a car.

- Describe circumstances in which the average velocity of a segment of a trip appears to totally contradict reason. Explain why.

## Instantaneous Velocity

When you apply the definition of average velocity to points on a graph that are far apart, sometimes the resulting value is extremely unreasonable as you discovered in the example of Adrienne's trip. When you bring the points on the graph closer and closer together, the calculated value of the velocity is nearly always very reasonable. In the Quick Lab, you brought the points so close together that they merged into one point. To perform a calculation of velocity, you had to draw a tangent line and use two points on that line. The value that you obtain in this way is called the **instantaneous velocity**. It might seem strange to define a velocity at one instant in time when velocity was originally defined as a *change* in position over a time *interval*. However, as you saw in the Quick Lab, you can make the time interval smaller and smaller, until the two points actually meet and become one point.

• **Think It Through**

- A jet-ski is able to maintain a constant speed as it turns a corner. Describe how the instantaneous speed of the jet-ski will differ from its instantaneous velocity during the turning process?

- The concept organizer on page 50 illustrates how to calculate the average velocity from a position versus time graph.
  1. Sketch a position time graph with a smooth curve having increasing slope. Select and mark two points on the curve. Draw in a dotted line to represent the average velocity between those two points.
  2. Now mark a point directly between the first two points. How would the average velocity for the very small time interval represented by the single dot look? Sketch it.

- Draw a concept organizer to show how position, displacement, average velocity, instantaneous velocity, and time are related.

The straight lines that you drew between points on the curve are called *chords* of the curve. When the straight line finally touches only one point on the curve, it becomes a **tangent** line. The magnitude of the velocity of an object at the point where the tangent line touches the graph is the slope of the tangent. You now have the tools to do a detailed analysis of position-time data.

**Math Link**

One, and only one, tangent line exists at any one point on a curve. Notice in the diagram that if the slope of the tangent line is changed, either increased or decreased, the line then cuts two points on the curve. It is no longer a tangent line but is now a *secant* line. A secant line intersects a curve at two points and continues beyond those points. How is the tangent line related to the trigonometric function named tangent?

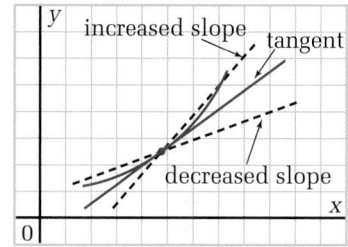

## INSTANTANEOUS VELOCITY

The *instantaneous velocity* of an object, at a specific point in time, is the *slope of the tangent* to the curve of the position-time graph of the object's motion at that specific time.

**Note:** Although average velocity is symbolized as $\vec{v}_{ave}$, a subscript is not typically used to indicate instantaneous velocity. When a subscript is present, it usually refers to the time or circumstances represented by that specific instantaneous velocity.

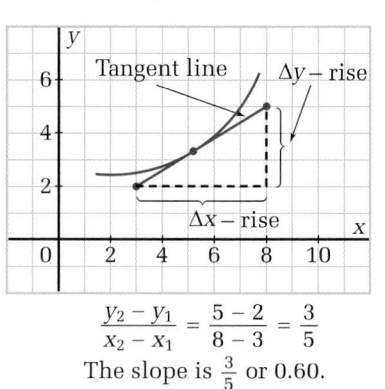

$$\frac{y_2 - y_1}{x_2 - x_1} = \frac{5 - 2}{8 - 3} = \frac{3}{5}$$

The slope is $\frac{3}{5}$ or 0.60.

## MODEL PROBLEM

### Determining Instantaneous Velocity

**The plot shown here is a position-time graph of someone riding a bicycle. Assume that position zero is the cyclist's home. Find the instantaneous velocity for at least nine points on the curve. Use the calculated values of velocity to draw a velocity-time graph. In your own words, describe the bicycle ride.**

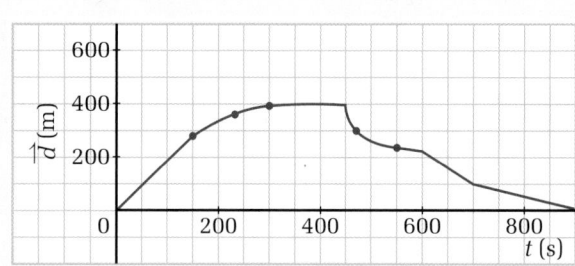

### Frame the Problem

- Between 0 s and 100 s, the graph is a straight line. Therefore, the velocity for that period of time is constant. Label the segment of the graph "A."

- Between 100 s and 350 s, the graph is curved, indicating that the velocity is changing. Label the segment of the graph "B."

- Between 350 s and 450 s, the graph is horizontal. There is no change in position, so the velocity is zero. Label the segment of the graph "C."

- At 450 s, the cyclist turns around and starts toward home. Up to 600 s, the graph is a curve, so the velocity is changing. Label the segment "D."

*continued* ▶

Describing Motion • MHR **55**

*continued from previous page*

- Between 600 s and 700 s and again between 700 s and 900 s, the graph forms straight lines. The velocity is constant during each period. Label those sections "E" and "F."

- At 900 s, the cyclist is back home.

- The motion is in one dimension so denote direction by positive and negative values.

## Identify the Goal

Find the value of the instantaneous velocity at nine points in time.

Draw a velocity-time graph.

## Variables and Constants

| Involved in the problem | Known | Unknown |
|---|---|---|
| $\vec{d}_n$    $\Delta \vec{d}_n$ | $\vec{d}_n$ | $\vec{v}_n$ |
| $t_n$ | $t_n$ | |
| $\vec{v}_n$ | | |

For all points (n) from $t = 0$ s to $t = 900$ s

For all points (n) from $t = 0$ s to $t = 900$ s

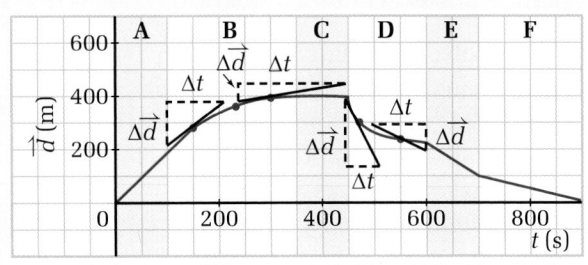

## Strategy

Redraw the graph with enough space below it to draw the velocity-time graph on the same time scale.

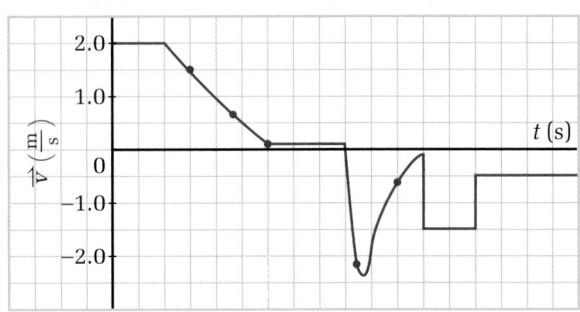

**Time period or point**

Identify the linear segments of the position-time graph. Select at least five points on the non-linear segments. Draw lines that are tangent to the graph at these points.

$t_0$ to $t_{100}$

Calculate the velocity for each linear segment of the graph and the points at which you have drawn tangent lines. Record the velocities in a table.

$t_{150}$

$t_{225}$

Plot the points on the velocity-time graph. Connect the points with a smooth curve where the points do not form a straight line.

$t_{300}$

$t_{350}$ to $t_{450}$

$t_{475}$

$$\frac{\vec{d}_2 - \vec{d}_1}{t_2 - t_1} = \vec{v}$$

$$\frac{200 \text{ m} - 0 \text{ m}}{100 \text{ s} - 0 \text{ s}} = 2.0 \ \frac{\text{m}}{\text{s}}$$

$$\frac{350 \text{ m} - 200 \text{ m}}{200 \text{ s} - 100 \text{ s}} = 1.5 \ \frac{\text{m}}{\text{s}}$$

$$(\text{not shown}) = 0.80 \ \frac{\text{m}}{\text{s}}$$

$$\frac{450 \text{ m} - 360 \text{ m}}{450 \text{ s} - 250 \text{ s}} = 0.45 \ \frac{\text{m}}{\text{s}}$$

slope is zero, $\vec{v} = 0.0 \ \dfrac{\text{m}}{\text{s}}$

$$\frac{210 \text{ m} - 400 \text{ m}}{530 \text{ s} - 450 \text{ s}} = -2.4 \ \frac{\text{m}}{\text{s}}$$

| $t_{550}$ | $\dfrac{180 \text{ m} - 300 \text{ m}}{700 \text{ s} - 490 \text{ s}} = -0.57 \dfrac{\text{m}}{\text{s}}$ |
|---|---|
| $t_{600}$ to $t_{700}$ | $\dfrac{100 \text{ m} - 250 \text{ m}}{700 \text{ s} - 600 \text{ s}} = -1.5 \dfrac{\text{m}}{\text{s}}$ |
| $t_{700}$ to $t_{900}$ | $\dfrac{0.0 \text{ m} - 100 \text{ m}}{900 \text{ s} - 700 \text{ s}} = -0.50 \dfrac{\text{m}}{\text{s}}$ |

**A:** The cyclist is riding at a constant velocity, away from home.

**B:** The cyclist slows down and, at the end of the segment, stops.

**C:** The cyclist is not moving.

**D:** The cyclist starts toward home at a high velocity, then slows. The position vector is positive, because the cyclist is at a positive position in relation to home. However, the velocity is negative because the cyclist is moving in a negative direction, toward home.

**E:** The cyclist is still heading toward home, but at a constant velocity.

**F:** The cyclist slows even more but is still at a positive position and a negative velocity.

## Validate

The slopes of the curve in A and B are positive (line and tangents go up to the right); therefore, the velocities should all be positive. They are.

The slope in C is zero so the velocity should be zero. It is.

The slopes of the curve in D, E, and F are all negative (lines and tangents go down to the right); therefore, the velocities should be negative. They are.

### PRACTICE PROBLEMS

4. Redraw the position-time graph shown here. Determine the velocity in each of the linear segments and for at least three points along the curved section. Use the calculated velocities to draw a velocity-time graph of the motion. Explain the circumstances that make the position vector negative and the velocity vector positive.

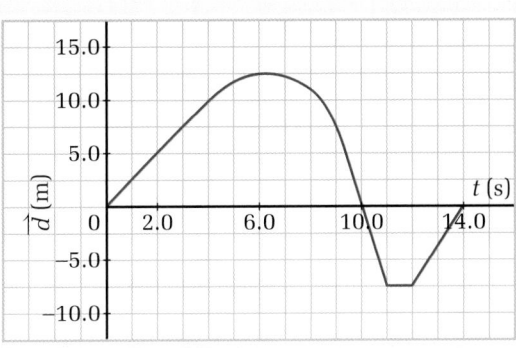

*continued from previous page*

5. Using the data table, draw a position-time graph. For points that do not lie on a straight line, connect the points with a smooth curve. Calculate the velocity for a sufficient number of points so that you can draw a good velocity-time graph.

| Time (s) | Position (m) | Time (s) | Position (m) |
|---|---|---|---|
| 0.0 | 0.0 | 16.0 | 0.0 |
| 2.0 | −10.0 | 17.0 | 10.0 |
| 4.0 | −20.0 | 18.0 | 20.0 |
| 6.0 | −30.0 | 20.0 | 25.0 |
| 8.0 | −36.0 | 22.0 | 30.0 |
| 10.0 | −38.0 | 24.0 | 26.6 |
| 12.0 | −32.0 | 26.0 | 23.3 |
| 13.0 | −27.0 | 28.0 | 20.0 |
| 14.0 | −10.0 | 30.0 | 0.0 |

## Concept Organizer

What type of velocity is the police officer measuring with the radar gun? A radar gun takes data points that are so close together, that it measures instantaneous velocity. If the car is moving with a constant velocity, the instantaneous velocity is the same as the constant velocity. If the car's velocity is changing, the radar gun will not measure average velocity. How would you measure a car's average velocity? To understand and report data correctly, you need to know how a measuring instrument works as well as knowing the precise meaning of specific terms.

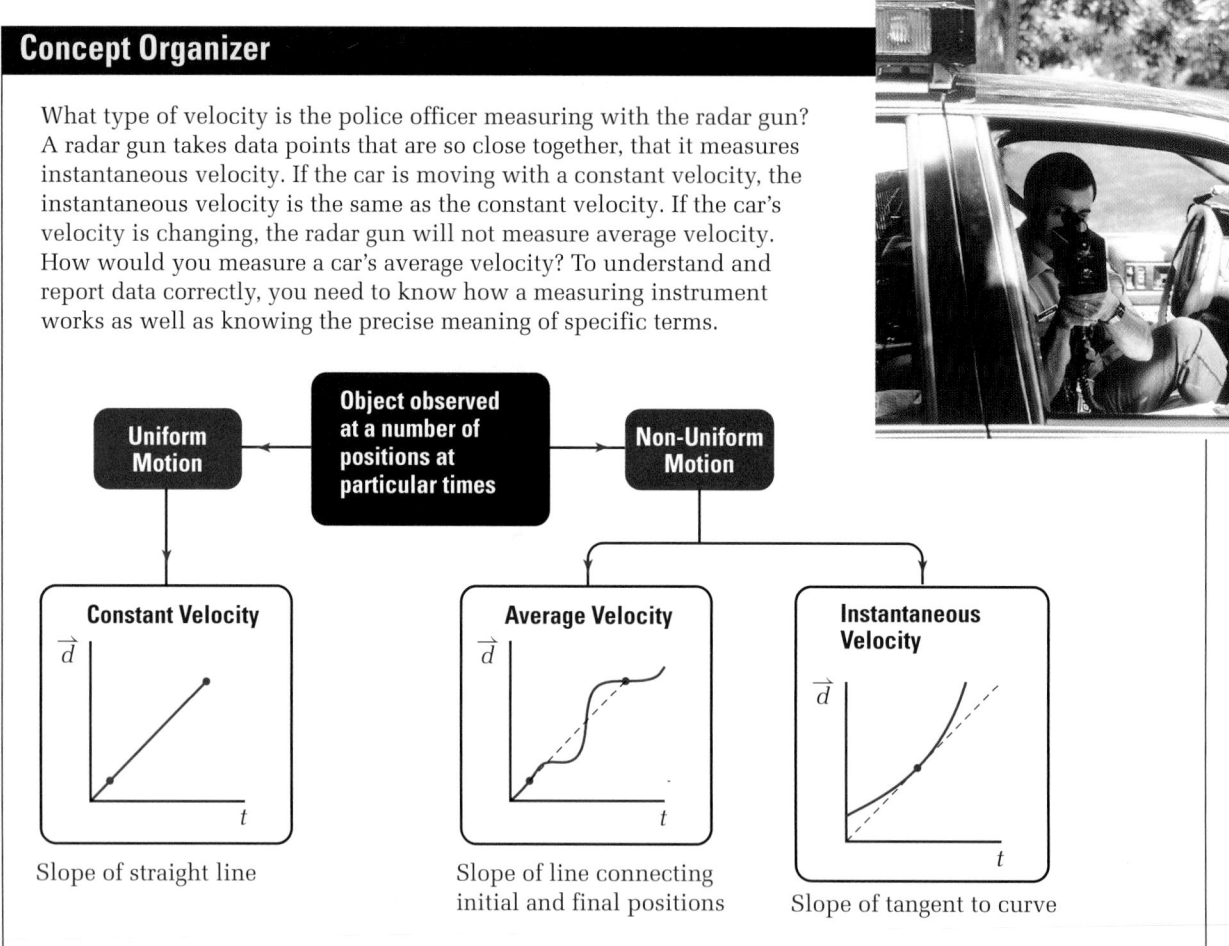

**Figure 2.18** This concept organizer will help you understand and remember three ways in which you can describe velocity.

# Rocket Motion

Imagine that you fire a toy rocket straight up into the air. Its engine burns for 8.0 s before it runs out of fuel. The rocket continues to climb for 4.0 s, then stops and begins to fall back down. After falling freely for 4.0 s, a parachute opens and slows the descent. The rocket then reaches a terminal velocity of 6.0 m/s[down]. Using videotape footage, you determined the rocket's altitude, $h$, at 2 s intervals. Your data are shown in the table.

| Phase | Time (seconds) | Position (metres[up]) |
|---|---|---|
| 1 engine on | 0 | 0 |
| | 2 | 10 |
| | 4 | 40 |
| | 6 | 90 |
| | 8 | 160 |
| 2 engine off (rising) | 10 | 220 |
| | 12 | 240 |
| 3 engine off (falling) | 14 | 220 |
| | 16 | 160 |
| 4 parachute opens | 18 | 92 |
| | 20 | 48 |
| | 22 | 28 |
| 5 terminal velocity | 24 | 20 |
| | 26 | 12 |
| | 28 | 4 |
| | 30 | 0 |

Make a motion analysis table like the one shown here, but add three more columns and label them: Time interval, Displacement, and Average velocity. Perform the indicated calculations for all intervals between the points listed, then complete the table.

Plot an average velocity-time graph. Remember, when you plot *average* velocity, you plot the point that lies at the midpoint of the time interval. For example, when you plot the average velocity for the interval from 2 s to 4 s, you plot the point at 3 s. Draw a smooth curve through the points.

Determine the instantaneous velocity at $t = 7$ s, $t = 13$ s, $t = 21$ s, and $t = 25$ s.

## Analyze and Conclude

1. In your own words, describe the motion of the rocket during each of the five stages.

2. What is the condition of the position-time graph when the velocity-time graph passes through zero? Explain the meaning of this point.

3. Under what specific conditions is the velocity-time graph a straight, horizontal line?

4. Compare the instantaneous velocities that you calculated for times 7 s, 13 s, 21 s, and 25 s, with the average velocities for the intervals that included those times. In which cases are the instantaneous and average velocities nearly the same? Quite different? Explain why.

5. Explain why it is reasonable to draw the line connecting the points on the position-time graph as a smooth curve rather than connecting the dots with a straight line.

1. **C** Explain why the following situations do not represent uniform motion.

   (a) driving through downtown at rush hour

   (b) start and stop sport drills that are executed at top speed

   (c) pendulum swinging with a constant frequency

   (d) a ball rolling down a ramp

   (e) standing on a merry go round that rotates at a constant number of revolutions per minute

2. **I** Analyze the following position time graph and sketch the velocity time graph of the same data.

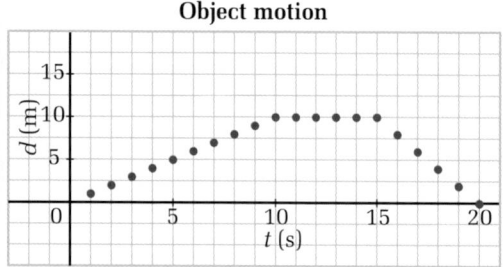

Object motion

3. **C** Explain the relationship between:

   (a) slope and position time graphs.

   (b) slope and velocity time graphs.

   (c) average velocity, constant velocity, and instantaneous velocity.

   (d) tangent line on a position time graph, and time interval.

   (e) negative time and a position time graph.

   (f) velocity, acceleration, and terminal velocity.

   (g) m/s and km/h.

4. **MC** Both physicists and mathematicians use observations from the physical world to create theories. Suggest criteria that separates physicists from mathematicians.

5. **MC** A boy and a girl are going to race twice around a track. The girl's strategy is to run each lap with the same speed. A boy is going to run the first lap slower so he can run the final lap faster. Suppose that the girl's speed is $x$ m/s for both laps and the boy's speed for the first lap is $(x - 2)$ m/s, and for the second lap is $(x + 2)$ m/s. Decide who will win and justify your response. If you need help, try using real numbers for the speeds.

6. **K/U** By what factor does velocity change if the time interval is increased by a factor of 3 and the displacement is decreased by a factor of two?

Many times in the last section, you read that an object's velocity was increasing, decreasing, or that it was changing direction. Once again, physicists have a precise way of stating the changes in velocity. **Acceleration** is a vector quantity that describes the rate of change of velocity.

---

### ACCELERATION

Acceleration is the quotient of the change in velocity and the time interval over which the change takes place.

$$\vec{a} = \frac{\Delta \vec{v}}{\Delta t}$$

| Quantity | Symbol | SI unit |
|---|---|---|
| acceleration | $\vec{a}$ | $\frac{\text{m}}{\text{s}^2}$ (metres per second squared) |
| change in velocity | $\Delta \vec{v}$ | $\frac{\text{m}}{\text{s}}$ (metres per second) |
| time interval | $\Delta t$ | s (seconds) |

**Unit Analysis**

$$\frac{\frac{\text{metres}}{\text{second}}}{\text{second}} = \frac{\frac{\text{m}}{\text{s}}}{\text{s}} = \frac{\text{m}}{\text{s}^2}$$

---

The units of acceleration — metres per second squared — do not have an obvious meaning. If you think about the basic definition of acceleration, however, the meaning becomes clear. The velocity of an object changes by a certain number of metres per second *every second*. For example, analyze the statement, "A truck is travelling at a constant velocity of 20 m/s[E], then accelerates at 1.5 m/s²[E]." This acceleration means that the truck's velocity increases by 1.5 m/s[E] every second. One second after it starts accelerating, it will be travelling at 21.5 m/s[E]. One second later, it will be travelling at 23 m/s[E]. The truck's velocity increases by 1.5 m/s[E] *every second*, as long as it is accelerating.

## Direction of Acceleration Vectors

The direction of the acceleration vector is the direction of the *change* in the velocity and not the direction of the velocity itself. To determine the direction of the acceleration vector, it is helpful to visualize the direction in which you would have to push on an object to *cause* a particular change in velocity.

**SECTION EXPECTATIONS**

- Define and describe the concept of acceleration.
- Design and conduct an experiment to determine variables that effect the acceleration due to gravity.
- Relate the direction of acceleration to the direction of a change in velocity.
- Interpret patterns and trends in motion data by hand or computer drawn graphs.

**KEY TERMS**

- acceleration
- constant (uniform) acceleration
- non-uniform acceleration
- average acceleration
- instantaneous acceleration

**ELECTRONIC LEARNING PARTNER**

To enhance your understanding of the language of acceleration go to the Electronic Learning Partner for an interactive activity.

Figure 2.19 shows the motion of a van that starts from rest, speeds up, travels at a constant velocity, slows down, and then stops. The frame of reference shows the origin at the left, with the x-axis pointing in a positive direction to the right. When the van is speeding up, the average velocity vectors and the average acceleration vector point in the same direction (+). When the van is travelling at a constant speed, the average acceleration is zero. When the van is slowing down, the average velocity vectors (+) and the average acceleration vector (−) are in opposite directions.

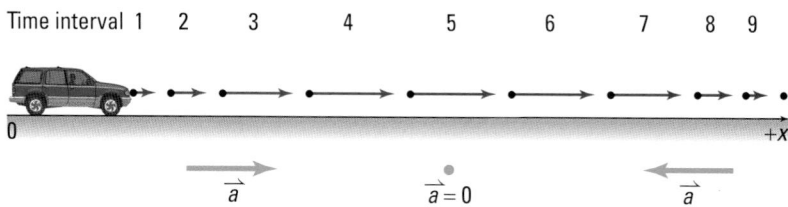

**Figure 2.19** When the van is moving in a positive direction but slowing down, the direction of the acceleration is negative.

Consider the directions that the average velocity and average acceleration vectors point if the van turns around and travels back to its starting point. As shown in Figure 2.20, when the van is speeding up in a negative direction, both the average velocity vectors and the acceleration vector point in the negative direction. While the van travels at constant velocity, the average velocity vectors are negative and the acceleration vector is zero. As the van slows down to stop, the average velocity vectors are pointing in the negative direction and the average acceleration vector is pointing in the positive direction.

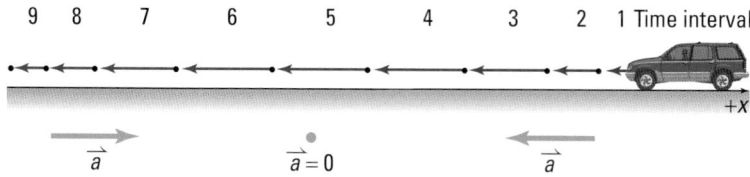

**Figure 2.20** When the van is moving in a negative direction and slowing down, the direction of acceleration is positive.

An object can accelerate without either speeding up or slowing down. If the magnitude of the velocity does not change but the direction does change, the object is accelerating. To visualize the direction of the acceleration vector in such cases, study Figure 2.21. Imagine the direction that you would have to push on the tip of the initial velocity vector to make it overlap with the final velocity vector. The direction of the acceleration vector is from the tip of the initial velocity vector toward the tip of the final velocity vector.

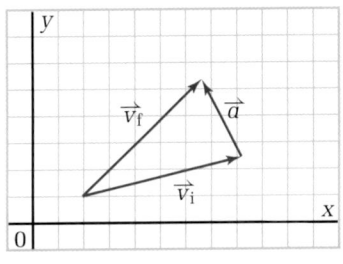

**Figure 2.21** Envision pushing on the tip of $\vec{v}_1$, until it overlaps with $\vec{v}_2$.

## • *Think It Through*

- The following charts refer to the van's journeys in Figures 2.19 and 2.20. Redraw the charts below and, using as examples the two rows that have been completed, fill in the remaining rows.

| Images in figure | Direction of velocity vector | Direction of acceleration vector | Description of motion |
|---|---|---|---|
| **Figure 2.19 Van is moving in the positive direction.** | | | |
| 1-2-3 | positive | positive | speeding up in positive direction |
| 4-5-6 | | | |
| 7-8-9 | | | |
| **Figure 2.20 Van is moving in the negative direction** | | | |
| 1-2-3 | | | |
| 4-5-6 | | | |
| 7-8-9 | negative | positive | slowing down in negative direction |

- Sketch each of the combinations of initial and final velocity vectors, and add to your sketch another vector showing the direction of the acceleration vector.

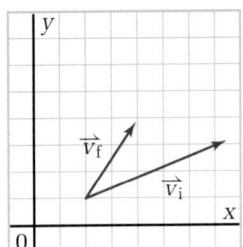 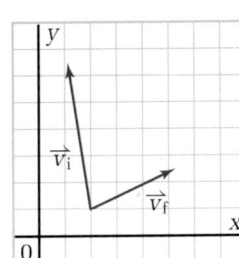 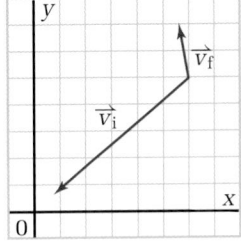

Physicists often use the term "uniform motion" to apply to motion with a constant velocity, and "uniformly accelerated motion" to apply to motion with a constant acceleration.

# Uniform and Non-Uniform Acceleration

Have you noticed the similarity in the mathematical expressions for velocity and acceleration?

$$\vec{v} = \frac{\Delta \vec{d}}{\Delta t} = \frac{\vec{d}_2 - \vec{d}_1}{t_2 - t_1} \qquad \vec{a} = \frac{\Delta \vec{v}}{\Delta t} = \frac{\vec{v}_2 - \vec{v}_1}{t_2 - t_1}$$

The mathematical operations you performed on position vectors to find velocity are nearly the same as those you will perform on velocity vectors to find acceleration. The similarity applies to both equations and graphs. For example, the slope of a velocity-time graph is the acceleration. If the velocity graph is curved, the slope of the tangent to the velocity-time graph at a specific time is the acceleration of the object at that time. The terms applied to velocity also apply to acceleration. **Constant** or **uniform acceleration** means that the acceleration does not change throughout specified time intervals. As well, **non-uniform acceleration** means that the acceleration is changing with time.

The terms, "average," "constant," and "instantaneous" apply to acceleration in the very same way that they apply to velocity. **Average acceleration** is an acceleration calculated from initial and final velocities and the time interval. Constant acceleration means that the acceleration is not changing over a certain interval of time. The velocity-time graph for the time interval is a straight line. **Instantaneous acceleration** is the acceleration found at one moment in time, and is equal to the slope of the tangent to velocity-time graph at that point in time.

To see the connections among time, position, velocity, and acceleration of a moving object, consider the example of a ball that is thrown straight up in the air with an initial velocity of 20.0 m/s. The position-time data are listed in Table 2.5 and the graphs of position, velocity, and acceleration are shown in Figure 2.22.

The data for the velocity-time graph were determined by the slope of the position-time graph. (Only four tangent lines are shown.) The velocity-time graph of data taken from the slopes of the position-time graph is a straight line with a negative slope that is the same everywhere. Since the slope of the velocity-time graph is the acceleration, the acceleration has the same negative value throughout the motion. (The value is −9.81 m/s².)

**Table 2.5** Position-Time Data

| $t$ (s) | $\vec{d}$ (m) |
|---------|---------------|
| 0.0 | 0.0 |
| 0.5 | 8.8 |
| 1.0 | 15.1 |
| 1.5 | 19.0 |
| 2.0 | 20.4 |
| 2.5 | 19.4 |
| 3.0 | 15.9 |
| 3.5 | 10.0 |
| 4.0 | 1.6 |

**Note:** Since the motion is in one dimension, direction is indicated by plus (+) or minus (−).

**Figure 2.22** Position, velocity, and acceleration graphs for Table 2.6

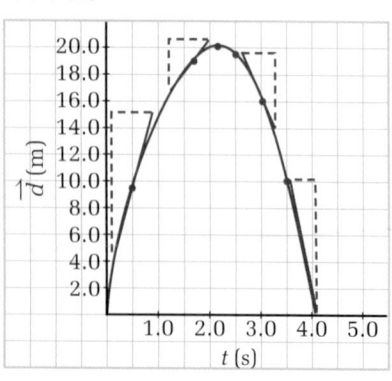
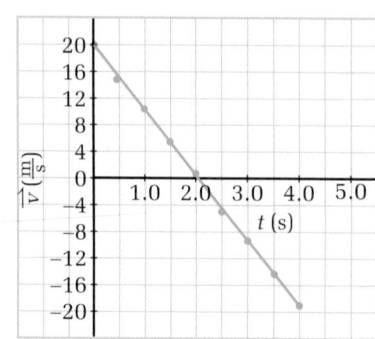
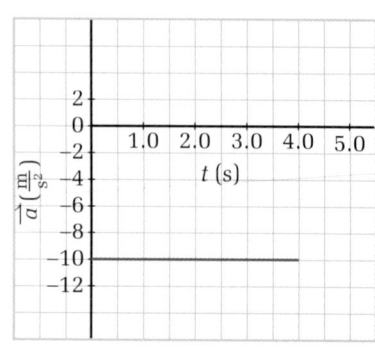

# Graphing the Motion of a Van

You qualitatively analyzed the motion of a van earlier. Now, using the example of the ball thrown into the air, you can do a more detailed analysis of the van's motion. The table shown here includes the time and position data, with one worked example for finding acceleration.

| Time $t$ (s) | Position $\vec{d}$ (m) | Velocity $\frac{\Delta \vec{d}}{\Delta t}$ (m/s) | Acceleration $\frac{\Delta \vec{v}}{\Delta t}$ (m/s$^2$) |
|---|---|---|---|
| 0.0 | 0.0 | | |
| | | 6.0 | |
| 2.0 | 12 | | +3.0 |
| | | 12 | |
| 4.0 | 36 | | |
| 6.0 | 48 | | |
| 8.0 | 96 | | |
| 10.0 | 142 | | |
| 12.0 | 190 | | |
| 14.0 | 226 | | |
| 16.0 | 250 | | |
| 18.0 | 262 | | |

## Sample Calculation

Notice that the velocity that will be plotted at $t = 1.0$ s is the average velocity between $t = 0.0$ s and $t = 2.0$ s. The velocity that will be plotted at $t = 3.0$ s is the average velocity between $t = 2.0$ s and 4.0 s. The acceleration that will be plotted at t = 2.0 s is the average acceleration between $t = 1.0$ s and $t = 3.0$ s.

$$\vec{v}_1 = \frac{\Delta \vec{d}_{0 \to 2}}{\Delta t_{0 \to 2}} = \frac{\vec{d}_2 - \vec{d}_0}{t_2 - t_0} = \frac{12 \text{ m} - 0.0 \text{ m}}{2.0 \text{ s} - 0.0 \text{ s}}$$

$$= \frac{12 \text{ m}}{2.0 \text{ s}}$$

$$= 6.0 \frac{\text{m}}{\text{s}}$$

$$\vec{v}_3 = \frac{\Delta \vec{d}_{2 \to 4}}{\Delta t_{2 \to 4}} = \frac{\vec{d}_4 - \vec{d}_2}{t_4 - t_2} = \frac{36 \text{ m} - 12 \text{ m}}{4.0 \text{ s} - 2.0 \text{ s}}$$

$$= \frac{24 \text{ m}}{2.0 \text{ s}}$$

$$= 12 \frac{\text{m}}{\text{s}}$$

$$\vec{a}_2 = \frac{\Delta \vec{v}_{1 \to 3}}{\Delta t_{1 \to 3}} = \frac{\vec{v}_3 - \vec{v}_1}{t_3 - t_1} = \frac{12 \frac{\text{m}}{\text{s}} - 6.0 \frac{\text{m}}{\text{s}}}{3.0 \text{ s} - 1.0 \text{ s}}$$

$$= \frac{6.0 \frac{\text{m}}{\text{s}}}{2.0 \text{ s}}$$

$$= 3.0 \frac{\text{m}}{\text{s}^2}$$

Complete the table for all average velocities and average accelerations. Then plot position-time, velocity-time, and acceleration-time graphs. On the position-time graph, select one point between 0 and 4 s and one point between 6 and 10 s. Draw tangents to the curve and determine their slopes.

## Analyze and Conclude

1. How well do the average and instantaneous velocities that you calculated agree with each other?

2. Separate the graphs into three sections: (a) 0 s to 8s, (b) 8 s to 12 s, and (c) 12 s to 20 s. For each of these three time periods, compare all three graphs in the following ways.

   (a) How do the shapes of the graphs (curved, straight, horizontal) relate to each other?

   (b) How do the signs of the values (positive, zero, or negative) relate to each other?

3. Under what circumstances can the van be moving but have a zero acceleration?

4. Under what circumstances is the sign of the velocity the same as the sign of the acceleration?

5. What general statement can you make about the motion of the van when the direction of the acceleration vector is opposite to the direction of the velocity vector?

- How do intervals of constant acceleration appear on an acceleration-time graph?

- How do intervals of constant acceleration appear on a velocity-time graph?

- What does a straight-line slope indicate on an acceleration-time graph?

- What would a curved line indicate on an acceleration-time graph?

- Explain circumstances in which an object would be accelerating but have an instantaneous velocity of zero?

- How does uniform acceleration differ from uniform motion?

## Concept Organizer

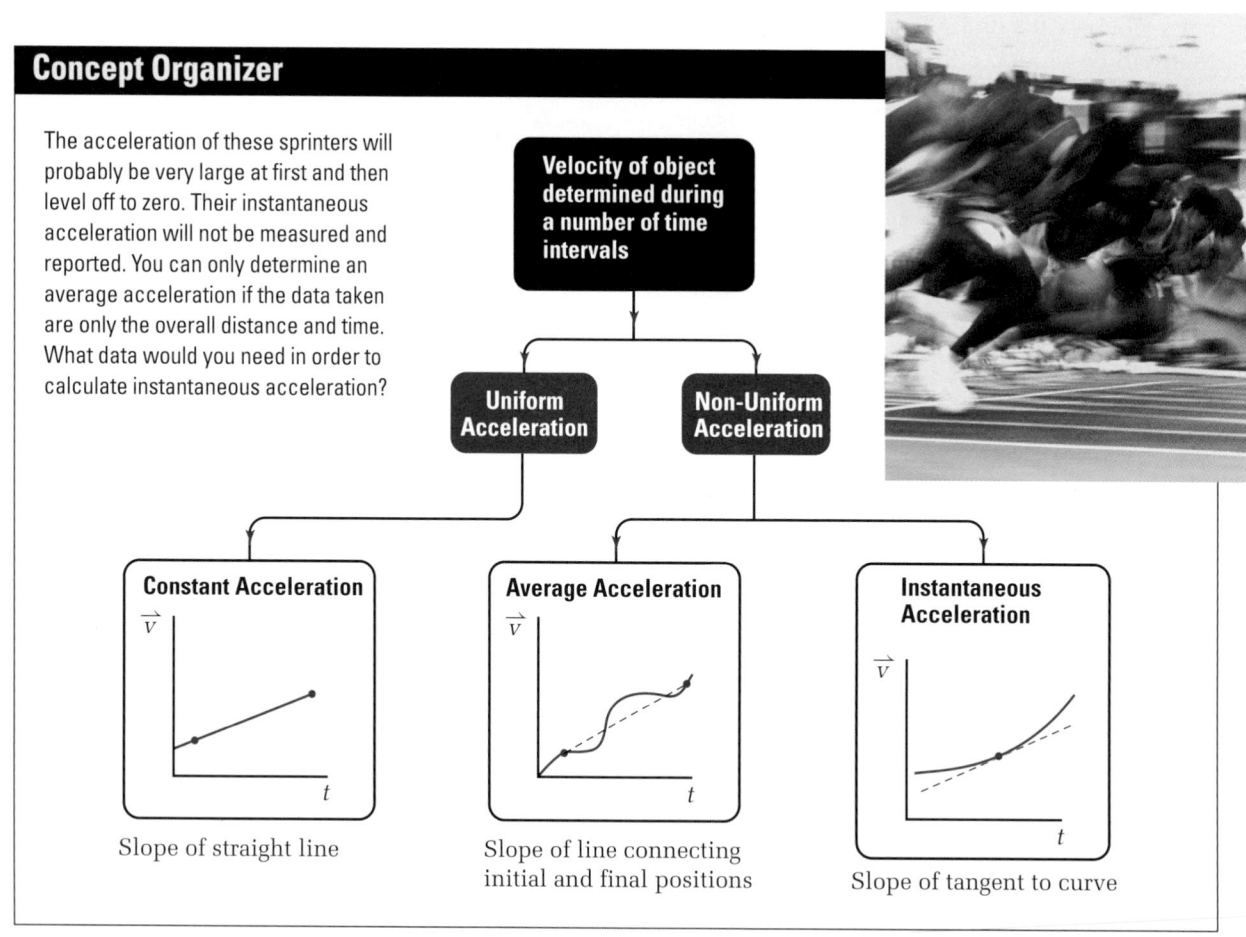

The acceleration of these sprinters will probably be very large at first and then level off to zero. Their instantaneous acceleration will not be measured and reported. You can only determine an average acceleration if the data taken are only the overall distance and time. What data would you need in order to calculate instantaneous acceleration?

**Velocity of object determined during a number of time intervals**

**Uniform Acceleration**

**Non-Uniform Acceleration**

**Constant Acceleration**

Slope of straight line

**Average Acceleration**

Slope of line connecting initial and final positions

**Instantaneous Acceleration**

Slope of tangent to curve

**Figure 2.23** The three ways in which you can describe acceleration are very similar to the ways of describing velocity.

## Balancing Forces in Structural Engineering

Dr. Jane Thorburn knows about the importance of balancing forces. In her work as a structural engineer, she designed highway bridges for the New Brunswick Department of Transportation. Now a professor at Dalhousie University in Halifax, Nova Scotia, she teaches courses on structural engineering and conducts research on the behaviour of structural steel members.

According to Newton's laws, any stable structure — such as a building, bridge, or tower — must produce internal reactions equal in magnitude and opposite in direction to all of the forces acting on it. Structures will be inadequate unless they can balance these external forces, also called "loads."

Structural engineers like Dr. Thorburn determine the right design and materials that will allow a structure to support all possible loads. For example, all structures must be able to support their own mass. They also might have to bear temporary loads related to their use, such as traffic, people, or furniture. At times, a structure might need to balance environmental forces caused by temperature changes, wind, snow accumulation, or earthquakes.

Dr. Thorburn worked on the Hammond River bridge in New Brunswick. When designing this bridge and choosing the materials for its construction, she took several factors into consideration. The bridge had to support the weight of two lanes of traffic. It also had to support its own weight, so the building materials had to be as light as possible. Dr. Thorburn also took environmental factors into account. For example, since the Hammond River is the site of an important salmon run, she wanted to minimize the number of concrete bridge supports, or piers, that were embedded in the riverbed, to reduce any effects that the piers might have on the salmon fishery. In her design, she was able to use only three piers.

Dr. Thorburn also had to determine how the bridge materials could be moved into the construction area, how they would connect together, and how to balance forces during the actual construction of the bridge.

Like most structural engineers, Dr. Thorburn uses special computer software that enables her to create mathematical models of her bridge designs and to determine the impact of external loads on these designs. Her calculation of forces and her understanding of building materials ensure the safety and stability of highway bridges and other structures.

# INVESTIGATION 2-A

## Acceleration Due to Gravity

One of the most famous stories of physics tells of how Galileo dropped cannonballs of various masses off the leaning Tower of Pisa to disprove the accepted theories of free-fall. Until Galileo's time, natural philosophers (the name for scientists of the time) thought that heavy objects fell "faster" than light objects. However, based on Galileo's thinking and experimentation, scientists now agree that acceleration due to gravity *in a vacuum* is

- uniform
- independent of the mass of the falling object

This investigation challenges you to verify Galileo's model experimentally and to determine the numerical value of the acceleration due to gravity. Since you will be working in air and not a vacuum, your results will provide some information about the effect of air resistance on the acceleration of falling objects.

### Problem

Verify that acceleration due to gravity is uniform and independent of mass.
Determine the numerical value of acceleration due to gravity.

### Equipment

- spark timer
- recording tape
- variety of small objects (rubber stoppers, steel balls, wooden beads, film canisters filled with different amounts of sand) ( CAUTION Do not open canisters)
- cellophane tape
- retort stand
- clamp

### Procedure

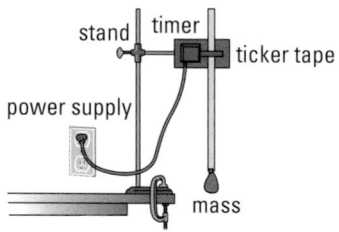

1. Set your spark timer at 10 Hz so the time between dots will be 0.10 s.

2. Clamp a spark timer to a retort stand. Secure the retort stand close to the edge of a desk or lab bench so that an object pulling the recording tape through the timer can fall to the floor.

3. Attach a small object to a piece of recording tape that is 1 m long.

4. Thread the recording tape through the timer.

5. Hold the object in place. Turn on the timer and release the object.

6. Repeat step 5 for at least three objects of different masses. Collect enough tapes so that each member of your lab group has one tape to analyze.

7. While the object is falling, the timer will record a series of dots on the tape that will look like the diagram. Locate the first clear dot that marks the beginning of a series of at least 10 time intervals. Label this dot "$\vec{d}_0$" to designate it as the origin of your frame of reference. Label the next 10 dots "$\vec{d}_1$," "$\vec{d}_2$," ..., "$\vec{d}_{10}$" to mark the position of the object at the end of each of 10 time intervals.

8. Make a table with the following headings: Position, Time, Time interval, Displacement, Average velocity, Change in velocity, Acceleration.

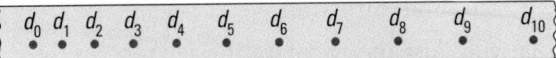

9. Use the label "$t_0$" for the instant in time at which the object is at position $\vec{d}_0$. Record the time and position of the object for the sequence of 10 positions following your designation of $\vec{d}_0$.

10. Complete the table by performing the indicated calculations.

11. Construct position-time, average velocity-time, and average acceleration-time graphs.

## Analyze and Conclude

1. Using the slope of your average velocity-time graph, calculate the value of the acceleration.

2. Compare your value of acceleration calculated in the above step to the values in your table that you calculated for individual intervals. Do they agree or is there a significant difference among them? If they differ, which values do you think are more accurate?

3. Compare the average acceleration determined from your velocity-time graph with the values determined by (a) other members of your group and (b) other groups in your class. Do the masses of the objects appear to have any influence on the value calculated for acceleration? If so, what effect does mass appear to have?

4. Considering all of the comparisons you have just made, does your class data support the model that says that acceleration due to gravity is constant and independent of mass? If not, explain how your class data contradict the model. How might you account for any discrepancies?

5. The accepted value for acceleration due to gravity is 9.81 m/s$^2$. Calculate the percent deviation of your own calculated value from the accepted value. If you need to review the method for determining percent deviation, go to Skill Set 1.

6. Calculate an average of the class data for the acceleration due to gravity. Omit any data points that are extremely different from the majority of the values. Calculate the percent deviation of the class average value to the accepted value. How might you account for any discrepancies?

7. Discuss how well (or poorly) your class data support the accepted value of 9.81 m/s$^2$ for the acceleration due to gravity.

8. Describe any factors that might be affecting the free-fall of each object as it pulls the recording tape through the timer.

9. How might air friction affect your data?

10. Discuss any other possible reasons for a deviation from the accepted value for acceleration due to gravity in a vacuum.

11. Identify possible errors that could have arisen during your experiment and suggest refinements to your procedure to minimize these errors.

12. Identify and discuss any evidence that the shape of an object affects its free-fall acceleration.

## Apply and Extend

13. Design and conduct an investigation to determine how the shape of an object affects its acceleration due to gravity in air. Determine what shape is the most aerodynamic; that is, determine what shape allows the object to accelerate downward with an acceleration as close to 9.81 m/s$^2$ as possible? (**Note:** To correctly test for the effect of shape, each object must have the same mass.)

**CAUTION** Get your teachers approval.

1. **I** The following graphs represent the motion of two students, Al and Barb, walking back and forth in front of the school, waiting to meet friends.

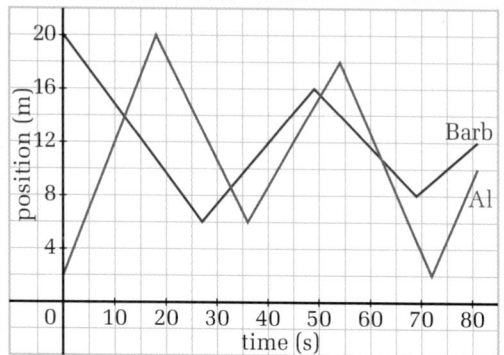

(a) During what periods of time are Al and Barb walking in the same direction?

(b) At what points do Al and Barb meet?

(c) During what periods of time are Al and Barb facing each other?

(d) Which student is, on the average, walking faster than the other? Explain your reasoning.

2. **K/U** Describe the similarities and differences between:

(a) constant acceleration and non-uniform acceleration.

(b) average acceleration and instantaneous acceleration.

3. **C** Explain the relationship between:

(a) tangent line on a velocity time graph, time interval, and acceleration.

(b) negative acceleration and deceleration.

(c) m/s and m/s$^2$

4. **K/U** How is the direction of an acceleration vector determined?

5. **K/U** Describe a motion when:

(a) velocity and acceleration vectors are in the same direction.

(b) velocity and acceleration vectors are in opposite directions.

6. **I** Throw a ball up into the air (from rest) and catch it. Sketch the path of the ball. Label points where the vertical velocity is zero. Label points where the acceleration is zero.

7. **I** Draw conclusions about the acceleration of the motion represented by the following graphs.

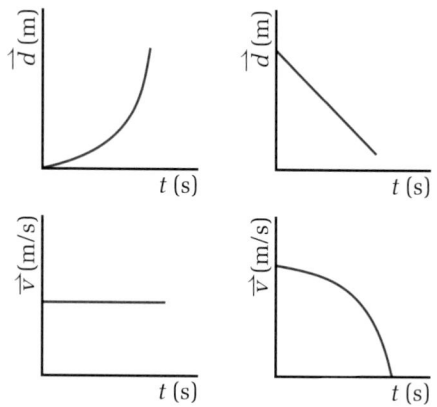

8. **I** Design a simulation on interactive physics software to verify that acceleration due to gravity is uniform and independent of mass.

### UNIT PROJECT PREP

Vehicles and objects are commonly seen speeding up, slowing down, and changing direction in motion pictures.

- What do you feel when a vehicle starts or stops suddenly? How can you use this in your "virtual reality"?

- An object accelerating towards the ground can be a dramatic situation. What effects can you use to simulate this?

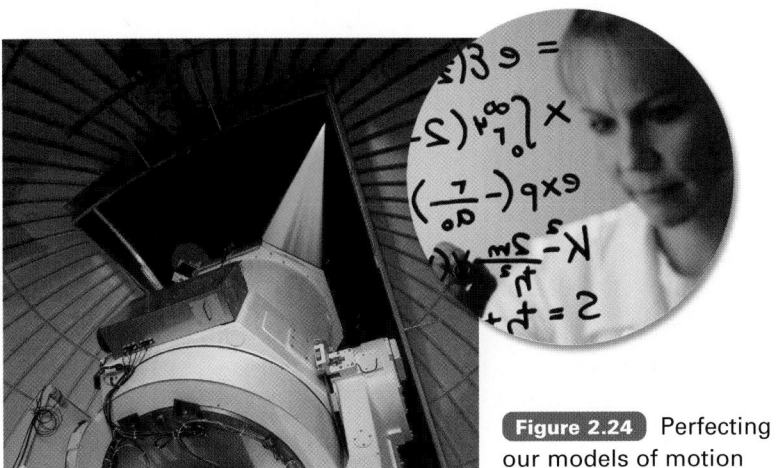

**Figure 2.24** Perfecting our models of motion

- Identify the variables and equations used to mathematically model the motion of an object.

- Apply quantitative relationships among displacement, velocity, and acceleration.

- Develop problem-solving skills and strategies through the analysis and synthesis of information.

Throughout this chapter, you have been developing models of motion. Your models have taken the form of stick figures, mathematical definitions of displacement, velocity, and acceleration, and graphs. In this section, you will call on all of those models to build a set of mathematical equations called the **equations of motion** (or of kinematics) for uniform acceleration. As the name implies, these equations apply *only* to situations in which the *acceleration is constant*.

A very important feature of the equations of motion is that they apply independently to each dimension, so you will use them to analyze motion in one direction at a time. For example, you will analyze only north-south motion or only vertical (up and down) motion. Consequently, the variables in the equations represent only the parts, or components, of the vector quantities, position, displacement, velocity, and acceleration.

Figure 2.25 shows a position and displacement vector separated into components. Since components of vectors apply to only one dimension, they are not vectors themselves. Therefore, vector notations will not be used in the equations of motion.

## Deriving the Kinematic Equations

The fundamental definitions of displacement, velocity, and acceleration form the basis of the set of equations you will develop and apply. Start with the definition of acceleration in one dimension.

$$a = \frac{\Delta v}{\Delta t}$$

- equations of motion

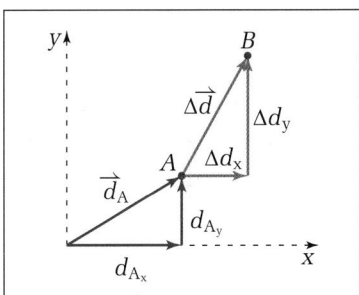

**Figure 2.25** The displacement of an object that moved from point A to point B is represented by the vector $\vec{\Delta d}$. The object moved a distance, $\Delta d_x$, in the $x$ direction and a distance, $\Delta d_y$ in the $y$ direction. Therefore, $\Delta d_x$ is called the $x$-component of the displacement vector and $\Delta d_y$ is the $y$-component. Any vector can be divided into components. The $x$- and $y$-components of position vector $\vec{d}_A$ are also shown, and are labelled as "$d_{Ax}$" and "$d_{Ay}$".

In some cases, you will know the initial and final velocities for a certain time interval and want to determine the acceleration. So, you will use the expanded form of the mathematical definition for the change in velocity: $\Delta v = v_f - v_i$. Substituting this expression into the original equation for acceleration, you obtain a useful equation.

$$a = \frac{v_f - v_i}{\Delta t}$$

In many cases, you will know the initial velocity and acceleration and want to find the final velocity for a time interval. Algebraically rearranging the above equation will give you another useful form.

- Mulitply both sides of the equation by $\Delta t$ and simplify.

$$a\Delta t = \left(\frac{v_f - v_i}{\Delta t}\right)\Delta t$$
$$v_f - v_i = a\Delta t$$

- Add $v_i$ to both sides of the equation and simplify.

$$v_f - v_i + v_i = a\Delta t + v_i$$
$$v_f = v_i + a\Delta t$$

## MODEL PROBLEMS

### Changing Velocities

1. **A slight earth tremor causes a large boulder to break free and start rolling down the mountainside with a constant acceleration of 5.2 m/s². What was the boulder's velocity after 8.5 s?**

### Frame the Problem

- Sketch and label a diagram of the motion.

- Choose the *direction of the motion* of the boulder as the *positive x* direction so the displacement will be positive.

- The boulder was *stationary* before the tremor, so its *initial velocity* was zero.

- The boulder's *acceleration* was constant, so the *equations of motion* apply to the problem.

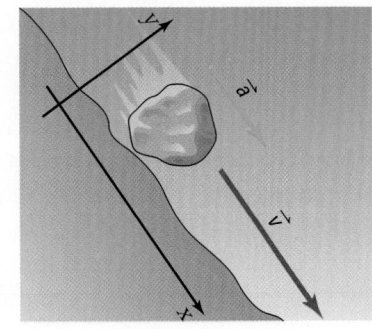

### Identify the Goal

The velocity (in one dimension), $v$, of the boulder 8.5 s after it started rolling

### Variables and Constants

| Involved in the problem | | Known | Implied | Unknown |
|---|---|---|---|---|
| $v_i$ | $a$ | $a = 5.2$ m/s² | $v_i = 0$ m/s | $v_f$ |
| $v_f$ | $\Delta t$ | $\Delta t = 8.5$ s | | |

## Strategy

Select the equation that relates the final velocity to the initial velocity, acceleration and time interval.

All of the needed quantities are known so substitute them into the equation.

Simplify.

The final velocity of the boulder was 43 m/s.

## Calculations

$$v_f = v_i + a\Delta t$$

$$v_f = 0.0 \ \frac{m}{s} + \left(5.2 \ \frac{m}{s^2}\right)(8.3 \ s)$$

$$v_f = 43.16 \ \frac{m}{s}$$

## Validate

The units cancel to give metres per second, which is correct for velocity. The product of 5 x 8 is 40, so you would expect the answer to be slightly larger than 40 m/s.

2. **A skier is going 8.2 m/s when she falls and starts sliding down the ski run. After 3.0 s, her velocity is 3.1 m/s. How long after she fell did she finally come to a stop? (Assume that her acceleration was constant.)**

## Frame the Problem

- Sketch and label the situation.
- Choose a coordinate system that places the skier at the origin when she falls and places her motion in the positive x direction.
- When the skier falls and begins to slide, her *initial velocity* is the same as her skiing velocity.
- Friction begins to *slow* her down and, eventually, she will come to a *stop*.
- Her *final velocity* will be *zero*.
- Since her *acceleration is constant*, the *equation of motion* relating initial and final velocities to acceleration and time applies to this problem.

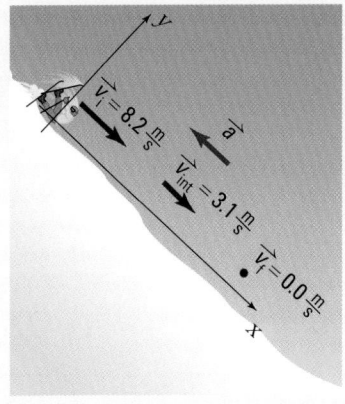

## Identify the Goal

The total time, $\Delta t$, it takes for the skier to stop sliding

*continued* ▶

*continued from previous page*

## Variables and Constants

| Involved in the problem | Known | Implied | Unknown |
|---|---|---|---|
| $v_i$          $a$ | $v_i = 8.2 \ \frac{m}{s}$ | $v_f = 0.0 \ \frac{m}{s}$ | $a$ |
| $v_{intermediate}$          $\Delta t$ | $v_{int} = 3.1 \ \frac{m}{s}$ | | $\Delta t$ |
| $v_f$ | | | |

## Strategy

You know the initial and final velocities but you need to find the acceleration in order to solve for the time interval. In this case, it is best to use the definition for acceleration.

Use the information about the intermediate velocity to find the acceleration. At the intermediate time, the "final" velocity is 3.1 m/s. Substitute values into the equation.

Knowing the acceleration, you can use it to find the length of the entire time interval from the initial fall to the time the skier stopped. Use the same form of the equation, but use the calculated acceleration. Also, in this part of the problem, the final velocity is zero.

## Calculations

$$a = \frac{v_f - v_i}{\Delta t}$$

$$a = \frac{3.1 \ \frac{m}{s} - 8.2 \ \frac{m}{s}}{3.0 \ s}$$

$$a = \frac{-5.1 \ \frac{m}{s}}{3.0 \ s}$$

$$a = -1.7 \ \frac{m}{s^2}$$

**Substitute first**

$$a = \frac{v_f - v_i}{\Delta t}$$

$$-1.7 \ \frac{m}{s^2} = \frac{0.0 \ \frac{m}{s} - 8.2 \ \frac{m}{s}}{\Delta t}$$

$$\left(-1.7 \ \frac{m}{s^2}\right)\Delta t = \frac{0.0 \ \frac{m}{s} - 8.2 \ \frac{m}{s}}{\cancel{\Delta t}}\cancel{\Delta t}$$

$$\frac{\cancel{\left(-1.7 \ \frac{m}{s^2}\right)}\Delta t}{\cancel{\left(-1.7 \ \frac{m}{s^2}\right)}} = \frac{-8.2 \ \frac{m}{s}}{\left(-1.7 \ \frac{m}{s^2}\right)}$$

$$\Delta t = 4.823 \ s$$

**Solve for $\Delta t$ first**

$$a = \frac{v_f - v_i}{\Delta t}$$

$$a\Delta t = \frac{v_f - v_i}{\cancel{\Delta t}}\cancel{\Delta t}$$

$$\frac{\cancel{a}\Delta t}{\cancel{a}} = \frac{v_f - v_i}{a}$$

$$\Delta t = \frac{0.0 \ \frac{m}{s} - 8.2 \ \frac{m}{s}}{-1.7 \ \frac{m}{s^2}}$$

$$\Delta t = \frac{-8.2 \ \frac{m}{\cancel{s}}}{-1.7 \ \frac{\cancel{m}}{s^{\cancel{2}}}}$$

$$\Delta t = 4.823 \ s$$

It took 4.8 s for the skier to come to a stop.

## Validate

The units cancelled to give seconds, which is correct for a time interval. Eight seconds is a reasonable time period for a skier to slide to a stop.

6. An Indy 500 race car's velocity increases from +6.0 m/s to +38 m/s over a 4.0 s time interval. What is its average acceleration?

7. A stalled car starts to roll backward down a hill. At the instant that it has a velocity of 4.0 m/s down the hill, the driver is able to start the car and start accelerating back up. After accelerating for 3.0 s, the car is travelling uphill at 3.5 m/s. Determine the car's

acceleration once the driver got it started. (Assume that the acceleration was constant.)

8. A bus is travelling along a street at a constant velocity when the driver steps on the brakes and brings the bus to a stop in 3.0 s. If the brakes cause the bus to accelerate at $-8.0 \text{ m/s}^2$, at what velocity was the bus travelling when the brakes were applied?

## Building Equations

The next logical step in building a set of equations would be to rearrange the equation that defines velocity. However, a problem arises when you try to use the equation. Can you see why?

$$v = \frac{\Delta d}{\Delta t}$$

$$v\Delta t = \left(\frac{\Delta d}{\Delta t}\right)\Delta t$$

$$\Delta d = v\Delta t$$

This equation is valid *only if the velocity is constant*, that is, if the motion is uniform. The equations of motions are developed for *constant acceleration*. So, unless that constant acceleration is zero, the velocity will be changing.

To find a relationship between displacement and velocity for a changing velocity, turn, once again, to graphs. First, consider an object moving at a constant velocity. The velocity-time graph is simply a horizontal line, as shown in Figure 2.26. As well, displacement is the product of the constant velocity and the time interval. Notice on the graph that velocity and time interval form the sides of a rectangle. Since the area of a rectangle is the product of its sides, velocity times time must be the same as the area of the rectangle.

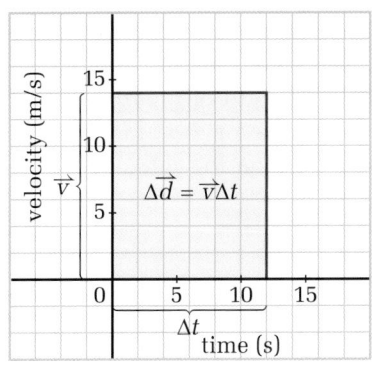

**Figure 2.26** The area under the curve of a velocity-time graph is the displacement.

In fact, the displacement of an object is always the same as the area under the velocity-time graph. When the graph is a curve, you can approximate displacement by estimating the area under the curve. In Figure 2.27 (A), the area of each of the small squares is 5.0 m/s times 1.0 s or 5.0 m. Counting the number of squares and multiplying by 5.0 m gives a good estimate of displacement. You can make your estimate more accurate by dividing up the area into small rectangles as shown in Figure 2.27 (B). Notice that the corner of each rectangle above the curve on the left is nearly the same area as the space between the rectangle and the curve on the right. So the area of the rectangle is very nearly the same as the area under the curve. When you add the areas of all of the rectangles, you have a very close approximation of the displacement.

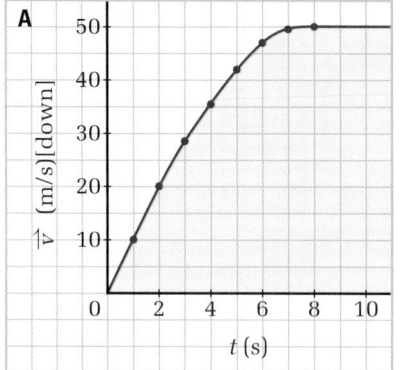
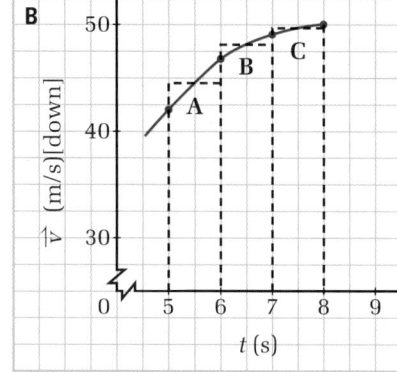

**Figure 2.27** **(A)** The area under the curve is the displacement. **(B)** You can increase the accuracy of the area determination by making the columns narrower. (Why?)

How does the knowledge that the area under the curve is the same as the displacement help you to develop precise equations for displacement under constant acceleration? Consider the shape of the velocity-time graph for an object travelling with uniform acceleration. The graph is a straight line, as shown in Figure 2.28. If you draw a rectangle so that the line forming the top is precisely at the midpoint between the initial velocity and the final velocity, you will find that the line intersects the velocity curve exactly at the midpoint of the time interval. The top of the rectangle and the velocity line create a congruent triangle. If you cut out the triangle on the right (below the graph), it would fit perfectly into the triangle on the left (above the graph). The area of the rectangle is *exactly* the same as the displacement. The height of the rectangle is the average of the initial and final velocities for the time interval or $v_{ave} = \frac{v_i + v_f}{2}$. You can now use this expression for velocity in the equation developed for displacement, above.

$$\Delta d = v \Delta t$$

$$\Delta d = \frac{v_i + v_f}{2} \Delta t$$

You have just developed another useful equation of motion for uniform acceleration.

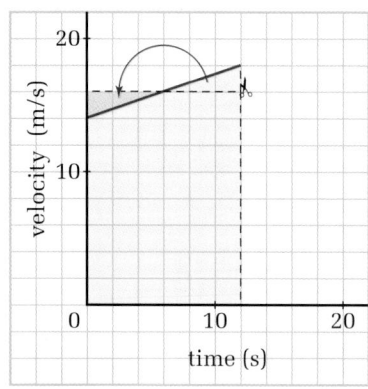

**Figure 2.28** Confirm the results of the analysis by using the formula for the area of a trapazoid.
$$A = \frac{(side\ 1 + side\ 2)}{2}\ (width)$$

Often, you know the initial velocity of an object and its acceleration but not the final velocity. You can develop an expression for displacement that does not include final velocity.

- Start with the equation above.
$$\Delta d = \frac{v_i + v_f}{2} \Delta t$$

- Recall the expression you developed for final velocity.
$$v_f = v_i + a\Delta t$$

- Substitute this value into the first equation.
$$\Delta d = \left( \frac{v_i + (v_i + a\Delta t)}{2} \right) \Delta t$$

- Combine like terms.
$$\Delta d = \left( \frac{2v_i + a\Delta t}{2} \right) \Delta t$$

- Multiply through by $\Delta t$.
$$\Delta d = \frac{2v_i \Delta t}{2} + \frac{a\Delta t^2}{2}$$

- Simplify.
$$\Delta d = v_i \Delta t + \frac{1}{2} a\Delta t^2$$

Table 2.6 summarizes the equations of motion and indicates the variables that are related by each equation. Notice that, in every case, the equation relates four of the five variables. Therefore, if you know three of the variables, you can find the other two. First, use an equation that relates the three known variables to a fourth. Then, find an equation that relates any three of the four you now know to the fifth.

**Table 2.6** Equations of Motion under Uniform Acceleration

| Equation | Variables | | | | |
|---|---|---|---|---|---|
| | $\Delta d$ | $v_i$ | $v_f$ | $a$ | $\Delta t$ |
| $a = \dfrac{v_f - v_i}{\Delta t}$ | | x | x | x | x |
| $v_f = v_i + a\Delta t$ | | x | x | x | x |
| $\Delta d = \dfrac{v_i + v_f}{2} \Delta t$ | x | x | x | | x |
| $\Delta d = v_i \Delta t + \dfrac{1}{2} a\Delta t^2$ | x | x | | x | x |

**History Link**

Research how Galileo Galilei used the equation $\Delta d = v_i \Delta t + \frac{1}{2} a\Delta t^2$ to calculate the acceleration due to gravity. Given the fact that he could not use a stopwatch to measure time, how close was his value for $a$?

**ELECTRONIC LEARNING PARTNER**

Your Electronic Learning Partner provides interactive activities exploring linear, parabolic and hyperbolic relationships.

## Think It Through

- Derive an equation that relates $v_i$, $v_f$, $\Delta d$, and $a$. (Hint: Notice that $\Delta t$ is not involved.) Solve for $\Delta t$ in the first equation. Substitute that value into $\Delta t$ in the third equation. Solve for $v_f^2$. Can you prove that $v_f^2 = v_i^2 + 2ad$?

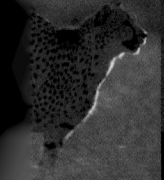

# INVESTIGATION 2-B

## Stop on a Dime

**TARGET SKILLS**
- Initiate and planning
- Hypothesizing
- Analyzing and interpreting

In this investigation, you will be challenged to build a vehicle that, when launched from a ramp, will travel a horizontal distance of 3.0 m and come to rest on a dime!

### Problem

Design, build, and test a vehicle. Enter it into a competition.

### Equipment

- flat 1.0 m ramp
- dime
- materials of your choice for building a vehicle

### Procedure

#### Designing and Building

1. With a partner or a small group, discuss potential designs for your vehicle, according to the following criteria.

   (a) Prefabricated kit and prefabricated wheels are not allowed.

   (b) Propulsion may come only from the energy gained by rolling the vehicle down a 1.0 m ramp.

   (c) You may adjust the angle of the ramp.

   (d) The vehicle must be self-contained. No external guidance systems, such as tracks, guide wires, or strings, are permitted.

   Be creative. Do not limit your thinking to a traditional four-wheeled vehicle.

2. Collect materials and build your vehicle according to your design. Get your teacher's approval before testing.

3. Test your vehicle and make adjustments until you are satisfied with its performance. Collect data on at least three trial runs.

#### Entering the Competition

4. As a class, establish criteria for being allowed to enter the competition. For

example, the vehicle must stop within 10 cm of the dime in at least one test run.

5. Submit a written application for entry into the competition. The application must include the following.

   (a) Description of design features

   (b) Outline of any major problems encountered in testing the vehicle, accompanied by a discussion of the solutions you discovered

   (b) Data from trial runs, including position-time, velocity-time, and acceleration-time graphs. Data must include at least three time intervals while accelerating down the ramp and five time intervals after the vehicle begins its horizontal motion.

#### The Competition

6. As a class, decide on the scoring system for the competition. Decide how many points will be given for such results as coming within 6.0 cm of the dime. Decide on other possible criteria for points. For example, points could be given for sturdiness, creative use of materials, originality, or aesthetic appeal.

7. Hold the competition.

### Analyze and Conclude

1. Analyze the performance of your own vehicle in comparison with your own criteria.

2. Analyze your vehicle in comparison with the vehicle that won the competition.

3. Considering what you learned from the competition, how would you design your vehicle differently if you were to begin again?

4. Summarize what you have learned about motion from this challenge.

## Applying the Equations of Motion

1. **You throw a rock off a cliff, giving it a velocity of 8.3 m/s, straight down. At the instant you released the rock, your hiking buddy started a stopwatch. You heard the splash when the rock hit the river below, exactly 6.9 s after you threw the rock. How high is the cliff above the river?**

## Frame the Problem

- Make a sketch of the problem and assign a coordinate system.

- The rock had an *initial velocity downward*. Since you chose downward as negative, the initial velocity is *negative*.

- The rock is *accelerating* due to gravity.

- The acceleration due to gravity is *constant*. Therefore, the equations of motion for uniform acceleration apply to the problem.

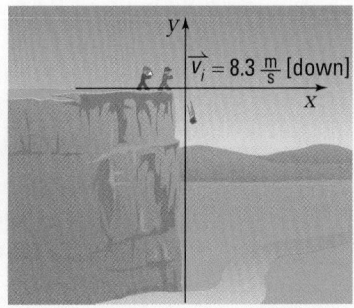

## Identify the Goal

The displacement, $\Delta d$, from the top of the cliff to the river below

## Variables and Constants

| Involved in the problem | | Known | Implied | Unknown |
|---|---|---|---|---|
| $v_i$ | $\Delta t$ | $v_i = -8.3$ m/s | $a = -9.81$ m/s$^2$ | $\Delta d$ |
| $a$ | $\Delta d$ | $\Delta t = 6.9$ s | | |

## Strategy

Use the equation of motion that relates the unknown variable, $\Delta d$, to the three known variables, $v_i$, $a$, and $\Delta t$.

Substitute the known variables.

Simplify.

## Calculations

$$\Delta d = v_i \Delta t + \frac{1}{2} a \Delta t^2$$

$$\Delta d = \left(-8.3 \ \frac{m}{s}\right)(6.9 \ s) + \frac{1}{2}\left(-9.81 \ \frac{m}{s^2}\right)(6.9 \ s)^2$$

$$\Delta d = -57.27 \ m + \left(-4.905 \ \frac{m}{s^2}\right)\left(47.61 \ s^2\right)$$

$$\Delta d = -57.27 \ m - 233.53 \ m$$

$$\Delta d = -290.8 \ m$$

The cliff was $2.9 \times 10^2$ m above the river. The negative sign indicates that the distance is in the negative direction, or down, from the origin, the point from which you threw the rock.

*continued* ▶

*continued from previous page*

## Validate

All of the units cancel to give metres, which is the correct unit. The
sign is negative, which you would expect for a rock going down.
The rock fell a long distance (more than a quarter of a kilometre), so
6.9 s is a reasonable length of time for the rock's fall to have taken.

---

2. **A car travels east along a straight road at a
   constant velocity of 18 m/s. After 5.0 s, it
   accelerates uniformly for 4.0 s. When it
   reaches a velocity of 24 m/s, the car proceeds
   with uniform motion for 6.0 s. Determine the
   car's total displacement during the trip.**

> **PROBLEM TIP**
>
> When the type of motion of an object changes, the prob-
> lem must be split into phases. Each phase is treated as a
> separate problem. The "final" conditions of one phase
> become the "initial" conditions of the next phase.

## Frame the Problem

- Make a diagram of the motion of the car that
  includes the known variables during each
  phase of the car's motion.

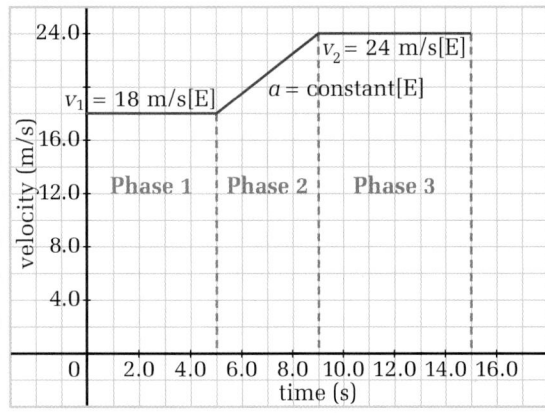

- During phase 1, the car is moving with
  *uniform motion*. Since the *acceleration* is
  *zero*, the equations of motion with uniform
  acceleration are not needed. The equation
  *defining velocity* applies to this phase.

- During phase 2 of the motion, the car is
  *accelerating*. Therefore, use the equation
  of motion with *uniform acceleration* that
  relates time, initial velocity, and final
  velocity to displacement.

- During phase 3, the car is again moving with
  *uniform motion*.

- The *total displacement* of the car is the *sum*
  of the displacements for all three phases.

## Identify the Goal

The total displacement, $\Delta d$, of the car for the duration of the motion

## Variables and Constants

| Involved in the problem | | Known | | Unknown |
|---|---|---|---|---|
| $v_1$ | $\Delta t_3$ | $v_1 = 18$ m/s | $\Delta t_2 = 4.0$ s | $\Delta d_{total}$ |
| $v_2$ | $\Delta d_1$ | $v_2 = 24$ m/s | $\Delta t_3 = 6.0$ s | $\Delta d_1$ |
| $\Delta t_1$ | $\Delta d_2$ | $\Delta t_1 = 5.0$ s | | $\Delta d_2$ |
| $\Delta t_2$ | $\Delta d_3$ | | | $\Delta d_3$ |

## Strategy

Use the equation that defines velocity.

Substitute values.

Simplify.

The displacement for phase 1 was 90 m east.

Use the equation of motion that relates
time, initial velocity, and final velocity
to displacement.

Substitute values.

Simplify.

The displacement during phase 2 was 84 m east.

Use the equation that defines velocity.

Substitute values.

Simplify.

The displacement during phase 3 was 144 m east.

Find the sum of the displacements for all
three phases.

The total displacement for the trip was $3.2 \times 10^2$ m east.

## Calculations

$\Delta d = v \Delta t$

$\Delta d_1 = v_1 \Delta t_1$

$\Delta d_1 = \left( 18 \, \frac{m}{s} [E] \right) (5 \, s)$

$\Delta d_1 = 90 \text{ m[E]}$

$\Delta d = \frac{v_i + v_f}{2} \Delta t$

$\Delta d_2 = \frac{v_1 + v_2}{2} \Delta t_2$

$\Delta d_2 = \frac{18 \, \frac{m}{s} [E] + 24 \, \frac{m}{s} [E]}{2} (4 \, s)$

$\Delta d_2 = 21 \, \frac{m}{s} [E] \, 4 \, s$

$\Delta d_2 = 84 \text{ m[E]}$

$\Delta d = v \Delta t$

$\Delta d_3 = v_2 \Delta t_3$

$\Delta d_3 = \left( 24 \, \frac{m}{s} [E] \right) 6 \, (s)$

$\Delta d_3 = 144 \text{ m[E]}$

$\Delta d_{total} = \Delta d_1 + \Delta d_2 + \Delta d_3$

$\Delta d_{total} = 90 \text{ m[E]} + 84 \text{ m[E]} + 144 \text{ m[E]}$

$\Delta d_{total} = 318 \text{ m[E]}$

## Validate

In every case, the units cancelled to give metres, which is the correct unit for displacement. The duration of the trip was short (15 s),
so the displacement cannot be expected to be very long. The answer
of 318 m[E] is very reasonable.

*continued* ▶

3. **A truck is travelling at a constant velocity of 22 m/s north. The driver sees a traffic light turn from red to green soon enough, so he does not have to alter his speed. Meanwhile, a woman in a sports car is stopped at the red light. At the moment the light turns green and the truck passes her, she begins to accelerate at 4.8 m/s². How far have both vehicles travelled when the sports car catches up with the truck? How long did it take for the sports car to catch up with the truck?**

## Frame the Problem

- The truck and the sports car leave the traffic signal at the *same time.* Define this time as $t = 0.0$ s.

- The truck passes the sports car at the traffic light. Let this point be $d = 0.0$ m.

- The truck travels with *uniform motion*, which means *constant velocity.* The truck's motion can therefore be described using the *equation that defines velocity.*

- The car's *initial velocity is zero.* Then, the car travels with *uniform acceleration.* The *equation of motion* that relates displacement, initial velocity, acceleration, and time interval describes the car's motion.

- When the sports car catches up with the truck, both vehicles have travelled for the *same length of time* and the *same distance.*

## Identify the Goal

The displacement, $\Delta d$, that the sports car and truck travel from the traffic light to the point where the sports car catches up with the truck

The time interval, $\Delta t$, that it takes for the sports car to catch up with the truck

## Variables and Constants

| Involved in the problem | | Known | Implied | Unknown | |
|---|---|---|---|---|---|
| $a_{car}$ | $v_{truck}$ | $v_{truck} = 22$ m/s | $v_{i(car)} = 0.0$ m/s | $\Delta d_{car}$ | $\Delta t_{car}$ |
| $v_{i(car)}$ | $\Delta t_{truck}$ | $a_{car} = 4.8$ m/s² | | $\Delta d_{truck}$ | $\Delta t_{truck}$ |
| $\Delta t_{car}$ | $\Delta d_{truck}$ | | | | |
| $\Delta d_{car}$ | | | | | |

## Strategy

Write the equation that defines velocity for the motion of the truck.

Write the equation of motion for the sports car.

## Calculations

$\Delta d = v\Delta t$

$\Delta d_{truck} = 22 \, \frac{m}{s} \Delta t_{truck}$

$\Delta d = v_i\Delta t + \frac{1}{2}a\Delta t^2$

$\Delta d_{car} = 0.0 \, \frac{m}{s}\Delta t_{car} + \frac{1}{2}4.8 \, \frac{m}{s^2}\Delta t_{car}^2$

$\Delta d_{car} = 2.4 \, \frac{m}{s^2}\Delta t_{car}^2$

## Strategy

The displacement for the sports car and the truck are the same. Call them both $\Delta d$.

The time interval is the same for the sports car and the truck. Call them both $\Delta t$.

You now have two equations and two unknowns. Solve for $\Delta t$ in the equation for the sports car. Then substitute that expression into $\Delta t$ for the truck. This will give you one equation with only one unknown, $\Delta d$.

## Calculations

$$\Delta d_{car} = \Delta d_{truck} = \Delta d$$

$$\Delta t_{car} = \Delta t_{truck} = \Delta t$$

$$\Delta d = 22\ \frac{m}{s}\Delta t$$

$$\frac{\Delta d}{22\ \frac{m}{s}} = \frac{\cancel{22}\ \frac{m}{s}}{\cancel{22}\ \frac{m}{s}}\Delta t$$

$$\Delta t = \frac{\Delta d}{22\ \frac{m}{s}}$$

$$\Delta d = 2.4\ \frac{m}{s^2}\Delta t^2$$

$$\Delta d = 2.4\ \frac{m}{s^2}\left(\frac{\Delta d}{22\ \frac{m}{s}}\right)^2$$

Solve for displacement.

$$\Delta d\left(22^2\ \frac{m^2}{s^2}\right) = 2.4\ \frac{m}{s^2}\frac{\Delta d^2}{\cancel{22\ \frac{m^2}{s^2}}}\cancel{22^2\ \frac{m^2}{s^2}}$$

$$484\ \frac{m^2}{s^2}\Delta d = 2.4\ \frac{m}{s^2}\Delta d^2$$

Subtract $484\ \frac{m^2}{s^2}\Delta d$ from both sides of the equation.

$$484\ \frac{m^2}{s^2}\Delta d - 484\ \frac{m^2}{s^2}\Delta d = 2.4\ \frac{m}{s^2}\Delta d^2 - 484\ \frac{m^2}{s^2}\Delta d$$

$$0.0 = 2.4\ \frac{m}{s^2}\Delta d^2 - 484\ \frac{m^2}{s^2}\Delta d$$

Factor out the $\Delta d$.

$$0.0 = \left(2.4\ \frac{m}{s^2}\Delta d - 484\ \frac{m^2}{s^2}\right)\Delta d$$

Set each of the factors equal to 0.

$$2.4\ \frac{m}{s^2}\Delta d - 484\ \frac{m^2}{s^2} = 0 \quad \text{or} \quad \Delta d = 0.0\ \frac{m}{s}$$

Solve for $\Delta d$.

$$2.4\ \frac{m}{s^2}\Delta d = 484\ \frac{m^2}{s^2}$$

$$\frac{\cancel{2.4\ \frac{m}{s^2}}\Delta d}{\cancel{2.4\ \frac{m}{s^2}}} = \frac{484\ \frac{m^2}{s^2}}{2.4\ \frac{m}{s^2}}$$

$$\Delta d = 201.67\ m$$

You found two solutions for displacement of the sports car and truck when setting the time intervals and displacements of the two vehicles equal to each other. The value of zero for displacement simply means that they had the same displacement (zero) at time zero. The displacement of the sports car and truck was $2.0 \times 10^2$ m when the sports car caught up with the truck.

*continued* ▶

## Strategy

To find the time it took for the sports car to catch up with the truck, substitute the displacement into the equation relating displacement of the truck and time interval.

## Calculations

$$\Delta t = \frac{\Delta d}{22\frac{m}{s}}$$

$$\Delta t = \frac{201.66 \text{ m}}{22\frac{m}{s}}$$

$$\Delta t = 9.167 \text{ s}$$

It took 9.2 s for the sports car to catch up with the truck.

## Validate

The units cancelled to give metres for displacement and seconds for time interval, which is correct. A second equation exists for calculating time interval from displacement. Substitute the displacement into the equation relating time and displacement for the truck, and solve for time interval. It should give the same value, 9.167 s.

The values are in agreement.

$$\Delta d_{car} = 2.4\frac{m}{s^2}\Delta t_{car}^2$$

$$201.67 \text{ m} = 2.4\frac{m}{s^2}\Delta t_{car}^2$$

$$\frac{201.67 \text{ m}}{2.4\frac{m}{s^2}} = \frac{2.4\frac{m}{s^2}}{2.4\frac{m}{s^2}}\Delta t_{car}^2$$

$$\Delta t_{car}^2 = 84.027 \text{ s}^2$$

$$\Delta t = 9.167 \text{ s}$$

## PRACTICE PROBLEMS

9. A field hockey player starts from rest and accelerates uniformly to a speed of 4.0 m/s in 2.5 s

   (a) Determine the distance she travelled.

   (b) What is her acceleration?

10. In a long distance race, Michael is running at 3.8 m/s and is 75 m behind Robert, who is running at a constant velocity of 4.2 m/s. If Michael accelerates at 0.15 m/s², how long will it take him to catch Robert?

11. A race car accelerates at 5.0 m/s². If its initial velocity is 200 km/h, how far has it travelled after 8.0 s?

12. A motorist is travelling at 20 m/s when she observes that a traffic light 150 m ahead of her turns red. The traffic light is timed to stay red for 10 seconds. If the motorist wishes to pass the light without stopping just as it turns green again, what will be the speed of her car just as it passes the light?

## 2.5    Section Review

1. **K/U** Define kinematics.

2. **MC** Refer to model problem #1 on page 77. Given the explanation of timing, is 6.9 s likely to be exact? Explain the types of errors that could lead to an over measurement of the time interval. How might the hiking buddies minimize their error?

3. **C** Develop an appropriate problem for each of the following formulas.

   (a) $a = \frac{(v_f - v_i)}{\Delta t}$

   (b) $\Delta d = \frac{(v_i + v_f)}{2}\Delta t$

   (c) $\Delta d = v_i\Delta t + \frac{1}{2}a\Delta t^2$

## REFLECTING ON CHAPTER 2

- An object is in motion if it changes position in a particular frame of reference or coordinate system.
- A motion diagram documents an object's position in a frame of reference at particular instants in time.
- Vector quantities are described in terms of their magnitude and direction.
- A position vector locates an object with a magnitude and direction from the origin of a frame of reference.
- A displacement vector designates the change in position of an object.
$$\Delta \vec{d} = \vec{d_f} - \vec{d_i}$$
- A time interval $t$ is the time elapsed between two instants in time.
$$\Delta t = t_f - t_i$$
- Velocity is the rate of change of position or the displacement of an object over a time interval.
$$\vec{v}_{ave} = \frac{\Delta \vec{d}}{\Delta t}$$
- Position-time graphs reveal patterns of uniform and non-uniform motion. Uniform motion or constant velocity appears as a straight slope. The average velocity during a time interval is determined by finding the slope of the line connecting the initial and the final positions of the object for the time interval. The instantaneous velocity is calculated by finding the slope of the tangent to the line of the graph at a particular instant in time.

- Acceleration is the rate of change of velocity of an object over a time interval.
$$\vec{a}_{ave} = \frac{\Delta \vec{v}}{\Delta t}$$
- Velocity-time graphs reveal patterns of uniform and non-uniform acceleration. Uniform or constant acceleration appears as a straight slope. The average acceleration during a time interval is determined by finding the slope of the line connecting the initial and the final velocities of the object for the time interval. The instantaneous acceleration is calculated by finding the slope of the tangent to the line of the graph at a particular instant in time.
- The displacement of an object during a particular time interval can be found by determining the area under the curve of a velocity-time graph.
- The mathematical equations which are used to analyze the motion of an object undergoing constant acceleration relate various combinations of five variables: the object's initial velocity $v_i$, its final velocity $v_f$, its acceleration $a$, a time interval $\Delta t$, the object's displacement $\Delta d$ during the time interval.
$$a = \frac{(v_f - v_i)}{\Delta t}$$
$$\Delta d = v_i \Delta t + \frac{1}{2} a \Delta t^2$$
$$\Delta d = \frac{1}{2}(v_i + v_f)\Delta t$$
$$v_f = v_i + a\Delta t$$

### Knowledge/Understanding

1. Define the following: **(a)** kinematics **(b)** dynamics **(c)** mechanics **(d)** velocity **(e)** acceleration **(f)** frame of reference **(g)** center of mass **(h)** vector **(i)** scalar

2. What is meant by the illusion of motion while watching a movie or video?

3. In terms of graphing distinguish between the following:
   **(a)** average velocity and instantaneous velocity

   **(b)** average acceleration and instantaneous acceleration

4. Describe, using dynamics, how one could produce a non-uniform acceleration of an object?

5. Describe how to determine the area under a velocity-time curve with a non-uniform acceleration (i.e. increasing or decreasing slope).

6. Distinguish between position and displacement.

7. Identify the quantity that is changing every second when an object is accelerating.

**8.** Describe some practical applications of acceleration.

**9.** What does a negative area calculation under a velocity-time graph mean?

## Inquiry

**10.** You are given the results of a 60 Hz recording ticker tape timer experiment of a cart rolling down a ramp onto a level lab bench. By analyzing the dots on the ticker tape describe how you know the cart was
**(a)** moving with a positive uniform acceleration
**(b)** moving with a negative uniform acceleration
**(c)** moving with constant velocity
**(d)** at rest.

**11.** A student's school is directly north of her home. While she roller blades to school she accelerates uniformly from rest to a modest velocity and then maintains this velocity, until she meets a friend half way to school at which point she stops for a while. She then continues to move at the same constant velocity until she gets near the school, where she slows uniformly until she stops.
**(a)** Sketch a position-time graph for her motion
**(b)** Sketch a velocity-time graph for her motion

## Communication

**12.** Explain what physical quantity is measured by a car's speedometer.

## Making Connections

**13.** Refer to kinematics principles to suggest solutions to improve road safety. Consider such aspects as: vehicle design and safety features, driver training, reaction time, roadway design, construction and maintenance, road signs, maximum driving speed, and road safety enforcement.

**14.** Make a list of the of the advantages and disadvantages of the following means of transportation: **(a)** train **(b)** plane **(c)** car **(d)** ship. Which means of transportation do you feel is adapting the most and least to meet the needs of our Canadian society.

## Problems for Understanding

**15.** A truck is transporting new cars to a car dealership. There are 8 cars on the truck's trailer. Describe a frame of reference in which a car is:
**(a)** moving.
**(b)** at rest.

**16.** Draw a dot diagram to illustrate the motion of a car travelling from one traffic light to the next. When the traffic light turns green the car's speed increases, it then travels at a constant speed, and then brakes to slow down to a stop at the next traffic light.

**17.** A girl is taking her dog for a walk. They walk 5.0 km[N] and then turn around and walk 12 km[S].
**(a)** What is the total distance that they travelled?
**(b)** What is their displacement?
**(c)** What displacement would they have to walk to get back to their starting point?

**18.** A cyclist is travelling with an average velocity of 5.9 m/s[W]. What will be their displacement after 1.2 h?

**19.** A canoeist paddles 1.6 km downstream and then turns around and paddles back upstream for 1.2 km. The entire trip takes 45 minutes.
**(a)** What is the displacement of the canoeist?
**(b)** Calculate the average velocity of the canoeist.

**20.** The closest star to our solar system is Alpha Centauri, which is $4.12 \times 10^{16}$ m away. How long would it take light from Alpha Centauri to reach our solar system if the speed of light is $3.00 \times 10^8$ m/s? (Provide an answer in both seconds and in years.)

**21.** Vectorville and Scalartown are 20.0 km apart. A cyclist leaves Vectorville and heads for Scalartown at 20.0 km/h. A second cyclist leaves Scalartown for Vectorville at exactly the same time at a speed of 15.0 km/h.
**(a)** Where will the two cyclists meet between the two towns?
**(b)** How much time passes before they meet (in minutes)?

22. Describe each of the situations below as either uniform motion or non-uniform motion.
    (a) a car on the highway travels with its cruise control set at 90 km/h
    (b) a skydiver jumps from a plane and falls faster and faster through the air
    (c) a piece of paper that is dropped, flutters to the ground
    (d) a satellite travels in a circular orbit above the earth at a constant speed
    (e) you sit quietly, enjoying an autumn day.

23. Graph the following data. Find the slope at different intervals and sketch a speed-time graph of the motion.

| t (s) | d (m) | t (s) | d (m) |
|-------|-------|-------|-------|
| 0 | 0 | 11 | 41 |
| 1 | 4 | 12 | 43 |
| 2 | 8 | 13 | 44.5 |
| 3 | 12 | 14 | 45.5 |
| 4 | 16 | 15 | 46 |
| 5 | 20 | 16 | 46 |
| 6 | 24 | 17 | 46 |
| 7 | 28 | 18 | 46 |
| 8 | 32 | 19 | 46 |
| 9 | 35.5 | 20 | 46 |
| 10 | 38.5 | | |

24. A car is travelling at 14 m/s when the traffic light ahead turns red. The car brakes and comes to a stop in 5.0 s. Calculate the acceleration of the car.

25. At the very end of their race, a runner accelerates at 0.3 m/s² for 12 s to attain a speed of 6.4 m/s. Determine the initial velocity of the runner.

26. The acceleration due to gravity on the moon is 1.6 m/s²[down]. If a baseball was thrown with an initial velocity of 4.5 m/s[up], what would its velocity be after 4.0 s?

27. When the traffic light turns green the car's speed increases, it then travels at a constant speed, and then brakes to slow down to a stop at the next traffic light.
    (a) Sketch a position-time graph to represent the car's motion.
    (b) Sketch a velocity-time graph to represent the car's motion.

28. A car that starts from rest can travel a distance of $5.0 \times 10^1$ m in a time of 6.0 s.
    (a) What is the final velocity of the car at this time?
    (b) What is the acceleration of the car?

29. A cyclist is travelling at 5.6 m/s when she starts to accelerate at $0.60 \, \text{m/s}^2$ for a time interval of 4.0 s.
    (a) How far did she travel during this time interval?
    (b) What velocity did she attain?

30. A truck is travelling at 22 m/s when the driver notices a speed limit sign for the town ahead. He slows down to a speed of 14 m/s. He travels a distance of 125 m while he is slowing down.
    (a) Calculate the acceleration of the truck.
    (b) How long did it take the truck driver to change his speed?

31. A skydiver falling towards the ground with his parachute open accelerates at 3.2 m/s². Calculate his displacement if after 8.0 s he attained a velocity of 15 m/s[down].

32. A car is travelling on the highway at a constant speed of 24 m/s. The driver misses the posted speed limit sign for a small town she is passing through. The police car accelerates from rest at 2.1 m/s². From the time that the speeder passes the police car:
    (a) How long will it take the police car to catch up to the speeder?
    (b) What distance will the cars travel in that time?

**Answers to Numerical Problems**

1. −1.0 m/s  2. 1.2 m/s[N57°W]  3. (a) 0.29 m/s (b) 75 m or 175 m (c) 75 m (d) 0.87 m/s  4. for linear segments: 2.5 m/s, −7.5 m/s, 0.0 m/s, 3.8 m/s  6. 8.0 m/s²  7. 2.5 m/s²[up]  8. 24 m/s  9. (a) 5.0 m  (b) 1.6 m/s²  10. 34 s  11. $6 \times 10^2$ m  12. 10 m/s

**W**hy do some people find water skiing so exhilarating? Certainly, travelling at high speeds is thrilling, but for many, the challenge of sports has another dimension — the desire to gain and maintain control over the motion of your body or that of the vehicle you are operating.

Although a water-skier depends on the towboat for propulsion, the skier can travel in a different direction and at a different speed than the boat. Similarly, windsurfers rely on the wind, but learn how to harness its energy to sail in directions other than that of the wind. The challenge is to understand what affects the direction of the motion and use that knowledge to create the desired motion. The water-skier and windsurfer develop a sense or "feel" for the techniques that they need to control their motion. However, to put a satellite into orbit or to build a guidance system for a rocket, the knowledge of motion and its causes must be much more precise and based on calculations.

You began your study of motion in the last chapter by developing the concepts of displacement, velocity, and acceleration. You performed detailed analyses of these quantities in one dimension, or a straight line. In this chapter, you will expand your knowledge to two dimensions and learn more about the vector nature of the quantities that describe motion.

# Pushed Around

## Changing Course

How is the *direction of acceleration* related to the *change in velocity* of a marble rolling in a plane? Try this activity to explore the answer to this question. Draw an *x–y*-coordinate system on a piece of easel paper, and place the paper flat on your desk or lab bench. Draw vectors A and B on your chart, as shown in the diagram.

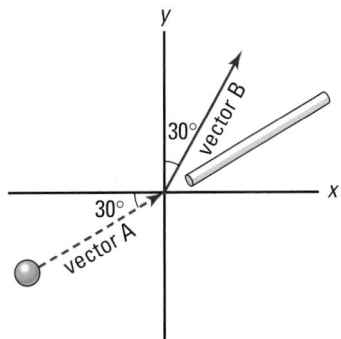

Roll a marble so that it travels along vector A at a reasonably slow speed. As the marble approaches the origin, gently blow on the marble through a straw, causing the marble to leave the origin, travelling along vector B. Practise this procedure until you have perfected your technique. On the easel paper, sketch the direction in which you needed to blow in order to create the desired change in direction of the motion of the marble. Repeat the procedure several more times with other pairs of vectors. Predict the direction you will need to blow to cause the change in velocity. Test your predictions.

### Analyze and Conclude

1. Analyze your sketches and look for a pattern that relates the direction of the push you exerted on the marble by blowing on it and the change in direction of the motion of the marble.

2. Describe the marble's motion by using the concepts of constant velocity, changes in velocity, and acceleration. Explain your reasoning.

## Watch Those Curves!

Draw a curved line on the easel paper. This time, investigate how you can cause the marble to follow the curve. Set the marble rolling toward the curve and then either blow or tap on it gently so that the marble follows the curved line. Carefully describe what you must do to keep the marble curving.

### Analyze and Conclude

1. Describe the marble's *speed* throughout its journey along the curved path.

2. Does the marble maintain a *constant velocity* during this experiment? Explain your reasoning.

3. Summarize the pattern you have found between the direction in which you blow or tap on the marble and its change in velocity as it follows the curve.

- Draw vector scale diagrams to visualize and analyze the nature of motion in a plane.

- Analyze motion by using scale diagrams to add and subtract vectors.

- Solve problems involving motion in a horizontal plane.

Imagine describing the motion of an expert water-skier to someone who had not watched the skier demonstrate his technique. You would probably do a lot of pointing in different directions. In a sense, you would be using vectors to describe the skier's motion. You have probably used vectors and even **vector diagrams** many times without realizing it. Did you ever draw a "treasure map" as a child? Have you ever gone hiking and drawn routes on a topographical map? If so, you already have a sense of the clarity with which vector diagrams help you to describe, analyze, and plan motion.

## Representing Vectors

You will represent all vector quantities with arrows that point in the direction of the quantity. The length of the arrow is proportional to the magnitude of the quantity you are representing. Vector quantities have direction, so you need a frame of reference or **coordinate system** to represent a direction. Vectors also have magnitude, so you need a scale to indicate and calculate magnitude.

Since you are probably comfortable with maps, they provide a good frame of reference to familiarize yourself with the rules for drawing vectors. For example, the map in Figure 3.1 shows displacement vectors between your home and school and between the fitness club and the library. The vectors are drawn to scale. On the map, 1.0 cm represents 2.0 km. Have you noticed that two of the displacement vectors are identical? The direction of displacement vectors B and C is 27 degrees north of east and the magnitude of each is 4.5 km. The beginning and ending points of a vector do not define the vector, only the length and direction. You can move a vector in its frame of reference as long as you do not change the length or the direction in which it points.

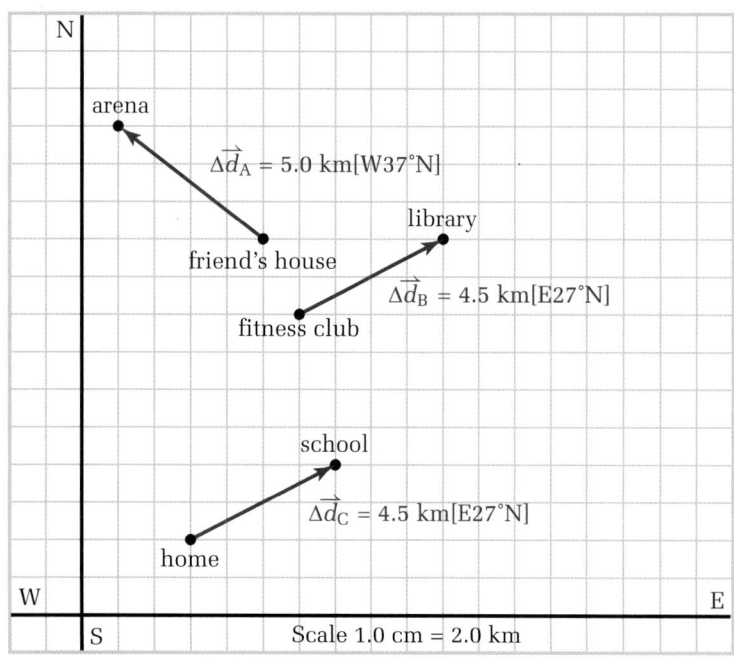

**Figure 3.1** The three displacement vectors are drawn to a scale of 1.0 cm to 2.0 km. To draw the directions from the symbols used here, start by pointing in the first compass direction. Then, rotate the stated number of degrees toward the last compass direction.

## Adding Vectors Graphically

Addition of vectors starts with some basic rules of arithmetic and then includes a few more rules. You have known for a long time that you cannot add "apples and oranges" or centimetres and metres. Similarly, you can add only vectors that represent the same quantity and are drawn to the same scale. Follow the steps listed in Table 3.1 to learn the graphical method for adding vectors. Any two vectors having the same units can be added according to the procedure in Table 3.1. The vector representing the sum is often called the **resultant vector.**

**Table 3.1** Graphical Vector Addition $(\vec{R} = \vec{A} + \vec{B})$

| Procedural step | Graph |
| --- | --- |
| Establish a coordinate system and choose a scale. | *(coordinate grid with x and y axes marked at −4, −2, 2, 4)* |
| Place the first vector $(\vec{A})$ in a coordinate system. | *(vector $\vec{A}$ drawn from origin)* |
| Place the tail of the second vector $(\vec{B})$ at the tip of the first vector. | *(vectors $\vec{A}$ and $\vec{B}$ drawn)* |
| Draw a vector from the tail of the first vector $(\vec{A})$ to the tip of the second $(\vec{B})$. Label it "$\vec{R}$." | *(vectors $\vec{A}$, $\vec{B}$, and $\vec{R}$ drawn)* |
| With a ruler, measure the length of $\vec{R}$. | *(vectors with ruler measuring $\vec{R}$)* |
| With a protractor, measure the angle between $\vec{R}$ and the horizontal axis. | *(vectors with protractor measuring angle)* |

**ELECTRONIC LEARNING PARTNER**

Go to your Electronic Learning Partner to visualize the process of adding vectors.

## Adding Vectors

A kayaker sets out for a paddle on a broad stretch of water. She heads toward the west, but is blown off course by a strong wind. After an hour of hard paddling, she arrives at a lighthouse that she knows is 12 km southwest of her starting point. She lands and waits for the wind to die down. She then paddles toward the setting sun and lands on a small island that is 8 km west of the lighthouse. In the calm of the evening, the kayaker plans to paddle straight back to her starting point. Use a vector diagram to determine her displacement from her starting point to the island. In which direction should she now head and how far will she have to paddle to go directly to the point from which she originally started paddling?

## Frame the Problem

- Make a scale diagram of the problem from the following sketch (not to scale). Your scale should be 1.0 cm : 2.0 km.

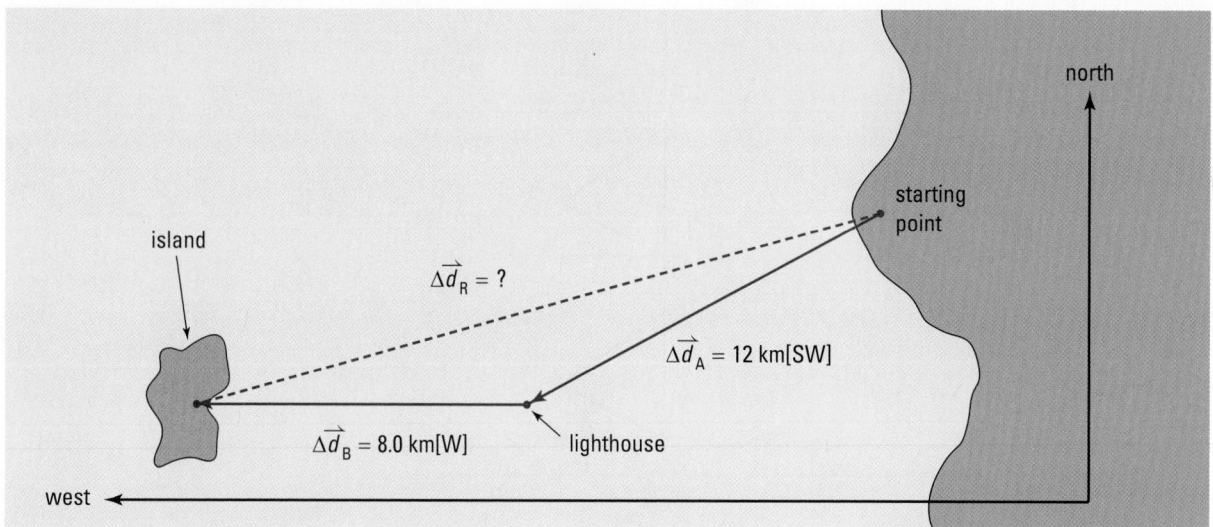

- The kayaker's journey consisted of two separate steps represented by the *two displacement vectors* in the diagram.

- The *vector sum* of the two vectors yields one *resultant vector* that shows the kayaker's final displacement.

- To return to the point she left earlier in the day, the kayaker will have to paddle a *displacement* that is *equal in magnitude* to her *resultant vector* and *opposite* in *direction*.

## Identify the Goal

(a) The displacement, $\Delta \vec{d}_R$, of the first two legs of the kayaker's trip

(b) The displacement, $\Delta \vec{d}$, needed to return to the point from which she originally started

## Variables and Constants

| Involved in the problem | | Known | Unknown |
|---|---|---|---|
| $\Delta \vec{d}_A$ | $\Delta \vec{d}_R$ | $\Delta \vec{d}_A = 12$ km[SW] | $\Delta \vec{d}_R$ |
| $\Delta \vec{d}_B$ | $\Delta \vec{d}$ | $\Delta \vec{d}_B = 8.0$ km[W] | $\Delta \vec{d}$ |

## Strategy

Measure the length of the resultant displacement vector in the scale diagram.

Multiply the length of the vector by the scale factor.

With a protractor, measure the angle between the horizontal axis and the resultant vector.

## Calculations

$\Delta \vec{d}_R = 4.75$ cm

$\Delta \vec{d}_R = 4.75 \text{ cm} \left( \dfrac{4.0 \text{ km}}{1.0 \text{ cm}} \right)$

$\Delta \vec{d}_R = 19$ km

$\theta = 27°$

(a) The resultant displacement is 19 km[W27°S].

(b) To return to the point from which she originally started her trip, the kayaker would have to paddle 19 km in the direction opposite to [W27°S], or 19 km[E27°N].

## Validate

The total distance that the kayaker paddled was 20 km (8 km + 12 km). However, her path was not straight. In paddling back to shore, her trip formed a triangle. Since any side of a triangle must be shorter than the sum of the other two, you expect that her direct return trip will be shorter than 20 km. In fact, it was 1.0 km shorter.

## PRACTICE PROBLEMS

1. An airplane flies with a heading of [N50.0°W] from Toronto to Sault Ste. Marie, a distance of 500 km ($5.0 \times 10^2$ km). The airplane then flies 750 km on a heading of [E10.0°S] to Ottawa.

   (a) Determine the airplane's displacement for the trip.

   (b) In what direction will the plane have to fly in order to return directly to Toronto?

2. A canoeist starts from her campsite, paddles 3.0 km due north, and then 4.0 km due west.

   (a) Determine her displacement for the trip.

   (b) In what direction would she have to head her canoe in order to paddle straight home?

*continued* ▶

3. From a lookout point, a hiker sees a small lake ahead of her. In order to get around it, she walks 4.5 km in a straight line toward the end of the lake. She makes a 60.0° turn to the right and walks to a campsite that is 6.4 km in the new direction. Determine her displacement from the lookout point when she has reached the campsite. (See the map on the right.)

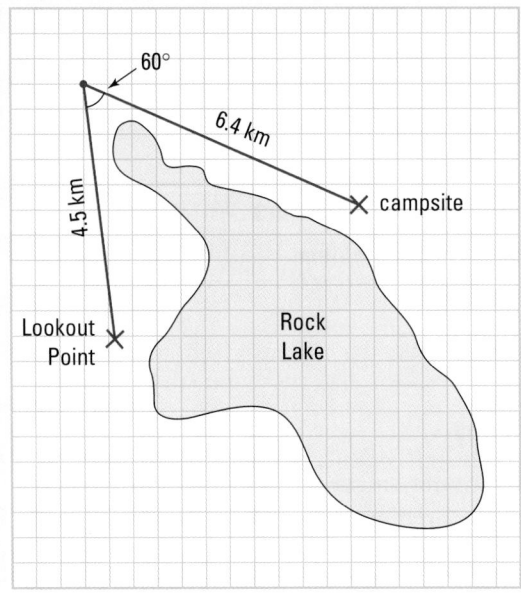

4. A boat heads out from port for a day's fishing. It travels 21.0 km due north to the first fishing spot. It then travels 30.0 km[W30.0°S] to a second spot. Finally, it turns and heads [W10.0°N] for 36.0 km.

   (a) Determine the boat's displacement for the entire journey.

   (b) In what direction should the boat point so as to head straight to its home port?

## Subtracting Vectors Graphically

The equations of motion with uniform acceleration include several cases in which you must subtract vector quantities. For example, displacement is defined as the difference of position vectors ($\Delta\vec{d} = \vec{d_2} - \vec{d_1}$). To find acceleration, you must first find a change in velocity, which is a difference in vector quantities ($\Delta\vec{v} = \vec{v_2} - \vec{v_1}$). In Chapter 2, you applied these equations to one dimension, so subtracting these quantities involved only arithmetic subtraction. However, when you work in two (or three) dimensions, you must account for the vector nature of these quantities.

Fortunately, vector subtraction is very similar to vector addition. Recall from basic math the expression, "A − B is equivalent to A + (−B)." The only additional piece of information that you need is the definition of the negative of a vector. In Figure 3.2, you can see that the negative of a vector is the same in magnitude and opposite in direction. Now, follow the directions in Table 3.2 for the first method for subtracting vectors graphically.

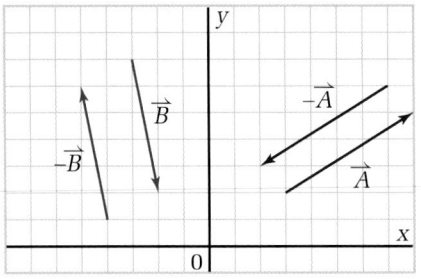

**Figure 3.2** To draw the negative of a vector, draw a line parallel to the positive vector and with an identical length. Then put an arrow head on the opposite end relative to the original vector.

**Table 3.2** Graphical Vector Subtraction ($\vec{R} = \vec{A} - \vec{B}$): Method 1

| Procedural step | Graph |
|---|---|
| Establish a coordinate system and choose a scale. | |
| Place the first vector ($\vec{A}$) in the coordinate system. | |
| Draw the negative of vector $\vec{B}$. | |
| Place the tail of $-\vec{B}$ at the tip of $\vec{A}$. | |
| Draw a vector from the tail of $\vec{A}$ to the tip of $-\vec{B}$. Label it "$\vec{R}$." | |
| With a ruler, measure the length of $\vec{R}$. | |
| With a protractor, measure the angle between $\vec{R}$ and the horizontal axis. | |

When subtracting vectors, you have an option of two different methods. Table 3.3 describes the second procedure.

**Table 3.3** Graphical Vector Subtraction ($\vec{R} = \vec{A} - \vec{B}$): Method Two

| Procedural step | Graph |
|---|---|
| Establish a coordinate system and choose a scale. | |
| Place the first vector ($\vec{A}$) in the coordinate system. | |
| Place the tail of the second vector ($\vec{B}$) at the *tail* of the first vector ($\vec{A}$). | |
| Draw a vector from the *tip* of the second vector ($\vec{B}$) to the *tip* of the first vector ($\vec{A}$). Label this vector "$\vec{R}$." | |
| With a ruler, measure the length of $\vec{R}$. | |
| With a protractor, measure the angle between $\vec{R}$ and the horizontal axis. | |

## Subtracting Vectors

A water-skier begins his ride by being pulled straight behind the boat. Initially, he has the same velocity as the boat ($50 \frac{km}{h}$[N]). Once up, the water-skier takes control and cuts out to the side. In cutting out to the side, the water-skier changes his velocity in both magnitude and direction. His new velocity is $60 \frac{km}{h}$[N60°E]. Find the water-skier's *change* in velocity.

## Frame the Problem

- Make a scale diagram of the water-skier's initial and final velocities.

- The water-skier follows directly behind the boat, while rising to the surface.

- He then *changes* the *magnitude* and *direction* of his motion.

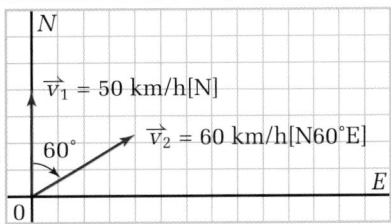

## Identify the Goal

The change in the velocity, $\Delta \vec{v}$, of the water-skier

## Variables and Constants

| Involved in the problem | Known | Unknown |
|---|---|---|
| $\vec{v_1}$ | $\vec{v_1} = 50 \ \frac{km}{h}$[N] | $\Delta \vec{v}$ |
| $\vec{v_2}$ | $\vec{v_2} = 60 \ \frac{km}{h}$[N60°E] | |
| $\Delta \vec{v}$ | | |

## Strategy

Write the mathematical definition for the change in velocity.

Draw a coordinate system and choose a scale. Put $\vec{v_2}$ in the coordinate system.

## Calculations

$$\Delta \vec{v} = \vec{v_2} - \vec{v_1}$$

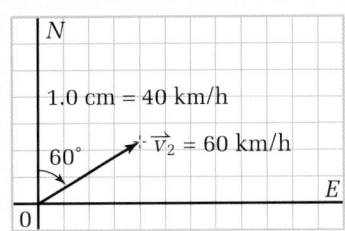

*continued* ▶

Place the tail of vector $-\vec{v}_1$ at the tip of $\vec{v}_2$ and draw $\Delta\vec{v}$.

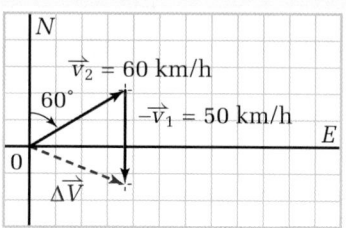

Measure the magnitude and direction of $\Delta\vec{v}$.

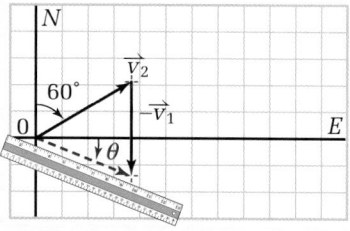

$$|\Delta\vec{v}| = 1.4 \text{ cm}, \ \theta = 21°$$

Multiply the magnitude of the vector by the scale factor 1.0 cm = 40 km/h.

$$|\Delta\vec{v}| = 1.4 \text{ cm} \left( \frac{40 \ \frac{\text{km}}{\text{h}}}{1.0 \text{ cm}} \right)$$

The water-skier's change in velocity was 56 km/h[E21°S].

$$|\Delta\vec{v}| = 56 \ \frac{\text{km}}{\text{h}}$$

## Validate

The magnitude of the change in the water-skier's velocity is smaller than the sum of the two velocities, which it should be.

Use the alternative method to check the answer. Draw a coordinate system to the same scale. Draw vectors $\vec{v}_1$ and $\vec{v}_2$ with their tails together. Draw a vector from the tip of $\vec{v}_1$ to the tip of $\vec{v}_2$. Measure the magnitude and direction of the resultant vector. As you can see in the diagram, these values are identical to those obtained using the first method.

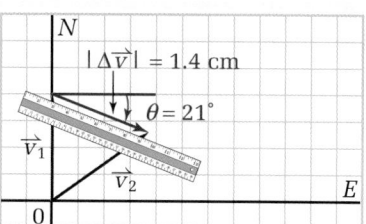

## PRACTICE PROBLEMS

Use both methods of vector subtraction to solve the following problems.

5. Given: $\vec{P} = 12$ km[N], $\vec{Q} = 15$ km[S], $\vec{R} = 10$ km[N30°E]

   (a) Use method 1 of graphical vector subtraction to solve each of the following:
   (i) $\vec{P} - \vec{Q}$  (ii) $\vec{R} - \vec{Q}$  (iii) $\vec{Q} - \vec{R}$

   (b) Use method 2 of graphical vector subtraction to solve each of the following:
   (i) $\vec{P} - \vec{Q}$  (ii) $\vec{R} - \vec{Q}$  (iii) $\vec{P} - \vec{R}$

6. A car is travelling east at 45 km/h. It then heads north at 50 km/h ($5.0 \times 10^1$ km/h). Determine its change in velocity.

7. An airplane is flying at $2.00 \times 10^2$ km/h [S30.0°W]. It makes a smooth wide turn and heads east at $2.00 \times 10^2$ km/h. Find its change in velocity.

8. A hockey puck hits the boards at a velocity of 12 m/s at an angle of 30° to the boards. It is deflected with a velocity of 10 m/s at an angle of 25° to the boards. Determine the puck's change in velocity.

9. A runner's velocity is recorded at four different points along the route, $\vec{v_1} = 3.5$ m/s[S], $\vec{v_2} = 5.0$ m/s[N12°W], $\vec{v_3} = 4.2$ [W], and $\vec{v_4} = 2.0$ m/s[S76°E].

(a) Graphically determine the change in velocity between $\vec{v_1}$ and $\vec{v_2}$.

(b) Calculate the change in velocity between $\vec{v_1}$ and $\vec{v_3}$.

(c) Graphically determine the change in velocity between $\vec{v_1}$ and $\vec{v_4}$.

## Think It Through

- Review your findings in the "Changing Course" lab at the beginning of this chapter. Assume that the marble rolls with a constant speed of 25 cm/s. Use method 2 for subtracting vectors to show $\Delta\vec{v}$ for the rolling marble when it changes direction. Do you see a relationship between the direction you find for $\Delta\vec{v}$ and the direction in which you blew on the marble to make it change directions?

- A marble is rolling across a lab bench. In what direction would you need to give the marble a sharp poke so that it would turn and travel at an angle of 90° to its original direction? Illustrate your thinking by drawing a sketch to show the original direction of the marble, the new direction of the marble, and the direction of the poke. Verify your thinking by conducting a mini-experiment.

- Two identical cars, travelling at the same speed, approach an intersection at right angles to each other on a day when the streets are very icy. Unfortunately, neither car is able to stop. During the collision, the cars are jammed together and the combined wreckage slides off the street. Draw a sketch of the accident and illustrate the direction in which the wreckage will travel.

## Multiplying and Dividing Vectors by Scalars

A brief review of the equations of motion will reveal one more important type of mathematical operation on vector quantities. These quantities are often multiplied or divided by the scalar quantity, $\Delta t$, the time interval. What happens to a vector when it is multiplied or divided by a scalar? To help answer that question, review the definition of average velocity: $\vec{v}_{ave} = \frac{\Delta\vec{d}}{\Delta t}$. When the displacement vector is divided by the time interval, the magnitude changes and the units change. The only thing that does *not* change is the direction of the vector. In summary, when a vector is multiplied or divided by a scalar, a different quantity is created, but the direction remains the same.

## Dividing Vectors by Scalars

1. A hiker sets out for a trek along a mountain trail. After 2 h, she checks her global positioning system and finds that she is 8 km[W40°N] from her starting point. What was her average velocity for the trek?

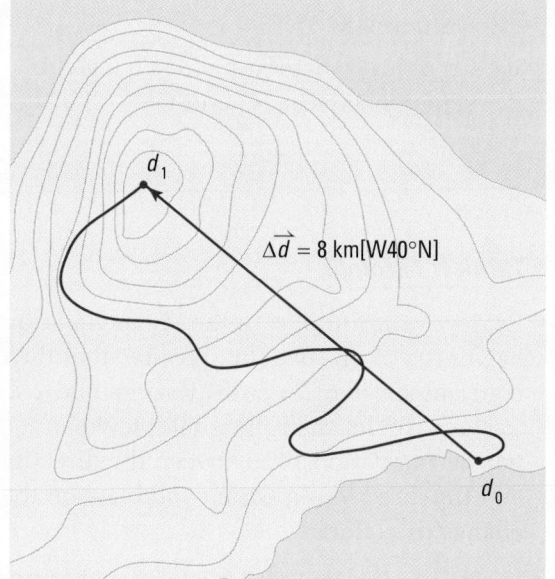

$\vec{\Delta d} = 8$ km[W40°N]

## Frame the Problem

- The hiker follows a meandering path; however, *displacement* is a *vector* that depends only on the *initial* and *final positions*.

- *Velocity* is the vector quotient of *displacement* and the *time interval*.

- The *direction* of the hiker's *velocity* is therefore the same as the *direction* of her *displacement*.

## Identify the Goal

The hiker's average velocity, $\vec{v}_{ave}$, for the trip

## Variables and Constants

| Involved in the problem | Known | Unknown |
|---|---|---|
| $\Delta t$ | $\Delta t = 2$ h | $\vec{v}_{ave}$ |
| $\vec{\Delta d}$ | $\vec{\Delta d} = 8$ km[W40°N] | |
| $\vec{v}_{ave}$ | | |

## Strategy

Use the equation that defines average velocity to calculate the magnitude of the average velocity.

The direction of the velocity is the same as the direction of the displacement.

## Calculations

$$\vec{v}_{ave} = \frac{\vec{\Delta d}}{\Delta t}$$

$$\vec{v}_{ave} = \frac{8 \text{ km[W40°N]}}{2 \text{ h}}$$

$$\vec{v}_{ave} = 4 \frac{\text{km}}{\text{h}}[\text{W40°N}]$$

The average velocity for the trip was 4 km/h[W40°N].

## Validate

The magnitude, 4 km/h, is quite reasonable for a hike. The direction can only be the same as the direction of the displacement.

**2.** A hot-air balloon rises into the air and drifts with the wind at a rate of 24 km/h[E40°N] for 2 h. The wind shifts, so the balloon changes direction and drifts south at a rate of 40 km/h for 1.5 h before landing. Determine the balloon's displacement for the flight.

## Frame the Problem

- Make a diagram of the problem.

- The balloon trip takes place in *two stages* (Phase A and Phase B).

- Data for the trip is given in terms of *velocities* and *time intervals*.

- The *total displacement* is the *vector sum* of the two *displacement vectors*.

$\vec{v}_A = 24 \text{ km/h[E40°N]}$

$\vec{v}_B = 40 \text{ km/h[S]}$

**PROBLEM TIP**

Average velocity is the *displacement divided by time interval.* You should *never* attempt to find an overall average velocity by taking the vector sum of two velocities.

## Identify the Goal

Total displacement, $\Delta \vec{d}_{total}$, for the balloon trip

## Variables and Constants

| Involved in the problem | | Known | Unknown |
|---|---|---|---|
| $\Delta t_A$ | $\Delta \vec{d}_A$ | $\Delta t_A = 2.0 \text{ h}$ | $\Delta \vec{d}_A$ |
| $\Delta t_B$ | $\Delta \vec{d}_B$ | $\Delta t_B = 1.5 \text{ h}$ | $\Delta \vec{d}_B$ |
| $\vec{v}_{A \ ave}$ | $\Delta \vec{d}_{total}$ | $\vec{v}_{A \ ave} = 24 \ \dfrac{\text{km}}{\text{h}}$ | $\Delta \vec{d}_{total}$ |
| $\vec{v}_{B \ ave}$ | | $\vec{v}_{B \ ave} = 40 \ \dfrac{\text{km}}{\text{h}}$ | |

## Strategy

Use the velocity for Phase A to calculate the displacement for Phase A.

Select a scale and draw scale diagrams for Phase A.

## Calculations

$\Delta \vec{d}_A = \vec{v}_A \Delta t_A$

$\Delta \vec{d}_A = 24 \ \dfrac{\text{km}}{\text{h}} \text{[E40°N]}(2.0 \text{ h})$

$\Delta \vec{d}_A = 48 \text{ km[E40°N]}$

**Phase A**

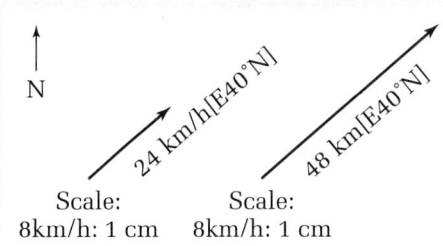

N

24 km/h[E40°N]

48 km[E40°N]

Scale: 8km/h: 1 cm    Scale: 8km/h: 1 cm

Velocity Vector × 2 = Displacement Vector

*continued* ▶

*continued from previous page*

## Strategy

Use the velocity for Phase B to calculate the displacement for Phase B.

Draw diagrams for Phase B using the scale you selected for Phase A.

On a graph, draw displacement vector A with its tail at the origin.

Draw displacement vector B with its tail at the tip of vector A.

Draw the resultant velocity vector from the tail of A to the tip of B. Label it "$\Delta\vec{d}_{total}$."

Measure the magnitude and direction of the total displacement vector.

Multiply the magnitude of the vector by the scale factor (1 cm = 20 km)

The balloon trip had a displacement of 47 km[E38°S].

## Calculations

$\Delta\vec{d}_B = \vec{v}_B \Delta t_B$

$\Delta\vec{d}_B = 40\ \dfrac{km}{h}[S](1.5\ h)$

$\Delta\vec{d}_B = 60\ km[S]$

$|\Delta\vec{d}_{total}| = 2.35\ cm \times \left(\dfrac{20\ km}{cm}\right)$

$|\Delta\vec{d}_{total}| = 47\ km$

$\theta = 38°$

## Phase B

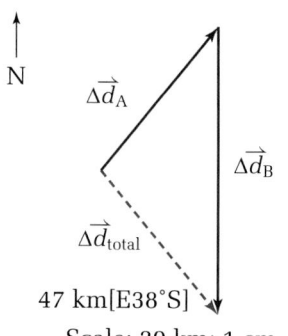

Scale: 8km/h: 1 cm     Scale: 8km/h: 1 cm

Velocity Vector × 1.5 = Displacement Vector

**Total Displacement**

47 km[E38°S]

Scale: 20 km: 1 cm

## Validate

The wind caused a significant change in course. Since the balloon made a fairly sharp turn, you would expect that the total displacement would be shorter than the sum of the two legs of the trip. This was, in fact, the case (47 km is much shorter than 48 km + 60 km).

### PRACTICE PROBLEMS

10. A hang-glider launches herself from a high cliff and drifts 6.0 km south. A wind begins to blow from the southeast, which pushes her 4.0 km northwest before she lands in a field. Calculate her average velocity for the trip if she is in the air for 45 min.

**11.** A light plane leaves Buttonville Airport, north of Toronto, and flies 80.0 km[W30.0°S] to Kitchener. After picking up a passenger, it flies to Owen Sound, which is 104 km[N20.0°W] of Kitchener. The entire trip took 2.5 h.

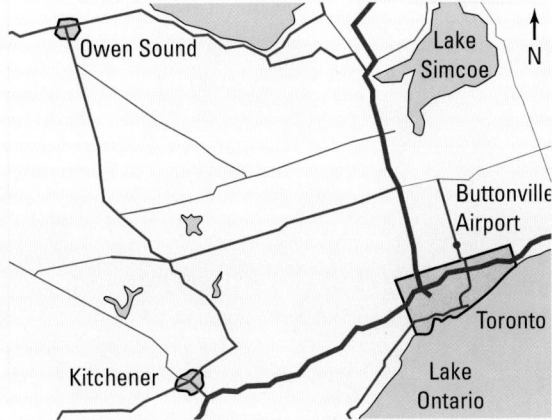

**(a)** Calculate the average velocity of the plane for the entire trip from Toronto to Owen Sound.

**(b)** If the pilot wants to fly straight back to Buttonville from Owen Sound in 1.0 h, with what velocity will she need to fly? Assume that there is no wind.

**12.** A canoeist paddles across a calm lake with a velocity of 3.0 m/s north for 30.0 min. He then paddles with a velocity of 2.5 m/s west for 15.0 min. Determine his displacement from where he began paddling.

**13.** A hiker sets out on a trek heading [N35°E] at a pace of 5.0 km/h for 48.0 min. He then heads west at 4.5 km/h for 40.0 min. Finally, he heads [N30°W] for 6.0 km, until he reaches a campground 1.5 hours later.

**(a)** Draw a displacement vector diagram to determine his total displacement.

**(b)** Determine his average velocity for the trip.

(Hint: Remember the definition of average velocity.)

# 3.1    Section Review

**1.** **C** Describe the difference between

**(a)** vectors and scalars

**(b)** positive and negative vectors of the same magnitude

**(c)** a resultant vector for displacement during of trip and the vector representing the return trip

**(d)** vector addition and vector subtraction

**(e)** a coordinate system and a frame of reference

**2.** **K/U** A map has a scale where 50 km is equal to 1 cm. If two towns are 7 cm apart on the map, what is their actual separation in km?

**3.** **I** Investigate how the resultant vector will change when vector A is added to:

**(a)** vector $\vec{A}$

**(b)** vector $-\vec{A}$

**(c)** vector $2\vec{A}$

**(d)** vector $-5\vec{A}$

**4.** **K/U** Define average velocity.

**5.** **K/U** A hiker sets out down the trail at a pace of 5 km/h for one hour and then returns to her camp at the same pace to pick up her trail map. What is her average velocity upon arriving back at camp?

## SECTION EXPECTATIONS

- Describe the motion of an object that is in a moving medium using velocity vectors.

- Analyze quantitatively, the motion of an object relative to different reference frames.

## KEY TERM

- relative velocity

**Figure 3.3** The dog is moving relative to the water, and the water is moving relative to the river bank. How would you describe the dog's velocity relative to the river bank?

Have you ever been stopped at a stoplight and suddenly felt that you were starting to roll backward? Your immediate instinct was to slam on the brakes, but then you realized that you were not moving after all. You sheepishly realized that, in fact, the car beside you was creeping forward. Your mind had been tricked. Subconsciously, you assumed that the car beside you had remained stopped and that your car was moving backward. Indeed, you *were* moving, but not in the way in which your instincts were telling you. You were moving backward relative to the car beside you, because the other car was moving forward relative to the ground. Relative motion can be deceiving. The dog in the photograph may think that the river bank is moving sideways while he is swimming directly across the river. Is it?

## Relative Velocity

Vector addition is a critical tool in calculating **relative velocities**. You have discovered that it is necessary to define a reference frame to describe any velocity. How do you relate velocities in two different reference frames? For example, an aviator must use the ground as a frame of reference to plot an airplane trip. However, when the plane is airborne, the air itself is moving relative to the ground, carrying the plane with it. So the aviator must account for both the motion of the plane relative to the air and the air relative to the ground. By defining velocity vectors for the plane relative to

**COURSE CHALLENGE**

**Staying in Orbit**

Investigate the question, "Are geosynchronous satellites ever in the shade?" A quick lab idea to help you solve this problem is presented on the following web site (follow the **Science Resources** and **Physics 11** links), and on the Electronic Learning Partner. *www.school. mcgrawhill.ca/resources/*

the air and for the air relative to the ground, the aviator can use vector addition to calculate the velocity of the plane relative to the ground, the critical piece of information.

An understanding of relative velocities is not a new problem. Imagine sailing on the high seas on a ship such as the one in Figure 3.4, in the days before communication technology was highly developed. Understanding wind and ocean currents was critical for navigation. The following model problems show you how to perform such calculations.

**PHYSICS FILE**

Nearly everyone has heard of Einstein's theory of relativity, but very few people understand it. According to the theory, relative velocities, as well as lengths of objects and time intervals, would appear to be very strange if you could travel close to the speed of light. For example, imagine that a train travelling close to the speed of light is passing through a short tunnel. An observer, who is stationary relative to the ground, would see the last car disappear into the tunnel before seeing the engine come out the other end. An observer on the train would perceive that the engine was leaving one end of the tunnel before the last car entered the other end. Now that's relativity!

**Figure 3.4** The lives of early sailors depended on their ability to accurately predict and control the motion of their ships, without any modern technology.

## MODEL PROBLEMS

### Calculating Relative Velocities

1. A canoeist is planning to paddle to a campsite directly across a river that is 624 m wide. The velocity of the river is 2.0 m/s[S]. In still water, the canoeist can paddle at a speed of 3.0 m/s. If the canoeist points her canoe straight across the river, toward the east: (a) How long will it take her to reach the other river bank? (b) Where will she land relative to the campsite? (c) What is the velocity of the canoe relative to the point on the river bank, where she left?

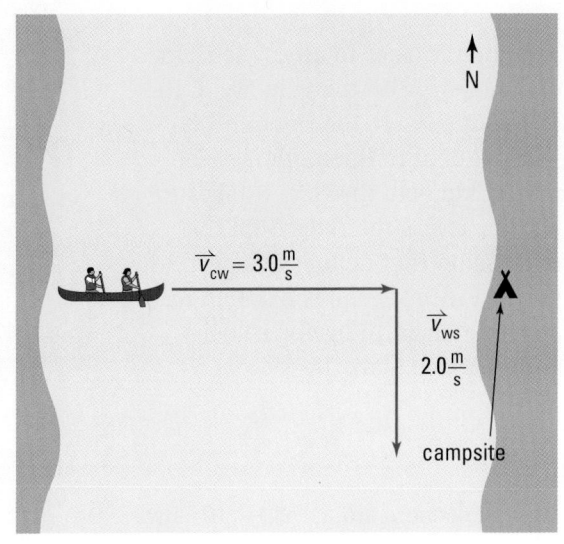

### Frame the Problem

■ Make a sketch of the situation described in the problem.

■ The *canoe* is *moving* relative to the *river*.

*continued* ▶

*continued from previous page*

- The *river* is *moving* relative to the *shore*.

- The *motion* of the *river* toward the south will *not* affect the *eastward motion* of the canoe.

- While the canoeist is *paddling east* across the river, the river is *carrying the canoe* downstream [S] with the current.

- The vector sum of the velocity of the *canoe* *relative* to the *water* and the *water* relative to the *shore* is the velocity of the *canoe* relative to the *shore*.

## Identify the Goal

(a) The time, $\Delta t_{SE}$, it takes for the canoe to reach the far bank

(b) The displacement, $\Delta \vec{d}_S$, of the canoe from the campsite at the point where the canoe comes ashore

(c) The velocity, $\vec{v}_{cs}$, of the canoe relative to the shore

## Variables and Constants

| Involved in the problem | Known | Unknown |
|---|---|---|
| $\Delta \vec{d}_E$ (width of river) | $\Delta \vec{d}_E = 624$ m | $\Delta \vec{d}_S$ |
| $\Delta \vec{d}_s$ (distance from campsite point where canoe went ashore) | $\vec{v}_{ws} = 2.0 \, \frac{m}{s}$ [S] | $\vec{v}_{ws}$ |
| | $\vec{v}_{cw} = 3.0 \, \frac{m}{s}$ [E] | $\vec{v}_{cw}$ |
| $\vec{v}_{ws} \quad \vec{v}_{cs} \quad \Delta t_E$ | | $\vec{v}_{cs}$ |
| $\vec{v}_{cw} \quad \theta \quad \Delta t_{SE}$ | | |

## Strategy

The time it takes to cross the river depends *only* on the velocity of the canoe relative to the river and is independent of the motion of the river. Calculate the time it takes to paddle across the river from the expression that defines average velocity. Since time is a scalar, use absolute magnitudes for velocity and displacement.

## Calculations

### Substitute first

$$|\vec{v}_{cw}| = \frac{|\Delta \vec{d}_E|}{\Delta t_{SE}}$$

$$3.0 \, \frac{m}{s} = \frac{624 \text{ m}}{\Delta t_{SE}}$$

$$3.0 \, \frac{m}{s} \Delta t_{SE} = \frac{624 \text{ m}}{\cancel{\Delta t_{SE}}} \cancel{\Delta t_{SE}}$$

$$3.0 \, \frac{m}{s} \Delta t_{SE} = 624 \text{ m}$$

$$\frac{\cancel{3.0 \, \frac{m}{s}} \Delta t_{SE}}{\cancel{3.0 \, \frac{m}{s}}} = \frac{624 \, \cancel{m}}{3.0 \, \frac{\cancel{m}}{s}}$$

$$\Delta t_{SE} = 208 \text{ s}$$

### Solve for $\Delta t$ first

$$|\vec{v}_{cw}| = \frac{|\Delta \vec{d}_E|}{\Delta t_{SE}}$$

$$|\vec{v}_{cw}| \Delta t_{SE} = \frac{|\Delta \vec{d}_E|}{\cancel{\Delta t_{SE}}} \cancel{\Delta t_{SE}}$$

$$\frac{\cancel{|\vec{v}_{cw}|} \Delta t_{SE}}{\cancel{|\vec{v}_{cw}|}} = \frac{|\Delta \vec{d}_E|}{|\vec{v}_{cw}|}$$

$$\Delta t_{SE} = \frac{|\Delta \vec{d}_E|}{|\vec{v}_{cw}|}$$

$$\Delta t_{SE} = \frac{624 \, \cancel{m}}{3.0 \, \frac{\cancel{m}}{s}}$$

$$\Delta t_{SE} = 208 \text{ s}$$

(a) It took the canoeist $2.1 \times 10^2$ s (or 3.5 min) to paddle across the river when her canoe was pointed directly east across the river.

## Strategy

During the 208 s that the canoeist was paddling, the river current was carrying her south, down the river. To find the distance down river that she landed, simply find the distance that she would travel at the velocity of the current. Use the equation for the definition of average velocity. (**Note:** When using an answer to a previous part of a problem in another calculation, use the unrounded value.)

## Calculations

### Substitute first

$$\Delta \vec{d}_S = \vec{v}_{ws}\, \Delta t_{SE}$$

$$\Delta \vec{d}_S = 2.0\ \tfrac{m}{s}\,[S](208\ \cancel{s})$$

$$\Delta \vec{d}_S = 416\ m[S]$$

### Solve for $\Delta t$ first

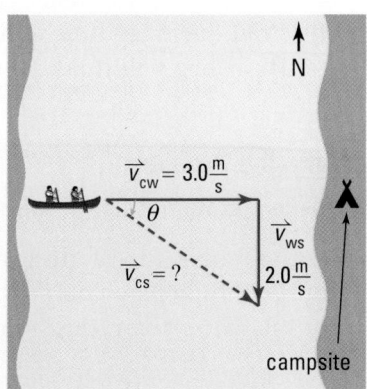

**(b)** The river carried the canoeist $4.2 \times 10^2$ m down river from the campsite.

To find the velocity of the canoe relative to the shore, find the vector sum of the velocity of the canoe relative to the water and the velocity of the water relative to the shore. Since the two vectors that you are adding are perpendicular to each other, the resultant is the hypotenuse of a right triangle. You can use the Pythagorean theorem to find the magnitude of the velocity.

$$\left|\vec{v}_{cs}\right|^2 = \left|\vec{v}_{cw}\right|^2 + \left|\vec{v}_{ws}\right|^2$$

$$\left|\vec{v}_{cs}\right|^2 = \left(3.0\ \tfrac{m}{s}\right)^2 + \left(2.0\ \tfrac{m}{s}\right)^2$$

$$\left|\vec{v}_{cs}\right|^2 = 9.0\left(\tfrac{m}{s}\right)^2 + 4.0\left(\tfrac{m}{s}\right)^2$$

$$\left|\vec{v}_{cs}\right|^2 = 13.0\left(\tfrac{m}{s}\right)^2$$

$$\left|\vec{v}_{cs}\right| = 3.6\ \tfrac{m}{s}$$

From the diagram, you can see that the tangent of the angle, $\theta$, is the velocity of the water relative to the shore divided by the velocity of the canoe relative to the water. Use this expression to find $\theta$.

$$\tan\theta = \frac{\left|\vec{v}_{ws}\right|}{\left|\vec{v}_{cw}\right|}$$

$$\tan\theta = \frac{2.0\ \cancel{\tfrac{m}{s}}}{3.0\ \cancel{\tfrac{m}{s}}}$$

$$\tan\theta = 0.6666$$

$$\theta = \tan^{-1} 0.6666$$

$$\theta = 33.69°$$

**(c)** The velocity of the canoe relative to the shore was 3.6 m/s[E34°S].

---

## Validate

In every case, the units cancelled to give the correct unit, (a) time in seconds, (b) displacement in metres and direction, and (c) velocity in metres per second and direction.

You would expect the velocity of the canoe relative to the shore to be larger than either the canoe relative to water and the water relative to the shore. You would also expect it to be less than the sum of the two magnitudes. All of these values were observed. The answers were all reasonable.

---

*continued* ▶

2. The canoeist in question 1 wants to head her canoe in such a direction that she will actually travel straight across the river to the campsite. (a) In what direction must she point the canoe? (b) Find the magnitude of her velocity relative to the shore. (c) How long will it take the canoeist to paddle to the campsite?

## Frame the Problem

- Make a sketch of the problem.

- As the canoeist is paddling across the river, she must continuously *paddle upstream* to make up for the *current* carrying her *downstream*.

- Her *velocity relative* to the *shore* will have a *direction of east*, but the *magnitude* will be *less* than the *3.0 m/s* that she paddles *relative* to the *water*.

- The effective *distance* that she *paddles* will be *greater* than the *width* of the river, because she is, in a sense, going upstream.

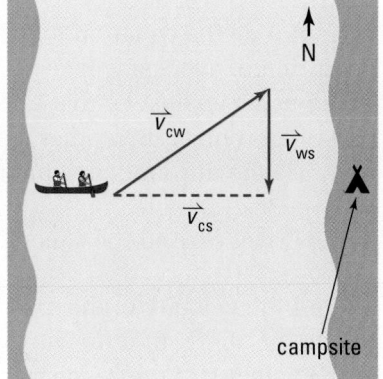

## Identify the Goal

(a) The direction, $\theta$, in which the canoe must point

(b) The magnitude of the velocity, $|\vec{v}_{cs}|$, of the canoe relative to the shore

(c) The time interval, $\Delta t$, to paddle to the campsite

## Variables and Constants

| Involved in the problem | Known | Unknown |
|---|---|---|
| $\vec{v}_{ws}$  $\vec{v}_{cs}$ | $\vec{v}_{ws} = 2.0\ \dfrac{m}{s}\,[S]$ | $|\vec{v}_{cs}|$ |
| $\vec{v}_{cw}$  $\theta$ | $|\vec{v}_{cw}| = 3.0\ \dfrac{m}{s}$ | $\theta$ |
| $\Delta t$ | | $\Delta t$ |

## Strategy

In this case, the magnitudes of the hypotenuse and opposite side of the triangle are known. Find the angle.

## Calculations

$$\sin \theta = \frac{2.0\ \frac{m}{s}}{3.0\ \frac{m}{s}}$$

$$\sin \theta = 0.6667$$

$$\theta = \sin^{-1} 0.6667$$

$$\theta = 41.8°$$

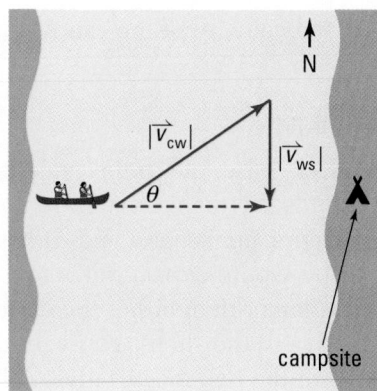

(a) The canoeist must point the canoe 42° upstream in order to paddle directly across the river to the campsite.

## Strategy

Use the Pythagorean theorem to find the magnitude of the velocity of the canoe relative to the shore.

## Calculations

### Substitute first

$$\left|\vec{v}_{cw}\right|^2 = \left|\vec{v}_{cs}\right|^2 + \left|\vec{v}_{ws}\right|^2$$

$$\left(3.0\ \frac{m}{s}\right)^2 = \left|\vec{v}_{cs}\right|^2 + \left(2.0\ \frac{m}{s}\right)^2$$

$$9.0\left(\frac{m}{s}\right)^2 - 4.0\left(\frac{m}{s}\right)^2 = \left|\vec{v}_{cs}\right|^2 + 4.0\left(\frac{m}{s}\right)^2 - 4.0\left(\frac{m}{s}\right)^2$$

$$9.0\left(\frac{m}{s}\right)^2 - 4.0\left(\frac{m}{s}\right)^2 = \left|\vec{v}_{cs}\right|^2$$

$$5.0\left(\frac{m}{s}\right)^2 = \left|\vec{v}_{cs}\right|^2$$

$$\left|\vec{v}_{cs}\right| = 2.24\left(\frac{m}{s}\right)$$

### Solve for $\left|\vec{v}_{cs}\right|$ first

$$\left|\vec{v}_{cw}\right|^2 = \left|\vec{v}_{cs}\right|^2 + \left|\vec{v}_{ws}\right|^2$$

$$\left|\vec{v}_{cw}\right|^2 - \left|\vec{v}_{ws}\right|^2 = \left|\vec{v}_{cs}\right|^2 + \left|\vec{v}_{ws}\right|^2 - \left|\vec{v}_{ws}\right|^2$$

$$\left|\vec{v}_{cs}\right|^2 = \left|\vec{v}_{cw}\right|^2 - \left|\vec{v}_{ws}\right|^2$$

$$\left|\vec{v}_{cs}\right| = \sqrt{\left|\vec{v}_{cw}\right|^2 - \left|\vec{v}_{ws}\right|^2}$$

$$\left|\vec{v}_{cs}\right| = \sqrt{\left(3.0\ \frac{m}{s}\right)^2 - \left(2.0\ \frac{m}{s}\right)^2}$$

$$\left|\vec{v}_{cs}\right| = \sqrt{9.0\left(\frac{m}{s}\right)^2 - 4.0\left(\frac{m}{s}\right)^2}$$

$$\left|\vec{v}_{cs}\right| = \sqrt{5.0\left(\frac{m}{s}\right)^2}$$

$$\left|\vec{v}_{cs}\right| = 2.24\ \frac{m}{s}$$

**(b)** The magnitude of the velocity of the canoe relative to the shore was 2.2 m/s.

Use the calculated velocity of the canoe relative to the shore and the known distance across the river to find the time interval for crossing the river.

### Substitute first

$$\vec{v}_{cs} = \frac{\Delta \vec{d}}{\Delta t}$$

$$2.23\ \frac{m}{s}[E] = \frac{624\ m[E]}{\Delta t}$$

$$2.23\ \frac{m}{s}[E]\Delta t = \frac{624\ m[E]}{\cancel{\Delta t}}\cancel{\Delta t}$$

$$\frac{2.23\ \cancel{\frac{m}{s}}\ \cancel{[E]}\Delta t}{2.23\ \cancel{\frac{m}{s}}\ \cancel{[E]}} = \frac{624\ \cancel{m[E]}}{2.23\ \frac{m}{s}\ \cancel{[E]}}$$

$$\Delta t = 279\ s$$

### Solve for $\Delta t$ first

$$\vec{v}_{cs} = \frac{\Delta \vec{d}}{\Delta t}$$

$$\vec{v}_{cs}\Delta t = \frac{\Delta \vec{d}}{\cancel{\Delta t}}\cancel{\Delta t}$$

$$\frac{\cancel{\vec{v}_{cs}}\ \Delta t}{\cancel{\vec{v}_{cs}}} = \frac{\Delta \vec{d}}{\vec{v}_{cs}}$$

$$\Delta t = \frac{\Delta \vec{d}}{\vec{v}_{cs}}$$

$$= \frac{624\ \cancel{m}[E]}{2.23\ \frac{m}{s}\ \cancel{[E]}}$$

$$= 279\ s$$

**(c)** It took $2.8 \times 10^2$ s (or 4.7 min) for the canoe to cross the river and reach shore at the campsite.

*continued* ▶

*continued from previous page*

## Validate

In every case, the units cancelled to give the correct units for the desired quantity.

You would expect that it would take longer to paddle across the river when taking the current into account than it would to paddle directly across relative to the water. It took about 70 s, or more than a minute, longer.

---

### PRACTICE PROBLEMS

**14.** A kayaker paddles upstream in a river at 3.5 m/s relative to the water. Observers on shore note that he is moving at only 1.7 m/s upstream. Determine the velocity of the current in the river.

**15.** A jet-ski speeds across a river at 11 m/s relative to the water. The jet ski's heading is due south. The river is flowing west at a rate of 5.0 m/s. Determine the jet-ski's velocity relative to the shore.

**16.** A bush pilot wants to fly her plane to a lake that is 250.0 km [N30.0°E] from her starting point. The plane has an air speed of 210.0 km/h, and a wind is blowing from the west at 40.0 km/h.

   **(a)** In what direction should she head the plane to fly directly to the lake?

   **(b)** If she uses the heading determined in (a), what will be her velocity relative to the ground?

   **(c)** How long will it take her to reach her destination?

**17.** An airplane travels due north for $1.0 \times 10^2$ km, then due west for $1.5 \times 10^2$ km, and then due south for $5.0 \times 10^2$ km.

   **(a)** Use vectors to find the total displacement of the airplane.

   **(b)** The time the airplane takes to fly the three different parts of the trip are as follows: 20.0 minutes, 40.0 minutes, and 12.0 minutes. Calculate the velocities for each of the three segments of the trip.

   **(c)** Calculate the average velocity for the total trip. (Hint: this is not the same as the average speed.)

**18.** A swimmer is standing on the south shore of a river that is 120 m wide. He wants to swim straight across and knows that he can swim 1.9 m/s in still water. He drops a stick in the water and finds that it floats with the current to a point 24 m west in 30.0 s.

   **(a)** Determine the direction in which the swimmer should head so that he lands directly across the river on the north bank.

   **(b)** If he follows your advice, determine how long it will take him to reach the far shore.

**19.** A hiker heads [N40.0°W] and walks in a straight line for 4.0 km. She then heads [E10.0°N] and walks in a straight line for 3.0 km. Finally, she heads [S40.0°W] and walks in a straight line for 2.5 km.

   **(a)** Determine the hiker's total displacement for the trip.

   **(b)** In what direction would she have to head in order to walk straight back to her starting position?

   **(c)** If her average walking speed was 4.0 km/h, how long did the total trip take?

**20.** A lone canoeist paddles from Tobermory heading directly east. When there is no wind, the velocity of the canoe is 1.5 m/s. However, a strong wind is blowing from the north, and the canoe is pushed southwards at a rate of 0.50 m/s.

   **(a)** Use vectors to calculate the resultant velocity of the canoe relative to the shore.

   **(b)** Check your solution by using the Pythagorean theorem.

# Go with the Flow

**TARGET SKILLS**

- Initiating and planning
- Predicting
- Performing and recording
- Communicating results

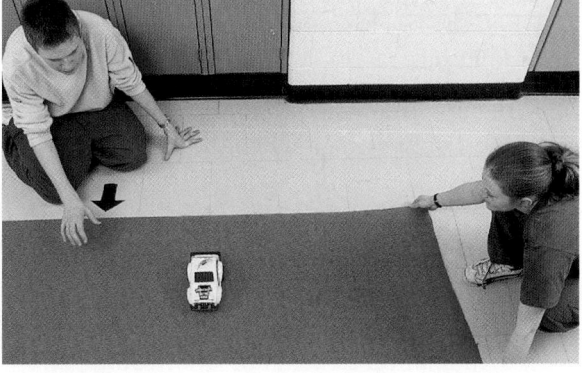

In this investigation, you will simulate the motion of a canoe travelling across a river. This process will help you to sharpen your skills of working with relative velocities and clarify your understanding of these concepts.

## Prediction

Predict the point at which your "canoe" will come ashore on the "river bank" under several different conditions.

## Problem

Test your predictions about the point where your canoe will come ashore on the river bank under several different conditions.

Determine the direction in which the canoe must head in order to reach the river bank directly across the river from where it started.

## Equipment

- battery-powered toy car (or physics bulldozer)
- 2 retort stands
- stopwatch
- protractor
- metre stick
- newspaper
- masking tape
- string

## Procedure

1. Make a paper river by taping together six sheets of newspaper. Newspaper sizes vary, but your river should end up being approximately 1 m wide and 3 m long. (A piece of brown wrapping paper can be used as an alternative to the newspaper.) Measure and record the exact width of your river.

2. A toy car will serve as the canoe. Design and carry out a procedure to determine

   (a) the canoe's speed

   (b) whether the canoe travels at a constant velocity

   Record your procedure, data, calculations, and conclusions.

3. Have one member of your group pull the river along at a constant velocity. Develop a technique for ensuring that the river "flows" at the same constant velocity throughout the investigation.

4. Design and carry out a procedure for determining the velocity of the river.

   Record your procedure, data, calculations, and conclusions.

*continued* ▶

*continued from previous page*

5. Predict how long it will take the canoe to cross the river from one edge to the other when the river is *not* flowing. Test your prediction.

6. Make the following predictions about the motion of the canoe when the river is flowing.

   (a) Predict whether the motion of the river will affect the time it takes for the canoe to travel from one bank to the other. Explain your reasoning. Include in your explanation a sketch of what you think will happen, and the frames of reference that you considered.

   (b) Assume that the canoe is pointed directly across the river. Predict where the canoe will come ashore on the opposite river bank.

   (c) Predict the direction in which the canoe must be pointed for it to travel directly across the river. Draw a vector diagram to support your predictions.

   (d) Predict the time it will take for the canoe to cross the river when pointed in the direction you determined in part (c).

7. Test your predictions according to the following procedures.

   (a) Measure the time it takes for the canoe to cross the river when the river is not flowing.

   (b) Based on your prediction in step 6(b), mark the starting point and the predicted ending point of the river-crossing when the river is flowing. Place a retort stand at each of the two positions and tie a string from one to the other along the predicted path. Start the river flowing. Start a stopwatch when you start the canoe, and time the crossing. Observe the crossing to see how well the canoe followed the predicted path.

   (c) Develop a method to ensure that you can align the canoe in the direction predicted in step 6(c). Start the river flowing, then start the canoe crossing at the predetermined angle. Time the crossing with a stopwatch. Observe the motion of the canoe and the point where it goes ashore on the far side of the river.

8. If the direction that you predicted in step 6(c) and tested in step 7(c) did not result in the canoe going ashore directly across the river, re-evaluate your velocities and calculations. Refine and repeat your experiment several times until your observations match your predictions as closely as is reasonable.

## Analyze and Conclude

1. What was your prediction about the effect of the motion of the river and the time it took for the canoe to cross the river when the canoe was pointed directly across the river? How well did your observations support your prediction? Explain.

2. Use the concept of frames of reference to answer the following question in two different (opposite) ways. Does the canoe move in the direction in which it is pointing?

3. How well did your observations support your predictions about the time interval of the crossing when the canoe was pointed in a direction that resulted in its moving directly across the flowing river? Explain.

4. How well were you able to predict the correct direction in which to point the canoe in order to cause it to move directly across the flowing river? If you had to make adjustments to your prediction and re-test the procedure, explain why this was necessary.

5. Discuss any problems you encountered in making and testing predictions involving relative velocities.

1. **C** Explain why the concept of relative velocity is useful to pilots and canoeists.

2. **C** Discuss whether all velocities can be considered relative velocities.

3. **K/U** When you are in a car moving at 50 km/h,

   **(a)** what is your velocity relative to the car?

   **(b)** what is your velocity relative to the ground?

4. **K/U** On a moving train, you walk to the dining car, which is forward of your own car. Draw velocity vectors of the train relative to the ground, you relative to the train, and you relative to the ground for a point on your trip towards and away from the dining car.

5. **K/U** State the object and reference frame of the resultant vector when the following velocity vectors are added.

   **(a)** plane relative to air + air relative ground

   **(b)** canoe relative to water + water relative ground

   **(c)** balloon relative to ground + ground relative to air

   **(d)** swimmer relative to ground + ground relative to water

6. **C** Use common language to communicate

   **(a)** air speed relative to the ground

   **(b)** the heading of someone in a kayak

   **(c)** water speed relative to the ground

7. **K/U** A canoe is headed directly across a river that is 200 m wide. Instead of moving with constant velocity, the canoe moves with a constant acceleration of $4.0 \times 10^{-2}$ m/s$^2$. If the river is flowing with a constant velocity of 2.0 m/s, how long will it take for the canoe to reach the other side? How far down the river will it land? Sketch the shape of the path the canoe will follow.

8. **I** A boy heads north across a river at a speed of $x$ m/s. A current of $\frac{x}{2}$ m/s heads west.

   **(a)** Develop a vector diagram to indicate where he will land on the opposite beach.

   **(b)** Develop a vector diagram to indicate the direction that he should head in order to land at the dock directly north of his starting position.

**ELECTRONIC LEARNING PARTNER**

Your Electronic Learning Partner has more information about motion in a plane.

**UNIT PROJECT PREP**

Simple visual effects can be impressive when they play with the viewer's expectations.

- How can other objects and changing backgrounds be used to create the illusion of motion?
- Can relative velocities be used to create comic or dramatic situations?
- Try using a strobe light to create interesting velocity effects.

  **CAUTION** Strobe lights can cause seizures in people with certain medical conditions.

- Apply vector definitions of position, displacement, velocity and acceleration to motion in a plane.

- Solve quantitative problems of motion in a plane.

- resolved

- components

When solving vector addition and subtraction problems graphically, you probably noticed that the method is very imprecise. Measurements with rulers and protractors create a large uncertainty. So you will not be surprised to learn that there is a more precise method. If you want more precision, you can choose to use the method presented in this section.

An important clue to the more precise method lies in the canoe investigation and other problems in which you added vectors that were at an angle of 90°, or right angles, relative to each other. When you determined the vector sum of the canoe's velocity and the river's velocity (pages 105–107), you were able to use the Pythagorean theorem to calculate precisely. You did not have to measure with a ruler. The precise method for adding and subtracting vectors is based on right triangles and the rules of trigonometry.

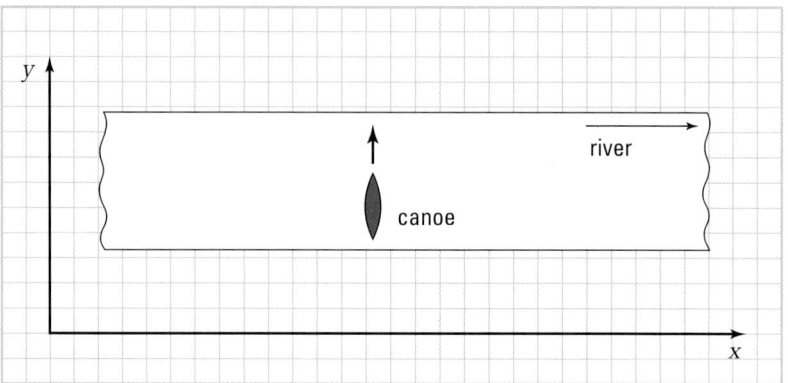

**Figure 3.5** Use the Pythagorean theorem to calculate vectors precisely.

## Vector Components

**ELECTRONIC LEARNING PARTNER**

To learn more about vector components, go to your Electronic Learning Partner.

The vectors that you want to add or subtract are rarely at right angles to each other; however, any vector can be separated, or **resolved**, into components that *are* at right angles to each other. **Components** are parts of a vector that lie on the axes of a coordinate system. Since components are confined to one direction, they are scalar quantities, not vectors.

When working with vector components, the *x–y*-coordinate system is much more convenient to use than a system based on compass directions. Follow the steps in Table 3.4 to learn how to resolve a vector into components. If you need to review the definitions of the trigonometric functions, sine, cosine, and tangent, turn to Skill Set 4.

**Table 3.4** Resolving Vectors into x- and y-Components

| Procedural Step | Graph |
|---|---|
| Draw the vector with its tail at the origin of the coordinate system. | (graph showing vector $\vec{A}$ from origin) |
| Identify the angle that the vector makes with the x-axis and label it "$\theta$." | (graph showing vector $\vec{A}$ with angle $\theta$) |
| Draw a vertical line from the tip of the vector to the x-axis. The line from the origin to the base of this vertical line is the x-component of the vector. | (graph showing $\vec{A}$ with x-component $A_x$) |
| Draw a horizontal line from the tip of the vector to the y-axis. The line from the origin to the base of this line is the y-component. | (graph showing $\vec{v}$ with y-component $A_y$) |
| Write the equation that defines $\sin \theta$. Notice that the y-component $(A_y)$ of the vector is identical to the line from the tip of the vector to the x-axis. | $\sin \theta = \dfrac{A_y}{|\vec{A}|}$ |
| Solve for the y-component, $A_y$. | $A_y = |\vec{A}| \sin \theta$ |
| Write the equation that defines $\cos \theta$. | $\cos \theta = \dfrac{A_x}{|\vec{A}|}$ |
| Solve for the x-component, $A_x$. | $A_x = |\vec{A}| \sin \theta$ |

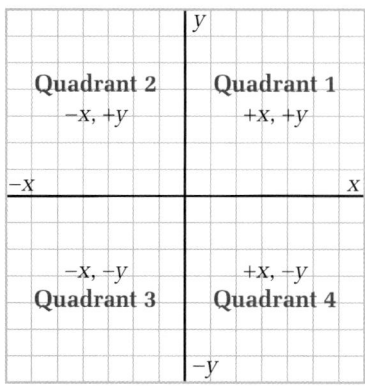

**Figure 3.6** Vector components are scalars, but the sign, positive or negative, is very important.

### PHYSICS FILE

When working in an x–y-coordinate system, mathematicians and physicists report the direction of a vector by giving the angle that the vector makes with the positive x-axis. You find the angle by starting at the positive x-axis and rotating until you reach the location of the vector. If a vector has an angle greater than 90°, it lies in a quadrant other than the first. Vectors in the second, third, or fourth quadrants have at least one component that is negative. Figure 3.6 summarizes the signs of the x- and y-components in the four quadrants. When you use an angle to calculate the magnitude of the components, you would use the angle the vector makes with the nearest x-axis. The model problems show you how to find the components of a vector.

## Resolving Vectors

1. **Find the *x*- and *y*-components of vector $\Delta \vec{d}$, which has a magnitude of 64 m at an angle of 120°.**

## Frame the Problem

- The angle is between 90° and 180°, so it is in the *second quadrant.* Therefore, the *x-component* is *negative* and the *y-component* is *positive.*

- Use *trigonometric functions* to find the components of the vector.

## Identify the Goal

The components, $\Delta d_x$ and $\Delta d_y$, of vector $\Delta \vec{d}$

## Variables and Constants

| Involved in the problem | | Known | Unknown |
|---|---|---|---|
| $\Delta d_x$ | $\Delta \vec{d}$ | $\Delta \vec{d} = 64$ m | $\Delta d_x$ |
| $\Delta d_y$ | $\theta$ | $\theta = 120°$ | $\Delta d_y$ |

## Strategy

Draw the vector with its tail at the origin of an *x*–*y*-coordinate system.

## Calculations

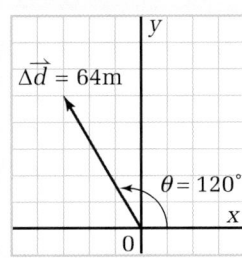

Identify the angle with the closest *x*-axis. Label it "$\theta_R$".

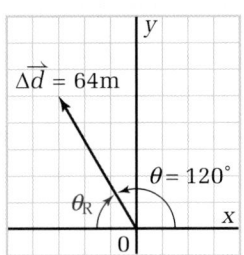

$\theta_R = 180° - 120°$
$\theta_R = 60°$

## Strategy

Draw lines from the tip of the vector to each axis, so that they are parallel to the axes.

Calculate the components according to the directions in Table 3.4.

Determine signs of the components.

The x-component of the vector is −32 m and the y-component is +55 m.

## Calculations

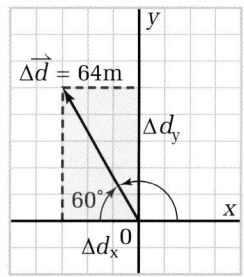

$$\Delta d_x = |\vec{\Delta d}| \cos \theta \qquad \Delta d_y = |\vec{\Delta d}| \sin \theta$$

$$\Delta d_x = 64 \text{ m } \cos 60° \qquad \Delta d_y = 64 \text{ m } \sin 60°$$

$$\Delta d_x = 64 \text{ m } (0.5000) \qquad \Delta d_y = 64 \text{ m } (0.8660)$$

$$\Delta d_x = 32 \text{ m} \qquad \Delta d_y = 55.4 \text{ m}$$

The x-component lies on the negative x-axis so it is negative. The y-component lies on the positive y-axis so it is positive.

## Validate

Use the Pythagorean theorem to check your answers.

$$|\vec{\Delta d}|^2 = \Delta d_x{}^2 + \Delta d_y{}^2$$
$$|\vec{\Delta d}|^2 = (32 \text{ m})^2 + (55.4 \text{ m})^2$$
$$|\vec{\Delta d}|^2 = 1024 \text{ m}^2 + 3069.2 \text{ m}^2$$
$$|\vec{\Delta d}|^2 = 4093.2 \text{ m}^2$$
$$|\vec{\Delta d}| = 64 \text{ m}$$

The value agrees with the original vector.

---

**2. Resolve the vector, $\vec{v} = 56 \frac{\text{km}}{\text{h}} [\text{N50°E}]$, into components.**

---

## Frame the Problem

- Make a sketch of the vector.

- The vector is described in compass directions.

- *Converting* to an *x–y-coordinate* system will simplify the process of finding components. Let +y be north, making −y south. East will become +x and west will be −x.

- You will need to find the *angle* with the closest *x-axis*.

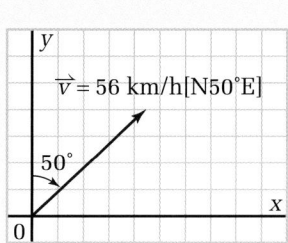

*continued* ▶

## Identify the Goal

The $x$- and $y$-components of the vector $\vec{v} = 56 \, \frac{km}{h} \, [N50°E]$

## Variables and Constants

| Involved in the problem | Known | Unknown |
|---|---|---|
| $\vec{v}$ $\quad$ $v_x$ | $\vec{v} = 56 \, \frac{km}{h} \, [N50°E]$ | $v_x$ |
| $\theta_R$ $\quad$ $v_y$ | | $v_y$ |
| | | $\theta_R$ |

## Strategy

Draw an $x$–$y$-coordinate system and indicate that the axes also represent the compass directions. Draw the vector with its tail at the origin of the coordinate system. Identify the angle, $\theta_R$.

## Calculations

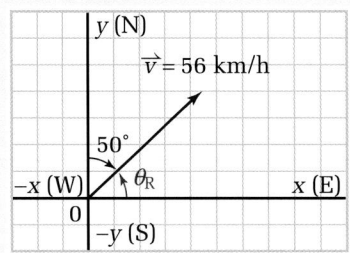

$\theta_R = 90° - 50°$
$\theta_R = 40°$

Since the 50° is the angle made by the vector and the $y$-axis [N], use it to find $\theta_R$, the angle the vector makes with the $x$-axis.

Draw lines from the tip of the vector to each axis, so that they are parallel to the axes.

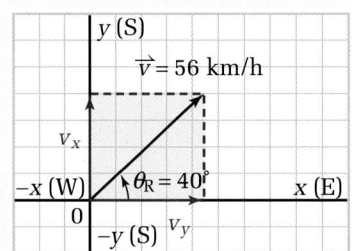

Calculate the components according to the directions in Table 3.4.

$v_x = |\vec{v}| \cos \theta$

$v_x = 56 \, \frac{km}{h} \, \cos 40°$

$v_x = 56 \, \frac{km}{h} \, (0.7660)$

$v_x = 42.9 \, \frac{km}{h}$

$v_y = |\vec{v}| \sin \theta$

$v_y = 56 \, m \, \sin 40°$

$v_y = 56 \, \frac{km}{h} \, (0.6428)$

$v_y = 36 \, \frac{km}{h}$

Determine the signs of the components.

The vector lies in the first quadrant, so all of the components are positive.

The $x$-component of the vector is 43 km/h and the $y$-component is 36 km/h.

## Validate

Apply the Pythagorean theorem to the components in their unrounded form.

$$|\vec{v}|^2 = v_x^2 + v_y^2$$

$$|\vec{v}|^2 = \left(42.9\ \frac{km}{h}\right)^2 + \left(36\ \frac{km}{h}\right)^2$$

$$|\vec{v}|^2 = 3136.28\left(\frac{km}{h}\right)^2$$

$$|\vec{v}| = 56\ \frac{km}{h}$$

The value agrees with the magnitude of the original vector.

### PRACTICE PROBLEMS

**21.** Resolve the following vectors into their components.

  **(a)** a position of 16 m at an angle of 75°

  **(b)** an acceleration of 8.1 m/s² at an angle of 145°

  **(c)** a velocity of 16.0 m/s at an angle of 225°

**22.** Resolve the following vectors into their components.

  **(a)** a displacement of 20.0 km[N20.0°E]

  **(b)** a velocity of 3.0 m/s[E30.0°S]

  **(c)** a velocity of 6.8 m/s[W70.0°N]

**23.** A hot-air balloon has drifted 60.0 km [E60.0°N] from its launch point. It lands in a field beside a road that runs in a north-south direction. The balloonists radio back to their ground crew to come and pick them up. The ground crew can travel only on roads that run north-south or east-west. The roads are laid out in a grid pattern, with intersections every 2.0 km. How far east and then how far north will the pickup van need to travel in order to reach the balloon?

## Vector Addition and Subtraction by Using Components

Examine Figure 3.7 to begin to see how resolving vectors into their components will allow you to add or subtract vectors in a precise yet uncomplicated way. In the figure, $\vec{R}$ is the resultant vector for the addition of $\vec{A}$, $\vec{B}$, and $\vec{C}$. You can also see that $R_x$ is equal in length to $A_x + B_x + C_x$. The same is true for the $y$-components.

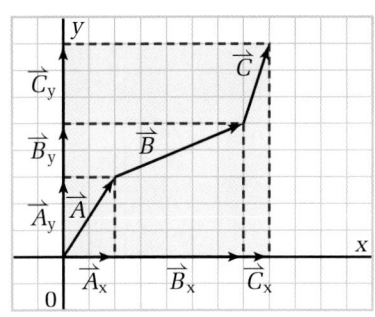

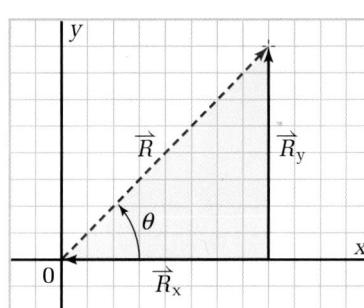

**Figure 3.7** The projection of each vector on the $x$- and $y$-axes are the components of the vector.

When you want to add or subtract vectors, you can separate the vectors into their components, add or subtract the components, and then find the resultant vector by using the Pythagorean theorem.

$$R_x = A_x + B_x + C_x + \cdots \qquad R_y = A_y + B_y + C_y + \cdots \qquad |\vec{R}|^2 = R_x{}^2 + R_y{}^2$$

You can find the angle, $\theta$, from the components of the resultant vector, because they make the sides of a right triangle. Find the ratio of $R_y$ to $R_x$, then use your calculator to find the angle for which the tangent is the ratio.

$$\tan \theta = \frac{R_y}{R_x}$$

## MODEL PROBLEMS

### Using Vector Components

1. **A sailboat sailed [N60°E] for 20.0 km. A strong wind began to blow, causing the boat to travel an additional 12.0 km[W25°N]. Determine the boat's displacement for the entire trip.**

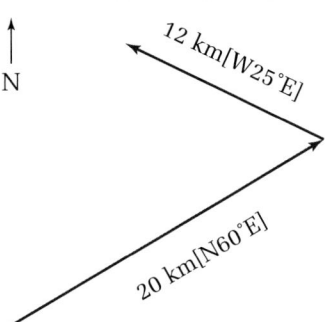

### Frame the Problem

- Make a sketch of the problem.

- The trip consisted of *two displacements* in two *different directions*.

- The *vector sum* of the *displacements* is the total displacement.

### Identify the Goal

The total displacement, $\Delta\vec{d}_T$, for the trip

### Variables and Constants

| Involved in the problem | | | Known | Unknown | | |
|---|---|---|---|---|---|---|
| $\Delta\vec{d}_1$ | $\Delta d_{1x}$ | $\Delta d_{1y}$ | $\Delta\vec{d}_1 = 20$ km[N60°E] | $\Delta\vec{d}_T$ | $\Delta d_{1x}$ | $\Delta d_{1y}$ |
| $\Delta\vec{d}_2$ | $\Delta d_{2x}$ | $\Delta d_{2y}$ | $\Delta\vec{d}_2 = 12$ km[W25°N] | $\theta_1$ | $\Delta d_{2x}$ | $\Delta d_{2y}$ |
| $\Delta\vec{d}_T$ | $\Delta d_{Tx}$ | $\Delta d_{Ty}$ | | $\theta_2$ | $\Delta d_{Tx}$ | $\Delta d_{Ty}$ |
| $\theta_1$ | $\theta_2$ | $\theta$ | | $\theta$ | | |

### Strategy

Draw vectors $\Delta\vec{d}_1$ and $\Delta\vec{d}_2$ on x–y-coordinate systems, where the positive y-axis coincides with north. Identify $\theta_1$ and $\theta_2$, the angles that the vectors make with the x-axis.

### Calculations

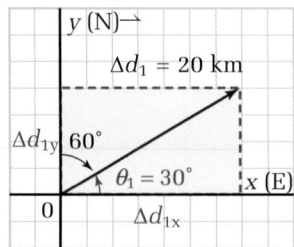

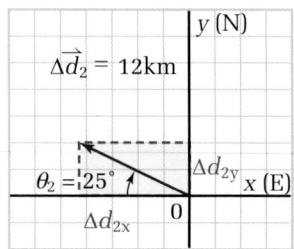

## Strategy

The angle [N60°E] refers to the angle that the vector makes with the $y$-axis. Find the size of the angle the vector makes with the $x$-axis.

Draw the lines from the tips of the vectors to the axes that are parallel to the axes.

Calculate the $x$- and $y$-components of the vectors:

$\vec{\Delta d_1} = 20$ km[N60°E]

$\vec{\Delta d_2} = 12$ km[W25°N]

Identify the signs of the components.

Make a table in which to list the $x$- and $y$-components. Add them to find the components of the resultant vector.

Use the Pythagorean theorem to find the magnitude of the resultant vector.

Calculate the angle, $\theta$, using the components of the resultant vector.

## Calculations

$\theta_2 = 25°$

$\theta_1 = 90° - 60°$

$\theta_1 = 30°$

$\Delta d_{1x} = |\vec{\Delta d_1}| \cos \theta_1$

$\Delta d_{1x} = 20$ km $\cos 30°$

$\Delta d_{1x} = 20$ km $(0.8660)$

$\Delta d_{1x} = 17.3$ km

$\Delta d_{1y} = |\vec{\Delta d_1}| \sin \theta_1$

$\Delta d_{1y} = 20$ km $\sin 30°$

$\Delta d_{1y} = 20$ km $(0.5000)$

$\Delta d_{1y} = 10$ km

$\Delta d_{2x} = |\vec{\Delta d_2}| \cos \theta_2$

$\Delta d_{2x} = 12$ km $\cos 25°$

$\Delta d_{2x} = 12$ km $(0.9063)$

$\Delta d_{2x} = 10.88$ km

$\Delta d_{2y} = |\vec{\Delta d_2}| \sin \theta_2$

$\Delta d_{2y} = 12$ km $\sin 25°$

$\Delta d_{2y} = 12$ km $(0.4226)$

$\Delta d_{2y} = 5.07$ km

$\vec{\Delta d_1}$ is in the first quadrant so both of the components are positive.

$\vec{\Delta d_2}$ is in the second quadrant, so the $x$-component is negative and the $y$-component is positive.

| Vector | x-component | y-component |
|---|---|---|
| $\vec{\Delta d_1}$ | 17.3 km | 10.0 km |
| $\vec{\Delta d_2}$ | −10.9 km | 5.07 km |
| $\vec{\Delta d_T}$ | 6.4 km | 15.07 km |

$|\vec{\Delta d_T}|^2 = (\Delta d_{Tx})^2 + (\Delta d_{Ty})^2$

$|\vec{\Delta d_T}|^2 = (6.4 \text{ km})^2 + (15.07 \text{ km})^2$

$|\vec{\Delta d_T}|^2 = 40.96 \text{ km}^2 + 227.1 \text{ km}^2$

$|\vec{\Delta d_T}|^2 = 248.06 \text{ km}^2$

$|\vec{\Delta d_T}| = 16.4$ km

$\tan \theta = \dfrac{\Delta d_{Ty}}{\Delta d_{Tx}}$

$\tan \theta = \dfrac{15.07 \text{ km}}{6.4 \text{ km}}$

$\tan \theta = 2.35$

$\theta = \tan^{-1} 2.35$

$\theta = 66.98°$

The total displacement was 16 km at an angle of 67°, or $\vec{\Delta d_T} = 16$km[E67°N].

*continued* ▶

*continued from previous page*

## Validate

In the original diagram of the trip, you can see that the sailboat makes a sharp turn. Therefore, you would expect that the total displacement (16 km) would be much smaller than the total distance (32 km) that the boat travelled. The answer is quite reasonable. You can also use the Pythagorean theorem to show that the original components were calculated correctly.

$$|\Delta \vec{d_1}|^2 = (17.3 \text{ km})^2 + (10.0 \text{ km})^2 = 299.29 \text{ km}^2 + 100 \text{ km}^2 = 399.29 \text{ km}^2$$
$$|\Delta \vec{d_1}| = 20 \text{ km}$$
$$|\Delta \vec{d_2}|^2 = (-10.9 \text{ km})^2 + (5.07 \text{ km})^2 = 118.81 \text{ km}^2 + 25.7 \text{ km}^2 = 144.5 \text{ km}^2$$
$$|\Delta \vec{d_2}| = 12 \text{ km}$$

---

2. **You are the pilot of a small plane and want to reach an airport, 600 km due south, in 4.0 h. A wind is blowing at 50 km/h[S35°E]. With what heading and airspeed should you fly to reach the airport on time?**

---

## Frame the Problem

- Your destination is directly *south*.

- A strong wind is blowing *east of south*. The *heading* of the plane will have to account for the wind.

- The *vector sum* of the *velocity* of the plane in relation to the air and the *velocity* of the air

in relation to the ground, must be the same as the needed *total velocity* of the plane in relation to the ground.

- You must use *vector addition*.

---

## Identify the Goal

The velocity, $\vec{v}_{pg}$, of the plane
(**Note:** A velocity has not been reported until both the magnitude and direction are given.)

## Variables and Constants

| Involved in the problem | | | Known | Unknown | | |
|---|---|---|---|---|---|---|
| $\vec{v}_{pa}$ | $v_{pax}$ | $v_{pay}$ | $\vec{v}_{ag} = 50 \dfrac{\text{km}}{\text{h}}[\text{S35°E}]$ | $\vec{v}_{pa}$ | $v_{pax}$ | $v_{pay}$ |
| $\vec{v}_{ag}$ | $v_{agx}$ | $v_{agy}$ | $\Delta \vec{d}_{pg} = 600 \text{ km}[\text{S}]$ | $\theta_{ag}$ | $v_{agx}$ | $v_{agy}$ |
| $\vec{v}_{pg}$ | $v_{pgx}$ | $v_{pgy}$ | $\Delta t = 4.0 \text{ h}$ | $\vec{v}_{pg}$ | $v_{pgx}$ | $v_{pgy}$ |
| $\theta_{ag}$ | $\Delta \vec{d}_{pg}$ | $\Delta t_{pg}$ | | $\theta_{pg}$ | | |
| $\theta_{pg}$ | | | | | | |

## Strategy

You can find the ground speed and direction that the plane must attain to arrive on time by using the mathematical definition for velocity.

Since you now know the wind velocity and the necessary velocity of the plane in relation to the ground, you can use the expression for the vector sum of the velocities to find the velocity of the plane in relation to the air. This quantity *is* the heading and airspeed of the plane. First, solve for $\vec{v}_{pa}$.

Draw the two known vectors on an $x$–$y$-coordinate system (+$y$ coincides with north), with their tails at the origin.

Identify the angles they make with the $x$-axis.

Define and draw the vector $-\vec{v}_{ag}$.

Find the $x$- and $y$-components of the vectors $\vec{v}_{pg}$ *and* $-\vec{v}_{ag}$.

## Calculations

$$\vec{v} = \frac{\Delta \vec{d}}{\Delta t}$$

$$\vec{v}_{pg} = \frac{\Delta \vec{d}_{pg}}{\Delta t_{pg}}$$

$$\vec{v}_{pg} = \frac{600 \text{ km[S]}}{4.0 \text{ h}}$$

$$\vec{v}_{pg} = 125 \frac{\text{km}}{\text{h}} \text{[S]}$$

$$\vec{v}_{pg} = \vec{v}_{pa} + \vec{v}_{ag}$$

$$\vec{v}_{pg} - \vec{v}_{ag} = \vec{v}_{pa} + \vec{v}_{ag} - \vec{v}_{ag}$$

$$\vec{v}_{pg} - \vec{v}_{ag} = \vec{v}_{pa}$$

$$\vec{v}_{pa} = \vec{v}_{pg} - \vec{v}_{ag}$$

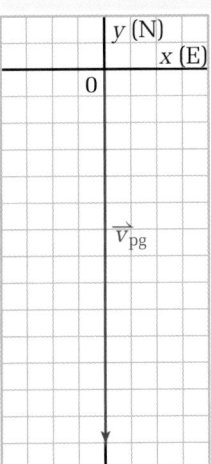

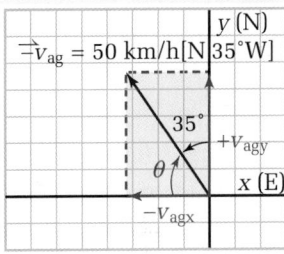

$\vec{v}_{pg}$ is pointed directly south, or along the negative $y$-axis. Therefore, it has no $x$-component. Its $y$-component is the same as the vector.

$$v_{pg\,y} = |\vec{v}_{pg}|$$

$$v_{pg\,y} = -125 \frac{\text{km}}{\text{h}}$$

$$\theta_{ag} = 90° - 35° = 55°$$

$$v_{ag\,x} = |\vec{v}_{ag}| \cos \theta_{ag}$$

$$v_{ag\,x} = 50 \frac{\text{km}}{\text{h}} \cos 55°$$

$$v_{ag\,x} = 50 \frac{\text{km}}{\text{h}} (0.5736)$$

$$v_{ag\,x} = 28.68 \frac{\text{km}}{\text{h}}$$

$$v_{ag\,y} = |\vec{v}_{ag}| \sin \theta_{ag}$$

$$v_{ag\,y} = 50 \frac{\text{km}}{\text{h}} \sin 55°$$

$$v_{ag\,y} = 50 \frac{\text{km}}{\text{h}} (0.8191)$$

$$v_{ag\,y} = 40.96 \frac{\text{km}}{\text{h}}$$

*continued* ▶

## Strategy

Determine the signs of the components.

Add the components of $\vec{v}_{pg}$ and $-\vec{v}_{ag}$ to obtain the components of $\vec{v}_{pa}$.

Use the Pythagorean theorem to find the magnitude of $\vec{v}_{pa}$.

Find the angle the resultant makes with the $x$-axis.

Since the components are both negative, the vector lies in the third quadrant. However, use positive values to find the tangent. The result will give the angle from the negative $x$-axis into the fourth quadrant.

The plane's airspeed must be 89 km/h and it must fly at a heading of [W71°S].

## Calculations

$\vec{v}_{pg}$ has no $x$-component and the $y$-component is negative.

$-\vec{v}_{ag}$ is in the second quadrant, so the $x$-component is negative and the $y$-component is positive.

| Vector | $x$-component | $y$-component |
|---|---|---|
| $\vec{v}_{pg}$ | $0.0\ \dfrac{\text{km}}{\text{h}}$ | $-125\ \dfrac{\text{km}}{\text{h}}$ |
| $\vec{v}_{ag}$ | $-28.68\ \dfrac{\text{km}}{\text{h}}$ | $40.96\ \dfrac{\text{km}}{\text{h}}$ |
| $\vec{v}_{pa}$ | $-28.68\ \dfrac{\text{km}}{\text{h}}$ | $-84.04\ \dfrac{\text{km}}{\text{h}}$ |

$$|\vec{v}_{pa}|^2 = v_{pa\ x}{}^2 + v_{pa\ y}{}^2$$

$$|\vec{v}_{pa}|^2 = \left(-28.68\ \frac{\text{km}}{\text{h}}\right)^2 + \left(-84.04\ \frac{\text{km}}{\text{h}}\right)^2$$

$$|\vec{v}_{pa}|^2 = 822.54\left(\frac{\text{km}}{\text{h}}\right)^2 + 7062.7\left(\frac{\text{km}}{\text{h}}\right)^2$$

$$|\vec{v}_{pa}|^2 = 7885.3(c)^2$$

$$|\vec{v}_{pa}| = 88.8\ \frac{\text{km}}{\text{h}}$$

$$\tan \theta_{pa} = \frac{84.04\ \frac{\text{km}}{\text{h}}}{28.68\ \frac{\text{km}}{\text{h}}} = 2.93$$

$$\theta_{pa} = \tan^{-1} 2.93$$

$$\theta_{pa} = 71.1°$$

## Validate

The wind is blowing toward the southeast and the pilot wants to fly directly south. The component of the wind blowing south will help the plane to get there faster, but the component of the wind blowing east will blow the plane off course if the pilot does not compensate. The pilot must head slightly west to make up for the wind blowing east, so you would expect that the pilot would have to fly slightly west of south. (Note that [W71°S] is the same as [S19°W].) This is in perfect agreement with the calculations.

24. Use components to verify any three of your scale-diagram solutions for Practice Problems 12 through 20 in Section 3.2.

25. A pleasure boat heads out of a marina for sightseeing. It travels 2.7 km due south to a small island. Then it travels 3.4 km[S26°E] to another island. Finally, it turns and heads [E12°N] for 1.9 km to a third island.

    (a) Determine the boat's displacement for the entire journey.

    (b) In what direction should the boat be pointed to head straight home?

26. A jet-ski driver wants to head to an island in the St. Lawrence River that is 5.0 km [W20.0°S] away. If he is travelling at a speed of 40.0 km/h and the St. Lawrence is flowing 6.0 km/h[E]

    (a) in what direction should he head the jet-ski?

    (b) how long will it take him to reach the island?

27. A space shuttle is approaching the Alpha International Space Station at a velocity of 12 m/s relative to the space station. A landing cable is fired toward Alpha with a velocity of 3.0 m/s, at an angle of 25° relative to the shuttle. What velocity will the cable appear to have to an observer looking out of a window in the space station?

## Acceleration Vectors in a Plane

Imagine yourself enjoying the thrill of making a smooth turn on a jet-ski, as shown in Figure 3.8. Feel the forces on your body as you lean into the turn. Now, think about the meaning of acceleration in a way you might not have considered it before.

Suppose that the jet-ski was originally travelling north with a speed of 60 km/h. The driver makes a sharp turn so that the jet-ski is travelling west at the same speed. If the turn took 12 s, what was the average acceleration of the jet-ski?

Your first thought might be that the jet-ski does not accelerate at all, because its speed did not change, but remember that acceleration is the rate of *change* of *velocity*.

**Figure 3.8** No matter how smooth you make the turn, you still must accelerate.

$$\vec{a}_{ave} = \frac{\Delta \vec{v}}{\Delta t}$$

The velocity of an object is a vector quantity that includes a magnitude and a direction. If either magnitude or direction changes, the velocity changes. Therefore, the object has accelerated. In the current example, the magnitude of the jet-ski's velocity did not change, but its direction did. The jet-ski was accelerating for 12 s, the time interval during which the direction was changing. To calculate this average acceleration, you must use vector subtraction, as well as division by a scalar. The direction of acceleration is the same as the direction of the *change* in velocity.

You can qualitatively determine the direction of the acceleration by drawing the initial and final velocity vectors on a coordinate system, with both of their tails at the origin. Then, determine the direction in which you would have to push on the initial velocity vector to convert it into the final velocity vector.

## Acceleration on a Curve

**Calculate the acceleration of the jet-ski as described in the text.**

### Frame the Problem

- Make a sketch of the motion in the problem.

- The *speed* of the jet-ski does not change; however, the *direction* does.

- The jet-ski accelerated.

- The equations of motion for constant acceleration do *not* apply.

- To calculate *acceleration* in *two dimensions*, you must use *vector* subtraction.

- Calculating *acceleration* in two dimensions also involves *dividing* a vector by a *scalar*.

- The *direction* of the acceleration will be the *same* as the direction of the *change in velocity*.

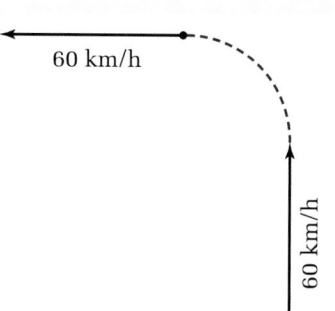

### Identify the Goal

The average acceleration, $\vec{a}_{ave}$, of the jet-ski during the turn

### Variables and Constants

| Involved in the problem | | Known | Unknown | |
|---|---|---|---|---|
| $\vec{v}_i$ | $v_{ix}$ | $\vec{v}_i = 60 \ \dfrac{km}{h} \ [N]$ | $\vec{a}_{ave}$ | $v_{ix}$ |
| $\vec{v}_f$ | $v_{fx}$ | $\vec{v}_f = 60 \ \dfrac{km}{h} \ [W]$ | $\theta$ | $v_{fx}$ |
| $\vec{a}_{ave}$ | $v_{iy}$ | $\Delta t = 12 \ s$ | | $v_{iy}$ |
| $\Delta t$ | $v_{fy}$ | | | $v_{fy}$ |
| $\theta$ | $\Delta\vec{v}$ | | | |

## Strategy

Make a sketch of the initial and final velocity vectors on an x–y-coordinate system, where +y coincides with north. Define and sketch the negative of the initial velocity vector.

## Calculations

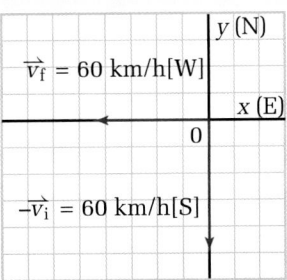

Determine the values of the x- and y-components of $-\vec{v}_i$ and $\vec{v}_f$. Add these components to obtain the components of $\Delta\vec{v}$.

| Vector | x-component | y-component |
|---|---|---|
| $-\vec{v}_i$ | $0.0\ \dfrac{\text{km}}{\text{h}}$ | $-60\ \dfrac{\text{km}}{\text{h}}$ |
| $\vec{v}_f +$ | $-60\ \dfrac{\text{km}}{\text{h}} +$ | $0.0\ \dfrac{\text{km}}{\text{h}} +$ |
| $\Delta\vec{v}$ | $-60\ \dfrac{\text{km}}{\text{h}}$ | $-60\ \dfrac{\text{km}}{\text{h}}$ |

Use the Pythagorean theorem to calculate the magnitude of $\Delta\vec{v}$.

$$|\vec{v}|^2 = v_x^2 + v_y^2$$

$$|\Delta\vec{v}|^2 = \left(-60\ \frac{\text{km}}{\text{h}}\right)^2 + \left(-60\ \frac{\text{km}}{\text{h}}\right)^2$$

$$|\Delta\vec{v}|^2 = 3600\left(\frac{\text{km}}{\text{h}}\right)^2 + 3600\left(\frac{\text{km}}{\text{h}}\right)^2$$

$$|\Delta\vec{v}|^2 = 7200\left(\frac{\text{km}}{\text{h}}\right)^2$$

$$|\Delta\vec{v}| = 84.85\ \frac{\text{km}}{\text{h}}$$

Convert velocity to SI units, so that you can use it to calculate acceleration.

$$\frac{84.85\ \cancel{\text{km}}}{\cancel{\text{h}}} \times \frac{\cancel{\text{h}}}{3600\ \text{s}} \times \frac{1000\ \text{m}}{\cancel{\text{km}}} = 23.57\ \frac{\text{m}}{\text{s}}$$

Calculate the direction of $\Delta\vec{v}$ by determining the angle, $\theta$. Since both of the components are negative, the vector lies in the third quadrant.

$$\tan\theta = \frac{-60\ \frac{\text{km}}{\text{h}}}{-60\ \frac{\text{km}}{\text{h}}} = 1$$

$$\theta = \tan^{-1} 1$$

$$\theta = 45°$$

Since both components are negative, the angle is in the third quadrant.

$$\theta = 45°[\text{W}45°\text{S}]$$

Calculate the acceleration, $\vec{a}$.

$$\vec{a} = \frac{\Delta\vec{v}}{\Delta t}$$

$$\vec{a} = \frac{23.57\ \frac{\text{m}}{\text{s}}[\text{W}45°\text{S}]}{12\ \text{s}}$$

$$\vec{a} = 1.96\ \frac{\text{m}}{\text{s}^2}[\text{W}45°\text{S}]$$

The acceleration of the jet-ski was 2.0 m/s²[W45°S].

*continued* ▶

## Validate

The direction of the acceleration [SW] is correct because, as the diagram shows, that is the direction in which you would have to push on the initial velocity vector in order to change its direction to match the final velocity vector.

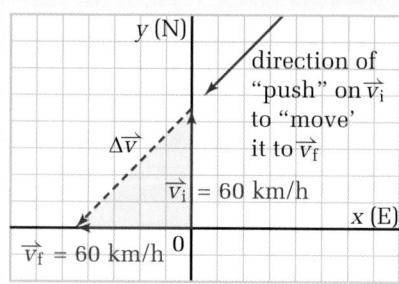

---

### PRACTICE PROBLEMS

**28.** You are jogging south down the street at a pace of 2.5 m/s. You come to an intersection and maintain your pace as you turn and head east down a side street. It takes you 4 s to make the turn. Determine your average acceleration during the turn.

**29.** A soccer player is running down the field with the ball at a speed of 4.0 m/s. He cuts to the right at an angle of 30.0° to his original direction to receive a pass. If it takes him 3.0 s to change his direction, what is his average acceleration during the turn?

**30.** A marble is rolling across the table with a velocity of 21 cm/s. You tap on the marble with a ruler at 90.0° to its direction of motion. The marble accelerates in the direction of the tap for 0.50 s at a rate of 56 cm/s². Determine the new velocity of the marble.

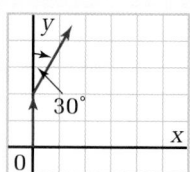

---

# 3.3    Section Review

**1.** **C** Explain how to resolve a vector.

**2.** **C** What use is the Pythagorean theorem in resolving vectors?

**3.** **K/U** Draw examples of velocity vectors for the following cases:

(a) the $x$- and $y$-components are both positive

(b) the $x$-component is positive and the $y$-component is negative

(c) the $x$-component is negative and the $y$-component is positive

(d) the $x$-component is zero and the $y$-component is negative

**4.** **C** Explain whether it is possible to drive around a curve with zero acceleration.

**5.** **C** Describe situations where the average acceleration is zero, positive, and negative.

**6.** **K/U** If a car travels at 50 km/h east and then turns to travel 50 km/h north, what direction is the acceleration?

**7.** **K/U** Consider a standard $x$–$y$-coordinate system. In which quadrant(s) does a vector have:

(a) two positive components?

(b) two negative components?

(c) one positive and one negative component?

- Vectors are represented by arrows that point in the direction of the quantity with respect to a frame of reference or coordinate system. The length of the arrow is proportional to the magnitude.

- To add vectors graphically, place the first vector in a coordinate system. Place the tail of the second vector at the tip of the first. The vector formed along the third side of this triangle by connecting the tail of the first vector to the tip of the second represents the sum of both vectors, and is called the resultant vector.

- Two methods for subtracting vectors graphically are provided. In method 1, place the first vector in a coordinate system. Draw the negative of the second vector, and place its tail at the tip of the first. The vector along the third side of the triangle represents the difference between the two vectors. For method 2, place the two vectors on a coordinate system, *tail* to *tail*. The difference of the two vectors is represented by the vector drawn from the tip of the first vector to the tip of the second vector.

- The direction of a vector does not change when multiplied or divided by a scalar quantity. The magnitude and the units are affected.

- Relative velocity describes motion with respect to a specific coordinate system. A dog's velocity swimming directly across a fast-flowing river could be given relative to the moving water or to the ground.

- Vector components are parts of a vector that are at right angles to each other. They lie on the axes of a coordinate system. Since components are confined to one direction, they are scalar quantities.

- Resolving a vector involves separating it into its components.

- Turning a corner at constant speed involves a change in velocity (the direction is changing) and therefore is associated with an acceleration given by: $\vec{a}_{ave} = \frac{\Delta \vec{v}}{\Delta t}$

## Knowledge/Understanding

1. Define or describe the following:
   - (a) horizontal plane
   - (b) scale vector diagram
   - (c) coordinate system
   - (d) frame of reference
   - (e) resultant vector
   - (f) vector components

2. Why are vectors so useful in solving physics problems?

3. When an airplane travels in a series of straight line segments, you can find the displacement for the whole trip by adding the various displacement vectors.
   - (a) Why can't you add up the various velocity vectors for the different segments of the trip, to get the average velocity?
   - (b) Describe how you would obtain the average velocity for the trip?

4. Suppose you swim across a river, heading towards a position on the far bank directly opposite from where you started from. If there is a strong current in the river, and you end up downstream from the position you were headed for, were you moving faster than you would be moving if there were no current? Explain.

5. A pilot wants to fly due north. However, a strong wind is blowing from the west. Therefore the pilot maintains a heading of a few degrees west of north so that her final ground speed will be in a direction due north. Is the airplane accelerating in this situation? Why or why not?

## Inquiry

6. In the Multi-Lab at the beginning of the chapter, a marble rolled along straight line A, then changed direction and rolled along straight line B. The lines made an angle of 30 degrees with each other. If the marble's speed, both before and after changing direction was 24 cm/s, calculate its *change in velocity*.

## Communication

7. Summarize the ideas about vectors and motion presented in this chapter, by using one or both of the following organizers:

    Make a table listing the vector quantities introduced in this chapter (i.e. displacement, velocity, acceleration) and the various rules used to work with the quantities (i.e., sum, difference, product, components, etc.). In the first column, list the quantity or rule. In the second column, give a definition. In the third column, illustrate with a small diagram.

    Organize the vector quantities and rules you learned in this chapter into a *concept map*. This map should show the connection between the various concepts, quantities, and rules you learned.

8. Brainstorm as a class, or in a small group, the common mistakes students make when working with vectors. Make a list and give a concrete example for each item in the list (i.e., adding the speed of various flight segments to get the average speed).

## Making Connections

9. Why do airplane pilots usually take off and land at airports so that they are facing the wind?

10. Imagine that you are a transportation consultant. You are hired by the province of Ontario, in the year 2015, to help plan a series of high-speed train lines across the province that connect the 20 most populous cities in the province. On the one hand, you want to minimize the time it takes for business people to travel between any two Ontario cities. On the other hand, you want to minimize the total amount of high speed track that will be constructed, as the construction, maintenance, and environmental impact of high speed lines is very high.

    Work in a group and brainstorm a list of the different things you would want to know in order to design the network of high speed railways.

    Critique the list of items that you came up

with, and decide which items are the most important to know with certainty.

    Your options include setting up straight lines between two cities, or "crooked" lines that connect a series of cities more or less in the same direction. For which cities would it make sense to set up a single high-speed line?

## Problems for Understanding

11. An airplane, going from Toronto to Ottawa, flies due east for 290 km and then due north for 190 km. Compare this trip with the trip the airplane makes if it flies in a straight line from Toronto to Ottawa. What is the same about both trips? What is different? When might a pilot want to fly along the component paths rather than in a direct line from one place to another?

12. A boat travels 10.0 km in a direction [N20.0°W] over still water. What are the components of its displacement in each of the four directions of the compass: N, S, E, and W?

13. A car travelling at 50.0 km/h due north, turns a corner and continues due west at 50.0 km/h. If the turn takes 5.0 s to complete, calculate the car's (a) change in velocity and (b) acceleration during the turn.

14. A person walks 3.0 km[S] and then 2.0 km[W], to go to the movie theatre.
    (a) Draw a vector diagram to illustrate the displacement.
    (b) What is the total displacement?

15. A person in a canoe paddles 5.6 km[N] across a calm lake in a time of 1.0 h. He then turns west and paddles 3.4 km in 30.0 minutes.
    (a) Calculate the displacement of the canoeist from his starting point.
    (b) Determine the average velocity for the trip.

16. A cyclist is moving with a constant velocity of 5.6 m/s[E]. He turns a corner and continues cycling at 5.6 m/s[N].
    (a) Draw a vector diagram to represent the change in velocity.
    (b) Calculate the change in velocity.

17. A cyclist travels with a velocity of 6.0 m/s[W] for 45 minutes. She then heads south with a speed of 4.0 m/s for 30.0 minutes.

**(a)** Calculate the displacement of the cyclist from her starting point.

**(b)** Determine the average velocity for the trip.

18. Thao can swim with a speed of 2.5 m/s if there is no current in the water. The current in a river has a velocity of 1.2 m/s[S]. Calculate Thao's velocity relative to the shore if
    **(a)** she swims upstream
    **(b)** she swims downstream

19. A physics teacher is on the west side of a small lake and wants to swim across and end up at a point directly across from his starting point. He notices that there is a current in the lake and that a leaf floating by him travels 4.2 m[S] in 5.0 s. He is able to swim 1.9 m/s in calm water.

    **(a)** What direction will he have to swim in order to arrive at a point directly across from his position?

    **(b)** Calculate his velocity relative to the shore.

    **(c)** If the lake is 4.8 km wide, how long will it take him to cross?

20. Resolve the following vectors into their components:
    **(a)** a displacement of 20.0 m[N25°E]
    **(b)** a displacement of 48 km[S35°E]
    **(c)** a velocity of 15 m/s[S55°W]
    **(d)** an acceleration of 24 m/s$^2$[N30.0°W]

21. A canoeist is paddling across a lake with a velocity of 3.2 m/s[N]. A wind with a velocity of 1.2 m/s[N20.0°E] starts and alters the path of the canoeist. What will be the velocity of the canoeist relative to the shore?

22. A person is jogging with a velocity of 2.8 m/s[W] for 50.0 minutes, and then runs at 3.2 m/s[N30.0°W] for 30.0 minutes. Calculate the displacement of the runner (answer in km).

23. A jogger runs 15 km[N35°E], and then runs 7.5 km[N25°W]. It takes a total of 2.0 hours to run.

    **(a)** Determine the displacement of the jogger.

    **(b)** Calculate the jogger's average velocity.

24. A sailboat is moving with velocity of 11.0 m/s[E] when it makes a turn to continue at a velocity of 12.0 m/s[S40.0°E]. The turn takes 45.0 seconds to execute. Calculate the acceleration of the sailboat.

25. An airplane heads due north from Toronto towards North Bay, 300 km away with a velocity relative to the wind of 400 km/h. There is a strong wind of 100 km/h blowing due south.

    **(a)** What is the velocity of the airplane with respect to the ground?

    **(b)** How long will it take the airplane to fly from Toronto to North Bay?

    **(c)** From North Bay back to Toronto (assuming the same wind velocity)?

    **(d)** Would the total trip take the same time if the wind velocity was 200 km/h?

26. A canoeist wants to travel straight across a river that is 0.10 km wide. However, there is a strong current moving downstream with a velocity of 3.0 km/hr. The canoeist can maintain a velocity relative to the water of 4.0 km/hr.

    **(a)** In what direction should the canoeist head to arrive at a position on the other shore directly opposite to his starting position?

    **(b)** How long will the trip take him?

---

**Numerical Answers to Practice Problems**

**1. (a)** $4.0 \times 10^2$ km[E28°N] **(b)** W28°S **2. (a)** 5.0 km **(b)** E37°S
**3.** 9.5 km, 36° to the right of the lookout **4. (a)** 62.6 km [W11.3°S] **(b)** E12°S **5. (a) (i)** 27 km[N] **(ii)** 24 km[N12°E] **(iii)** 24 km[S12°W] **(b) (i)** 27 km[N] **(ii)** 24 km[N12°E] **(iii)** 6.0 km [W34°S] **6.** 67 km/h [W48°N] **7.** 346 km/h[E30.0°N] **8.** 10 m/s in direction 7° from the normal to the boards, towards the puck's initial direction **9. (a)** 8.4 m/s[N7.1°W] **(b)** 5.5 m/s [W4.0 × $10^1$°N] **(c)** 3.6 m/s[E57°N] **10.** 5.7 km/h[S42°W]
**11. (a)** 48 km/h[W29°N] **(b)** 1.2 × $10^2$ km/h[E29°S]
**12.** $5.8 \times 10^3$ m[N23°W] **13. (a)** 9.2 km[N24°W] **(b)** 3.1 km/h [N24°W] **14.** −1.8 m/s **15.** 12 m/s[S24°W] **16. (a)** N20.5°E
**(b)** 227 km/h[N30.0°E] **(c)** 1.10 h **17. (a)** 1.6 × $10^2$ km[W18°N]
**(b)** 3.0 × $10^2$ km/h[N], 2.2 × $10^2$ km/h[W], 2.5 × $10^2$ km/h[S]
**(c)** 1.3 × $10^2$ km/h[W18°N] **18. (a)** N25°E **(b)** 69 s **19. (a)** 2.1 km [W54°N] **(b)** S54°E **(c)** 2.4 h **20. (a)** 1.6 m/s[E18°] **21. (a)** 4.1 m, 15 m **(b)** −6.6 m/s$^2$, 4.6 m/s$^2$ **(c)** −11.3 m/s, −11.3 m/s
**22. (a)** 6.84 km, 18.8 km **(b)** 2.6 m/s, −1.5 m/s **(c)** −2.3 m/s, 6.4 m/s **23.** 3.0 × $10^1$ km[E], 5.2 × $10^1$ km[N] **25. (a)** 5.9 km[E34°]
**(b)** [W56°N] **26. (a)** W17°S **(b)** 8.7 min **27.** 15 m/s in a direction 4.9° to the shuttle **28.** 0.9 m/s$^2$[NE] **29.** 0.7 m/s$^2$ at 104° to the right of his original direction **30.** 35 cm/s in a direction 53° away from its original direction

The forest of steel and cables shown here manipulates enormous amounts of mass by exerting huge forces with intricate precision. Rebuilding a city after a natural or political disaster requires hauling, lifting, and fastening millions of kilograms of concrete and steel. The technology that enables societies to build and rebuild structures at a breathtaking rate is rooted in an understanding of forces that dates back more than 400 years.

The inset photograph provides a close look at one of the hundreds of cranes used to build the high-rise buildings of a large city. The crane must be relatively easy to assemble and take apart. At the same time, it must be capable of lifting great masses and moving them to specific places within a new building's perimeter. The crane's slender arm belies its tremendous strength. By controlling the movement of a series of massive counterweights, the crane operator can guide the crane to every corner of the growing structure.

The ability to lift, balance, and move objects from feather-light sheet music to massive steel girders requires forces. Chapter 4 explores the development of the concept of force. Why do stationary objects remain at rest and why do moving objects tend to continue to move? You will study Newton's laws of physics and learn how to use them to predict the motion of macroscopic objects. You will also discover how Newton's model fits into the current theories of motion of subatomic particles and Einstein's theories of relativity. You will also survey the fundamental forces of nature, which are governed by the properties of the universe that emerged in the first few seconds after the Big Bang.

# Predicting Motion

## Marbles

What will happen when you roll a marble horizontally across the floor? First, predict what will happen to the marble after it leaves your hand. Using a diagram, explain your prediction to your partner. Now, try the experiment.

### Analyze and Conclude

1. Was your prediction correct?

2. Provide a reasonable explanation for the results.

## Thinking about Space

Imagine that you are in a spaceship out in intergalactic space, very far away from any stars. You fire the ship's rockets for a while and then shut them down.

### Analyze and Conclude

1. Describe the motion of your spacecraft after you turn the engines off.

2. Explain your answer to your partner using diagrams and a written explanation.

3. Discuss the possibilities with your partner and try to agree on a reasonable answer.

## Tossing a Coin

Toss a coin vertically up into the air with a quick motion of your hand. Predict the motion of the coin from the moment it leaves your fingers. Draw a diagram that illustrates your prediction and explain it to your partner. If you and your partner do not agree, make separate predictions. Try the experiment and carefully observe the outcome.

### Analyze and Conclude

1. Attempt to explain why the coin follows the path it does.

2. If you and your partner do not agree on an explanation, try to find a flaw in one of the explanations.

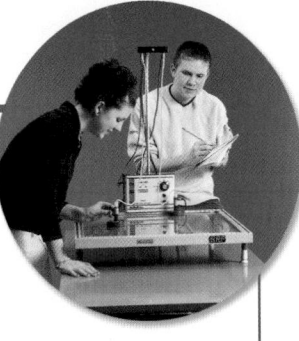

## Air Table

What will happen if you tap on a puck that is lying on an air table or on a cart on an air track? Make a prediction then justify your answer to your partner using a diagram. Carry out the experiment.

### Analyze and Conclude

1. Was your prediction correct?

2. Explain the results.

• Describe and assess Galileo's contribution to the study of dynamics.

• Design and perform simple experiments to verify Galileo's predictions.

**KEY TERMS**

- inertia
- mechanics
- kinematics
- force
- dynamics

**Figure 4.1** This photograph shows how an object is affected by forces. Forces sometimes stop, sometimes propel, and sometimes restrain an object.

The crash-test dummy in the photograph was in motion before the car abruptly stopped. The dummy continued to move until it experienced a stopping force. Because the dummy was not restrained by a seatbelt, the windshield was the first object to make contact and thus provide a stopping force. The seatbelt would have been a far better option.

If you have ever ridden on a public transit bus or a subway, you may recall being flung backward as the vehicle accelerated away from the stop. Later, upon arriving at the next stop, you were thrown forward when the vehicle came to a halt. In this example, as in the above photo, you see evidence of a property that is shared by all matter — the tendency of an object to resist any change in its motion. This property is called **inertia**.

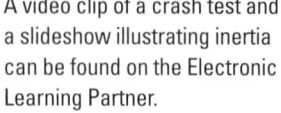

**ELECTRONIC LEARNING PARTNER**

A video clip of a crash test and a slideshow illustrating inertia can be found on the Electronic Learning Partner.

**DEFINITION OF INERTIA**

Inertia is the natural tendency of an object to remain in its current state of motion. The amount of an object's inertia is directly related to its mass.

# Galileo's Perception of Inertia

Throughout history, inquisitive people have attempted to understand why an object moves or remains at rest. Aristotle's (384–322 B.C.E.) observations led him to conclude that a constant force will yield a constant speed. His idea went unchallenged for nearly 2000 years, but it is, in fact, false. French philosopher, Jean Buridan (1300–1358) believed that objects stayed in motion because they possessed "impetus" — something inside that makes them continue to move.

Galileo carefully considered both ideas. He conducted a three-part thought experiment in an attempt to understand why objects move the way they do.

**Table 4.1** Galileo's Thought Experiment on Motion

| Observations | Predictions | Assumptions | Diagram |
|---|---|---|---|
| A ball rolling down a slope speeds up. A ball rolling up a slope slows down. | Therefore, a ball rolling on a horizontal surface should continue without speeding up or slowing down | The reason objects do slow down on horizontal surfaces is the result of the force of friction. | released at this height · reaches this height before stopping |
| A ball rolling up a slope that is not as steep as the slope it rolled down will continue farther along the shallower slope. | The ball will continue up the shallow slope until it has reached the height from which it was originally released. | The reason objects do not quite reach the same height is due to the force of friction. | same result with less slope |
| When the second slope is zero (horizontal), the ball will continue to roll. | The ball would continue forever. | Again, the force of friction will prevent this from occurring naturally. | continues forever without stopping |

**Summary:** An object will naturally remain at rest or in uniform motion unless acted on by an external force.

Galileo's thought experiment challenged the commonly held belief that an object's uniform motion was the result of continued force. Instead, he viewed uniform motion as being a state just as natural as rest. The experiment also contradicted the "impetus" theory proposed by Buridan. According to Galileo, an object's movement remains unchanged, not because of something inside of it, but because there is no force resisting the motion.

 **Language Link**

*Impetus* is commonly used in the English language to describe a stimulus that incites action. For example, "The energetic audience gave the actors the *impetus* to give their very best performances." Notice that, even in this context, the impetus — the motivating force — is external. That is, the impetus comes from the energetic audience, not from within the actors. What other common terms are derived from the word *impetus*?

Test Galileo's ideas using a Hot Wheel's™ track and a marble. Although friction is not completely removed, it will be sufficiently reduced to observe what Galileo envisioned in his thought experiments.

Galileo had identified the natural tendency of mass to continue doing what it is already doing — that is, to continue in uniform motion or remain at rest. Galileo was defining inertia. His ingenious ideas were not immediately accepted, but his concept of inertia has stood the test of time and is the only classical law incorporated into Einstein's theories about the nature of our universe.

## Examples of Inertia

Revisit the Multi Lab, Thinking about Space. Your spacecraft will continue to move forever, at a constant speed in one direction, unless something causes it to stop or change direction. This motion is natural, in the same way that it is natural for the rocks of Stonehenge to remain stationary. There is no actual force involved in inertia. In fact, motion remains constant due to the absence of any force.

## How Forces Affect Motion

As you learned in Chapter 2, predicting and describing an object's motion in terms of its displacement, velocity, and acceleration, are aspects of a branch of physics called **kinematics**. The branch of physics that explains *why* objects move the way they do is called **dynamics**. Together, kinematics and dynamics form a branch of physics called **mechanics**. The study of dynamics involves **forces**, which you can regard as a push or a pull on an object. Forces cause *changes* in motion.

Table 4.2 lists examples of moving objects that could be studied within the field of dynamics.

**Figure 4.2** These stones were moved without the aid of powerful machines. Once placed, not even a tornado could exert enough force to move them.

**Table 4.2** Objects and Their Motion

| Object | Motion | Theoretical Explanation |
|--------|--------|-------------------------|
| electron | remains in motion near the nucleus of its atom | attracted by positively charged protons in the nucleus |
| snowflake | drifts toward the ground | Earth's gravity |
| baseball | flies off after contact with the bat | contact with bat |
| skydiver | reaches terminal speed while falling to Earth | air friction |
| Earth | orbits around the Sun | Sun's gravity |

1. **K/U** A marble is fired into a circular tube that is anchored onto a frictionless table-top. Which of the five paths will the ball take as it exits the tube and moves across the tabletop? Justify your answer.

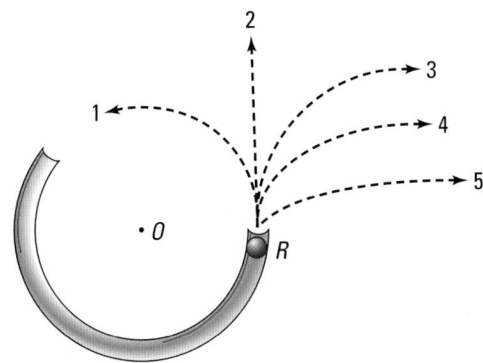

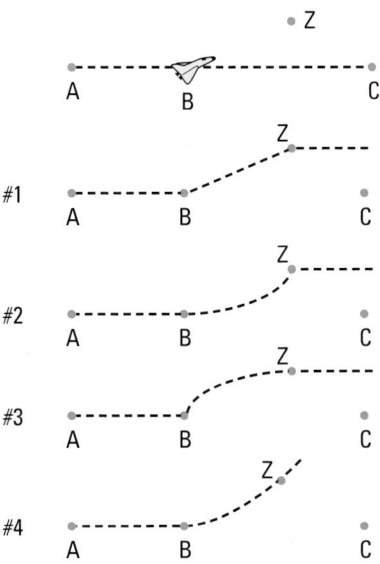

2. **MC** Imagine the following scenario. In a sudden burst of energy, you cleaned your room. A few moments later, your mother saw your room and appeared surprised, but you were not surprised. About an hour later, you went back into your room to find that the books were all on the floor. Your bed was on the wrong side of the room and all of the dresser drawers were on the bed. You were shocked! Use technical terms, including inertia, to explain why you were so surprised.

3. **K/U** A spacecraft is lost in deep space, far from any objects, and is drifting along from point A toward point C. The crew fires the on-board rockets that exert a constant force exactly perpendicular to the direction of drift. If the constant thrust from the rockets is maintained from point B until point Z is reached, which diagram best illustrates the path of the spacecraft?

4. **C** Decide whether each of the following statements is true or false. If the statement is false, rewrite it to make it true.

   (a) Inertia is the result of stationary mass.

   (b) An object will be at rest or slowing down if no force is acting on it.

5. **C** Galileo thought deeply about motion and its causes.

   (a) Describe his thought experiments, including any assumptions that he made.

   (b) How did Galileo's conclusions challenge current beliefs of his time?

6. **C**
   (a) Define kinematics, dynamics, and mechanics.

   (b) Produce a table similar to Table 4.2 listing the motion and a theoretical explanation of three different objects.

- contact force
- non-contact force
- weight
- acceleration due to gravity
- static frictional force
- kinetic frictional force
- coefficient of friction
- normal force
- net force
- free body diagram

Objects interact with other objects by exerting forces on each other. You move your pencil across your page, the desk at which you are sitting supports your books, and the air you breathe circulates around the room. Your desk is in direct contact with your books, so this interaction is an example of a **contact force**. Conversely, **non-contact forces** act over a distance, as is the case when two magnets attract or repel each other without touching.

## Gravitational Force

**Figure 4.3** Vince Carter is able to exert a large force in a short period of time, allowing him to jump extremely high, as he demonstrated during the 2000 Olympics in Sydney.

In Section 4.1, you studied an important property of matter — inertia. You learned that the inertia of an object was directly related to its mass. A second and equally important property of matter is gravity. Any two masses exert a mutual gravitational attractive force on each other. If you hold a tennis ball in each hand, you do not notice any force acting between them because gravitational forces are very weak. It is only because Earth has such an enormous mass that you are aware that it is exerting a strong force on you. Did you know, however, that you are exerting an equal force on Earth? Gravitational forces always act in pairs.

According to Newton's law of universal gravitation, the mutual attractive force between any two masses acts along a line joining their centres. Therefore, the force of gravity between Earth and any object is directed along a line between the centre of the object and the centre of Earth. Newton's law also states that the strength of the gravitational attractive force diminishes as the distance between their centres increases. The lengths of the arrows in Figure 4.4 represent the relative strength of the gravitational attractive force at different distances from Earth's centre. Although the attractive force decreases with distance it never truly reaches zero. The force of gravity, although relatively weak, has an infinite range. It is, therefore, a non-contact force.

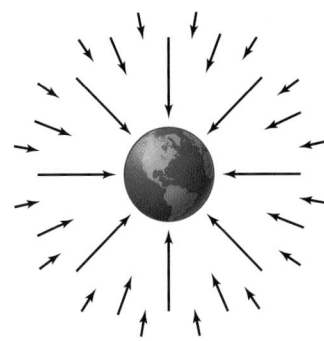

**Figure 4.4** The gravitational force exerted by Earth on an object diminishes as the object's distance from Earth increases.

### ● *Think It Through*

- Consider the last sentence in the preceding paragraph. If it is true, then compare, qualitatively, the *force of gravity* acting on you due to the mass *of Earth* when you are on Earth's surface to

  **(a)** the force of gravity acting on you when you are aboard the International Space Station,

  **(b)** the force of gravity acting on you when you are out past the orbit of Pluto.

## Weight

You have probably seen pictures of astronauts bouncing along the surface of the Moon. Even in their bulky space suits and oxygen tanks, they can jump significantly higher and drop back down more slowly than they could on Earth. What makes the difference? If an astronaut had a mass of 60 kg on Earth, he or she would still have a mass 60 kg when arriving on the Moon. However, astronauts weigh much less on the Moon than on Earth. The distinction between mass and weight becomes clear when you compare the effects of Earth's gravity to the Moon's gravity. You have a specific mass regardless of where you are located — on Earth, the Moon, or in intergalactic space. Your weight, however, is influenced by the force of gravity. In fact, **weight** is defined as the force of gravity acting on a mass. Therefore, your weight would be much lower on the Moon than it is on Earth. On the other hand, if Jupiter had a surface that astronauts could walk on, they would find themselves pinned to the ground weighing 2.5 times more than they weigh on Earth.

### TRY THIS...

Poke a small hole in the side of a plastic cup near the bottom. Predict what will happen when you fill the cup with water and then allow it to fall toward the ground in the upright position. Record your prediction with the aid of a diagram. Test your prediction. Explain the results.

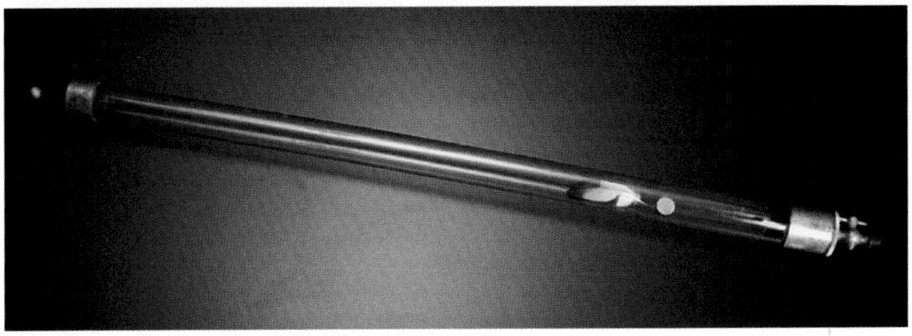

**Figure 4.5** Acceleration due to gravity does not depend on mass, as demonstrated by the example of the feather and the coin in a vacuum tube.

As you know, the force of gravity is influenced by the masses of the two interacting objects as well as the distance between their centres. Thus you might expect that the relationship between an object's mass and the force of gravity acting on it would be very complex. Fortunately, it is not. There is a common factor that was also a topic of contemplation for Aristotle and Galileo. Aristotle believed that more massive objects fall faster than less massive objects. He predicted that a mass ten times greater than another mass would fall ten times faster. Galileo reasoned that a large mass would have more inertia than a small mass and therefore a greater force would be required to change the motion of larger mass than the smaller mass. Since the gravitational force on a large mass is greater than the gravitational force on a small mass, the masses should move in the same way under the influence of gravity. He conducted the experiment and found that all objects fell at almost exactly the same rate. He attributed the slight difference in falling time to air resistance. Galileo concluded, correctly, that, at any given location, and in the absence of air resistance, all objects will fall with the same acceleration. Physicists now call this the **acceleration due to gravity** and give it the symbol, $g$. You may have seen a demonstration of a coin and a feather falling at exactly the same rate when enclosed in an evacuated tube like the one shown in Figure 4.5 on the previous page.

Since the value of $g$ is influenced by both the mass of Earth and the distance from Earth's centre, the value of $g$ varies with location. On Earth's surface, $g$ is approximately 9.81 m/s$^2$, but actually ranges from 9.7805 m/s$^2$ at the equator to 9.8322 m/s$^2$ at the poles because Earth is not perfectly round. It is really a flattened sphere that bulges in the middle. Consequently, objects at the poles are closer to the centre of Earth than are objects located at the equator. Some values of $g$ are listed in Table 4.3 for comparison.

**Table 4.3** Free-Fall Accelerations Due to Gravity on Earth

| Location | Acceleration due to gravity (m/s$^2$) | Altitude (m) | Distance from Earth's centre (km) |
|---|---|---|---|
| North Pole | 9.8322 | 0 (sea level) | 6357 |
| equator | 9.7805 | 0 (sea level) | 6378 |
| Mt. Everest (peak) | 9.7647 | 8850 | 6387 |
| Mariana Ocean Trench* (bottom) | 9.8331 | 11 034 (below sea level) | 6367 |
| International Space Station* | 9.0795 | 250 000 | 6628 |

*These values are calculated.

Because the mass and radius of the planets vary significantly, the acceleration due to gravity is quite different from planet to planet. Values of $g$ for the Moon and a few planets are listed in Table 4.4.

**Table 4.4** Free-Fall Accelerations Due to Gravity in the Solar System

| Location | Acceleration due to gravity (m/s²) |
|---|---|
| Earth | 9.81 |
| Moon | 1.64 |
| Mars | 3.72 |
| Jupiter | 25.9 |

To summarize, you have discovered that weight is the force of gravity acting on a mass. You have also seen that the acceleration due to gravity incorporates all of the properties of the gravitational attractive force, except the mass of the object, that affect the strength of the force of gravity — mass of the planet and the distance between the centre of the object and the planet. Now you are ready to put them together in the form of a mathematical equation.

---

**WEIGHT**

An object's weight, $F_g$, is the product of its mass, $m$, and the acceleration due to gravity, $g$.

$$\vec{F_g} = m\vec{g}$$

| Quantity | Symbol | SI unit |
|---|---|---|
| force of gravity (weight) | $\vec{F_g}$ | N (newton) |
| mass | $m$ | kg (kilogram) |
| acceleration due to gravity | $\vec{g}$ | m/s² (metres per second squared) |

**Unit Analysis**

(mass) (acceleration) $= \text{kg}\dfrac{\text{m}}{\text{s}^2} = \text{N}$

**Note:** The symbol $g$ is reserved for acceleration due to gravity on Earth. In this textbook, $g$ with an appropriate subscript will denote acceleration due to gravity on a celestial object other than Earth, for example, $g_{\text{Moon}}$.

---

**Web Link**

**www.school.mcgrawhill.ca/resources/**
Do you have a dramatic flair? Check out the Internet site above to read about — and perhaps even test — Galileo's arguments of logic refuting Aristotle's teachings concerning falling objects. Galileo actually wrote the words! He presented the arguments using two fictitious characters. Salviati voiced the beliefs of Galileo, while Aristotle's ideas were embodied in Simplicio. If you enjoy a good debate, this English translation will captivate you. Go to the Internet site and follow the links for **Science Resources** and **Physics 11** to find out where to go next.

**MISCONCEPTION**

**Mass is not Weight**

Everyday language often confuses an object's mass with its weight. You may have a *mass* of 78 kg, but you do not *weigh* 78 kg. Your weight would be $\vec{F_g} = m\vec{g} = (78 \text{ kg})(9.81 \text{ m/s}^2) = 765 \text{ N[down]}$. Many scales and balances convert the newton reading to a mass equivalent on Earth. A spring scale designed for use on Earth would give incorrect results on the Moon.

The language of forces can sometimes seem to complicate a concept that you already understand. Attempt to *describe* the interactions between the objects listed below in terms of forces.

## Problem

What interactions exist between objects and how would things change if some interactions were removed?

## Hypothesis

Form a hypothesis about which interactions will always exist between these common objects, regardless of where you might go to test them, including intergalactic space.

## Equipment

- small mass (for example, a pop can)
- feather
- hockey puck
- magnets (2)
- string

## Procedure

1. Place the pop can on the lab bench. How are the pop can and the lab bench interacting?
   (a) Describe how the pop can interacts with the lab bench.
   (b) Describe how the lab bench interacts with the pop can.
2. Drop the feather from 2.0 m above the floor.
   (a) During its descent, is the feather interacting with any objects?
   (b) What interaction causes the feather to fall?
   (c) Describe what would happen if the feather did not interact with any objects during its descent.

3. Slide the hockey puck across the lab bench.
   (a) Describe three interactions affecting the hockey puck as it slides.
   (b) Describe one interaction between the puck and the lab bench.

4. Attach two magnets to strings. Hold the magnets by the strings and allow them to approach each other but do not allow them to touch.
   (a) Describe the interaction between the magnets.
   (b) Describe the interaction between the strings and the magnets.
   (c) What would happen if the interaction between the strings and magnets was removed?

## Analyze and Conclude

1. List the interactions that are contact forces.

2. List the interactions that are non-contact forces.

3. Did your hypothesis include each interaction that you discovered during the activity?

4. Did you describe any interactions that could not be classified as a force?

## Apply and Extend

5. (a) Describe the forces acting on a skydiver as she falls to Earth with the parachute not yet deployed.
   (b) Is the skydiver accelerating during the entire descent before deploying the parachute?
   (c) Describe what force acts on the skydiver when the parachute is deployed.

6. A baseball is thrown high into the air. Describe the forces acting on the ball: (a) as it rises, (b) at the highest point, and (c) during the fall back to Earth.

## Weight and Mass Calculations

1. **Calculate the weight of a 4.0 kg mass on the surface of the Moon.**

---

### Frame the Problem

- The object is on the surface of the *Moon*.

- Its *weight* is the *force* of *gravity* acting on it.

- *Weight* is related to *mass* through the *acceleration* due to *gravity*.

- The *acceleration* due to *gravity* on the *Moon* is given in Table 4.4.

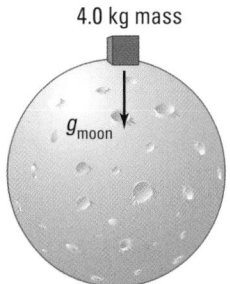

4.0 kg mass

$g_{moon}$

---

### Identify the Goal

Weight, or force of gravity, $F_g$, acting on a mass on the Moon

### Variables and Constants

| Involved in the problem | | Known | Implied | Unknown |
|---|---|---|---|---|
| $m$ | $\vec{F}_g$ | $m = 4.0$ kg | $\vec{g}_{\text{Moon}} = 1.64$ m/s$^2$ [down] | $\vec{F}_g$ |
| $\vec{g}_{\text{Moon}}$ | | | | |

### Strategy

The acceleration due to gravity is known for the surface of the Moon.

Use the equation for weight.

Substitute in the variables and solve.

1 kg$\frac{\text{m}}{\text{s}^2}$ is equivalent to 1 N.
Convert to the appropriate number of significant digits.

The 4.0 kg mass would weigh 6.6 N[down] on the surface of the Moon.

### Calculations

$$\vec{F}_{g\text{ Moon}} = m\vec{g}_{\text{Moon}}$$

$$\vec{F}_{g\text{ Moon}} = (4.0 \text{ kg})\left(1.64\frac{\text{m}}{\text{s}^2}\right)[\text{down}]$$

$$\vec{F}_{g\text{ Moon}} = 6.56 \text{ kg}\frac{\text{m}}{\text{s}^2}[\text{down}]$$

$$\vec{F}_{g\text{ Moon}} = 6.6 \text{ N}[\text{down}]$$

---

### Validate

Weight is a force and, therefore, should have units of newtons, N.

*continued* ▶

**2. A student standing on a scientific spring scale on Earth finds that he weighs 825 N. Find his mass.**

## Frame the Problem

- *Weight* is defined as the *force of gravity* acting on a *mass*.

- If you know the *weight* and the *acceleration* due to *gravity*, you can find the *mass*.

- The *acceleration* due to *gravity on Earth* is given in Table 4.4.

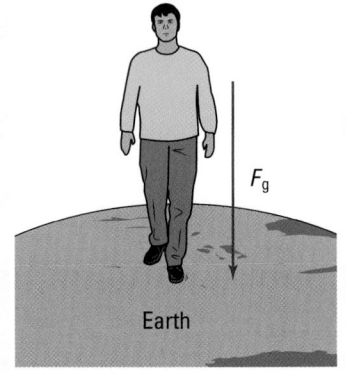

Earth

## Identify the Goal

The mass, $m$, of the student

## Variables and Constants

| Involved in the problem | Known | Implied | Unknown |
|---|---|---|---|
| $m$ $\qquad$ $\vec{F}_g$ | $\vec{F}_g = 825$ N[down] | $\vec{g} = 9.81$ m/s²[down] | $m$ |
| $\vec{g}$ | | | |

## Strategy

Use the equation for the force of gravity.

Simplify.

Express N as $kg\frac{m}{s^2}$ in order to cancel units.

## Calculations

$$\vec{F}_g = m\vec{g}$$

**Substitute first**

$$825 \text{ N[down]} = m\ 9.81\frac{m}{s^2}\text{[down]}$$

$$\frac{825 \text{ N}}{9.81\frac{m}{s^2}} = \frac{m\ \cancel{9.81\frac{m}{s^2}\text{[down]}}}{\cancel{9.81\frac{m}{s^2}\text{[down]}}}$$

$$m = 84.1\frac{kg\frac{m}{s^2}}{\frac{m}{s^2}}$$

$$m = 84.1 \text{ kg}$$

**Solve for $m$ first**

$$\frac{\vec{F}}{\cancel{\vec{g}}} = \frac{m\cancel{\vec{g}}}{\cancel{\vec{g}}}$$

$$m = \frac{\vec{F}}{\vec{g}}$$

$$m = \frac{825 \text{ N}\cancel{\text{[down]}}}{9.81\frac{m}{s^2}\cancel{\text{[down]}}}$$

$$m = 84.1\ \frac{kg\frac{m}{s^2}}{\frac{m}{s^2}}$$

$$m = 84.1 \text{ kg}$$

The student has a mass of 84.1 kg.

## Validate

Force has units of newtons and mass has units of kilograms.
The result, 84.1 kg, is a reasonable value for a mass of a person.

1. Find the weight of a 2.3 kg bowling ball on Earth.

2. You have a weight of 652.58 N[down] while standing on a spring scale on Earth near the equator.

   **(a)** Calculate your mass.

   **(b)** Determine your weight on Earth near the North Pole.

   **(c)** Determine your weight on the International Space Station. Why would this value be impossible to verify experimentally?

3. The lunar roving vehicle (LRV) pictured here has a mass of 209 kg regardless of where it is, but its weight is much less on the surface of the Moon than on Earth. Calculate the LRV's weight on Earth and on the Moon.

4. A 1.00 kg mass is used to determine the acceleration due to gravity of a distant, city-sized asteroid. Calculate the acceleration due to gravity if the mass has a weight of $3.25 \times 10^{-2}$ N[down] on the surface of the asteroid.

## Friction

When Galileo developed his principles of mechanics, he attributed many phenomena to frictional forces. Hundreds of years of observations have supported his conclusions. Since frictional forces are involved in essentially all mechanical movements, a more detailed understanding of these forces is critical to making predictions and describing motion. Unlike gravity and magnetism, frictional forces are contact forces.

Frictional forces inhibit relative motion between objects in contact with each other. In this section, you will focus on friction between surfaces. The frictional forces that act on an object moving through a fluid such as air or water are much more complex and you will not deal with them quantitatively in this chapter.

Two types of frictional forces are involved when you slide an object over a surface. A **static frictional force** exists when you start to move an object from rest. A **kinetic frictional force** exists while the object is moving. You probably discovered that the static frictional force that you must overcome to start an object moving is larger the kinetic frictional force. Figure 4.6 on page 146 shows a graph of actual experimental data obtained by slowly increasing

### TRY THIS...

Push a large mass horizontally across the lab bench with only one finger. Start pushing gently. Gradually increase the force you exert on the mass until it begins moving, then try to keep it moving with a constant velocity. Compare the force required to start the mass moving to the force that you must exert to keep it moving. Repeat the process until you can conclude whether there is any difference in these two forces. (You may have already experienced a similar effect if you have ever tried to slide a large box across a room.) Comment on and attempt to explain any difference in the force needed to start an object sliding and the force needed to keep the object in motion.

an applied force on an object. Once the object started to move, it maintained a constant speed. The graph clearly demonstrates how the static frictional force increases to a maximum before the object gives way and begins to move. The object then requires a smaller force to keep it moving at a constant speed.

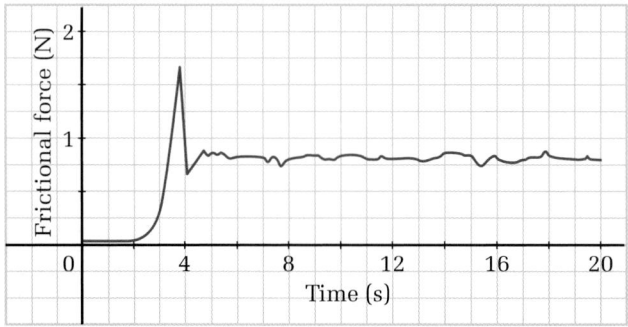

**Figure 4.6** Sensitive equipment and careful experimentation yield results that clearly show how a static frictional force increases until the object begins moving. The kinetic frictional force is less than the maximum static frictional force.

## QUICK LAB — How Sticky Is Your Shoe?

**TARGET SKILLS**

- Initiating and planning
- Analyzing and interpreting
- Communicating results

Using a force probe and a computer, determine the maximum static frictional force that you can cause your shoe to exert. Predict what factors might affect your shoe's "stickiness." Conduct an experiment to test your predictions on a variety of surfaces. (This experiment can also be conducted using a Newton spring scale.)

### Analyze and Conclude

1. Describe the nature of the surfaces on which you tested your shoe.

2. Did some surfaces cause your shoe to be "stickier" than others? Offer an explanation about the differences in these surfaces.

3. Compare the static frictional force with the kinetic frictional force. Describe the results of your comparison.

4. What steps might you take to ensure that your shoe is as sticky as it possibly can be?

The strength of a frictional force between two surfaces depends on the nature of the surfaces. Although some surfaces create far more friction than others, all surfaces create some friction. The appearance of smooth surfaces can be deceiving. The photograph in Figure 4.7 shows a magnified cross section of a highly polished steel surface. The jagged peaks are far too small to see with the unaided eye or to feel with your hand, yet they are high enough to interact with objects that slide over them.

The force of friction is actually an electromagnetic force acting between the surface atoms of one object and those of another. In fact, if two blocks of highly polished steel were cleaned and placed together in a perfect vacuum, they would weld themselves together and become one block of steel. In reality, small amounts of air, moisture, and contaminants accumulate on surfaces and prevent such "ideal" interactions.

When two surfaces are at rest and in contact, the surface atoms interact to form relative strong attractive forces. When you push on one object, static friction "pushes back" with exactly the same magnitude as an applied force until the applied force is great enough to break the attractive force between the surface atoms. When the object begins to move, new "bonds" are continually being formed and broken in what you could call a stick-and-slip process. Once the object is in motion, the stick-and-slip process repeats itself over and over in rapid succession. This process is responsible for the noise produced when two objects slide past one another. The squealing of tires on dry pavement and the music created by passing a bow over a violin string are examples of such noises.

Figure 4.8 illustrates how surfaces appear to make contact. The amount of contact and the types of atoms and molecules making up the materials passing over one another play a significant role in determining how large a frictional force will be. Sliding a 5.0 kg block of ice on a sheet of ice requires much less force than sliding the same block across rubber. Experimentation yields a "stickiness" value called a **coefficient of friction** for specific combinations of surfaces. Table 4.5 lists *coefficients of static friction* for objects at rest and *coefficients of kinetic friction* for objects in motion. Coefficients of friction are experimentally obtained and depend entirely on the *two* interacting surfaces.

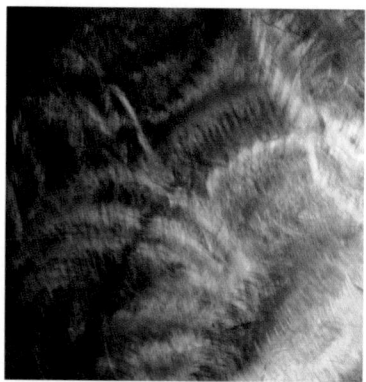

**Figure 4.7** Highly polished steel that feels very smooth still has bumps and valleys that will collide with other surface imperfections when rubbed together.

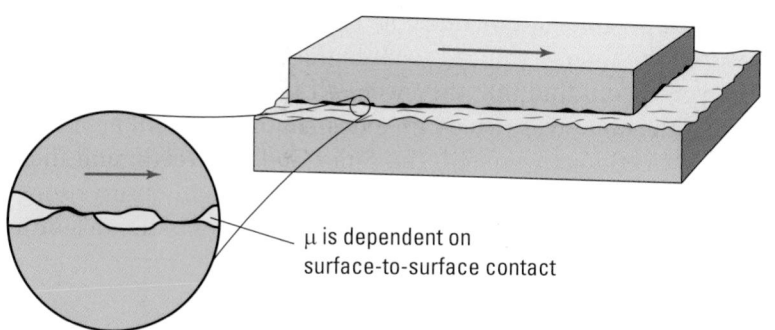

μ is dependent on surface-to-surface contact

**Figure 4.8** The coefficient of friction depends on each of the two surfaces in contact and must be experimentally obtained.

Hold a book against a flat wall by holding one hand under the book. With your other hand, gently press the book against the wall. Move your hand that is under the book. Catch the book if it begins to slip. Repeat the process but exert a little more pressure on the book toward the wall until the book does not fall when you remove your hand from beneath it. Summarize what your observations tell you about frictional forces.

**Table 4.5** Coefficients of Friction

| Surfaces | Coefficient of Static Friction $\mu_s$ | Coefficient of Kinetic Friction $\mu_k$ |
|---|---|---|
| rubber on dry solid surfaces | 1 – 4 | 1 |
| rubber on dry concrete | 1.00 | 0.80 |
| rubber on wet concrete | 0.70 | 0.50 |
| glass on glass | 0.94 | 0.40 |
| steel on steel (unlubricated) | 0.74 | 0.57 |
| steel on steel (lubricated) | 0.15 | 0.06 |
| wood on wood | 0.40 | 0.20 |
| ice on ice | 0.10 | 0.03 |
| Teflon™ on steel in air | 0.04 | 0.04 |
| lubricated ball bearings | < 0.01 | < 0.01 |
| synovial joint in humans | 0.01 | 0.003 |

direction of motion

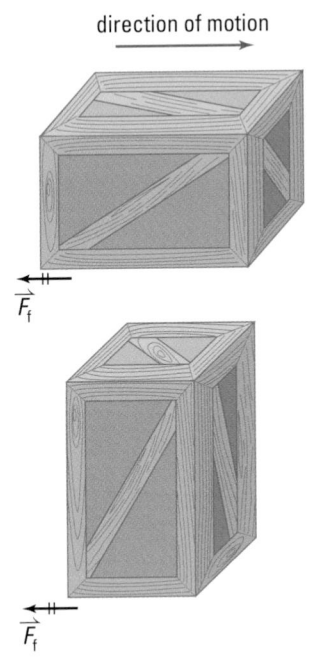

**Figure 4.9** The uniform distribution of mass will yield approximately the same frictional force regardless of the side in contact with the floor.

The force of friction depends not only on the types of surfaces that are in contact, but also on the magnitude of the forces that are pressing the two surfaces together. Whenever any object exerts a force on a flat surface such as a wall, floor, or road surface, that surface will exert a force back on the object in a direction perpendicular to the surface. Such a force is called a **normal force**. If you have ever attempted to slide a dresser full of clothes across a carpeted floor, you will know that by removing the drawers as well as the clothes the job gets much easier. A full dresser weighs significantly more than an empty one. The carpeted floor must support the weight of the dresser, and does so with a normal force. By reducing the weight of the dresser, you are also reducing the normal force.

An observation that might be surprising is that the force of surface friction is independent of velocity. Fluid friction, on the other hand, is affected by velocity in a complex way. A second, possibly surprising observation is that the force of friction is independent of the area of contact. Consider the crate in Figure 4.9. If you measured the forces required to slide the crate along any of its sides, you would find that they were all the same.

When the crate was placed on sides having different areas, the materials in contact were still the same, and the weight and therefore normal forces were the same. Experiments and observations have shown that these two factors, alone, determine the magnitude of the frictional force between surfaces.

Friction is caused by a large variety of atomic and molecular interactions. These reactions are so varied that firm "laws" do not apply. The relationships that have been developed are consistent but can be applied only under the following conditions. If the conditions are met, you can consider the results of calculations to be very good approximations, but not exact predictions.

- The force of friction is independent of surface area *only* if the mass of the object is evenly distributed.

- Certain plastics and rubbers have natural properties that often do not fit the standard model of friction, (for example, adhesive tape, "ice-gripping" tires).

- The two interacting surfaces must be flat. If such things as spikes or ridges are present that penetrate the opposite surface, the principles discussed above no longer apply (See Figure 4.10).

**Figure 4.10** The ridges on the bulldozer tracks are penetrating the soil, creating interactions that cannot be considered as simple surface friction.

---

**SURFACE FRICTION**

The magnitude of the force of surface friction is the product of the coefficient of friction and the magnitude of the normal force. The direction of the force of friction is always opposite to the direction of the motion.

$$F_f = \mu F_N$$

| Quantity | Symbol | SI unit |
|---|---|---|
| force of friction | $F_f$ | N (newton) |
| coefficient of friction | $\mu$ | none (coefficients of friction are unitless) |
| normal force | $F_N$ | N (newton) |

**Note:** Since the direction of the normal force is perpendicular to the direction of the force of friction, vector notations are omitted.

---

**MODEL PROBLEMS**

### Working with Friction

1. **During the winter, owners of pickup trucks often place sandbags in the rear of their vehicles. Calculate the increased static force of friction between the rubber tires and wet concrete resulting from the addition of 200 kg ($2.00 \times 10^2$ kg) of sandbags in the back of the truck.**

---

*continued* ▶

## Frame the Problem

- Sketch the problem.

- The addition of the sandbags will not change the *coefficient of friction* between the tires and the wet road.

- The sandbags will increase the weight of the truck, thereby increasing the *normal* force.

- The equation relating *frictional force* to the coefficient of friction and the normal force applies to this problem.

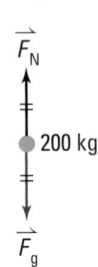

$\vec{F}_N$

200 kg

$\vec{F}_g$

## Identify the Goal

The increase in the frictional force, $F_f$, resulting from placing sandbags in the back of the truck

## Variables and Constants

| Involved in the problem | | | Known | Implied | Unknown |
|---|---|---|---|---|---|
| $m_{sandbags}$ | $g$ | $F_N$ | $m_{sandbags} = 2.00 \times 10^2$ kg | $g = 9.81$ m/s$^2$ | $F_{g\ sandbags}$ |
| $F_{g\ sandbags}$ | $\mu_s$ | | | $\mu_s = 0.70$ | $F_N$ |

### Strategy

In this case, the additional normal force is equal to the weight of the sandbags.

Use the equation for weight to find the weight and thus the normal force.

Substitute and solve.

$\dfrac{\text{kg} \cdot \text{m}}{\text{s}^2}$ is equivalent to N.

All of the values needed to find the additional frictional force are known so substitute into the equation for frictional forces.

### Calculations

$$F_N = F_g$$

$$F_g = mg$$

$$F_g = (2.00 \times 10^2 \text{ kg})\left(9.81\frac{\text{m}}{\text{s}^2}\right)$$

$$F_g = 1.962 \times 10^3\ \frac{\text{kg} \cdot \text{m}}{\text{s}^2}$$

$$F_N = 1.962 \times 10^3 \text{ N}$$

$$F_f = \mu_s F_N$$

$$F_f = (0.70)(1.962 \times 10^3 \text{ N})$$

$$F_f = 1.373 \times 10^3 \text{ N}$$

The sandbags increased the force of friction of the tires on the road by $1.4 \times 10^3$ N.

## Validate

The force of friction should increase with the addition of weight, which it did.

**2.** A horizontal force of 85 N is required to pull a child in a sled at constant speed over dry snow to overcome the force of friction. The child and sled have a combined mass of 52 kg. Calculate the coefficient of kinetic friction between the sled and the snow.

## Frame the Problem

- Sketch the forces acting on the child and sled.

- The applied force just overcomes the force of friction, therefore, the *applied force* must be *equal* to the *frictional force*.

- The sled is neither sinking into the snow, nor is it rising off of the snow; therefore, the *weight* of the sled must be exactly equal to the *normal force* supporting it.

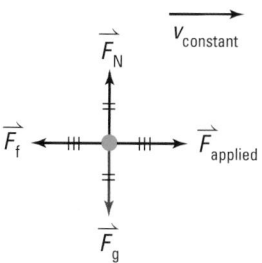

## Identify the Goal

The coefficient of kinetic friction, $\mu_k$

## Variables and Constants

| Involved in the problem | | Known | Implied | Unknown |
|---|---|---|---|---|
| $m$ | $F_f$ | $m = 52$ kg | $g = 9.81$ m/s$^2$ | $F_g$ |
| $F_{applied}$ | $F_N$ | $F_{applied} = 85$ N | | $F_f$ |
| $F_g$ | $\mu_k$ | | | $F_N$ |
| $g$ | | | | $\mu_k$ |

## Strategy

The conditions for the equation describing surface friction are met, so use the equation.

Since the normal force is equal to the weight, use the equation for weight.

Solve.

$\dfrac{\text{kg} \cdot \text{m}}{\text{s}^2}$ is equivalent to N.

Apply the equation for a frictional force.

## Calculations

$$F_f = \mu_k F_N$$

$$F_N = Fg$$

$$F_g = mg$$

$$F_N = mg$$

$$F_N = (52 \text{ kg})\left(9.81 \frac{\text{m}}{\text{s}^2}\right)$$

$$F_N = 510.12 \frac{\text{kg} \cdot \text{m}}{\text{s}^2}$$

$$F_N = 510.12 \text{ N}$$

$$F_f = \mu_k F_N$$

*continued* ▶

*continued from previous page*

## Strategy

## Calculations

### Substitute first

$$F_f = \mu_k F_N$$

$$85 \text{ N} = \mu_k 510.12 \text{ N}$$

$$\frac{85 \text{ N}}{510.12 \text{ N}} = \frac{\mu_k \cancel{510.12 \text{ N}}}{\cancel{510.12 \text{ N}}}$$

$$\mu_k = 0.1666$$

### Solve for $\mu_k$ first

$$\frac{F_f}{F_N} = \frac{\mu_k \cancel{F_N}}{\cancel{F_N}}$$

$$\mu_k = \frac{85 \text{ N}}{510.12 \text{ N}}$$

$$\mu_k = 0.1666$$

Therefore, the coefficient of kinetic friction between the sled and the snow is 0.17.

## Validate

The coefficient of friction between a sled and snow should be relatively small, which it is.

## PRACTICE PROBLEMS

**5.** A friend pushes a 600 g ($6.00 \times 10^2$ g) text-book along a lab bench at constant velocity with 3.50 N of force.

**(a)** Determine the normal force supporting the textbook.

**(b)** Calculate the force of friction and coefficient of friction between the book and the bench.

**(c)** Which coefficient of friction have you found, $\mu_s$ or $\mu_k$?

**6.** A 125 kg crate full of produce is to be slid across a barn floor.

**(a)** Calculate the normal force supporting the crate.

**(b)** Calculate the minimum force required to start the crate moving if the coefficient of static friction between the crate and the floor is 0.430.

**(c)** Calculate the minimum force required to start the crate moving if half of the mass is removed from the crate before attempting to slide it.

**7.** Avalanches often result when the top layer of a snow pack behaves like a piece of glass, and begins sliding over the underneath layer. Calculate the force of static friction between two layers of horizontal ice on the top of Mount Everest, if the top layer has a mass of $2.00 \times 10^2$ kg. (Refer to Table 4.5 for the coefficient of friction.)

**8.** Assume that, in the "Try This" experiment on page 148, you discovered that you had to push the book against the wall with a force of 63 N in order to prevent it from falling. Assume the mass of the book to be 2.2 kg. What is the coefficient of static friction between the book and the wall? (Hint: Be careful to correctly identify the source of the normal force and the role of the frictional force in this situation.)

## Concept Organizer

The types of surfaces in contact directly affect the magnitude of the force of friction.

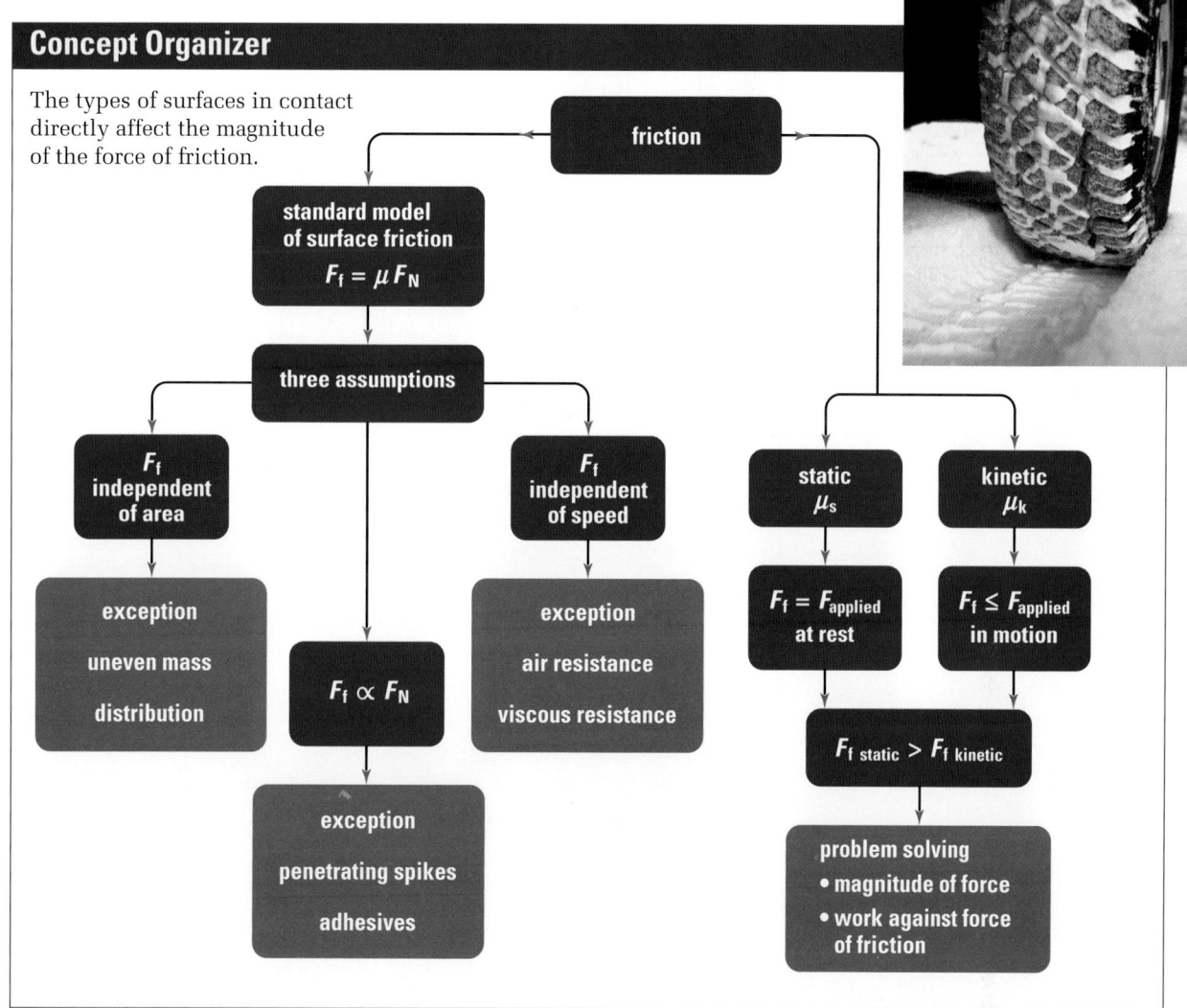

friction

standard model of surface friction
$F_f = \mu F_N$

three assumptions

$F_f$ independent of area

exception
uneven mass
distribution

$F_f \propto F_N$

exception
penetrating spikes
adhesives

$F_f$ independent of speed

exception
air resistance
viscous resistance

static
$\mu_s$

$F_f = F_{applied}$
at rest

kinetic
$\mu_k$

$F_f \leq F_{applied}$
in motion

$F_{f\ static} > F_{f\ kinetic}$

problem solving
• magnitude of force
• work against force of friction

**Figure 4.11** Understanding the standard model of friction and the exceptions.

# Free Body Diagrams

You have seen many cases in which more than one force was acting on an object. You also learned that, according to Galileo's concepts of inertia, the motion of a body will change only if an external force is applied to it. When several forces are acting, how do you know which force might change an object's state of motion? The answer is, "All of them acting together." You must look at every force that is acting on an object and find the vector sum of all of the forces before you can predict the motion of an object. This vector sum is the **net force** on the object. A free body diagram is an excellent tool to help you keep track of all of the forces.

**Free body diagrams** represent all forces *acting on* one object, and only the forces acting on the object. Forces that the object exerts on other objects do not appear in free body diagrams because they have no effect on the motion of the object itself.

In drawing a free body diagram, you represent the object as a single dot to help focus interest on the forces involved and not on the creator's artistic flair. You will represent each force *acting* on the object with an arrow. The arrow's direction shows the direction of the force and the arrow's relative length provides information about the magnitude of the force. Forces that have the same magnitude should be sketched with approximately the same length, forces that are larger should be longer, and smaller forces should be shorter. Study the examples in Figure 4.12.

**Figure 4.12** Free body diagrams for some everyday objects.

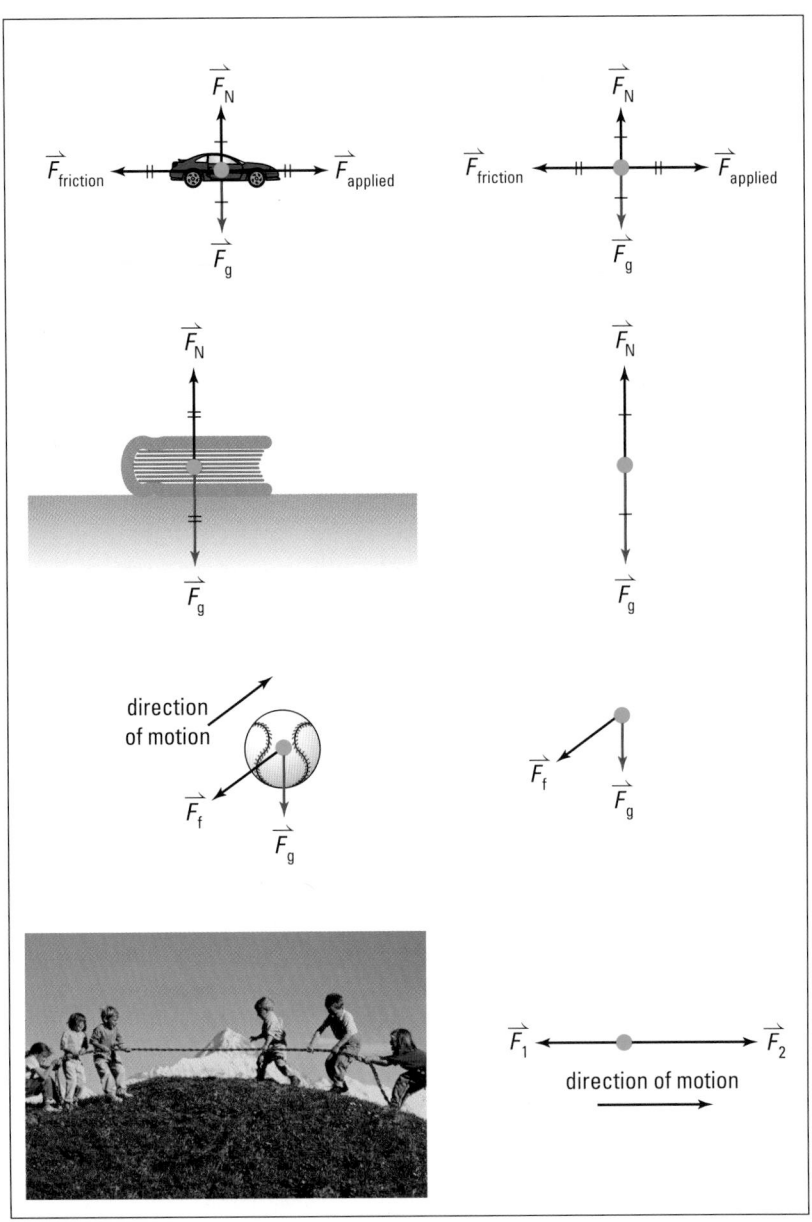

1. **K/U** Describe how and why acceleration due to gravity varies around the globe.

2. **K/U** A Ping Pong™ ball is struck with a paddle, sails over the net, bounces off of the table, and continues to the floor. Describe the forces acting on the ball throughout the trip.

3. **C** Explain the difference between contact and non-contact forces and provide examples of each.

4. **C** Explain, using force arguments, how it is possible for a feather and a coin in a vacuum to fall toward Earth with the same acceleration.

5. **C** A news reporter states that the winning entry in a giant pumpkin-growing contest "had a weight of 354 kg." Explain the error in this statement and provide values for both the weight and mass of the winning pumpkin.

6. **C** Describe the forces acting on a bag of chips resting on a tabletop.

7. **C** Explain why the coefficient of static friction is greater than the coefficient of kinetic friction.

8. **K/U** Imagine that all of the objects pictured are on the Moon, where there is no atmosphere.

(a) Rank the objects in order of weight, putting the heaviest object first.

(b) Rank the objects from largest to smallest according to the magnitude of the force of gravity acting on each of them.

(c) If each object was dropped simultaneously, which would hit the Moon's surface first?

9. **C**

(a) State three assumptions implied by the friction equation: $F_f = \mu F_N$.

(b) Discuss whether or not these assumptions are valid for each situation shown in the following pictures.

10. **K/U** Predict the motion that each object would undergo based on the free body diagrams illustrated.

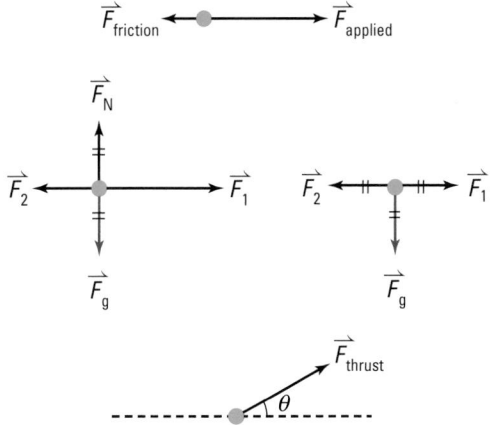

*continued* ▶

**11.** (K/U)

(a) Why is a dot used in a free body diagram?

(b) Why are the lengths of the force vector sketches important?

(c) Describe the purpose of a free body diagram.

**12.** (K/U) Draw a free body diagram showing the forces acting on a monkey that is hanging at rest

(a) from a vine

(b) from a spring

**13.** (K/U)

(a) Draw a free body diagram showing the forces acting on the ball during its flight.

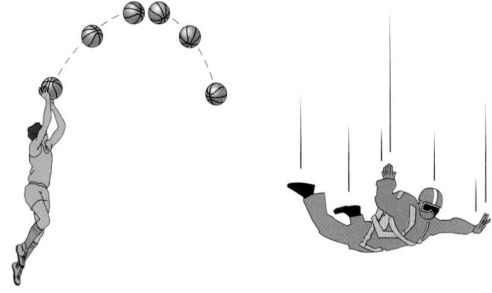

(b) Draw a free body diagram showing the forces acting on the skydiver before he deploys his parachute.

**14.** (K/U) Draw a free body diagram showing the forces acting on the pop can.

**15.** (K/U) Draw a free body diagram for each of the following situations.

(a) A submarine moves horizontally with constant velocity through deep water.

(b) A car accelerates from a stoplight.

(c) A pail is lifted from a deep well at constant velocity using a rope.

(d) A neon sign hangs motionless, suspended by two cables. One cable runs horizontally, connecting the sign to a wall; the other cable runs up and away from the wall, connecting the sign to the ceiling.

**UNIT PROJECT PREP**

Friction and inertia are often involved in comic situations in the movies.

- Have you ever tried running across a skating rink and then tried to stop yourself?
- Have you ever pushed or rolled a heavy object and then had difficulty stopping it?

The understanding of force and motion progressed slowly from the time of Aristotle through to Galileo and then to Newton. Today, we use concepts summarized by Sir Isaac Newton in his book called *Principia Mathematica Philosophiae Naturalis* published in 1686. Newton's three laws of motion are currently used to predict force and motion interactions for macroscopic objects. The laws are over 400 years old. Has no progress been made in terms of understanding force and motion in the past 400 years? Yes, in fact, there has been significant progress.

The physics of force and motion is currently divided into two very different categories, classical mechanics and quantum mechanics. This textbook deals with **classical mechanics**, sometimes called **Newtonian mechanics**. Newtonian mechanics treats energy and matter as separate entities and uses Newton's laws of motion to predict the results of interactions between objects. The principles of Newtonian mechanics, although formulated 400 years ago, accurately predict and describe the behaviour of large-scale objects such as baseballs, cars, and buildings. Newtonian mechanics provides a connection between the acceleration of a body and the forces acting on it. It deals with objects that are large in comparison to the size of an atom, and with speeds that are much less than the speed of light ($c = 3.0 \times 10^8$ m/s). **Quantum mechanics**, on the other hand, attempts to explain the motion and energy of atoms and subatomic particles.

In the early part of the twentieth century, Einstein developed two ingenious theories of relativity, which, in part, deal with objects travelling close to the speed of light. Also included in his theories is the proposal that mass and energy are, in fact, different manifestations of the same entity. His famous equation, $E = mc^2$, describes how these two quantities are related. Einstein's theories predict everything that Newtonian mechanics is able to, plus much more. It is important to understand that Einstein's relativistic mechanics is an extension of Newtonian mechanics, not a replacement. Developing a conceptual framework to understand force and motion in terms of Newton's laws is necessary before attempting to grasp these more advanced theories of physics.

## Newton's First Law

Newton originally considered Aristotle's ideas, but came to adopt Galileo's perspective that a body will tend to stay at rest or in uniform motion unless acted on by an external force. Clearly, many objects are at rest even though all objects on Earth are subjected to the force of gravity. So what is the meaning of "external force?"

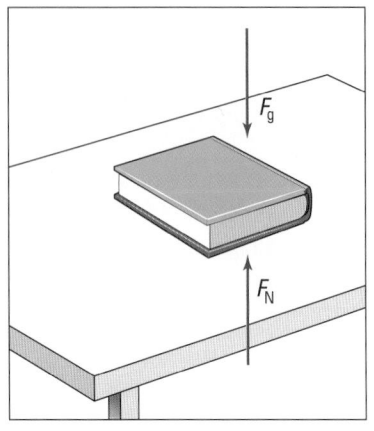

**Figure 4.13** The forces on the book are balanced so the book remains at rest.

Consider the book on the desk in Figure 4.13. The book does not fall under the force of gravity because it is supported by the normal force of the desk pushing up on it. The key to understanding the meaning of "external force" is to consider all forces acting on an object. If the vector sum of all of the forces acting on an object is zero, then there is no *net force* acting on the object and its motion will not change. You would say that the forces are balanced and the object is in equilibrium.

### NEWTON'S FIRST LAW — THE LAW OF INERTIA

An object at rest or in uniform motion will remain at rest or in uniform motion unless acted on by an external force.

Hockey provides concrete examples of Newton's first law. A hockey puck at rest on the ice will not spontaneously start to move. Likewise, a puck given some velocity will continue to slide in a straight line at a constant speed. If the ice was truly frictionless and the ice surface was infinitely large, the puck would continue to slide forever.

You can apply Newton's laws to each dimension, independently. Consider the case in which a cart is pulled across a smooth tabletop with constant velocity. During the trip, a steel ball bearing is fired directly upward out of the cart. Magically, the ball will travel up and then back down, all the while remaining directly above the cart. In fact it is not magic, but Newton's first law, that is demonstrated here. The ball has the same horizontal velocity as the cart before it is launched. The ball maintains a horizontal velocity equal to that of the cart as it moves through the air. The forces that caused the ball to rise and fall were in the vertical dimension and had no effect on its horizontal motion.

**Figure 4.14** Hockey provides a nearly frictionless surface for the puck to slide on and plenty of opportunity to observe how forces change the speed and direction of objects.

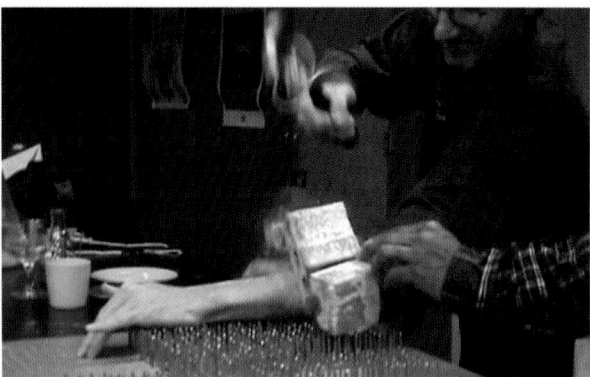

**Figure 4.15** This performer's arm is resting on a bed of nails, yet he is able to break a brick over it without suffering any injury. Use Newton's law of inertia to explain how this is possible.

# Inertial Reference Frames

Situations exist where the law of inertia seems to be invalid. Imagine a car speeding away from a stoplight. A passenger appears to be abruptly pushed back against the seat. To the pedestrian observing the car from the street corner, Newton's law of inertia seems to be completely accurate. The passenger, under the influence of the unbalanced force of the seat pushing on her, is accelerating *relative to the ground*. The observer is in an **inertial frame of reference**.

The passenger sitting in a car feels the seat pushing on her as the car accelerates away from a stoplight. The seat is exerting a strong force onto her back, and yet she is not moving *relative to the car*. To the passenger, an unbalanced force is acting on a mass (her body) and yet there is no apparent motion. This appears to violate the law of inertia. However, Newton's laws apply to inertial reference frames. The passenger is in an *accelerating* frame of reference, which is a **non-inertial reference frame**. Notice that, when the car reaches a constant velocity, the passenger no longer feels the pressure of the back of the seat pushing on her. A reference frame moving at a constant velocity is an inertial reference frame.

---

## DEFINITION OF AN INERTIAL REFERENCE FRAME

An inertial reference frame is one in which Newton's law of inertia is valid. An inertial reference frame must not be accelerating.

---

### Think It Through

- Consider the example shown below. The passenger in a car travelling with constant velocity tosses and then catches a tennis ball. In the passenger's frame of reference, the ball goes straight up and comes straight back down under the influence of gravity. A roadside observer sees the tennis ball trace a different trajectory. Explain how different observers can see two different trajectories of the ball and yet both observers perceive that Newton's first law is validated by the observation.

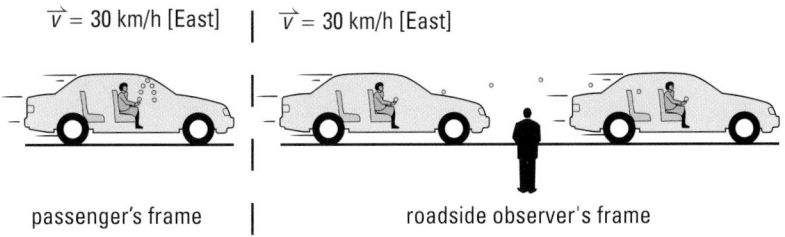

$\vec{v}$ = 30 km/h [East]  |  $\vec{v}$ = 30 km/h [East]

passenger's frame  |  roadside observer's frame

Both frames are Inertial

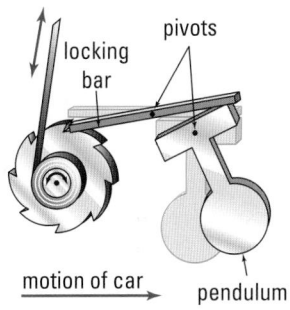

## Newton's Second Law

Imagine three young friends playing in the snow with a toboggan on flat ground. Assume that (a) each friend has the same mass, $m$, (b) can pull with the same force, $F$, (c) the toboggan's mass is so small that it can be ignored, and (d) the toboggan glides so easily on top of the snow that friction can be ignored. The friends take turns pulling and being pulled on the toboggan. A comparison of the resulting accelerations is shown in the chart below. Look for a pattern that relates force, mass, and acceleration.

| | | | |
|---|---|---|---|
| **Net force** | $|\vec{F}|$ | $|\vec{F}|$ | $2|\vec{F}|$ |
| **Mass on toboggan** | $2m$ | $m$ | $m$ |
| **Acceleration** | $\frac{1}{2}|\vec{a}|$ | $|\vec{a}|$ | $2|\vec{a}|$ |

**COURSE CHALLENGE**

**The Cost of Altitude**

Research current costs associated with getting objects into space. Later, you may use this data to help estimate the cost of delivering electric energy through a space-based power system.

Learn more from the **Science Resources** section of the following web site: **www.school.mcgrawhill.ca/ resources/** and find the *Physics 11 Course Challenge.*

Newton observed motions as simple as those of the toboggan described above and as complex as the motion of planets around the sun. From these observations, he developed his second law of motion. As you may have deduced from the data in the chart above, when a net force, $F$, acts on a mass, $m$, the resulting acceleration of the mass, $a$, is proportional to the magnitude of the force and inversely proportional to the amount of mass. The direction of the acceleration is the same as that of the net force. The form of the law described here and the more familiar form are detailed in the box on the next page.

### Think It Through

- Picture this. You and your family are moving. There are boxes everywhere. You just carried a very heavy box out to the truck and have come back for another one. You reach for a box that you believe to be full and very heavy. However, it is empty. What do you think will happen when you start to lift it? Explain, in terms of forces and acceleration, what happens when anyone starts to lift an object that they believe to be much heavier than it actually is.

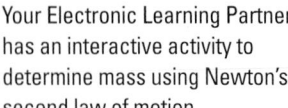

**ELECTRONIC LEARNING PARTNER**

Your Electronic Learning Partner has an interactive activity to determine mass using Newton's second law of motion.

## NEWTON'S SECOND LAW

Force is the product of mass and acceleration, or, acceleration is the quotient of the force and the mass.

$$\vec{F} = m\vec{a} \quad \text{or} \quad \vec{a} = \frac{\vec{F}}{m}$$

| Quantity | Symbol | SI unit |
|---|---|---|
| acceleration | $\vec{a}$ | $m/s^2$ (metre per second squared) |
| force | $\vec{F}$ | N (newton) |
| mass | $m$ | kg (kilogram) |

**Unit Analysis**

(mass) (acceleration) = $kg\ m/s^2$ = N

### MODEL PROBLEM

## Applying Newton's Second Law

1. A man is riding in an elevator. The combined mass of the man and the elevator is $7.00 \times 10^2$ kg. Calculate the magnitude and direction of the elevator's acceleration if the tension ($T$) in the supporting cable is $7.50 \times 10^3$ N ($T$ is the applied force).

## Frame the Problem

- Sketch a free body diagram of the man and elevator.

- The cable exerts an *upward force* on the man and elevator.

- *Gravity* exerts a *downward force* on the man and elevator.

- The net force on the man and elevator will determine the *acceleration* according to Newton's second law.

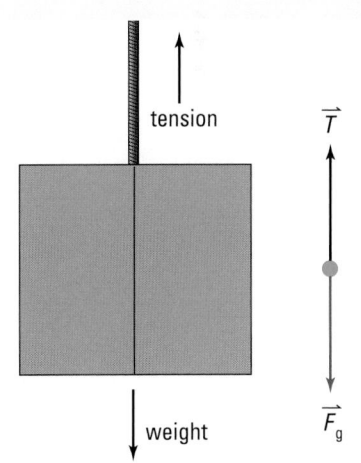

## Identify the Goal

The acceleration, $\vec{a}$, of the elevator

*continued* ▶

## Variables and Constants

| Involved in the problem | | Known | Implied | Unknown |
|---|---|---|---|---|
| $\vec{T}$ | $\vec{a}$ | $\vec{T} = 7.50 \times 10^3$ N | $\vec{g} = 9.81 \frac{m}{s^2}$ [down] | $\vec{F}_g$ |
| $\vec{F}_g$ | $m$ | $m = 7.00 \times 10^2$ kg | | $\vec{F}_{net}$ |
| $\vec{F}_{net}$ | $\vec{g}$ | | | $\vec{a}$ |

## Strategy

Since the motion is all along one line, up and down, denote direction with signs only. Let up be positive and down be negative.

Find the force of gravity acting on the man and elevator using the equation for weight. Since "down" was chosen as negative, the acceleration due to gravity becomes negative.

Find the net force acting on the man and elevator by finding the vector sum of the tension and force of gravity acting on the elevator and man.

Apply Newton's second law in terms of acceleration and solve.

Write N as $\frac{kg \cdot m}{s^2}$ so you can cancel units.

## Calculations

$$\vec{F}_g = m\vec{g}$$

$$\vec{F}_g = (7.00 \times 10^2 \text{ kg})\left(-9.81 \frac{m}{s^2}\right)$$

$$\vec{F}_g = -6.867 \times 10^3 \frac{kg \cdot m}{s^2}$$

$$\vec{F}_g = -6.87 \text{x} 10^3 \text{ N}$$

$$\vec{F}_{net} = \vec{T} + \vec{F}_g$$

$$\vec{F}_{net} = +7.50 \times 10^3 \text{ N} - 6.867 \times 10^3 \text{ N}$$

$$\vec{F}_{net} = +6.33 \times 10^2 \text{ N}$$

$$\vec{a} = \frac{\vec{F}}{m}$$

$$\vec{a} = \frac{+6.33 \times 10^2 \text{ N}}{7.00 \times 10^2 \text{ kg}}$$

$$\vec{a} = +9.043 \times 10^{-1} \frac{\frac{kg \cdot m}{s^2}}{kg}$$

$$\vec{a} = +9.043 \times 10^{-1} \frac{m}{s^2}$$

The elevator was accelerating upward at 0.904 m/s$^2$.

## Validate

The tension was greater than the weight, causing a net upward force to exist. A net force will cause an object to accelerate upward. The units cancelled to give metres per square second, which is correct for acceleration.

2. **A curler exerts an average force of 9.50 N[S] on a 20.0 kg stone. (Assume that the ice is frictionless.) The stone started from rest and was in contact with the girl's hand for 1.86 s.**

   **(a) Determine the average acceleration of the stone.**

   **(b) Determine the velocity of the stone when the curler releases it.**

## Frame the Problem

- Draw a free body diagram of the problem.

- The *downward force of gravity* is balanced by the *upward normal force*. Therefore there is *no net force* in the vertical direction. These forces do not affect the acceleration of the stone.

- The only *horizontal force* on the stone is the *force exerted* by the curler. Therefore it is the *net force* on the stone.

- The *net force* determines the *acceleration* of the stone according to *Newton's second law* of motion.

- After the stone leaves the curler's hand, there is *no* longer a horizontal *force* on the stone and thus, it is *no* longer *accelerating*.

- The *equations of motion* for uniform acceleration apply to the motion of the stone.

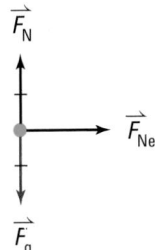

## Identify the Goal

(a) The acceleration, $\vec{a}$, of the curling stone

(b) The final velocity, $\vec{v}$, of the stone as it leaves her hand

## Variables and Constants

| Involved in the problem | | Known | Implied | Unknown |
|---|---|---|---|---|
| $\vec{F}_{applied}$ | $\vec{v_1}$ | $\vec{F}_{applied} = 9.5$ N[S] | $\vec{v_1} = 0.0 \ \frac{m}{s}$ [S] | $\vec{a}$ |
| $m$ | $\vec{v_2}$ | $m = 20.0$ kg | | $\vec{v_2}$ |
| $\vec{a}$ | $\Delta t$ | $\Delta t = 1.86$ s | | |

*continued* ▶

*continued from previous page*

## Strategy

You know the net force so use Newton's second law in terms of acceleration.

## Calculations

$$\vec{a} = \frac{\vec{F}}{m}$$

$$\vec{a} = \frac{9.50 \text{ N[S]}}{20.0 \text{ kg}}$$

$$\vec{a} = 0.475 \frac{\frac{\text{kg} \cdot \text{m}}{\text{s}^2}}{\text{kg}} \text{[S]}$$

$$\vec{a} = 0.475 \frac{\text{m}}{\text{s}^2} \text{[S]}$$

(a) The average acceleration of the stone was 0.475[S].

Recall the equations of motion from Chapter 2. Use the equation that relates initial and final velocities, acceleration, and the time interval.

Substitute in the known variables and solve for $\vec{v_2}$.

$$\vec{v_2} = \vec{v_1} + \vec{a}\Delta t$$

$$\vec{v_2} = 0.0 \frac{\text{m}}{\text{s}^2}\text{[S]} + \left(0.475 \frac{\text{m}}{\text{s}^2}\text{[S]}\right)(1.86 \text{ s})$$

$$\vec{v_2} = 0.8835 \frac{\text{m}}{\text{s}}\text{[S]}$$

(b) The velocity of the stone when it left the curlers hand was $0.884\frac{\text{m}}{\text{s}}\text{[S]}$.

## Validate

The curling stone experienced a net force acting toward the south. A net force causes acceleration in the direction of the force. The stone accelerated in the direction of the force, gaining speed as it went. The units cancelled to give $\frac{\text{m}}{\text{s}^2}\text{[S]}$ for acceleration and $\frac{\text{m}}{\text{s}}\text{[S]}$ for velocity. These are the correct units.

## PRACTICE PROBLEMS

9. A 4.0 kg object experiences a net force of 2.2 N[E]. Calculate the acceleration of the object.

10. A 6.0 kg object experiences an applied force of 4.4 N[E] and an opposing frictional force of 1.2 N[W]. Calculate the acceleration of the object.

11. A 15 kg object experiences an applied force of 5.5 N[N] and an opposing frictional force of 2.5 N[S]. If the object starts from rest, how far will it have travelled after 4.0 s?

12. A 45 kg student rides his 4.0 kg bicycle, exerting an applied force of 325 N[E].

(a) Calculate the acceleration of the cyclist if frictional resistance sums to 50.0 N[W].

(b) How far will the student have travelled if he started with a velocity of at 3.0 m/s[E] and accelerated for 8.0 s?

13. A stretched elastic exerts a force of 2.5 N[E] on a wheeled cart, causing it to accelerate at 1.5 m/s$^2$[E]. Calculate the mass of the cart, ignoring frictional effects.

14. The driver of a $1.2 \times 10^3$ kg car travelling 45 km/h[W] on a slippery road applies the brakes, skidding to a stop in 35 m. Determine the coefficient of friction between the road and the car tires.

# INVESTIGATION 4-B

## Force and Acceleration

**TARGET SKILLS**
- Performing and recording
- Analyzing and interpreting

Children playing with a toboggan in the snow will gain an intuitive sense that force, mass, and acceleration are related. Early scientists developed the same intuition, but only through experimentation could they formulate mathematical models that accurately predicted observed results. In this investigation, you will determine the validity of Newton's second law.

### Problem
Obtain experimental evidence to support Newton's second law.

### Hypothesis
State Newton's second law in the form of an hypothesis.

### Equipment
- elastic bands
- dynamics cart
- metre stick
- ticker-tape timer or motion sensor

### Procedure
1. Tape the elastic band onto the end of the metre stick as shown. Determine the length of an unstretched elastic band. Mark the length on the ruler with a piece of tape.

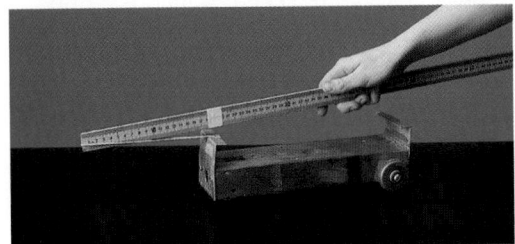

2. Attach the free end of the elastic to the dynamics cart.

3. Set up your equipment to collect distance versus time data on the cart as you pull it along with the elastic.

4. Obtain data with the elastic stretched to 1.0 cm, 2.0 cm, 3.0 cm, and 4.0 cm. It is crucial that you ensure that the stretch in the elastic remains exactly the same for the entire trip.

**CAUTION** As you run trials with more stretch in the elastic, the cart's final speed will increase dramatically. Have a partner waiting to catch the moving cart. Ensure that your path is free of obstacles or doors that could swing open.

5. Generate velocity versus time graphs for each length trial.

### Analyze and Conclude
1. What was the purpose of ensuring that the amount of stretch in the elastic band remained constant?

2. Use the velocity versus time graphs to obtain an average acceleration for each trial. Ensure that you select a time interval that is the same length for each trail when determining the average acceleration. (A larger time interval will yield better results.) Use the results to generate a force versus acceleration graph. You can express the force as centimetres of stretch of the elastic band.

3. Find the slope of the best-fit line from the force versus acceleration graph. What does this slope represent?

4. From your results, develop a mathematical model that relates force, mass, and acceleration.

5. Use your mathematical model to predict how the slope of the line in the force versus acceleration graph would appear if (a) two carts and (b) three carts were pulled using the same elastic. If time allows, test your prediction.

### The Physics of a Car Crash

A car accident has occurred at a busy intersection. A passenger in one of the cars is seriously injured and both vehicles are extensively damaged.

One of the drivers says she was stopped at a green light, waiting to make a left-hand turn, when an oncoming car swerved and hit her. The driver of the other car says he hit her car because she began to turn as he was passing through the intersection. The police officers investigating the accident also hear conflicting stories from witnesses. Nobody seems to know exactly what happened. It is time to bring in an accident investigator with expertise in physics and motion to determine how the crash actually occurred.

Accident investigators use the principles of physics, such as work and the conservation of energy, to determine the cause of car accidents. The investigators consider a number of factors and make detailed measurements at the scene of an accident. They might consider road conditions, damage to vehicles, the pre- and post-accident positions of vehicles, and vehicle characteristics such as weight and size. An investigator might be asked to determine the speed of each vehicle on impact by considering their masses and the distance they travelled after impact. For example, in the accident described above, the investigator would need to determine if the driver of the car in the left-turn lane was, in fact, stationary at the time of impact.

Several different career options are available for accident investigators. Police officers with specialized training are involved in accident investigation and reconstruction, while other investigators are consultants hired on a contractual basis by police departments, insurance companies, and individual citizens. Still others might be full-time employees of insurance companies or legal firms. Accident investigators are also often called on to serve as expert witnesses in criminal or civil law cases.

The training required to become an accident investigator varies. Some investigators have earned degrees in civil or traffic engineering. An excellent understanding of physics and the ability to perform detailed tasks accurately and without bias are important requirements. A knowledge of computers is becoming increasingly important as collision analysis becomes more computerized.

For information about where and how law enforcement accident investigators receive their training, contact the Community Services branch of a municipal, provincial, or federal law enforcement agency in your area.

### Going Further

1. Investigators are not always able to make actual measurements at the scene of an accident. Often they must reconstruct a scene or determine the cause of an accident from photographs and reports alone. List the factors that an investigator would need to know about the accident scenario just described.

2. How could an investigator determine these factors?

### Web Link

*www.school.mcgrawhill.ca/resources/*
In addition to motor vehicle accidents, some investigators work on air, marine, and rail accident cases. For example, investigators for the Canadian Transportation Safety Board worked on the Swissair Flight 111 disaster near Peggy's Cove in Nova Scotia. To learn about this investigation, go to the above Internet site and follow the links for **Science Resources** and **Physics 11** to find out where to go next.

# The Vector Nature of Force

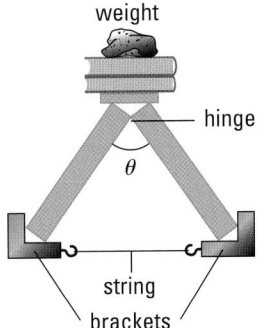

By using the apparatus shown, you can investigate the vector nature of force. Cut five lengths of thread to connect the base of the apparatus. Shorter lengths will create a taller isosceles triangle; longer lengths will create a shorter, wider isosceles triangle. Measure the angle at the top of the triangle and then pile small masses onto the top of the apparatus until the thread snaps. Record the angle and the amount of mass. Repeat this procedure for each length of thread.

## Analyze and Conclude

1. Find the weight of the total amount of mass required to break each thread.

2. The following diagram illustrates how the weight of the supported mass can be used to determine the tension in the thread. Find the breaking tension for each length of thread.

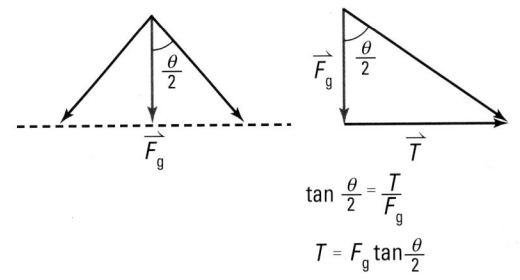

$$\tan \frac{\theta}{2} = \frac{T}{F_g}$$

$$T = F_g \tan \frac{\theta}{2}$$

For a given amount of mass, the weight remains constant. The weight is the vertical component of the force compressing the wooden post. For the mass to be supported, the weight (vertical component) and the tension in the thread (horizontal component) must add "tip to tail" to form a closed triangle with the resultant force in the post. Notice that the weight, for a given amount of mass, does not change; therefore, as the angle is increased, the tension force must get progressively larger to keep the system in equilibrium. The tension force, T, can be found using simple trigonometry.

3. Organize your data to compare breaking weight to interior angle.

4. Draw conclusions about the two-dimensional nature of force.

## Think It Through

- Which clothesline will be under the greatest tension, assuming that both pairs of pants are identical? Why? [Hint: Review your results from the Quick Lab.]

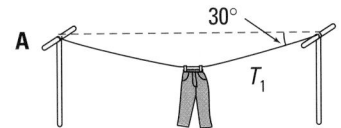

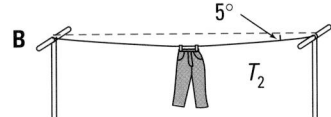

- If the tension in the ropes remains the same, what angle, $\theta$, between the two ropes would result in the greatest force in the direction of motion?

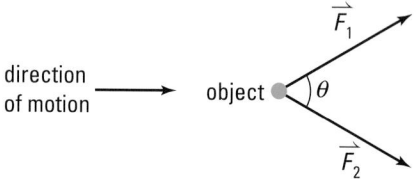

Thus far, you have performed force calculation for cases in which the direction of the net force was obvious. However, in many cases, the direction of the net force is not clear. Before you solve the problem, you must first find the magnitude and direction of the net force. Fortunately, the methods for adding and subtracting vectors involved in motion can be applied to any vector quantity.

Young children provide wonderful examples of the vector nature of forces. If you have ever had two children pulling each of your arms in different but not opposite directions, then you have witnessed the vector nature of forces. One child pulls you in one direction, the other pulls you in another direction, and you end up moving in a third direction that is actually determined by the vector sum of the original two forces.

**Figure 4.16** Which way does the net force act?

## MODEL PROBLEMS

### Forces in Two Dimensions

1. **Three children are each pulling on their older sibling, who has a mass of 65 kg. The forces exerted by each child are listed here. Use a scale diagram to determine the resultant acceleration of the older sibling.**

$\vec{F}_1 = 45$ N[E]
$\vec{F}_2 = 65$ N[S40°W]
$\vec{F}_3 = 20$ N[N75°W]

### Frame the Problem

- The force of gravity on the older sibling is balanced by the normal force of the ground. Therefore, you can neglect vertical forces because there is no motion in the vertical plane.

- Draw a free body diagram representing horizontal forces on the older sibling.

- The *net force* in the horizontal plane will determine the magnitude and direction of the *acceleration* of the older sibling.

- *Newton's second law* applies to this problem.

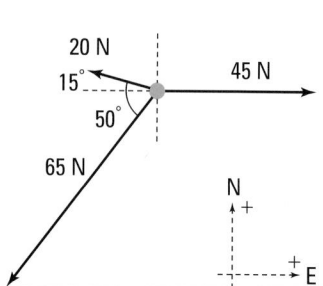

## Identify the Goal

The acceleration, $\vec{a}$, of the older sibling

## Variables and Constants

| Involved in the problem | | Known | Unknown |
|---|---|---|---|
| $\vec{F}_1$ | $\vec{a}$ | $\vec{F}_1 = 45$ N[E] | $\vec{a}$ |
| $\vec{F}_2$ | $\theta$ | $\vec{F}_2 = 65$ N[S40°W] | $\theta$ |
| $\vec{F}_3$ | | $\vec{F}_3 = 20$ N[N75°W] | |

## Strategy

Draw a scale diagram, adding the vectors "tip to tail."
If you need review, turn to Table 3.1 on page 91.

Measure the length of the resultant force vector.

Use the scale factor to determine the magnitude of the force.

Use a protractor to measure the angle.

Use Newton's second law in terms of acceleration.

## Calculations

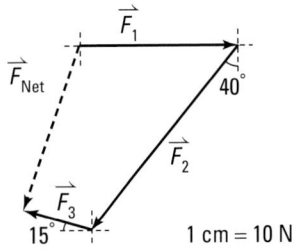

1 cm = 10 N

$|\vec{F}_{net}| = 4.8$ cm

$|\vec{F}_{net}| = (4.8 \text{ cm})\left(\dfrac{10 \text{ N}}{\text{cm}}\right)$

$|\vec{F}_{net}| = 48$ N

$\theta = [\text{S20°W}]$

$\vec{a} = \dfrac{\vec{F}}{m}$

$\vec{a} = \dfrac{48 \text{ N[S20°W]}}{65 \text{ kg}}$

$\vec{a} = 0.7385 \dfrac{\frac{\text{kg} \cdot \text{m}}{\text{s}^2} \text{ [S20°W]}}{\text{kg}}$

$\vec{a} = 0.7385 \dfrac{\text{m}}{\text{s}^2} [\text{S20°W}]$

The older sibling will have an acceleration of 0.74[S20°W].

## Validate

The acceleration value is reasonable. Units cancelled to give $\frac{\text{m}}{\text{s}^2}$ which is correct for acceleration.

---

2. **Solve the same problem using the method of components rather than a scale diagram.**

---

*continued*▶

*continued from previous page*

## Frame the Problem

- The force of gravity on the older sibling is balanced by the normal force of the ground. Therefore, you can neglect vertical forces because there is no motion in the vertical plane.

- Draw a free body diagram representing horizontal forces on the older sibling.

- The *net force* in the horizontal plane will determine the magnitude and direction of the *acceleration* of the older sibling,

- *Newton's second law* applies to this problem.

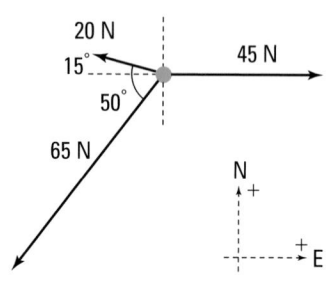

## Identify the Goal

The acceleration, $\vec{a}$, of the older sibling

## Strategy and Calculations

Draw each vector with its tail at the origin of an *x*–*y*-coordinate system where +*y* coincides with north and +*x* coincides with east.

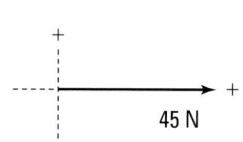
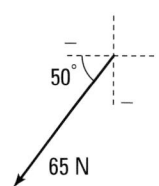
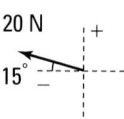

Find the angle with the nearest *x*-axis.

| | | |
|---|---|---|
| East coincides with the *x*-axis so the angle is 0° | In the angle [S40°W], the 40° angle is with the −*y*-axis. The angle with the −*x*-axis is 90° − 40° = 50° | In the angle [N75°W] the 75° angle is with the +*y*-axis. The angle with the −*x*-axis is 90° − 75° = 15° |

Find the *x*-component of each force vectors.

| | | |
|---|---|---|
| $F_{1x} = \lvert\vec{F_1}\rvert \cos 0°$ | $F_{2x} = -\lvert\vec{F_2}\rvert \cos 50°$ | $F_{3x} = -\lvert\vec{F_3}\rvert \cos 15°$ |
| $F_{1x} = (45\text{ N})(1.000)$ | $F_{2x} = -(65\text{ N})(0.6428)$ | $F_{3x} = -(20\text{ N})(0.9659)$ |
| $F_{1x} = 45$ N | $F_{2x} = -41.78$ N | $F_{3x} = -19.32$ N |
| | The angle is in the third quadrant so *x* is negative. | The angle is in the second quadrant so *x* is negative |

Find the *y*-components of each force vector.

| | | |
|---|---|---|
| $F_{1y} = \lvert\vec{F_1}\rvert \sin 0°$ | $F_{2y} = -\lvert\vec{F_2}\rvert \sin 50°$ | $F_{3y} = \lvert\vec{F_3}\rvert \sin 15°$ |
| $F_{1y} = (45\text{ N})(0.0)$ | $F_{2y} = -(65\text{ N})(0.7660)$ | $F_{3y} = (20\text{ N})(0.2588)$ |
| $F_{1y} = 0.0$ N | $F_{2y} = -47.79$ N | $F_{3y} = 5.176$ N |
| | The angle is in the third quadrant so *y* is negative. | |

Make a table in which to list the $x$- and $y$-components. Add them to find the components of the resultant vector.

| Vector | x-component | y-component |
|--------|-------------|-------------|
| $\vec{F}_1$ | 45 N | 0.0 N |
| $\vec{F}_2$ | −41.78 N | −47.79 N |
| $\vec{F}_3$ | −19.32 N | 5.176 N |
| $\vec{F}_{net}$ | −16.1 N | −42.614 N |

Use the Pythagorean Theorem to find the magnitude of the net force.

$$|\vec{F}_{net}|^2 = (F_{x\ net})^2 + (F_{y\ net})^2$$

$$|\vec{F}_{net}|^2 = (-16.1\ \text{N})^2 + (-42.614\ \text{N})^2$$

$$|\vec{F}_{net}|^2 = 259.21\ \text{N}^2 + 1815.9\ \text{N}^2$$

$$|\vec{F}_{net}|^2 = 2075.163\ \text{N}^2$$

$$|\vec{F}_{net}| = 45.55\ \text{N}$$

Use trigonometry to find the angle $\theta$.

Since both the $x$- and $y$-components are negative, the angle is in the third quadrant.

$$\tan \theta = \frac{-41.614\ \text{N}}{-16.1\ \text{N}}$$

$$\tan \theta = 2.619$$

$$\theta = \tan^{-1} 2.619$$

$$\theta = 69.1°$$

The net force on the older sibling is 46 N at an angle of 69° from the $x$-axis in the third quadrant. This result is equivalent to 46 N[W69°S] or 46 N[S21°W].

Apply Newton's second law in terms of acceleration to find the older sibling's acceleration.

$$\vec{a} = \frac{\vec{F}}{m}$$

$$\vec{a} = \frac{45.55\ \text{N[S21°W]}}{65\ \text{kg}}$$

$$\vec{a} = 0.70077\ \frac{\frac{\text{kg} \cdot \text{m}}{\text{s}^2}}{\text{kg}} \text{[S21°W]}$$

$$\vec{a} = 0.70077\ \frac{\text{m}}{\text{s}^2} \text{[S21°W]}$$

The acceleration of the older sibling is 0.70 $\frac{\text{m}}{\text{s}^2}$[S21°W].

## Validate

Using components gives nearly the same answer as the scale diagram method. You would expect the method of components to yield more accurate results. Also, the units cancelled to give m/s² which is correct for acceleration.

## PRACTICE PROBLEMS

**15.** A swimmer is propelled directly north by a force of 35.0 N. Moving water exerts a second force of 20 N[E]. Use both a scale diagram and a mathematical solution to determine the net force acting on the swimmer.

**16.** Find the resultant force acting on each object pictured. Obtain values by measuring the vectors.

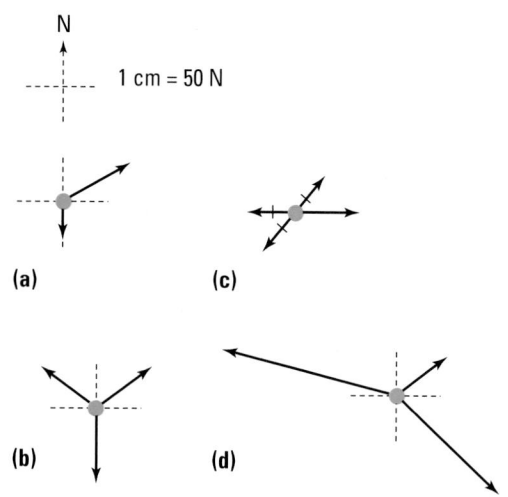

**17.** A train car is pulled along the tracks by a force of 1500 N from a pickup truck driving beside the tracks. The rope connecting the truck and the train car makes an angle of 15° to the direction of travel.

**(a)** Find the component of the pulling force in the direction of travel.

**(b)** Find the component of the pulling force perpendicular to the direction of travel.

**18.** A student pushes a 25 kg lawn mower with a force of 150 N. The handle makes an angle of 35° to the horizontal.

**(a)** Find the vertical and horizontal components of the applied force.

**(b)** Calculate the normal force supporting the lawn mower while it is being pushed.

**(c)** Calculate the net force propelling the mower if a frictional force of 85 N exists.

**(d)** Calculate the horizontal acceleration of the lawn mower. (Remember: Only part of the $F_{applied}$ is parallel to the direction of horizontal acceleration.)

---

### TRY THIS...

Perform each of the following.
- Stretch an elastic band between your hands.
- Gently push a toy across the lab bench.
- Push with all your might on a concrete wall.

Describe what you felt in each situation. Describe the forces that you applied and the forces that you felt applied to you. Draw free body diagrams for each object. Draw free body diagrams of one of your hands for each scenario. Make a general statement about action forces (the forces you applied) and reaction forces (the ones you felt).

## Newton's Third Law

For Newton's first and second laws, you focussed on individual objects and all of the forces acting on one specific object. The net force determined the change, or lack of change in the motion of that object. In Newton's third law, you will consider the interactions between two objects. You will not even consider all of the forces acting on the two objects but instead concentrate on only the force involved in the interaction between those two objects.

Newton realized that every time an object exerted a force on a second object, that object exerted a force back on the first.

---

### NEWTON'S THIRD LAW

For every action force on object B due to object A, there is a reaction force, equal in magnitude but opposite in direction, due to object B acting back on object A.

$$\vec{F}_{A \text{ on } B} = -\vec{F}_{B \text{ on } A}$$

---

Newton's third law states that forces always act in pairs. An object cannot experience a force without also exerting an equal and opposite force. There are always *two forces* acting and *two objects* involved. To develop an understanding of Newton's third law, consider something as simple as walking across the room.

If someone asked you what force caused you to start moving across the room after standing still, you might say, "I push on the floor with my feet." Think about that statement. You *push* on the floor. According to Newton's second law, when you exert a force on an object, that object should move. So, by pushing on the floor, you should cause the floor to move. However, many other objects, such as the walls, the subfloor, and other structures are also pushing on the floor, making the sum of all of the forces equal to zero. Therefore, according to Newton's first law, the floor does not move. You cannot explain why you can walk across the room without calling on Newton's third law. According to Newton's third law, when you exert a frictional force on the floor, it exerts an equal and opposite frictional force on you. In reality, the floor pushes on you, propelling you across the floor as shown in Figure 4.17.

### Think It Through

- Draw force diagrams for each of the situations illustrated here.

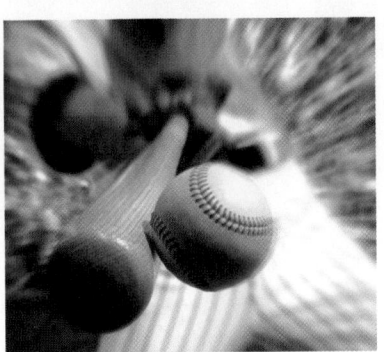

**PHYSICS FILE**

An American physicist, Robert Goddard, published a paper in 1919 that suggested rockets could be used to attain altitudes higher than the atmosphere. Editors of the New York Times ridiculed Goddard, claiming a rocket would not work outside of the atmosphere.

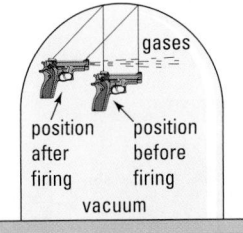

Goddard used the demonstration pictured to show the editors their error.

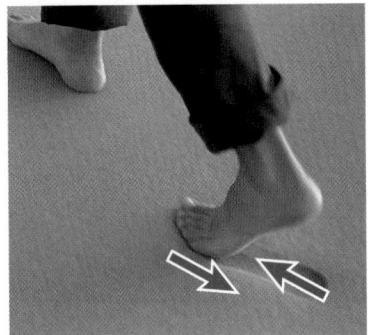

**Figure 4.17** Newton's third law explains how the floor pushes you across the room.

- Look at the illustration of (A) two identical football players
colliding and (B) one of the football players travelling with the
exact same speed colliding with a wall. Use Newton's third
law and compare the forces exerted on each player in both
situations. Explain your answer.

A                    B

## Concept Organizer

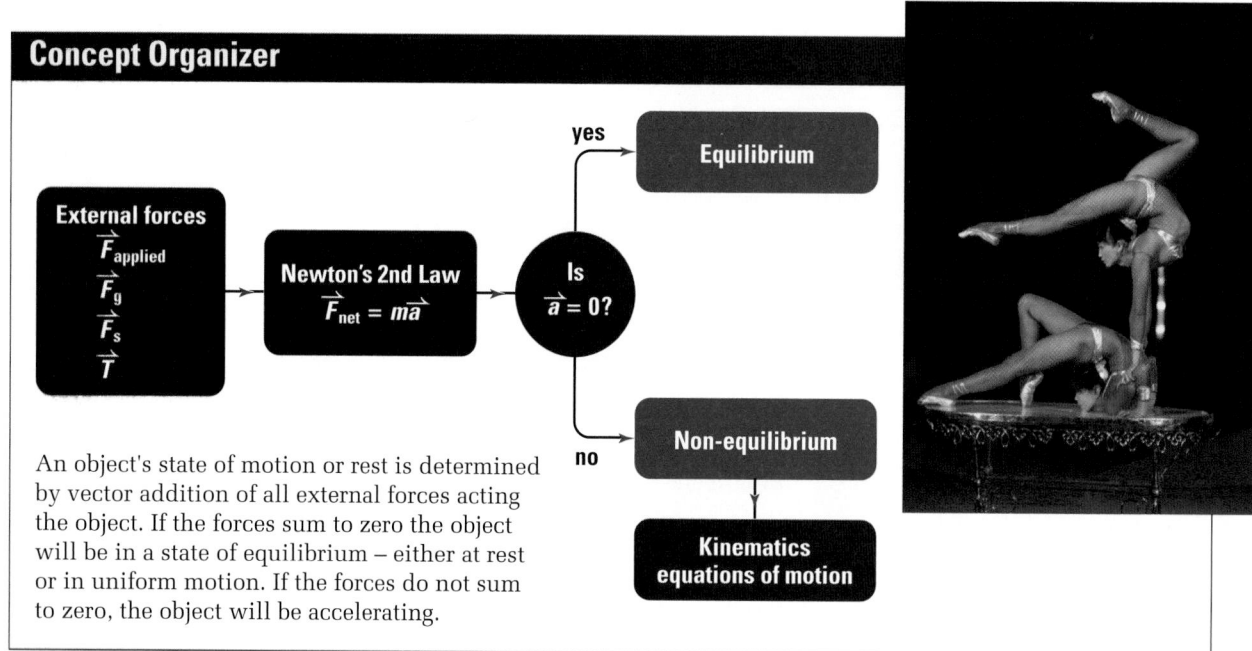

**External forces**
$\vec{F}_{applied}$
$\vec{F}_{g}$
$\vec{F}_{s}$
$\vec{T}$

**Newton's 2nd Law**
$\vec{F}_{net} = m\vec{a}$

**Is** $\vec{a} = 0$?

yes → **Equilibrium**

no → **Non-equilibrium** → **Kinematics equations of motion**

An object's state of motion or rest is determined
by vector addition of all external forces acting
the object. If the forces sum to zero the object
will be in a state of equilibrium – either at rest
or in uniform motion. If the forces do not sum
to zero, the object will be accelerating.

**Figure 4.18** Understanding
Newton's second law of force
and motion.

Classical mechanics is a powerful tool for analyzing and predicting the behaviour of macroscopic objects. Remember, this term
applied to objects large enough to see, such as a chair, a scooter,
an ironing board, or a clothesline. Solving problems relating to
such objects may require the application of one, two, or even all
three of the laws to reach a solution. The laws are valid in all
inertial reference frames.

1. **K/U** State Newton's first law and give two examples.

2. **K/U** State Newton's second law and give two examples.

3. **K/U** State Newton's third law and give two examples.

4. **K/U** By how much will an object's acceleration change if
   (a) the force is doubled?
   (b) the mass of the object is halved?
   (c) the mass is doubled and the force is halved?

5. **K/U** State the equal-and-opposite force pairs in each of the following situations:
   (a) kicking a soccer ball
   (b) a pencil resting on a desk
   (c) stretching an elastic band

6. **K/U** A block of pure gold is perfectly balanced by a 2.0 kg lead mass. Describe what would happen to this setup if it was placed on the Moon.

7. **K/U** Consider a football player throwing a football. The chart represents action-reaction force pairs from the situation.

| Type of force (action-reaction pairs) | Objects involved | | |
|---|---|---|---|
| | Player | Ball | Earth |
| gravitational | ✓ | | ✓ |
| normal | ✓ | | ✓ |
| friction | ✓ | | ✓ |
| normal | ✓ | ✓ | |
| gravitational | | ✓ | ✓ |

Construct similar charts for each of the following situations.
   (a) A hockey player strikes a puck resting on the ice.

   (b) A person pours water from a bucket toward the ground.

8. **K/U** Determine the net force in each of the following situations:
   (a) A race car travels at 185 km/h[W].
   (b) Two tug-of-war teams are at a standoff, each pulling with 1200 N of force.
   (c) The Voyager 1 space probe moves at 25 000 km/h in deep space beyond our solar system.

9. **K/U** Apply Newton's third law to each situation in order to determine the reaction force, magnitude, and direction.
   (a) A soccer ball is kicked with 85 N[W].
   (b) A bulldozer pushes a concrete slab directly south with 45 000 N of force.
   (c) A 450 N physics student stands on the floor.

10. **C** How would you describe the vector nature of Newton's second law to a friend? Provide at least one example.

### UNIT PROJECT PREP

Cartoons are famous for situations in which characters "forget" that forces, such as gravity, exist.  When the character "remembers," the consequences are predictably disastrous.

- Can you create a situation where a character forgets that a force or reaction force exists and suffers the consequences?
- How can you use exaggeration to give the illusion of defying Newton's laws? For example, a person running leaps over a two-storey house, or a child moves a refrigerator with one finger.
- Can you reverse an action and its reaction?

- Identify the four fundamental forces of nature.
- Describe scientific models for the fundamental forces.

KEY
TERMS

- exchange particle
- strong nuclear force
- electromagnetic force
- weak nuclear force
- force of gravity
- Grand Unification Theory
- Big Bang
- super force
- Big Crunch

Throughout this chapter, you have been learning how to describe the motion of objects under the influence of forces. You have focussed on the motion resulting from the forces and not on the forces themselves. To some extent, you explored the force of gravity, a non-contact force, and friction, a contact force. You may have wondered about two questions. First, what really happens when two objects come into "contact" with each other? And second, how do non-contact forces extend their influence through apparently empty space? These questions are not trivial, and demand a deeper look at the nature of force in general. In fact, scientists have identified four fundamental forces of nature — electromagnetic, strong nuclear, weak nuclear, and gravitational forces. Physicists have classified all forces that exist as one of these four fundamental forces.

## Four Fundamental Forces

What do physicists know about the four forces today and how have they learned about these properties? One important source of information is quantum theory and the study of elementary particles, those particles that make up the familiar proton, neutron, and electron. Current theory holds that material objects interact with each other, exerting forces on each other through **exchange particles**. Specific elementary particles, much less massive than a proton, travel from one object to another "carrying" the force. In this way, each force is "carried" or *mediated* by the exchange of a particle. The properties of these exchange particles then determines the characteristics of the four forces.

### Strong Nuclear Force

The **strong nuclear force** is the strongest of the fundamental forces. It is able to overcome the repulsion of positively charged protons, keeping them tightly packed in the nucleus of an atom. The exchange particles of the force are called pions, with other heavy particles also being involved. The strong nuclear force has a very short range, not much longer than the diameter of a proton itself.

### Electromagnetic Force

Electric forces and magnetic forces were considered to be separate forces until the 1860s when James Clerk Maxwell was able to demonstrate that they were different manifestations of the same force — the **electromagnetic force**. The electromagnetic force is mediated by a massless particle known as a photon. The massless nature of the photon makes the effective range infinite, even though the strength of the force decreases rapidly as the distance between the objects increases. As described earlier in this chapter, so-called contact forces actually belong to the electromagnetic force category. Two objects may appear to come into contact on a macroscopic scale, but, in fact, it is the repulsion of each material's electrons that are interacting. Therefore, most "everyday" forces (other than gravity) are really examples of the electromagnetic force.

## Weak Nuclear Force

Understanding the function of the **weak nuclear force** has been particularly challenging. The weak nuclear force is very weak, 10 000 times weaker than the strong nuclear force. This force acts over the shortest range of any of the fundamental forces. Despite these meagre statistics, the weak nuclear force plays a major role in the structure of the universe. It is an exchange force mediated by the exchange of three different particles called vector bosons. The weak nuclear force is responsible for radioactive decay. Specifically, the weak force changes the flavour (type) of an elementary particle called a quark. When this process occurs, a neutron in the nucleus transforms into a proton.

## Gravitational Force

The **force of gravity** is the most familiar to all forces. You have experienced it from the moment you tried to take your first step. Nevertheless, it is the least understood of the four fundamental forces. Newton's model of gravity allows us to carry out the complex calculations required to get humans to the Moon. Einstein's model of gravity actually removes the concept that gravity is a force at all, suggesting instead that it is the result of large masses bending the fabric of space-time.

Both models serve their respective purpose, but fail to paint a clear picture as to what gravity really is. Gravity is theorized to be an exchange force with a massless mediating particle called a graviton. The massless nature of the graviton allows gravity to have infinite range similar to the electromagnetic force. However, the graviton is the only exchange particle never to have been detected. Gravity is by far the weakest of the four fundamental forces

## Grand Unification Theory

Having four separate and quite different classes of forces is unsettling to many physicists. Since James Clerk Maxwell was able to find a theory that unified the electric and magnetic forces, physicists have searched for a **Grand Unification Theory** to show that all of the observed forces are four manifestations of one single force of nature. Albert Einstein spent the last 20 years of his life searching for a unification theory and failed. Nevertheless, scientists have not given up. With advances in technology, they have been able to dig deeper into the depths of the atom and have gathered a wealth of data. In 1967, three physicists introduced a theory that seemed to successfully unite the weak nuclear force and the electromagnetic force. The search is still on.

Table 4.6 summarizes the properties of the four fundamental forces of nature. The table includes the exchange particles that have been found (using particle accelerators), as well as those that are suspected to exist.

**Table 4.6** Fundamental Forces Summary

| Fundamental Force | Function | Relative strength | Range | Exchange particle(s) |
|---|---|---|---|---|
| **strong nuclear** | It holds the nucleus of each atom together. | 1 | $10^{-15}$ m (diameter of a proton) | $\pi$ (pions), others |
| **electromagnetic** | Like charges repel, unlike charges attract. | 1/137 | infinite | photon (massless) |
| **weak nuclear** | It is involved in radioactive decay. | $10^{-5}$ | $10^{-17}$ m (1% of the diameter of a proton) | $W^+$, $W^-$, $Z_0$ (vector bosons) |
| **gravity** | Matter attracts matter. | $6 \times 10^{-39}$ | infinite | graviton? (massless) |

While elementary particle physicists were looking inside of the atom, even inside protons and electrons, cosmologists were looking out across the universe and back in time to learn more about the four forces of nature.

## The Big Bang

Innovative experiments and technological advances continually add to the accumulating data that currently form the basis used to understand the universe. Albert Einstein gave us the now well-tested and widely accepted theory of general relativity, which establishes the relationships among matter, energy, time, and space. Einstein assumed, without discussion, that the universe was static (that is, it remains the same size). A Russian theorist, Aleksander Friedmann, predicted that Einstein's model of the universe would be incredibly unstable. According to the theory, the universe would either collapse in on itself or expand outward to infinity, depending on a relatively small difference in the total mass of the universe.

At the same time, physicists were collecting data that showed that galaxies were moving apart. The data supported the expanding universe idea. If the universe is expanding, then it must have had a beginning or starting point from which the expansion started. As it turns out, Einstein's theory could be adapted to fit an expanding universe. An English physicist, Fred Hoyle, completely discounted the expanding universe theory and nicknamed it the **Big Bang** in an attempt to discredit it. Well, the catchy name stuck, and the theory has held up in the face of masses of data collected over the past 70 years.

**Figure 4.19** The Big Bang theory of the origin of our universe pictured as a function of time

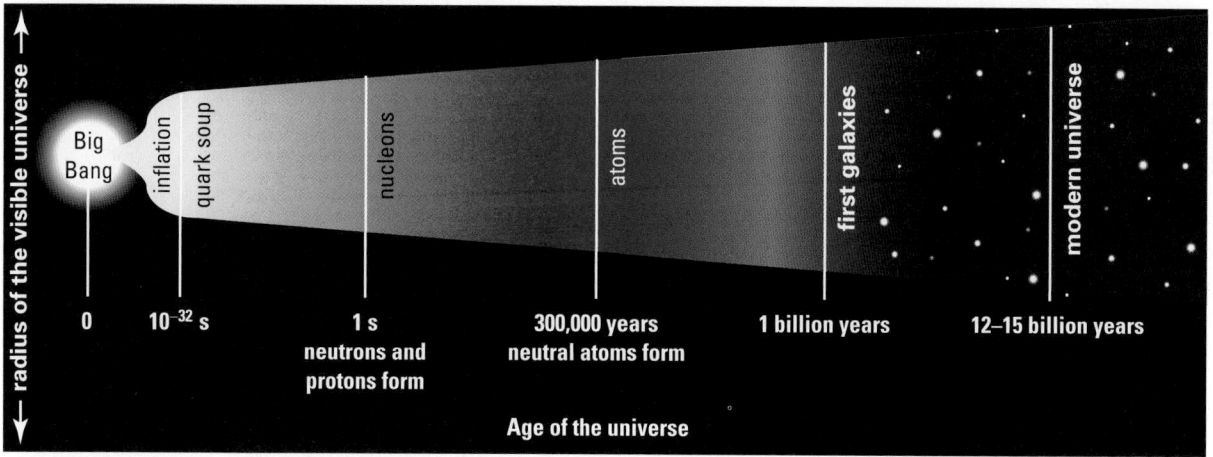

Pondering the origins of the universe is fodder for theologians, mystics, philosophers, and scientists. In science, the slow methodical approach is adopted. Only information obtained by experimentation or observation is accepted. Understanding the characteristics of the four fundamental forces of nature is a direct result of this process. Approximately 10 to 20 billion years ago, all of the matter and energy that can be observed in the universe was concentrated in an area smaller than the point of a needle. The minute universe explosively began to expand and cool at an incredible rate. During the first few nanoseconds of the explosion, at the incomprehensibly high temperatures near $10^{32}$ K, the strengths of the four forces were the same. This unified force is nicknamed the **super force**.

Within about $10^{-4}$ s, the universe cooled to a temperature of only 100 millions times the temperature of the Sun's core and the four fundamental forces of nature had acquired their present-day characteristics. The Big Bang model of the universe predicts the existence and characteristics of the four fundamental forces. The model also predicts how the formation of matter continued as the energy cooled out into elementary particles, then atomic nuclei, finally forming the host of elements that exist on today's periodic table. Observation and experimentation from several fields of physics have provided the evidence that is helping to unravel the mysteries of our universe.

### Astronomy

Giant ground-based telescopes and the Hubble telescope allow astronomers to view distant galaxies billions of light-years away (Figure 4.20), and to see what the universe was like when it was very young.

### Particle Physics

Particle accelerators probe into the world of high-energy physics, re-creating environments similar to moments after the Big Bang. Figure 4.21 shows the amazing patterns they find.

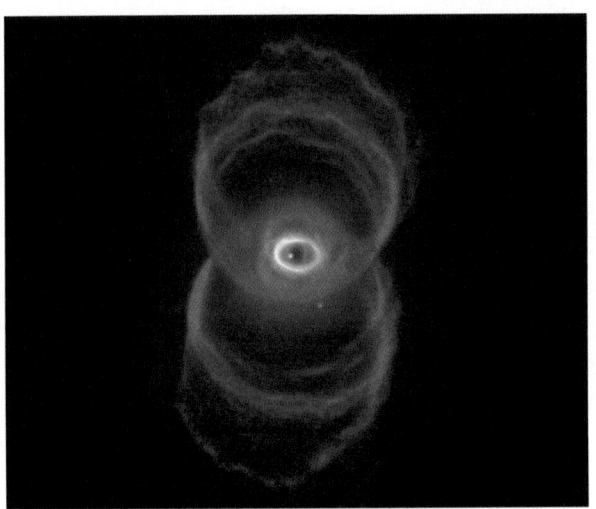

**Figure 4.20** The light from some stars has been travelling toward Earth for billions of years.

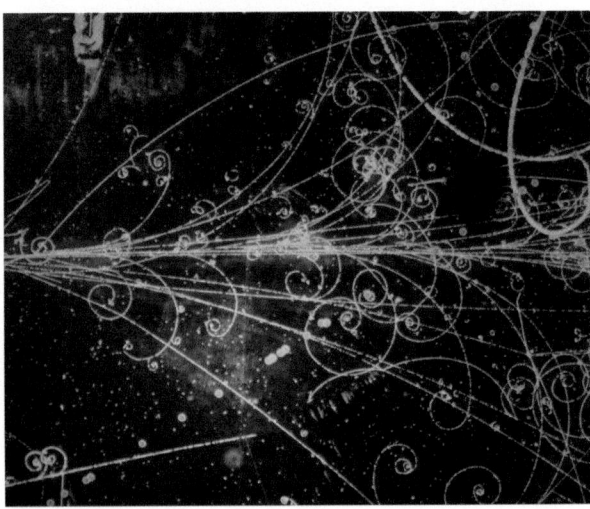

**Figure 4.21** Particles of different mass and charge produce different trajectories. You cannot see the particles but you can see their tracks.

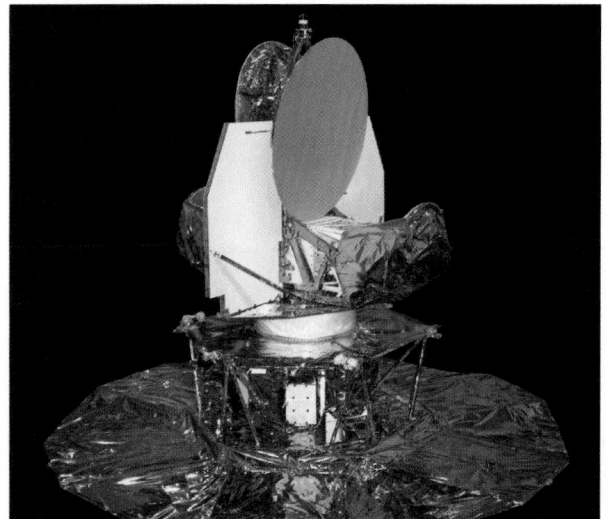

**Figure 4.22** This second-generation satellite named MAP, short for Microwave Anisotropy Probe, will continue the analysis of its predecessor COBE (Cosmic Background Explorer). It will be parked at the Second Lagrange point where Earth's gravitational attraction will be exactly matched by the Sun's pull.

■ **REPORT CARD FOR MAJOR THEORIES**

| Concept | Grade | Comments |
|---|---|---|
| The universe evolved from a hotter, denser state | A+ | Compelling evidence drawn from many corners of astronomy and physics |
| The universe expands as the general theory of relativity predicts | A− | Passes the tests so far, but few of the tests have been tight |
| Dark matter made of exotic particles dominates galaxies | B+ | Many lines of indirect evidence, but the particles have yet to be found |

**Figure 4.23** The Big Bang origin theory of our universe actually refers to several competing explanations as to how the universe came to its present form. This table compares some of the competing ideas and highlights the major feature of each.

## Quantum Mechanics

Quantum mechanics, a branch of modern physics that deals with matter and energy on atomic scales, predicts that the four forces of nature were unified until approximately $1 \times 10^{-43}$ s after the Big Bang.

## Satellites

Sophisticated technology is allowing astronomers to peer back to the moment when time began by looking at the background radiation left over from the Big Bang (Figure 4.22). The picture generated from the background radiation data provides the largest-scale view of the universe that is possible.

## Theoretical Physics

Working within the framework of the Big Bang theory, several component theories are attempting to answer some intriguing questions, as Figure 4.23 shows.

Big Bang cosmology, or the Standard Cosmological Model as it is also known, is evolving as new theories add to and change certain aspects of it, but it may always leave some questions unanswered. For instance, what was the universe like before the Big Bang? What does the distant future hold for the universe once the last stars exhaust their fuel? What was the nature of the super force? (The super force is the name given to the state when all four fundamental forces, electromagnetic, weak nuclear, strong nuclear,

**PHYSICS FILE**

The background microwave radiation generated during the Big Bang bathes us on Earth even today. If you have ever used a television set that operates with an antenna rather than cable or satellite, part of the snow observed when a channel is not being received is due to the Big Bang background radiation.

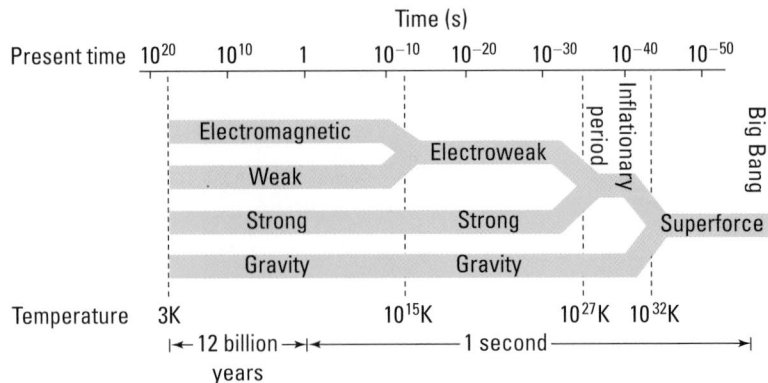

**Figure 4.24** Current theory time line of the creation of our universe

and gravity, were united.) How could anything exist with the power to expand our universe to what it is today? Understanding the characteristics of each fundamental force that exists today is the first step in a long road to understanding the unification of the forces.

Figure 4.24 provides a time line detailing when each of the four forces is believed to have obtained its current properties.

Although it is incredibly weak and the least understood of the fundamental forces, gravity has the honour of determining the ultimate fate of the universe — if enough mass exists in the universe, gravity could ultimately stop and reverse the expansion of our universe, leading to what is sometimes referred to as the **Big Crunch**.

## 4.4    Section Review

1. **K/U** What evidence first suggested the concept of the Big Bang?

2. **K/U** List and briefly describe three fields in physics that have aided in providing knowledge about the origin of the universe.

3. **K/U** Describe each of the four fundamental forces as they exist today.

4. **C** Prepare a short presentation that you could use to teach a Grade 5 class about the four fundamental forces.

5. **K/U** It has been suggested that "the weakest of all forces will eventually win out, causing the end of the universe as we know it." Describe what this statement means.

6. **K/U** According to current theory, how do each of the four fundamental forces exert their influence through empty space?

7. **C** Summarize the contributions of astronomy, particle physics, quantum mechanics, and mathematical theoretical physicists to our current understanding of the universe.

- Inertia is the natural tendency of an object to remain at rest or in uniform motion.
- The amount of inertia depends on the amount of mass of an object.
- The force of gravity acting on an object near a celestial object, such as Earth, is called weight. An object's weight is given by $\vec{F}_g = m\vec{g}$.
- The normal force acts perpendicular to the plane of the surfaces in contact.
- The force of friction is the product of the *coefficient* of friction between the contact surfaces and the *normal force* pressing the objects together ($F_f = \mu F_N$). The force of friction always acts in a direction to oppose motion. The coefficient of friction, $\mu$, is dependent on the types of materials in contact.
- Application of the standard model of friction assumes that
  - **(a)** the force of friction is independent of area of contact
  - **(b)** the force of friction is proportional to the normal force
  - **(c)** the force of friction is independent of the velocity of motion
- Free body diagrams represent all forces *acting on* one object (and only those forces).
- Forces that the object exerts on other objects are *not* shown in a free body diagram.

- The object is represented as a single dot and an arrow is used to represent each force *acting on* the object.
- The direction of each force arrow represents the direction of the force and the arrow's relative length provides information about the magnitude of the force.
- **Newton's 1$^{st}$ law:** An object will stay at rest or in straight-line motion at a constant speed unless acted on by an external force.
- **Newton's 2$^{nd}$ law:** An object will accelerate in the direction of the unbalanced net force. The magnitude of the acceleration will be proportional to the magnitude of the force and inversely proportional to the mass, $\vec{a} = \dfrac{\vec{F}}{m}$.
- **Newton's 3$^{rd}$ law:** For every action force on object B, there is an equal in magnitude but opposite in direction reaction force acting back on object A. Thus, forces always act in equal and opposite pairs.
- Our current model of the universe categorizes all known forces as one of four distinct types: **strong nuclear**, **weak nuclear**, **electromagnetic**, and **gravitation**. The Standard Cosmological Model suggests that each of these forces took on their physical characteristics moments after the Big Bang.

## Knowledge and Understanding

1. Describe three examples of inertia.
2. A small stuffed animal hangs from the rear view mirror of a car turning a corner. Sketch the position of the stuffed animal relative to the mirror during the turn as seen by a passenger. Explain the reason for the perceived movement of the stuffed animal.
3. Astronauts working outside of the space shuttle in Earth orbit are able to move large satellites. One astronaut referred to the satellites not as heavy, but as massive. Explain the astronaut's comment.
4. You are a passenger in a car that is driving on the highway at 100 km/h. Explain, in terms of inertia, what happens to you if the driver brakes suddenly?
5. Consider the motion of an object. State Aristotle's, Buridan's, and Galileo's understanding of what is now termed inertia.
6. If gold were sold by weight, at which of the two locations would you prefer to buy it: At a location on the equator at sea level, or at the North Pole at sea level? If it were sold by mass, where would you prefer to buy it? Explain.

7. Describe how the normal force acting between a block and a board changes when the board-block combination are (a) horizontal and (b) at an angle to the horizontal.

8. State Newton's three laws and provide two examples for each.

9. Scientists have classified all forces that exist as one of four fundamental forces. List and describe the four fundamental forces.

## Inquiry

10. Design and carry out an experiment to compare the values of the maximum acceleration of a bus or subway train when speeding up and slowing down. Is the magnitude of the acceleration greater when speeding up or when slowing down?

11. Sketch a graph that shows the velocity of a sky diver through three phases:
    (a) from the time she jumps from the plane to the time when she opens her parachute
    (b) from the time when she opens her parachute to the time when she reaches terminal velocity
    (c) from the time when she reaches terminal velocity, with her parachute fully deployed, until the time when she lands
    Use the graph to make conclusions about the forces acting on the sky diver during each of the three phases of her fall.

12. A sky diver uses a GPS system to measure his velocity every second during a free fall. The recorded velocities at the end of each of the first 5 seconds are 9.5 m/s, 18 m/s, 25 m/s, 30 m/s, and 32 m/s. Plot a velocity-time graph of this free fall, assuming his velocity is 0 m/s at time 0 s. Is the graph linear? What does the shape of the graph imply about the sky diver's acceleration and the forces acting on him during these 5 seconds?

13. In Investigation 4-B on page 165, you varied the force on a dynamics cart and observed the change in the cart's acceleration. Design an experiment to test what happens to the acceleration of the cart when its mass is varied, instead of the force.

14. If it is winter time, design and carry out an experiment to test the success of different kinds of wax in reducing friction on snow skis.

## Communications

15. Use a labeled diagram to show the principal forces acting on a car that is slowly braking as it moves towards a stop light. The length of the force vectors should represent their magnitude. Make reference to this diagram to explain why accidents occur when the road is covered with ice.

16. Summarize Newton's three laws of motion in a table with three columns, headed Law, Example, and Catch phrase. In the third column, create a personalized version of each law, such as "You get what you give."

17. Describe three benefits of using free-body diagrams in engineering problems involving the design of bridges.

## Making Connections

18. Make a concept map or flow chart showing how Newton's contributions to mechanics enabled scientific discovery and technological application to advance in the centuries following.

19. The work of Newton was so revolutionary for European society that it inspired several contemporary poets to compose poems reflecting on his work. From your perspective, living in the 21$^{st}$ century, create a free-verse poem describing the impact of the scientific revolution of Galileo, Newton, and others in changing our view of nature, society, and technology.

20. List all the sports you can think of that involve an object being propelled at high velocity. Choose one of these sports and describe how the sport has been changed, because of technological advances in the equipment or safety concerns in the use of the equipment.

21. Propose a course of action to a government committee looking into building a high-speed rail link between Montreal, Ottawa, and Toronto. The proposal should consider economic, environmental, political, and safety issues.

## Problems for Understanding

22. A piano is to be slid across the floor. It has a mass of 450 kg.
    (a) Calculate the normal force supporting the piano.
    (b) If the coefficient of static friction between the floor and the piano is 0.35, calculate the minimum amount of force needed to get the piano to move.
    (c) Once the piano is moving, a horizontal force of $1.1 \times 10^3$ N is necessary to keep it moving at a constant speed. Determine the coefficient of kinetic friction.

23. Draw a free-body diagram to show the magnitude of the forces acting in the following situations:
    (a) A person on a scooter uses one of her feet to accelerate forward.
    (b) The person glides briefly at a constant speed.
    (c) She slows down as she continues to glide.

24. As it moves through the water a 400 kg boat experiences a resistance force of 2 500 N from the air and 3 200 N force of resistance from the water. If the motor provides a forward force of 6 000 N:
    (a) Determine the net force.
    (b) Calculate the acceleration of the boat.

25. A croquet ball with a mass of 300 g is thrown with an initial velocity of 6.0 m/s[right]. A force of friction of 0.45 N causes the ball to come to a stop. How long did it take the ball to roll to a stop?

26. A cyclist is travelling at 21 km/h when she sees a stop sign ahead. She applies the brakes and comes to a stop in 15 m. The mass of the cyclist and the bike is 73 kg.
    (a) Calculate the acceleration of the cyclist.
    (b) Determine the coefficient of friction between the road and the bike tires.

27. A water-skier is being pulled behind a boat. The skier leaves the wake behind the boat and moves so that the rope makes an angle of 48° with the back of the boat. Calculate the horizontal and vertical components if the tension in the rope is 620 N.

28. A toboggan with a mass of 15 kg is being pulled with an applied force of 45 N at an angle of 40° to the horizontal. What is the acceleration if the force of friction opposing the motion is 28 N?

29. A grocery cart is being pushed with a force of 450 N at an angle of 30.0° to the horizontal. If the mass of the cart and the groceries is 42 kg,
    (a) Calculate the force of friction if the coefficient of friction is 0.60.
    (b) Determine the acceleration of the cart.

30. Calculate the net force if the following three forces are all being applied at the same time: 40 N[S], 60 N[N], and 30 N[N35°E].

31. Two boxes are side by side on a frictionless surface. A horizontal force is applied to move both boxes.
    (a) Calculate the acceleration of both boxes.
    (b) Determine the force that the $4.0 \times 10^1$ kg box applies to the $2.0 \times 10^1$ kg box.

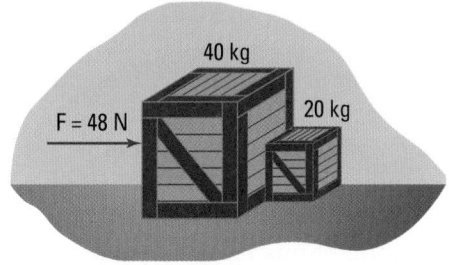

### Numerical Answers to Practice Problems

1. 23 N  2. (a) 66.722 kg (b) 656.03 N (c) 605.81 N
3. $W_{Earth} = 2.05 \times 10^3$ N, $W_{Moon} = 3.43 \times 10^2$ N
4. $3.25 \times 10^{-2}$ m/s²  5. (a) 5.89 N (b) 3.50 N; 0.595 (c) $\mu_k$
6. (a) $1.23 \times 10^3$ N (b) 527 N (c) 264 N  7. $1.95 \times 10^2$ N  8. 0.34
9. 0.55 m/s²[E]  10. 0.53 m/s²[E]  11. 1.6 m[N]  12. (a) 5.6 m/s²
(b) $2.0 \times 10^2$ m [E]  13. 1.7 kg  14. 0.23  15. 40 N[N30°E]
16. (a) 43 N[E] (b) 7.4 N[N] (c) 15 N[E] (d) 15 N[W28°S]
17. (a) $1.4 \times 10^3$ N (b) $3.9 \times 10^2$ N  18. (a) $F_x = 120$ N, $F_y = -86$ N
(b) $3.3 \times 10^2$ N (c) 38 N in direction 2.3° below the horizontal (d) 1.5 m/s²

The zoetrope—an early device for creating "motion pictures."

## Background

In this unit, you have studied both how to describe different kinds of motion and how physicists provide explanations of why objects move the way they do. In developing your understanding, you highlighted that objects move in relation to a frame of reference. The manipulation of this relationship between an object and its frame of reference allows film-makers to create different "realities."

## Challenge

In groups of three or four, you are to research, write a storyline for, and produce a short animated film sequence, video, or special effects show that demonstrates an aspect of how film-makers use physics to create a virtual reality. Think about how filmmakers make use of the following. In Chapter 2, you examined how an object that is speeding up will have an increasing displacement between each consecutive time frame. In Chapter 3, you explored how objects appear to be moving differently relative to different frames of reference. In Chapter 4, you considered how natural forces cause an object to change speed or direction, although you do not see any physical force being applied. You can also draw on Units 3 and 4 for researching the physics involved in creating optical illusions and virtual realities. Unit 5 will stimulate your thinking about "magic tricks."

Here are the components you are required to include in the project.

**A** Entertainment Technology
Decide on which era of the entertainment industry you wish to study. You could decide to make one of the early film technologies such as a magic lantern to present your finished product. Or, you may wish to explore software for creating three-dimensional, computer-generated objects.

**B** The Physics of Motion
Decide what aspect of physics you are going to demonstrate. How do you create the illusion of an object changing speed or direction? How can frames of reference be manipulated to portray a rocket flying off into space when in reality the whole film is created inside a studio? How do special sound or light effects help create a desired virtual reality?

**C** Plot and Setting
Decide on a situation and short story as a context for your cartoon or film. Your special effects or virtual reality will have more impact if it is based on a believable story.

## Web Link

**www.school.mcgrawhill.ca/resources/**
For more information about computer animation, special effects, and virtual reality, go to the above Internet site. Follow the links for **Science Resources** and **Physics 11** to find out where to go next.

## Design Criteria

**A.** Brainstorm ideas and conduct preliminary research to enhance them.

**B.** For further ideas, consider some of the following:

- Interview the technical staff of a theatre group about how they create special effects.

- Visit a theme park and note carefully how virtual realities are created. For example, how do they use movable chairs to simulate a thrilling auto race?

- View a digital videodisc (DVD) that contains additional segments that explain how special effects were created. Make notes on the physics involved. Bear in mind that your own effects will be much simpler than those in big-budget movies.

**C.** Write a proposal for a *manageable* amount of work. Include a flow chart with time lines for tasks for each member of the group.

- If your project is to build an early piece of technology, you will develop relatively few frames. If you are using computer simulations, your animation or video will be longer.

- Part of the evaluation of your project will be the evaluation of your ability to develop a proposal for a reasonable amount of work.

**D.** Prepare a written presentation of your project, including

- a title page with the names of your group members

- a summary of several pages that lists the technology you used, the aspect of physics that you dealt with, and the reality that you created in your animation sequence, video, or special effects show

how similar the final project is to your proposal
- **assess your project based on how clearly the physical concepts are conveyed**
- **assess your project using the rubric designed in class**

- a log of your work, including your proposal, outline of initial research, storyboard, a description of any problems encountered and how you solved them, and a list of references and resources.

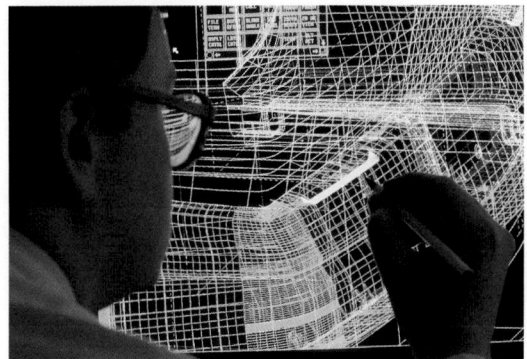

Three-dimensional modelling used by engineers is one example of an application of virtual reality.

### Action Plan

**1.** Construct the technology based on self-made or ready-made plans, and become familiar with the operation of the technology.

**2.** Develop a storyboard (a shot-by-shot sequence of your storyline).

**3.** Outline the physics that you are using to guide the development of your storyline.

**4.** Complete production: complete storyboard artwork, block scene-by-scene staging, and film the video or rehearse the live show.

**5.** Present the completed project.

### Evaluate

**1.** As a class, design a rubric to evaluate the projects, including proposal and presentation.

**2.** Use your rubric to evaluate your project.

## Knowledge and Understanding

### True and False

In your notebook, indicate whether each statement is true or false. Correct each false statement.

1. Average velocity equals distance over elapsed time.

2. On a velocity-time graph the slope equals the acceleration.

3. One walks 10 km north and then 10 km south. His displacement is 20 km.

4. An object falling freely experiences non-uniform acceleration.

5. The order of adding vectors does not matter in determining the resultant.

6. Vectors representing different kinds of quantities can be added and subtracted together.

7. The downstream velocity of a river that flows north has no effect on a boat's westward velocity.

8. An object must experience a force to keep it in motion.

9. Free body diagrams show all forces acting on an object.

10. At any one location, all objects, no matter how massive, fall with the same acceleration if air resistance is ignored.

11. Newtonian Physics is used to solve all force and motion interactions on both the microscopic and macroscopic level.

### Multiple Choice:

In your notebook, write the letter of the best answer for each of the following questions.

12. The speed of a vehicle travelling at 90 km/h is equal to:
    (a) 324 m/s    (c) 129.6 m/s    (e) none of these
    (b) 0.4 m/s    (d) 25 m/s

13. A woman walks 15 km[N], 4 km[W], 2 km[S] and 4 km[E]. The resultant displacement is
    (a) 17 km    (c) 25 km[N]    (e) 13 km[N]
    (b) 17 km[N]    (d) 8 km[N]

14. If action and reaction forces are always equal and opposite then why do objects move at all?

(a) one object has more mass than the other object

(b) the forces act on different objects

(c) the reaction forces take over since the action forces acted first

(d) the reaction force is slower to react because of inertia

(e) the action and reaction forces are not exactly equal

15. A cable on an elevator exerts a 6 KN upward force. The downward force of gravity on the elevator is also 4 KN. The elevator could be
    (a) moving upward with constant speed.
    (b) moving downward with decreasing speed.
    (c) moving upward with decreasing speed.
    (d) moving downward with increasing speed.
    (e) moving upward with increasing speed.

16. An airplane is moving at constant velocity in a straight level flight. What is the <u>net</u> force acting on the plane?
    (a) zero
    (b) upward
    (c) downward
    (d) in the direction of motion
    (e) opposite to the direction of motion

### Short Answer

In your notebook, write a sentence or short paragraph to answer each of the following questions.

17. What could happen in each of the following situations?
    (a) A vehicle tries to come to a stop at a traffic light on an icy street.
    (b) A passenger in a car does not have their seat belt on when the driver must make a quick stop.
    (c) A student has placed their textbooks in the back window of a car. While driving they have to make a sudden stop to avoid driving through a stop sign.

18. Draw a properly labelled graph that illustrates uniformly decreasing acceleration.

19. Distinguish between the average velocity and the instantaneous velocity.

20. Which of the four fundamental forces is the weakest?

21. What is the normal force?

22. What is the difference between static and kinetic frictional forces?

23. (a) What does the slope of the tangent to the curve on a position-time graph represent?

    (b) What does the slope of the tangent to the curve on a velocity-time graph represent?

    (c) What does the area under the curve on a velocity-time graph represent?

## Inquiry

24. Design a simple experiment involving the dropping of a coin to test Galileo's ideas of inertia on a train or subway moving with constant velocity. State your hypothesis, and write out a simple procedure. What variables will you need to control? What variable will you measure? If possible, carry out the experiment. What do you conclude?

25. When you stand on your bathroom scales you measure your weight. But what will your scales show if you stand on them in an elevator at the moment when it begins to accelerate up? Down? Design and carry out an experiment to test your hypothesis.

26. You and four of your friends live at five different locations in a large city. You want to find a central meeting place that minimizes the total travel of all five of you. Can you think of a way of solving this problem using elastic bands? If you have found a solution, try it out using a map of the locations of your residences. Can you explain the physics behind how this works?

## Communication

27. A jet ski moving at 30 km/h on water can make a 90° turn in a very short distance, while a supertanker moving with the same velocity needs a distance of up to 20 kilometers to turn through the same angle. Explain this phenomenon in terms of Newton's First Law.

28. Create a concept map or graphic organizer that links the "derived" SI units for force, velocity, and acceleration, to the "basic" SI units for distance, mass, and time.

29. Design a section of a new roller coaster and create a schematic diagram of a roller coaster car, with people in it, at several locations on this section. For each location, draw a free body diagram of the forces acting on the car. Based on this, make some safety recommendations for the structural engineer to consider.

30. Create a computer program (i.e., spreadsheet, programming language, etc.) to solve for the magnitude and direction of the resultant force acting on a point, given the magnitudes and directions of the individual forces acting on that point.

## Making Connections

31. Propose a course of action to a government committee looking into building a high-speed rail link between Montreal, Ottawa, and Toronto. The proposal should consider economic, environmental, political, and safety issues.

32. How has tire manufacturing technology impacted transportation?

## Problems for Understanding

Show complete solutions for all problems that involve equations and numbers.

33. A delivery truck travels 15 km north, then 13 km east and finally heads south for 18 km. Determine the truck's displacement.

34. A car is traveling $5.0 \times 10^1$ km/h[N]. It turns a corner and heads down a side street at $4.0 \times 10^1$ km/h[E]. Determine the car's change in velocity.

35. A tourist is travelling south to Toronto late at night and has her car set on cruise control. On the highway she sees a sign that says "Toronto 165 km" and notices that it is 10:30 p.m. At 11:00 p.m. she sees a second sign that says "Toronto 110 km".

    (a) How much time passed from when the tourist saw the first sign to when she saw the second sign?

    (b) What is her displacement for the time interval?

    (c) Calculate the velocity of the tourist.

**36.** Use the position-time graphs below to answer the following questions:

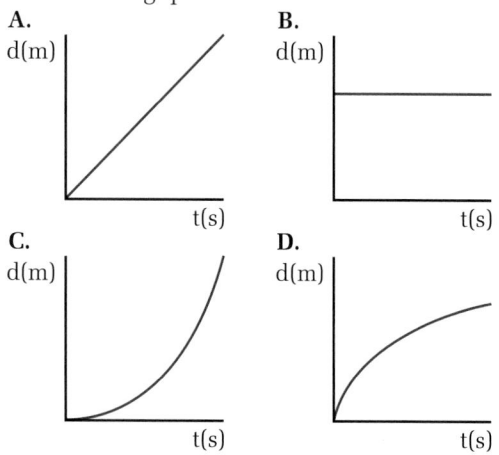

Which graph best describes each situation:
  **(i)** A car is stopped at a stoplight.
  **(ii)** The light turns green, so the car gradually increases in speed.
  **(iii)** A car is travelling on a highway at a constant speed.
  **(iv)** A car slows down as they approach a school zone.

**37.** A student is late for school. She runs out the door and starts down the street at 8.0 km/h. Three minutes later, her Mom notices that she left her physics homework on the table and immediately runs after her at 12.0 km/h.
  **(a)** How far does she get in 3 minutes
  **(b)** How long does it take her mom to catch her (in minutes)?
  **(c)** How far away from home does her mom catch her?

**38.** On a highway a car is travelling at 28 m/s[N] when it increases its speed to pass another car. Calculate the acceleration of the car if it reaches a speed of 33 m/s in 2.0 s.

**39.** How far does a car travel if it accelerates from 22 m/s[W] to 28 m/s[W] at a rate of 3.0 m/s²?

**40.** How long does it take a race car, accelerating from a velocity of 6.0 m/s at 4.0 m/s² to travel a distance of 216 m?

**41.** A sprinter is running the 100 m dash. For the first 1.75 s of the race she accelerates from rest to a speed of 5.80 m/s. For the rest of the 100 m she continues at a constant speed. What time did the sprinter achieve for the race?

**42.** Two vectors X and Y are shown:

Choose from the choices below to answer the following questions:

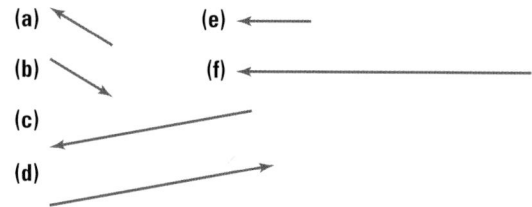

  **(i)** Which most accurately represents the sum of the vectors X and Y?
  **(ii)** Which most accurately represents the difference of the vectors X and Y, (X − Y)?
  **(iii)** Which most accurately represents the product $(\frac{1}{2}X)$?

**43.** A person in a kayak paddles across a calm lake at 2.5 m/s[S] for $3.0 \times 10^1$ minutes. He then heads east at 2.0 m/s for $2.0 \times 10^1$ min.
  **(a)** Calculate the displacement of the kayak from its starting point.
  **(b)** Determine the average velocity for the trip.

**44.** Jerry watches a stick float downstream in a river and notes that it moves 12 m[E] in $2.0 \times 10^1$ s. His friend Ben is starting on the south side of the river and is going to swim across. In still water Ben knows that he can swim with a speed of 1.7 m/s. What is Ben's velocity relative to the shore? If the river is 1.5 km wide, how long will it take Ben to cross the river? How far downstream will he land?

**45.** A passenger climbs aboard a northbound bus and walks toward the back at a rate of 1.8 m/s. The bus starts off up the street at 9.2 m/s. What velocity will the passenger appear to be walking relative to

  **(a)** a person standing on the sidewalk.
  **(b)** a person who is walking 2.1 m/s south along the sidewalk.
  **(c)** a person who is walking 2.1 m/s north along the sidewalk.

46. A speedboat is towing a water skier. At the beginning of the ride the skier is following straight along behind the boat which is traveling 68 km/h due north. The skier cuts out to the right side so that he is heading at an angle of 48° to the direction of the boat. Calculate the skier's change in velocity if his speed while cutting to the side is 76 km/h.

47. A sailboat is using its motor to travel with a velocity of 42 km/h[E 40.0°S] when a wind from the north starts blowing at 5.0 km/h, what will be the velocity of the sailboat relative to the shore?

48. A pilot wants to land at a small lake that is [N30.0°W] of the airport that she is starting from. The wind has a velocity of 25.0 m/s[W] and the air speed of the plane is $1.90 \times 10^2$ m/s. What direction will the plane have to fly to get to its destination? What will be the velocity of the plane relative to the ground?

49. A pitcher throws a baseball with a velocity of 26 m/s[S]. It strikes a player's bat and the velocity changes to $3.0 \times 10^1$ m/s[N]. If the player's bat was in contact with the ball for $3.0 \times 10^{-3}$ s, determine the acceleration of the ball.

50. A cyclist is travelling at 12 m/s[E] when she turns a corner and continues at a velocity of 12 m/s[N]. If the cyclist took 2.5 s to complete the turn, calculate her acceleration.

51. A basketball player is running down the court at 3.0 m/s[N]. It takes him 2.0 s to change his velocity to receive a pass. His acceleration is 1.4 m/s²[E 5.0° S]. Calculate the new velocity of the player.

52. A toboggan is being pulled with an applied force that is at an angle of 30° to the horizontal.
    (a) Draw a free-body diagram to illustrate all the forces acting on the toboggan.
    (b) Predict the motion of the toboggan if the horizontal component of the applied force is greater than the frictional force.
    (c) Predict the motion of the toboggan if the horizontal component of the applied force is equal to the frictional force.
    (d) Predict the motion of the toboggan if the horizontal component of the applied force is less than the frictional force.

53. A grocery cart has a mass of 32.0 kg. An applied force of $4.00 \times 10^2$ N[E] is used to move the cart. The cart starts from rest and the force is applied for 5.0 s.
    (a) Calculate the force of friction acting on the grocery cart if the coefficient of friction between the cart and the asphalt is 0.87.
    (b) Calculate the acceleration of the grocery cart.
    (c) How far does the cart move in 5.0 s?

54. A wagon is used to help deliver papers. There is a force applied at an angle of 25° to the horizontal that causes the wagon to move a distance of 15 m[N] in 10.0 s from rest. There is a force of friction of 3.1 N acting on the 27 kg wagon.
    (a) What is the acceleration of the wagon?
    (b) What is the net force acting on the wagon?
    (c) Calculate the applied force.

55. A tug-of-war has started over a popular toy. One child pulls with a force of $2.0 \times 10^1$ N[W], a second child pulls with a force of 15 N[N] and a third child pulls with a force of $4.0 \times 10^1$ N[E30°S]. Calculate the net force on the toy.

56. A parachutist jumps from a plane and falls faster and faster through the air. At one point in time her acceleration is 8.0 m/s² [down]. If she has a mass of 65 kg, calculate the force of air resistance that is acting opposite to her motion.

**COURSE CHALLENGE**

**Space-Based Power**

Consider the following as you begin gathering information for your end-of-course project:

- Analyze the contents of this unit and begin recording concepts, diagrams, and formulas that might be useful.

- As you collect ideas attempt to collect information in a variety of ways, including conceptual organizers, useful web sites, experimental data, and perhaps unanswered questions to help you create your final assessment product.

- Scan magazines, newspapers, and the Internet for interesting information to enhance your project.

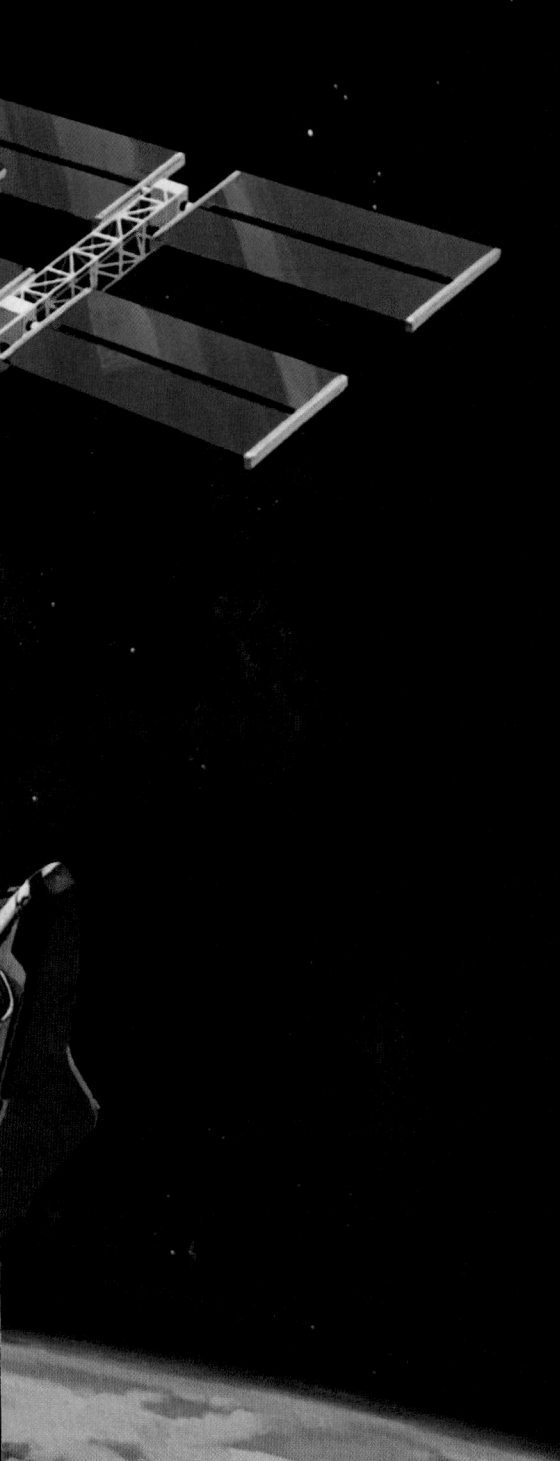

On November 20, 1998, the first piece of the International Space Station lifted off aboard a Russian proton rocket. This launch marked the dawn of a new era in both space science and international co-operation. The collaborative expertise of nearly 100 000 people from 16 different countries is making this immense project possible. In all, 46 separate launches will be required to haul all of the components to this construction site orbiting 400 km above Earth's surface.

Hauling these delicate objects into orbit is a technological feat in itself, requiring millions of tonnes of rocket fuel. Once in orbit, the pieces must be assembled with precision. Much of the delicate work is performed by a new-generation Canadarm called the "Special Purpose Dexterous Manipulator". Larger jobs require the 17 m Canadarm. Both "arms" are part of the Canadian Mobile Services System that is permanently attached to the Space Station.

Once the station is assembled, what will keep the robotic arms functioning, the air filters operating, and the inhabitants cozy and warm? The answers to these questions lie in our ability to understand and then manipulate energy. You will begin to appreciate the real significance and accomplishment that the International Space Station represents as you learn to see the world as a physicist does. Understanding energy is the first step to understanding the technological world in which you live.

### UNIT PROJECT PREP

In the unit project on pages 318–319 you will analyze a sport or sports equipment. Consider:

- How do skateboarders "jump" their boards?
- What energy transformations are designed into hockey helmets for player protection?

**A** football flies through the air. The kicker's foot has just done work on the football, causing it to seemingly defy gravity as it soars high above the field. The kick returner anxiously waits as the ball falls faster and faster toward his arms. The opposing team bears down on him as rapidly as the ball descends. Catching the ball, he runs about six yards. Then you hear the clash of helmets as the opposing tackles bring him to the ground. Although you cannot actually see the energy that has been transferred to the ball and among the players, you most certainly can see and even hear its effects. By simply using your own five senses, you can witness the effects of energy and energy transformations.

Although it may not be obvious, every object you see has some form of energy. When you observe people walking, curtains blowing in the breeze, a jet plane in the distance, or hear the quiet humming of a computer fan, you are detecting evidence of energy transformations. In this chapter, you will learn to understand and describe, both conceptually and mathematically, some important types of energy transformations.

# nergy Transformations

## Hit a Block

Suspend a mass on a string as shown in the photo. Keeping the string taut, pull the mass upward away from the block. Release the mass so it is free to swing down and strike the stationary block. Predict how varying the height of the mass will affect the motion of the block after it is hit by the mass. Repeat the procedure several times, holding the mass at different heights.

### Analyze and Conclude

1. What force causes the mass to swing?

2. Using technical terms, write a statement that describes the relationship between the height of the block and the resulting motion of the mass.

## Wind-Up Toy

Your task is to develop a relationship between the number of turns used to wind a toy car and the distance the toy travels. Devise a method to ensure that the toy travels in a straight line.

### Analyze and Conclude

1. How is energy stored in the toy when you wind it up?

2. What causes the toy to move?

3. What force causes the toy to stop?

4. Make a general statement about the number of winding turns and energy.

5. Make a general statement about the number of winding turns and the distance travelled.

6. What happened to the stored energy?

## Come-Back Can

Obtain a hammer, two nails, one elastic band, a coffee can, tape, and items to act as weights. Attach your weights to *one* side of the elastic band with a string. Punch a hole in the centre of the plastic lid and the bottom of the can. Slip the elastic band through each hole. Ensure that the weights are in the centre of the elastic band. Put a nail through the loop of the elastic band and securely tape everything in place.

### Analyze and Conclude

1. On a smooth, flat surface, gently roll the can away from yourself.

2. Release the can and describe what happens.

3. Suggest an explanation for what you observed.

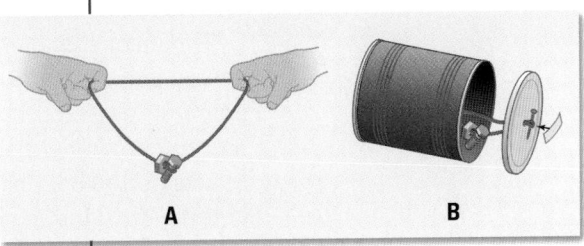

A                    B

- Describe the requirements for doing work.

- Analyze and interpret experimental data representing work done on an object.

- Identify the relationship between work done and displacement along the line of force.

KEY
TERMS

- kinetic energy

- potential energy

- mechanical energy

- work

- joule

Each morning, people throughout Canada perform the same basic activities as they prepare for the day ahead. The ritual might begin by swatting the alarm clock, turning on a light, and heading for a warm shower. Following a quick breakfast of food taken from the refrigerator, they hurry on their way, travelling by family car, bus, subway, train, bicycle, or on foot. This ritual, repeated the world over, demands energy. Electrical energy sounds the alarm, lights the hallways, heats the water, and refrigerates and then cooks the food. Fossil fuels provide the energy for the engines that propel our vehicles. Energy is involved in everything that happens and, in fact, is the reason that everything *can* happen.

**Figure 5.1** One thousand cars could be driven about 350 km, the distance from Ottawa to Hamilton, with the amount of energy it takes to launch a space shuttle into orbit. That requires $2.5 \times 10^{12}$ J of energy.

## Types of Energy

Physicists classify energy into two fundamental types — **kinetic energy** (the energy of motion) and **potential energy** (energy that is stored). The many different forms of energy, such as light energy, electrical energy, and sound energy that you will study in this and other units, all fit into one of these two categories. In this chapter, you will focus on one form of energy called **mechanical energy**.

The mechanical energy of an object is a combination of kinetic energy and potential energy. For example, the football in the photograph on page 194 has kinetic energy because it is moving. It also has potential energy because it is high in the air. The force of gravity acts on the ball, causing it to fall. As it falls, its speed increases and it gains kinetic energy. The best way to begin to understand energy is to study the relationship between energy and work.

# Defining Work

If you have ever helped someone to move, you will understand that lifting heavy boxes or sliding furniture along a rough floor or carpet requires a lot of energy and is hard work. You may also feel that solving difficult physics problems requires energy and is also hard work. These two activities require very different types of work and are examples of how, in science, we need to be very precise about the terminology we use.

In physics, a force does work on an object if it causes the object to move. Work is always done *on* an object and results in a change in the object. **Work** is not energy itself, but rather it is a transfer of mechanical energy. A pitcher does work on a softball when she throws it. A bicycle rider does work on the pedals, which then cause the bicycle to move along the road. You do work on your physics textbook each day when you lift it into your locker. Each of these examples demonstrates the two essential elements of work as defined in physics. There is always a *force* acting on an object, causing the object to move a certain *distance*.

You know from experience that it takes more work to move a heavy table than to move a light chair. It also takes more work to move the table to a friend's house than to move it to the other side of the room. In fact, the amount of work depends directly on the magnitude of the *force* and the *displacement* of the object along the line of the force.

---

**WORK**

Work is the product of the force and the displacement when the force and displacement vectors are parallel and pointing in the same direction.

$$W = F_{\parallel}\Delta d$$

| Quantity | Symbol | SI unit |
|---|---|---|
| magnitude of the force (parallel to displacement) | $F_{\parallel}$ | N (newton) |
| magnitude of the displacement | $\Delta d$ | m (metre) |
| work done | $W$ | J (joule) |

**Unit Analysis**

(force)(displacement) = N · m = J

**Note:** Both force and displacement are vector quantities, but their product, work, is a scalar quantity. For this reason, vector notations will not be used. Instead, a subscript on the symbol for force will indicate that the force is parallel to the displacement.

**Figure 5.2** The joule was chosen as the unit of work in honour of a nineteenth-century physicist, James Prescott Joule.

The derived unit of work, or newton metre (N · m), is called a **joule** (J). One joule of work is accomplished by exerting exactly one newton of force on an object, causing it to move exactly one metre.

The definition for work applies to an individual force, not the net force, acting on an object. As shown in Figure 5.3, two forces are acting on the box. Both forces — the applied force and the force of friction — are doing work. You can calculate the work done by the applied force or the work done by the frictional force.

**Figure 5.3** When you were determining the motion of objects in Chapter 4, you used the net force acting on the object. The net force is really the vector sum of all of the forces acting on the object. When calculating work, you determine the work done by one specific force, not the net force.

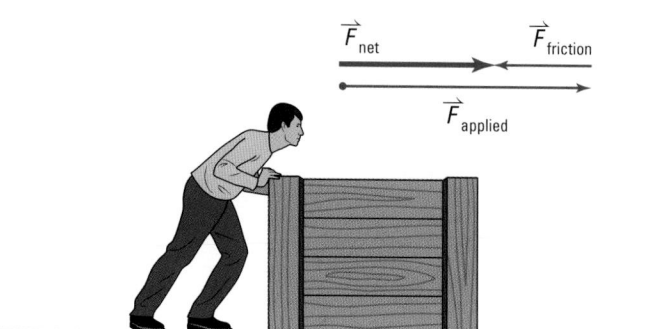

## MODEL PROBLEM

### Determining the Amount of Work Done

**A physics student is rearranging her room. She decides to move her desk across the room, a total distance of 3.00 m. She moves the desk at a constant velocity by exerting a horizontal force of $2.00 \times 10^2$ N. Calculate the amount of work the student did on the desk in moving it across the room.**

### Frame the Problem

- The student *applies a force* to the desk.

- The *applied force* causes the desk to move.

- The *constant velocity* of the desk means that the *acceleration is zero*; thus, the net force on the desk is zero. Therefore, a *frictional force* must be balancing the applied force.

- Since the *force applied* by the student acts in the *same direction* as the *displacement* of the desk, the student is *doing work* on the desk.

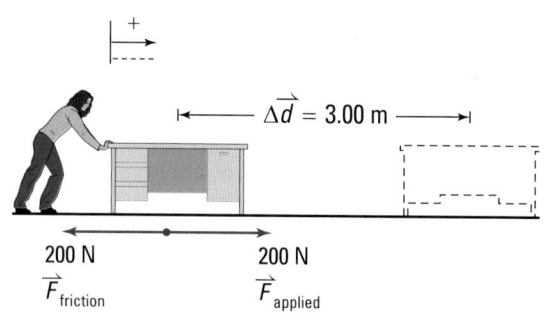

### Identify the Goal

Amount of work, $W$, done by the girl on the desk while moving it across the room

## Variables and Constants

**Involved in the problem**

$\Delta d$

$F_{applied}$ (in the direction of the motion)

$W$

**Known**

$\Delta d = 3.00$ m

$F_{applied} = 2.00 \times 10^2$ N

**Unknown**

$W$

**PROBLEM TIP**

When asked to calculate work done, be sure to identify the force specified by the problem. Then, consider only that force when setting up the calculation. Other forces may be doing work on the object, but you should consider only the work done by the force identified.

### Strategy

Use the formula for work done by a force acting in the same direction as the motion.

All of the needed variables are given, so substitute into the formula.

Multiply.

An N · m is equivalent to a J, therefore,

### Calculations

$$W = F_{\parallel}\Delta d$$

$$W = (2.00 \times 10^2 \text{ N})(3.00 \text{ m})$$

$$W = 6.00 \times 10^2 \text{ N} \cdot \text{m}$$

$$W = 6.00 \times 10^2 \text{ J}$$

The student did $6.00 \times 10^2$ J of work while moving the desk.

### Validate

The applied force was used, because that is the force identified by the problem. The frictional force also did work, but the problem statement did not include work done by friction. Work has units of energy or joules, which is correct.

### PRACTICE PROBLEMS

1. A weight lifter, Paul Anderson, used a circular platform attached to a harness to lift a class of 30 children and their teacher. While the children and teacher sat on the platform, Paul lifted them. The total weight of the platform plus people was $1.1 \times 10^4$ N. When he lifted them a distance of 52 cm, at a constant velocity, how much work did he do? How high would you have to lift one child, weighing 135 N, in order to do the same amount of work that Paul did?

2. A 75 kg boulder rolled off a cliff and fell to the ground below. If the force of gravity did $6.0 \times 10^4$ J of work on the boulder, how far did it fall?

3. A student in physics lab pushed a 0.100 kg cart on an air track over a distance of 10.0 cm, doing 0.0230 J of work. Calculate the acceleration of the cart. (Hint: Since the cart was on an air track, you can assume that there was no friction.)

## When Work Done Is Zero

Physicists define work very precisely. Work done on an object is calculated by multiplying the *force* times the *displacement* of the object when the two vectors are *parallel*. This very precise definition of work can be illustrated by considering three cases where intuition suggests that work has been done, but in reality, it has not.

### Case 1: Applying a Force That Does Not Cause Motion

Consider the energy that you could expend trying to move a house. Although you are pushing on the house with a great deal of force, it does not move. Therefore, the work done on the house, according to the equation for work, is zero (see Figure 5.4). In this case, your muscles feel as though they did work; however, they did no work on the house. The work equation describes work done by a force that moves the object on which the force is applied. Recall that work is a transfer of energy to an object. In this example, the *condition* of the house has not changed; therefore, no work could have been done on the house.

### Language Link

"The *condition* of the house has not changed." In this sentence, the term "condition" is used to represent a measurable and obvious change in the total energy of the house. A change in condition would mean that the house gained some form of kinetic or potential energy from the work done by pushing on it. How does kicking a stationary soccer ball change the condition of the ball?

$\Delta d = 0$
$W = 0$

**Figure 5.4**   $W = F_{\parallel}\Delta d$   $W = F_{\parallel} \times 0$   $W = 0$

### Case 2: Uniform Motion in the Absence of a Force

Recall from Chapter 4 that Newton's first law of motion predicts that an object in motion will continue in motion unless acted on by an *external* force. A hockey puck sliding on a frictionless surface at constant speed is moving and yet the work done is still zero (see Figure 5.5). Work was done to start the puck moving, but because the surface is frictionless, a force is not required to keep it moving; therefore, no work is done on the puck to keep it moving.

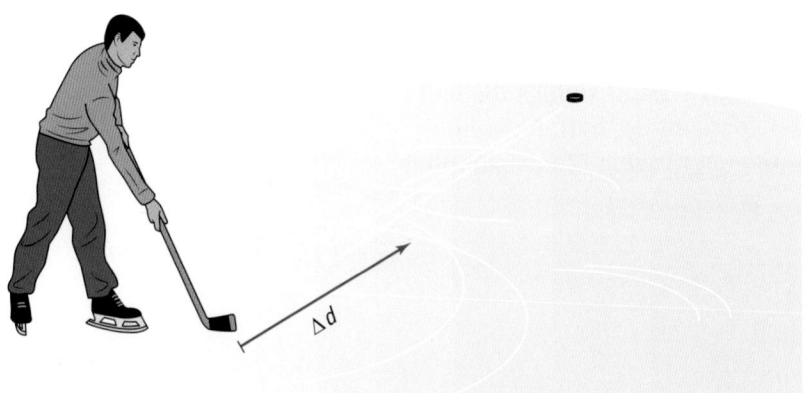

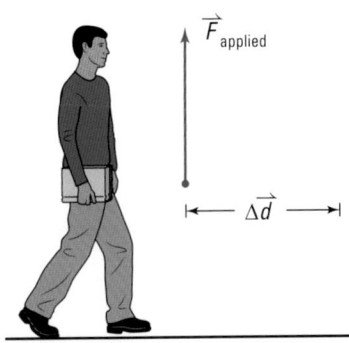

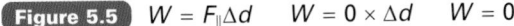

**Figure 5.5** $W = F_{\parallel}\Delta d$ $\quad W = 0 \times \Delta d$ $\quad W = 0$

**Figure 5.6** You are exerting an upward force (against gravity) on your book to prevent it from falling. However, since this force is perpendicular to the motion of the book, it does no work on the book.

## Case 3: Applying a Force That Is Perpendicular to the Motion

Assume that you are carrying your physics textbook down the hall, at constant velocity, on your way to class. Your hand applies a force directly upward to your textbook as you move along the hallway. When considering the work done on the textbook by your hand, you can see that the upward force is perpendicular (i.e., at 90°) to the displacement. In this case, the work done by your hand on the textbook is zero (see Figure 5.6). It is important to note that your hand does do work on the textbook to accelerate it when you begin to move, but once you and the textbook are moving at a constant velocity, you are no longer doing work on the book.

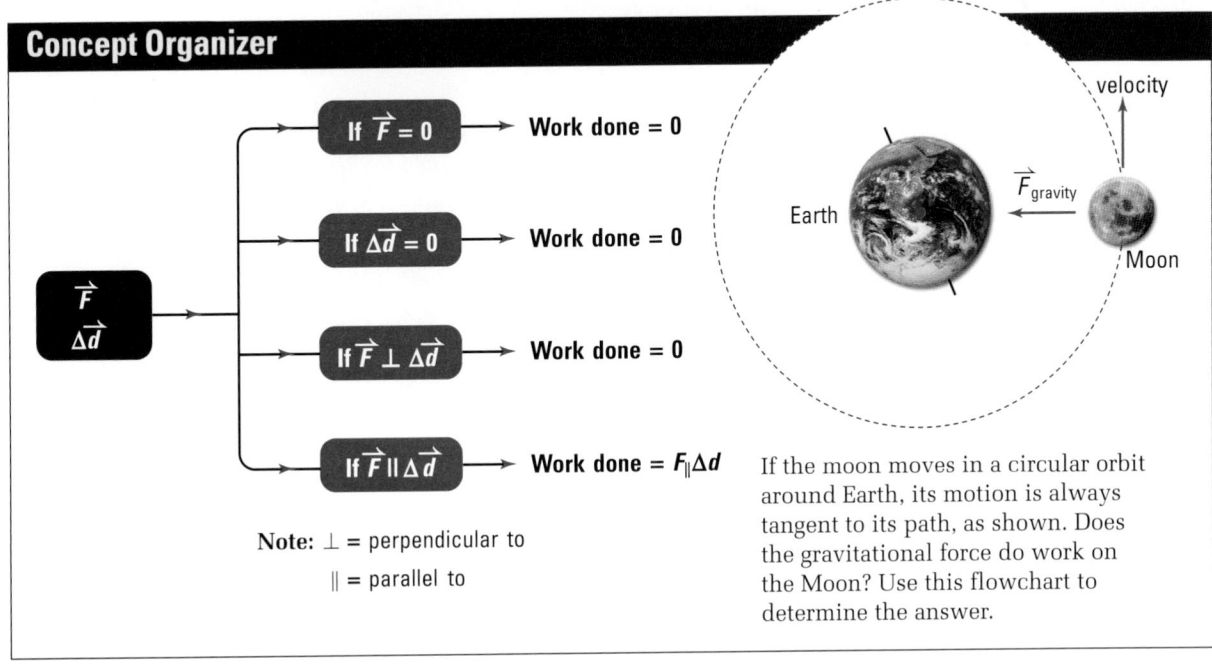

**Figure 5.7** Making decisions about work done

## Work Done by Swinging a Mass on a String

A child ties a ball to the end of a 1.0 m string and swings the ball in a circle. If the string exerts a 10 N force on the ball, how much work does the string do on the ball during a swing of one complete circle?

### Frame the Problem

- The string *applies a force* to the ball.

- The ball is *moving*.

- The direction of the motion of the ball is *perpendicular* to the direction of the force.

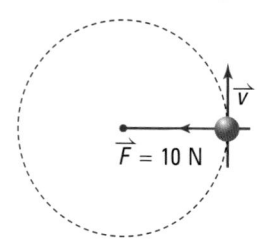

### Identify the Goal

Amount of work, $W$, done by the string on the ball

### Variables and Constants

| Involved in the problem | Known | Unknown |
|---|---|---|
| $\Delta d$ | $\Delta d = 2\pi\ (1.0\ \text{m})$ | $W$ |
| $F_{\text{applied}}$ | (circumference of a circle of radius 1.0 m) | $F_{\parallel}$ |
| $W$ | $F_{\text{applied}} = 10\ \text{N}$ | |
| $F_{\parallel}$ | (perpendicular to the direction of the motion) | |

**PROBLEM TIP**

When solving problems involving work, consider the following questions.

- Is a force acting on the object to be moved?

- Does the force *cause* displacement of the object?

- Is the force acting in the same direction as the displacement?

If the answer to each question is yes, then you can safely apply the equation for work done. If the answer to any of the questions is no, compare the problem to the cases discussed on pages 200 to 201. The work done might be zero.

### Strategy

The force that the string applies on the ball is not in the direction of the motion; therefore, it cannot do work on the ball. No information is given about any possible force that is parallel to the direction of the motion. However, the problem did not ask for work done by any force other than that exerted by the string.

The string does no work on the ball as it swings around on the end of the string.

### Calculations

$W = 0\ \text{J}$

### Validate

The orientations of the force and displacement vectors fit the conditions described in Case 3. It is not possible for the force to do work. Therefore the work must be zero.

4. With a $3.00 \times 10^2$ N force, a mover pushes a heavy box down a hall. If the work done on the box by the mover is $1.90 \times 10^3$ J, find the length of the hallway.

5. A large piano is moved 12.0 m across a room. Find the average horizontal force that must be exerted on the piano if the amount of work done by this force is $2.70 \times 10^3$ J.

6. A crane lifts a 487 kg beam vertically at a constant velocity. If the crane does $5.20 \times 10^4$ J of work on the beam, find the vertical distance that it lifted the beam.

7. A teacher carries his briefcase 20.0 m down the hall to the staff room. The teacher's hand exerts a 30.0 N force upward as he moves down the hall at constant velocity.

   (a) Calculate the work done by the teacher's hand on the briefcase.

   (b) Explain the results obtained in part (a).

8. A $2.00 \times 10^2$ N force acts horizontally on a bowling ball over a displacement of 1.50 m. Calculate the work done on the bowling ball by this force.

9. The *Voyager* space probe has left our solar system and is travelling through deep space, which can be considered to be void of all matter. Assume that gravitational effects may be considered negligible when *Voyager* is far from our solar system.

   (a) How much work is done on the probe if it covers $1.00 \times 10^6$ km travelling at $3.00 \times 10^4$ m/s?

   (b) Explain the results obtained in part (a).

10. An energetic group of students attempts to remove an old tree stump for use as firewood during a party. The students apply an average upward force of 650 N. The 865 kg tree stump does not move after 15.0 min of continuous effort, and the group gives up.

    (a) How much work did the students do on the tree stump?

    (b) Explain the results obtained in part (a).

## Work Done by Changing Forces

We have restricted our discussion so far to forces that remain constant throughout the motion. However, the definition of work as given by $W = F_{\parallel}\Delta d$ applies to all cases, including situations where the force changes. Mathematically, solving problems with changing forces goes beyond the scope of this course. However, you can use a graph to approximate a solution without using complex mathematics. A force-versus-position graph allows you to determine the work done, whether or not the force remains constant. Examine Figure 5.8, in which the graph shows a constant force of 10 N acting over a displacement of 4.0 m. The area under the force-versus-position line is given by the shaded rectangle.

$$\text{Area under the curve} = \text{area of the shaded rectangle}$$
$$= \text{length} \times \text{width}$$
$$= (10 \text{ N})(4.0 \text{ m})$$
$$= 40 \text{ N} \cdot \text{m}$$
$$= 40 \text{ J}$$

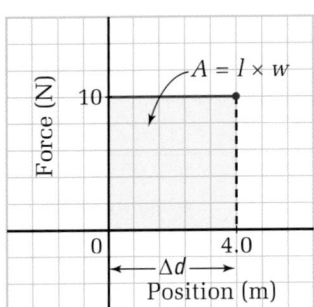

**Figure 5.8** The length of the shaded box is $F_{\parallel}$ = 10 N. The width is $\Delta d$ = 4.0 m. Therefore, the area is $F_{\parallel} \times \Delta d$ = 40 N · m. The area under the curve is the same as the work done by the force.

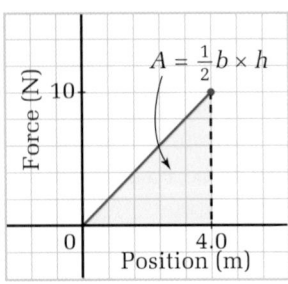

**Figure 5.9** When the force increases linearly, the average force can be calculated by using the equation

$$F_{ave} = \tfrac{1}{2}(F_{initial} + F_{final})$$

The area under the curve is in the shape of a triangle. Thus, the work done can be calculated by calculating the area of the triangle.

This result is identical to the one obtained by applying the equation for the work done: $W = F_{\parallel}\Delta d$.

$$W = F_{\parallel}\Delta d$$
$$= (10 \text{ N})(4.0 \text{ m})$$
$$= 40 \text{ N} \cdot \text{m}$$
$$= 40 \text{ J}$$

A force-versus-position graph can be used to determine work done even when the applied force does not remain constant. Consider the force-versus-position graph in Figure 5.9. It shows a force that starts at zero and increases steadily to 10 N over a displacement of 4.0 m. In this case, the area under the curve forms a triangle. Even though the force does not remain constant, it is still possible to calculate the work done by finding the area of the triangle.

Work done = area under the force-versus-position curve

= area of the triangle

= $\tfrac{1}{2}$ base × height

= (0.5)(4.0 m)(10 N)

= 20 N · m

= 20 J

Work done = average force multiplied by the displacement

= $\tfrac{1}{2}$(0 N + 10 N)(4.0 m)

= (0.5)(10 N)(4.0 m)

= 20 N · m

= 20 J

Notice that calculating the area of the triangle yields the same results as the equation for work, if the average force is used. In many situations, however, the applied force changes in a way that makes it difficult to obtain an average force. Solving problems that involve such a changing applied force can be solved using calculus. Since this advanced mathematical technique is not a requirement of this course, you can estimate solutions to problems involving changing forces by estimating the area under a curve, such as the one shown in Figure 5.10.

The force-versus-position curve in Figure 5.10 represents the force exerted on a golf ball by the club when the golfer tees off. Notice how the force changes, reaching a maximum and then falling back to zero. To calculate the area under the curve and, therefore, the work done on the ball by the club, you must count the squares and estimate the area in the partial squares. This method yields a result that is a close approximation to the numerical answer that could be obtained using calculus.

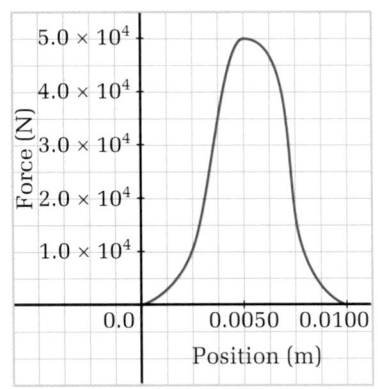

**Figure 5.10** Even when a force-versus-position curve is irregular, the area under the curve gives the work done. This curve represents the force of a golf club on the ball when the club strikes the ball.

## Estimating Work from a Graph

**Determine the amount of work done by the changing force represented in the force-versus-position plot shown here.**

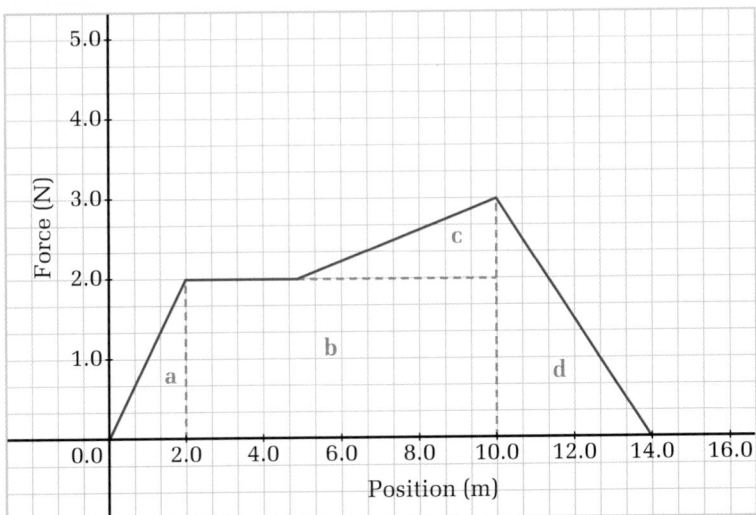

## Frame the Problem

- A *changing force* is acting on an object.

- Since the force is not constant, the formula for work does not apply.

- The *area under a force-position curve* is equal to the work done when the units of force and displacement on the graph are used correctly.

- You can divide the area into segments that have simple geometric shapes. Then, use the formulas for the shapes to find the areas. The sum of the areas for each shape gives the total area and, thus, the work done.

## Identify the Goal

The work, $W$, done by the forces represented on the force-versus-position plot

## Variables and Constants

| Involved in the problem | Known | Unknown |
|---|---|---|
| Scale for force and displacement on graph | Scale for force and displacement on graph | $A_a$ |
| | | $A_b$ |
| $A_a$  $A_b$ | | $A_c$ |
| $A_c$  $A_d$ | | $A_d$ |
| $W$ | | $W$ |

*continued* ▶

## Strategy

Divide the area under the curve into simple triangles and rectangles, as shown on the plot.

Calculate $A_b$, the area of the rectangle b.

Calculate $A_a$, the area of triangle a.

Calculate $A_c$, the area of triangle c.

Calculate $A_d$, the area of triangle d.

Find $A_T$, the total area.

Work is equal to the total area under the curve.

## Calculations

Area = base × height

base = 10.0 m − 2.0 m = 8.0 m

height = 2.0 N − 0.0 N = 2 N

$A = b \times h$
$A_b = (8.0 \text{ m})(2.0 \text{ N})$
$A_b = 16.0 \text{ N} \cdot \text{m}$
$A_b = 16 \text{ J}$

Area = $\frac{1}{2}$base × height

base = 2.0 m − 0.0 m = 2.0 m

height = 2.0 N − 0.0 N = 2 N

$A = \frac{1}{2}b \times h$
$A_a = \frac{1}{2}(2.0 \text{ m})(2.0 \text{ N})$

$A_a = 2.0 \text{ N} \cdot \text{m}$

$A_a = 2.0 \text{ J}$

Area = $\frac{1}{2}$base × height

base = 10.0 m − 5.0 m = 5.0 m

height = 3.0 N − 2.0 N = 1.0 N

$A = \frac{1}{2}b \times h$
$A_c = \frac{1}{2}(5.0 \text{ m})(1.0 \text{ N})$

$A_c = 5.0 \text{ N} \cdot \text{m}$

$A_c = 5.0 \text{ J}$

Area = $\frac{1}{2}$base × height

base = 14.0 m − 10.0 m = 4.0 m

height = 3.0 N − 0.0 N = 3.0 N

$A = \frac{1}{2}b \times h$
$A_d = \frac{1}{2}(4.0 \text{ m})(3.0 \text{ N})$

$A_d = 6.0 \text{ N} \cdot \text{m}$

$A_d = 6.0 \text{ J}$

$A_T = A_a + A_b + A_c + A_d$
$\quad = 2.0 \text{ J} + 16.0 \text{ J} + 5.0 \text{ J} + 6.0 \text{ J}$
$\quad = 29.0 \text{ J}$

$W = 29 \text{ J}$

The force represented by the graph did 29 J of work.

---

## Validate

By looking at the graph, you can estimate that the average force is close to 2 N. Therefore, a rough estimate of the work would be

$$\text{Work} \approx 2 \text{ N} \times 14 \text{ m} = 28 \text{ J}$$

This is close enough to give you confidence in the value of 29 J.

**11.** Determine the amount of work done by the forces represented in the four force-versus-position plots that follow.

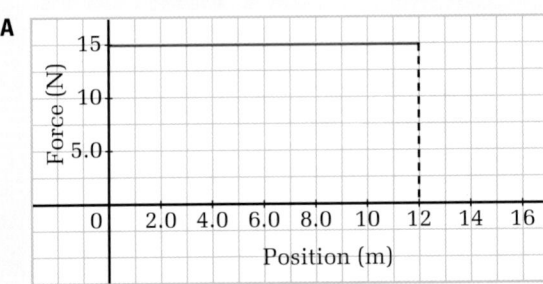

A

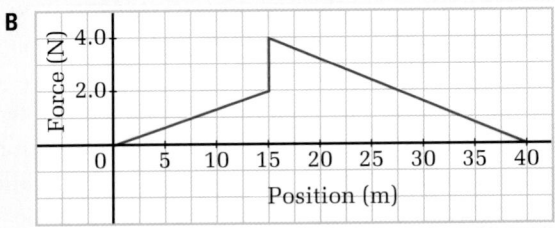

B

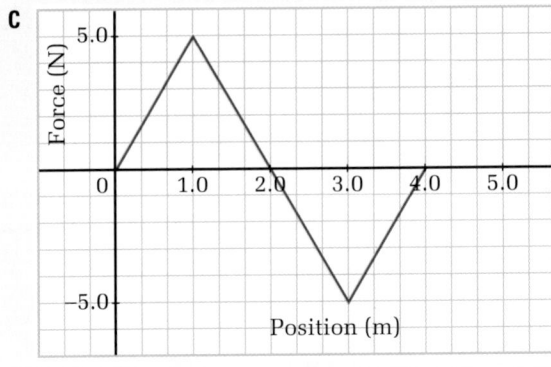

C

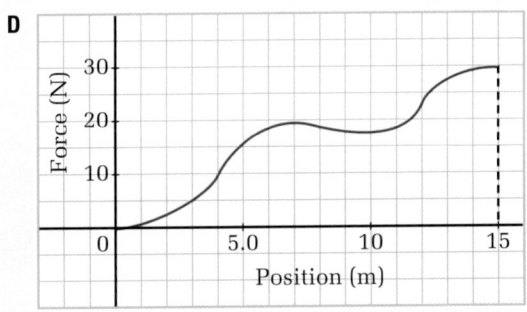

D

**12.** Draw a force-versus-position plot that represents a constant force of 60 N exerted on a Frisbee™ over a distance of 80.0 cm. Show the work done on the Frisbee™ by appropriately shading the graph.

**13.** Stretch a rubber band and estimate the amount of force you are using to stretch it. (Hint: A 100 g mass weighs approximately 1N.) Notice how the force you must exert increases as you stretch the rubber band. Draw a force-versus-position graph of the force you used to stretch the rubber band for a displacement of 15 cm. Use the graph to estimate the amount of work you did on the rubber band.

## Constant Force at an Angle

Everyday experience rarely provides situations in which forces act precisely parallel or perpendicular to the motion of an object. For example, in Figure 5.11 on the next page, the applied force exerted by the child pulling the wagon is at an angle relative to the direction of the wagon's displacement. In cases such as this, only part of the force vector or a component of the force does work on the wagon.

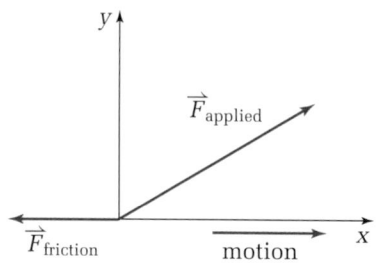

**Figure 5.11** To calculate the work done on the wagon, use only the component of the force that is in the direction of the displacement.

To determine the work done by the child on the wagon, you must first find the component of the force that is parallel to the direction of motion of the wagon. The first step in performing this calculation is choosing a coordinate system with one axis along the direction of motion of the wagon. Then, use simple trigonometry to *resolve* the force into its components — one that is parallel to the direction of motion and one that is perpendicular to the

## QUICK LAB

# Cart Bungee

**TARGET SKILLS**

- Predicting
- Identifying variables
- Performing and recording

Connect one elastic band between a cart and a force sensor on an inclined slope. Set up a motion sensor to track the displacement of the cart. Have the computer generate a force-versus-displacement graph as you release the cart and it moves down the incline until the elastic band stops it. Use the computer software to calculate the area under the force-versus-displacement curve.

### Analyze and Conclude

1. What does the area under the curve on your graph represent?

2. Predict how changing the number of elastic bands will affect the area under the curve.

3. Predict how changing the mass of the cart will affect the area under the curve.

4. What other variables could you change? Predict how they would affect the area under the curve.

5. If you have the opportunity, test your predictions.

direction of motion. As shown in Figure 5.12, the parallel component of the force vector is $F_x = |\vec{F}| \cos \theta$. Note that the angle, $\theta$, is always the angle between the force vector and the displacement vector.

To summarize, when the applied force, $\vec{F}$, acts at an angle to the displacement of the object, use the component of the force parallel to the direction of the motion, $F_x = |\vec{F}| \cos \theta$, to calculate the work done. In these cases, the equation for work is derived as follows.

- This equation applies to cases where the force and displacement vectors are in the same direction and, thus, vector notations are not used.

$$W = F_{\parallel} \Delta d$$

- This equation applies to cases where the x-component is parallel to the direction of the motion, and thus, vector notations are not used.

$$W = F_x \Delta d$$

- The component of the original force, $\vec{F}$, that is parallel to the displacement

$$F_x = |\vec{F}| \cos \theta$$

- Substitute the expression for $F_x$ into the expression for work.

$$\therefore W = |\vec{F}| \cos \theta \, \Delta d$$

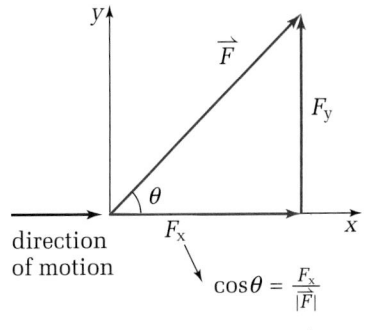

$$\cos\theta = \frac{F_x}{|\vec{F}|}$$

$$F_x = |\vec{F}| \cos\theta$$

**Figure 5.12** To find the horizontal component of the force, start with the definition of the cosine of an angle. Then solve for $F_x$.

---

**WORK**

Work done when the force and displacement are *not* parallel and pointing in the same direction

$$W = F\Delta d \cos \theta$$

$\theta$ is the angle between the force and displacement vectors.

**Note:** Since work is a scalar quantity and only the magnitudes of the force and displacement affect the value of the work done, vector notations have been omitted.

---

The example of the child pulling the wagon demonstrates some subtle aspects of the definition of work. Consider the question "Does the force of gravity do work on the wagon?" To answer this, follow the reasoning process outlined in Figure 5.7, the Concept Organizer, on page 201. You can see that the force of gravity is not zero, but that it is perpendicular to the displacement. Therefore, we may conclude that the work done by gravity on the wagon in this case is zero. Note that $F \cos \theta = F \cos 90° = F \times 0 = 0$.

## Think It Through

- A child is pulling a wagon up a ramp. The applied force is parallel to the ramp. Is gravity doing any work on the wagon? Explain your reasoning.

- Examine the figure below. Assume that, in both case A and case B, the mass of the crate, the frictional force, and the constant velocity of the crate are the same. In which case, A or B, is the worker exerting a greater force on the crate by pulling on the rope? Explain your reasoning.

A          B

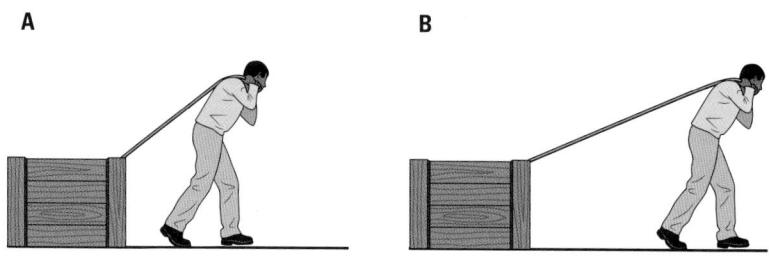

# Elevator Versus a Ramp

What forces are acting on the mass? How will these forces affect its motion?

Set up an inclined plane. Carefully measure both its length and height. Using a Newton spring scale or a force sensor, carefully pull a cart up the incline at a constant velocity. Observe and record the force required. Now lift the cart at a constant velocity through the same height as the incline. Again, record the force required. Predict which method will require more work to accomplish. Calculate the work done in each case.

### Analyze and Conclude

1. Compare the work done in each case. How accurate was your prediction?

2. Discuss the experimental variables that most directly affect the work done.

3. Repeat the experiment using a large, rough mass that will slide up the incline rather than roll as the cart did. Predict which method of pulling the large mass will require more work. Was your prediction correct? Why or why not?

## Positive and Negative Work

Does the force of friction do work on the mass in Figure 5.13? It is parallel to the displacement, but it is acting in the opposite direction. This means that the angle between the displacement and the force of friction is 180°. Applying the revised equation for work, we find the following results.

Apply the equation for work to this case $\quad W = F\Delta d \cos 180°$
Because $\cos 180° = (-1)$ $\qquad\qquad\qquad W = F\Delta d(-1)$
Therefore $\qquad\qquad\qquad\qquad\qquad\quad W = -F\Delta d$

**Figure 5.13** In what direction is the force of friction acting?

The work done by the frictional force is non-zero and negative. *Negative work* done by an external force *reduces* the energy of a mass. The energy does not disappear, but is, instead, lost to the surroundings in the form of heat or thermal energy. If the person stopped pulling the mass, its motion would quickly stop, as the frictional force would reduce the energy of motion to zero.

*Positive* work adds energy to an object; *negative* work removes energy from an object. In many situations, such as the one shown in Figure 5.14, two different forces are doing work on the same object. One force is doing positive work and the other is doing negative work.

**Figure 5.14** The hammer does positive work on the nail — the applied force and the displacement are in the same direction as the nail moves into the wood. The force of friction does negative work on the nail — the force of friction is opposite to the displacement as the nail moves into the wood. The nail also does negative work on the hammer — the applied force is opposite to the direction of the displacement. The hammer stops moving.

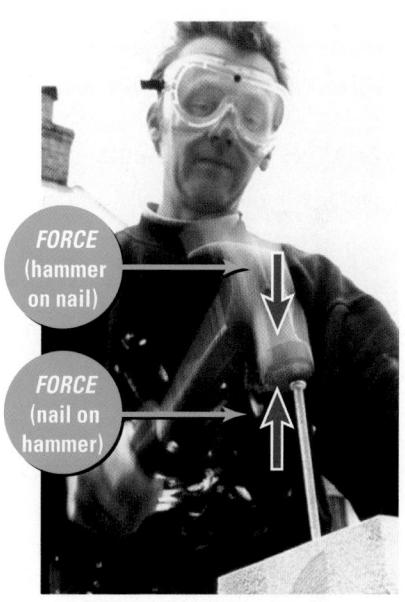

FORCE (hammer on nail)

FORCE (nail on hammer)

### MODEL PROBLEM

### Doing Positive and Negative Work

Consider a weight lifter bench-pressing a barbell weighing $6.50 \times 10^2$ N through a height of 0.55 m. There are two distinct motions: (1) when the barbell is lifted up and (2) when the barbell is lowered back down. Calculate the work done on the barbell during each of the two motions.

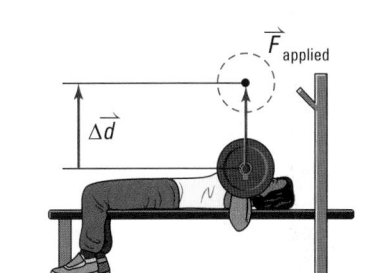

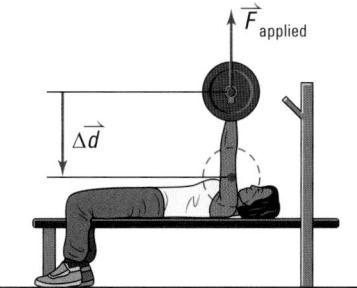

*continued* ▶

## Frame the Problem

- The weight lifter *lifts* and *lowers* the barbell at a *constant velocity*. Therefore, the force *she exerts* is equal to the *weight* of the barbell.

- Use the formula for work when the direction of the force is at an angle with the direction of the displacement.

- Positive work adds energy to the object. Negative work removes energy from the object.

- The angle between the force acting on the barbell and its displacement is
  0° while lifting
  180° while lowering

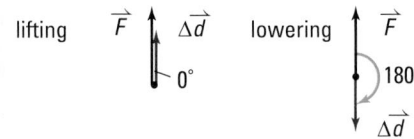

## Identify the Goal

Work, $W$, done while lifting the barbell

Work, $W$, done while lowering the barbell

## Variables and Constants

| Involved in the problem | Known | Unknown |
|---|---|---|
| $F$ | $F = 6.50 \times 10^2\,[\text{up}]$ | $W$ |
| $\Delta d$ | $\Delta d = 0.55\text{ m}\,[\text{up}]$ while lifting | |
| $\theta$ | $\Delta d = 0.55\text{ m}\,[\text{down}]$ while lowering | |
| $W$ | $\theta = 0°$ while lifting | |
| | $\theta = 180°$ while lowering | |

## Strategy

Use the formula for work when the force is not the same direction as the displacement.

Substitute in the values.

Multiply.

## Calculations

$$W = F\Delta d \cos\theta$$

**Lifting**

$W = (6.5 \times 10^2\text{ N})(0.55\text{ m})\cos 0°$

$W = 3.65 \times 10^2\text{ N} \cdot \text{m } (+1)$

$W = 3.7 \times 10^2\text{ J}$

**Lowering**

$W = (6.5 \times 10^2\text{ N})(0.55\text{ m})\cos 180°$

$W = 3.65 \times 10^2\text{ N} \cdot \text{m } (-1)$

$W = -3.7 \times 10^2\text{ J}$

The weight lifter did $3.7 \times 10^2$ J of work to lift the barbell and $-3.7 \times 10^2$ J of work to lower it.

## Validate

The barbell gained energy when it was raised. The energy that the barbell gained would become very obvious if it was dropped from the elevated position — it would accelerate downward, onto the weight lifter!

The weight lifter does negative work on the bar to lower it. She is removing the energy that she previously had added to the bar by lifting it. If the weight lifter did not do negative work on the bar, it would accelerate downward.

14. A large statue, with a mass of 180 kg, is lifted through a height of 2.33 m onto a display pedestal. It is later lifted from the pedestal back to the ground for cleaning.

    (a) Calculate the work done by the applied force on the statue when it is being lifted onto the pedestal.

    (b) Calculate the work done by the applied force on the statue when it is lowered down from the display pedestal.

    (c) State all of the forces that are doing work on the statue during each motion.

15. A mechanic exerts a force of 45.0 N to raise the hood of a car 2.80 m. After checking the engine, the mechanic lowers the hood. Find the amount of work done by the mechanic on the hood during each of the two motions.

16. A father is pushing a baby carriage down the street. Find the total amount of work done by the father on the baby carriage if he applies a 172.0 N force at an angle of 47° with the horizontal, while pushing the carriage 16.0 m along the level sidewalk.

17. While shopping for her weekly groceries, a woman does 2690 J of work to push her shopping cart 23.0 m down an aisle. Find the magnitude of the force she exerts if she pushes the cart at an angle of 32° with the horizontal.

18. A farmer pushes a wheelbarrow with an applied force of 124 N. If the farmer does 7314 J of work on the wheelbarrow while pushing it a horizontal distance of 77.0 m, find the angle between the direction of the force and the horizontal.

## 5.1    Section Review

1. **K/U** In each of the following cases, state whether you are doing work on your textbook. Explain your reasoning.

   (a) You are walking down the hall in your school, carrying your textbook.

   (b) Your textbook is in your backpack on your back. You walk down a flight of stairs.

   (c) You are holding your textbook while riding up an escalator.

2. **K/U** Your lab partner does the same amount of work on two different objects, A and B. After she stops doing work, object A moves away at a greater velocity than object B. Give two possible reasons for the difference in the velocities of A and B.

3. **C** Describe two different scenarios in which you are exerting a force on a box but you are doing no work on the box.

4. **K/U** State all of the conditions necessary for doing positive work on an object.

5. **I** A student used a force meter to pull a heavy block a total distance of 4.5 m along a floor. Part of the floor was wood, another part was carpeted, and a third part was tiled. In each case, the force required to pull the block was different. The table below lists the distance and the force recorded for the three parts of the trip.

| Floor surface | Distance pulled | Force measured |
|---|---|---|
| wooden floor | 1.5 m | 3.5 N |
| carpet | 2.5 m | 6.0 N |
| tiled floor | 0.5 m | 4.5 N |

   (a) Use these data to construct a force-versus-distance graph for the motion.

   (b) Calculate the total work performed on the heavy block throughout the 4.5 m.

- Analyze the factors that determine an object's kinetic energy.

- Communicate the relationships between work done to an object and the object's change in kinetic energy.

- Use equations to analyze quantitative relationships between motion and kinetic energy.

**KEY**
**TERMS**

- work-kinetic energy theorem

- work-energy theorem

**Figure 5.15** Olympic triathlon winner Simon Whitfield crossing the finish line ahead of all of the other competitors.

Simon Whitfield won a gold medal in the triathlon at the Sydney Olympics. Chemical reactions in his muscles caused them to shorten. This shortening of his muscles did work on the bones of his skeleton by exerting a force that caused them to move. The resulting motion of his bones allowed him to run faster than all of his competitors. How can you mathematically describe the motion resulting from the work his muscles did? In this section, you will investigate the relationship between doing work on an object and the resulting motion of the object.

## Kinetic Energy

A baseball moves when you throw it. A stalled car moves when you push it. The subsequent motion of each object is a result of the work done on it. The energy of motion is called kinetic energy. By doing some simple "thought experiments" you can begin to develop a method to quantify kinetic energy. First, imagine a bowling ball and a golf ball rolling toward you with the same velocity. Which ball would you try hardest to avoid? The bowling ball would, of course, do more "work" on you, such as crushing your toe. Since both balls have the same velocity, the mass must be contributing to the kinetic energy of the balls. Now, imagine two golf balls flying toward you, one coming slowly and one rapidly. Which one would you try hardest to avoid? The faster one, of course. Using the same reasoning, an object's velocity must contribute to its kinetic energy. Could kinetic energy be a mathematical combination of an object's mass and velocity?

Dutch mathematician and physicist Christian Huygens (1629–1695) looked for a quantity involving mass and velocity that was characteristic of an object's motion. Huygens experimented with collisions of rigid balls (similar to billiard balls). He discovered that if he calculated the product of the mass and the square of the velocity (i.e. $mv^2$) for each ball and then added those products together, the totals were the same before and after the collisions. German mathematician Gottfried Wilhelm Leibniz (1646–1716), Huygens' student, called the quantity *vis viva* for "living force." It was many years later, after numerous, detailed observations and calculations, that physicists realized that the correct expression for the kinetic energy of an object, resulting from the work done on it, is actually *one half* of the quantity that Leibniz came up with.

**Figure 5.16** When billiard balls collide, the work each ball does on another gives the other ball kinetic energy only.

---

### KINETIC ENERGY

Kinetic energy is one half the product of an object's mass and the square of its velocity.

$$E_k = \tfrac{1}{2}mv^2$$

| Quantity | Symbol | SI unit |
|---|---|---|
| kinetic energy | $E_k$ | J (joule) |
| mass | $m$ | kg (kilogram) |
| velocity | $v$ | $\dfrac{m}{s}$ (metres per second) |

**Unit Analysis**

$(mass)(velocity)^2 = kg(\frac{m}{s})^2 = kg\frac{m^2}{s^2} = kg\frac{m}{s^2}m = N \cdot m = J$

**Note:** When velocity is squared, it is no longer a vector. Therefore, vector notation is not used in the expression for kinetic energy.

---

### MODEL PROBLEM

#### Calculating Kinetic Energy

A 0.200 kg hockey puck, initially at rest, is accelerated to 27.0 m/s. Calculate the kinetic energy of the hockey puck **(a)** at rest and **(b)** in motion.

hockey puck at rest

hockey puck in motion

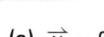

**(a)** $\vec{v_1} = 0$

**(b)** $\vec{v_2} = 27.0\ \frac{m}{s}$

#### Frame the Problem

- A hockey puck was *at rest* and was then *accelerated*.

- A moving object has *kinetic energy*.

- The amount of *kinetic energy* possessed by an object is related to its *mass* and *velocity*.

*continued* ▶

*continued from previous page*

## Identify the Goal

The kinetic energy, $E_k$, of the hockey puck
at rest
at a velocity of 27.0 $\frac{m}{s}$

## Variables and Constants

| Involved in the problem | Known | Implied | Unknown |
|---|---|---|---|
| $m$ | $m = 0.200$ kg | $v_1 = 0.0 \frac{m}{s}$ (at rest) | $E_k$ |
| $v_1$ (at rest) | $v_2 = 27.0 \frac{m}{s}$ (moving) | | |
| $v_2$ (moving) | | | |
| $E_k$ | | | |

## Strategy

Use the equation for kinetic energy.

All of the needed values are known, so substitute into the formula.

Multiply.

1 kg $\frac{m}{s^2}$ is equivalent to 1 N.

1 N·m is equivalent to 1 J.

## Calculations

$$E_k = \tfrac{1}{2}mv^2$$

| At rest | Moving |
|---|---|
| $E_k = \tfrac{1}{2}(0.200 \text{ kg})\left(0 \frac{m}{s}\right)^2$ | $E_k = \tfrac{1}{2}(0.200 \text{ kg})\left(27.0 \frac{m}{s}\right)^2$ |
| $E_k = 0 \text{ kg} \frac{m^2}{s^2}$ | $E_k = 72.9 \text{ kg} \frac{m^2}{s^2}$ |
| $E_k = 0 \text{ kg} \frac{m}{s^2} \text{ m}$ | $E_k = 72.9 \text{ kg} \frac{m}{s^2} \text{ m}$ |
| $E_k = 0 \text{ N} \cdot \text{m}$ | $E_k = 72.9 \text{ N} \cdot \text{m}$ |
| $E_k = 0 \text{ J}$ | $E_k = 72.9 \text{ J}$ |

The puck had zero kinetic energy while at rest, and 72.9 J of kinetic energy when moving.

## Validate

A moving object has kinetic energy, while an object at rest has none.
The units are expressed in joules, which is correct for energy.

**PRACTICE PROBLEMS**

19. A 0.100 kg tennis ball is travelling at 145 km/h. What is its kinetic energy?

20. A bowling ball, travelling at 0.95 m/s, has 4.5 J of kinetic energy. What is its mass?

21. A 69.0 kg skier reaches the bottom of a ski hill with a velocity of 7.25 m/s. Find the kinetic energy of the skier at the bottom of the hill.

**Table 5.1** Examples of Mechanical Kinetic Energies

| Item | Information | | |
|---|---|---|---|
| | **Mass** | **Velocity** | **Kinetic energy** |
| Meteor Crater meteor | $3.0 \times 10^4$ kg | $9.3 \times 10^6 \frac{m}{s}$ | $1.3 \times 10^{18}$ J |
| International Space Station | $4.44 \times 10^5$ kg | $2.74 \times 10^5 \frac{km}{h}$ | $1.29 \times 10^{15}$ J |
| aircraft carrier | $9.86 \times 10^7$ kg | $40 \frac{km}{h}$ | $6.1 \times 10^9$ J |
| tractor trailer | $1.8 \times 10^4$ kg | $50 \frac{km}{h}$ | $1.7 \times 10^6$ J |
| hockey player | 120 kg | $12 \frac{m}{s}$ | $8.6 \times 10^3$ J |
| pitched baseball | $2.50 \times 10^2$ g | $1.00 \times 10^2 \frac{km}{h}$ | 9.65 J |
| person | 75 kg | $0.5 \frac{m}{s}$ | 9 J |
| housefly | 2.0 mg | $7.2 \frac{km}{h}$ | $4.0 \times 10^{-3}$ J |
| electron in computer monitor | $9.11 \times 10^{-31}$ kg | $2.0 \times 10^8 \frac{m}{s}$ | $1.8 \times 10^{-14}$ J |

*Note that all velocities were converted to m/s and all mass values to kg before calculating the $E_k$.

## Think It Through

• As a car leaves a small town and enters a highway, the driver presses on the accelerator until the speed doubles. By what factor did the car's kinetic energy increase when its speed doubled?

• If two cars were travelling at the same speed but one car had twice the mass of the other, was the kinetic energy of the larger car double, triple, or quadruple the kinetic energy of the smaller car?

• Examine all of the moving objects in the photographs in Figure 5.17, "Visualizing kinetic energy," on the following page, and answer the following questions.

   **1.** Which moving object probably has the greatest velocity? Explain why you chose that object.

   **2.** Work is being done on which of the objects in the photos? What is the force doing the work in each case?

   **3.** Which objects are probably losing kinetic energy?

   **4.** Which objects are probably gaining kinetic energy?

   **5.** Which object has the greatest amount of kinetic energy?

**Figure 5.17** Visualizing kinetic energy

# Work and Kinetic Energy

The special relationship between doing work on an object and the resulting kinetic energy of the object is called the **work-kinetic energy theorem**. Everyday experience supports this theorem. If you saw a hockey puck at rest on the ice and a moment later saw it hurtling through the air, you would conclude that someone did work on the puck, by exerting a large force over a short distance, to make it move. This correct conclusion illustrates how doing work on an object gives the object increased velocity or kinetic energy.

**Figure 5.18** Find out how the work done on the hockey puck is related to the puck's kinetic energy.

To develop a mathematical expression that relates work to the energy of motion, assume that all of the work done on a system gives the system kinetic energy only. Start with the definition of work and then apply Newton's second law. To avoid dealing with advanced mathematics, assume that a constant force gives the system a constant acceleration so that you can use the equations of motion from Unit 1. Since work and kinetic energy are scalar quantities, vector notation will be omitted from the derivation. This is valid as long as the directions of the force and displacement are parallel and the object is moving in a straight line.

### PHYSICS FILE

Deriving or generating a mathematical equation to describe what you observe in the world is one of the things that theoretical physicists do. The challenge is to make appropriate substitutions from information that you already know and develop a useful relationship.

### Math Link

Try to derive the work-kinetic energy theorem using the work equation, $W = F_\parallel \Delta d$, and the equation of motion, $v_2^2 = v_1^2 + 2a\Delta d$, from Chapter 2. You may prefer this derivation rather than the one illustrated. (Hint: Solve the motion equation for acceleration.)

- Assume that the force is constant and write the equation for work.

$$W = F_\parallel \Delta d$$

- Recall Newton's second law, $\vec{F} = m\vec{a}$. Assume the force and the acceleration are parallel to the direction of the displacement and motion is in one direction. Then omit vector symbols and use $F = ma$.

$$F = ma$$

- Substitute $ma$ for $F$ in the equation for work.

$$W = ma\Delta d$$

- Recall, from Chapter 2, the definition of acceleration for uniformly accelerated motion.

$$a = \frac{v_2 - v_1}{\Delta t}$$

- Rewrite the equation for work in terms of initial and final velocities by substituting the definition for acceleration into $a$.

$$W = m\frac{(v_2 - v_1)}{\Delta t}\Delta d$$

- Also, from Chapter 2, recall the equation for displacement for uniformly accelerated motion.

$$\Delta d = \frac{(v_1 + v_2)}{2}\Delta t$$

- Divide both sides of the equation by $\Delta t$ to obtain an equation for $\frac{\Delta d}{\Delta t}$.

$$\frac{\Delta d}{\Delta t} = \frac{(v_1 + v_2)}{2}$$

- Rewrite the equation for work by substituting the value for $\frac{\Delta d}{\Delta t}$.

$$W = m\frac{(v_2 - v_1)(v_1 + v_2)}{2}$$

- Expand the brackets (FOIL).

$$W = \frac{1}{2}m(v_1 v_2 + v_2^2 - v_1^2 - v_1 v_2)$$

- Simplify by combining like terms.

$$W = \frac{1}{2}m(v_2^2 - v_1^2)$$

- Expand. Notice that the result is in the form of initial and final kinetic energies.

$$W = \frac{1}{2}mv_2^2 - \frac{1}{2}mv_1^2$$

- Conclude that work done on an object results in a change in the kinetic energy of the object.

$$W = E_{k2} - E_{k1}$$
$$W = \Delta E_k$$

The delta symbol, $\Delta$, denotes change. The expressions

$$W = \Delta E_k$$
$$W = E_{k2} - E_{k1}$$

are mathematical representations of the work-kinetic energy theorem which describes how doing work on an object can change the object's kinetic energy (energy of motion).

The work-kinetic energy theorem is part of the broader **work-energy theorem**. The work-energy theorem includes the concept that work can change an object's potential energy, thermal energy, or other forms of energy.

### Think It Through

- Raj cannot understand why his answer to a problem is wrong. He was trying to calculate the amount of work done by a hockey stick on a hockey puck. The puck was moving slowly when the stick hit it, making it move faster. First, Raj found the difference of the final and initial velocities, and squared the value. He then multiplied his value by the mass of the puck. Finally, he divided by two. Explain to Raj why his answer was wrong. Tell him how to solve the problem correctly.

## QUICK LAB

## Pulling a Cart

### TARGET SKILLS

- **Performing and recording**
- **Analyzing and interpreting**
- **Communicating results**

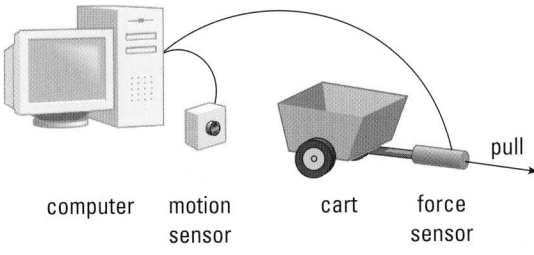

computer   motion     cart    force
           sensor              sensor

A baseball bat does a lot of work on a baseball in a very short period of time. If all goes as the batter plans, the result may be a long fly ball. The work-kinetic energy theorem can predict and quantify the work done on the baseball and its resulting motion. In this investigation, you will test the work-kinetic energy theorem using a slightly more controlled environment than a baseball diamond. Set up the force and motion sensors using one interface, as shown in the

illustration. (Alternatively, use a Newton scale and ticker tape.) Pull on the cart with the force meter and collect data that will allow you to

- generate a force-versus-position graph
- determine the final velocity of the cart

### Analyze and Conclude

1. Determine the total work that you did on the cart.

2. Using your value for work, predict the final speed of the cart.

3. Determine the actual final speed of the cart.

4. Discuss whether your data support the work-kinetic energy theorem.

5. Give possible reasons for any observed discrepancies.

### Science of the Sole

When the world's athletes take a run at the gold in the Olympics, few people know that a Canadian physicist is running in spirit right beside them. Dr. Benno Nigg, a professor at the University of Calgary and founder and chairperson of its Human Performance Laboratory, is a leading expert in biomechanics. Biomechanics is the science of how living things move. With biomechanics, scientists can determine how dinosaurs really walked, tens of millions of years ago. Biomechanics can also shed light on how humans walk, or, more importantly for athletes, how humans run.

It may sound strange, but there is a lot of physics in the design of running shoes. Dr. Nigg's work demands a precise application of forces, energy, and thermal physics. For example, the hardness of the shoe affects which kinds of muscle fibres are activated. This, in turn, will have an effect on the runner's fatigue. So an in-depth understanding of how the mechanical energy of the impact of the shoe is transmitted to the runner is vital. These discoveries are even now being used by athletics-wear companies to design the next generation of running shoe.

Dr. Nigg's interest in this kind of physics has taken him far. He has won awards from around the world, and is a member of the Olympic Order and the International Olympic Committee Medical Commission. He has consulted for all of the largest sportswear and equipment companies. If you look at the qualifications for researchers and technicians at any of these companies, you will see very quickly that almost every position requires a knowledge of biomechanics. This is truly a career that can bring an interest in sports and an interest in science together.

Dr. Nigg will tell you that, when he first entered the field, his main qualifications were a broad educational background and a deep

Dr. Benno Nigg

interest in physics. He says these enabled him to go anywhere: "When you study physics, you open doors."

### Going Further

According to Dr. Nigg, the best way to learn about biomechanics and the careers available is to get hands-on experience. Many students don't realize that very often university and corporate laboratories offer summer programs specifically aimed at high school students. Call a local university or an athletics-equipment manufacturer and see what they have to offer.

### Web Link

**www.school.mcgrawhill.ca/resources/**

The Internet has a number of excellent resources that can help to explain some of the more difficult concepts and considerations involved in determining how runners run and what we can do to help them run faster. For example, Dr. Nigg's *Biomechanigg* page illustrates a number of these concepts. Go to the web site shown above and click on **Science Resources** and then on **Physics 11** to find out where to go next. Try to create your own method for measuring the forces involved in walking and running.

## Applying the Work-Kinetic Energy Theorem

1. A physics student does work on a 2.5 kg curling stone by exerting $4.0 \times 10^1$ N of force horizontally over a distance of 1.5 m.

    (a) Calculate the work done by the student on the curling stone.

    (b) Assuming that the stone started from rest, calculate the velocity of the stone at the point of release. (Consider the ice surface to be effectively frictionless.)

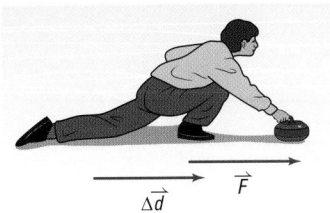

### Frame the Problem

- The curling stone was *initially at rest*; therefore, it had *no kinetic energy*.

- The student did *work* on the stone, giving it *kinetic energy*.

- The *force* exerted by the student was in the *same direction* as the *displacement* of the stone; thus, the equation for work applies.

- Since an ice surface has so little friction, we can *ignore* any effects of *friction*.

- The ice surface is level; therefore, there was no change in the height of the stone.

- The *work-kinetic energy theorem* applies to the problem.

### Identify the Goal

Work done by the student on the stone
Velocity of the stone at release

### Variables and Constants

| Involved in the problem | | Known | Implied | Unknown |
|---|---|---|---|---|
| $F$ | $v_{initial}$ | $F = 4.0 \times 10^1$ N | $v_{initial} = 0 \dfrac{m}{s}$ | $W$ |
| $\Delta d$ | $E_{k(final)}$ | $\Delta d = 1.5$ m | $E_{k(initial)} = 0$ J | $E_{k(final)}$ |
| $W$ | $v_{final}$ | $m = 2.5$ kg | | $v_{final}$ |
| $E_{k(initial)}$ | $m$ | | | |

### Strategy

The formula for work done by a force parallel to the displacement applies.

The values are known, so substitute.

Multiply.

An N·m is equivalent to a J.

### Calculations

$$W = F_{\parallel}\Delta d$$

$$W = (4.0 \times 10^1 \text{ N})(1.5 \text{ m})$$

$$W = 6.0 \times 10^1 \text{ N} \cdot \text{m}$$

$$W = 6.0 \times 10^1 \text{ J}$$

The student did $6.0 \times 10^1$ J of work on the curling stone.

*continued* ▶

## Strategy

Knowing the work, you can use the work-kinetic energy theorem to find final kinetic energy.

## Calculations

$$W = E_{k(final)} - E_{k(initial)}$$

| **Substitute first** | **Solve for $E_{k(final)}$ first** |
|---|---|
| $W = E_{k(final)} - E_{k(initial)}$ | $W = E_{k(final)} - E_{k(initial)}$ |
| $6.0 \times 10^1 \text{ J} = E_{k(final)} - 0 \text{ J}$ | $W + E_{k(initial)} = E_{k(final)}$ |
| $6.0 \times 10^1 \text{ J} + 0 \text{ J} = E_{k(final)}$ | $6.0 \times 10^1 \text{ J} + 0 \text{ J} = E_{k(final)}$ |
| $E_{k(final)} = 6.0 \times 10^1 \text{ J}$ | $E_{k(final)} = 6.0 \times 10^1 \text{ J}$ |

Knowing the final kinetic energy, you can use the formula for kinetic energy to find the final velocity.

$$E_k = \frac{1}{2}mv^2$$

Divide both sides of the equation by the terms beside $v^2$.

| **Substitute first** | **Solve for $v_{final}$ first** |
|---|---|
| $6.0 \times 10^1 \text{ J} = \frac{1}{2}(2.5 \text{ kg})v^2$ | $E = \frac{1}{2}mv^2$ |
| $\dfrac{6.0 \times 10^1 \text{ J}}{\frac{1}{2}(2.5 \text{ kg})} = \dfrac{\frac{1}{2}(2.5 \text{ kg})}{\frac{1}{2}(2.5 \text{ kg})}v^2$ | $\dfrac{E}{\frac{1}{2}m} = v^2$ |
| $48\dfrac{\text{kg·m}^2}{\text{kg·s}^2} = v^2$ | $\dfrac{2E}{m} = v^2$ |

Take the square root of both sides of the equation.

$$\sqrt{48\frac{\text{m}^2}{\text{s}^2}} = v \qquad\qquad \sqrt{\frac{2E}{m}} = v$$

Simplify.

$$v = \pm 6.928 \, \frac{\text{m}}{\text{s}}$$

$$\sqrt{\frac{2(6.0 \times 10^1 \text{ J})}{2.5 \text{ kg}}} = v$$

$$\sqrt{48\frac{\text{kg m}^2}{\text{kg s}^2}} = v$$

$$v = \pm 6.928 \, \frac{\text{m}}{\text{s}}$$

The velocity of the stone, on release, was 6.9 m/s.

## Validate

The student did work on the stationary stone, transferring energy to the stone. The stone's kinetic energy increased and, therefore, so did its velocity, as a result of the work done on it.

Notice that *directional information* (±) is *lost* when the *square root* is taken. You must deter-mine which answer (positive or negative) is correct by choosing the one that logically agrees with the situation. In this case, the direction of the applied force was considered positive; there-fore, any motion resulting from the application of this force will also be positive.

**2.** A 75 kg skateboarder (including the board), initially moving at 8.0 m/s, exerts an average force of $2.0 \times 10^2$ N by pushing on the ground, over a distance of 5.0 m. Find the new kinetic energy of the skateboarder if the trip is completely horizontal.

## Frame the Problem

Make a sketch of the force and motion vectors.

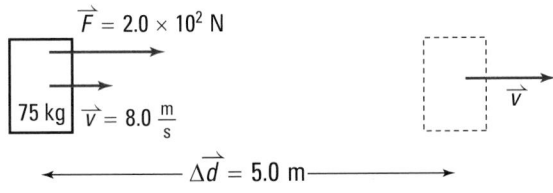

- The skateboarder has an *initial velocity* and, therefore, an *initial kinetic energy.*

- He exerts a force by pushing along the ground. As a result, the *ground exerts a force* on him in the *direction of his motion.*

- The ground *does work* on the skateboarder, thus changing his *kinetic energy.*

- Assuming that the friction is negligible, *all of the work* goes into *kinetic energy.* Therefore, the *work-kinetic energy theorem* applies to the problem.

## Identify the Goal

The final kinetic energy, $E_{k2}$, of the skateboarder

## Variables and Constants

| Involved in the problem | | Known | Unknown |
|---|---|---|---|
| $E_{k1}$ | $W$ | $v = 8.0 \; \dfrac{\text{m}}{\text{s}}$ | $E_{k1}$ |
| $E_{k2}$ | $F$ | $m = 75$ kg | $E_{k2}$ |
| $v$ | $\Delta d$ | $\Delta d = 5.0$ m | $W$ |
| $m$ | | $F = 2.0 \times 10^2$ N | |

A skateboarder does work by pushing on the ground. The work gives the skateboarder kinetic energy.

## Strategy

A tree diagram showing relationships among the variables is often helpful when several steps are involved in obtaining a solution.

You can calculate final kinetic energy if you can find the initial kinetic energy and the work done.

You can calculate the initial kinetic energy if you know the mass and the initial velocity.

The mass and initial velocity are known.

You can calculate the work done if you can find the force and the displacement.

The force and displacement are known.

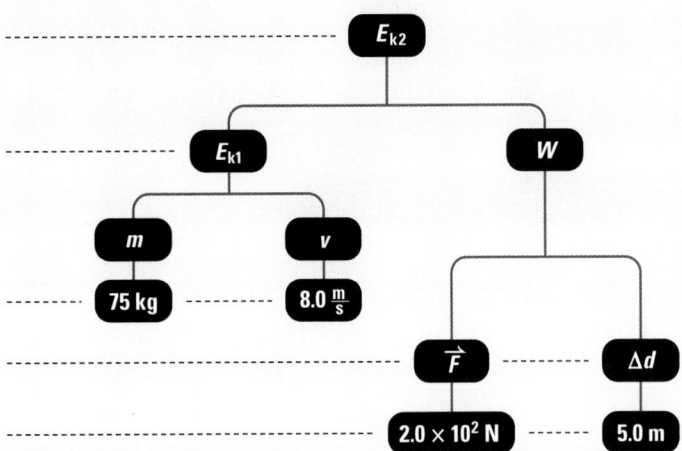

The problem is essentially solved. All that remains is the math.

*continued ▶*

*continued from previous page*

| **Strategy** | **Calculations** |
|---|---|
| Find the initial kinetic energy by using the formula. | $E_k = \frac{1}{2}mv^2$ |
| Substitute. | $E_{k1} = \frac{1}{2}(75 \text{ kg})(8.0 \frac{\text{m}}{\text{s}})^2$ |
| Multiply. | $E_{k1} = 2.4 \times 10^3 \text{ kg} \frac{\text{m}^2}{\text{s}^2}$ |
| Simplify. | $E_{k1} = 2.4 \times 10^3 \text{ kg} \frac{\text{m}}{\text{s}^2} \text{ m}$ |
| 1 N·m is equivalent to 1 J. | $E_{k1} = 2.4 \times 10^3 \text{ N} \cdot \text{m}$ |
| | $E_{k1} = 2.4 \times 10^3 \text{ J}$ |
| Find work by using the formula. | $W = F_{\parallel}\Delta d$ |
| Substitute. | $W = (2.0 \times 10^2 \text{ N})(5.0 \text{ m})$ |
| Multiply. | $W = 1.0 \times 10^3 \text{ N} \cdot \text{m}$ |
| 1 N·m is equivalent to 1 J. | $W = 1.0 \times 10^3 \text{ J}$ |
| Find the final kinetic energy by using the work-kinetic energy theorem. | $W = E_{k2} - E_{k1}$ |

|  | **Substitute first** | **Solve for $E_{k2}$ first** |
|---|---|---|
| Add $E_{k1}$ or its value to both sides of the equation. | $1.0 \times 10^3 \text{ J} = E_{k2} - 2.4 \times 10^3 \text{ J}$ <br> $1.0 \times 10^3 \text{ J} + 2.4 \times 10^3 \text{ J} = E_{k2}$ | $W + E_{k1} = E_{k2}$ <br> $1.0 \times 10^3 \text{ J} + 2.4 \times 10^3 \text{ J} = E_{k2}$ |
| Simplify. | $E_{k2} = 3.4 \times 10^3 \text{ J}$ | $E_{k2} = 3.4 \times 10^3 \text{ J}$ |

The skateboarder's final kinetic energy was $3.4 \times 10^3$ J.

## Validate

The ground did positive work on the skateboarder, exerting a force in the direction of the motion. Intuitively, this work should increase the velocity and by definition, the kinetic energy, which it did.

### PRACTICE PROBLEMS

**22.** A 6.30 kg rock is pushed horizontally across a 20.0 m frozen pond with a force of 30.0 N. Find the velocity of the rock once it has travelled 13.9 m. (Assume there is no friction.)

**23.** The mass of an electron is $9.1 \times 10^{-31}$ kg. At what speed does the electron travel if it possesses $7.6 \times 10^{-18}$ J of kinetic energy?

**24.** A small cart with a mass of 500 g is accelerated, uniformly, from rest to a velocity of 1.2 m/s along a level, frictionless track. Find the kinetic energy of the cart once it has reached a velocity of 1.2 m/s. Calculate the force that was exerted on the cart over a distance of 0.1 m in order to cause this change in kinetic energy.

25. A child's toy race car travels across the floor with a constant velocity of 2.10 m/s. If the car possesses 14.0 J of kinetic energy, find the mass of the car.

26. A 1250 kg car is travelling 25 km/h when the driver puts on the brakes. The car comes to a stop after going another 10 m. What was the average frictional force that caused the car to stop? If the same car was travelling at 50 km/h when the driver put on the brakes and the car experienced the same average stopping force, how far would it go before coming to a complete stop? Repeat the calculation for 100 km/h. Make a graph of stopping distance versus speed. Write a statement that describes the relationship between speed and stopping distance.

## 5.2 Section Review

1. **K/U** State the work-kinetic energy theorem and list three common examples that effectively support the theorem.

2. **C** Discuss how the following supports the work-kinetic energy theorem. A cue ball is at rest on a pool table, and then moves after being struck by a pool cue.

3. **K/U** A pitcher does work, $W$, on a baseball when he pitches it. How much more work would he have to do to pitch the ball three times as fast?

4. **K/U** Two identical cars are moving down a highway. Car X is travelling twice as fast as car Y. Both drivers see deer on the road ahead and apply the brakes. The forces of friction that are stopping the cars are the same. What is the ratio of the stopping distance of car X compared to car Y?

5. **K/U** Two cars, A and B, are moving. B's mass is half that of A and B is moving with twice the velocity of A. Is B's kinetic energy four times as great, twice as great, the same, or half as great as A's kinetic energy?

6. **I** How much force do the tires of a bicycle apply to the pavement when they are braking and/or skidding to a stop? Design and perform an experiment that will help you determine the average braking force indirectly. (It would be very difficult to measure the braking force of skidding tires directly.) Use the work-kinetic energy theorem. It implies that the initial kinetic energy of the bicycle and rider, before the brakes are applied, will be equal to the work done by the brakes and tires in stopping the bicycle.

   (a) What equipment, materials, and tools will you need to determine the initial kinetic energy of the bicycle and rider before applying the brakes? What will you need to determine the stopping distance?

   (b) Develop a procedure that lists all the steps you will follow.

   (c) Repeat the experiment several times with the same rider. Next, repeat it with different riders. What do you conclude?

SECTION
EXPECTATIONS

• Describe the characteristics common to all forms of potential energy.

• Communicate the relationships between work done to an object and the object's change in gravitational potential energy.

• Use equations to analyze situations that change an object's gravitational potential energy.

KEY
TERM

• gravitational potential energy

When you do work on an object, will the object always gain kinetic energy? Are there situations where you do work on an object but leave the object at rest? The work done by the student in the photograph is a clear demonstration that an object may remain motionless after work is done on it. In this section, you will consider how doing work on an object can result in a change in potential energy, rather than in kinetic energy. Potential energy is sometimes described as the energy stored by an object due to its *position* or *condition*.

**Figure 5.19** When you lift groceries onto a shelf, you have exerted a force on the groceries. However, when they are on the shelf, they have no kinetic energy. What form of energy have the groceries gained?

## Potential Energy

Consider the work you do on your physics textbook when you lift it from the floor and place it on the top shelf of your locker. You have exerted a force over a distance. Therefore, you have done work on the textbook and yet it is not speeding off out of sight. The work you did on your textbook is now stored in the book by virtue of its position. Your book has gained potential energy. By doing work against the force of gravity, you have given your book a special form of potential energy called **gravitational potential energy**.

Gravitational potential energy is only one of several forms of potential energy. For example, chemical potential energy is stored in the food you eat. Doing work on an elastic band by stretching it stores elastic potential energy in the elastic band. A battery contains both chemical and electrical potential energy.

# Gravitational Potential Energy

For hundreds of years, people have been using the gravitational potential energy stored in water. Many years ago, people built water wheels like the one shown in Figure 5.20. Today, we create huge reservoirs and dams that convert the potential energy of water into electricity.

**Figure 5.20** Gravitational potential energy is stored in the water. When the water begins to fall, it gains kinetic energy. As it falls, it turns the wheel, giving the wheel kinetic energy.

To determine the factors that contribute to gravitational potential energy, try another "thought experiment." Ask yourself the following questions.

- If a golf ball and a Ping Pong™ ball were dropped from the same distance, which one might you try to catch and which one would you avoid?

- If one golf ball was dropped a distance of 10 cm and another a distance of 10 m, which one would hit the ground harder?

- Which golf ball would hit the surface with the greatest impact, one dropped a distance of 1.0 m on Earth or 1.0 m on the Moon?

Everyday experience tells you that mass and height (vertical distance between the two positions) contribute to an object's gravitational potential energy. Your knowledge of gravity also helps you to understand that $g$, the acceleration due to gravity, affects gravitational potential energy as well.

An important characteristic of all forms of potential energy is that there is no absolute zero position or condition. We measure only changes in potential energy, not absolute potential energy. Physicists must always assign a reference position and compare the potential energy of an object to that position. Gravitational potential energy depends on the difference in height between two positions. Therefore, the zero or reference level can be assigned to any convenient position. We typically choose the reference position as the solid surface toward which an object is falling or might fall.

**PHYSICS FILE**

The constant $g$, the acceleration due to gravity, affects objects even when they are not moving. When something is preventing an object from falling, such as your desk holding up your book, $g$ influences its weight. The weight of an object is its mass times $g$. If nothing were preventing it from falling, your book, or any other object, would accelerate at $9.81 \text{m/s}^2$, the value of $g$. The value of $g$ varies with the size and mass of the planet, moon, or star. On the Moon, for example, your book would weigh less and, if falling, would accelerate at a lower rate ($1.62 \text{m/s}^2$).

The equation for gravitational potential energy in the box on the right is an example of what physicists call a "special case." The numerical value of $g$, $9.81\,\text{m/s}^2$, applies only to cases near Earth's surface. The value would be different out in space or on a different planet. Therefore, because the equation contains $g$, the equation itself applies only to cases close to Earth's surface. For example, you could not use the equation to find the gravitational potential energy of an astronaut in the International Space Station.

## GRAVITATIONAL POTENTIAL ENERGY

Gravitational potential energy is the product of mass, the acceleration due to gravity, and the change in height.

$$E_g = mg\Delta h$$

| Quantity | Symbol | SI unit |
|---|---|---|
| gravitational potential energy | $E_g$ | J (joule) |
| mass | $m$ | kg (kilogram) |
| acceleration due to gravity | $g$ | $\frac{\text{m}}{\text{s}^2}$ (metres per second squared) |
| change in height (from reference position) | $\Delta h$ | m (metre) |

**Unit Analysis**

$(\text{mass})(\text{acceleration})(\text{height}) = \text{kg}\frac{\text{m}}{\text{s}^2}\text{m} = \text{N} \cdot \text{m} = \text{J}$

## MODEL PROBLEM

### Calculating Gravitational Potential Energy

You are about to drop a 3.0 kg rock onto a tent peg. Calculate the gravitational potential energy of the rock after you lift it to a height of 0.68 m above the tent peg.

### Frame the Problem

Make a sketch of the situation.

- You do *work* against gravity when you *lift* the rock.

- All of the work gives *gravitational potential energy* to the rock.

- The expression for *gravitational potential energy* applies.

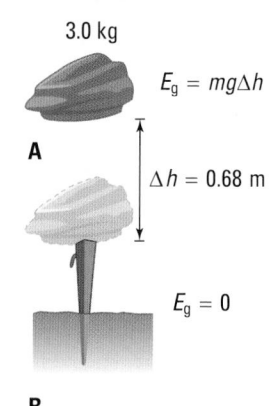

3.0 kg

$E_g = mg\Delta h$

A

$\Delta h = 0.68$ m

$E_g = 0$

B

## Identify the Goal

The gravitational potential energy, $E_g$, of the rock

## Variables and Constants

| Involved in the problem | Known | Implied | Unknown |
|---|---|---|---|
| $E_g$ | $m = 3.0$ kg | $g = 9.81 \frac{m}{s^2}$ | $E_g$ |
| $m$ | $\Delta h = 0.68$ m | | |
| $\Delta h$ | | | |
| $g$ | | | |

| Strategy | Calculations |
|---|---|
| Use the formula for gravitational potential energy. | $E_g = mg\Delta h$ |
| Substitute. | $E_g = (3.0 \text{ kg})(9.81 \frac{m}{s^2})(0.68 \text{ m})$ |
| Multiply. | $E_g = 2.0 \times 10^1 \text{ kg} \frac{m}{s^2} \text{ m}$ |
| | $E_g = 2.0 \times 10^1 \text{ N} \cdot \text{m}$ |
| 1 N · m is equivalent to 1 J. | $E_g = 2.0 \times 10^1 \text{ J}$ |

The rock has $2.0 \times 10^1$ J of gravitational potential energy.

## Validate

Doing work on the rock resulted in a change of position of the rock relative to the tent peg. The work done is now stored by the rock as gravitational potential energy.

### PRACTICE PROBLEMS

27. A framed picture that is to be hung on the wall is lifted vertically through a distance of 2.0 m. If the picture has a mass of 4.45 kg, calculate its gravitational potential energy with respect to the ground.

28. The water level in a reservoir is 250 m above the water in front of the dam. What is the potential energy of each cubic metre of surface water behind the dam? (Take the density of water to be 1.00 kg/L.)

29. How high would you have to raise a 0.300 kg baseball in order to give it 12.0 J of gravitational potential energy?

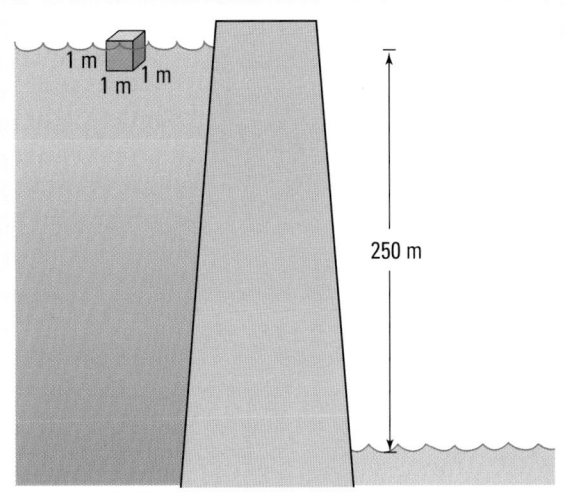

# Work and Gravitational Potential Energy

To develop a mathematical relationship between work and gravitational potential energy, start with the equation for work.

- Work is the product of the force that is parallel to the direction of the motion and the distance that the force caused the object to move.

$$W = F_{\parallel}\Delta d$$

- Recall from Chapter 4 that the force of gravity on a mass near Earth's surface is given by $\vec{F} = m\vec{g}$, where $g = 9.81$ m/s$^2$. Since the force of gravity and the acceleration due to gravity are always downward, and since work is a scalar quantity, we will omit vector notations.

$$F = mg$$

- Substitute $mg$ for $F$ into the expression for work.

$$\therefore W = mg\Delta d$$

- Substitute $\Delta h$ for height in place of $\Delta d$ to emphasize that the displacement vector is vertical.

$$W = mg\Delta h$$

- This is the equation for work done to lift an object to height $\Delta h$, relative to its original position.

$$W = mg\Delta h$$

- The work, $W$, done on the object has become gravitational potential energy stored in the object by virtue of its position.

$$E_g = mg\Delta h$$

Depending on your choice of a reference level, an object may have some gravitational potential energy before you do work on it. For example, choose the floor as your reference. If your book was on the desk, it would have an amount of gravitational potential energy, $mg\Delta h_1$, in relation to the floor, where $\Delta h_1$ is the height of the desk. Then you do work against gravity to lift it to the shelf, where it has gravitational potential energy $mg\Delta h_2$, where $\Delta h_2$ is the height of the shelf. The work you did *changed* the book's gravitational potential. You can describe this change mathematically as

$$W = mg\Delta h_2 - mg\Delta h_1$$
$$W = E_{g2} - E_{g1}$$
$$W = \Delta E_g$$

The mathematical expression above is a representation of the work-energy theorem in terms of gravitational potential energy.

**Figure 5.21** Visualizing potential energy

## • *Think It Through*

- Object A has twice the mass of object B. If object B is 4.0 m above the floor and object A is 2.0 m above the floor, which one has the greater gravitational potential energy?

- If both objects in the question above were lowered 1.0 m, would they still have the same ratio of gravitational potential energies that they had in their original positions? Explain your reasoning.

- You carry a heavy box up a flight of stairs. Your friend carries an identical box on an elevator to reach the same floor as you. Which one, you or your friend, did the greatest amount of work on a box against gravity? Explain your reasoning.

- Examine the photographs in Figure 5.21 on the previous page, "Visualizing potential energy," then answer the following questions.

  1. Name all of the objects in the photographs that clearly illustrate gravitational potential energy that might soon be converted into kinetic energy. Explain why you made your choices.

  2. From your list for the previous question, which object do you think has the most gravitational potential energy? Explain.

  3. List as many examples as you can of forms of potential energy other than gravitational potential energy.

## MODEL PROBLEM

### Applying the Work-Energy Theorem

**A 65.0 kg rock climber did $1.60 \times 10^4$ J of work against gravity to reach a ledge. How high did the rock climber ascend?**

### Frame the Problem

- The rock climber did *work* against *gravity*.

- Work done against gravity *increased* the rock climber's *gravitational potential energy*.

- The *work-energy theorem* that applies to potential energy is appropriate for this situation.

## Identify the Goal

The vertical height, $\Delta h$, that the climber ascended

## Variables and Constants

| Involved in the problem | | Known | Implied | Unknown |
|---|---|---|---|---|
| $W$ | $E_{g1}$ | $W = 1.60 \times 10^4$ J | $g = 9.81 \frac{\text{m}}{\text{s}^2}$ | $E_{g2}$ |
| $E_{g2}$ | $g$ | $m = 65.0$ kg | $E_{g1} = 0$ J | $\Delta h$ |
| $m$ | $\Delta h$ | | | |

## Strategy

Use the work-energy theorem to find the climber's gravitational potential energy from the amount of work done.

Choose the starting point as your reference for gravitational potential energy, so that $E_{g1}$ will be zero. Solve.

Use the value for gravitational potential energy to find the height.

Divide both sides of the equation by the value in front of $\Delta h$.

Simplify.

Convert J to $\text{kg} \frac{\text{m}^2}{\text{s}^2}$, so that you can cancel units.

## Calculations

$$W = E_{g2} - E_{g1}$$

**Substitute first**

$$W = E_{g2} - E_{g1}$$

$$1.6 \times 10^4 \text{ J} = E_{g2} - 0 \text{ J}$$

$$E_{g2} = 1.6 \times 10^4 \text{ J}$$

$$E_{g2} = mg\Delta h$$

$$1.6 \times 10^4 \text{ J} = 65 \text{ kg } 9.81 \frac{\text{m}}{\text{s}^2}\Delta h$$

$$\frac{1.6 \times 10^4 \text{ J}}{65 \text{ kg } 9.81 \frac{\text{m}}{\text{s}^2}} = \frac{65 \text{ kg } 9.81 \frac{\text{m}}{\text{s}^2}\Delta h}{65 \text{ kg } 9.81 \frac{\text{m}}{\text{s}^2}}$$

$$\frac{1.6 \times 10^4 \text{ J}}{65 \text{ kg } 9.81 \frac{\text{m}}{\text{s}^2}} = \Delta h$$

$$\Delta h = 25.09 \frac{\frac{\text{kg} \cdot \text{m}^2}{\text{s}^2}}{\text{kg} \frac{\text{m}}{\text{s}^2}}$$

$$\Delta h = 25.09 \text{ m}$$

**Solve for $E_{g2}$ first**

$$W = E_{g2} - E_{g1}$$

$$W + E_{g1} = E_{g2}$$

$$1.6 \times 10^4 \text{ J} + 0 \text{ J} = E_{g2}$$

$$1.6 \times 10^4 \text{ J} = E_{g2}$$

$$E_{g2} = mg\Delta h$$

$$\frac{E_{g2}}{mg} = \frac{mg\Delta h}{mg}$$

$$\Delta h = \frac{E_{g2}}{mg}$$

$$\Delta h = \frac{1.6 \times 10^4 \text{ J}}{65 \text{ kg}}$$

$$\Delta h = 25.09 \frac{\frac{\text{kg} \times \text{m}^2}{\text{s}^2}}{\frac{\text{kg} \times \text{m}}{\text{s}^2}}$$

$$\Delta h = 25.09 \text{ m}$$

The rock climber ascended 25.1 m.

---

## Validate

The climber did a large amount of work, so you would expect that the climb was quite high.

The units canceled to give m, which is correct for height.

*continued* ▶

## PRACTICE PROBLEMS

**30.** A student lifts her 2.20 kg pile of textbooks into her locker from where they rest on the ground. She must do 25.0 J of work in order to lift the books. Calculate the height that the student must lift the books.

**31.** A 46.0 kg child cycles up a large hill to a point that is a vertical distance of 5.25 m above the starting position. Find

 **(a)** the change in the child's gravitational potential energy

 **(b)** the amount of work done by the child against gravity

**32.** A 2.50 kg pendulum is raised vertically 65.2 cm from its rest position. Find the gravitational potential energy of the pendulum.

**33.** A roller-coaster train lifts its passengers up vertically through a height of 39.4 m from its

starting position. Find the change in gravitational potential energy if the mass of the train and its passengers is $3.90 \times 10^3$ kg.

**34.** The distance between the sixth and the eleventh floors of a building is 30.0 m. The combined mass of the elevator and its contents is $1.35 \times 10^3$ kg.

 **(a)** Find the gravitational potential energy of the elevator when it stops at the eighth floor, relative to the sixth floor.

 **(b)** Find the gravitational potential energy of the elevator when it pauses at the eleventh floor, relative to the eighth floor.

 **(c)** Find the gravitational potential energy of the elevator when it stops at the eleventh floor, relative to the sixth floor.

## 5.3 Section Review

**1.** **K/U** List three forms of potential energy other than gravitational potential energy and give an example of each.

**2.** **K/U** Is gravitational potential energy always measured from one specific reference point? Explain.

**3.** **K/U** Define the term "potential" as it applies to "gravitational potential energy."

**4.** **C** Describe what happens to the gravitational potential energy of a stone dropped from a bridge into a river below. How has the amount of gravitational potential energy changed when the stone is (a) halfway down, (b) three quarters of the way down, and (c) all of the way down?

**5.** **C** Your physics textbook is sitting on a shelf above your desk. Explain what is wrong with the statement, "The gravitational potential energy of the book is 20 J."

**6.** **C** The following is the derivation of the relationship between work and gravitational potential energy.

$$W = F_\parallel \Delta d$$
$$W = mg\Delta d$$
$$W = mg\Delta h$$

 **(a)** Explain why $mg$ could be substituted for force in this derivation but not in the derivation for the relationship between work and kinetic energy.

 **(b)** Explain why $\Delta h$ was substituted for $\Delta d$.

**7.** **K/U** An amount of work, $W$, was done on one ball to raise it to a height $h$. In terms of $W$, how much work must you do on four balls, all identical to the first, to raise them to twice the height $h$?

The word conservation, as it is often used, means "saving" or "taking care of" something. You frequently hear about the need for "conservation of fossil fuels" or "conservation of the wilderness." These two examples share a meaning for conservation that is different from the one used by physicists. As used in physics, conservation means that something remains *constant*. "Conservation of energy" implies that the total energy of an isolated system remains constant. The energy however, may change from one form to another, or it may be transferred from one object to another *within* the system.

## Conservation of Mechanical Energy

In some processes, not only total energy but total *mechanical* energy is conserved, that is, remains constant. The example of a falling rock in Figure 5.22 illustrates such a process. The rock, initially at rest, falls from a height, *h*, above the ground. The rock's gravitational potential energy is being transformed into kinetic energy as it falls farther and faster. Its mechanical energy, however, is being conserved.

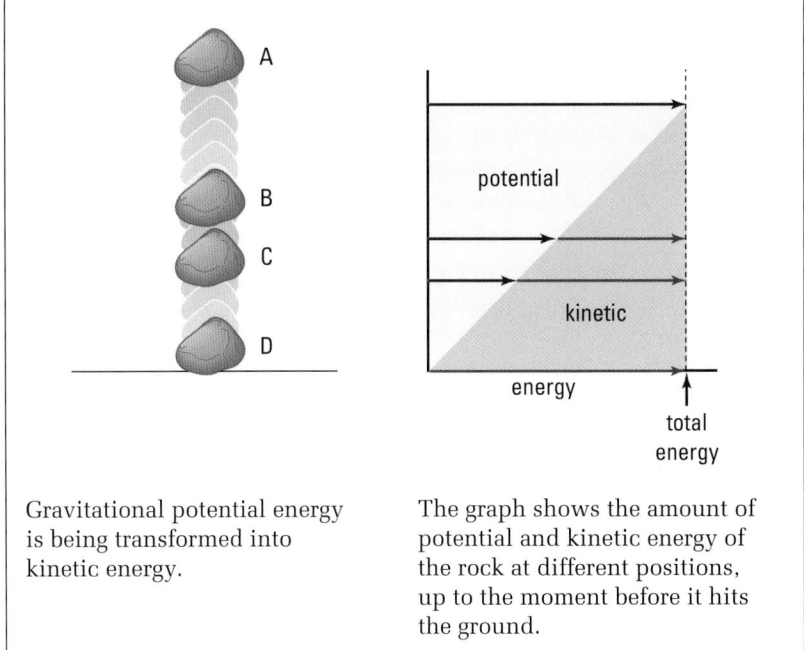

Gravitational potential energy is being transformed into kinetic energy.

The graph shows the amount of potential and kinetic energy of the rock at different positions, up to the moment before it hits the ground.

**Figure 5.22** Although gravitational potential energy is being transformed into kinetic energy, the total mechanical energy of the rock remains the same as it falls.

**SECTION EXPECTATIONS**

- Analyze the relationship between kinetic energy and potential energy.

- Design an experiment to investigate the conservation of mechanical energy.

- Communicate observations that would lead you to conclude that mechanical energy is not conserved.

**KEY TERMS**

- conservative force
- non-conservative force
- law of conservation of mechanical energy
- law of conservation of energy

**PHYSICS FILE**

In physics, a system may be defined in any way that you choose to define it. A system may be one object, such as a rock. It may be a combination of objects. Once you have chosen your system, you must not change your definition while you are making energy calculations, or the rules of physics will no longer apply.

- Imagine an amusement park that has rides like those illustrated here. The masses of all of the cars are identical and the same four people go on each ride. The wheels and track are effectively frictionless. Each car starts from rest at level A. What are the relative speeds of the cars when they reach level B?

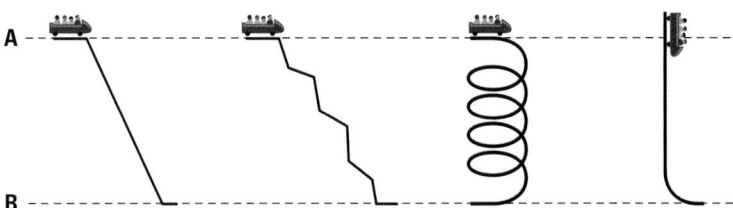

- While playing catch by yourself, you throw a ball straight up as hard as you can throw it. Neglecting air friction, how does the speed of the ball, when it returns to your hand, compare to the speed with which it left your hand?

You can say, intuitively, that the kinetic energy of an object, the instant before it hits a solid surface, is the same as the gravitational potential energy at the point from which it fell from rest. However, physicists try to show such relationships mathematically. Then, they can make calculations based on the principles. Start with the general expression for kinetic energy, $E_k = \frac{1}{2}mv^2$, and show how the kinetic energy at the end of a free fall relates to the gravitational potential energy, $E_g = mg\Delta h$, at a distance $\Delta h$ above the surface.

The first step in deriving this relationship is to find the velocity of an object after it falls a distance, $\Delta h$.

- Recall from Chapter 2 the equation of motion for uniform acceleration.

$$v_2^2 = v_1^2 + 2a\Delta d$$

- The object was initially at rest.

$$v_1 = 0$$
$$v_2^2 = 0 + 2a\Delta d$$

- Substitute $\Delta h$, representing height, for $\Delta d$.

$$\Delta d = \Delta h$$
$$v_2^2 = 0 + 2a\Delta h$$

- Since the only force acting on the object is gravity, the object's acceleration is $g$, the acceleration due to gravity.

$$a = g$$

- This is the square of the velocity of the object, just before hitting a surface, after falling through height $\Delta h$.

$$v_2^2 = 2g\Delta h$$

- Substitute the previous expression for $v^2$ into the general expression for kinetic energy.

$$E_k = \frac{1}{2}mv^2$$

- The kinetic energy of the object just before it hits the surface becomes

$$E_k = \frac{1}{2}m(2g\Delta h)$$

- Notice that the kinetic energy just before the object hits the solid surface is identical to

$$E_k = mg\Delta h$$

- The equation for gravitational potential energy a distance $\Delta h$ above the surface is

$$E_g = mg\Delta h$$

The total energy of the object is not only the same at the top and at the bottom, but throughout the entire trip. The mathematical process used above could be applied to any point throughout the fall. You could show that, for any point within the fall, the gravitational potential energy that was lost up to that point has been transformed into kinetic energy. Therefore, the sum of the kinetic and potential energies remains constant at all times throughout the fall.

Obviously, when an object hits a solid surface and stops, the mechanical energy is no longer conserved. Also, any effects of air friction were neglected in the discussion. Although the total energy of an isolated system is always conserved, mechanical energy is not necessarily conserved. Is there any specific characteristic of a process that you could use to determine whether or not mechanical energy will be conserved?

## Conservative and Non-Conservative Forces

If you lift your book one metre above a table and release it, it will drop back onto the table, gaining kinetic energy as it falls. If you push your book across the table, will it automatically return to its original spot, gaining kinetic energy as it moves? Of course not. In the first case, you were doing work against the gravitational force. In the second case, you were doing work against a frictional force.

These two forces, gravity and friction, represent two important classes of forces. Before defining these classes of forces, consider another property of doing work against them. If you lift your book to a certain height, then carry it across the room, you have done the same amount of work that you would do if you simply lifted it straight up. However, if you push your book from side to side as you move it from one end of the table to the other, you have done more work than you would if you pushed it in a straight line. The amount of work you do against friction depends on the path through which you push the object.

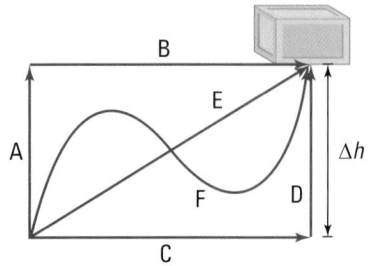

**Figure 5.23** Regardless of the path taken, the work done to lift the box to height $\Delta h$ is identical. Work done against a conservative force is *independent* of the path.

A **conservative force** is one that does work on an object so that the amount of work done is *independent* of the path taken. The force of gravity is an example of a conservative force because it takes the same amount of work to lift a mass to height $\Delta h$, regardless of the path. The work done, and therefore the gravitational potential energy, depend only on the height $\Delta h$.

Friction, on the other hand, is not a conservative force. The work done against a frictional force when a crate is pushed across a rough floor depends on whether the path is straight, or curved, or zigzagged. The work done *is* path-dependent, and therefore the force of friction is **non-conservative**.

Work done by conservative forces results in energy changes that are independent of the path and are therefore reversible. Work done by non-conservative forces results in energy changes that are dependent on the path, and therefore may not be reversed. We can now state the conditions under which mechanical energy is conserved. This statement is called the **law of conservation of mechanical energy**.

---

**LAW OF CONSERVATION OF MECHANICAL ENERGY**

The *total mechanical energy* of a system always remains constant if work is done by conservative forces.

$$E_T = E_g + E_k$$

where $\begin{cases} E_T \text{ is the total mechanical energy of the system.} \\ E_g \text{ is the gravitational potential energy of the system.} \\ E_k \text{ is the mechanical kinetic energy of the system.} \end{cases}$

---

**Table 5.2** Examples of Conservative and Non-Conservative Forces

| Conservative forces | Non-conservative forces |
|---|---|
| gravity | friction |
| electric | air resistance (a specific example of a frictional force) |
| magnetic | |
| nuclear | any applied thrust force (e.g. a rocket or a motor) |
| elastic* | |

*Items such as springs and elastic bands are not "perfectly elastic"; therefore, the forces they exert are not truly conservative.

### Think It Through

- A marble oscillates back and forth in a U-shaped track, repeatedly transferring gravitational potential energy to kinetic energy, and back to gravitational potential energy. Does the law of conservation of mechanical energy apply? Defend your reasoning by discussing conservative and non-conservative forces.

## Applying the Law of Conservation of Mechanical Energy

A crane lifts a car, with a mass of $1.5 \times 10^3$ kg, at a constant velocity, to a height of 14 m from the ground. It turns and drops the car, which then falls freely back to the ground. Neglecting air friction, find

(a) the work done by the crane in lifting the car

(b) the gravitational potential energy of the car at its highest point, in relation to the ground

(c) the velocity of the car just before it strikes the ground after falling freely for 14 m

**PROBLEM TIP**

**Strategy for Solving Energy Transformation Problems**

1. Always draw a diagram of the system you are analyzing.

2. Label the initial and final positions.

3. Write an equation to represent the total energy at each position.

4. Describe and write equations to link energy totals at each point.

5. Solve for the unknowns.

## Frame the Problem

Make a sketch of all of the actions in the problem.

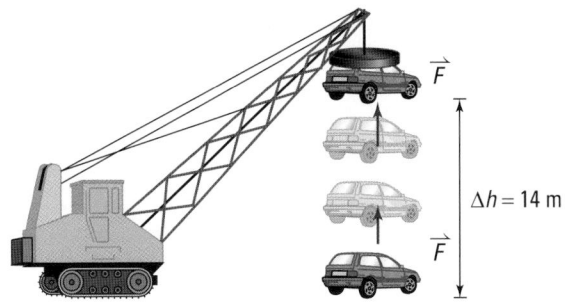
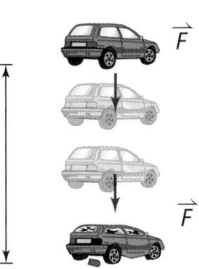

- As the crane lifts the car, it is *doing work* on the car, against the force of gravity. Since the crane is moving the car at a constant velocity, the *forces* are *balanced*. The force exerted by the crane is equal in *magnitude* and *opposite in direction* to the force of gravity.

- The work done by the crane gives *gravitational potential energy* to the car.

- When the car is *falling*, the force of *gravity is doing work* on the car.

*continued* ▶

- The *work done* on the car by gravity gives the car *kinetic energy*.

- Since we are neglecting air friction, the expression for *conservation of mechanical energy* applies to this problem.

- The *total mechanical energy* of the car is the same as the *work done* on the car by the crane. The work done on the car by the crane gave the car *mechanical energy*.

## Identify the Goal

(a) Work, $W$, done by the crane

(b) Gravitational potential energy, $E_g$, of the car at the highest point

(c) Kinetic energy, $E_k$, of the car just before it hits the ground

## Variables and Constants

| Involved in the problem | | Known | Unknown |
|---|---|---|---|
| $W$ | $v$ | $m = 1.5 \times 10^3$ kg | $W$ |
| $F_{\text{crane}}$ | $m$ | $\Delta h = 14$ m | $F_{\text{crane}}$ |
| $\Delta h$ | $F_g$ | **Implied** | $F_g$ |
| $E_{g(\text{top})}$ | $E_{k(\text{top})}$ | $g = 9.81 \frac{\text{m}}{\text{s}^2}$ | $E_{g(\text{top})}$ |
| $E_{g(\text{bottom})}$ | $E_{k(\text{bottom})}$ | $E_{k(\text{top})} = 0$ J | $E_\text{T}$ |
| $E_\text{T}$ | | $E_{g(\text{bottom})} = 0$ J | $E_{k(\text{bottom})}$ |
| $g$ | | | $v$ |

> **PROBLEM TIP**
>
> Often when you are solving conservation of mechanical energy problems, the mass variable can easily be eliminated, as in the following example.
>
> | $E_\text{T} = E_g$ | at the top |
> |---|---|
> | $E_\text{T} = E_k$ | just before impact |
>
> $$E_{g(\text{top})} = E_{k(\text{bottom})}$$
> $$\cancel{m}g\Delta h = \frac{1}{2}\cancel{m}v^2 \quad \text{each } m \text{ cancels}$$
>
> Solving for one of the other variables, $\Delta h$ or $v$, is now easier.

## Strategy

Find the work done by the crane on the car by using the equation for work done by a force parallel to the direction of the displacement.

The force exerted by the crane on the car is $F_{\text{crane}}$. It is equal to the weight of the car, $F_g$.

All of the values are known, so substitute into the equation.

Multiply.

1 kg $\frac{\text{m}}{\text{s}^2}$ is equivalent to 1 N.

1 N · m is equivalent to 1 J.

(a) The crane did $2.1 \times 10^5$ J of work on the car.

## Calculations

$$W = F_\parallel \Delta d$$

$$W = F_{\text{crane}} \, \Delta h$$
$$W = F_g \, \Delta h$$
$$W = mg\Delta h$$

$$W = (1.5 \times 10^3 \text{ kg})(9.81 \tfrac{\text{m}}{\text{s}^2})(14 \text{ m})$$

$$W = 2.06 \times 10^5 \text{ kg} \tfrac{\text{m}}{\text{s}^2}\text{m}$$

$$W = 2.06 \times 10^5 \text{ N} \cdot \text{m}$$

$$W = 2.06 \times 10^5 \text{ J}$$

## Strategy

Find the gravitational potential energy of the car at its highest point by using the expression for gravitational potential energy.

All of the values are known, so substitute into the equation.

Multiply.

$1 \text{ kg} \frac{\text{m}}{\text{s}^2}$ is equivalent to 1 N.

$1 \text{ N} \cdot \text{m}$ is equivalent to 1 J.

**(b)** The car had $2.1 \times 10^5$ J of gravitational potential energy at its highest point..

The values needed to calculate the velocity of the car just before it hits the ground are not all known. In cases such as this, a tree diagram is often helpful in determining the values that you need

You can calculate the velocity of the car just before it hits the ground if you know the car's mass and kinetic energy just before it hits the ground.

The mass is known.

You can calculate the kinetic energy of the car just before it hits the ground if you know the total mechanical energy and the gravitational potential energy of the car at that point.

The gravitational potential energy at the bottom is zero.

You can calculate the total energy of the car if you know both the gravitational potential energy and the kinetic energy at the same point.

Since the car was not in motion at the top, the kinetic energy at that point was zero. You found the gravitational potential energy at the top in the second part of the problem.

## Calculations

$$E_g = mg\Delta h$$

$$E_{g(\text{top})} = \left(1.5 \times 10^3 \text{ kg}\right)\left(9.81 \, \frac{\text{m}}{\text{s}^2}\right)\left(14 \text{ m}\right)$$

$$E_{g(\text{top})} = 2.06 \times 10^5 \text{ kg} \, \frac{\text{m}}{\text{s}^2}\text{m}$$

$$E_{g(\text{top})} = 2.06 \times 10^5 \text{ N} \cdot \text{m}$$

$$E_{g(\text{top})} = 2.06 \times 10^5 \text{ J}$$

> ### PROBLEM TIP
>
> When you report an answer to any part of a problem, you must report it using the correct number of significant digits. However, when you use that answer in a subsequent part of the problem, use the value you calculated before rounding to the proper number of significant digits. Only round a number when you have completed the calculation and are reporting a solution. This will lead to a more accurate final solution.

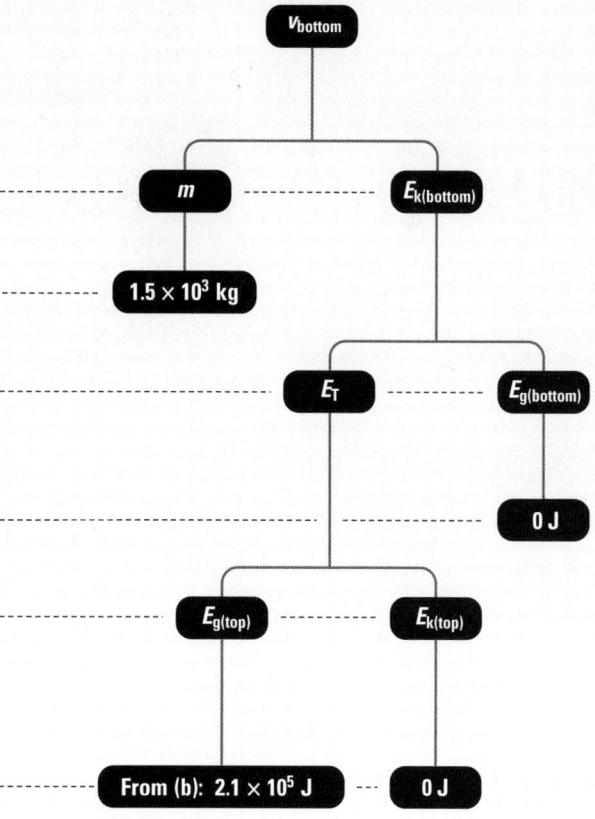

*continued* ▶

*continued from previous page*

All of the needed values are known. All that remains is to do the calculations.

| **Strategy** | **Calculations** |
|---|---|

Find $E_T$ using the equation for conservation of mechanical energy.

$$E_T = E_{g(top)} + E_{k(top)}$$

All of the needed values are known, so substitute into the equation.

$$E_T = 2.06 \times 10^5 \text{ J} + 0 \text{ J}$$
$$E_T = 2.06 \times 10^5 \text{ J}$$

Knowing $E_T$, use the equation for conservation of mechanical energy to find $E_{k(bottom)}$.

$$E_T = E_{g(bottom)} + E_{k(bottom)}$$

Substitute.

$$2.06 \times 10^5 \text{ J} = 0 \text{ J} + E_{k(bottom)}$$
$$E_{k(bottom)} = 2.06 \times 10^5 \text{ J}$$

The kinetic energy of the car just before it hit the ground was $2.1 \times 10^5$ J.

Use $E_{k(bottom)}$ and the equation for kinetic energy to find $v$ at the bottom of the fall, just before the car hit the ground.

$$E_k = \frac{1}{2}mv^2$$

**Substitute first**

**Solve for *v* first**

All of the needed values are known, so you can solve the problem.

$$2.06 \times 10^5 \text{ J} = \frac{1}{2}(1.5 \times 10^3 \text{ kg})v^2$$

$$\frac{2E_{k(bottom)}}{m} = v^2$$

Isolate $v^2$.

$$\frac{2.06 \times 10^5 \text{ J}}{\frac{1}{2}(1.5 \times 10^3 \text{ kg})} = \frac{\frac{1}{2}\cancel{(1.5 \times 10^3 \text{ kg})}\,v^2}{\cancel{\frac{1}{2}(1.5 \times 10^3 \text{ kg})}}$$

$$\sqrt{\frac{2E_{k(bottom)}}{m}} = v$$

$$\sqrt{\frac{2(2.06 \times 10^5 \text{ J})}{1.5 \times 10^3 \text{ kg}}} = v$$

Take the square root of both sides of the equation.

$$v^2 = 2.8 \times 10^2 \; \frac{\text{kg m}^2}{\text{s}^2 \, \cancel{\text{kg}}}$$

$$v = \sqrt{2.8 \times 10^2 \; \frac{\text{m}^2}{\text{s}^2}}$$

$$v = \pm 16.7 \; \frac{\text{m}}{\text{s}}$$

$$v = \sqrt{2.8 \times 10^2 \; \frac{\text{N} \cdot \text{m}}{\text{kg}}}$$

$$v = \sqrt{2.8 \times 10^2 \; \frac{\cancel{\text{kg}} \cdot \text{m} \, \text{m}}{\cancel{\text{kg}} \, \text{s}^2}}$$

$$v = \pm 16.7 \; \frac{\text{m}}{\text{s}}$$

By taking the square root, directional information was lost. Since you know that the car was falling, select the negative sign for the answer.

**(c)** The velocity of the car just before it hit the ground was $-1.7 \times 10^1$ J.

## Validate

As the crane lifts the car, the energy total of the system increases until it reaches a maximum at the highest point (14 m), because work is being done on the system (car). Since the work done on the car transfers energy to the car, you would expect the gravitational potential energy of the car to be the same as the work done by the crane on the car. It is.

In every case, the units cancelled to give the correct units: joules for energy and metres per second for velocity.

**PRACTICE PROBLEMS**

**35.** You throw a ball directly upward, giving it an initial velocity of 10.0 m/s. Neglecting friction, what would be the maximum height of the ball? (Explain why you do not need to know the mass of the ball.)

**36.** A $4.0 \times 10^4$ kg roller coaster starts from rest at point A. Neglecting friction, calculate its potential energy relative to the ground, its kinetic energy, and its speed at points B, C, and D.

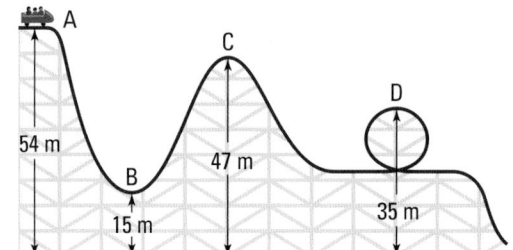

**37.** A wrecking ball, with a mass of 315 kg, hangs from a crane on 10.0 m of cable. If the crane swings the wrecking ball so that the angle that the cable makes with the vertical is 30.0°, what is the potential energy of the wrecking ball in relation to its lowest position? What will be the kinetic energy of the wrecking ball when it falls back to the vertical position? What will be the speed of the wrecking ball?

**38.** A 2.5 kg lead ball and a 55 g piece of lead shot are both dropped from a height of 25 m. Neglecting air friction, what is the kinetic energy of each object just before it hits the ground? What is the velocity of each object just before it hits the ground?

**39.** A 32 kg crate was pushed down a frictionless ramp. Its initial velocity at the top of the ramp was 3.2 m/s. Its velocity when it reached the bottom of the ramp was 9.7 m/s. The ramp makes an angle of 25° with the horizontal. How long is the ramp?

**40.** A worker on the roof of a building that is under construction dropped a 0.125 kg wrench over the edge. Another workman on the eighth floor saw the wrench go by and determined that its speed at that level was 33.1 m/s. The first floor of the building is 12.0 m high and each successive floor is 8.00 m high. Neglecting air friction, how many floors does the building have? How fast was the wrench falling just before it hit the ground? What was its kinetic energy just before it hit the ground?

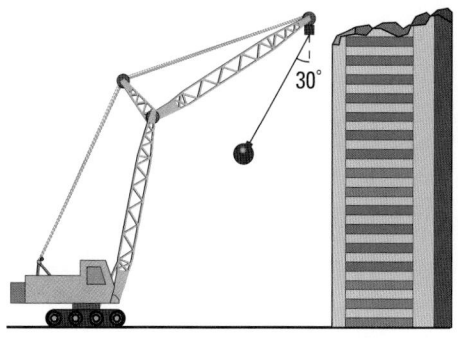

**ELECTRONIC LEARNING PARTNER**

A video clip of a wrecking ball in action can be found on the Electronic Learning Partner.

## Pole-Vaulting: The Art of Energy Transfer

Athlete Stacy Dragila won a gold medal for pole-vaulting at the 2000 Olympic Games in Sydney, the first games to feature women's vaulting. During her winning vaults she demonstrated a fundamental principle of physics — the conservation of energy. Dragila has speed, power, and near-perfect technique. When she vaults, she uses and transfers energy as efficiently as possible.

Before a pole-vaulter begins to sprint toward the bar, she is storing chemical energy in her muscles. If the vaulter is well trained, she will be able to use the chemical energy as efficiently as possible. This chemical energy is converted into kinetic energy as she sprints toward the bar. Near the end of the sprint, the vaulter places her pole in a socket. Kinetic energy from the sprint bends the pole and the pole gains elastic potential energy. When the pole begins to extend from its bend, it converts the elastic energy into gravitational potential energy, lifting the vaulter into the air.

Ideally, a vaulter would be able to convert all of her kinetic energy into elastic energy. But not even Dragila has perfect technique. Even if she did, some of the energy would still be converted to forms that do not perform useful work. Some kinetic energy is converted to thermal and sound energy during the sprint, and some elastic energy is absorbed by the pole as thermal energy during the lift.

The important thing to remember is that the amount of energy involved in the pole vault remains constant. The type of energy and the efficiency with which the athlete uses the energy make the difference between champions such as Dragila and other pole-vaulters.

Olympic gold medallist Stacy Dragila in action

### Analyze

1. Sporting events are sometimes held at high altitudes, such as the 1968 Olympic Games in Mexico City. The thinner air poses challenges for some athletes, but pole-vaulters, and also long-jumpers and high-jumpers, tend to benefit. Why?

2. Pole-vaulting began as an inventive way to get across rivers and other barriers. The first poles were made of bamboo. Today's poles are much more sophisticated. What principles need to be considered when designing a pole?

## Conservation of Mechanical Energy

**TARGET SKILLS**

- Initiating and planning
- Predicting
- Performing and recording
- Analyzing and interpreting

You have learned about mechanical energy and worked problems based on the conservation of energy. Now you will design an experiment that will test the law of conservation of mechanical energy. A well-designed pendulum has such a small amount of friction that it can be neglected.

### Problem

How can you verify the conservation of mechanical energy for a system in which all of the work is done by conservative forces?

### Prediction

Make predictions about the type of energy transformations that will occur and when, during the cycle, they will occur.

### Equipment

- parts for assembling a pendulum
- motion sensor
- metre stick or other measuring device

### Procedure

Write your own procedure for testing the law of conservation of energy using a pendulum. Clearly describe the required materials and the proposed procedure.

Verify, with the teacher, that the experiment is safe before proceeding.

Make detailed predictions identifying the positions at which the mechanical energy of the pendulum will be (a) all gravitational potential, (b) all kinetic, and (c) half gravitational potential and half kinetic.

### Analyze and Conclude

1. Which variables did you control in the experiment?

2. How might you improve your experiment?

3. How well did your results verify the law of conservation of energy? Was any discrepancy that you found within limits that are acceptable to you?

4. If mechanical energy was lost, explain how and why it might have been lost.

### Apply and Extend

5. The transformation of energy from one form to another is never completely efficient. For example, predict what will happen to the chemical potential energy of a car's fuel as it is transferred to the kinetic energy of the motion of the car.

# Conservation of Total Energy

The law of conservation of mechanical energy was generated using Newton's laws of motion. Evidence also confirms the derived results and, in fact, allows the law of conservation of energy to be expanded to include work done by non-conservative forces.

Once again, consider the car in the model problem. This time do not neglect air resistance. As the car falls through the atmosphere, it collides with air molecules. The resulting friction produces heat or thermal energy, similar to the results observed when you rub your hands together. This frictional force acts through the entire distance that the car falls. Since the frictional force is in the opposite direction to the force of gravity, it does negative work on the car. Therefore, some of the gravitational potential energy is transferred into thermal energy of the surroundings. This loss of potential energy to thermal energy reduces the amount of kinetic energy gained by the car. In this case, since friction, a non-conservative force, has done work, mechanical energy is not conserved. Careful experimentation with sensitive equipment has verified that if all forms of energy within the system, including the thermal energy of air molecules and the increased thermal energy of the car, are added together, the total energy of the system remains constant. These observations led to a fundamental principle of physics, the **law of conservation of energy**.

## LAW OF CONSERVATION OF ENERGY

Energy is neither created nor destroyed. It simply changes form or is transferred from one body to another. The total energy of an isolated system, including all forms of energy, always remains constant.

In summary, mechanical energy is conserved if work done by non-conservative forces is zero ($W_{nc} = 0$). In this case, total mechanical energy will remain constant and be the sum of the kinetic and potential energies. If the work done by *non-conservative forces is not zero*, then mechanical energy is *not conserved*.

ELECTRONIC
LEARNING PARTNER

Your Electronic Learning Partner has an interactive simulation to show how energy conservation relates to pendulums and roller coasters.

## WORK AND NON-CONSERVATIVE FORCES

The work done by non-conservative forces is the difference of the final mechanical energy of a system and the initial mechanical energy of a system.

$$W_{nc} = E_{final} - E_{initial}$$

where $\begin{cases} W_{nc} \text{ is the work done by non-conservative forces.} \\ E_{initial} \text{ is the initial mechanical energy of the system.} \\ E_{final} \text{ is the final mechanical energy of the system.} \end{cases}$

## Concept Organizer

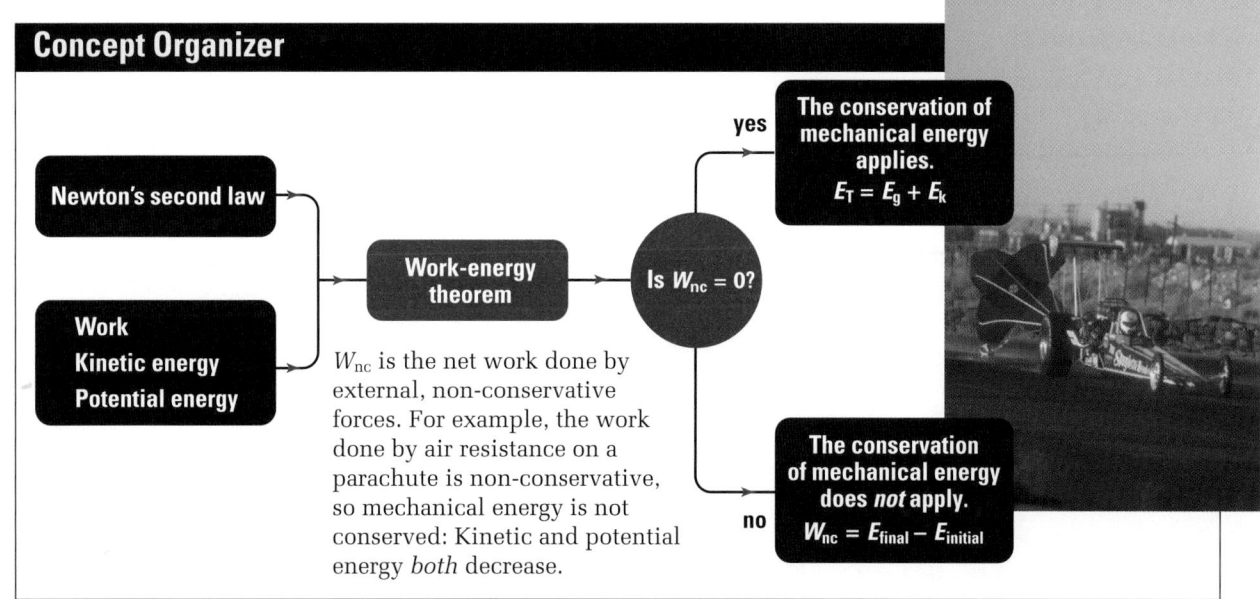

**Figure 5.24** Applying the work-energy theorem

---

MODEL PROBLEM

### Work Done by Air Friction

A 65.0 kg skydiver steps out from a hot air balloon that is $5.00 \times 10^2$ m above the ground. After free-falling a short distance, she deploys her parachute, finally reaching the ground with a velocity of 8.00 m/s (approximately the speed with which you would hit the ground after having fallen a distance of 3.00 m).

(a) Find the gravitational potential energy of the skydiver, relative to the ground, before she jumps.

(b) Find the kinetic energy of the skydiver just before she lands on the ground.

(c) How much work did the non-conservative frictional force do?

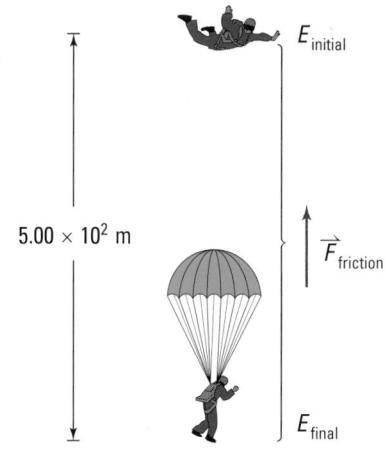

---

### Frame the Problem

- Make a sketch of the problem and label it.

- The hot air balloon can be considered to be at rest, so the vertical velocity of the skydiver initially is *zero*.

- As she starts to fall, the *force of air friction* does *negative* work on her, converting much

of her *gravitational potential energy* into *thermal energy*.

- The rest of her *gravitational potential energy* was converted into *kinetic energy*.

*continued* ▶

*continued from previous page*

## Identify the Goal

Initial gravitational potential energy, $E_g$, of the skydiver relative to the ground

Final kinetic energy, $E_k$, of the skydiver just before touchdown
Work done by non-conservative force, $W_{nc}$

## Variables and Constants

| Involved in the problem | | Known | Unknown |
|---|---|---|---|
| $E_{initial}$ | $W_{nc}$ | $m = 65.0$ kg | $E_{initial}$ |
| $E_g$ | $E_{final}$ | $g = 9.81 \frac{m}{s^2}$ | $E_g$ |
| $m$ | $E_k$ | $\Delta h = 5.00 \times 10^2$ m | $W_{nc}$ |
| $g$ | $v$ | $v = 8.00 \frac{m}{s^2}$ | $E_{final}$ |
| $\Delta h$ | | $E_k$ | |

## Strategy

Use the equation for gravitational potential energy. All of the needed variables are known, so substitute into the equation.

Multiply.

1 kg $\frac{m}{s^2}$ is equivalent to 1 N.

1 N · m is equivalent to 1 J.

(a) The skydiver's initial gravitational energy, before starting the descent, was $3.19 \times 10^5$ J.

Use the skydiver's velocity just before touchdown to find the kinetic energy before touchdown. All of the needed values are known, so substitute into the equation for kinetic energy, then multiply.

1 kg $\frac{m^2}{s^2}$ is equivalent to 1 J.

(b) The skydiver's kinetic energy just before landing was $2.08 \times 10^3$ J.

Use the equation for the conservation of total energy to find the work done by air friction. The final mechanical energy is entirely kinetic energy and the initial mechanical energy was entirely gravitational potential energy. These values were just calculated.

(c) Air friction did $3.17 \times 10^5$ J of negative work on the skydiver.

## Calculations

$$E_g = mg\Delta h$$

$$E_g = (65.0 \text{ kg})(9.81 \tfrac{m}{s^2})(5.00 \times 10^2 \text{ m})$$

$$E_g = 3.188 \times 10^5 \text{ kg} \tfrac{m}{s^2} \text{ m}$$

$$E_g = 3.188 \times 10^5 \text{ N} \cdot \text{m}$$

$$E_g = 3.188 \times 10^5 \text{ J}$$

$$E_k = \tfrac{1}{2}mv^2$$

$$E_k = \tfrac{1}{2}(65.0 \text{ kg})(8.00 \tfrac{m}{s})^2$$

$$E_k = 2.080 \times 10^3 \text{ kg} \tfrac{m^2}{s^2}$$

$$E_k = 2.080 \times 10^3 \text{ J}$$

$$W_{nc} = E_{final} - E_{initial}$$
$$W_{nc} = 2.080 \times 10^3 \text{ J} - 3.188 \times 10^5 \text{ J}$$
$$W_{nc} = -3.167 \times 10^5 \text{ J}$$

## Validate

The $3.17 \times 10^5$ J of work done by air resistance is very large. This is fortunate for the skydiver, as it allows her to land softly on the ground, having lost 99.3 percent of her mechanical energy. This large loss of mechanical energy is what a parachute is designed to do.

How fast would the skydiver in the model problem be travelling just before hitting the ground if the work done by air resistance was ignored?

## PRACTICE PROBLEMS

**41.** A 0.50 kg basketball falls from a 2.3 m shelf onto the floor, then bounces up to a height of 1.4 m before you catch it.

**(a)** Calculate the gravitational potential energy of the ball before it falls.

**(b)** Ignoring frictional effects, determine the speed of the ball as it strikes the floor, assuming that it fell from rest.

**(c)** How fast is the ball moving just before you catch it?

**42.** A 2.0 g bullet initially moving with a velocity of 87 m/s passes through a block of wood. On exiting the block of wood, the bullet's velocity is 12 m/s. How much work did the force of friction do on the bullet as it passed through the wood? If the wood block was 4.0 cm thick, what was the average force that the wood exerted on the block?

**43.** The Millennium Force, the tallest roller coaster in North America, is 94.5 m high at its highest point. What is the maximum pos-

sible speed of the roller coaster? The roller coaster's actual maximum speed is 41.1 m/s. What percentage of its total mechanical energy is lost to thermal energy due to friction?

**44.** A 15 kg child slides, from rest, down a playground slide that is 4.0 m long, as shown in the figure. The slide makes a 40° angle with the horizontal. The child's speed at the bottom is 3.2 m/s. What was the force of friction that the slide was exerting on the child?

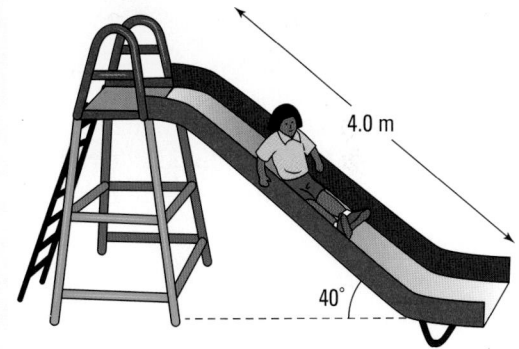

In this chapter, you have focussed on mechanical energy. As you continue to study other forms of energy, you will discover that many of the principles that you learned here will apply to these other forms of energy. For example, in the next chapter, you will learn how to quantify the energy that is "lost" from a system when work is done by non-conservative forces such as friction. It may seem obvious to you now that friction generates heat or thermal energy. However, the nature of heat was hotly debated by scientists for many years.

1. **K/U** How are conservation of mechanical energy and conservation of energy different?

2. **C** An extreme sport in-line skating competition often involves aerial tricks at the top of a curved ramp that has the shape of a cylinder cut in half. Describe how this activity demonstrates the concept of conservation of mechanical energy.

3. **C** How is the conservation of energy demonstrated if the in-line skater falls during the competition mentioned in question 2?

4. **K/U** A skier starts from rest at the top of a ski jump, skis down, and leaves the bottom of the jump with a velocity $v$. Assuming that the ski jump is frictionless, how many times higher would the jump have to be for the skier to jump with twice the velocity?

5. **K/U** A bullet is fired into a block of wood and penetrates to a distance $\Delta d$. An identical bullet is fired into a block of wood with a velocity three times that of the first. How much farther, in terms of $\Delta d$, does the second bullet penetrate the wood?

6. **MC** Some Olympic sports transform kinetic energy into gravitational potential energy of the athlete's body to aid performance. These sports include:

- bob sledding
- high diving off a diving board
- pole vaulting
- snow ski racing
- snow ski jumping
- trampoline jumping

(a) In which of these sports can you increase your gravitational potential energy by using better sports equipment? (for example, a better diving board will allow you to jump higher).

(b) Suppose you can use better materials to produce equipment with greater elastic potential energy. Which sports will be affected?

**ELECTRONIC LEARNING PARTNER**

Go to the Electronic Learning Partner to quiz yourself on these concepts.

**UNIT PROJECT PREP**

Athletic activities from walking to skydiving involve work being done by conservative and non-conservative forces.

- Identify conservative and non-conservative forces involved with your project topic.
- How are non-conservative forces used in the design of sports equipment to increase safety?

## REFLECTING ON CHAPTER 5

- Work is the transfer of energy from one system to another, or from one form to another. The equation $W = F_{\parallel}\Delta d$ applies to work done by a constant force that is parallel to the direction of the motion.

- When the force is not constant, the work done can be estimated from the area under the curve of a force-versus-position graph.

- When the force is not parallel to the displacement, only the component of the force that is in the direction of the displacement does work: $W = F\Delta d\cos\theta$, where $\theta$ is the angle between the direction of the force and displacement vectors.

- Positive work on a system adds energy to the system.

- Negative work on a system removes energy from the system.

- Kinetic energy is the energy of motion.
$$E_k = \tfrac{1}{2}mv^2$$

- Gravitational potential energy of a system is its energy stored due to its position above a reference level. $E_g = mg\Delta h$, where $\Delta h$ is the vertical distance between the system and the reference level.

- Mechanical energy of a system is conserved when work is done by conservative forces.

- Total energy is conserved even when work is done by non-conservative forces. The work done by non-conservative forces decreases the mechanical energy of the system.
$$W_{nc} = E_{final} - E_{initial}$$

- The law of conservation of energy states that the total energy of an isolated system is *always* conserved.

### Knowledge/Understanding

1. What is energy?
2. How would you describe a physicist's concept of work to a non-physicist?
3. Describe a scenario where there is an applied force and motion and yet no work is done.
4. A bat hitting a baseball is difficult to analyze because the applied force is continuously changing throughout the collision. Describe how such a situation may be investigated without the use of calculus.
5. Describe the two general types of energy into which all forms of energy can be classified.
6. How much more kinetic energy would a baseball have if
   (a) its speed was doubled?
   (b) its mass was doubled?
7. Differentiate between potential energy and gravitational potential energy.

### Inquiry

8. Design an experiment to test the following hypothesis: *The work required to pull a smooth wood block a distance of 1.0 m along a smooth 45° slope is three quarters of the work required to lift the block vertically a distance of 1.0 m.*
   (a) List the equipment and materials you will need to test this hypothesis.
   (b) Develop a procedure that lists all the steps you will follow in your investigation.
   (c) What independent variables are you changing? What dependent variables will you be measuring? What variables are you controlling?
   (d) Suggest some sample observations that might confirm the hypothesis.
   (e) Suggest some sample observations that might refute the hypothesis.
   (f) Carry out the investigation. Record and analyze your observations, and form a conclsion. Include the percentage deviation of your result from the predicted one.

## Communication

**9.** In this chapter, you have learned many new concepts about work and energy. These concepts include the following terms and concrete examples: mechanical work, gravitational potential energy, kinetic energy, law of conservation of energy, energy transfer, conservative versus non-conservative forces, friction, force-distance graphs, motion sensor or photogate, force sensor or force meter, cars on inclines, building elevator, braking car, skidding bicycle, bowling balls, bow and arrow in archery, sports performance, careers, rockets and space stations, pendulums, wind-up toys, weight lifting, pushing boxes.

Create a graphic organizer, such as a concept map, mind map, or flow chart, to show the logical connections between the concepts, events, and objects in the list.

**10.** Draw a force-versus-displacement plot that represents a constant force of 60 N exerted on a Frisbee™ over a distance of 80.0 cm. Show the work done on the Frisbee™ by shading the graph.

## Making Connections

**11.** A coin is dropped from a cliff, 553 m down into the ocean. While in free fall, it eventually reaches terminal velocity (see Chapter 2). Discuss with the aid of a free-body diagram:
  **(a)** What force does work on the coin, causing it to fall?
  **(b)** What force does work on the coin, reducing its acceleration to zero when terminal velocity is reached?
  **(c)** Is the coin doing work on any objects during the fall? Explain.
  **(d)** How fast would the coin be falling if there was no force doing negative work on it?

**12.** Make a list of Olympic sports that involve exerting a force over a distance. For example, running involves pushing with a force over a distance. Your foot pushes backwards on the track through a distance as you move forward.

**13.** Choose *one* of these sports. Analyze the way a force is applied through a distance. Suggest a possible improvement so that less work (either less force or less distance) is needed to produce the same velocity or distance.

## Problems for Understanding

**14.** Calculate the velocity a 1.0 g raindrop would reach if it fell from a height of 1.0 km. Explain why raindrops do not reach such damaging speeds.

**15.** A car travels at a constant velocity of 27.0 m/s for $1.00 \times 10^2$ m.
  **(a)** Name all of the forces that act on the car.
  **(b)** Which, if any, of these forces are doing work on the car? How much work are these forces doing?

**16.** In order to start her computer, a student pushes in the button to turn on the monitor. This action requires her to do 0.20 J of work. Find the average force that must be applied if she pushes the button a total distance of 0.450 cm.

**17.** A young girl pushes her toy box at a constant velocity with a force of 50.0 N. Calculate the work done by the girl if she moves the box 7.00 m.

**18.** A total of 684 J of work are done on a couch by moving it 3.00 m at a constant velocity against a frictional force of 80.0 N. Find the net force exerted on the couch.

**19.** A horse pulls a wagon that was initially at rest. The horse exerts a horizontal force of 525 N, moving the wagon 18.3 m. The applied force then changes to 345 N and acts for an additional distance of 13.8 m. Calculate the total work done by the horse on the wagon during the trip.

**20.** A man drags a small boat 6.00 m across the dock with a rope attached to the boat. Find the amount of work done if the man exerts a 112.0 N force on the rope at an angle of 23° with the horizontal.

**21.** A 65.0 kg rock is moved 12.0 m across a frozen lake. If it is accelerated at a constant rate of 0.561 m/s$^2$ and the force of friction is ignored, calculate the work done.

22. The following force-versus-position graph shows the horizontal force on a cart as it moves along a frictionless track. If the cart has a mass of 1.25 kg, find the kinetic energy and velocity of the cart at the following positions: 5.0 m, 15.0 m, and 25.0 m.

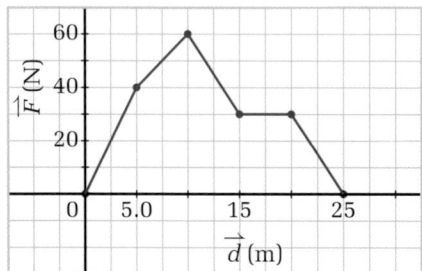

23. The diagram provided shows a man pulling a box across the floor. Assume that the force of friction can be ignored and that the acceleration of the box is 1.27 m/s². Find the angle to the horizontal that the man must pull.

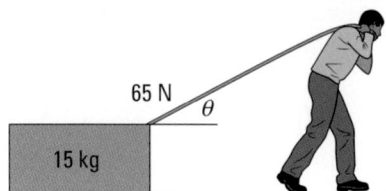

24. A 55 kg running back travelling at 6.3 m/s moves toward a 95 kg linebacker running at 4.2 m/s. Which athlete has more kinetic energy?

25. A 68 kg in-line skater starts from rest and accelerates at 0.21 m/s² for 15 s.
   (a) Find her final velocity and total kinetic energy after the 15 s of travel.
   (b) If she exerts a breaking frictional force of 280 N, find her stopping distance.

26. A girl climbs a long flight of stairs. She travels a horizontal distance of 30.0 m and a vertical distance of 14.0 m. If her change in gravitational potential energy relative to the ground is 6800 J, find the mass of the girl.

27. A 2.00 kg mass is attached to a 3.00 m string and is raised at an angle of 45° relative to the rest position, as shown. Calculate the gravitational potential energy of the pendulum relative to its rest position. If the mass is released, determine its velocity when it reaches its rest position.

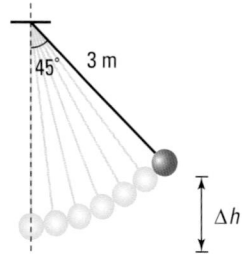

28. A 5.50 kg pendulum is vertically raised 6.25 m and released. Find the maximum velocity of the pendulum and discuss, with the aid of a diagram, where the maximum velocity occurs.

29. A roller coaster at a popular amusement park has a portion of the track that is similar to the diagram provided. Assuming that the roller coaster is frictionless, find its velocity at the top of the loop.

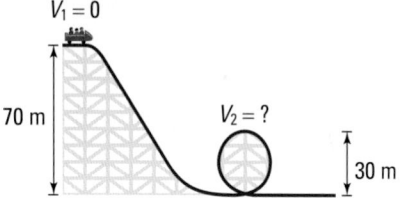

**Numerical Answers to Practice Problems**

**1.** $5.7 \times 10^3$ J; 42 m **2.** 82 m **3.** 2.30 m/s² **4.** 6.33 m **5.** 225 N **6.** 10.9 m **7. (a)** 0 J **(b)** force is perpendicular to direction of motion **8.** $3.00 \times 10^2$ J **9. (a)** 0 J **(b)** no forces are acting so no work is done (velocity is constant) **10. (a)** 0 J **(b)** the tree didn't move, so $\Delta d$ is zero **11. A.** 180 J **B.** 65 J **C.** 0 J **D.** ~ 240 J **14. (a)** $4.1 \times 10^3$ J **(b)** $-4.1 \times 10^3$ J **(c)** gravity and applied force **15.** raising: +126 J; lowering: −126 J **16.** $1.9 \times 10^3$ J **17.** $1.4 \times 10^2$ N **18.** 40.0° **19.** 81.1 J **20.** $1.0 \times 10^1$ kg **21.** 1810 J **22.** 11.5 m/s **23.** $4.1 \times 10^6$ m/s **24.** 0.4 J; 4 N **25.** 6.35 kg **26.** 3000 N; 40 m; 160 m; d ∝ v² **27.** 87 J **28.** $2.4 \times 10^6$ J **29.** 4.1 m **30.** 1.16 m **31. (a)** 2370 J **(b)** 2370 J **32.** 16 J **33.** $1.51 \times 10^6$ J **34. (a)** $1.59 \times 10^5$ J **(b)** $2.38 \times 10^5$ J **(c)** $3.97 \times 10^5$ J **35.** 5.1 m **36. B:** $E_g = 5.9 \times 10^6$ J; $E_K = 1.5 \times 10^7$ J; $v = 28$ m/s **C.** $E_g = 1.8 \times 10^7$ J; $E_K = 2.4 \times 10^6$ J; 11 m/s **D.** $E_g = 1.4 \times 10^7$ J; $E_K = 7.2 \times 10^6$ J; 19 m/s **37.** $E_g = 4140$ J; $E_K = 4140$ J; $v = 5.12$ m/s **38.** ball: 610 J, 22 m/s; shot: 13 J, 22 m/s **39.** $1.0 \times 10^1$ m **40.** 15 floors; 49.3 m/s 152 J **41. (a)** 11 J **(b)** 6.7 m/s **(c)** 4.2 m/s **42.** −7.4 J; −180 N **43.** 43.1 m/s; 8.9% **44.** 75 N

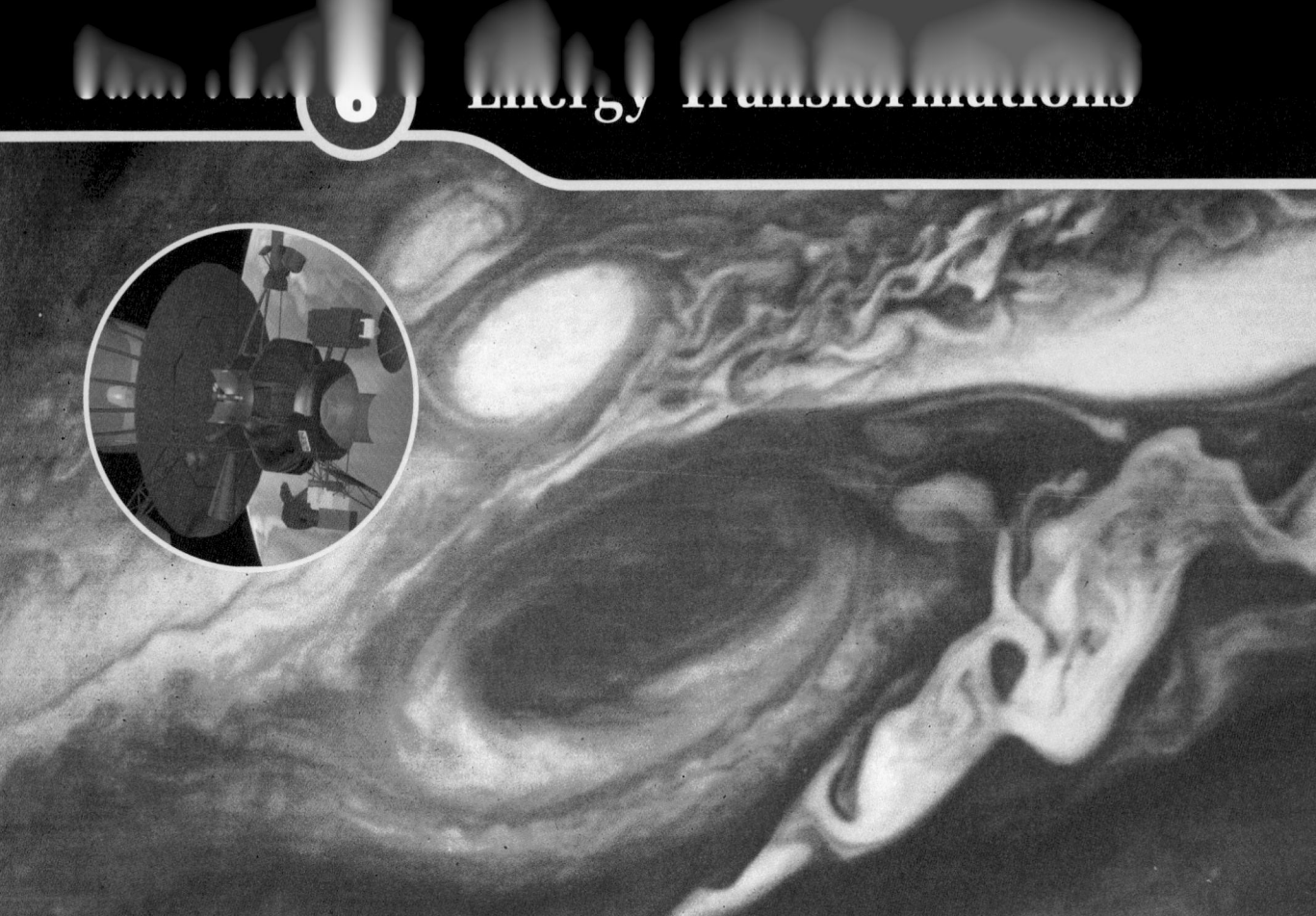

# Energy Transformations

The planet Jupiter poses many challenging questions to space scientists. What powers Jupiter's giant storms? Jupiter radiates twice as much energy as it receives from the Sun. What is the source of this excess energy? In search of answers, NASA's spacecraft Galileo, dropped a probe into Jupiter's swirling atmosphere. Data revealed that Jupiter undergoes a phenomenon called the Kelvin-Helmholtz contraction. The planet's own gravitational attraction is causing it to shrink. In so doing, it converts gravitational potential energy into heat. This heat drives the violent weather patterns. The probe descended a mere 0.1 percent into the dense atmosphere before being crushed by extremely high pressures and temperatures. Data collected during the descent, however, support a controversial theory that Jupiter experiences helium rain deep within its nearly bottomless atmosphere. There, atmospheric pressures, a million times greater than Earth's, cause gaseous helium to condense into a liquid.

Understanding how energy is transformed is yet another window into understanding the universe. Space scientists are learning more about a variety of subjects, from the formation of planets to the mechanisms that drive our own climate patterns.

Whether or not you realize it, you have been conducting informal investigations all of your life. Possibly you learned from experience the difference between hot and cold. This investigation allows you to evaluate the theories you developed from childhood experiences.

## Problem

Develop a theory that can explain observed differences between hot and cold objects. The best theory should allow you to accurately predict how heated objects will behave.

## Hypothesis

Form a hypothesis about what happens to an object as it is heated. What observable changes occur?

## Equipment

- pencil and paper

## Procedure

1. Working in a small group, generate a list of observations from memory that suggest an object is becoming hotter. To get started, make a list of the differences between hot and cold objects.

2. List three different ways that heat might be transferred.

3. Within your small group, attempt to create a theory that explains what "heat" is. Write explanations of how your theory relates to as many as possible of the observations you and your group generated in step 1. For example, your theory should be able to explain why heated objects expand. The theory does not have to be based on current scientific knowledge, but instead, must — as accurately as possible — explain and predict observed phenomena.

## Analyze and Conclude

1. Test your theory's completeness. Use your theory to predict answers to each of the following questions.

   (a) Why do objects expand when heated?

   (b) What keeps a hot-air balloon aloft?

   (c) Why is heat generated when you rub your hands together?

2. Develop a question that your theory is able to explain, and then pass it to another group to see if its theory can also provide an explanation.

Perhaps a panting dog on a hot day can help you formulate your theory of "heat."

- Describe and compare temperature, thermal energy, and heat.

- Analyze situations involving the transfer of thermal energy.

- Identify the relationship between specific heat capacity and thermal energy.

- Compare observations to predictions made by the current scientific model of thermal energy.

**KEY TERMS**

- kinetic molecular theory

- heat

- thermal energy

- absolute zero

- thermal equilibrium

- thermosphere

- specific heat capacity

- latent heat of fusion

- latent heat of vaporization

In the development of your own theory of heat, you have asked questions similar to those asked by physicists for hundreds of years. Over the centuries, many misconceptions about the nature of heat arose. These were eventually proved incorrect by observation and experiment. How well does your concept of heat and thermal energy coincide with the current theory of heat and thermal energy?

### Think It Through

- Write a criticism of each of the following statements about heat and thermal energy.

  1. "The terms 'heat,' 'temperature,' and 'thermal energy' each describes essentially the same thing."

  2. "Heat is a type of substance that resides within an object."

  3. "A piece of wood and a piece of steel, both removed from a pot of boiling water and placed in a single sealed container, will cool to different temperatures."

  4. "Increasing thermal energy will automatically increase temperature."

- As you study this chapter, revisit your criticisms and, if necessary, modify them.

## Phlogiston Theory

Possibly the earliest concept of heat was proposed by the early Greek philosopher Empedocles (492–435 B.C.E.). He asserted that matter consisted of four elements: earth, air, fire, and water. According to this theory, a piece of wood contained a solid mass (earth), a great deal of moisture (water), and empty spaces filled with gas (air), and could be made to *release* its fire. It was hundreds of years before scientists began doing detailed experiments and determined that there were many types of gases, not just "air." Some gases supported burning and others did not.

Some 400 years ago, German scientist Georg Ernest Stahl (1660–1734) presented his theory that heat was a fluid that he called "phlogiston." Stahl proposed that phlogiston flowed from, into, or out of a substance during burning. For example, when wood burned, the phlogiston flowed out and the ash left behind was the pure substance. The phlogiston theory explained many observed phenomena, including the smelting of ore. However, about 100 years later, French scientist Antoine Lavoisier (1743–1794) demonstrated that combustion (burning) required oxygen. The phlogiston theory was set aside, but the concept of heat as a substance continued beyond the life of the phlogiston theory.

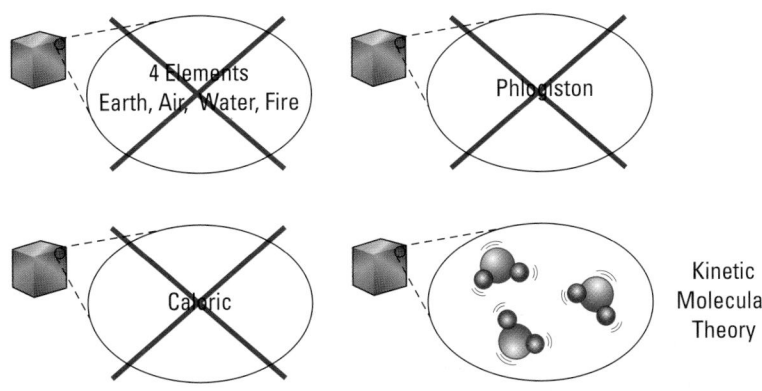

Benjamin Thompson (1753–1814), who later became Count Rumford, Imperial Count of the Holy Roman Empire, used science to solve a wide variety of practical problems. When he was superintendent of an arsenal in Bavaria, he observed that, when boring cannons, the metal became extremely hot. He made measurements and showed that the amount of heat produced was proportional to the work done by the boring process. He was one of the first scientists to seriously challenge the caloric theory. It was many years after his death, however, that the scientific community finally accepted the concept that mechanical work can be transformed into thermal energy.

**Figure 6.1** The kinetic molecular theory of heat dismisses the earlier false assumption that heat was a substance.

## Caloric Theory

British chemist and physicist Joseph Black (1728–1799) performed extensive experiments in an effort to understand heat. His discoveries led to yet another theory of heat, based on a substance called "caloric." Caloric was an invisible, "massless" substance that existed in all materials. It could not be created or destroyed, and exhibited "self-repulsive" forces that made it flow from high concentrations to low concentrations (hot to cold). Everyday experience, as well as most scientific observations, seemed to support this caloric theory. However, one exception eventually brought an end to the theory. Friction, as simple as rubbing your hands together, creates heat. If caloric could not be created or destroyed, how could the act of rubbing two cool objects together — doing mechanical work — generate heat? Many scientists, such as James Joule, began to search for the "mechanical equivalent of heat."

## Kinetic Molecular Theory

The current theory of heat evolved along with the discovery that all matter was made up of particles. The **kinetic molecular theory** is based on findings of scientists of the mid-nineteenth century. The findings showed that all matter is composed of particles that are always in motion. Particles in hot objects move more rapidly, thus have more kinetic energy, than particles in cooler objects. **Heat** is now defined as the *transfer* of thermal energy. Heat is not a substance. It does not "flow." **Thermal energy** is the kinetic energy of the particles of a substance due to their constant, random

**ELECTRONIC LEARNING PARTNER**

Go to your Electronic Learning Partner to learn more about the theories discussed here.

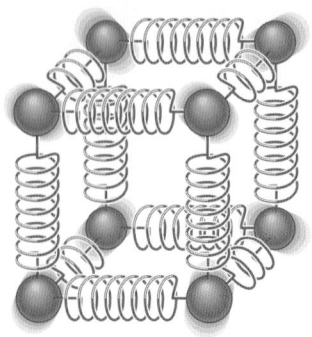

**Figure 6.2** Although the molecules in a solid are held together, they can still move and vibrate, as though connected by springs.

### Biology Link

When your skin's surface temperature falls too low, your nervous system activates certain muscles. The muscle fibres slide and rub against one another, generating thermal energy. As more clusters of muscles become involved, more heat is produced. Your bloodstream then carries this heat to surrounding tissues, which helps raise your temperature back to normal. Shivering is a very efficient method of generating thermal energy in your body. What are some ways that other organisms use to generate thermal energy? Consider lizards, dogs, and butterflies.

motion. The transfer of thermal energy is the result of fast-moving particles colliding with slower-moving ones, and in the process, transferring energy — thermal energy.

Work and heat are the two methods by which energy can be transferred between a system and its surroundings. These concepts make it possible to express the first law of thermodynamics in quantitative form, as shown in the box below.

---

### FIRST LAW OF THERMODYNAMICS

The change in the energy of a system is the sum of the work and heat exchanged between a system and its surroundings.

$$\Delta E = W + Q$$

| Quantity | Symbol | SI unit |
| --- | --- | --- |
| change in energy of a system | $\Delta E$ | J (joule) |
| work | $W$ | J (joule) |
| heat | $Q$ | J (joule) |

When using the quantitative form of the first law of thermodynamics, remember the following points.

- When a system does work on its surroundings, it loses energy, so $W$ is negative.

- When the surroundings do work on the system, the energy of the system increases, so $W$ is positive.

- When heat transfers thermal energy out of a system, the system loses energy, so $Q$ is negative.

- When heat transfers thermal energy into a system, the system gains energy, so $Q$ is positive.

---

Currently, all observed phenomena associated with the heating and cooling of objects support the kinetic molecular theory of heat and thermal energy. Like all forms of energy, thermal energy can do work or can be the result of work. So, the next time you are waiting for the bus on a cold winter day and you rub your hands together to warm them up, remember that you are doing work, transferring muscle energy of movement into thermal energy of warmth, and in the process, disproving the caloric theory of heat.

**Figure 6.3** Rubbing your hands together transforms mechanical work into thermal energy.

**Table 6.1** The Changing Theories of Heat

| Observations | Caloric theory | Kinetic molecular theory |
|---|---|---|
| Objects expand when heated. | More heat means more caloric has flowed in, causing an increase in volume. | More energetic random motions cause the particles to occupy more space. |
| Thermal energy lost by one object will equal the heat gained by another. | The heat or "caloric" flowed from one object to the other. | Energy is transferred from one object to another by collisions. |
| States of matter (solid, liquid, gas) | The more caloric contained within a substance determined its state. For example, a gas occupies a lot of space; therefore, it must contain a lot of caloric. | As the kinetic energy of the molecules increases, they are able to break bonds with neighbouring molecules and change from a solid to a liquid or a liquid to a gas. |
| Temperature increases are caused by friction; for example, by rubbing hands together. | The act of rubbing two objects of the same temperature together and getting heat means that caloric was created — which does not fit the theory.<br><br>The caloric theory *fails* here. | Mechanical work can be transformed into thermal energy. The work done on the particles increases their random motion. |

# Temperature

The concept of temperature is also fraught with misconceptions that stem from the original error of assuming heat was a substance. Recall that the kinetic molecular theory defines thermal energy as the random motion of the particles that make up an object. Although all particles of a substance are not moving at the same speed, the *average* speed or kinetic energy is a useful property of the substance. The temperature of a substance is a measure of the *average* kinetic energy of its atoms or molecules.

An analogy may clarify this concept. Consider the average height of a Grade 9 versus a Grade 12 class. The average height of the Grade 9 class will be much less than that of the Grade 12 class, even though there may very well be some Grade 9 students who are taller than most of those in Grade 12, or some Grade 12 students who are shorter than most of those in Grade 9. In a hot object, a greater percentage of particles will have higher kinetic energies than they will in a cool object.

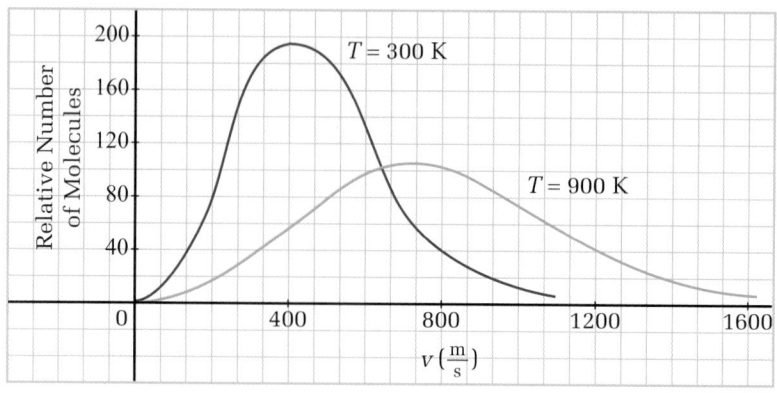

The temperature of an object does not depend on its mass, but the total thermal energy of an object does. If a pot of boiling water and a cup of boiling water both have a temperature of 100°C, the amount of thermal energy in the pot of water will be much greater than that in the cup. If the hot water heater was broken and you had to take a bath in cold water, you would obviously choose to pour a full pot of hot water into the tub rather than a full cup of hot water.

**Figure 6.5** A full kettle of hot water has more thermal energy than a full cup of the same water.

## The Celsius Temperature Scale

Normal human body temperature is 37°C (degrees Celsius). Comfortable room temperature is usually 20°C. You know that your body temperature is warmer than room temperature, but do you know how these values were determined? An old thermometer may measure body temperature in °F (degrees Fahrenheit). Normal body temperature using the Fahrenheit scale is 98.6°F. The fact that normal body temperature is readily represented by two different values highlights the difficulties in developing a theory of heat and defining temperature. Scientists attempted to assign numerical values to temperature before understanding the kinetic molecular theory of matter. Several temperature scales were proposed, but the most popular scale today was established by the Swedish physicist, Anders Celsius (1704–1744). Celsius chose the freezing point of water at standard atmospheric pressure (101.3 kPa) to be 0°C and the boiling point to be 100°C. Therefore, one degree Celsius is one one-hundredth of the temperature difference between the freezing and boiling points of water.

---

**PHYSICS FILE**

When Celsius devised his temperature scale in 1742, he assigned values in reverse to what are used today (freezing point was 100°C and the boiling point was 0°C).

---

## The Kelvin Scale and Absolute Zero

What would it mean for the temperature of a substance to be absolute zero? Since temperature is directly related to the average kinetic energy of the particles in a substance, then **absolute zero** would have to be the temperature at which the substance has zero kinetic energy. There would be no movement of particles.

If particles did not move, they could not collide with each other or with the walls of a container. No collisions would mean no pressure.

In 1848, William Thomson, who became Lord Kelvin, applied concepts from the newly developing field of thermodynamics (movement of thermal energy) to theoretically devise an absolute temperature scale. At about the same time, several scientists were attempting to experimentally determine the coldest possible temperature. The agreement between the theoretical and the experimental approach was excellent. Absolute zero was determined to be −273.15°C.

Absolute zero was experimentally determined by supercooling low-density gases, such as hydrogen or helium, at constant volume. A graph of pressure versus temperature was plotted. The resulting plot (see Figure 6.6) formed a straight line. When extrapolated to zero pressure the line crossed the temperature axis at −273.15°C. Further cooling would result in a gas that exerted a negative pressure and that cannot and does not exist.

The size of one unit on the Kelvin scale is the same as one degree on the Celsius scale. However, the unit is not called a degree. Scientists have agreed on the convention that temperatures using the Kelvin scale will be reported without the word "degree." For example, water freezes at 273.15 kelvins, not 273.15 *degrees* kelvin. The SI unit for temperature is the kelvin (K). To convert from Celsius to kelvin, use the relationship shown in the box below.

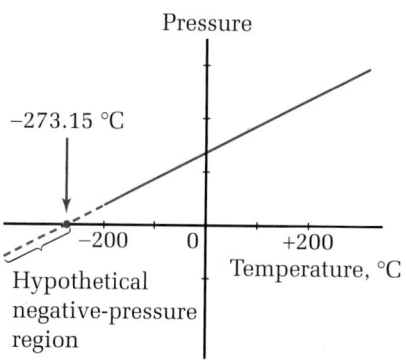

**Figure 6.6** Absolute zero temperature is unattainable in the laboratory because, as quantum theory asserts, the lowest possible energy level for atoms and molecules is not zero energy. Nevertheless, temperatures very close to absolute zero can be obtained and you can extrapolate a graph to zero pressure (thus zero energy) and determine the position of absolute zero on a temperature scale.

---

## CELSIUS TO KELVIN SCALE CONVERSION

The temperature in kelvins is the sum of the Celsius temperature and 273.15.

$$T = T_C + 273.15$$

| Quantity | Symbol | SI unit |
|---|---|---|
| temperature on the Kelvin scale | $T$ | K (kelvin) |
| temperature on the Celsius scale | $T_C$ | none (°C is not an SI unit) |

---

### • *Think It Through*

• Extrapolating the line on the pressure-versus-temperature plot provided the value of absolute zero.

1. Predict what a volume-versus-temperature plot might look like when cooling a low-density gas.

2. At what point would the line cross the horizontal axis?

3. Describe the significance of this intercept.

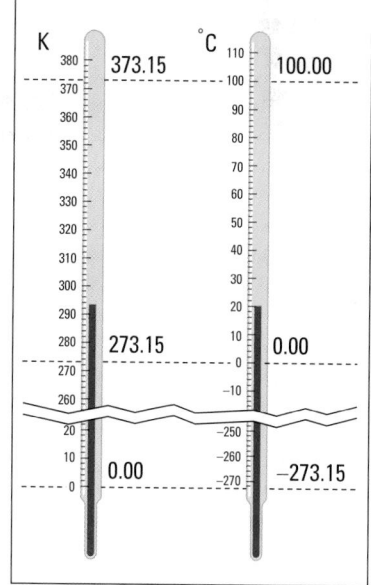

**Figure 6.7** A comparison of the temperature scales shows that one Celsius degree is the same size as its kelvin counterpart.

Human temperature receptors can readily sense temperature differences but, surprisingly, estimating the actual temperature is not always that simple. Obtain three large beakers. Place hot water (but not so hot that you cannot put your hand in it) in one beaker, room temperature water in another, and ice water in the last. Simultaneously submerge one hand in the hot water and one in the ice water for 60 s. Then remove both hands and place them in the beaker filled with water at room temperature. Describe how the room temperature water feels to each hand. In terms of the kinetic molecular theory, explain why the room temperature water feels different to each of your hands.

# Thermal Equilibrium

Place an ice cube in a glass of pop. The ice cube will start to melt and the pop will start to get cooler. This is the natural result of placing two objects of different temperatures together. According to the kinetic molecular theory, energy is being transferred from the warmer pop to the cooler ice cube by millions of collisions. Eventually, the melted ice cube and the pop will reach the same temperature — or **thermal equilibrium**. Although the second law of thermodynamics can be quite complex, one way to express it in a simple form is to state that "thermal energy is always transferred from an object at a higher temperature to an object at a lower temperature" or "an isolated system will always progress toward thermal equilibrium."

**Before thermal equilibrium**

Hot body

Hot body     Cold body

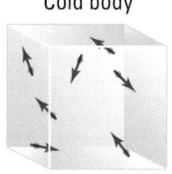

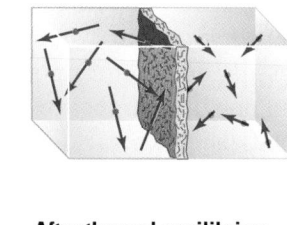

Cold body

**After thermal equilibrium**

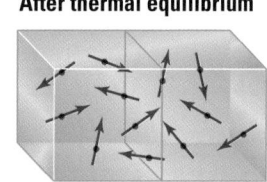

**Figure 6.8** **(left)** Particles in a hot object have greater speeds and therefore, greater kinetic energy, than do particles in cold objects; **(right)** thermal energy is transferred from a hot object to a colder object until thermal equilibrium is reached.

**Space Link**

The top portion of our atmosphere, called the *thermosphere*, is approximately where the space shuttle flies. There the temperature reaches extraordinarily high values, up to 1500°C during the daytime. When the astronauts leave the shuttle to repair satellites or construct portions of the International Space Station, they require a very special suit. The suit protects them from debris (moving faster than the speed of sound), from harmful ultraviolet radiation, and from freezing to death. How could they freeze to death if the temperature is 1500°C? The answer lies in the difference between thermal energy and temperature. The atmosphere at that altitude is at an extremely low pressure. When a molecule interacts with a high-energy ultraviolet ray, it gains a large amount of kinetic energy and starts moving extremely fast. Since temperature is a measure of how fast the molecules are moving, the temperature is very high. However, there is very little thermal energy at this altitude because there are so few molecules moving about. If an astronaut held out a thermometer to take the temperature, the chance that it would be struck by a fast-moving molecule is slim. Although the temperature is high, the total amount of thermal energy is exceedingly small.

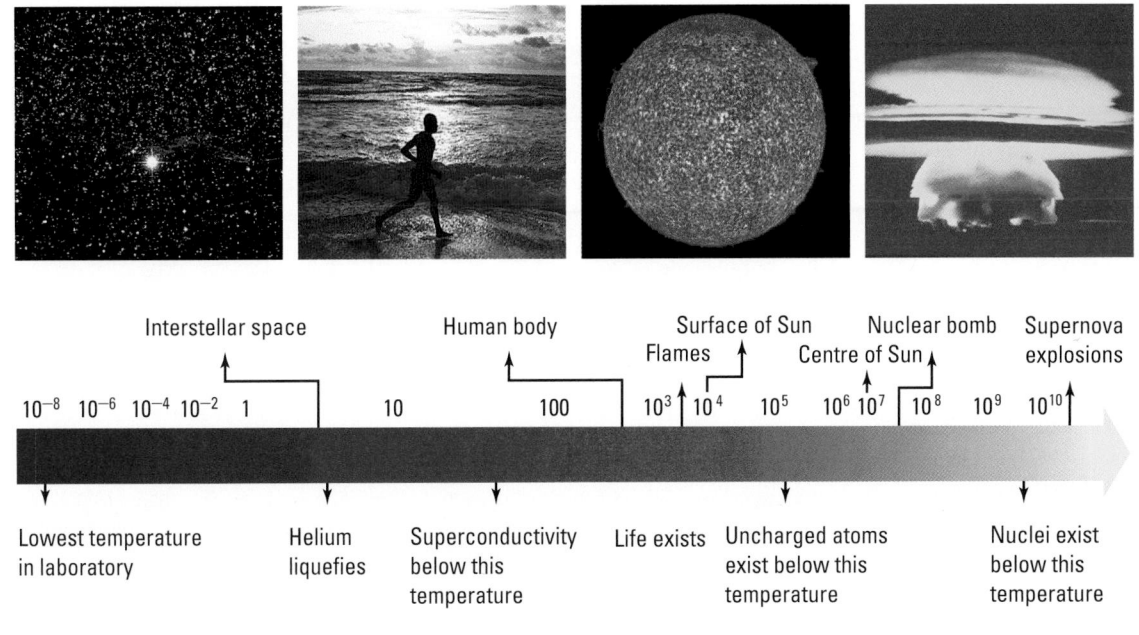

Figure 6.9 An extremely wide range of temperatures can be found throughout the universe.

• **Think It Through**

• Will putting pop cans in a mixture of ice and water cool the pop more or less quickly than placing them in ice alone?

## Specific Heat Capacity

Just knowing that thermal energy is always transferred from an object at a high temperature to an object at a lower temperature cannot tell you what the equilibrium temperature will be. The amount of mass of an object and the nature of the material determine the extent of the temperature change. Clearly, for any two objects made of the same material, the object with more mass will require more heat to achieve a specific rise in temperature. The nature of the material plays an equally important role in determining the amount of temperature increase. Different materials have varying capacities to absorb heat for a given temperature change. The specific heat capacity of a substance depends on the type of material and must be obtained experimentally. The **specific heat capacity** of a substance is the amount of energy that must be added to raise 1.0 kg of the substance by 1.0 K. Table 6.2 on the following page provides specific heat capacities for some common substances.

 **Web Link**

*www.school.mcgrawhill.ca/ resources/*
The Sun's surface temperature is about 6000°C, but the corona (the Sun's atmosphere) is much hotter — millions of degrees hotter. How does all of that energy get into the corona without heating up the surface? Try to answer this question, then visit the above web site to compare your answer with that of the experts. Follow the links for **Science Resources** and **Physics 11**.

The extraordinarily large specific heat capacity of water is one of the reasons why you can go outside in the middle of winter or on the hottest day of the summer without a significant, possibly dangerous, change in your body temperature. A large specific heat capacity means that a great deal of heat transfer must occur before the temperature of water will change significantly. Since the composition of the human body includes a very large percentage of water, a large amount of heat must be added or removed to change the body temperature. The human body also has very capable thermoregulation systems that maintain our body temperature within a couple of degrees Celsius.

**Table 6.2**

Specific Heat Capacities of Some Representative Solids and Liquids

| Substance | Specific Heat Capacity J/kg · °C |
|---|---|
| **Solids** | |
| aluminum | 900 |
| copper | 387 |
| glass | 840 |
| human body (37°C) | 3500 |
| ice (−15° C) | 2000 |
| steel | 452 |
| lead | 128 |
| silver | 235 |
| **Liquids** | |
| ethyl alcohol | 2450 |
| glycerin | 2410 |
| mercury | 139 |
| water (15° C) | 4186 |

## HEAT REQUIRED FOR A TEMPERATURE CHANGE

The amount of heat ($Q$) required to raise the temperature of a quantity ($m$) of a substance by an amount $\Delta T$ is the product of the mass, specific heat capacity, and temperature change.

$$Q = mc\Delta T$$

| Quantity | Symbol | SI unit |
|---|---|---|
| amount of heat transferred | $Q$ | J (joule) |
| mass | $m$ | kg (kilogram) |
| specific heat capacity | $c$ | J/kg · K (joules per kilogram per kelvin) |
| change in temperature | $\Delta T$ | K (kelvins) |

**Note:** The size of a K and a °C are the same. Therefore, when temperature *changes* are used, the two scales will give the same result. Most tables report $c$ in J/kg · °C.

The specific heat capacity of a *gas* is dependent not only on the type of gas, but also on the variations in pressure and volume. Different values are obtained for specific heat capacities when the measurements are made under conditions of constant pressure compared to those obtained under constant volume. Table 6.3 lists the different values for specific heat capacity that are obtained for selected gases.

**Table 6.3** Specific Heat Capacity of Gases

| Gas | Specific heat capacity J/kg · °C | |
|---|---|---|
| | Constant pressure, $c_p$ | Constant volume, $c_v$ |
| ammonia | 2190 | 1670 |
| carbon dioxide | 833 | 638 |
| nitrogen | 1040 | 739 |
| oxygen | 912 | 651 |
| water vapour (100°C) | 2020 | 1520 |

All values are for 15°C and 101.3 kPa of pressure, except for water vapour.

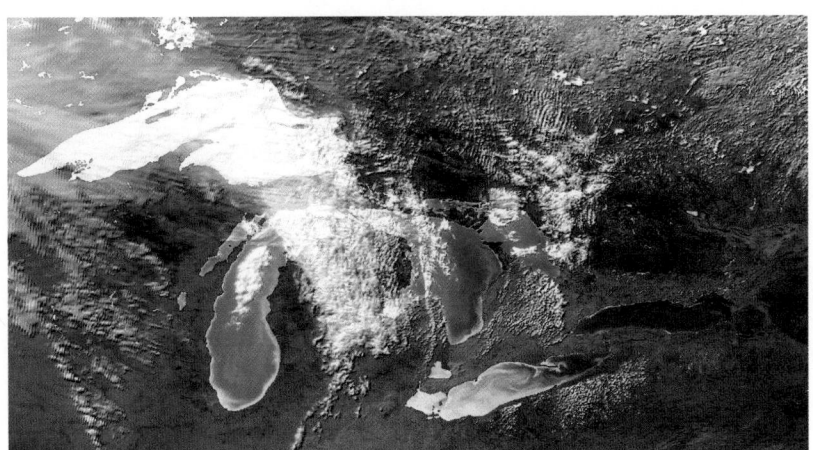

**Earth Link**

The large specific heat capacity of water plays an important role in weather and climate patterns. For example, Lake Ontario is very deep, containing much more water than Lake Erie. In the winter, Lake Ontario stays much warmer than the cold air masses that move over it. The result is that the warmer water heats the moving air masses above it and loads the air with moisture. When the air mass moves back over the much colder land, the air mass cools and the excess moisture is released as snow. This type of snowfall is termed "lake-effect snow." The city of Buffalo receives significantly more snow than Toronto because of lake-effect snow. Attemp to determine the time of year when the image in Figure 6.10 was captured. Do a web search to find a thermograph of an area of interest to you. Are you able to identify in what season the thermograph was taken?

**Figure 6.10** This thermograph of the Great Lakes shows dramatic temperature differences in the water. Warmer water is yellow (about 25°C), cooling through dark blue (about 5°C). Cloud cover appears white.

## MODEL PROBLEMS

### Calculating a Temperature Change

1. A 55.0 kg person going for an hour-long, brisk walk generates approximately $6.50 \times 10^5$ J of energy. If the body's temperature-regulating systems did not remove this thermal energy, by how much would the walker's body temperature increase?

### Frame the Problem

- *Thermal energy* generated by muscle action is normally removed by body thermoregulation systems.

- If the thermal energy was to remain in the body, the *body temperature would increase*.

- The thermal energy generated by the muscles, $Q$, is transferred to all parts of the body by the bloodstream, distributing the heat throughout the body.

*continued* ▶

## Identify the Goal

Change in body temperature, due to walking briskly for one hour, if no heat was released from the body

## Variables and Constants

| Involved in the problem | Known | Implied | Unknown |
|---|---|---|---|
| $Q$ $c$ | $Q = 6.50 \times 10^5$ J | $c = 3500 \dfrac{\text{J}}{\text{kg} \cdot {}^\circ\text{C}}$ | $\Delta T$ |
| $m$ $\Delta T$ | $m = 55.0$ kg | (Obtained from Table 6.2) | |

## Strategy

The equation for heat transfer causing a temperature change applies.

Substitute in given variables and solve.

## Calculations

$$Q = mc\Delta T$$

**Substitute first**

$$6.50 \times 10^5 \text{ J} = (55.0 \,\cancel{\text{kg}})\left(3500 \,\frac{\text{J}}{\cancel{\text{kg}} \cdot {}^\circ\text{C}}\right)\Delta T$$

$$6.50 \times 10^5 \text{ J} = \left(1.925 \times 10^5 \frac{\text{J}}{{}^\circ\text{C}}\right)\Delta T$$

$$\frac{6.50 \times 10^5 \,\cancel{\text{J}}}{1.925 \times 10^5 \frac{\cancel{\text{J}}}{{}^\circ\text{C}}} = \frac{\left(\cancel{1.925 \times 10^5 \frac{\text{J}}{{}^\circ\text{C}}}\right)}{\cancel{1.925 \times 10^5 \frac{\text{J}}{{}^\circ\text{C}}}}\Delta T$$

$$\Delta T = 3.376\,{}^\circ\text{C}$$

**Solve for $\Delta T$ first**

$$\frac{Q}{mc} = \frac{\cancel{mc}\Delta T}{\cancel{mc}}$$

$$\frac{Q}{mc} = \Delta T$$

$$\Delta T = \frac{6.50 \times 10^5 \,\cancel{\text{J}}}{(55.0 \,\cancel{\text{kg}})\left(3500 \frac{\cancel{\text{J}}}{\cancel{\text{kg}} \cdot {}^\circ\text{C}}\right)}$$

$$\Delta T = 3.376\,{}^\circ\text{C}$$

The walker's body temperature would increase by 3.38°C (resulting in a body temperature of 40.3°C) if the heat generated by walking was not removed. A temperature increase of 3.38°C is considered to be a serious fever, not something that an evening walk should cause.

## Validate

Specific heat capacity of an average human body was taken from Table 6.2. Temperature has units of °C, which is correct.

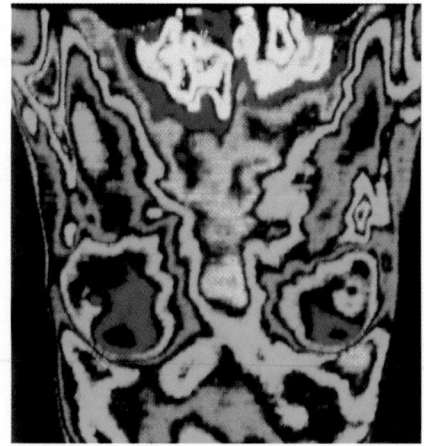 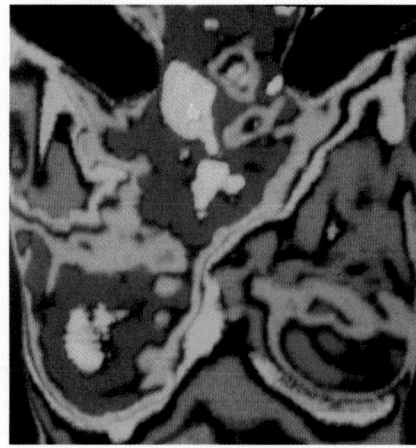

**Figure 6.11** Changes in body temperature in different body organs can be detected using thermographs. Physical activity speeds up certain cellular processes, and so does cell division, as shown here. **(left)** Healthy breasts register a bluish colour in thermographs, indicating cooler temperatures. **(right)** The red to white colour shown here indicates higher temperatures, associated with accelerated cell division of malignant tissue.

2. A typical warm-water shower (without an energy-saving showerhead) consumes 130 kg ($1.30 \times 10^2$ kg) of water at a temperature of 65.0°C.

    (a) Calculate how much energy is required to heat the water if it begins at a temperature of 15.0°C.

    (b) Calculate the electrical cost of the shower if the utility company's charge is $0.15 per kilowatt-hour (kW · h). A kilowatt-hour is a convenient way for electrical utility companies to track the amount of energy that your home consumes in a month.

## Frame the Problem

- The *temperature* of water is to be *increased*.

- An amount of heat, $Q$, must be added to raise the water temperature.

- *Electric energy* will be *transformed* into *thermal energy* of the water.

## Identify the Goal

(a) The amount of heat, $Q$, required to raise the water temperature

(b) The cost of the electric energy required to heat the water

## Variables and Constants

| Involved in the problem | Known | Unknown |
|---|---|---|
| $Q$ | $m = 1.30 \times 10^2$ kg | $Q$ |
| $m$ | $c = 4186 \dfrac{J}{kg \cdot °C}$ | $\Delta T$ |
| $c$ | cost/kW · h = $0.15 per kW · h | |
| $\Delta T$ | $T_{initial} = 15.0°C$ | |
| $T_{initial}$ | $T_{final} = 65.0°C$ | |
| $T_{final}$ | | |

## Strategy

Find the change in temperature.

Find the energy required to cause the temperature change by using the formula for heat.

## Calculations

$\Delta T = T_{final} - T_{initial}$
$\Delta T = 65.0°C - 15.0°C$
$\Delta T = 50.0°C$

$Q = mc\Delta T$
$Q = (1.30 \times 10^2 \, \cancel{kg})(4186 \, \dfrac{J}{\cancel{kg} \cdot \cancel{°C}})(50.0°C)$
$Q = 2.721 \times 10^7$ J

(a) $2.72 \times 10^7$ J of energy must be added to raise the temperature of 130 kg of water by 50.0°C.

*continued* ▶

*continued from previous page*

## Strategy

Find the conversion factor from kilowatt hours to joules.

## Calculations

$$(1 \ \cancel{kW \cdot h}) \left( \frac{1000 \ \cancel{W}}{\cancel{kW}} \right) \left( \frac{1 \frac{J}{\cancel{s}}}{\cancel{W}} \right) \left( \frac{3600 \ \cancel{s}}{\cancel{h}} \right) = 3.6 \times 10^6 \ J$$

Use the conversion factor to convert energy, $Q$, to $kW \cdot h$.

$$Q = (2.72 \times 10^7 \ \cancel{J}) \left( \frac{1 \ kW \cdot h}{3.6 \times 10^6 \ \cancel{J}} \right) = 7.558 \ kW \cdot h$$

Determine the total cost.

$$(7.558 \ \cancel{kW \cdot h}) \left( \frac{\$0.15}{\cancel{kW \cdot h}} \right) = \$1.134$$

**(b)** Heating the water for an average shower costs $1.14.
(**Note:** Utility companies always round dollar figures up.)

## Validate

A joule is a small amount of energy, and a large amount of thermal energy is required to change the temperature of water significantly. Thus, $2.72 \times 10^7$ J is a reasonable amount of energy. The units cancel to give dollars in the final answer, which is the expected unit.

### PRACTICE PROBLEMS

1. A community pool holds $1.10 \times 10^6$ kg of water. First thing in the morning, the temperature of the pool was 20.0°C. When the temperature was checked later in the day, it was 27.0°C. Calculate how much energy was required to raise the temperature of the water.

2. A scientist added 1500 J of energy to each of two 1.0 kg samples of gas kept at constant volume. One gas is $CO_2$ and the other is $O_2$. Calculate the final temperature difference between the two samples if they both started at 20°C.

3. Washing the evening dishes requires 55.5 kg of water. Tap water is at a temperature of 10.0°C and the dishwasher's preferred water temperature is 45.0°C. Find the amount of energy that is required to heat the water. Calculate the electrical cost of washing the dishes if the local utility company charges $0.120 per kilowatt-hour.

4. A covered beaker of ethanol is sitting in a window in the sun. The temperature changes from 8.00°C to 16.00°C. If this requires $2.00 \times 10^4$ J of energy, find the amount of ethanol in the beaker in kilograms.

## Phase Change

The kinetic molecular theory neatly describes how energy can be added to a system without causing an increase in the temperature of the system. Consider a glass full of ice cubes. As thermal energy is added to the ice, the average kinetic energy of the particles increases, which is then reflected in a temperature increase. When the temperature reaches 0°C and energy continues to be added, the ice will melt into liquid water that also has a temperature of 0°C.

Adding more energy will melt more ice, until all of the ice is melted. Only after all of the ice has melted will the water temperature begin to increase once more. The kinetic molecular theory explains that as energy is added, the kinetic energy of the particles increases to a point where it is able to break the bonds that keep the ice cube solid. The temperature will not increase when energy is added during a phase change (from solid to liquid or liquid to gas). The added energy is being used to do work on the bonds that keep the object in a given state.

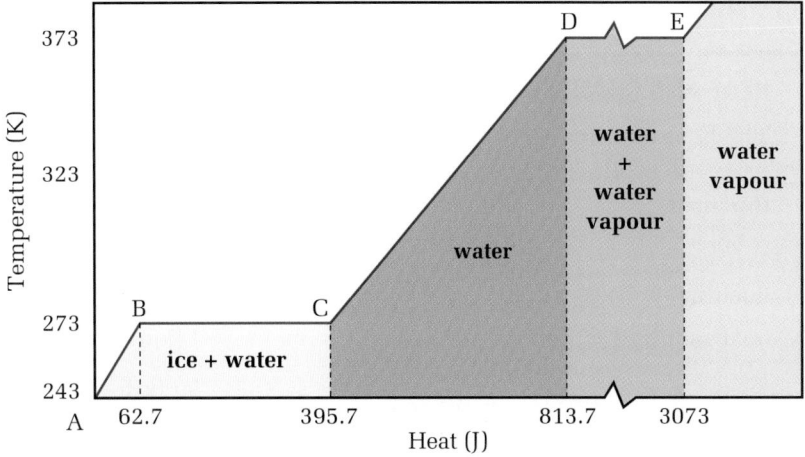

**Figure 6.12** This is the heating curve of water. The plateaus on this heating curve are associated with phase changes.

Phase changes require a change in internal energy. The amount of energy required for a phase change depends on the substance and whether the change of state is from solid to liquid (melting) or from liquid to gas (vaporization). The temperature remains constant during a phase change and, therefore, the "heat" remains hidden. For this reason, it is called the "latent heat" or hidden heat of a transformation of a substance. Table 6.4 lists the **latent heat of fusion** (melting) and **latent heat of vaporization** of several substances.

**Table 6.4**
Latent Heats of Fusion and Vaporization for Common Substances

| Substance | Latent heat of fusion $L_f$ (J/kg) | Latent heat of vaporization $L_v$ (J/kg) |
|---|---|---|
| helium | 5 230 | 20 900 |
| oxygen | 13 800 | 213 000 |
| ethyl alcohol | 104 000 | 854 000 |
| water | 334 000 | 2 260 000 |
| lead | 24 500 | 871 000 |
| gold | 64 500 | 1 578 000 |

Have you ever been outside on a calm winter's day when beautiful, large snowflakes flutter down from the sky? If you have, you may recall that the ambient temperature always seems a little warmer than it is on days without snow. There is some real physics behind that perception. When water vapour condenses into ice crystals, forming snowflakes, energy must be liberated. Looking at Table 6.4, you can see that $2.26 \times 10^6$ J of energy are released for every kilogram of snow formed. That is a lot of energy being released into the environment, which often results in a perceptible rise in air temperature.

## HEAT AND CHANGES OF STATE

The heat required to change the state of an amount of mass, $m$, is the product of the latent heat of the transformation and the mass.

$$Q = mL_f \quad \text{or} \quad Q = mL_v$$

| Quantity | Symbol | SI unit |
|---|---|---|
| heat required to change state | $Q$ | J (joule) |
| mass | $m$ | kg (kilogram) |
| latent heat of fusion | $L_f$ | J/kg (joules per kilogram) |
| latent heat of vaporization | $L_v$ | J/kg (joules per kilogram) |

## MODEL PROBLEM

### Heat Removed by Evaporation

When the ambient temperature is higher than your body temperature, the only way your body can rid itself of excess heat is by sweating. The sweat evaporates from the surface of your skin by absorbing energy from your body, thus lowering your temperature. Calculate the amount of energy that must be absorbed to evaporate 5.0 g of water.

### Frame the Problem

- *Thermal energy* in your skin is *absorbed* by the *water*.

- When the water *evaporates*, the thermal energy is *removed* from your body.

- The amount of energy required to evaporate the water is dependent on the *latent heat of vaporization* of water.

**PROBLEM TIP**

When using the equation containing latent heat of fusion or vaporization, always ensure that the mass and latent heat values have appropriate units. If the mass is in grams, then the latent heat must be in J/g. If the mass is in kg, then the latent heat must be in J/kg.

## Identify the Goal

Amount of energy absorbed by the water to evaporate 5.0 g

## Variables and Constants

| Involved in the problem | Known | Implied | Unknown |
|---|---|---|---|
| $Q$ | $m = 5.0$ g | $L_v = 2.26 \times 10^6$ J/kg | $Q$ |
| $m$ | | | |
| $L_v$ | | | |

## Strategy

The energy of vaporization formula applies to this situation.

Convert the mass to kilograms.

Substitute known values into formula and simplify.

Round to two significant figures.

## Calculations

$$Q = mL_v$$

$$(5.0 \text{ g}) \left( \frac{1 \text{ kg}}{1000 \text{ g}} \right) = 5.0 \times 10^{-3} \text{ kg}$$

$$Q = (5.0 \times 10^{-3} \text{ kg})(2.26 \times 10^6 \text{ J/kg})$$
$$Q = 1.13 \times 10^4 \text{ J}$$

$$Q = 1.1 \times 10^4 \text{ J}$$

In order to evaporate, 5.0 g of water needs to absorb $1.1 \times 10^4$ J of energy.

---

## Validate

Equivalent mass and latent heat values were used; therefore, the units of mass cancelled.

The water was evaporating; therefore, the latent heat of vaporization was used. Energy has units of joules, which is correct.

### PRACTICE PROBLEMS

**5.** Calculate the quantity of heat that must be removed by an ice-maker converting 4.0 kg of water to ice at 0.0°C.

**6.** Repeat the above calculation substituting ethyl alcohol for the water.

**7.** The Canadian Mint regularly produces gold coins. Calculate the amount of pure gold that the Mint could melt in one hour with a furnace capable of generating heat at a rate of 25 kJ/min.

(Read the Technology Link on the following page before attempting problems 8 and 9.)

**8.** Assuming that the snow is −4°C and that the water is dumped at +4°C, calculate the amount of heat energy required to operate a Metromelt for one hour.

**9.** Compare the heat energy required to operate a Metromelt for one day (eight hours) to the equivalent number of hot showers the energy could supply to the citizens of Toronto. Assume that an average morning shower uses 40 kg of water that has been heated from 15°C to 70°C.

**Why Steam Burns**

You must add $3.35 \times 10^5$ J of energy to 1.00 kg of ice to cause it to melt into liquid water. Remember, the energy being added is not causing the temperature to change, but rather causes the state to change from solid to liquid. It takes considerably more energy, $2.26 \times 10^6$ J, to turn 1.00 kg of 100°C water into 100°C steam. Likewise, to condense 100°C steam into 1.00 kg of 100°C water, $2.26 \times 10^6$ J must be removed. This is the reason that steam causes such terrible burns. Water at 100°C is not hot enough to cause very serious burns. The energy that is released from the steam condensing on the skin is what causes the injury.

### Technology Link

In some large cities across Canada, snow removal after a large storm poses a difficult problem. It must be cleared from roadways and from sidewalks. In the densely packed downtown city core, there is no place to put the snow. One solution to the problem is to melt the snow, using a specially designed machine. Toronto has five such machines, called "Metromelts." Each machine is 16 m long, 3 m wide, and 3.6 m tall. This giant boiler on wheels has a mass of 60 000 kg empty and can travel up to 24 km/h when it is not melting snow. The Metromelt is capable of melting 150 000 kg of snow in one hour. What are some other alternatives to Metromelts that other cities use?

## 6.1 Section Review

1. **MC** Why do some pots have copper bottoms?

2. **K/U** If you were to walk outside on a cold winter evening and touch a piece of wood, a metal fence pole, and a handful of snow, which would feel coldest? Justify your answer.

3. a) **K/U** Why is the specific heat capacity of the human body, although quite large, less than that of water?

   b) **MC** How does this large value help our survival?

4. **I** Extremely hot water is poured into two glasses. One glass is made of pure silver, the other of pure aluminum. Which glass will be hot to the touch first, the silver or aluminum glass? Explain.

5. **K/U** What happens to the work done when a bottle of lemonade is shaken?

6. **K/U** When wax freezes, is energy absorbed or released by the wax?

7. **C** A typical heating curve contains two plateaus. Describe why the plateaus exist.

8. **I** Why does rubbing alcohol at room temperature feel cool when a drop of it is placed on your skin?

### UNIT PROJECT PREP

Dramatic changes in temperature can affect the performance of athletes and sporting equipment.

■ Identify the role of temperature in your investigation topic.

■ How do materials used in sporting equipment compensate for changes in temperature?

A 400 t passenger jet, loaded with over 450 people and their luggage, waits at the end of a runway. In less than 60 s, the jet will speed down a 2 km long runway and lift into the air. Meanwhile, halfway around the world, in the Andes Mountains, an old converted school bus slowly bounces along a gravel road. Both the jet and the bus burn fuel to obtain energy for motion. The difference between these vehicles is, of course, the rate at which they convert the chemical energy into motion.

**SECTION**
**EXPECTATIONS**

- Define and describe power and efficiency.

- Analyze the factors that determine the amount of power generated.

- Apply quantitative relationships among power, energy, and time.

- Design and conduct experiments related to the transformation of energy.

- Consider relationships between heat and the law of conservation of energy.

**KEY**
**TERMS**

- power
- efficiency

**Figure 6.13** Jet engines do an incredible amount of work in a very short interval of time. They generate a large amount of *power*.

## Power

The engine in an old school bus could, over a long period of time, do as much work as jet engines do when a jet takes off. However, the school bus engine could not begin to do enough work fast enough to make a jet lift off. In this and many other applications, the rate at which work is done is more critical than the amount of work done. **Power** is the rate at which work is done. Since work is defined as a transfer of energy, power can also be defined as the rate at which energy is transferred.

The powerful machine shown here removed eighty million cubic metres of dirt from beneath the English Channel in less than six years. Two rail tunnels and a service tunnel, 52 km in length, now connect England to France. How important is the tunnel in stimulating trade between the United Kingdom and the European continent? What kinds of vehicles are allowed in the tunnel, and how are their exhaust gases controlled? You may find answers to these and other questions you would be interested in by investigating further.

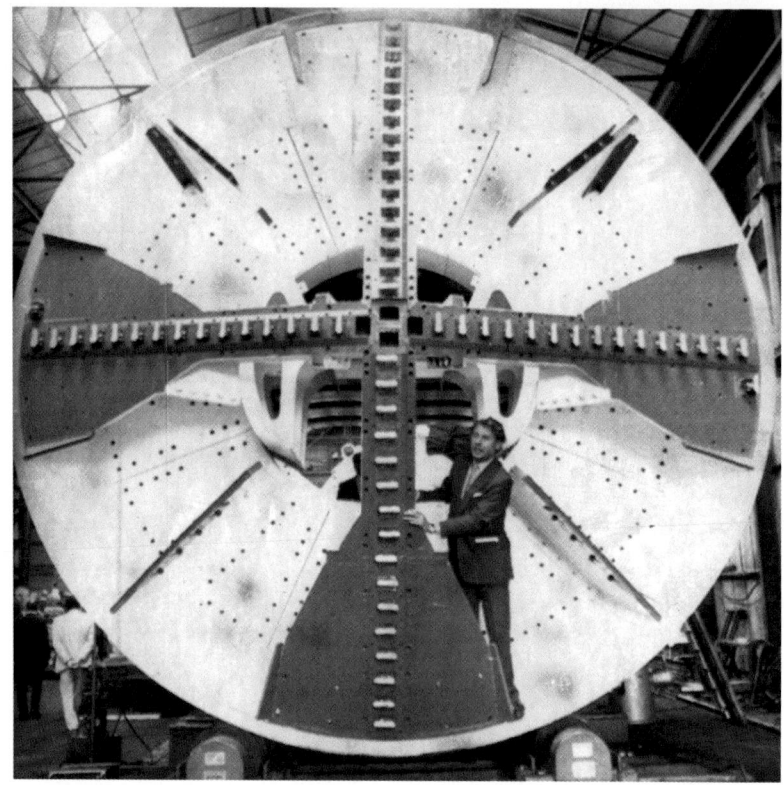

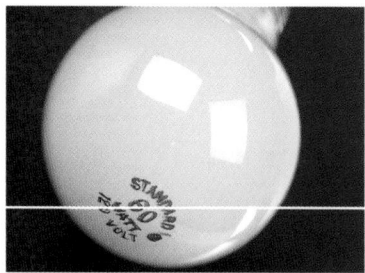

**Figure 6.14** Light bulbs and electric appliances are often labelled with a power rating.

---

### DEFINITION OF POWER

Power is the quotient of work and time interval.

$$P = \frac{W}{\Delta t} \quad \text{or} \quad P = \frac{E}{\Delta t}$$

| Quantity | Symbol | SI unit |
|---|---|---|
| power | $P$ | W (watt) |
| energy transferred | $E$ | J (joule) |
| work done | $W$ | J (joule) |
| time interval | $\Delta t$ | s (seconds) |

**Note:** A watt is equivalent to a joule per second: $\text{W} = \frac{\text{J}}{\text{s}}$

---

Any machine that does mechanical work or any device that transfers energy via heat can be described by its power rating; that is, the rate at which it can transfer energy. The SI unit of power, the watt, can be used to quantify the power of motors, rockets, or even dynamite, but it is most familiar as a power rating for a light bulb. A 60 W bulb transforms 60 J of electric energy into thermal energy and light in 1 s, as compared to a 100 W bulb that transforms 100 J of electric energy into light and thermal energy in 1 s.

The language of power is subtle and different from that of work. Recall that work is done *on* an object and results in a *transfer* of energy to that object. The *rate* of this energy transfer, or power, is often referred to as the power that is *generated* in doing the work. The term "power" not only applies to the rate at which energy is transferred from one object to another or transformed from one form to another, but also to the rate at which energy is transported from one location to another. For example, electric *power lines* carry electric energy across vast stretches of land.

### History Link

The unit, the watt, was named in honour of the Scottish engineer, James Watt, who made such great improvements in the steam engine that it hastened the Industrial Revolution. The ability to do work did not change, but the rate at which the work could be accomplished did. Watt did experiments with strong dray horses and determined that they could lift 550 pounds a distance of one foot in 1 s. He called this amount of power one horsepower (hp). Converting to SI units, 1 hp is equivalent to 746 W. What do you think your horsepower is? Design a simple experiment that you could use to determine your horsepower, and then read ahead in the Multi Lab in this section to see another way to determine your horsepower.

## MODEL PROBLEMS

### Calculating Power

1. A crane is capable of doing $1.50 \times 10^5$ J of work in 10.0 s. What is the power of the crane in watts?

### Frame the Problem

- The crane did *work* in a specified *time interval*.

- *Power* is defined as work done per unit time.

- Simply apply the power definition.

### PROBLEM TIP

Remember that a capital "*W*" is the variable that represents work done and "W" also represents the unit of power, the watt. Be careful not to confuse the two, as they are very different. To help distinguish the difference, in this text, symbols for quantities are in italics while units are in roman print.

*continued* ▶

*continued from page 23*

## Identify the Goal

Power generated by the crane

## Variables and Constants

| Involved in the problem | Known | Unknown |
|---|---|---|
| $W$ | $W = 1.50 \times 10^5$ J | $P$ |
| $\Delta t$ | $\Delta t = 10.0$ s | |
| $P$ | | |

| Strategy | Calculations |
|---|---|
| Use the formula for power. | $P = \dfrac{W}{\Delta t}$ |
| All needed variables are given, so substitute the variables into the formula. | $P = \dfrac{1.50 \times 10^5 \text{ J}}{10.0 \text{ s}}$ |
| Divide. | $P = 1.50 \times 10^4$ J/s<br>$P = 1.50 \times 10^4$ W |

The crane is able to generate $1.50 \times 10^4$ W of power.

## Validate

Work was given in joules and time in seconds.
Power is in J/s or W, which is correct.

---

2. **A cyclist and her mountain bike have a combined mass of 60.0 kg. She is able to cycle up a hill that changes her altitude by $4.00 \times 10^2$ m in 1.00 min.**

   (a) **How much work does she do against gravity in climbing the hill?**

   (b) **How much power is she able to generate?**

---

## Frame the Problem

- The cyclist is *doing work* against gravity by cycling uphill, thus *changing her altitude*.

- She is therefore *changing* her *gravitational potential energy*.

- Her *work done* will be equal to her change in *gravitational potential energy*.

- The *time interval* is given, and you can calculate the work done. Therefore, you can use the formula for *power*.

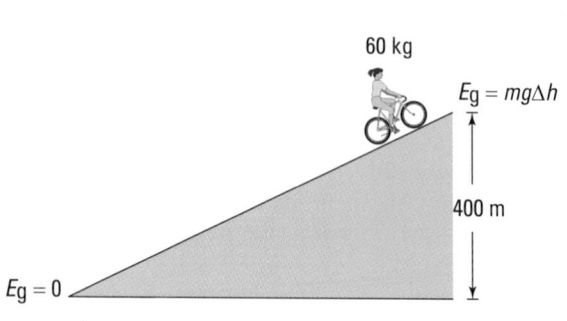

## Identify the Goal

(a) Her work done, $W$, in climbing the hill

(b) The power, $P$, that she generated

## Variables and Constants

| Involved in the problem | | Known | Implied | Unknown |
|---|---|---|---|---|
| $m$ | $g$ | $m = 60.0$ kg | $g = 9.81 \frac{m}{s^2}$ | $W$ |
| $\Delta d$ | $P$ | $d = 4.00 \times 10^2$ m [up] | | $E_g$ |
| $\Delta t$ | $E_g$ | $\Delta t = 1.00$ min | | $P$ |
| $W$ | | | | |

| Strategy | Calculations |
|---|---|
| Work done is equal to change in gravitational potential energy. | $W = E_g$ $E_g = mg\Delta h$ |
| Substitute known values. | $E_g = (60.0 \text{ kg})(9.81\frac{m}{s^2})(4.00 \times 10^2 \text{ m})$ |
| Multiply. | $E_g = 2.3544 \times 10^5 \frac{kg \cdot m}{s^2} \cdot m$ $E_g = 2.3544 \times 10^5 \text{ N} \cdot \text{m}$ $E_g = 2.3544 \times 10^5 \text{ J}$ $W = 2.3544 \times 10^5 \text{ J}$ |

(a) Therefore, the work done by the girl to cycle to the top of the 400 m hill is $2.35 \times 10^5$ J.

| | |
|---|---|
| Use the formula for power. | $P = \dfrac{W}{\Delta t}$ |
| Convert time to SI units. | $t = (1.00 \text{ min}) \left(\dfrac{60 \text{ s}}{\text{min}}\right)$ $t = 60.0$ s |
| The values are known, so substitute. Divide. | $P = \dfrac{2.3544 \times 10^5 \text{ J}}{60.0 \text{ s}}$ $P = 3.924 \times 10^3 \dfrac{\text{J}}{\text{s}}$ $P = 3.924 \times 10^3 \text{ W}$ |

(b) The cyclist generated $3.92 \times 10^3$ W of power. That amount of power is equivalent to 5.25 hp or sixty-five 60 W light bulbs.

---

## Validate

The power is in watts, which is correct.

*continued* ▶

continued from page 25

## PRACTICE PROBLEMS

10. A mover pushes a 25.5 kg box with a force of 85 N down a 15 m corridor. If it takes him 8.30 s to reach the other end of the hallway, find the power generated by the mover, in watts.

11. A chair lift carries skiers uphill to the top of the ski run. If the lift is able to do $1.85 \times 10^5$ J of work in 12.0 s, what is the power of the chair lift in both watts and horsepower?

12. A 75.0 kg student runs up two flights of stairs in order to reach her next class. The total height of the stairs is 5.75 m from the ground level. If the student can generate 200 W of power and has 20.0 s to reach her classroom at the top of the stairs, will the student be on time for class?

13. A small car travelling at 100 km/h has approximately $3.6 \times 10^5$ J of kinetic energy.
    (a) How much water in kg could that much energy warm from room temperature (20°C) to boiling (100°C)?
    (b) What amount of power would be generated if the total mass of water was warmed in 11 min?

14. A well-insulated shed, built to house a water pump, is heated by a single 100 W light bulb.
    (a) If the shed has 10.4 kg of air (assumed to be mostly nitrogen), how long, in minutes, would it take the light bulb to raise the air temperature from –8°C to +2°C? Assume that all of the bulb's energy is thermal energy.
    (b) Is your answer to part (a) reasonable? What assumptions could you improve?

15. A 2.0 kg bag of ice is used to keep a cooler of pop cold for 5.5 h. If the ice had an initial temperature of –4°C and, after 5 h, was liquid water with a temperature of 3°C, find the power of energy absorption.

# MULTI LAB Investigating Power

## Work, Power, and Gravity

You will need two marbles of different sizes, a golf ball, a stopwatch, and a board to act as a ramp. Set up the board so that it forms a ramp approximately 45° to the horizontal. Time the marbles and golf ball as they race to the bottom of the ramp. Verify that the race is fair by controlling necessary variables.

## Analyze and Conclude

1. What did you notice about the time required for the different-sized balls to reach the bottom of the ramp?

2. What did you notice about the relative speed of each ball when it reached the bottom of the ramp?

3. What effect did mass have on the time or speed?

4. What ball required the most power to be generated?

5. If gravity is generating the power, what limits exist on the amount of power that could be generated?

6. How does society make use of the power generation of gravity?

### TARGET SKILLS
- **Predicting**
- **Performing and recording**
- **Analyzing and interpreting**
- **Identifying variables**
- **Communicating results**

## What Is Your Horsepower?

How much horsepower can you develop? How important is speed? How important is mass? To determine your horsepower, all you need is your mass, a stopwatch, and a staircase of known height. Have a classmate record the time it takes you to climb a flight of stairs. Choose a relatively high flight of stairs and run more than one trial.

**CAUTION** If you have any respiratory or heart problems or any physical condition that could be compromised by running up stairs, do not actively participate in this investigation. Do the theory and calculations only.

### Analyze and Conclude

1. Make an educated guess as to what your horsepower will be. Refer to the History Link on page 277 to help you with your guess.

2. Calculate the work you did against gravity.

3. Use the amount of work you did to calculate the power you generated.

4. Report your answer in watts (W) and in horsepower (hp). **Note:** 746 W = 1 hp.

## Generating Power, Transforming Energy

Using a hand-held electric generator and electric winch assembly as shown, investigate what variables determine the amount of power generated. Try different masses and be very careful to observe the relative difficulty you experience when turning the hand crank.

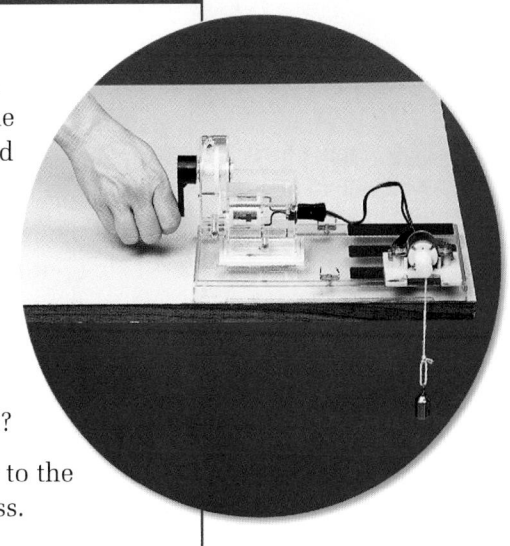

### Analyze and Conclude

1. What was generating the electric power?

2. What was generating the mechanical power?

3. What variables affect the amount of power generated?

4. Trace the energy path from your muscles all the way to the gravitational potential energy stored in the lifted mass.

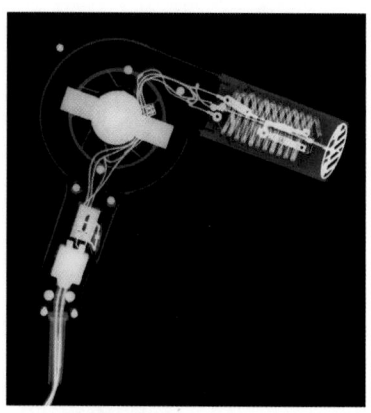

**Web Link**

*www.school.mcgrawhill.ca/ resources/*

Gravity is the force behind the generation of power in many examples in nature. Learn more about power generation in nature by searching the Internet. Follow the links for **Science Resources** and **Physics 11**.

# Efficiency

A light bulb is designed to convert electric energy into light energy. A car engine is designed to convert chemical potential energy stored in the fuel into kinetic energy for the car. However, both the light bulb and the car engine become extremely hot while they perform their designed function. Obviously, they have transformed much of the energy into thermal energy. While the light bulb and the car engine are transforming some of the potential energy into the desired form of energy, much energy is "lost."

As you well know, energy cannot be destroyed. However, it can be, and is, converted into forms that do no work or do not serve the intended purpose. Transforming energy from one form to another always involves some "loss" of useful energy. Often the lost energy is transformed into heat. The **efficiency** of a machine or device describes the extent to which it converts input energy or work into the intended type of output energy or work.

---

**DEFINITION OF EFFICIENCY**

Efficiency is the ratio of useful energy or work output to the total energy or work input.

$$\text{Efficiency} = \frac{E_o}{E_i} \times 100\% \quad \text{or} \quad \text{Efficiency} = \frac{W_o}{W_i} \times 100\%$$

| Quantity | Symbol | SI unit |
|---|---|---|
| useful output energy | $E_o$ | J (joule) |
| input energy | $E_i$ | J (joule) |
| useful output work | $W_o$ | J (joule) |
| input work | $W_i$ | J (joule) |
| efficiency | (none) | none; efficiency is a ratio; units cancel in ratios |

---

**Figure 6.15A** Even a hair dryer that is designed to produce thermal energy is not 100% efficient. What other "wasted" forms of energy does it produce?

In most machines and devices, as well as in natural systems, there is more than one energy transformation process. Energy path diagrams provide a visual way to represent how well the energy being put into a system is being transferred into useful work. They also show how and where the input energy is being lost — that is, being transformed into unwanted forms of energy. A few examples are shown in Figure 6.15B.

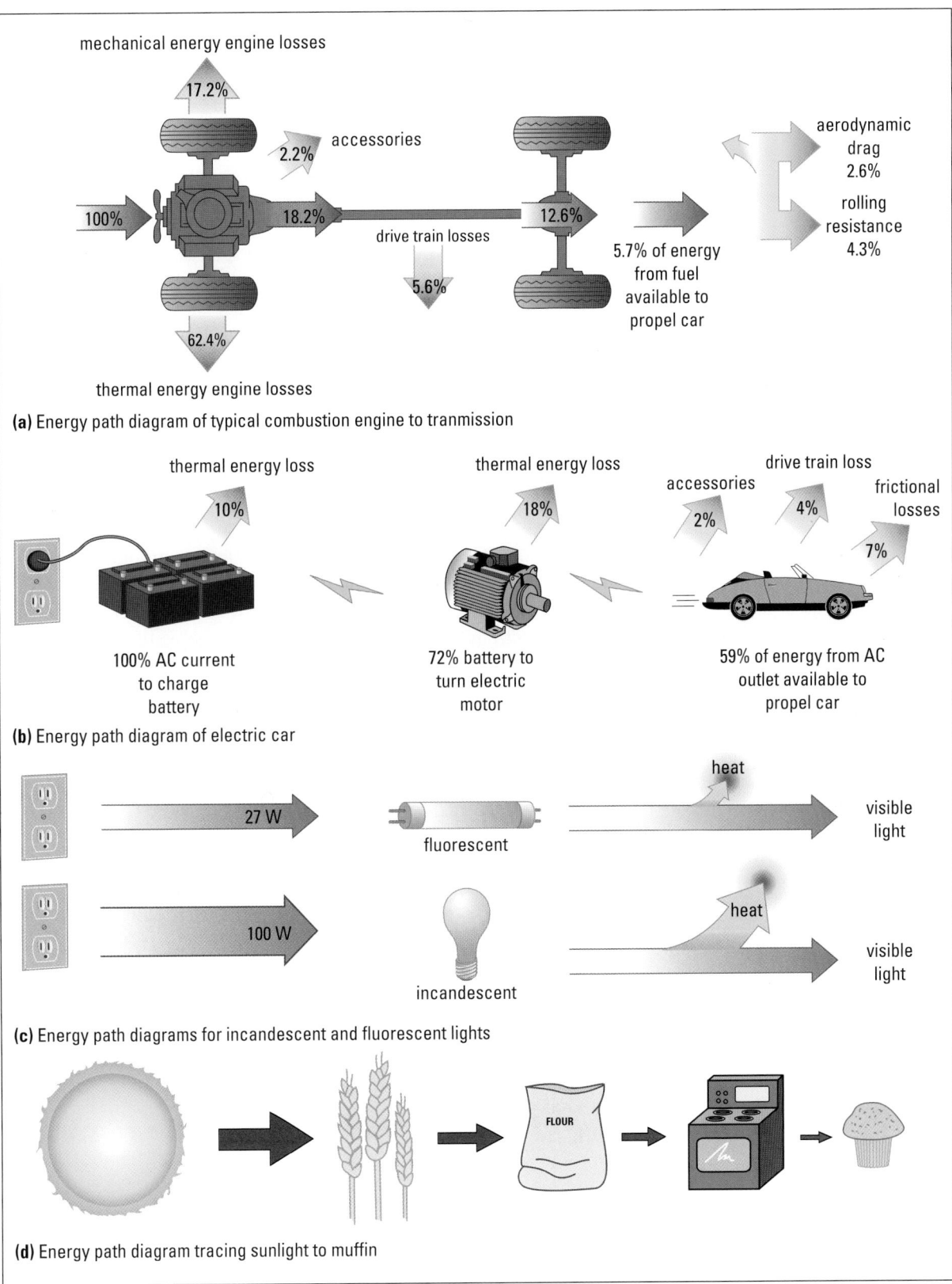

**(a)** Energy path diagram of typical combustion engine to tranmission

**(b)** Energy path diagram of electric car

**(c)** Energy path diagrams for incandescent and fluorescent lights

**(d)** Energy path diagram tracing sunlight to muffin

**Figure 6.15B** Energy path diagrams illustrate energy transformations into both useful and wasted forms.

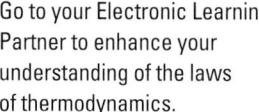

# Second Law of Thermodynamics

The three processes pictured in Figure 6.16 represent ones that do not violate the first law of thermodynamics, and yet they never occur spontaneously. The reason lies in the second law of thermodynamics, which (recall from Section 6.1) requires that thermal energy is always transferred from an object at a higher temperature to an object at a lower temperature. Part A of the figure can be compared to putting an ice cube in a drink to cool it, only to discover that the ice cube got colder and the drink got hotter. You would be shocked if this happened.

This concept can be applied to machines such as steam engines. While thermal energy is going from high- to low-temperature objects, some of the heat can do mechanical work. The second law of thermodynamics goes one step further by asserting that during the process of heat transfer and doing mechanical work, some thermal energy will leave the system without doing work. As illustrated in part B of the figure, it is not possible to use 100% of the thermal energy of a system to do work. No machine is 100% efficient.

Finally, part C of Figure 6.16 indicates that a random system will not spontaneously become ordered. Try to imagine that you are holding a glass of water and it suddenly feels warm, because all of the energetic molecules (hot) migrated to the walls of the glass, while the less energetic molecules (cool) moved to the centre.

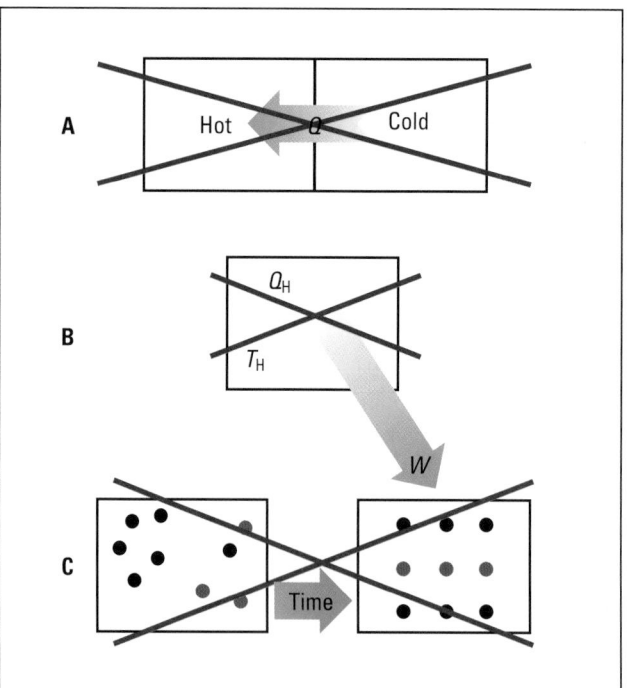

**Figure 6.16** The three processes shown here are allowed by the first law of thermodynamics but are forbidden by the second law.

## Calculating Efficiency

A model rocket engine contains explosives storing $3.50 \times 10^3$ J of chemical potential energy. The stored chemical energy is transformed into gravitational potential energy at the top of the rocket's flight path. Calculate how efficiently the rocket transforms stored chemical energy into gravitational potential energy if the 0.500 kg rocket is propelled to a height of $1.00 \times 10^2$ m.

### Frame the Problem

- The rocket engine transforms *chemical potential energy* first into kinetic energy that will propel the rocket *upward* against gravity.

- The kinetic energy is transformed into *gravitational potential energy*.

- If the energy transformation was 100% *efficient*, then the rocket would reach a height such that its *gravitational potential energy* would be equal to the chemical energy stored in the explosives.

- That will not happen, because a great deal of *energy is lost* as *thermal energy* created both by the combustion of the fuel and by friction with the rocket and the atmosphere.

**(a)** maximum height = maximum gravitational potential energy

**(b)** energy lost to atmosphere due to friction

**(c)** energy lost to environment as thermal energy from engine combustion

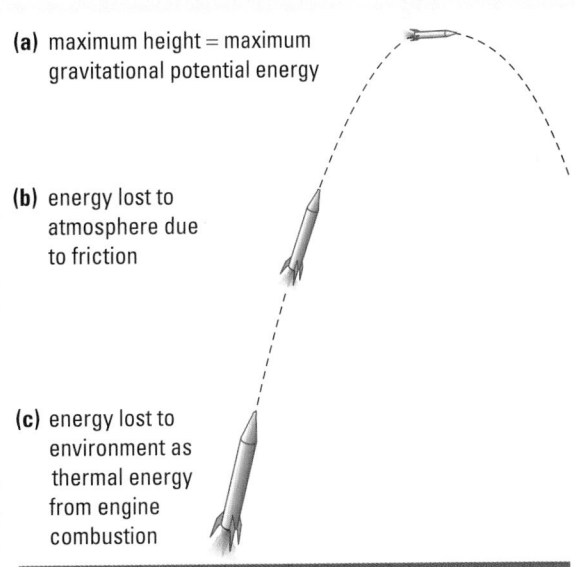

### Identify the Goal

Efficiency of the rocket and engine system in transforming chemical potential energy into gravitational potential energy

### Variables and Constants

| Involved in the problem | | Known | Implied | Unknown |
|---|---|---|---|---|
| $m$ | $W_i$ | $m = 0.500$ kg | $g = 9.81 \dfrac{\text{m}}{\text{s}^2}$ | $W_o$ |
| $\Delta h$ | $W_o$ | $\Delta h = 1.00 \times 10^2$ m | | efficiency |
| $g$ | efficiency | $W_i = 3.50 \times 10^3$ J | | |

*continued* ▶

## Strategy

The useful work output is the gravitational potential energy.

Calculate the gravitational potential energy that the rocket has at the top of its flight.

Substitute in the variables and multiply.

All the variables needed to calculate the efficiency have been determined.

Calculate the efficiency.

## Calculations

$W_o = E_g$

$E_g = mg\Delta h$

$W_o = E_g = (0.500 \text{ kg}) (9.81 \frac{\text{m}}{\text{s}^2})(100 \text{ m})$

$W_o = 490.5 \frac{\text{kg m m}}{\text{s}^2}$

$W_o = 490.5 \text{ N} \cdot \text{m}$

$W_o = 490.5 \text{ J}$

$\text{Efficiency} = \frac{W_o}{W_i} \times 100\%$

$\text{Efficiency} = \frac{490.5 \text{ J}}{3500 \text{ J}} \times 100\% = 14\%$

The energy stored in the rocket engine is transformed into gravitational potential energy (height) of the rocket with an efficiency of 14%. Most of the "lost" energy is transferred to the surroundings as thermal energy.

## Validate

The gravitational energy at the rocket's maximum height is correctly assumed to be the $W_o$.

Efficiency is given as a percentage, which is correct.

---

### PRACTICE PROBLEMS

**16.** A portable stereo requires 265 J of energy to operate the CD player, yielding 200 J of sound energy.

**(a)** How efficiently does the stereo generate sound energy?

**(b)** Where does the "lost" energy go?

**17.** A 49.0 kg child sits on the top of a slide that is located 1.80 m above the ground. After her descent, the child reaches a velocity of 3.00 m/s at the bottom of the slide. Calculate how efficiently the potential energy is converted to kinetic energy.

**18.** A machine requires 580 J of energy to do 110 J of useful work. How efficient is the machine?

**19.** An incandescent light bulb transforms 120 J of electric energy to produce 5 J of light energy. A florescent bulb requires 60 J of electrical energy to produce the same amount of light.

**(a)** Calculate the efficiency of each type of bulb.

**(b)** Why is the fluorescent bulb more efficient than the incandescent bulb?

**20.** A microwave oven transforms 345 J of radiant energy into 301 J of thermal energy in some food. Calculate the efficiency of this energy transformation.

21. A 125 g ball is thrown with a force of 85.0 N that acts through a distance of 78.0 cm. The ball's velocity just before it is caught is 9.84 m/s.

(a) Calculate the work done on the ball.

(b) Calculate the kinetic energy of the ball just before it is caught.

(c) What fraction of the energy transferred to the ball was lost to the atmosphere during flight?

22. Rubbing your hands together requires 450 J of energy and results in a thermal energy increase in your palms of 153 J. Calculate how efficiently the kinetic energy is converted to thermal energy.

23. A 1500 W hair dryer increases the thermal energy of 0.125 kg of carbon dioxide gas at constant pressure by $2.0 \times 10^3$ J in 2.0 s. Find the temperature rise of the carbon dioxide gas and the dryer's efficiency.

 **CANADIANS IN PHYSICS**

## Shifting Energy Sources: Piotr Drozdz and Azure Dynamics Inc.

For 100 years, combustion engines have supplied our cars with energy. This old-fashioned technology, however, is not efficient. Car and truck exhaust emissions are a major environmental problem, and owning a car is becoming more difficult with the rising cost of gasoline.

Electric cars have received a lot of attention as a possible alternative to combustion engines. They do not produce harmful emissions and they are more efficient, but the batteries that supply the electric energy need to be recharged frequently. Currently, there is little or no infra-structure to allow drivers to recharge their car batteries each day.

Piotr Drozdz, vice-president of technology at Azure Dynamics Inc. in Vancouver, is a pioneer in the field of hybrid electric vehicles (HEVs). HEVs combine the best of two worlds. They have combustion engines to provide the power needed when a car is travelling at high speeds, and electric motors for more efficient energy use at lower speeds. HEVs also do not have to be recharged. They generate their own energy through "regenerative braking." In HEVs, the brakes are coupled to the electrical system so that the kinetic energy of motion is converted to electrical energy.

Drozdz is the primary inventor of a system that acts as the "brains" and decides whether a hybrid electric vehicle uses its combustion engine or its more environmentally friendly electrical motor. The system uses sensors that ensure that energy use is as efficient as possible.

Vehicles powered by electricity is not a new idea. Electric cars like the one above were produced early in the last century. The electric motor was patented in 1821.

 **Web Link**

**www.school.mcgrawhill.ca/resources/**
For more information about hybrid electric vehicles and other types of alternatively powered cars, buses, and trucks, go to the above web site. Follow the links for **Science Resources** and **Physics 11**.

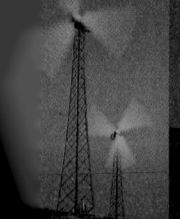

# INVESTIGATION 6-B

## Muscle Efficiency and Energy Consumption

**TARGET SKILLS**

- **Hypothesizing**
- **Performing and recording**
- **Analyzing and interpreting**
- **Communicating results**

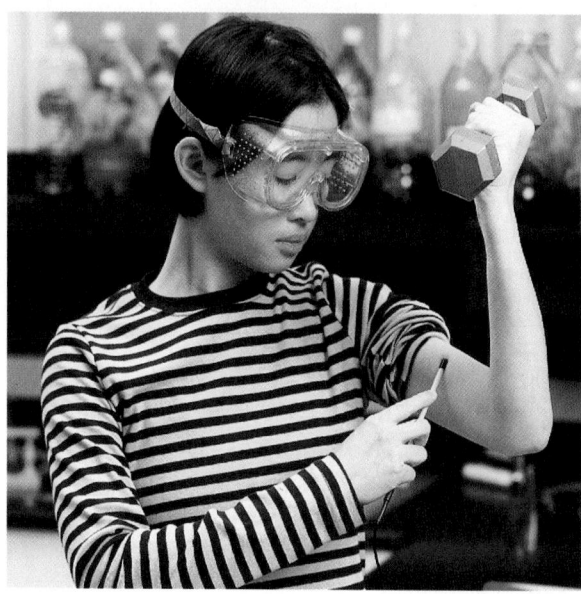

No real process is 100% efficient at transforming stored energy into either mechanical work or heat. Just as it is possible to calculate the efficiency of a light bulb or an automobile motor, with some basic assumptions, it is also possible to determine the approximate efficiency of your muscles.

## Problem

You can demonstrate quantitatively that your muscles lose energy when they work. How can the results be put in terms of work done, muscle efficiency, and food energy equivalence?

## Hypothesis

Form a hypothesis about the form the lost energy of the muscle will take. Predict the efficiency of your biceps muscle.

## Equipment

computer
data collection interface
temperature probe
dumbbell

## Procedure
### The Workout

1. Select a dumbbell with which you will just be able to do 10 biceps curls with one arm.

2. Connect and activate the computer interface and temperature probe.

3. Obtain a base temperature reading for your muscle by holding the probe firmly against your rested biceps for 60 s.

4. With the probe still firmly pressed against your biceps, perform 10 biceps curls with the dumbbell, using only one arm.

### Calculating Efficiency

5. Calculate the approximate volume of your biceps muscle. Assume that your arm is essentially a cylinder. Biceps curling involves the muscles on the front half of this cylinder. Therefore, you are able to approximate the volume of your biceps by measuring the length and circumference of your upper arm. Then, calculate the approximate volume of your biceps, using the method shown in the chart on the facing page.

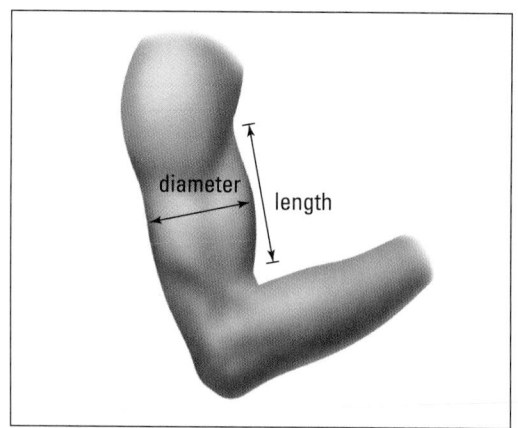

diameter    length

| | |
|---|---|
| Volume of a cylinder | $V_{cylinder} = \pi r^2 h$ |
| Circumference of a circle | $C = 2\pi r$ |
| Solve for $r$. | $r = \dfrac{C}{2\pi}$ |
| Substitute into the volume formula. | $V_{cylinder} = \pi \left(\dfrac{C}{2\pi}\right)^2 h$ |
| Expand. | $V_{cylinder} = \dfrac{\pi h C^2}{4\pi^2}$ |
| Simplify. | $V_{cylinder} = \dfrac{h C^2}{4\pi}$ |
| Divide by two for half of a cylinder. | $V_{half\ cylinder} = \dfrac{h C^2}{8\pi}$ |

6. Measure $C$ and $h$. If you take measurements in centimetres, your volume will be in cubic centimetres.

7. To approximate the mass of your biceps muscle, recall that a large percentage of your muscle fibres are water. Convert your biceps volume to the equivalent mass of water that it could contain by using the following relationship: 1 mL of water = 1 $cm^3$ = 1 g.

8. Calculate the quantity of heat generated. Recall:

   $Q = mc\Delta T$, where $m$ = mass of biceps converted to kg

   $c$ = 3500 J/°K kg (specific heat capacity for the human body)

   $\Delta T$ = change in temperature during workout

9. Determine the work done during the 10-repetition workout.

   Recall that $W = F_{\parallel}\Delta d$.

In this case, the force is equal in magnitude but opposite in direction to the force of gravity acting on the mass: $F = mg$.

The distance the mass is moved is equal to the length of the arm.

10. Calculate your biceps muscle's approximate efficiency.

   Recall: Efficiency = $\dfrac{W_o}{W_i} \times 100\%$,

   where $W_o$ = useful work output
   $W_i$ = total work output

   *In the case of curling a mass, the total work input is the sum of both the work done on the mass and the heat generated in the biceps. The useful work output is only the work done on the mass.*

## Analyze and Conclude

1. How did the biceps curls affect the temperature of your working arm?

2. What caused the observed changes?

3. What is the approximate efficiency of your biceps muscle?

4. List all of the assumptions that you made in determining efficiency. How could your estimate be improved?

5. What biological function(s) generated the heat?

6. A single chocolate chip cookie contains $1.34 \times 10^6$ J of energy. Calculate how many times you would have to curl the mass you used in the experiment to consume all of the energy supplied by the cookie.

# Keeping Warm

Birds and mammals need to maintain a nearly constant body temperature. Even with insulation such as fur or feathers, they still lose heat by radiation and conduction. Two Ontario high school students, Sandra Amicone and Laura Anderson, conducted an experiment to determine how an animal might maintain its body temperature as the ambient temperature (temperature of the surrounding air) dropped. The students placed a hamster in an environmental cage. They controlled the ambient air temperature surrounding the hamster and measured the amount of $CO_2$ that the animal exhaled. Their graph of $CO_2$ emission versus air temperature is shown here.

## Analyze and Conclude

1. Analyze the students' graph to determine how the hamster maintained a constant body temperature as the ambient temperature dropped.

2. Write a paragraph explaining the conclusion you drew from the data.

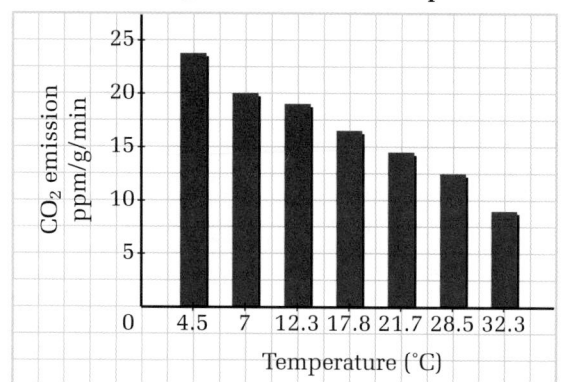

**Hamster $CO_2$ Emission versus Temperature**

---

## 6.2　　Section Review

1. **K/U** Describe the difference between work and power.

2. **C** Generate an energy-path diagram to show the electric energy consumed in your home.

3. **I** Develop an algebraic relationship for power in terms of force, $F$, and constant velocity, $v$. (Hint: Begin with the power formula and make substitutions for work, $W$.)

4. **MC** Based on the second law of thermodynamics, does an air conditioner pump "cold" in or "heat" out of a house?

5. **I** Consider Table 6.5. Draw an energy path diagram to suggest where energy is being consumed when travelling by (a) city bus and (b) ocean liner.

6. **MC** Using Table 6.5, compare cycling efficiency to driving, flying, and using a snowmobile.

**Table 6.5**
Transportation Energy Requirements

| Mode of transportation | Energy consumption (kJ/km) |
| --- | --- |
| bicycle | 52 |
| walking | 170 |
| city bus | 360 |
| car | 674 |
| jumbo jet | 2252 |
| snowmobile | 6743 |
| ocean liner | 8117 |

Near the beginning of this unit, you read a description of a typical Canadian's morning routine. The routine required the use of several different forms of energy, such as electricity to run the refrigerator and the microwave, gasoline to power the car, and even light for a solar calculator. Scientific and technological advances have led society from a world that required only food energy to be transformed into muscle power, to one that makes use of every imaginable form of energy. As well, you learned in the last section that, in every energy-transformation process, some useful energy is lost. Clearly, society cannot continue to demand more and more energy without consideration for the future generations.

The next challenge is to develop energy sources and processes that are sustainable. A sustainable resource is one that will not deplete over time and will not damage Earth's sensitive biosphere, while still being able to provide for the energy demands of society. To that end, the use of fossil fuels, which have significant detrimental effects on our biosphere, needs to be and is being replaced with the use of alternative fuel sources.

## Survey of Forms of Energy

A fundamental understanding of the physics of energy forms and transformations will help you to better evaluate the potential sources of energy for society. As you have learned, all forms of energy can be classified as one of two *types*, either as stored or potential energy or as moving or kinetic energy. Figure 6.17 on page 292 provides a survey of common forms of both potential and kinetic energies.

## Energy Sources for Today and Tomorrow

Following the review of scientific principles of energy, you will find a survey of technologies related to our energy sources. The short topics highlight current and alternative energy sources, their benefits, and their inadequacies. The dynamic nature of science guarantees that new discoveries will take future research into yet unimagined directions. The following descriptions provide background information and vocabulary to provide a starting point for meaningful discussion and research.

SECTION
EXPECTATIONS

• Analyze the economic, social, and environmental impacts of various energy sources.

• Analyze various energy-transformation technologies from social, economic, and environmental perspectives.

• Synthesize, organize, and communicate energy transformation concepts using diagrams.

KEY
TERMS

• fission
• fusion
• biogas
• wind power
• wave power
• tidal power

• geothermal energy
• photovoltaic cell
• fuel cell
• space-based power

**Technology Link**

Dutch researchers have developed carbon dioxide-trapping technologies, using both the ocean and underground rock deposits. These technologies may help to reduce the ever-increasing amount of greenhouse gas in the atmosphere. What are some technologies that are helping to reduce carbon dioxide emissions? Keep abreast of technological developments in this area as you progress through the course.

**Figure 6.17** Common forms of potential and kinetic energies

## POTENTIAL ENERGIES

### Energy Form

**Explanation**

**Chemical Potential**

Chemical potential energy is the energy contained within the bonds between atoms. These bonds can take many different forms, including energy derived from carbohydrates in food to energy stored in gasoline. Food energy allows us to do work and gasoline allows our vehicles to do work.

**Elastic Potential**

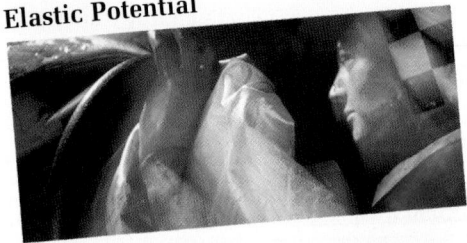

Elastic potential energy is the energy stored within a stretched or compressed object, such as an elastic band or a car bumper. Pole-vaulters store elastic energy in the pole and then use it to glide over the bar.

**Electric Potential**

Since opposite charges attract, electric potential energy can be stored by a separation of positive and negative charges. Electric potential energy does work as the electrons move from negatively to positively charged objects.

**Gravitational Potential**

Gravitational potential energy is the energy of position. A roller-coaster cart poised at the top of a large hill contains gravitational energy that will result in the cart's passengers' enjoyment of the ride.

**Nuclear**

Nuclear energy consists of the energy stored within the nucleus of an atom. When extremely large atoms, such as uranium or plutonium, undergo **fission**, or split in two, tremendous amounts of energy are released. Nuclear power stations use this energy to generate electric energy. Research continues in an attempt to harness energy released when two extremely small atoms undergo **fusion**, or join together.

## Energy Form

**Mechanical Kinetic**

## Explanation

Kinetic energy is the energy of motion. Any object that is moving has kinetic energy, from atoms and molecules to cars, to planets.

**Sound**

Sound energy travels through a substance as a wave. (You will study sound waves in Unit 3.) As a sound wave passes through, the atoms or molecules of the substance vibrate back and forth, colliding with the adjacent atoms or molecules. Sound vibrations cause a person's eardrums to vibrate.

**Thermal**

Thermal energy is the energy of random motion of the particles that make up an object or system. A bathtub filled with 65°C water has much more thermal energy than a teacup of water at the same temperature.

**Radiant**

Radiant energy travels as an electromagnetic wave. Although electromagnetic waves do not involve a moving mass, they do carry energy through space. In fact, in some applications, radiant energy is treated as massless particles or packets of energy called "photons." Radiant energy from the Sun supplies Earth with all of the energy required to sustain life. It also drives all weather systems.

### Think It Through

- List the nine forms of energy identified in Figure 6.17 and provide two examples of each that are different from those given there.

- A fraction of energy is always "lost" as energy is transferred from one form of energy to another. What form of energy does this "lost" quantity usually take?

### Web Link

**www.school.mcgrawhill.ca/resources/**
Check out the above web site to learn more about the latest in energy research and technology. Follow the links for **Science Resources** and **Physics 11**.

# Muscle Power

Canadian wrestler Daniel Igali,
Sydney Olympics gold medal winner

Hunting and gathering were a way of life for people of the Stone Age. With the beginning of agriculture, societies began to herd animals, as well as using animal muscle power to plow their fields. As human settlements increased in size, the energy needs required to circulate goods led to the domestication of animals not only for food but also for transport. Once the muscle power of animals was in use, the opportunity to build and pull heavy carts became evident. With the advent of the wheel, roads were required to travel on, and the construction of roadways demanded more human and animal muscle power than ever before.

Developing countries still depend on muscle power for the majority of their daily tasks, such as getting water from a distant well. First World countries still employ the use of muscle power, but often more for leisure than for sustaining life.

## Advantages

- Muscle power is 100% environmentally sustainable, there are no toxic by-products, and when the equipment (human or animal) expires, the remains fold neatly back into Earth's nutrient cycle.

## Disadvantages

- The amount of work that can be accomplished is limited to the physical strength and power of the work force.

- Human rights violations, such as child labour, occur, as well as unsafe working conditions.

| IN CANADA |  |
| --- | --- |
| **Energy Consumption** (kJ/passenger-kilometre) | |
| bicycle | 52.7 kJ |
| walking | 170 kJ |
| small car | 674 kJ |

## Muscle Power Energy Path

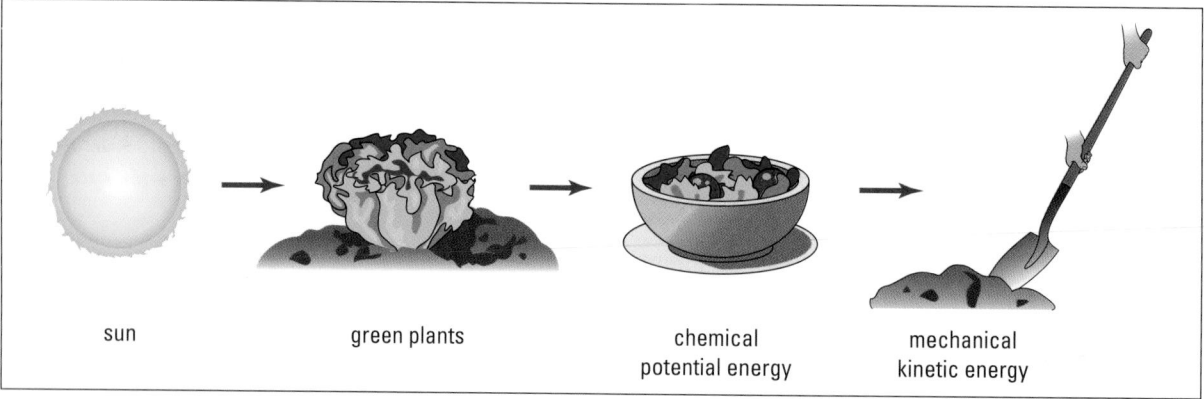

sun          green plants          chemical potential energy          mechanical kinetic energy

Steam-driven devices had been in existence for centuries before colonial times. However, they were inefficient and impractical. Then, James Watt made modifications that made steam engines so efficient that they sparked the Industrial Revolution. Soon, steam engines were being employed in every conceivable area of manufacturing and transport. The steam engine required energy — first wood, then coal was used to fuel society's appetite for production.

Fossil fuels are the remains of million-year-old plant life — now coal — or aquatic animal life — now gasoline and natural gas. Fossil fuels were first put to use by the Chinese and then the Romans, but it is the much more recent industrialization of so many societies that has made fossil fuels the primary fuel source of our planet. Approximately 1.4 trillion tonnes of coal are recovered annually.

Coal remained the primary fuel source until the mid-1950s, when oil-fired electricity generation and combustion engines became more economical. The consumption of oil increased rapidly, doubling every 15 years.

Today, 80% of available oil is consumed in North America, Western Europe, the former Soviet Union, and Japan. Two-thirds of the world's oil production comes from five Middle Eastern countries — Iran, Iraq, Kuwait, Saudi Arabia, and the United Arab Emirates.

### Advantages
- Fossil fuels are relatively inexpensive to process because the infrastructure to do so already exists.

### Disadvantages
- Combustion of coal and oil products releases harmful by-products into the air, reportedly contributing to 1000 deaths per year in Toronto alone.
- Combustion products are a source of greenhouse gases and contribute to global warming.
- The production of all types of plastics requires oil by-products, including materials required to create solar panels and other alternative energy transformation equipment.
- Supplies of fossil fuels are limited.

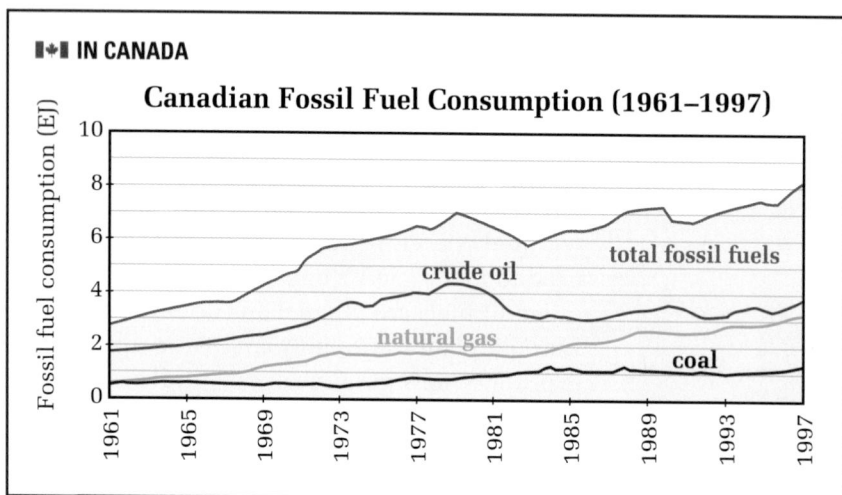

### Canadian Fossil Fuel Consumption (1961–1997)

Fossil fuel consumption (EJ)

total fossil fuels

crude oil

natural gas

coal

**Math Link**

Statistics Canada and the United Nations archives keep detailed records over a wide range of social, economic, and scientific data. Do research to obtain up-to-date data. Have the types of fuels Canadians use changed since the last data on this graph?

## Fossil Fuel Power Energy Path

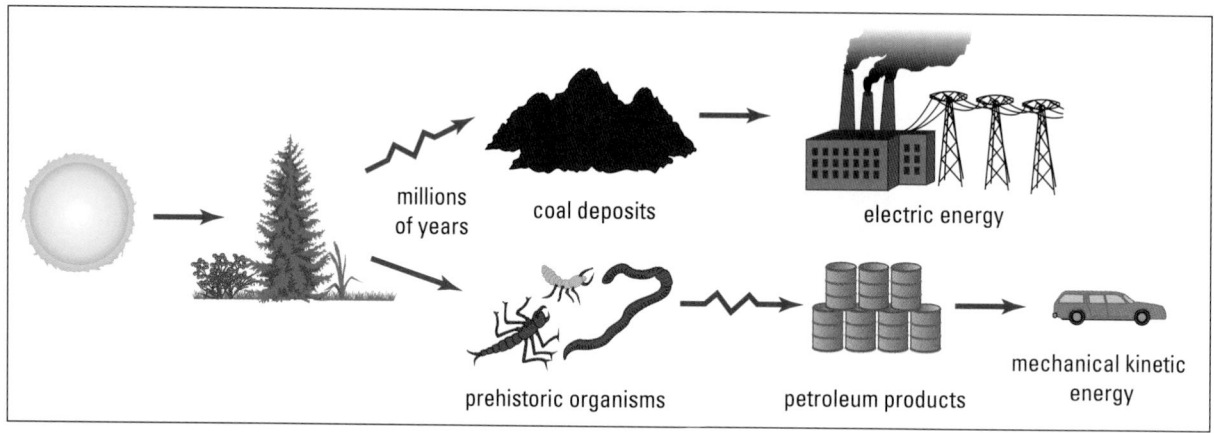

millions of years

coal deposits

electric energy

prehistoric organisms

petroleum products

mechanical kinetic energy

## Biomass

"Bio" means life, so bioenergy is energy from living things. The term "biomass" refers to the material from which we get bioenergy. Biomass is produced when the Sun's solar energy is converted into plant matter (carbohydrates) by the process of photosynthesis. Only green plants and photosynthetic algae, containing chlorophyll, are able to use solar energy that originated 150 million kilometres from Earth to synthesize biomass from which we get bioenergy. The simplest process employed to make use of this energy is eating. Every time you eat a fruit, a vegetable, or a processed version of either, you are taking advantage of the energy stored as biomass.

There are many methods currently used around the world to make the best possible use of biomass energy. Burning wood is becoming increasingly popular as a method of electricity generation. Combustion fuels that burn more cleanly are being developed. They use alcohol derived from corn as an additive. This fuel, while marginally more expensive, is available in Canada at agricultural co-operatives. This is an attempt to promote the environmental benefits as well as to provide another use for agricultural products.

Rotting plant matter breaks down under the action of bacteria in the same way that food energy is extracted in the human digestive tract.

One tonne of food waste can produce 85 $m^3$ of **biogas**, which is composed of methane, carbon dioxide, and hydrogen sulfide gas. The biogas is 60% methane and the rest is primarily carbon dioxide. It is an excellent source of fuel for heat- and power-generating plants. It is considered to be a $CO_2$ neutral fuel because the carbon dioxide that is released was only very recently removed from the atmosphere. Burning wood is also considered to be $CO_2$ neutral, unlike burning fossil fuel, which releases carbon dioxide that has not been in the atmosphere for millions of years. These types of systems are increasing dramatically in popularity in Scandinavian and European countries.

## Advantages

- Biogas systems are highly suitable for processing liquid manure and industrial wastes. The residues can be used as nutrient-rich fertilizers.

- Utilization of biogas systems for farm manure reduces nitrate pollution and the chance of water contamination from E-coli bacteria.

## Disadvantages

- Successful biogas systems currently rely on the co-operation of regional farmers and industrial sites to supply a centralized plant. This co-operation can be difficult to achieve.

### IN CANADA

- There is limited use of biomass energy generation.
- Most common use is combustion of wood for home heating.

## Biomass Power Energy Path

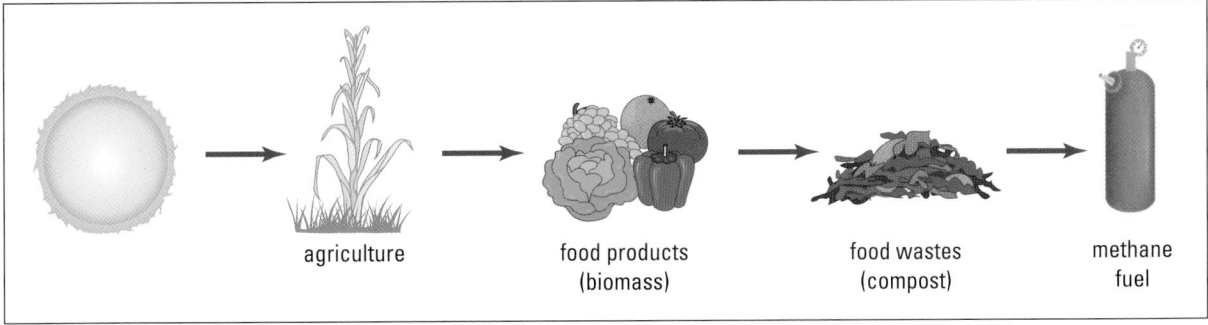

agriculture → food products (biomass) → food wastes (compost) → methane fuel

## Lake Ontario: Toronto's New Air Conditioner

Summertime in Toronto can be sweltering. Despite the summer heat, thousands of Torontonians go to work in downtown office buildings every day. Air-conditioning systems are used to keep the buildings cool, and the people inside them comfortable. These systems require massive amounts of energy, however, and they contribute to pollution and greenhouse gas production.

As a result, Enwave District Energy Ltd., a utility company, and the city of Toronto have decided to experiment with a more environmentally friendly air-conditioning system. The system makes use of the frigid temperature of the deepest water of Lake Ontario, and is called "deep-water cooling."

Office buildings are kept cool by circulating chilled water through air-conditioning systems. In traditional systems, the cooling of the water is electricity-driven and may require the burning of fuel, resulting in the emission of carbon dioxide. Older systems also use refrigerants, such as chlorofluorocarbons and hydrochlorofluorocarbons (CFCs and HCFCs), which damage the ozone layer.

The deep-lake cooling system in Toronto will use cold water from the depths of Lake Ontario near the Toronto Islands to cool the interiors of the city's buildings. Water will be drawn from about 70 m below the surface, where the temperature usually remains at 4°C. Onshore pumps will then suck the water from the lake and draw it into a heat exchanger through already existing pipes.

The heat exchanger transfers heat energy from the existing warm water to the cold lake water. As heat is transferred, the existing water becomes cooler. Finally, water distribution pipes circulate the new, chilled water throughout the building's air-conditioning system, cooling the office air.

Juri Pill, president of Enwave, says the deep-lake cooling project is in its final stages of design. In a few years, he claims, water from the depths of Lake Ontario will be cooling 46% of office buildings in the downtown Toronto core.

### Going further

1. Find out more about the construction costs of the project. How soon would the savings in electricity bills cover these costs?

2. The Deep-Lake Cooling Project uses 90 percent less electricity than regular air-conditioning systems, and will reduce carbon dioxide and refrigerant emissions. However, no new system is ever perfect, environmentally or otherwise. Consider the negative impacts or challenges of the project. Conduct a town hall meeting and address the concerns of possible opponents of the project.

How should Toronto keep its cool?

Tiverton, Ontario, wind generation station

The kinetic energy of the wind is thought to have carried Australian Aborigines from the mainland of southeast Asia to Australia 40 000 years ago. Travel by boat, using both muscle power and **wind power**, brought about the first contacts between distant people and created trade routes for merchandise and knowledge. Early civilizations harnessed wind power for sailing, but it was not until about 2000 years ago that the first windmills were constructed in China, Afghanistan, and Persia. The windmills were used mainly to pump water for irrigation of farmland. Over 1000 years ago, people of the country we now know as Holland used windmills to pump out large inland lakes that were prone to flooding. The land left behind became fertile farmland.

Wind power is currently used in many parts of the world to generate electricity. By 1996, the wind-turbine installation at California's windy Altamont Pass had a total generating capacity of 3000 MW. This generating capacity is the same as the capacity of one of the Bruce nuclear plants in Ontario. A wind-farm test site, capable of generating 6 MW, has also been installed in Tarifa in southern Spain, a location known as a popular windsurfing destination. The purpose of the test site was not only to determine the suitability of the site, but also to develop locally the required infrastructure of technologies, financing, and materials. Should the tests be successful, the site will provide the community of Tarifa with the ability to implement an 8000 MW facility by 2005. The 6 MW site has already contributed power to the local electrical grid and reduced the use of local oil-fired plants. During the first year of operation, the emissions of $CO_2$ were reduced by about 12 000 t, $SO_2$ by 5 t, and $NO_x$ gases by 4 t. Currently, Ontario operates a 0.6 MW facility in Tiverton.

### Advantages
- Electricity generation is possible with zero emissions of greenhouse gases.
- There is very limited potential for accidents that could cause widespread ecological damage.

### Disadvantages
- It is expensive to develop the needed infrastructure.
- Wind power use is viable only where there is a relatively constant wind of sufficient strength.
- Wind farms require a great deal of open space.
- Suitable sites are often migratory paths for birds. This can cause injury or death to some of the birds.

---

**I✦I IN CANADA**

- Canada's landscape and climate have the potential to provide 485 GW.
- Currently, 124 MW are produced — enough to power 34 000 homes.

---

## Wind Power Energy Path

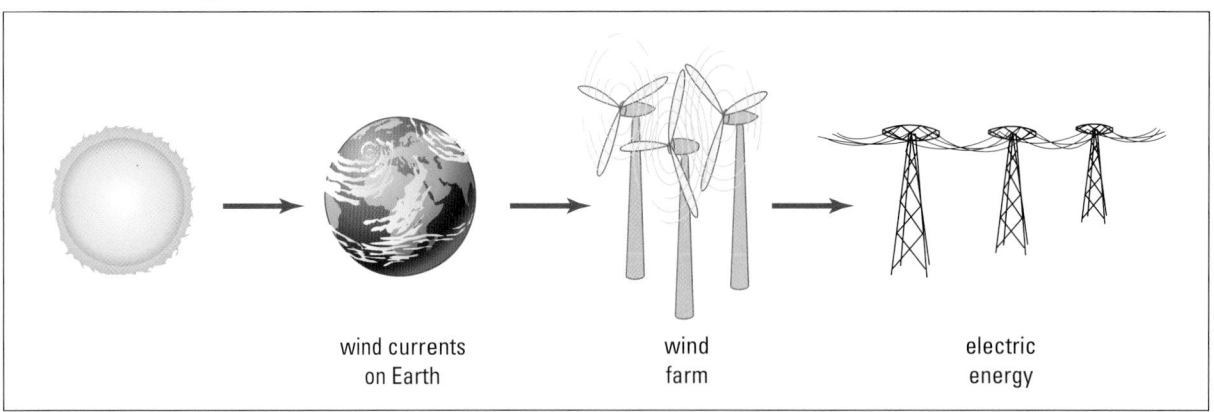

wind currents
on Earth

wind
farm

electric
energy

## Water Power

Canadian La Grande Complex in northern Québec

Electric energy generation using the gravitational potential energy of water is known as hydroelectric generation. In Ontario, seven separate facilities generate up to 2278 MW of power from the fast-moving waters approaching Niagara Falls. Ontarians are so familiar with the concept of generating power using water power that the term "hydro" is often used in place of electricity. The force of gravity does work on the water, pulling it down and providing it with a tremendous amount of kinetic energy. This kinetic energy is transformed into electric energy by very large turbines.

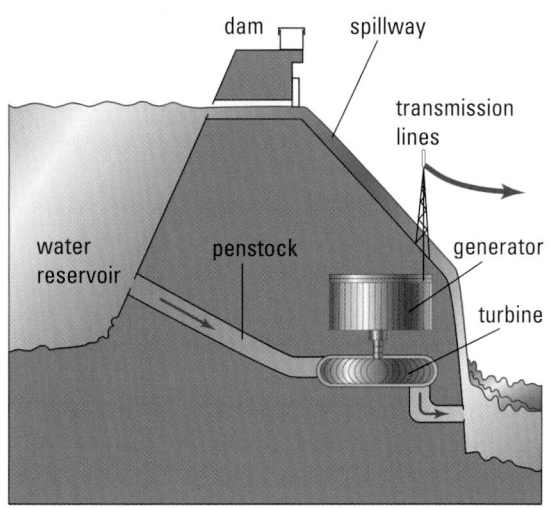

Energy is stored in the reservoir due to the tremendous mass of water and its height above the base of the dam. The turbine and generator convert this stored energy — gravitational potential energy — into electric energy.

The production of electric energy in this way appears to be a perfect solution, with minimal cost to the environment. This perception is false. In northern Québec, a massive hydro-electric generating dam project called the Canadian La Grande Complex was started in 1973. When completed, it will be the largest dam project in the world. The plan calls for the diversion of three rivers, reversing the flow of a fourth, and then channelling the water from all of those rivers into the La Grand-Rivière, which flows into James Bay. The Chinese government is currently constructing the Three Gorges Dam, which will be the largest single dam in the world when completed. For more details, read "Physics in the News" on page xxx.

Such large reservoirs sometimes flood thousands of hectares of farmland. In other cases, they drastically alter the ecosystem with unknown consequences. For many years, engineers did not have the technology to economically use smaller reservoirs for generating electricity. With improvements in technology, smaller generation facilities are becoming much more popular. Table 6.6 lists the number and capacity of small hydro-electric facilities in selected countries throughout the world.

**Table 6.6**
Number and Capacity of Small Hydro-Electric Facilities Worldwide

| Country | Number of small hydro plants | Power generation capacity (MW) |
| --- | --- | --- |
| Canada | 500 | 58 000 |
| China | 60 000 | 13 250 |
| France | 1 500 | 1 646 |
| Italy | 1 400 | 1 969 |
| Sweden | 1 350 | 8 400 |
| United States | 1 700 | 3 420 |

## Advantages
- Power generation is efficient.
- Once the facilities are operational, there is usually limited environmental impact.
- There are well-developed technologies and infrastructure.

## Disadvantages
- Water power requires fast-flowing water.
- Tremendous ecological damage often results from large generating facilities.

### ▐✦▌ IN CANADA
- 58 000 MW total hydro-electric generating capacity in Canada, including
  - 34 632 MW Québec
  - 8072 MW British Columbia
  - 7309 MW Ontario
- Canada exports power to the United States.

## Water Power Energy Path

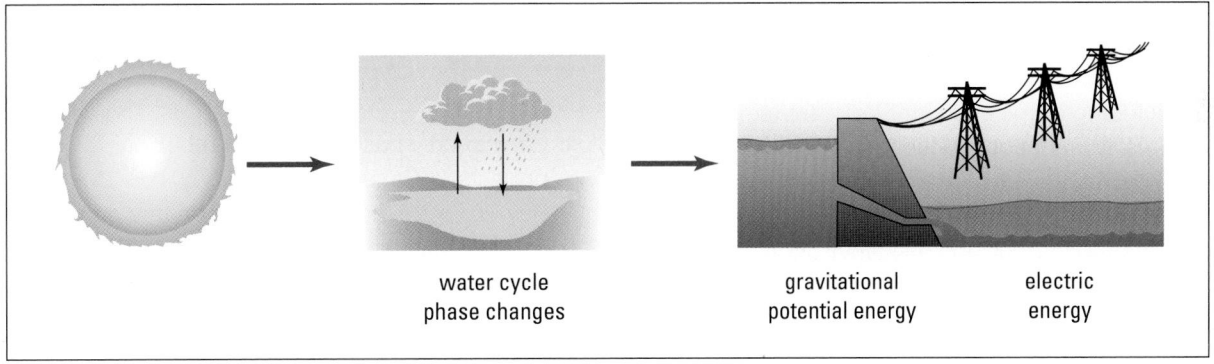

water cycle
phase changes

gravitational
potential energy

electric
energy

## China's Controversial Three Gorges Dam

The Three Gorges Dam on China's Yangtze River will be the largest dam ever built if it is completed as planned in 2008, but it may also be the most controversial construction project ever undertaken.

China's Three Gorges Dam

Almost 2 million Chinese citizens are being forced to leave their homes because a massive reservoir will be created upstream from the dam. About $120 \text{ km}^2$ of land will be flooded, including farms and significant archeological sites. Many people believe the dam will cause irreparable environmental damage and species extinction. Since the Chinese government began the project in 1994, environmentalists and other groups have tried to put an end to it.

Many people believe, however, that the future dam will benefit China and the environment. It will supply the country with needed energy in clean hydro-electric form. When water from the Yangtze River falls from the 185 m high dam, it will generate 16 750 MW of electrical power from 26 turbines — roughly the power output of 18 nuclear power plants. The dam will also cut back on the burning of fossil fuels, and thus reduce carbon dioxide emissions. Proponents also hope the dam and the newly available electrical energy will help boost China's economy. In addition to supplying energy, the dam will help control flooding of the Yangtze River, the third-largest river in the world. Hundreds of thousands of people have died in Yangtze River floods over the past 100 years.

In Canada, the dam has received mixed reviews. Canada's Export Development Corporation (EDC) has provided funds for the dam. The EDC reviewed the environmental and social impacts of the project and decided it met the requirements for financial support, says Rod Giles, EDC spokesperson.

Other Canadians remain steadfastly opposed to the dam and believe it comes at too high a price. "The Three Gorges Dam is unnecessary and costly," says Patricia Adams, executive director of Probe International, a Toronto-based watchdog group. Probe International has been trying to convince Chinese citizens and government officials in China and the Western world that the dam should not be supported.

### Going Further

1. Controversy has surrounded other dam projects. A Canadian example is the James Bay Project. Find out what made this dam controversial and compare the issues surrounding it with the issues surrounding the Three Gorges Dam.

2. Suggest key steps in the planning and approval of a project such as the Three Gorges Dam that would help resolve controversial issues more effectively.

Two thirds of Earth is covered by water, undoubtedly the largest solar collector available. The waves that ripple the ocean's surface only hint at the amount of energy collected and stored by the water. A better window for glimpsing the power that is stored as thermal energy in the oceans is to study the seasonal weather systems that form out in the middle of the Atlantic and Pacific Oceans — hurricanes and cyclones. The destructive power of hurricanes is demonstrated every year, as coastal communities are besieged by storms with winds of over 180 km/h. The ultimate source of the energy for these giant storms is the thermal energy of the ocean waters.

Tapping into the energy of the ocean in a reliable and predictable way has been approached in three basic ways — wave power, tidal power, and ocean thermal power. Tides were used as early as the eleventh century along the coast of present-day England to operate flour-grinding mills. Japan started researching the extraction of power from the ocean as a means to generate electricity as far back as 1945. Since then, engineers in several countries have studied numerous techniques for taking advantage of the power of the oceans. The invention of special turbines and hydraulic machines are opening the way toward practical uses of energy stored in the oceans.

---

### IN CANADA

**Wave Power**
Canada has open-water ocean coastlines capable of generating 70 MW/km of power on both the east and west coasts.

**Tidal Power**
The Bay of Fundy has more potential than anywhere else in the world.

**Ocean Thermal Power**
No development

---

## Wave Power

**Wave power** devices serve coastal communities in two important ways. One type of device, created in Japan and called "The Mighty Whale," is capable of extracting 60% of the waves' energy and, in doing so, reducing the waves' height by 80%. This provides both electric energy generation and shoreline protection. The sheltered region of shoreline may be suitable for aquaculture — farming of specific fish species. A Norwegian plant was successfully generating 0.5 MW of power until 1988 when 10 m high storm swells destroyed the facility.

### Advantages
- Countries that have access to an ocean have the potential to take advantage of wave power.
- There is a minimal threat to the environment.

### Disadvantages
- Currently few countries have invested in wave power research.
- Technology is expensive.

**Wave Power Energy Path**

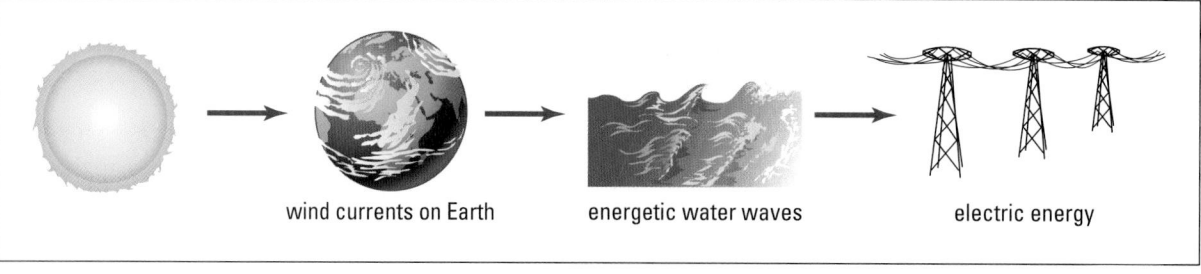

wind currents on Earth     energetic water waves     electric energy

## Tidal Power

The gravitational attraction of the moon as it orbits Earth causes bulges in the oceans nearest and farthest from the Moon. As our world rotates completely around on its axis once every 24 h, these two bulges become tides twice a day. Capturing high-tide waters, only to release them through turbines during low tide, is another method for generating electrical power from the ocean.

The first modern **tidal power** generation plant, with a capacity of 0.04 MW, was constructed in China in 1956. Since then, eight more tidal power stations have been built in China, with a total generating capacity of 6.2 MW. Canada built North America's first tidal power facility in 1994. The 17.8 MW plant was built on the Annapolis River west of Halifax in Nova Scotia. The Bay of Fundy Tidal Power Review studied several potential sites throughout Nova Scotia and New Brunswick. One of the most promising sites could generate an estimated 3800 MW of power. However, there are no current plans to construct tidal facilities because of the high production costs and the availability of several untapped sites that are suitable for hydro-electricity generation.

Physicists have calculated that rising and falling tides dissipate energy at a rate of two to three million megawatts. Unfortunately, only a small fraction of that energy is recoverable, approximately 23 000 MW worldwide, or about 1% of the available power from hydro

generation. This, and the fact that tidal facilities are viable in only a few locations around the world, means that tidal power will not become a global energy resource anytime soon.

| Tidal Power | |
|---|---|
| **Generation facility location** | **Power generation capacity (MW)** |
| France | 240.0 |
| Canada | 17.8 |
| China | 6.2 |

Bay of Fundy tidal power facility

## Advantages

- Zero greenhouse gas emissions are produced during the operation of the facility.
- There is a limited environmental impact at the power plant site.

## Disadvantages

- Few regions on Earth can take advantage of tidal power.
- Theoretical maximum power output per tidal cycle is only three times the power of one wind turbine.

**Web Link**

*www.school.mcgrawhill.ca/ resources/*
To learn more about tides and the potential of tidal power, go to the above web site. Follow the links for **Science Resources** and **Physics 11**.

**Tidal Power Energy Path**

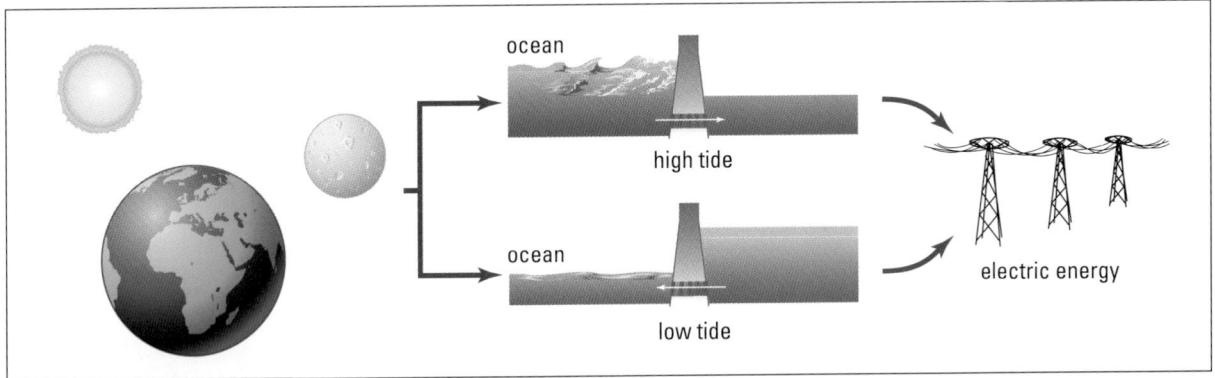

## Ocean Thermal Power

In 1881, a French physicist, Jacques d'Arsonval, proposed that the temperature differences between the surface and deep water of the oceans could be used to generate power. His system consisted of a large closed loop that would carry ammonia from the surface to the depths of the ocean and then back again. The ammonia would vaporize as it rose, surrounded by warm water, and drive electrical turbines. Then, it would condense again in the cold section of the loop at the bottom of the sea. In 1930, Georges Claude, a student of d'Arsonval and inventor of the neon light, built and tested an "open cycle" system in the waters off Cuba. The latest adaptation of both the open and closed systems uses temperature differences of at least 20°C between surface and deep waters to generate both electricity and fresh water. The system is known as the ocean thermal energy conversion (OTEC) system.

The Sun's energy warms the tropical ocean water to temperatures close to 30°C. The water,

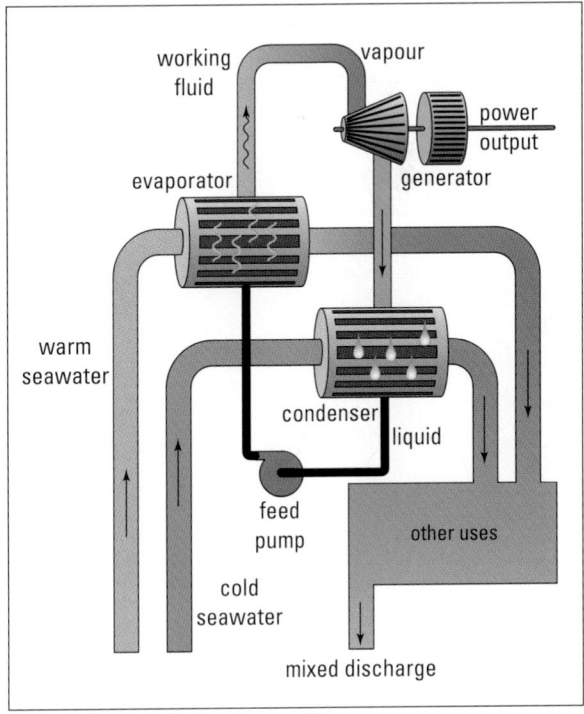

Closed-cycle OTEC system

due to its high specific heat capacity, maintains this temperature night and day, all year long. The water 1 km below the surface is significantly cooler, often around 4°C (water is most dense at 4°C). The electrical power generated by the OTEC system could be transmitted to a local power grid or used on site to produce hydrogen gas through the electrolysis of water. The electrical energy could also be used to make other valuable substances such as ammonia or methanol, which could be piped to other locations.

## Advantages
- OTEC operational systems would not contribute to greenhouse gas emissions.
- It is environmentally non-destructive at the site of the operation.

## Disadvantages
- It is suitable only in tropical ocean waters near the equator.
- Floating platform construction is expensive.

### Ocean Thermal Power Energy Path

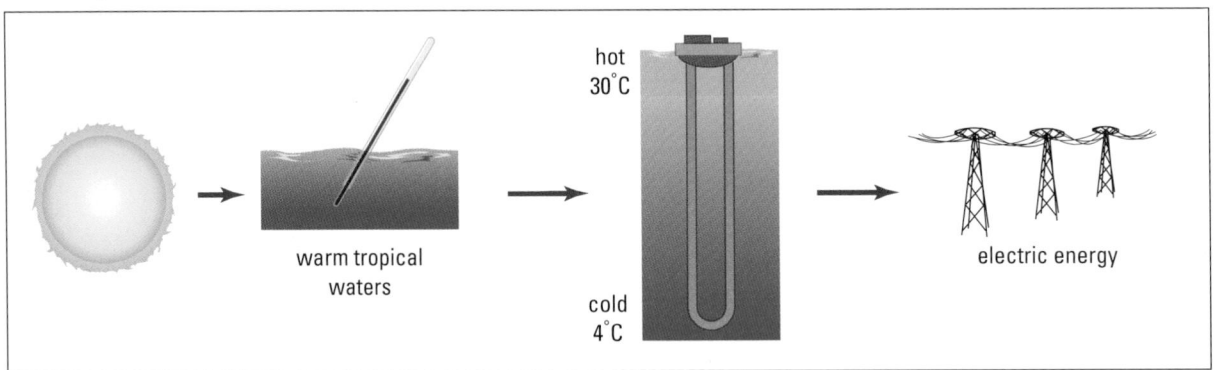

## Geothermal Power

**Geothermal energy** is the energy recovered from Earth's core. The thermal energy contained within Earth's core results from energy trapped almost 5 billion years ago during the formation of the planet, and from the heating effects of naturally decaying radioactive elements. The rate of heat flow through Earth's crust is 5000 times slower than the rate of energy arriving from the Sun. Therefore, Earth's surface temperature is the result of energy from the Sun, and not the heat flowing from its core.

Geothermal energy is not, in fact, a renewable energy source, because heat is extracted much more quickly than it is replaced. Although the total thermal energy is finite, the time it would take to deplete this resource would be measured in millennia because of the enormous size of Earth. For this reason, geothermal energy is considered to be renewable.

The four basic geologic formations that allow for the extraction of geothermal energy are hydrothermal; geopressurized; hot, dry rock; and magma. The distribution of favourable sites around the globe tends to be localized near regions of geologic instability that often experience active volcanoes or tectonic movement.

These regions are areas where Earth's crust is relatively thin, allowing economical access to the thermal energy beneath it. In Canada, British Columbia is the only area where it is currently feasible to use geothermal energy. California, Iceland, Italy, New Zealand, and Japan all have areas where geothermal energy is used on a significant scale. Geothermal energy is used to heat buildings, including homes, or to create steam that turns electrical generators.

### Advantages
- There are no greenhouse gas emissions.
- Geothermal energy is almost unlimited.
- It is a continuous power source that is not affected by weather or other factors.

### Disadvantages
- There are limited locations on Earth where it is economically viable.
- Naturally occurring, dissolved corrosive salts from the brine that is circulated cause problems for equipment and the environment.
- Hydrogen sulfide ($H_2S$) gas discharge does occur. It is toxic and even fatal in high concentrations.

---

**❖❖ IN CANADA**

- Only the west coast of Canada offers geothermal potential.
- Currently, a 60 MW test facility operates on British Columbia's Mount Meager.
- Mount Meager's wells turn water into 270°C steam.

---

### Geothermal Power Energy Path

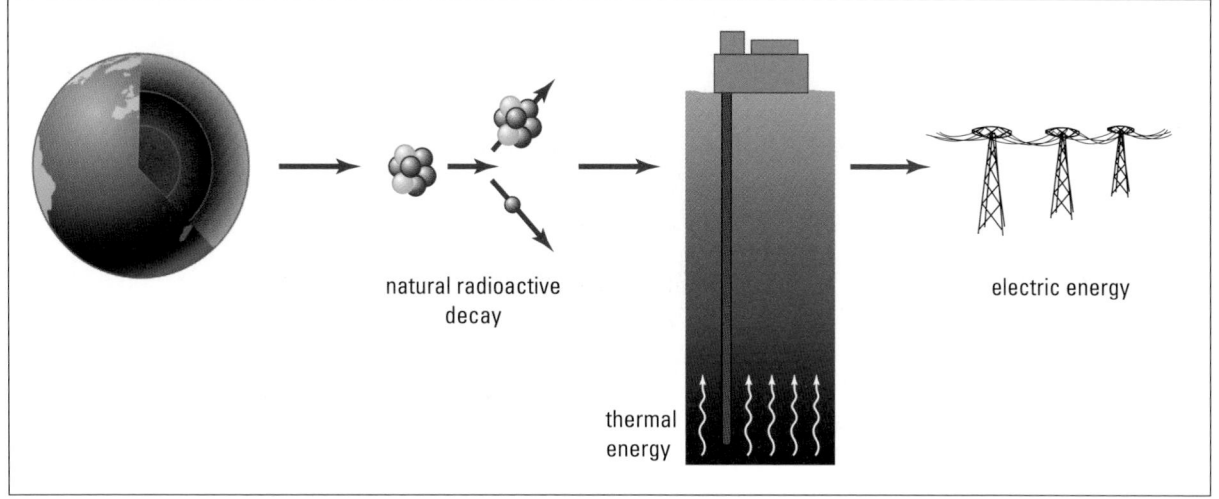

natural radioactive decay

thermal energy

electric energy

A solar-powered railroad crossing sign

In 1839, French physicist Edmund Becquerel (1820–1891) first discovered that generating electric current directly from sunlight was possible. Another 100 years would pass before commercial applications of the technology would appear. Today, solar cells are used to power everything from calculators and watches to small cities. Sunlight, the fuel required by solar cells, is 100% free and allows for electric energy generation that is completely free of greenhouse gas emissions.

**Photovoltaic cells** are composed of semiconductors, such as silicon. The sunlight knocks an electron from the crystal structure. Impurities added to the semiconductor do not allow the electron to fall directly back into place. The liberated electron will therefore follow the path of least resistance, which in the case of semiconductors is an external circuit. The flow of electrons in the external circuit can be used directly or stored in batteries for later use.

Several different materials and techniques are used to create photovoltaic cells. Very expensive cells are able to convert sunlight into electrical energy with efficiencies approaching 50 percent. Other procedures use thin films of amorphous silicon deposited on a variety of bases. These cells are much more economical to produce and have the advantage of being flexible and more durable. The weakness of amorphous silicon cells is that they degrade with exposure to sunlight, losing up to 50 percent of their efficiency over time.

Photovoltaic cells have been used very successfully in space, to provide power for satellites and space stations. The variety of different types of solar collectors has also allowed successful implementations throughout the world, from pole to pole. Small-scale power generation, for road signs or individual homes, is gaining popularity as storage battery systems improve. Large-scale solar power generation has been slower off the mark. This slow progress is due to the variability of sunlight. In addition, weather has a dramatic effect on a photovoltaic cell's ability to generate electric current. During nighttime hours, energy must be supplied by solar energy stored during the day. This type of intermittent power supply is a drawback of most renewable sources.

Solar cells are widely used in developing countries, where systematic power-delivery infrastructure does not exist. Individual homes consume the power to cook and cool their food and heat their homes. Industrialized countries demand significantly more power during the workday, when the sun is shining, than at night, so perhaps solar power generation will offer a supplement to help reduce the use of fossil fuels.

## Advantages

- Solar cells are 100% free of greenhouse gas emissions during operation.
- Silicon is the second most abundant element in Earth's crust.
- Reliable technology has already been successfully tested in various locations.

## Disadvantages

- Potentially toxic chemicals are released during manufacturing.
- Significant land area is required to produce significant amounts of electricity.

### Solar Power Energy Path

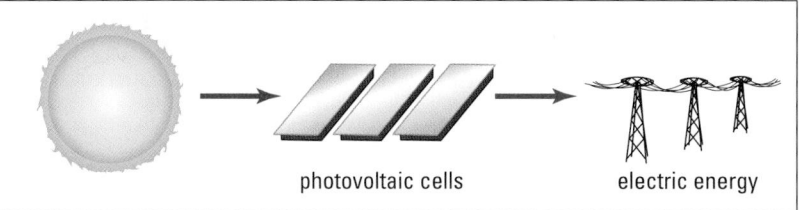

photovoltaic cells          electric energy

## Nuclear Power

After World War II, nuclear power was retooled for "peaceful" uses. In 1955, a U.S. Navy submarine, *Nautilus*, travelled almost 100 000 km powered by the controlled nuclear fission of a lump of uranium the size of a golf ball. Nuclear energy promised to be a clean, efficient energy source to meet rising global energy demands. Compared to coal, the amount of uranium that would need to be extracted from the ground was almost negligible. Further, electricity produced by nuclear fission did not involve the release of greenhouse gases into the environment.

Nuclear fission is the process of splitting extremely large atoms into two or more pieces, which releases an enormous amount of energy in the form of radiation or heat. The heat is used to boil water that eventually turns an electrical generator. Canada's CANDU reactor is a very popular choice worldwide because it has the capability of using unenriched uranium as fuel and has the fortunate record of never having had a catastrophic accident.

Nuclear power is currently used to generate approximately 16% of the global energy demand, which falls far short of the early visions of what nuclear power was to be. Nuclear power production faces two major problems: waste and safety. Some waste products of nuclear fission are extremely radioactive and will remain that way for thousands of years. Proponents of nuclear power note that the most radioactive by-products disintegrate within 100 years, leaving the remaining waste products much less hazardous to living creatures. They also point out that all of the nuclear waste ever stockpiled in Canada would barely fill two ice hockey rinks to the height of the boards. The safety issue is not as easily dismissed. Memories of the 1979 Three Mile Island accident in Pennsylvania and the 1986 disaster at Chernobyl in the former U.S.S.R. do not soon fade. The catastrophic consequences of a full-blown core meltdown similar to Chernobyl have led Sweden and Germany to legislate the reduction of nuclear generating facilities in their countries. Canada, although never having experienced a serious nuclear accident, has also scaled back reactor use because of both financial and safety reasons. Unfortunately, greenhouse-gas-belching plants fired by coal are now generating the electrical power that would have been generated by the nuclear plants.

Nuclear fusion could produce almost endless amounts of energy without greenhouse gas emissions or long-lived dangerous radioactive waste products. A 1000 MW fusion generator would have a yearly fuel consumption of only 150 kg of deuterium and 400 kg of lithium. However, scientists and engineers have yet to develop a method to sustain and control nuclear fusion reactions. Some researchers predict that within 40 to 50 years, fusion power will be available. Unfortunately, to date the only application of fusion is the hydrogen thermonuclear bomb. Will nuclear fusion, the energy that powers our Sun, become a mainstream power source?

 **Web Link**

www.school.mcgrawhill.ca/resources/
Current nuclear research is directed toward finding ways of using the spent fuel to generate more power and reduce nuclear waste. Learn more about the potential and problems associated with nuclear power by going to the above web site. Follow the links for **Science Resources** and **Physics 11**.

### Advantages
- There are no greenhouse gas emissions during electric energy generation.
- Small quantities of fuel generate enormous amounts of energy.

### Disadvantages
- There is the potential for catastrophic damage to human life and the environment that an accident — however unlikely — could cause.
- Operation requires continual monitoring by specially skilled individuals.
- Nuclear generating facilities are expensive to build and maintain.

| ▐•▌ IN CANADA |
| --- |
| • The CANDU reactor is the world's safest reactor. |
| • Canada is the world leader in uranium exports, holding 40% of the market share, totalling $1 billion per year. |

## Nuclear Power Energy Path

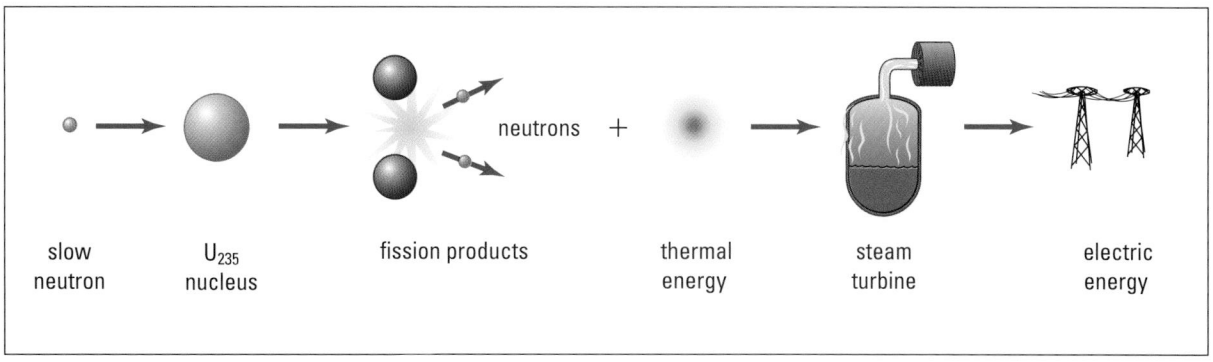

| slow neutron | U₂₃₅ nucleus | fission products | thermal energy | steam turbine | electric energy |

## Hydrogen Fuel Cells

One of Vancouver's fuel cell-powered buses

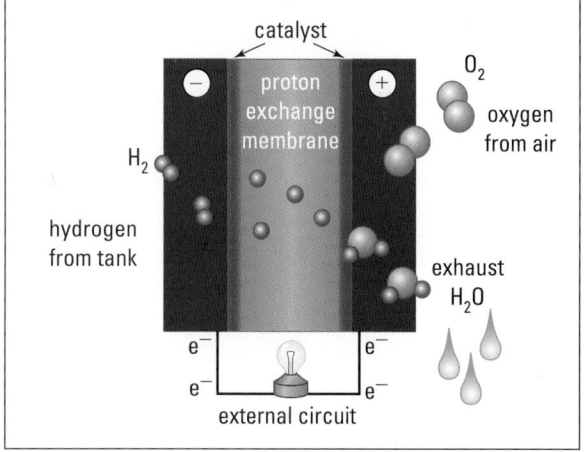

Fuel cell schematic

The most promising, and currently the most actively researched, method of non-fossil fuel power generation is the **fuel cell**. A fuel cell requires hydrogen as fuel at the anode and oxygen, or air, at the cathode. A reaction splits the hydrogen atom into an electron and a proton in the presence of platinum, which acts as a catalyst. The electrons take a different path from the anode to the cathode than do the protons. The electrons may therefore be used as electric current. The only waste product from the reaction is water vapour.

The technology to create fuel cells was first demonstrated at the University of Cambridge in 1839. It was not until the U.S. drive to put someone on the Moon that the technology was developed to a point of usefulness beyond a university laboratory. NASA first considered fuel-cell technology simply to produce potable water for the astronauts during the long trip to the Moon. The finished fuel cell versions that flew into space were used to produce both water and electric power on board the space capsule.

Fuel cell research and large-scale production have been limited due to the need for large amounts of very expensive platinum. In the past, platinum sheets were required, but new technology allows engineers to coat cheaper materials with a very thin layer of platinum. This method produces a large surface area of platinum, required for sustained reactions, at significantly reduced cost.

Applications of fuel cell technology have subsequently exploded in research popularity. Fuel cells are used to power automobiles, individual homes, and even remote villages. Vancouver has fuel cell-powered city buses operating on its downtown streets.

The real advantage of fuel cell technology is its ability to use common combustion fuels, such as methanol or even gasoline, as a source of hydrogen. Using gasoline in a fuel cell is more than twice as efficient as burning it in a combustion engine. Any technology that significantly reduces greenhouse gas emissions, while at the same time making use of existing infrastructure, such as the oil-refining systems, will be more readily adopted.

### Advantages

- The only emission is water vapour.
- Reformers allow current fuels to be utilized without combustion.
- There is scalable power generation. The cell produces only the required amount of electric current to power everything from laptops to whole towns.

### Disadvantages

- Hydrogen storage technology is new.
- Size and weight problems must be dealt with in smaller vehicles.

> **▮◆▮ IN CANADA**
>
> - Vancouver operates some city buses powered by fuel cells.
> - Vancouver trials have lead to a new generation of fuel cell bus that is 2000 kg lighter, capable of 205 kW or 275 hp.

### Hydrogen Power Energy Path

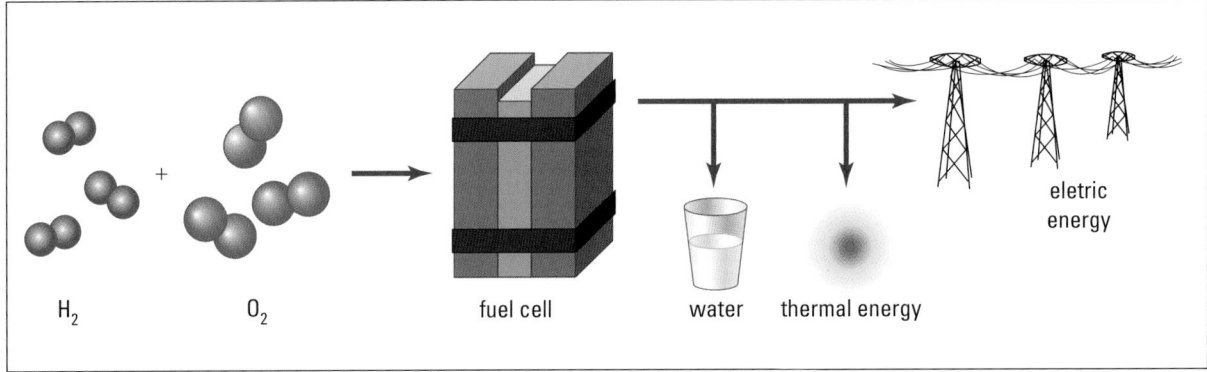

$H_2$     $O_2$          fuel cell          water     thermal energy          eletric energy

### Web Link

*www.school.mcgrawhill.ca/resources/*

Fuel cells may have the brightest future of any non-fossil fuel energy source. Billions of dollars are invested in research worldwide every year, resulting in exciting advances and innovative applications. Check out the above web site for more information about the latest in fuel-cell research and technology. Follow the links for **Science Resources** and **Physics 11**.

A microwave-receiving antenna needs to be very large, but because it allows up to 70% of visible light through, the land underneath can still be utilized.

Capturing solar energy on Earth is limited by weather conditions during the day and darkness at night. One solution to this inconsistency is to capture the solar energy in outer space and beam it to Earth.

The concept of transmitting power without the use of wires is not new. It was first tried in 1888. In 1908, U.S. engineer Nikola Tesla (1856–1943) attempted to transmit power around the globe by using a very tall tower in New York. His attempt failed, but the idea of transmitting power without wires did not die.

A Canadian experiment in 1987 demonstrated that a small aircraft could be kept flying indefinitely by using ground-based power that was beamed to the plane using microwaves. NASA demonstrated the ability to transmit 30 kW of power over 1.5 km with 82% efficiency. Efficiency is lost in large part because Earth's atmosphere attenuates, or weakens, energy beams by absorbing energy. Energy beams are radiant energy called "electromagnetic waves."

Electromagnetic waves include everything from very long TV and radio waves through microwaves, visible light, X rays, and very short gamma rays. Experiments show that microwaves with a frequency of 2.45 GHz pass through the atmosphere with little loss of energy —

2% during good weather conditions and up to 13% during storm conditions. The second advantage to using 2.45 GHz microwaves is that the magnetron tubes capable of producing them are already mass-produced for the more than 150 million microwave ovens now in operation.

The amount of solar radiation reaching the surface of Earth is about half of the power density, 1360 $W/m^2$, present in outer space. By placing satellites in geosynchronous orbit with very large solar collectors, approximately 50 $km^2$, we could take advantage of the increased intensity. The satellites would convert the solar power into microwaves and send it back to Earth in the same way communication satellites transmit data. A large rectifying antenna or rectenna, possibly 130 $km^2$, would capture the microwave energy and convert it into electrical power for distribution. The large rectennas could be placed on land or at sea, wherever power is required. The rectennas have the advantage of being largely transparent to visible light, and therefore the land beneath them could still be used for farming.

## COURSE CHALLENGE

### Does it Really Work?

Investigate the concept of space-based power from an environmental assessment perspective. Attempt to identify direct (e.g. launching the satellite) and indirect (e.g. mining the silicon for PV cells) impacts that implementing a space-based power system would have on our global environment. Present the information in an energy path diagram.

Learn more from the **Science Resources** section of the following web site: ***www.school.mcgrawhill.ca/ resources/*** to find *Physics 11 Course Challenge.*

## Advantages

- Solar energy is virtually 100% renewable without any $CO_2$ emissions.
- Power is readily deliverable to remote or developing areas.
- Solar energy is more sparing of land resources than other renewable sources of energy.

## Disadvantages

- The initial expense is large.
- Launching spacecraft releases $CO_2$ into the atmosphere.
- Communication bandwidth interference problems need more research.

> **🍁 IN CANADA**
>
> - Canada is a leader in space technology.
> - Canada successfully demonstrated wireless power transmission in 1987.

## Solar Energy Path in Space

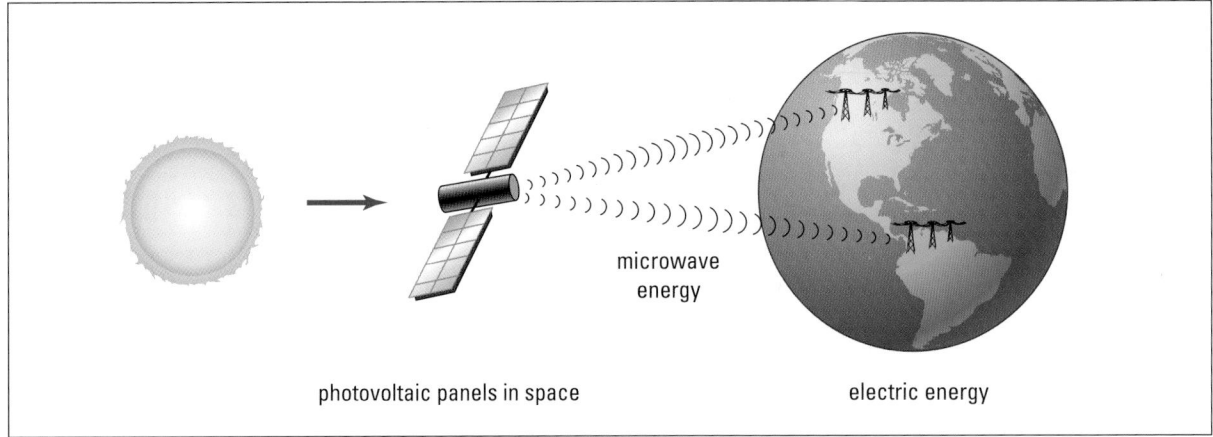

photovoltaic panels in space     microwave energy     electric energy

## 6.3     Section Review

1. **C** Most sources of energy are used to generate electric energy. Describe why the relative efficiency with which the energy is transformed is so important.

2. **K/U** Using microwaves to beam power to Earth will provide almost continuous power, although some of the beam power will be lost when transmitted through storm conditions on Earth. Why will stormy days cause the atmosphere to absorb more of the beam's power than days with clear skies will?

3. **MC** Hundreds of companies are investing millions of dollars into various ways of producing electric energy. Put yourself in the role of the president of one of those companies and research a source of energy that could provide sustainable power in Canada. Put together a presentation designed to convince the Canadian federal government to accept your plan for providing Canadians with electrical power.

## REFLECTING ON CHAPTER 6

- Thermal energy is a measure of the overall energy of an object, based on the random motions of the object's atoms and molecules.

- Temperature is a measure of the average kinetic energy of the atoms and molecules of a substance.

- Specific heat capacity of a substance is the amount of energy that must be added to raise 1.0 kg of material by 1.0 K. Specific heat capacity is used when calculating quantity of heat in joules.
$$Q = mc\Delta T$$

- Latent heat of a substance is the amount of energy required to cause the material to change state. Latent heat of fusion (melting or freezing) and latent heat of vaporization (vaporizing or condensing) are different for any one material. Latent heat capacity is used when calculating quantity of heat, in joules, required for a phase change.
$$Q = mL$$

- The first law of thermodynamics restates, and completes, the law of conservation of energy. Chapter 5 detailed how energy can be transferred from one system to another by doing work. This chapter details a second method of transferring energy, by heat.

- The second law of thermodynamics states that thermal energy moves from a system at a higher temperature to a system at a lower temperature. It includes the concept that no process is 100% efficient.

- Power, measured in watts, is defined as the amount of work done or energy transferred in a specific amount of time.
$$P = W/\Delta t$$

- Efficiency, provided as a percentage, is a comparison ratio between the useful work output of a system compared to the total work input.
$$\text{Efficiency} = \frac{W_o}{W_i} \times 100\%$$

- There are two types of energy — kinetic and potential. There are several forms of energy. Energy transformations change energy from one form to another. The transformation of energy from one form to another is never 100% efficient.

- Society requires energy to function, from simple methods of cooking our food to complex electrical grids that power our cities. Continued research and technological advancement will help meet our global energy demands in a sustainable way.

### Knowledge/Understanding

1. Describe the difference between thermal energy and temperature.

2. Differentiate between heat and thermal energy.

3. Describe the caloric theory of heat and its major flaw.

4. How does the kinetic molecular theory of heat explain the fact that rubbing your hands together generates thermal energy?

5. How is power different from work?

6. Define the two *types* of energy and provide four *forms* of energy for each type.

7. Define thermal equilibrium.

8. Describe the first law of thermodynamics and provide an example of the law in action.

9. Describe three processes that are allowed by the first law, but forbidden by the second law of thermodynamics.

10. Define specific heat capacity of a substance.

11. A news anchor reports that new batteries recently developed are able to generate twice the power of standard batteries for four times as long. How much more work can these new batteries do compare to their standard counterparts?

12. Does it take the addition of more energy to melt 1.0 kg of 0°C ice or boiling 1.0 kg of 100°C water? Explain.

13. Farmers often spray plants growing in their fields with water when unseasonably cold temperatures threaten to drop slightly below zero. How does this practice help keep the plants from freezing?

14. On a hot summer day, would a metal lawn chair have a higher temperature than a wooden lawn chair if they were both in the shade?

15. (a) Sketch and energy path diagram detailing a perfectly efficient bicycle-rider combination.
    (b) List several reasons why a bicycle-rider combination cannot be perfectly efficient.

16. Describe the fundamental limitation of muscle power.

17. Is the use of biomass through biogas generation considered to be $CO_2$ neutral? Explain.

18. How does a large body of water moderate the temperature (cooler summers, warmer winters) of a nearby area?

19. What provides the energy extracted in a tidal power electrical generating station? (Hint: What actually causes the tide?)

## Inquiry

20. Wind generation of electric power relies on almost continuously suitable wind conditions. These conditions are often present in areas that also serve a migration routes for birds. Devise a system that would allow the generation stations to co-exist with the birds.

21. Two friends are roasting marshmallows on an open fire. One friend uses a stick made from wet, green wood and the other uses a metal coat hanger. Which friend could hold their skewer nearer to the tip to remove the marshmallow after holding it in the fire?

22. Two young children are racing up a hill. Matt decides to run directly up the hill while Jen decides to zigzag to the top, exactly tripling her total distance as compared to Matt. They both reach the top of the hill at the same time.

(a) Who did the most work?
(b) Who was required to exert the most force?
(c) Who developed the most power?

## Communication

23. Beginning with the Phlogiston theory, describe the historical misconception about heat. Explain what experimental evidence eventually demonstrated the historical error.

24. A burn from boiling water is often much less serious than a burn caused by steam. Explain why steam causes so much more serious injury.

25. Describe the basic physical concepts behind the operation of a fuel cell. What advantages does fuel cell derived energy have over more historical energy sources?

26. Hydrogen fuel cell technology supplied energy and fresh water to astronauts in the 1960's. What factors, scientific, political and otherwise kept this technology from gaining widespread use in industrialized economies?

## Making Connections

27. The human body, composed largely of water, has a specific heat capacity of approximately 3500 J/kg · °C. How does this value help aid in human survival?

28. Discuss the process of societal evolution that caused humans to move from a society primarily based on muscle power to one that requires energy transformation technologies.

29. (a) Canadian demand for electric power was projected to be several times the current value in the early eighties. Why do you think that the projections overestimated the actual need for electric energy?
    (b) Suggest innovations or societal changes that you believe will help reduce electric energy demand in Canada over the next decade.

30. How has transportation energy transformation technologies impacted industrialized societies?

31. What advantages help make photovoltaic cell, electric systems popular in developing countries?

## Problems for Understanding

**32.** Calculate the amount of heat required to change the temperature of 2.0 kg of liquid water by 15°C.

**33.** A 21 kg aluminum block absorbs $1.5 \times 10^5$ J of energy. Calculate the change in temperature.

**34.** A perfectly insulating pail contains water that is heated, adding $4.5 \times 10^3$ J of energy while simultaneously being stirred which adds an additional $6.0 \times 10^2$ J of energy. According to the first law of thermodynamics, what is the total change in energy of the water?

**35.** Complete the table of temperature values.

| Degrees Celsius | Kelvin |
|---|---|
| 0°C | |
| | 373.15 K |
| 20 | |

**36.** Heating a home is substantially more costly if the air is humid (contains excess water vapour). Calculate the amount of energy required to raise the temperature of 5 kg of nitrogen gas and 5 kg of water vapour (100°C) by 15°C at constant volume.

**37.** During construction of the trans-Canada railway, each length of iron track needed to be welded to the next piece. This welding process was accomplished using a special chemical reaction, called the thermite reaction that generated incredibly hot temperatures and liquid iron. Calculate the amount of heat required to melt 5.0 kg of iron ($L_f$ = 289000 J/kg for iron).

**38.** Calculate the power developed by a runner able to do $7.0 \times 10^2$ J of work in 2.0 s.

**39.** Calculate the amount of energy required to operate each of the following devices for 30 min.
   **(a)** 150 W light bulb
   **(b)** 900 W hair dryer
   **(c)** 2000 W portable heater
   **(d)** $2.5 \times 10^6$ W electric motor

**40.** A 12 kg sled is pulled by a 15 N force at an angle of 35° to the horizontal along a frictionless surface.
   **(a)** Sketch the situation.
   **(b)** Calculate the acceleration of the sled.
   **(c)** Calculate the distance traveled by the sled in 3.0 s if it started from rest.
   **(d)** Calculate the work done on the sled in 3.0 s.
   **(e)** Calculate the power generated in pulling the sled.

**41.** A homemade go-cart has the following efficiencies:
   - Transformation of fuel energy to rotational energy of the axle – 15%
   - Transformation of the axle rotation to forward motion – 60%
   - Loss of forward motion due to air resistance – 16%

   Draw an energy path diagram detailing the efficiency of the go-cart.

**42.** A farmer is contemplating using a small water fall on his property for hydroelectric power generation. He collects data, and finds that 3000 kg of water fall 15.0 m every minute. Assuming the highest possible efficiency that he is able to achieve in transforming the water's gravitational potential energy to electric energy is 74%, what continuous power in Watts could he generate?

**43.** A wave power electric generating facility protects a shoreline from damage caused by large waves. During a storm 11.0 m swells bombard the facility and lose 80% of their height. Calculate the height of the waves reaching the shoreline.

---

**Numerical Answers to Practice Problems**

**1.** $3.22 \times 10^{10}$ J  **2.** 0.05°C  **3.** $8 \times 10^6$ J; 27 cents  **4.** 1.02 kg
**5.** $1.3 \times 10^6$ J  **6.** $4.2 \times 10^5$ J  **7.** 23 kg  **8.** $5 \times 10^{10}$ J  **9.** 47 000
**10.** $1.5 \times 10^2$ W  **11.** 15.4 kW; 20.7 hp  **12.** No, the student will be 1 second late  **13. (a)** 1.1 kg **(b)** 540 W  **14.** 12.8 min  **15.** 39 W
**16. (a)** 75% **(b)** into friction of moving parts  **17.** 26 %  **18.** 19%
**19. (a)** $\text{Eff}_{incand}$ = 4.2%, $\text{Eff}_{fl}$ = 8.4% **(b)** the fluorescent bulb heats up less than the incandescent bulb  **20.** 87.2%
**21. (a)** 6.24 J **(b)** 6.05 J **(c)** 3%  **22.** 34 %  **23.** 19°C

## Background

A 2 inch × 4 inch board (5 cm × 10 cm) and some roller-skate wheels started a phenomenon that is now a bona fide sport, the extreme sport of skateboarding. There is a lot of good physics behind both a good board and a good boarder. The move pictured, called an "ollie" after Alan "Ollie" Gelfand, was invented in the late 1970s. The board seems to stick to the rider's feet, but, in actuality, the rider and the board jump up by pressing down and then controlling specific rotational characteristics of the board.

Remarkable physics lies behind both the equipment and the tricks involved in skateboarding. The "ollie" pictured is a basic boarding manoeuver.

Boarding popularity slumped in the early 1970s, but with the advent of urethane wheels, boarders gained control, coupled with speed, and they have never looked back. Wheels need to cushion the ride, stick to the pavement enough to steer (but not so much as to stop a forced slide), and also allow the rider to move quickly without losing energy. Most wheels deform, or flatten, where they come into contact with the ground. The wheel must push back against the ground quickly, removing the flat spot before losing contact with the ground. If not, the energy that went into deforming the wheel will be "lost" and speed is compromised. A logical question

then comes to mind. Why not make the wheel rigid, so it does not flatten? In this case, the pavement would be forced to flatten and then even more energy would be lost — as elastic potential energy transferred to the pavement. Manufacturers are often faced with these questions in designing new equipment.

## Pre-lab Focus

In this project, you will systematically analyze sporting equipment of your choice (for example, a hockey helmet, roller blades, a snowboard, etc.) by using the physics knowledge that you have just learned. In Chapter 5, you learned that work is the transfer of energy from one system to another, or from one form to another. You also noted that energy is *always* conserved: it may transfer from one object to another, or from one form to another, but the total energy in the universe remains constant. In Chapter 6, the principle of conservation of energy was extended to include thermal energy and the first law of thermodynamics. Power was defined as the rate at which work is done. Understanding these and other concepts will allow you to examine the physics of sports and sports equipment.

## Materials

- Collect newspaper, magazine, and Internet articles on sports and sports equipment that are of interest to you. Attempt to gather information that will assist you in analyzing the physics concepts that make the sport possible.

- Review the materials as you collect them. You will need to decide on a theme, such as Safety, Durability, Sport-Specific "How'd They Do That," etc. Begin to focus your search for materials based on the theme you choose. Availability of resources may restrict your final choice, so be sure to collect enough materials to ensure success before locking yourself into any one specific theme.

## Initiate a Plan

A. Working with a partner, develop an investigation plan that will allow you to study, through research and experimentation, your selected sport or sporting equipment based on the energy, work, and power concepts that you have studied. Limit the equipment to be studied. Attempt to take one piece of sporting equipment through several rigorous tests, covering as much energy, work, and power content as possible, rather than testing several items in different areas.

B. Design a flowchart depicting the components of your plan (research, laboratory, presentation) and include tentative completion dates. Attempt to predict special needs (equipment, time, supervision) so that once the project is underway, you will not be sidetracked with unforeseen issues. Check with your teacher to ensure that your plan is appropriate for the allotted time.

C. Decide how the final information will be presented. You may work as a class, making decisions about how the final information needs to be presented and what the evaluation scheme will look like. For instance, you may decide as a class that you are each going to promote your selected sports equipment in a pamphlet or on an Internet site. Each promotion may be required to highlight the physics involved, the specific characteristics (safety, elasticity, durability), and the specific selling features. You may work as a class to build the assessment rubric together, ensuring that the criteria for the finished product are explicit and clear. Information presentation possibilities include a technical report; poster presentation; pamphlet or newsletter; multimedia presentation; and web site.

### ASSESSMENT

**After you complete this investigation**

- assess your procedure by having a classmate try to duplicate your results
- assess your presentation based on the clarity of the physical concepts conveyed

## Laboratory Testing

1. Use your plan to develop a suitable laboratory procedure, with equipment that is available to you, to systematically investigate specific characteristics of your selected sport or equipment.

2. Work with your teacher during this phase to ensure safety issues are not overlooked.

## Investigation Checklist

As your investigation proceeds, pay attention to the following checklist.

(a) Have you stated the purpose of the experiment (the question you want answered)?

(b) Have you written your hypothesis about what you expect the answer to be?

(c) Have you collected enough information from a variety of sources to design the experiment?

(d) Have you made a complete list of all the materials you will need?

(e) Have you identified the manipulated and controlled variables?

(f) Have you written a step-by-step procedure?

(g) Have you critically analyzed your procedure for possible errors or improvements?

(h) Have you repeated your experiment several times? Were the results similar each time?

## Analysis

Discuss your experimental results. Was your hypothesis shown to be correct? Have you evaluated the experimental errors?

## Assessing Your Experimental Design

List the successes and difficulties of your investigation. If you were to do it again, what changes would you make?

## Knowledge and Understanding

### True/False

In your notebook, indicate whether each statement is true or false. Correct each false statement.

1. Doing work on an object does not change the object's energy.
2. Mechanical kinetic energy is stored energy due to gravity.
3. Work done is proportional to the applied force and the square of the displacement in the same direction.
4. Work is done on an object by a force acting perpendicularly to the displacement.
5. It is possible to do negative work.
6. Gravitational energy is always measured from the same reference point.
7. Conservation of mechanical energy requires that a non-conservative force does the work.
8. The force of friction is a non-conservative force.
9. The first law of thermodynamics states that a change in the energy of a system is the sum of the work done and the heat exchanged between the system and its surroundings.
10. The caloric theory of heat is currently accepted as correct.
11. Temperature is a measure of the total thermal energy of a system.
12. During melting, an object releases energy.
13. Efficiency is a ratio of useful work input compared to the amount of work output.

### Multiple Choice

In your notebook, write the letter of the best answer for each of the following questions.

14. Work is not energy itself, but rather
    (a) it is a form of kinetic energy.
    (b) it is a form of gravitational potential energy.
    (c) it is a force.
    (d) it is a transfer of mechanical energy.
    (e) it is a result of parallel forces.
15. Which of the following are equivalent to a joule (J)?

(a) $N \cdot m^2$
(b) $kg \frac{m^2}{s^2}$
(c) $N \cdot m$
(d) $N \frac{m}{s}$
(e) both (b) and (c)

16. Work done is zero when
    (a) an applied force does not result in any motion.
    (b) uniform motion exists in the absence of a force.
    (c) the applied force is perpendicular to the displacement.
    (d) both (a) and (c).
    (e) all of the above.
17. A weight lifter lowers a barbell at constant speed. Down is assigned as positive. In doing so, the weight lifter
    (a) does positive work on the barbell.
    (b) allows gravity to do negative work on the barbell.
    (c) does not do any work on the barbell.
    (d) does negative work on the barbell.
    (e) allows the kinetic energy of the barbell to increase.
18. At midnight, you walk outside. The temperature of the air has been exactly –12°C for several hours. You touch a metal fence post, a block of wood, and some snow. Apply the concept of thermal equilibrium to identify the correct statement.
    (a) The fence post has the lowest temperature.
    (b) The brick will have the highest temperature.
    (c) The fence post, the block of wood, and the snow will all be at a temperature of –12°C.
    (d) The process of thermal equilibrium causes cold to flow from objects with low specific heat capacities.
    (e) Determining which object will have the lowest temperature is impossible without knowing the specific heat capacities.
19. Determine the only example of potential energy from the following forms of energy.
    (a) sound
    (b) thermal
    (c) radiant
    (d) nuclear
    (e) mechanical kinetic

**20.** A sustainable energy source derived from biological waste products is
   **(a)** solar power.
   **(b)** ocean thermal power.
   **(c)** biomass.
   **(d)** not feasible in any setting.
   **(e)** not considered to be $CO_2$ neutral.

**21.** Wind-based power provides some Canadians with reliable, cost-effective energy. Historians believe that wind energy was first utilized
   **(a)** by the Dutch to pump out inland lakes about 1000 years ago.
   **(b)** in regions of China 30 000 years ago.
   **(c)** about 2000 years ago by Persian sailors.
   **(d)** over 40 000 years ago to carry Australian Aborigines from mainland Asia to the island continent.
   **(e)** to generate electricity approximately 2000 years ago.

**22.** Humans have used solar energy, streaming from the Sun, for thousands of years. Which of the following energy sources do not ultimately trace back to the Sun?
   **(a)** fossil fuels          **(d)** nuclear
   **(b)** photovoltaic cells     **(e)** biomass
   **(c)** ocean thermal

**23.** Hydrogen fuel cells may be the single most effective sustainable power source alternative in the next 50 years. Which of the following statements concerning fuel cell technology is false?
   **(a)** The only waste product is water vapour.
   **(b)** Fuel cells are used to power automobiles, homes, and even small villages.
   **(c)** Hydrogen fuel cell technology was used to provide Apollo astronauts with fresh water and electric power.
   **(d)** Fuel cells require temperatures near $-72\,^\circ C$ to operate efficiently.
   **(e)** Fuel cells can utilize energy stored in gasoline through the use of reformers.

## Short Answer

**24.** Explain how work and a transformation of energy are related.

**25.** What type of energy does a wind-up toy contain after being wound just before release?

**26.** Does your arm, lifting completely vertically, do any work on your textbook as you carry it down the hall if the book's vertical position does not change?

**27.** Describe the work done by a nail on a hammer as the nail is driven into a wall. What evidence is there that the work done is negative?

**28.** What is the factor by which a javelin's kinetic energy is changed, if its velocity is increased to five times its initial velocity?

**29.** Draw a graphical representation showing total mechanical energy, gravitational energy, and mechanical kinetic energy versus time for the following processes.
   **(a)** A block of ice, initially at rest, slides down a frictionless slope.
   **(b)** A moving block of ice slides up a frictionless slope and instantaneously comes to rest.

**30.** What common misconception about heat may be the result of the phlogiston and caloric theories?

**31.** What aspects of the kinetic molecular theory are supported by observations that the caloric theory cannot explain?

**32.** How is temperature different from thermal energy?

**33.** Is it possible to cool an object to 0 K or absolute zero? Explain.

**34.** Describe how the specific heat capacity of a substance affects its ability to change temperature.

**35.** Thermal energy is being added to a substance, and yet the temperature is not increasing. Explain how this is possible.

**36.** What sustainable energy source opportunities exist in Canada?

**37.** **(a)** Wave power utilization provides a visual example of how energy is transformed from one form to another. Explain how this can be seen in the case of wave power.
   **(b)** Describe the process involved in extracting energy from the thermal energy of tropical oceans.

## Inquiry

**38.** Investigate the energy transformations that take place when an athlete is pole-vaulting.

**39.** Design a simple test that would allow you to differentiate between a substance with a very low specific heat capacity and one with a very high specific heat capacity.

**40.** How much energy do you use during the course of one week to heat water? Design a simple method of recording both the approximate quantities and the temperature changes of the hot water you use. Include everything from showers to hot coffee. Convert the energy required into a dollar value by using the price of electricity per kW · h in your area.

## Communication

**41.** Can the gravitational potential energy of a golf ball ever be negative? Explain without using a formula.

**42.** Design a tree diagram to illustrate the relationship between work, kinetic and potential energy, and related variables for each of the following cases.
  **(a)** A high jumper converts kinetic energy and work into height (or potential energy).
  **(b)** A curler causes a curling stone that starts at rest to slide down the rink, eventually stopping.
  **(c)** A grocery store employee restocks canned goods onto a top shelf.

**43.** Describe how the graph represents conservation of energy.

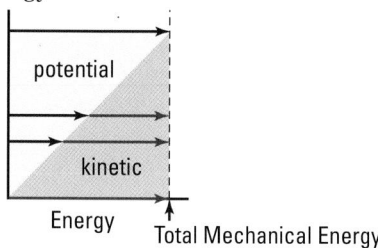

## Making Connections

**44.** Research an alternative source of energy that has not been discussed in the text. Determine the energy transformation efficiency and environmental impact.

**45.** A newspaper advertisement selling a digital daytimer claims it is 100% efficient. Is it possible that this advertisement is telling the truth? Explain your reasoning.

**46.** Describe which sustainable energy source you believe would be most effective, and the least environmentally harmful, in the tropics. Defend your choices with clear examples.

**47.** Is the temperature during a still winter day likely to increase or decrease slightly when large snowflakes begin to fall? Explain your reasoning.

**48.** How has the specific heat capacity of water helped to shape life on Earth?

**49.** A fuel cell-powered vehicle does not produce any emissions other than water vapour during its operation. Would you be willing to spend more to purchase a fuel cell vehicle rather than one with a combustion engine? Provide both a scientific and societal rationale for your answer.

## Problems for Understanding

**50.** The force-versus-position plots were created from data collected during low-speed automobile collisions with cars with different-style bumpers. Calculate the work done on the bumper in each case.

**(a)**

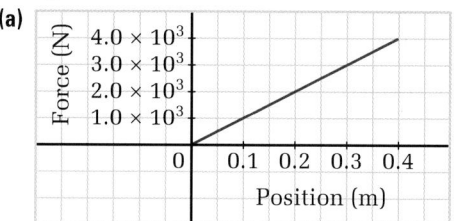

**(b)**

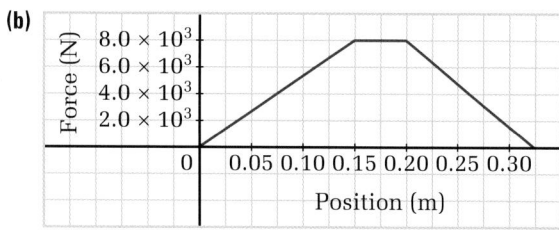

**51.** How fast will a 2.55 kg bowling ball be travelling if the 358 J of work done to the ball are transformed into kinetic energy?

**52.** Find the mass of an object if 250 J of work cause it to gain 3.2 m/s of velocity.

**53.** Calculate the gravitational potential energy of a 75 kg bag of salt that is 2.5 m above the ground.

**54.** A 250 kg roller coaster cart loaded with people has an initial velocity of 3.0 m/s. Find the velocity of the cart at A, B, and C.

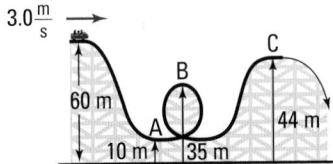

**55.** A raindrop reaches terminal velocity very quickly as it falls to Earth due to the work friction does during the descent.

(a) If a 0.500 g raindrop formed 1.50 km above Earth, calculate its speed as it struck the ground if air resistance is ignored. Answer in km/h.

(b) Determine the percentage of energy that is lost due to work done by friction if the 0.500 g raindrop reaches Earth travelling at 5.20 m/s.

**56.** A 45 kg cyclist travelling 15 m/s on a 7.0 kg bike brakes suddenly and slides to a stop in 3.2 m.

(a) Calculate the work done by friction to stop the cyclist.

(b) Calculate the coefficient of friction between the skidding tires and the ground.

(c) Are you able to determine if the tires were digging into the ground from your answer in part (b)? Explain.

**57.** A tow truck pulls a car by a cable that makes an angle of 21° to the horizontal. The tension in the cable is $6.5 \times 10^3$ N.

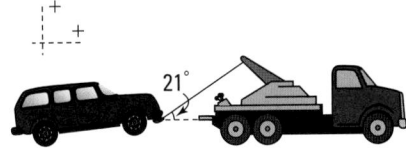

(a) How large is the force that causes the car to move horizontally?

(b) How much work has the tow truck done on the car after pulling it 3.0 km?

**58.** Glycerin, a major component of soap, is heated before being poured into moulds. Calculate the mass of glycerin if $5.50 \times 10^3$ J of energy causes a 4.20°C temperature increase.

**59.** The Sun heats $2.00 \times 10^2$ kg of ammonia gas that are kept at constant volume in a large tank. Assuming that solar radiation of $6.00 \times 10^2$ W caused the heating, determine the length of time it took to raise the ammonia's temperature by 12°C.

**60.** A garage-door opener uses 1200 W when operating. When the electric energy is transformed to kinetic energy, 20 percent of the power is lost. Friction between the chain and the support causes a loss of five percent. Friction between the rollers on the door and the track can range from 10 percent to 60 percent. The remaining power is used to actually move the door. Draw an energy path diagram to represent the power loss of the door opener.

**COURSE CHALLENGE**

**Space-Based Power**

Consider the following as you continue to build your physics research portfolio.

• Add important concepts, formulas, interesting and disputed facts, and diagrams from this unit.

• Review the information you have gathered in preparation for the end of course challenge. Consider any new findings to see if you want to change the focus of your project.

• Scan magazines, newspapers, and the Internet to find interesting information to enhance your project.

## OVERALL EXPECTATIONS

**DEMONSTRATE** an understanding of the properties of mechanical waves and sound.

**INVESTIGATE** the properties of mechanical waves and compare predicted results with experimental data.

**EVALUATE** the contributions to entertainment, health, and safety of technologies that make use of mechanical waves.

On May 22, 1960, in the early afternoon, an earthquake rocked the floor of the Pacific Ocean off the coast of south-central Chile. In fishing villages on the coast, many inhabitants took to their boats to escape the shaking. This was a mistake. About fifteen minutes later, the level of the ocean water dropped. Shallow harbours emptied of water, and boats thudded down onto the seabed. Then, the sea returned in a thunderous breaker, picking up the boats and pitching them onto the land.

The giant wave, a tsunami, was generated when a huge area of the ocean floor suddenly sank several metres during the earthquake. The westbound portion of the wave raced across the Pacific at speeds of up to 700 km/h, as fast as a small passenger jet. About fourteen hours later, a wave, three storeys high, swept to the shores of the Hawaiian Islands, 10 000 km from its starting point. The force of the debris-filled waters uprooted trees, bent parking meters, and pushed houses off their foundations.

Almost a day after the quake, the wave reached Japan, half a world away from its source. It was reported that 119 people were killed.

How do waves, like this tsunami, transfer their energy over such great distances? What keeps them going? What determines their speed and height? These, and other questions about wave action, can be answered through a study of waves, energy, and interestingly, the nature of sound.

### UNIT ISSUE PREP

Read ahead to pages 452–453. At the end of this unit, you will develop a noise policy document.
- How does the nature of sound waves affect how you address the issue of noise in your community?

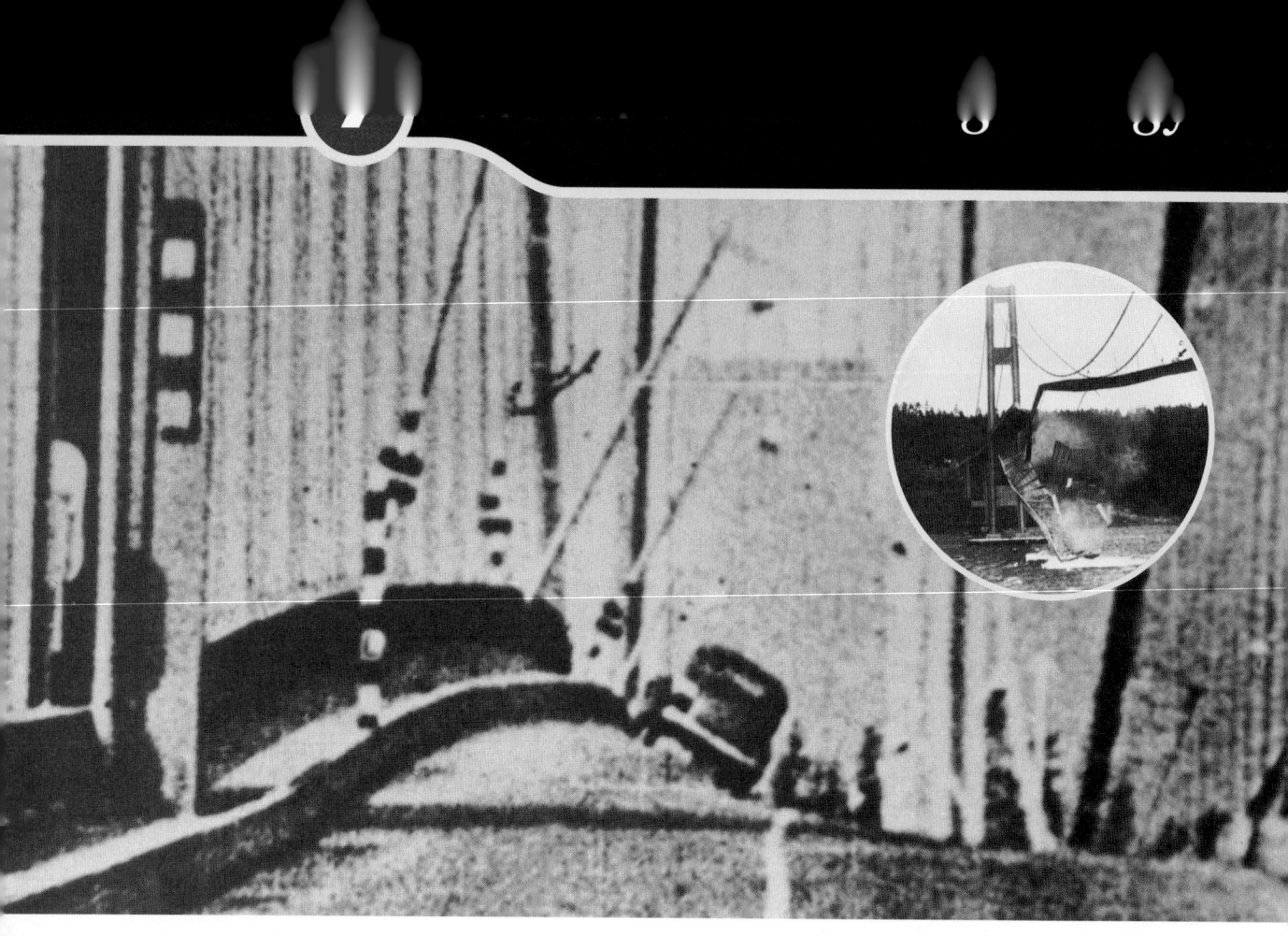

People living in coastal areas are very aware of the energy carried by tsunamis (large ocean waves) and the destruction that they can bring. The effects of small waves combining together can also lead to destruction. On November 5, 1940, many relatively small waves combined to produce such huge vibrations in the Tacoma Narrows Bridge (as shown in the large photograph) that the bridge was torn apart. The engineers who designed the suspension bridge, located near Tacoma, Washington, had no idea that the relatively moderate winds that blew through the Narrows could produce such destructive vibrations. Today, models of bridges such as this one are routinely tested in specially designed wind tunnels to ensure that small waves cannot combine to produce disasters like the Tacoma Narrows Bridge collapse.

The transfer of energy by waves can also be beneficial. Physiotherapists commonly use high frequency sound waves to reduce the pain and swelling of athletic injuries. High frequency sound waves (ultrasound) are also used for routine monitoring of the developing fetus in expectant mothers. Ultrasound does not present the dangers associated with X-ray imaging.

# INVESTIGATION 7-A

## Waves in a Spring

You have probably been observing waves of some sort all of your life. As a child, you might have dropped a pebble in water and watched the waves spread over the surface. Children often develop ideas about the nature of what they observe. These ideas can be helpful but, if they are incorrect, they can stand in the way of new learning. To see if your ideas about waves are correct, test them in this investigation.

### Problem

What affects the nature of a wave as it travels along a spring? What influences the speed of the wave?

### Prediction

Make predictions about the factors affecting the nature of a wave pulse as it travels down a spring. For example, predict whether such factors as the tension in the spring, the height of the pulse, and the distance travelled affect the speed of a wave pulse. How might these factors do so? In each case, explain your reasoning.

### Equipment

- large-diameter spring (such as a Slinky™)
- small-diameter spring
- stopwatch
- metre stick

### Procedure

1. Stretch the large spring out on the floor until you and your lab partner are about 8 m apart.

2. Move your hand rapidly to one side and back in order to send a single pulse down the spring.

3. Observe its speed, size, and anything else that you find significant.

4. Find a second way to produce a wave pulse. Experiment with different motions of the end of the spring.

5. Establish a method for measuring the speed of the wave pulse. Test the effect, if any, of the following on the speed of the wave pulse.

   (a) pulse size

   (b) distance travelled

   (c) tension in the spring

   (d) use of the small spring

(Hint: Be careful when you are testing one factor to control all of the other factors. When stretching the springs, be sure to use the same distance for timing the pulse.)

**CAUTION** Releasing one end of a stretched spring usually results in a tangled mess! When you are finished experimenting, do not *release either end*. Instead, walk toward your partner with your end of the spring.

### Analyze and Conclude

1. Compare your results with the predictions you made earlier about the factors that affect the speed of a wave pulse in a spring.

2. In each case, try to explain any discrepancies between your results and your predictions.

3. Describe any misconceptions you might have had about waves and how they travel.

• Describe and explain amplitude, frequency, and phase of vibration.

• Analyze and experiment with the components of, and conditions required for, resonance to occur in a vibrating object.

KEY
TERMS

• periodic motion
• cycle
• period
• rest position
• amplitude
• frequency
• hertz

• phase difference
• in phase
• out of phase
• natural frequency
• resonance

The motion of particles in a mechanical wave and the energy transmitted by a wave are quite different from the motion and energy that you studied in Units 1 and 2. After you sent a wave pulse down the spring in the investigation, all parts of the spring returned to their original positions. As the wave pulse passed through each section of the spring, that section moved from side to side or back and forth, but then returned to its initial position. Only the *energy* travelled down the spring. To initiate the wave, you had to move your hand back and forth. Similarly, most waves are started by vibrating objects such as the student's hand in Figure 7.1. Learning how to describe vibrations is an important first step in understanding and describing waves and their motion.

**Figure 7.1** You can transmit energy to the duck, making it bounce up and down, by creating water waves. Your hand, the water, and the duck vibrate only up and down, but the wave has transmitted energy from your hand to the duck.

## Amplitude, Period, Frequency, and Phase of Vibrations

When an object moves in a repeated pattern over regular time intervals, it is undergoing **periodic motion**. One complete repeat of the pattern is called a **cycle** or vibration. The time required to complete one cycle is the **period** ($T$). Figure 7.2 shows two types of periodic motion. A simple pendulum moves from side to side, perpendicular to its length, while a mass on a spring oscillates up and down, parallel to its length. When a pendulum or a mass on a spring is not in motion but is allowed to hang freely, the position it assumes is called its **rest position**. When in motion, the distance from the rest position to the maximum displacement is the **amplitude** ($A$) of the vibration.

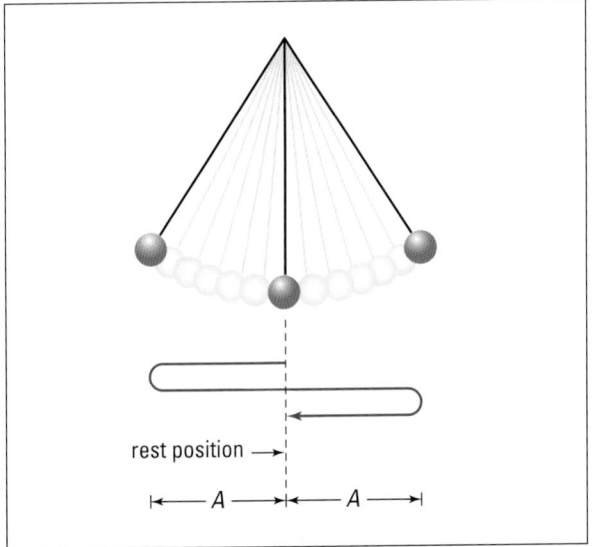

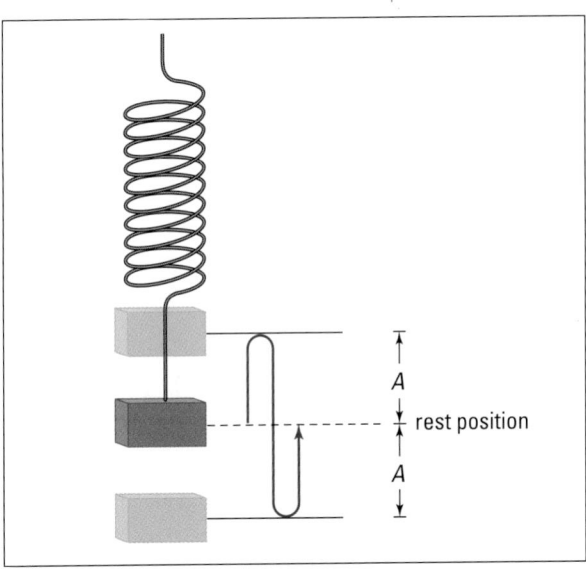

**Figure 7.2** **(A)** When a simple pendulum completes one full cycle of its motion, it is in its original position.

**(B)** One full cycle of the motion of the mass on a spring brings the mass back to the rest position.

One of the most common terms used to describe periodic motion is **frequency** ($f$), which is the number of cycles completed in a specific time interval. The frequency is the reciprocal, or inverse, of the period. The SI unit of frequency is $s^{-1}$ or $\frac{1}{s}$ (reciprocal seconds). This unit has been named the **hertz** (Hz) in honour of German scientist Heinrich Hertz (1857–1894), who discovered radio waves.

---

**PERIOD AND FREQUENCY**

The period is the quotient of the time interval and the number of cycles.

$$T = \frac{\Delta t}{N}$$

The frequency is the quotient of the number of cycles and the time interval.

$$f = \frac{N}{\Delta t}$$

The frequency is the reciprocal, or inverse, of the period.

$$f = \frac{1}{T}$$

| Quantity | Symbol | SI unit |
|---|---|---|
| period | $T$ | s (seconds) |
| frequency | $f$ | Hz (hertz) |
| time interval | $\Delta t$ | s (seconds) |
| number of cycles | $N$ | none (pure number) |

**Note:** $1\ \text{Hz} = \frac{1}{s} = 1\ s^{-1}$

---

The pendulum in the photograph is called a Foucault pendulum. It is named after French scientist Jean-Bernard-Leon Foucault (1819–1868), who built the first one in Paris. A Foucault pendulum has a very large mass suspended by a very long wire that does not constrain the pendulum to swing in a specific plane. Earth's rotation causes the plane of the swing of a Foucault pendulum to rotate very gradually. For example, the plane of Foucault's first pendulum in Paris made one complete rotation in about 32 h. This rotation was the first "laboratory" evidence that Earth rotates on its axis.

Even when vibrating objects have the same amplitude and frequency, they may not be at the same point in their cycles at the same time. When this occurs, we say that there is a **phase difference** between them. Two vibrating objects are **in phase** when they are always moving in the same direction at the same time. If, during any part of their cycles, the two objects are moving in opposite directions, they are vibrating **out of phase**. Figure 7.3 illustrates this.

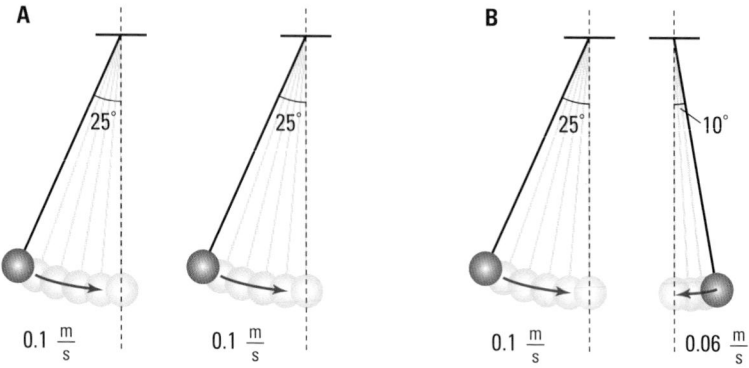

**Figure 7.3** **(A)** When two pendulums are moving in unison, they are said to be "in phase." **(B)** Since these two pendulums are moving in opposite directions, they are moving "out of phase."

## MODEL PROBLEM

### Period and Frequency

**A mass suspended from the end of a spring vibrates up and down 24 times in 36 s. What are the frequency and period of the vibration?**

### Frame the Problem

- The *mass* is undergoing *periodic motion*.
- The *period* is the time for one complete cycle.
- The *frequency* is the reciprocal of the *period*.

### Identify the Goal

**(a)** The period, $T$, of the motion

**(b)** The frequency, $f$, of the motion

## Variables and Constants

| Involved in the problem | | Known | Unknown |
|---|---|---|---|
| $T$ | $\Delta t$ | $N = 24$ | $T$ |
| $f$ | $N$ | $\Delta t = 36$ s | $f$ |

## Strategy

You can find the period because you know the total time interval and the number of cycles. Substitute these values into the equation.

Divide.

**(a)** The period of the vibrating mass is 1.5 s.

You can also find the frequency from the number of cycles and the total time interval.

Substitute.

Divide.

## Calculations

$$T = \frac{\Delta t}{N}$$

$$T = \frac{36 \text{ s}}{24}$$

$$T = 1.5 \text{ s}$$

$$f = \frac{N}{\Delta t}$$

$$f = \frac{24}{36 \text{ s}}$$

$$f = 0.67 \text{ s}^{-1}$$
$$f = 0.67 \text{ Hz}$$

**(b)** The frequency of the vibrating mass is 0.67 Hz.

---

## Validate

The unit of the period was seconds, which is correct.

The unit for the frequency was seconds to the negative one, or reciprocal seconds, which is equivalent to hertz. That is the correct unit of frequency.

The period and frequency were calculated individually, but they should be the reciprocal of each other. Check to see if they are.

$$f = \frac{1}{T}$$

$$f = \frac{1}{1.5 \text{ s}}$$

$$f = 0.67 \text{ s}^{-1}$$

$$f = 0.67 \text{ Hz}$$

The frequency is the reciprocal of the period.

## PRACTICE PROBLEMS

1. A metronome beats 54 times over a 55 s time interval. Determine the frequency and period of its motion.

2. Most butterflies beat their wings between 450 and 650 times per minute. Calculate in hertz the range of typical wing-beating frequencies for butterflies.

3. A watch spring oscillates with a frequency of 3.58 Hz. How long does it take to make 100 vibrations?

4. A child swings back and forth on a swing 12 times in 30.0 s. Determine the frequency and period of the swinging.

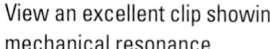

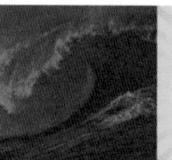

Assemble the apparatus shown here, using strings and masses tied to retort stands or other solid supports. Make two pendulums the same length, one longer than the pair, and one shorter. One at a time, pull each pendulum to the side and let it swing. Observe any response of the other pendulums.

**ELECTRONIC LEARNING PARTNER**

View an excellent clip showing mechanical resonance.

## Natural Frequencies and Resonance

When an object, like a simple pendulum or a mass on a spring, is allowed to vibrate freely, it vibrates at a specific frequency called its **natural frequency**. The natural frequency of a simple pendulum depends on its length — shorter pendulums have higher natural frequencies than longer ones. What determines the frequency with which a mass vibrates on the end of a spring?

When you push a child on a swing, you need not push very hard to make the child swing higher and higher. What you do have to do is to push at the right times, that is, with a frequency equal to the natural frequency of the swing and the child. As well, the cycle of pushing must be *in phase* with the motion of the child and the swing. This condition is true for all vibrating objects. If energy, no matter how small the amount, is added to a system during each cycle and none is removed, the amplitudes of the vibration will become very large. This phenomenon is called **resonance**.

---

QUICK LAB
## Natural Frequency of a Mass on a Spring

**TARGET SKILLS**
- **Identifying variables**
- **Performing and recording**

What factors affect the frequency of vibration of a mass on a spring? Does the frequency change with time? Does the amplitude affect the frequency? Does the amount of mass influence the frequency? Answer these questions by carrying out the following experiment. You will need two different-sized springs, three different masses, a stopwatch, a metre stick, a retort stand, and a clamp.

Securely attach a spring to a clamp on a retort stand. Hang a mass on the end of the spring. Stretch the spring a measured distance by pulling on the mass. Release the mass and determine the length of time it takes for the mass to complete five full cycles. Calculate the frequency. Carry out the procedure for three different amplitudes.

Repeat the experiment using a different mass. To determine if the frequency changes with time, after taking one measurement, allow the mass to continue vibrating on the spring and measure the time it takes for five more cycles.

Repeat the entire procedure for a different spring.

### Analyze and Conclude

1. Which factors affected the frequency of the mass on a spring?

2. Explain how you could determine that some factors did not affect the frequency.

3. What was the natural frequency of each of your combinations of spring, mass, and stretch distance?

You may have experienced another example of resonance when driving or riding in a car. If the wheels were not properly aligned and balanced, the entire car would start to vibrate severely at a certain speed. These vibrations can occur only when the vibrations produced by the spinning of the automobile's wheels are equal to a natural frequency of vibration of the automobile itself.

Mechanical resonance can cause serious problems for engineers constructing buildings, bridges, and aircraft. The Tacoma Narrows Bridge collapsed on November 7, 1940, because resonance was created in its centre span by relatively moderate winds of 60 to 70 km/h. Over a period of two hours, the vibrations of the centre span increased steadily, until they became so violent that the bridge collapsed into the river below (see Figure 7.4).

**History Link**

Galileo Galilei first investigated the constant frequency of a simple pendulum around 1600. Although he never constructed it, Galileo also proposed a design for a mechanical clock regulated by such a pendulum. It was not until around 1759 that John Harrison constructed an extremely accurate clock or chronometer. Finally, seafarers could determine their longitude on long ocean voyages with the aid of this accurate chronometer. This helped the Royal Navy to control the seas and Britain to dominate world trade during the nineteenth century.

**Web Link**

*www.school.mcgrawhill.ca/ resources/*
For more pictures and information about the Tacoma Narrows Bridge collapse, go to the above web site. Follow the links for **Science Resources** and **Physics 11** to find out where to go next.

**Figure 7.4** Resonance produced such violent vibrations in the original Tacoma Narrows Bridge that it collapsed.

# 7.1 Section Review

1. **K/U** Give three examples of periodic motion. What makes them periodic?

2. **C** Explain in your own words the meaning of the terms "cycle," "period," and "frequency."

3. **K/U** If a simple pendulum is lengthened, what happens to its frequency? To its period?

4. **K/U** Sketch diagrams illustrating
   (a) two masses on springs vibrating in phase
   (b) two pendulums vibrating out of phase

5. **K/U** Period is typically measured in units of seconds and frequency in units of hertz. How are these two units related to each other? Why are they related in this manner?

6. **C** Describe what happens when resonance occurs in an object. Explain how this is produced.

7. **MC** Provide two examples of resonance in everyday life.

- Describe and define the concepts of mechanical waves including medium and wavelength.

- Compare the relationships between frequency and wavelength to the speed of a wave in various mediums.

- Use scientific models to explain the behaviour of waves at barriers and interfaces between media.

- wave
- medium
- mechanical wave
- crests
- troughs
- wavelength
- frequency of a wave
- transverse wave
- longitudinal wave
- wave equation

What is a wave? When you first read the word, you probably imagine an ocean wave or a crowd in the stands at an athletic event doing "the wave." Soon you will know why the first image is a true wave in the scientific sense and the second image is not.

**Figure 7.5** **(A)** The water ski does work on the water, giving it energy. The wave transmits the energy across the surface of the water. **(B)** "The wave", in which sports fans wave their arms in unison, is not a wave at all. No energy is transmitted from one person to another. Each person is using his or her own energy to make "the wave."

A **wave** is a disturbance that transfers energy through a **medium**. While the disturbance, and the energy that it carries, moves through the medium, the matter does not experience net movement. Instead, each particle in the medium vibrates about some mean (or rest) position as the wave passes. The behaviour of many physical phenomena can be described as waves. Disturbances in ropes, springs, and water are easily recognized as waves in which energy moves through the medium. Although you cannot see sound waves, they also are true waves. You usually think of sound waves travelling through air. However, they can also travel through water, steel, or a variety of other materials.

All of these waves travel through matter and are known as **mechanical waves**. Their speed does not depend on their size or the amount of energy they carry. Neither does it decrease as they move through a medium. Rather, the speed of any wave in a particular medium is the same as the speed of any other wave in that same medium. The speed of a wave in a medium is a characteristic property of that medium, in much the same way that its density or boiling point is a property of the medium.

The forces between the particles of the medium and the mass of those particles determine the speed of a mechanical wave. The greater the force between those particles, the more rapidly each particle will return to its rest position; hence, the faster the wave will move along. However, the greater the mass (inertia) of a particle in the medium, the slower it will return to its rest position and the slower the wave will move along. Generally, there is friction in the medium. This friction acts to dampen or reduce the size or height of the wave. Unlike the friction acting on a material object that is moving, the friction does not affect the speed of the wave. Similarly, when a wave is given more energy, the shape of the wave is affected, but the speed with which it moves through the medium is unaffected.

## Describing Waves

All waves have several common characteristics that you can use to describe them. Figure 7.6 shows a periodic wave that looks as though it is frozen in time. The horizontal line through the centre is the rest or equilibrium position. The highest points on the wave are the **crests** and the lowest points are the **troughs**. The amplitude ($A$) is the distance between the rest position and a crest or trough. The **wavelength** ($\lambda$, the Greek letter lambda) is the shortest distance between two points in the medium that are in phase. Therefore, two adjacent crests are one wavelength apart. Similarly, two adjacent troughs are one wavelength apart.

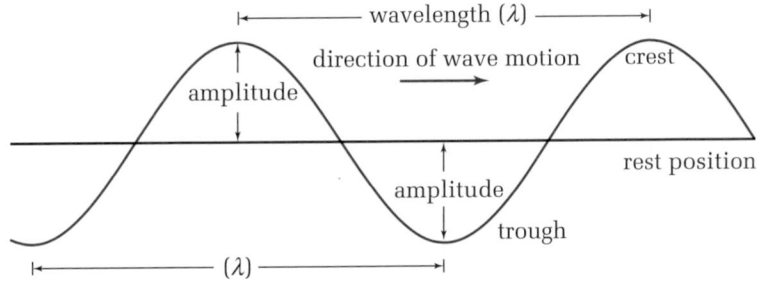

**Figure 7.6** This idealized wave illustrates the features that are common to all waves.

Now imagine that the wave is no longer frozen in time but is moving. The **frequency of a wave** ($f$) is the number of complete wavelengths that pass a point in a given amount of time. Similar to vibrating objects, the frequency of a wave is usually described in units of hertz. The frequency of a wave is the same as the frequency of the source producing it, so it does not depend on the medium. Also, similar to vibrating objects, the period of a wave is the time it takes for one full wavelength to pass a given point.

While a wave travels through a medium such as a spring, the particles do not need to vibrate in the same direction in which the wave is moving (see Figure 7.7). When the particles of a medium vibrate at right angles to the direction of the motion, the wave is called a **transverse wave**. Water waves and waves on a rope are examples of transverse waves.

**PHYSICS FILE**

Electromagnetic waves transmit energy just as mechanical waves do. However, electromagnetic waves do not require a medium but can travel through a vacuum. All forms of electromagnetic waves, from long radio waves, through visible light, up to gamma rays with very short wavelengths, travel at the same speed through a vacuum, $3 \times 10^8$ m/s. You could say that the speed of any electromagnetic wave is the same in the absence of a medium. However, when travelling through a medium, different wavelengths do, in fact, travel at different speeds. How fast is $3 \times 10^8$ m/s? Light can travel around the world 7.5 times in 1 s!

**COURSE CHALLENGE**

**Interference: Communication versus Energy Transmission**

Research current allocations of microwave bandwidths. How closely are these bandwidth slices packed? What kind of current research dealing with microwave interference is being done? Learn more from the **Science Resources** section of the following web site: **www.school. mcgrawhill.ca/resources/** and go to the *Physics 11 Course Challenge*.

**ELECTRONIC LEARNING PARTNER**

Learn more about wave terminology in your Electronic Learning Partner.

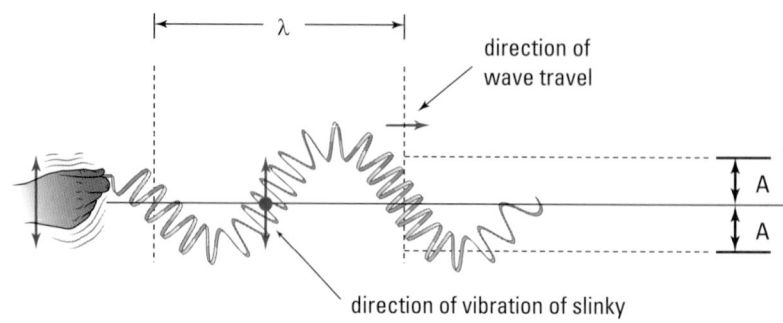

Figure 7.7  When a transverse wave travels along a spring, the segments of the spring vibrate from side to side, perpendicular to the direction of the wave motion.

When the particles of a medium vibrate parallel to the direction of the motion of the wave, it is called a **longitudinal wave** (see Figure 7.8). Once again, the speed of the wave is determined only by the medium. Sound waves, which you will study in more detail in Chapter 8, are examples of longitudinal waves.

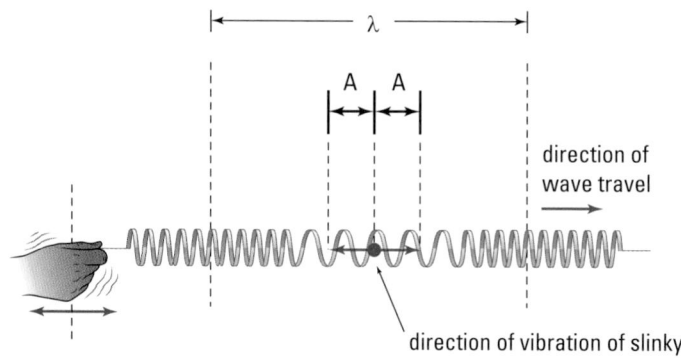

Figure 7.8  When a longitudinal wave travels along a spring, the segments of the spring vibrate parallel to the direction of the wave motion.

## Speed, Wavelength, and Frequency: A Universal Wave Equation

Knowing the speed of a wave provides critical information in many situations. For example, knowing the speed of waves in the deep ocean makes it possible to predict when a tsunami will hit a particular shore. Knowing the speeds of the different types of seismic waves from earthquakes enables geologists to locate the epicentre of an earthquake. As you have probably discovered, it is not easy to determine the speed of a wave pulse accurately. Fortunately, it is possible to determine this value from observable properties of waves travelling through a medium. Use the following problem-solving model to find the relationship.

### Earth Link

Earthquakes generate both transverse and longitudinal waves. Since the longitudinal waves travel faster through Earth and reach seismographs (earthquake detectors) first, they are called primary or P-waves. Transverse waves travel much slower and are called secondary or S-waves. From the difference in the time of arrival of the two waves, seismologists can estimate the distance from the detector to the quake. Do research for recent earthquake information. Especially focus on the recorded waves and time intervals. Attempt your own mathematical calculations for the distance between detectors and the earthquake's epicentre.

## Applying the Wave Equation

**1.** **A wave has an amplitude, $A$, frequency, $f$, and wavelength, $\lambda$. How can you find the speed of the wave using these variables?**

## Frame the Problem

- A source *vibrating* with a *frequency*, $f$, takes a time interval, $\Delta t = T$, to complete one cycle.

- During that *time*, the *wave* that it produced *moves* a *distance*, $\lambda$, along the medium.

- The average speed of any entity is the quotient of the distance it travelled and the time interval that it was travelling. This is true for waves as well as for moving objects.

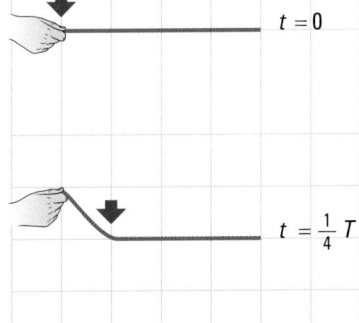

## Identify the Goal

The speed, $v$, of a wave

## Variables and Constants

| Involved in the problem | Known | Unknown |
|---|---|---|
| $\Delta d$    $A$ | $\Delta t = T = \dfrac{1}{f}$ | $v$ |
| $v$    $\Delta t$ | $\Delta d = \lambda$ | |
| | $A$ | |

| Strategy | Calculations |
|---|---|
| Use the formula for the velocity (or speed) of any entity. | $v = \dfrac{\Delta d}{\Delta t}$ |
| Substitute in known values. | $v = \dfrac{\lambda}{T}$ |
| Substitute $\dfrac{1}{f}$ for $T$. | $v = \dfrac{\lambda}{\frac{1}{f}}$ |
| Simplify. | $v = \lambda f$ |

The speed of a wave is the product of its wavelength and its frequency: $v = \lambda f$.

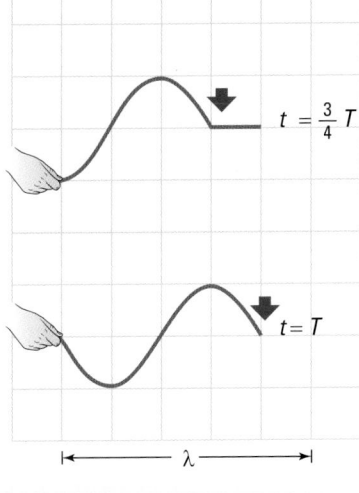

While the hand moves through one cycle, the wave moves one wavelength.

## Validate

The equation $v = f\lambda$ is called the **wave equation**. The wave equation can be seen simply as the form of the equation $v = \dfrac{\Delta d}{\Delta t}$ for waves, because we know that waves travel one wavelength in one period.

*continued* ▶

2. **A physics student vibrates the end of a spring at 2.8 Hz. This produces a wave with a wavelength of 0.36 m. Calculate the speed of the wave.**

## Frame the Problem

- *Vibrating* the end of a spring creates a *wave* in the spring.

- The *frequency* of the wave is the same as the *frequency* of the vibrating source.

- The *wave equation* applies to this situation.

## Identify the Goal

The speed, $v$, of the wave

## Variables and Constants

| Involved in the problem | Known | Unknown |
|---|---|---|
| $f$ | $f = 2.8$ Hz | $v$ |
| $\lambda$ | $\lambda = 0.36$ m | |
| $v$ | | |

## Strategy

Use the wave equation.

All needed variables are known, so substitute.

## Calculations

$v = f\lambda$

$v = (2.8 \text{ Hz})(0.36 \text{ m})$

$v = 1.0 \text{ ms}^{-1}$

$v = 1.0 \dfrac{\text{m}}{\text{s}}$

The speed of the wave is 1.0 m/s.

## Validate

The units reduced to m/s which is correct for speed.
A reasonable speed for a wave in a spring is 1.0 m/s.

3. **Water waves with wavelength 2.8 m, produced in a wave tank, travel with a speed of 3.80 m/s. What is the frequency of the straight vibrator that produced them?**

## Frame the Problem

- A vibrator in a water tank is producing waves.

- The *frequency* of a wave is the same as the *frequency* of the vibrating source.

- The *wave equation* applies to this situation.

## Identify the Goal

The frequency, $f$, of the vibrator

## Variables and Constants

| Involved in the problem | Known | Unknown |
|---|---|---|
| $f$ | $\lambda = 2.8$ m | $f$ |
| $\lambda$ | $v = 3.80$ m/s | |
| $v$ | | |

## Strategy

Use the wave equation to find the frequency of the wave.

All needed values are known, so substitute into the equation.

Divide by the coefficient of $f$.

Simplify.

## Calculations

$$v = f\lambda$$

**Substitute first**

$$3.80 \text{ m/s} = f(2.8 \text{ m})$$

$$\frac{3.80 \text{ m/s}}{2.8 \text{ m}} = f$$

$$f = 1.4 \text{ s}^{-1}$$

$$f = 1.4 \text{ Hz}$$

**Solve for $f$ first**

$$v = f\lambda$$

$$\frac{v}{\lambda} = \frac{f\lambda}{\lambda}$$

$$f = \frac{v}{\lambda}$$

$$f = \frac{3.80 \text{ m/s}}{2.8 \text{ m}}$$

$$f = 1.4 \text{ s}^{-1}$$

$$f = 1.4 \text{ Hz}$$

The frequency of the wave and, therefore, the frequency of the vibrator is 1.4 Hz.

## Validate

Since the wavelength (~3 m) is shorter than the distance the wave travels in 1 s (~4 m/s), you would expect that the period would be less than 1 s. If the period is less than 1 s, then the frequency should be more than 1 s, which it is.

## PRACTICE PROBLEMS

5. A longitudinal wave in a 6.0 m long spring has a frequency of 10.0 Hz and a wavelength of 0.75 m. Calculate the speed of the wave and the time that it would take to travel the length of the spring.

6. Interstellar hydrogen gas emits radio waves with a wavelength of 21 cm. Given that radio waves travel at $3.00 \times 10^8$ m/s, what is the frequency of this interstellar source of radiation?

*continued* ▶

7. Tsunamis are fast-moving ocean waves typically caused by underwater earthquakes. One particular tsunami travelled a distance of 3250 km in 4.6 h and its wavelength was determined to be 640 km. What was the frequency of this tsunami?

8. An earthquake wave has a wavelength of 523 m and travels with a speed of 4.60 km/s through a portion of Earth's crust.

    (a) What is its frequency?

    (b) If it travels into a different portion of Earth's crust, where its speed is 7.50 km/s, what is its new wavelength?

(c) What assumption(s) did you make to answer part (b)?

9. The speed of sound in air at room temperature is 343 m/s. The sound wave produced by striking middle C on a piano has a frequency of 256 Hz.

    (a) Calculate the wavelength of this sound.

    (b) Calculate the wavelength for the sound produced by high C, one octave higher than middle C, with a frequency of 512 Hz.

 **Web Link**

*www.school.mcgrawhill.ca/ resources/*
To learn more about tsunamis, go to the above web site. Follow the links for **Science Resources** and **Physics 11** to find out where to go next.

---

### THE WAVE EQUATION

The speed of a wave is the product of the wavelength and the frequency.

$$v = f\lambda$$

| Quantity | Symbol | SI unit |
|---|---|---|
| speed | $v$ | m/s (metres per second) |
| frequency | $f$ | Hz (or $s^{-1}$)(hertz) |
| wavelength | $\lambda$ | m (metres) |

**Unit Analysis**

(frequency)(wavelength) = Hz m = $s^{-1}$ m = m/s

---

### Think It Through

A student is holding one end of two different ropes, A and B. The other end of each rope is tied to a wall. Rope A is heavier than rope B, causing the speed of a wave in B to be twice as fast as the speed of a wave in A ($v_B = 2v_A$). The student initiates a wave pulse in both ropes by shaking the two ends at the same time. This gives the wave pulses identical amplitudes and time intervals. The diagram shows the shape and position of the wave pulse in rope A moments after the student initiated the pulses. Copy this figure into your notebook and, below it, sketch the appearance of rope B. Carefully consider the size, shape, and position of the wave pulse in rope B compared to rope A.

# Waves at Boundaries: Reflection and Transmission

When a wave moves from one medium into another, its *frequency remains the same* but its *speed changes*. As you have learned, the speed of a wave depends on the properties of the medium through which it is travelling.

Figure 7.9 shows what happens when a wave crosses the boundary between two springs joined end to end. The spring on the left is heavy and larger; thus, the wave pulse travels slowly. Once the wave pulse moves into the lighter, smaller spring on the right, it travels faster. You could call the heavy spring the "slow medium" and the light spring the "fast medium." Notice in the figure that, in addition to a transmitted wave pulse, there is also a reflected pulse. Notice, also, that the reflected pulse is on the same side of the spring as the incident and transmitted pulse. When a wave travels from a slow medium to a fast medium, the reflected wave is always on the same side of the rest position as the incoming wave.

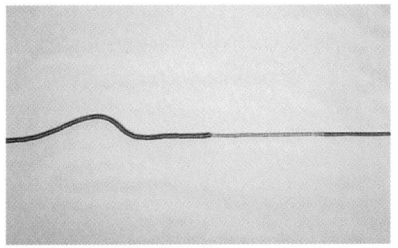 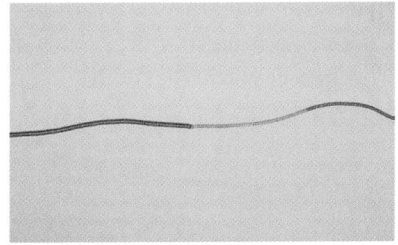

**Figure 7.9** At a slow-to-fast interface between two media, the transmitted and reflected pulses are on the same side of the spring.

When a wave pulse travels from a fast medium, such as a light spring, to a slow medium, such as a heavy spring, both transmitted and reflected wave pulses result. However, the reflected pulse is inverted, that is, on the opposite side of the spring, as shown in Figure 7.10.

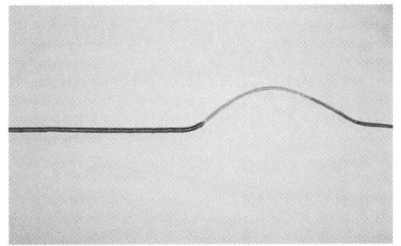 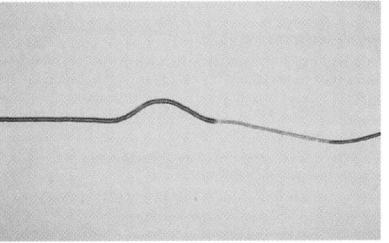

**Figure 7.10** At a fast-to-slow interface, the transmitted pulse is on the same side of the spring as the original pulse, but the reflected pulse is inverted.

When one end of a spring is attached firmly to a wall, for example, the reflected pulse is inverted as well (see Figure 7.11). You can consider reflection from an end attached to something solid to be a special case of transmission at a fast/slow boundary. The speed of the wave in the massive wall is much slower than in the spring.

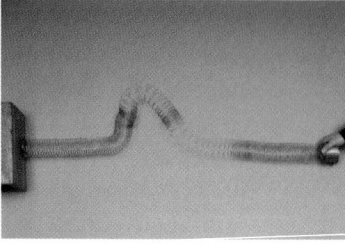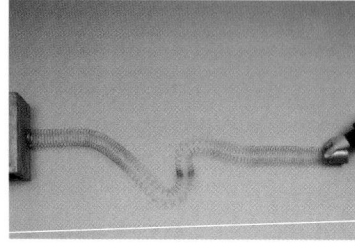

**Figure 7.11** When a wave pulse is reflected from a fixed end, the reflected wave pulse is inverted.

Similarly, when a pulse travels down a spring toward an end that is not attached to anything, the free end will reflect the pulse. As shown in Figure 7.12, the reflected pulse will not be inverted. You can consider reflection from a free end as a special case of transmission at a slow-to-fast boundary.

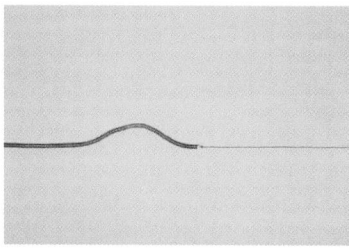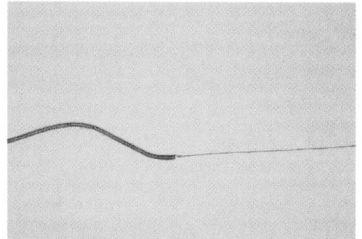

**Figure 7.12** When a wave pulse is reflected from a free end, the reflected pulse is on the same side as the original pulse.

## Think It Through

- A good model accurately predicts observations from the widest possible set of related phenomena. Assume that the wave model presented here is a good model and also applies to water waves.

  **1.** If a water wave travels through water, what changes might occur that will simulate a different medium?

  **2.** Discuss the types of boundaries that water waves might encounter.

  **3.** A water wave moves from the deep ocean to the much shallower waters of the Grand Banks off the coast of Newfoundland. Use the wave model to predict how the waves will behave as they cross this boundary.

1. **K/U** What are the essential characteristics of a wave?

2. **K/U** How do transverse and longitudinal waves differ? Give an example of each.

3. **C** Sketch a diagram of a transverse wave. Mark the amplitude and wavelength of the wave on your diagram. Also, mark two points, P1 and P2, that are in phase.

4. **C** Do the following:

   **(a)** State the wave equation relating the speed, frequency, and wavelength of a wave.

   **(b)** Explain how the wave equation can be derived from the fact that a wave travels a distance of one wavelength in a time of one period.

5. **K/U** A wave pulse is travelling down a spring from left to right as shown. Sketch what the spring would look like after the pulse had been reflected if

   **(a)** the opposite end of the spring was firmly held to the floor by another student.

   **(b)** the opposite end of the spring was held by a light thread so that it was free to move.

6. **K/U** After a transverse wave pulse has travelled 2.5 m through a medium, it has a speed of 0.80 m/s. How would this speed have differed if

   **(a)** the pulse had been twice the size?

   **(b)** the pulse had had twice the energy?

   **(c)** the pulse had travelled twice the distance?

7. **C** Explain the meaning of the statement, "The speed of a wave is a characteristic property of the medium through which it is travelling."

8. **K/U** A wave pulse is travelling down a spring with a speed of 2 m/s toward a second spring attached to its opposite end. Sketch what the two springs would look like after the pulse has passed into the second spring if

   **(a)** the speed of a wave in the second spring is 1 m/s.

   **(b)** the speed of a wave in the second spring is 4 m/s.

   **(c)** the speed of a wave in the second spring is 2 m/s.

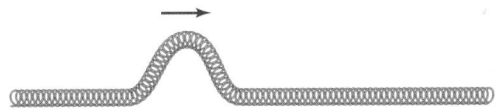

$2\ \frac{m}{s}$

- Explain and illustrate the principle of interference of waves.

- Communicate and graphically
illustrate the principle of
superposition.

- Analyze, measure, and interpret
the components of standing
waves.

### KEY TERMS

- resultant wave
- component wave
- constructive interference
- destructive interference
- node

- antinode
- standing wave
- natural frequency
- fundamental frequency
- fundamental mode
- overtone

When material objects such as billiard balls collide, they bounce off each other. One usually gains energy, while the other loses energy. In any case, they move off in different directions. In some cases, when an object has a large amount of energy and collides with another, the shape of the object undergoes a drastic change. It might break apart into many pieces or it might collapse or be crushed into an unrecognizable form. How do waves react when they meet?

The students in Figure 7.13 are sending wave pulses toward each other along the same spring. What will happen when the wave pulses meet? Will they collide and bounce off each other? Will the waves become distorted and unrecognizable? Will they simply pass through each other unchanged? Complete the following Quick Lab to find out for yourself.

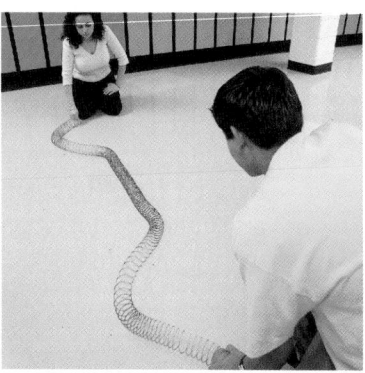

**Figure 7.13** How do collisions between wave pulses compare to collisions between material objects?

## QUICK LAB — Do Waves Pass Through or Bounce Off Each Other?

**TARGET SKILLS**
- Predicting
- Analyzing and interpreting

Predict what will happen when two wave pulses meet. Give the reasoning behind your predictions. With a partner, stretch a large spring out along the floor, to a length of about 8 m. Start wave pulses from each end at the same time and observe what happens. Test all of the following combinations.

- pulses of the same size on the same side of the spring
- pulses of different sizes on the same side of the spring
- pulses on the opposite sides of the spring

Discuss with your partner what you perceive to be happening. If you do not agree, design more experiments until you feel that you have a clear understanding of what happens when wave pulses meet.

### Analyze and Conclude

1. Did your final conclusion agree with your prediction? Explain any contradictions.

2. Describe your final conclusion about whether waves bounce off or pass through each other.

3. How did the design of your experiments help you to draw a firm conclusion?

# Superposition of Waves

You no doubt concluded from the Quick Lab that waves *do* pass through each other. During the time that the two waves overlap, they interact in a manner that temporarily produces a different-shaped wave. This **resultant wave** is quite unlike either of the two **component waves**. Each component wave affects the medium *independently*. Consequently, at any one time, the displacement of each point in the medium is the *sum* of the displacements of each component wave. Note that the displacements of the component waves can be either positive (+) or negative (–). These signs must be included when adding them. If one wave would have moved a particular point in the medium up three centimetres (+3 cm), and a second wave would have moved that point down six centimetres (–6 cm), then the resultant displacement would be three centimetres down (+3 cm + (–6 cm) = –3 cm). This behaviour of waves is known as the "principle of superposition."

When two waves displace the medium in the same direction, either up or down, the resultant displacement is larger than the displacement produced by either component wave alone. This interaction is called **constructive interference** (see Figure 7.15). As the wave pulses pass through each other and the peaks of each wave overlap, one point (A) in the medium will experience the maximum displacement.

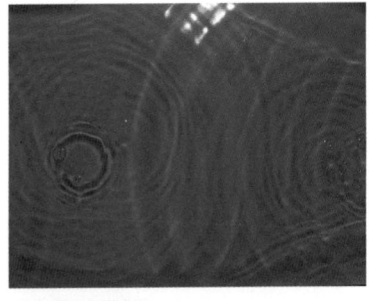

**Figure 7.14** As these water waves move through each other, you can readily see the details of each individual wave.

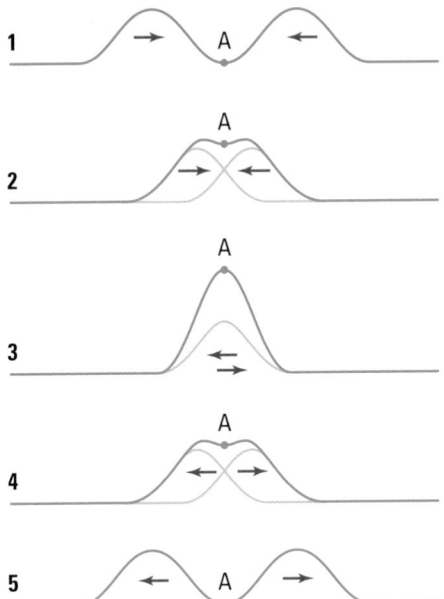

**Figure 7.15** Constructive interference results in a wave pulse that is larger than either individual pulse.

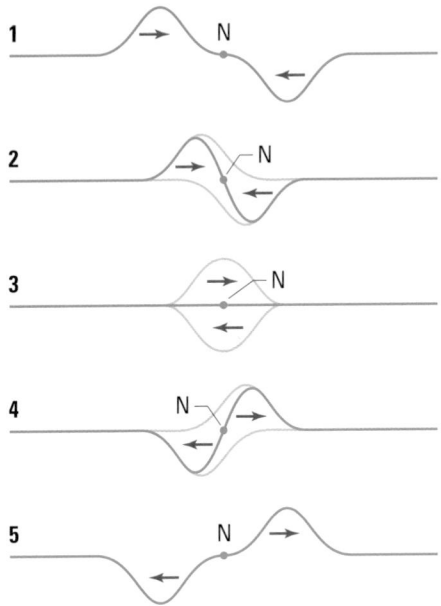

**Figure 7.16** Destructive interference results in a pulse that is smaller than the larger of the component waves. When the component pulses are identical in size but one is inverted, there is one moment when no pulse can be seen.

When two waves displace the medium in opposite directions, the resultant displacement is less than one, and sometimes both, of the component waves. This interaction is called **destructive interference**, and is illustrated in Figure 7.16 on the previous page. If the two pulses are identical in size, the point where they first meet (N) will not move at all as the waves interfere.

## Standing Waves

When periodic waves with the same shape, amplitude, and wavelength travel in opposite directions in a linear medium such as a rope or spring, they produce a distinct pattern in the medium that appears to be standing still. At intervals that are a half wavelength apart, the waves destructively interfere and create points, called **nodes**, that never move. Between each node, a point in the medium, called an **antinode**, vibrates maximally. Because the nodes do not move, the sense of movement of the two-component wave is lost and the resultant wave is called a **standing wave**.

Figure 7.17 illustrates how using the principle of superposition yields the pattern of fixed nodes, spaced half a wavelength apart, and points of maximum disturbance, or antinodes, also spaced half a wavelength apart. The antinodes are located at the midpoints between adjacent nodes. As you can see in part (A) of Figure 7.17, when the two identical component waves line up with troughs opposite crests, there is complete destructive interference and, momentarily, the medium is undisturbed.

A quarter of a period later, the yellow wave will have moved a quarter of a wavelength to the right and the blue wave will have moved a quarter of a wavelength to the left. Part (B) of Figure 7.17 shows how the two component waves are superimposed, producing constructive interference. This interference produces a resultant wave with an amplitude that is the sum of the component waves.

A quarter of a period after the situation depicted in part (B), the waves will again be lined up in opposition, so as to produce destructive interference in part (C). A quarter of a period later, the component waves will again be superimposed so as to produce constructive interference, as illustrated in part (D). This sequence will repeat over each period.

Part (E) of the figure represents the image you would see over a period of time. At the nodes, the medium does not move at all. In between the nodes, the standing wave appears as a blur, because the medium is moving up and down constantly. The resulting movement is characterized by a series of nodes spaced half a wavelength apart along the medium and a series of antinodes located at the midpoints between the nodes.

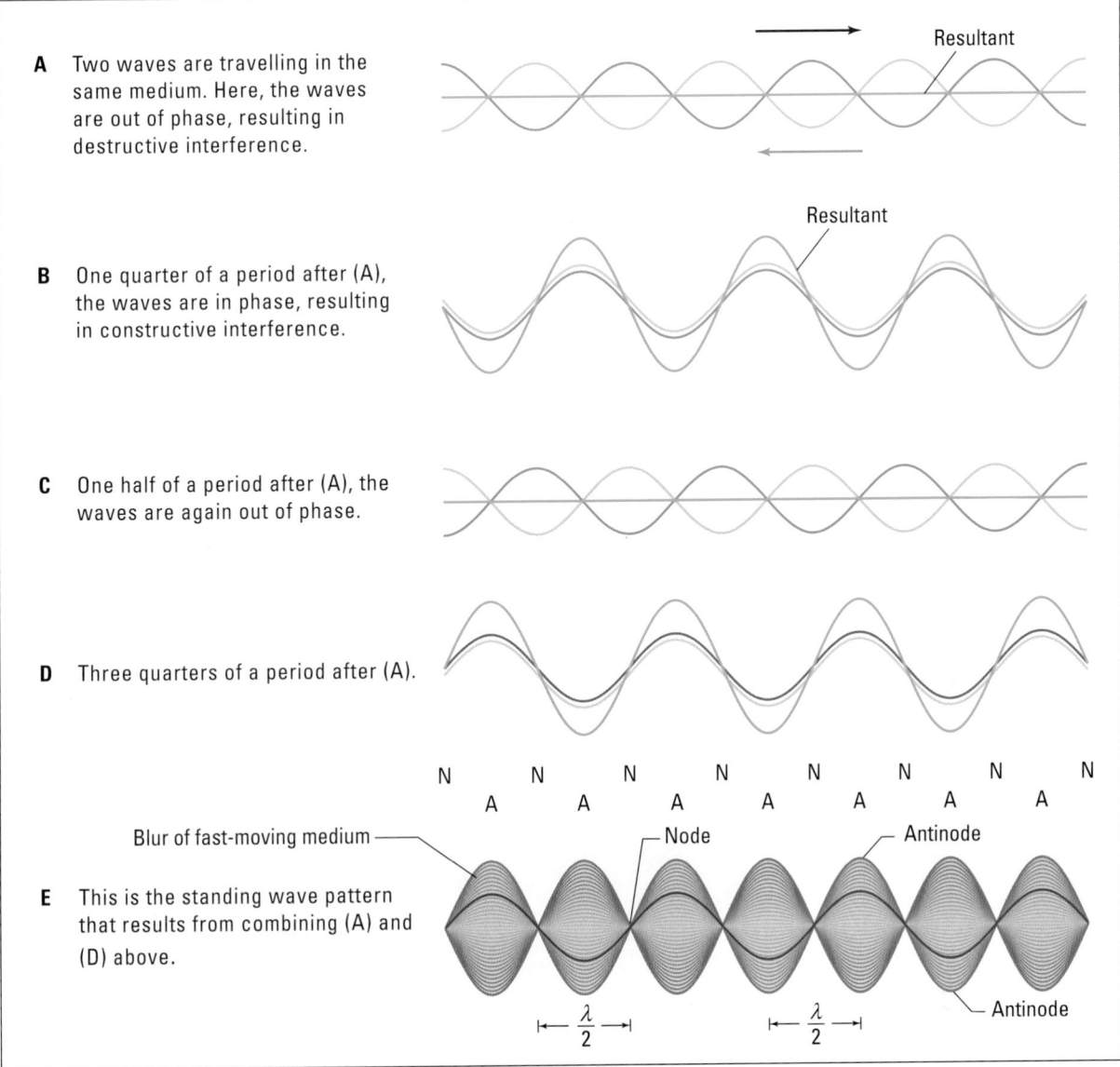

**A** Two waves are travelling in the same medium. Here, the waves are out of phase, resulting in destructive interference.

Resultant

**B** One quarter of a period after (A), the waves are in phase, resulting in constructive interference.

Resultant

**C** One half of a period after (A), the waves are again out of phase.

**D** Three quarters of a period after (A).

N    N    N    N    N    N    N    N
  A    A    A    A    A    A    A

Blur of fast-moving medium — — Node — Antinode

**E** This is the standing wave pattern that results from combining (A) and (D) above.

— Antinode

$\frac{\lambda}{2}$    $\frac{\lambda}{2}$

**Figure 7.17** As identical travelling waves proceed one quarter of a wavelength in opposite directions, the resultant wave goes from complete destructive interference to maximal constructive interference. Part (E) shows how the wave appears over a period of time. The red wave shows the wave at one instant in time.

A standing wave can also be produced in a linear medium using only one vibrating source. The second wave is produced by the reflection of the first wave from the far end of the medium. Because the second wave is a reflection, it will have essentially the same frequency, wavelength, and amplitude as the first. Whether or not resonance and, consequently, a standing wave will occur depends on the match between the frequency of vibration and the length of the linear medium.

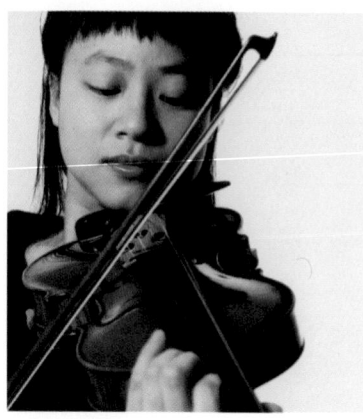

Figure 7.18 The position of a violinist's finger determines the effective length of the string and, therefore, determines which wavelengths will form a standing wave.

For example, consider a bow drawn across a violin string. The friction of the bow causes the string to vibrate at many different frequencies. Waves move in both directions away from the bow toward the fixed ends of the string. When they reach the ends, the waves reflect back. The propagated waves and the reflected waves interfere, sometimes constructively. Whether or not standing waves of a given frequency can form depends on the end points of the string where the string is fixed. Since the ends of the strings cannot move, standing waves can form only if nodes occur at the ends of the strings. When the string is vibrating at its resonance frequencies, it causes the body of the violin to vibrate and amplify the tone.

For every medium of a fixed length, there are many **natural frequencies** of vibration that produce resonance. Figure 7.19 shows a rope vibrating at three of its natural frequencies. The lowest natural frequency (corresponding to the longest wavelength) that will produce resonance on the rope is called the **fundamental frequency**. The standing wave pattern for a medium vibrating at its fundamental frequency displays the fewest number of nodes and antinodes and is called its **fundamental mode** of vibration. All natural frequencies higher than the fundamental frequency are called **overtones**. For example, the natural frequency that corresponds to a pattern with one node in the centre of the rope is called the "first overtone." The pattern continues with the addition of one node at a time. A rope or string may vibrate at several natural frequencies at the same time.

MISCONCEPTION

**You can't get something for nothing!**

Some people might think that constructive interference creates energy, or that destructive interference destroys energy. However, the law of conservation of energy still holds. Energy can neither be created nor destroyed. A standing wave simply redistributes the energy so that there is less energy near the nodes and more of it near the antinodes.

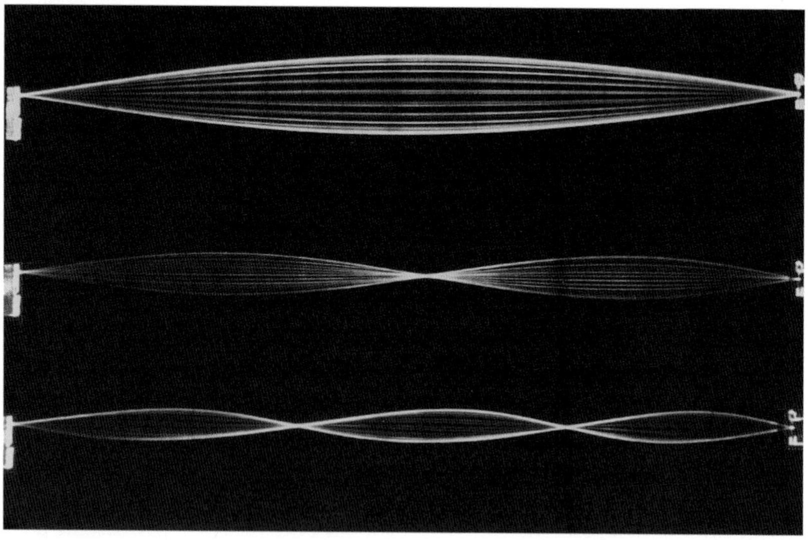

Figure 7.19 Resonance will occur in a vibrating rope for wavelengths that create nodes at the ends of the rope. There may be any number of nodes within the rope.

# INVESTIGATION 7-B

## Wave Speed in a Spring

**TARGET SKILLS**
- **Performing and recording**
- **Analyzing and interpreting**

If you know the speed of a wave, you can use it to determine the time it takes a wave to travel a given distance, or the distance a wave will travel in a given time. You can also use the speed to determine the wavelength of a wave when the frequency is known, or the frequency when the wavelength is known. The precision of any of these calculations depends on the precision with which the speed is known. Consequently, it is important to determine the speed as precisely and accurately as possible.

In this investigation, you will measure the speed of a wave using two different methods. Then, you will evaluate the methods and decide which is the more accurate. In order to make the best measurements, you will need at least three people in each group.

## Problem

To determine the speed of a wave in a stretched spring as precisely as possible, and to evaluate the accuracy of the result.

## Equipment

- long spring
- stopwatch
- metre stick or measuring tape

## Procedure
### Direct Measurement

1. Stretch the spring out between two partners. The third partner will carefully measure the length of the spring. Ensure that you maintain this length throughout the investigation.

2. Determine the optimum number of times that you can allow the pulse to reflect back and forth and still see the pulse clearly enough to make good time measurements. Send several test pulses down the spring to determine the optimum number of reflections to allow for one time measurement. (**Note:** You can

increase the precision of your measurements by allowing the pulse to travel longer distances. However, as the pulse reflects back and forth, friction causes the amplitude to decrease. The amplitude eventually becomes so small that it is hard to follow and thus decreases the precision of your measurements.)

3. Devise a method for determining the exact distance that a pulse has travelled when you make a measurement.

4. Prepare a data table with the following headings: Distance, Time, Speed. Allow enough rows for at least five trials.

5. Carry out at least five trials for measuring the time and distance data for a moving pulse.

6. Calculate the speed of the pulse and determine the average speed for the five or more trials that you performed.

### Indirect Measurement

7. To carry out an indirect measurement, you need to create standing waves in the spring. Practise the creation of standing waves by performing the following steps.

   (a) Stretch the spring out to about 8 m.

   (b) While one partner holds the end of the spring fixed, another should vibrate the opposite end back and forth.

   (c) Start with a very low frequency and try to get the spring to vibrate in its fundamental mode. You should see only one node at the end held in place and one more node very close to the end that is being vibrated.

   (d) Slowly increase the frequency of vibration until you can produce other standing wave patterns.

   (e) Determine the values of all of the natural frequencies that you were able to find.

*continued* ▶

*continued from previous page*

| time for 20 vibrations $\Delta t$ | $f = \dfrac{20}{\Delta t}$ | $\lambda$ = distance from fixed end to second node | $v = f\lambda$ |
| --- | --- | --- | --- |
|  |  |  |  |
|  |  |  |  |

**8.** Stretch the spring to exactly the same distance that you used when making the first set of measurements.

**9.** Prepare a data table like the one shown above. Allow for at least five trials.

**10.** One partner should hold one end of the spring firmly in place, while a second partner creates a standing wave by vibrating the other end of the spring. Find a frequency that creates at least two nodes in the spring.

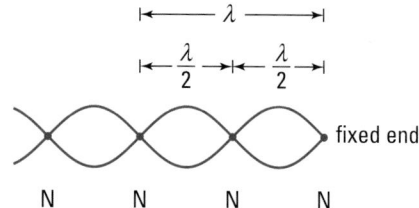

**11.** Let the third partner determine the wavelength by measuring the distance from the stationary end to the second node. Record the value in the data table.

**12.** Determine the time for 20 vibrations. Record the value in the data table.

**13.** Repeat steps 10 through 12 for at least five trials.

**14.** Calculate the speed of the wave for each trial. Determine the average of the speeds for all trials.

**Analyze and Conclude**

**1.** Compare the precision of measurement for the two methods for determining the speed of a wave. (**Note:** The range of the speeds in individual trials for each method is an indicator of precision. A narrow range of values indicates greater precision. If the calculated speeds were quite different from one trial to the next, the precision is low.) If you are unsure about the difference between precision and accuracy, review the meanings of these terms in Skill Set 2.

**2.** Compare the values of the speed of the wave for the two different methods. Do the ranges of values of speed for the two methods overlap? Are the values of average speed of the wave similar or quite different for the two different methods?

**3.** List the factors that might have contributed to any lack of compatibility of the two methods.

**4.** From your observations and analyses, which average speed do you think is the most accurate? Explain the reasoning on which you based your conclusion.

**5.** What is the relationship between the natural frequencies of the spring and its fundamental frequency?

**6.** How could you tell if you had missed one of the natural frequencies when you were finding natural frequencies above the fundamental?

Standing wave patterns can be set up in a variety of objects. If you carefully examine the photograph of the Tacoma Narrows Bridge collapse on page 326, you should see evidence of the standing wave that was set up in the bridge. You can also observe standing wave patterns in the radio antenna of a car as you travel at different speeds along a highway.

**Figure 7.20** The tones of a music box are created by the natural frequencies of tiny strips cut from a small sheet of metal. As the drum turns, pegs on the drum flip the metal strips and cause them to vibrate.

If you carefully examine the photograph of the Tacoma Narrows Bridge collapse on page 326,

## QUICK LAB

# Standing Waves in a Thin Piece of Wood

**TARGET SKILLS**
- **Analyzing and interpreting**
- **Communicating results**

Standing wave patterns can easily be set up in a long piece of wood moulding. Obtain a piece of quarter-round moulding 2 m to 3 m long and 0.50 cm thick. With the moulding oriented horizontally or vertically, vibrate one end of the moulding back and forth through a range of frequencies. You should be able to "feel" the resonance that is produced when you are vibrating the moulding at a natural frequency.

### Analyze and Conclude

1. How do the standing wave patterns produced in the moulding differ from those produced in the spring?

2. Describe the standing wave pattern produced by the fundamental frequency.

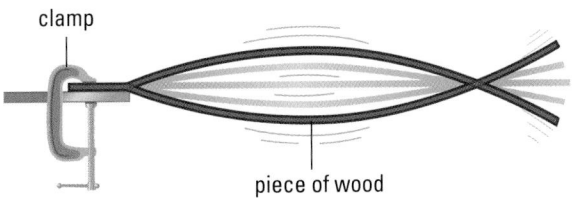

clamp

piece of wood

How is the wavelength associated with the fundamental frequency related to the length of the moulding?

1. **K/U** Two triangular pulses, each 2 cm high and 1 cm wide, were directed toward each other on a spring, as shown. Sketch the appearance of the spring at the instant that they met and completely overlapped. What kind of interference is this?

2. **K/U** Two triangular pulses, each 2 cm high and 1 cm wide, were directed toward each other along the same spring. However, the pulse approaching from the left was erect and the one approaching from the right was inverted. Sketch the appearance of the spring at the instant that the two pulses met and completely overlapped. What kind of interference is this?

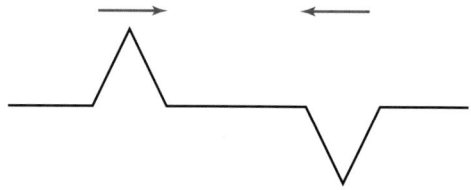

3. **K/U** An upright square pulse and an inverted triangular pulse were directed toward each other on a spring, as shown in the illustration. Sketch the appearance of the spring at the instant the two pulses met and completely overlapped. What principle did you use in constructing the shape of the spring for the instant at which the two pulses met? What does this principle state about how waves combine?

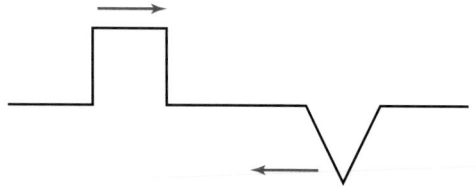

4. **C** Describe what you would see when a standing wave was set up in a spring. Why is it called a standing wave?

5. **K/U** What is a node? What is an antinode? Describe how the nodes and antinodes are distributed along the length of the standing wave pattern.

6. **C** Sketch the appearance of the standing wave pattern set up in a spring when it is fixed at one end and the other end is vibrated at (a) its fundamental frequency, (b) a frequency twice its fundamental frequency, and (c) a frequency three times its fundamental frequency.

7. **K/U** The figure shown here represents a spring vibrating at its second overtone. The points labelled (A), (B), and (C) represent the location of the central point of the string at various times.

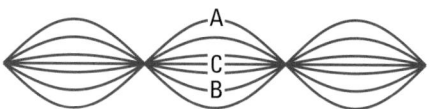

(a) At which location is the central point of the string moving at its maximum speed?

(b) At which location is its instantaneous speed zero?

(c) At which location is the point on the string moving with an intermediate speed?

(d) Explain the reasoning you used to answer the above questions.

Have you ever taken a stroll near a river, either in woodland or perhaps along a city street, and heard the splashing sounds of moving water before you could actually see the river?

**Figure 7.21** A river generates sound that often can be heard long before the river is in view. This phenomenon highlights some special properties of sound waves that result from their three-dimensional nature.

**SECTION EXPECTATIONS**

- Investigate the properties of mechanical waves through experimentation.

- Define and describe the concepts and units related to constructive and destructive interference.

- Draw and interpret interference of waves during transmission through a medium.

**KEY TERMS**

- wavefront
- ray
- normal line
- angle of incidence
- angle of reflection
- refraction
- diffraction
- nodal line
- antinodal line

So far in this chapter, you have explored the behaviour of waves in linear media such as springs and ropes. However, many wave phenomena that you will be studying, such as sound and light, are not confined to a single dimension. In fact, sound waves can travel around corners, as you can tell whenever you hear a sound before you can see its source. To understand such phenomena, you need to learn about waves in more than one dimension. In this section, you will briefly explore some wave behaviours that emerge when waves travel in two-dimensional media.

## Behaviour of Two-Dimensional Waves

The most visible two-dimensional waves are water waves. Observing and describing water waves will help you to understand sound and light waves, as well as many other types of waves. Water waves can take on a variety of shapes. Two examples, straight and circular waves, are shown in Figure 7.22 on the next page. The lines drawn across the crests are called **wavefronts**. Since the distance from one crest to the next is one wavelength, the distance between wavefronts is one wavelength. To indicate the direction of the motion of the wave, lines called **rays** are drawn perpendicular to the wavefronts. Rays are not physically *part of* the wave, but they do help to model it scientifically.

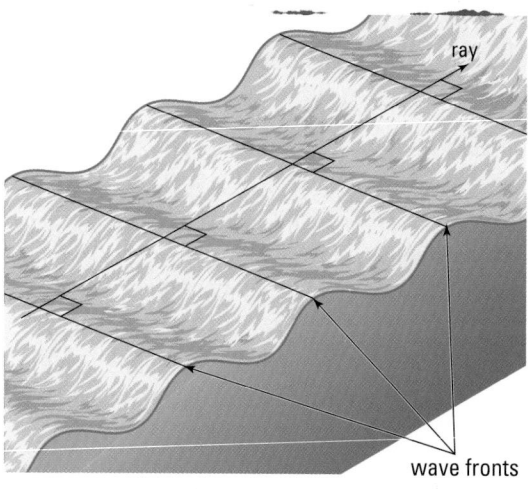

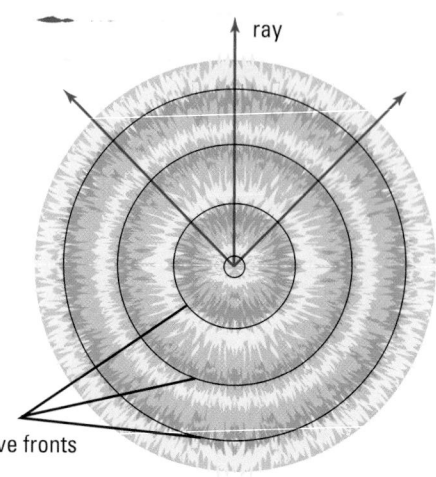

**Figure 7.22** (A) A straight object rocking back and forth can disturb the water's surface and create waves with straight wavefronts.

(B) A stone or pebble dropped into water can create circular waves. The crests move out in all directions from a central disturbance to water.

When a water wave encounters a solid barrier, it reflects in a way that is similar to a wave in a rope reflecting from an end that is firmly attached to a wall. However, water waves are not constrained to reflect directly backward. To quantitatively describe the way a two-dimensional wave reflects off a barrier, physicists define specific angles, as shown in Figure 7.23. At the point where a ray strikes the barrier, a line, called a **normal line**, is drawn perpendicular to the barrier surface. The **angle of incidence** is the angle between the normal line and the ray representing the incoming wave. The **angle of reflection** is the angle between the normal line and the ray representing the reflected wave.

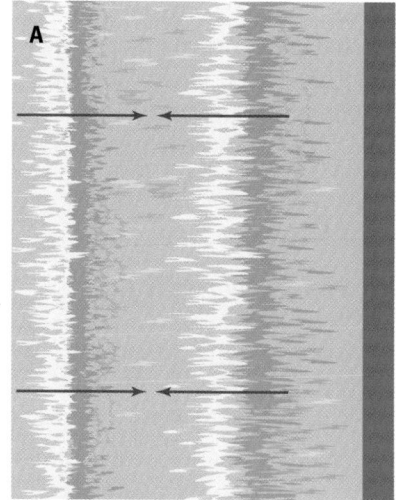

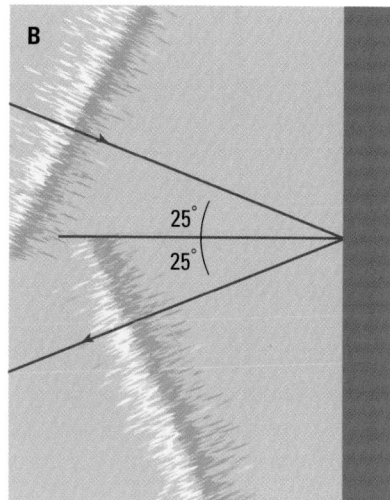

**Figure 7.23** (A) Waves travelling directly toward a straight barrier reflect straight back. (B) Waves arriving at a barrier at an angle reflect off the barrier at an angle.

# INVESTIGATION 7-C

## Waves on the Surface of Water

Patterns occur in a ripple tank because the crests of waves act as lenses (see Chapter 12) that focus the light and produce bright regions on the screen. The troughs act as lenses that spread the light out and produce dark regions.

## Problem

How can a ripple tank be used to study waves?

## Equipment

- ripple tank
- wooden dowel
- wave generator
- plastic or paraffin block (about 1 cm thick)

  **CAUTION**

Care must be taken with any electrical equipment near ripple tanks. Firmly attach lights and wave generators to the tank or lab bench, and keep all electrical wiring away from the water.

## Procedure

1. Assemble a ripple tank similar to the one shown below. Add water and level the tank so that the depth of the water is approximately 2 cm at all points in the tank.

2. Place a solid barrier in the tank. Use a dowel to generate single wave pulses, one at a time.

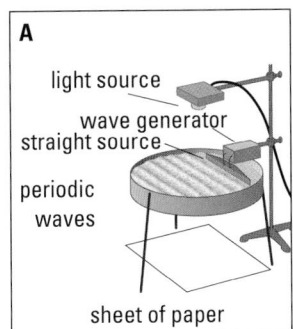 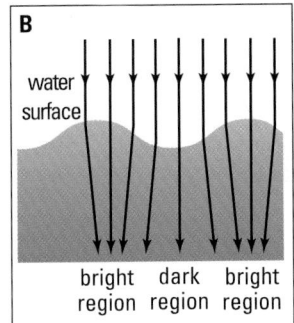

**(A)** You can see the details of water waves by using a ripple tank with a light shining directly onto the surface of the water. **(B)** The curves on the water's surface focus light, creating bright and dark regions.

3. First, send a straight wave pulse directly toward the barrier. Then send straight wave pulses toward the barrier at a variety of angles. Draw diagrams of the wavefronts, with rays showing their direction of motion.

4. Next, send straight wave pulses toward a concave barrier roughly the shape of a parabola. Adjust the shape of the barrier until the reflected wave appears to converge toward a point. Keeping the shape of the barrier the same, start a circular wave from the point you just found, using your finger. Observe what happens when these circular waves reflect from the parabolic barrier.

5. To record your observations, draw diagrams of the shape of the wavefronts and include rays to illustrate the path of their motion.

6. Place a block of plastic (or paraffin), approximately 1 cm thick, on the bottom of the tank at one end. With the dowel, make straight wave pulses. First, send the pulses directly toward the edge of the plastic. Then, send straight wave pulses toward the plastic at various angles with the edge of the plastic. Observe any changes in wavelength or direction after the waves pass over the plastic.

## Analyze and Conclude

1. Compare the angle of reflection to the angle of incidence when straight waves reflect from a straight barrier. State any general relationship that you observed.

2. How does a straight wave reflect from a parabolic barrier?

3. If a circular wave is started at the point where a parabola focusses a straight wave, what is the shape of the reflected wavefront?

4. How did the wavelength and direction of straight waves change when the waves passed from deeper water into more shallow water?

The equation $y = kx^2$ is the equation of a parabola. The vertex of this parabola is located at the origin (0,0) and it opens upward. The rate at which it opens is determined by the value of $k$. Radar antennas and satellite dishes for television are constructed with parabolic cross sections. They collect waves coming straight in and focus them into a small receiver. Carefully sketch a parabola with equation $y = x^2$, and use ray paths to form a hypothesis about the location of the focus.

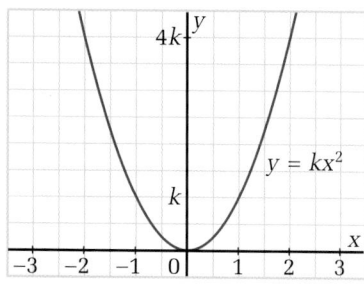

# Reflection and Refraction of Water Waves

In the previous Quick Lab, you probably discovered that, when a straight wave moves directly toward a straight barrier, the wave will be reflected directly backward. If a straight wave meets a barrier at an angle, it will reflect off at an angle. The angle of reflection will equal the angle of incidence.

Barriers with different shapes reflect waves in unique ways. For example, a concave barrier in the shape of a parabola will reflect straight waves so they become curved waves. The waves will converge toward, then pass through, a single point. As they continue through the point, they curve outward, or diverge.

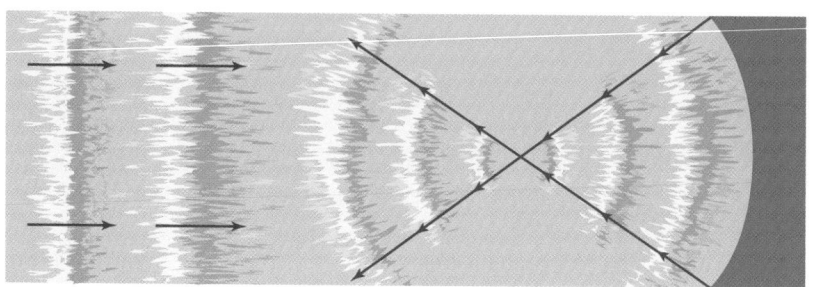

**Figure 7.24** A parabolic barrier will reflect straight waves through a single focal point.

# Diffracting Water Waves

**TARGET SKILLS**
- Identifying variables
- Analyzing and interpreting

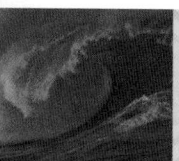  

**CAUTION**

Use extreme care when working with electrical equipment near ripple tanks. Ensure that lights and wave generators are firmly attached. Keep all electrical wiring away from the water.

Use a straight wave generator with a ripple tank to generate periodic water waves. Obtain a barrier with an opening that can be varied in size. Place the barrier in the tank, parallel to the straight wave generator. Position the barrier so that you have a good view of the wavefronts after they have passed through the opening. Observe the behaviour of the waves passing through the barrier under the following conditions.

- Vary the size, $D$, of the opening in the barrier.
- Vary the wavelength, $\lambda$, of the waves.

The property of waves that causes them to "bend around corners" is called "diffraction." Choose the size of opening that caused the greatest amount of diffraction and make two openings of that size that are quite close together. Observe the behaviour of straight waves as they reach and pass through the two openings.

## Analyze and Conclude

1. Do small or large openings in the barrier cause more diffraction of the water waves?

2. Are small or large wavelengths diffracted more?

3. Describe the pattern you observed that was caused by two openings located close together.

When waves travel from one medium into another, their speed changes. This phenomenon is called **refraction**. As a result of the change in speed, the direction of two-dimensional waves changes. You can demonstrate this effect in water waves without even changing from water to another medium. Refraction occurs in water because the speed of waves in water is influenced by the depth of the water. You probably observed refraction occurring in the ripple tank when you placed the thick sheet of plastic in the tank. Did the waves change direction when they went from the deeper water to the shallower water? You will study refraction of light waves in more depth in Chapter 11.

## Bending Around Corners: Diffraction of Waves

Waves and particles behave quite differently when they travel past the edge of a barrier or through one or more small openings in a barrier. Moving particles either reflect off the barrier or pass through the opening and continue in a straight line. What do waves do when they encounter the edge of a barrier or openings in a barrier?

In the previous Quick Lab, you probably discovered that when waves pass through small openings in barriers they do not continue straight through. Instead, they bend around the edges of the barrier. This results in circular waves that radiate outward. This behaviour of waves at barriers is called **diffraction**. The amount of diffraction is greatest when the size of the opening is nearly the same as the size of one wavelength.

**Biology Link**

Scientists who study bird vocalizations use parabolic reflectors to collect sounds from a distance and focus the sound waves to a point where they have placed a microphone. By aiming the reflector at a distant bird, they can "capture" the sound of that bird and nearly eliminate other sounds. Use print resources and the Internet to find out what researchers have discovered about how songs and calls are learned, and if they have "meaning."

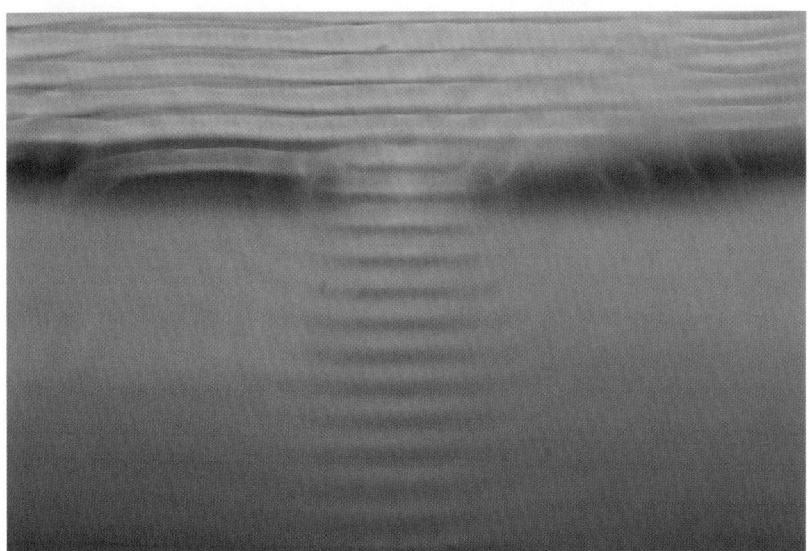

**Figure 7.25** When straight water waves reach a small opening in a barrier, the tiny part of the wave at the opening acts as a point source, similar to the effect of putting your finger in the water; thus, the waves move out in semicircles. This phenomenon is an example of the diffraction of waves.

## Sound Technology in the Movies

Today's movies aim to give moviegoers a total sensory experience. This includes not only stunning visual special effects, but spectacular sound as well. If a plane in the distance on the lower left of the screen zooms overhead, sound engineers want viewers to hear the sound pass overhead, too, and fade away behind their right shoulder. How is this done?

The invention of two-track stereo sound was a first step. The action was recorded with two separated microphones and played back on speakers at each side of the screen. This allowed the sound to follow a car moving across the screen. The effect could be enhanced by using four tracks, with two extra speakers at the back of the theatre.

In the 1980s, the producers of the film *Earthquake* wanted moviegoers to feel the ground shake. In theatres specially fitted for the film, many large speakers were placed around the walls and under the seats. When the earthquake began, these speakers pumped out loud, low frequency sound that caused the seats and floor to vibrate.

The newest advance in sound technology, three-dimensional sound, does more than just record and reproduce sound. Instead, audio engineers attempt to generate the sound you would have heard had you been in the scene. Computers analyze and reproduce the tiny delays and echoes that occur when you hear a sound. Imagine that someone near you drops a wineglass. The ear slightly closer to the event hears the sound first. Sound rebounding off the ceiling comes to you from above. Echoes arrive from the back of the room a fraction of a second later, although the time lag is so small that you do not actually distinguish these echoes from the original crash. Higher-pitched sounds bounce off you and rebound from the front. Lower-pitched sounds flow around you, and you hear them from behind once again. All of these audio clues help you to sense that the glass shattered in front of you, slightly to the right, and below.

The physical principles governing reflection, diffraction, and interference are all used to reproduce three-dimensional sound correctly. This advanced audio technology is in use in theatre sound systems and quality headsets.

### Analyze

1. What frequency might have been used in the movie *Earthquake* to produce a vibration of the floor but not make an audible note?

2. Explain how you are able to locate the source of a sound.

3. Describe how your life might be affected if you were unable to locate sound sources.

# Interference Patterns in Water Waves

If plane waves pass through two openings in a barrier that are close together, diffraction of the waves creates a unique pattern. A similar pattern is created by two point sources, located close together, generating circular waves that are in phase. As the waves from the two sources meet, interference creates the distinctive pattern seen in Figure 7.26. This pattern results from alternating nodes and antinodes radiating outward from the sources. Although the situation is much like standing waves on a string, the pattern spreads over the two-dimensional water surface.

You can analyze, even predict, the pattern by considering several lines radiating out from the centre and then determining how waves will interfere along these lines. Start with the perpendicular bisector of a line connecting the two sources. Then, pick any point on that line and draw lines from each source to the point, as shown in Figure 7.27 (A). Notice that these lines form congruent triangles because the bases are equal, the angles are equal, and they share a side. Therefore, the lines from the sources are equal in length. Since the two sources are emitting waves in phase, when the two waves pass through each other at point *P*, they will be at exactly the same phase in their cycles. Figure 7.27 (B) shows two crests meeting at point *P*. They will add constructively, making the amplitude at that point double that of each individual wave. A moment later, troughs will meet. This makes a trough that is double the size of each individual trough. This process occurs at every point along this central line. Thus, every point on the surface will be oscillating maximally.

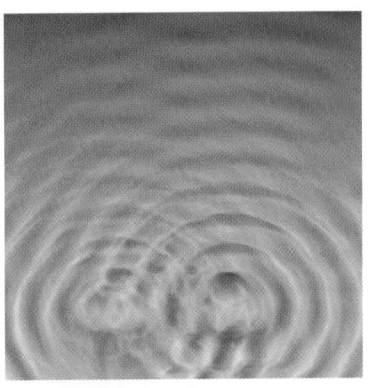

**Figure 7.26** A series of antinodes run along the vertical line up the centre. The lines spreading out beside the antinodes are formed by nodes.

## Biology Link

The phenomenon of diffraction provides an important tool in several fields of science. Rosalind Franklin obtained the pattern in the photograph shown here by passing X rays through a crystal of DNA. The pattern produced by the diffraction of X rays is shown on the film. This pattern provided James Watson and Francis Crick with key information that helped them discover the three-dimensional structure of DNA. Find out more about these scientists' amazing discoveries in your biology and chemistry courses, or go to print resources and the Internet for more.

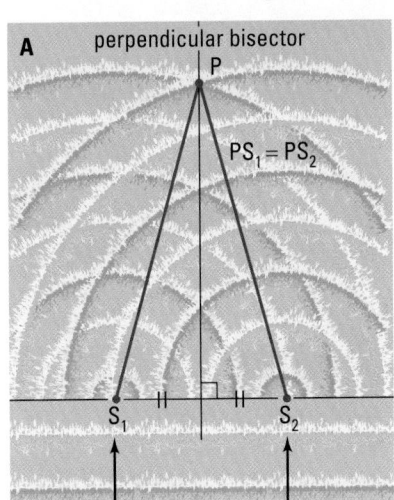

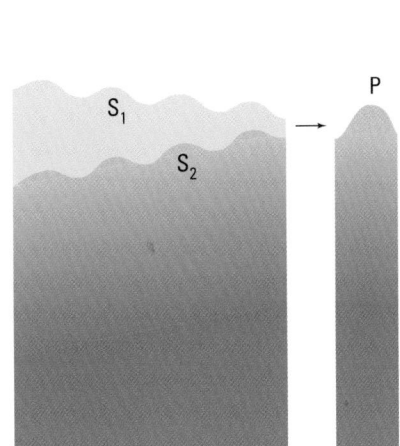

**Figure 7.27** **(A)** Constructive interference occurs for a point on the perpendicular bisector because it is an equal distance from both sources. Since the sources are creating waves in phase, they will still be in phase at point *P*. Crests are represented in blue and troughs in green. **(B)** Looking from the side, you can see crests superimposed at point *P*.

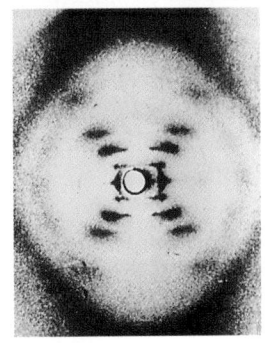

Chemists and biochemists use X-ray diffraction to determine the structure of many crystals and biological molecules.

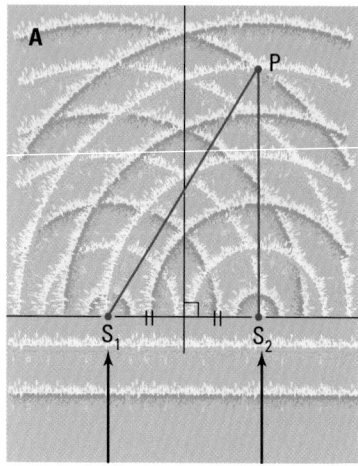

A

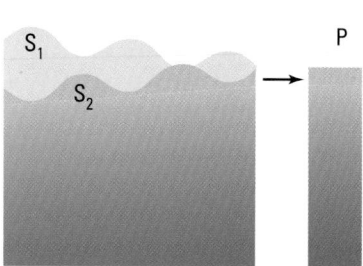

B

Next, consider a line radiating outward to the right (or left) of the centre line as shown in Figure 7.28 (A). This line is carefully chosen so that the difference in the distances from the sources to any point on this line is exactly one half of a wavelength. Again, Figure 7.28 (B) shows what you would see if you looked, from the side, at waves travelling along these lines. Notice that the waves that pass through each other at point $P_1$ on this line are exactly out of phase. No matter what stage of the cycle is passing this point, the waves will destructively interfere. The surface of the water along this line will not move. There are nodes all along the line, so it is called a **nodal line**. Although these lines appear straight in Figure 7.26, they are, in reality, slightly curved.

If you were to continue to draw lines radiating outward at greater and greater angles from the perpendicular bisector, you would reach points where the distance from one source would be one full wavelength longer than the other. Again, the waves would constructively interfere as they passed through each other, forming an **antinodal line**. When the distance to points on a line is 1.5 wavelengths, you would find another nodal line.

**Figure 7.28** **(A)** Destructive interference occurs at points that are a distance $\frac{1}{2}\lambda$ farther from one source than the other, because the waves will always be half a wavelength out of phase when they reach this line. **(B)** Looking from the side, you can see the result of a crest superimposed on a trough at point $P$.

---

## 7.4    Section Review

1. **C** Sketch the wave produced by dipping a finger into water in a ripple tank. Add rays to your diagram to illustrate the directions of wave movement.

2. **I** Two circular waves are sent out from points about 15 cm apart in a ripple tank.

   **(a)** Sketch their appearance a short period of time after they have met.

   **(b)** What does this tell you about how the two waves have moved?

3. **I** Sketch the appearance of a straight wave after it has passed through a small opening in a straight barrier. Add rays to your diagram to illustrate the directions of wavefront movement.

4. **C** Sketch a typical interference pattern produced by two point sources vibrating in phase.

> **UNIT ISSUE PREP**
>
> Diffraction is an important concept to consider in a noise policy document.
> - Have you ever heard sound waves diffract around barriers?
> - What is the relationship between the amount of diffraction and the sound wave frequency?

- Periodic motion occurs when an object moves in a repeated pattern (a cycle) over equal periods of time, $T$.

- The frequency of the motion, $f$, is the number of cycles completed in 1 s.
  $f = \dfrac{1}{T}$ and is measured in hertz (Hz):
  $$1 \text{ Hz } = \dfrac{1}{s} = s^{-1}$$

- The amplitude, $A$, of the vibration is the distance from the maximum displacement to the rest position.

- When an object is vibrated, even gently, at one of its natural frequencies, the amplitude of its vibration will increase, sometimes very dramatically. This phenomenon is known as resonance.

- A mechanical wave is a disturbance that transfers vibrational energy through a medium. A mechanical wave requires a medium.

- In a transverse wave, the vibration of the medium is at right angles to the direction of the wave. In a longitudinal wave, the vibration of the medium is parallel to the direction of the wave.

- The wavelength of a wave, $\lambda$, is the shortest distance between two points in the medium that are vibrating in phase; for example, the distance between two adjacent crests (or troughs).

- The vibrating source that produces the wave determines the frequency, $f$, of the wave. The frequency is equal to the number of wavelengths produced in 1 s.

- A wave travels with a constant speed in a homogeneous medium predicted by the wave equation $v = f\lambda$.

- When a wave passes from one medium into another, it is partially transmitted and partially reflected.

- If two waves are moving toward each other, they pass through each other without any permanent change in either wave.

- According to the principle of superposition, when two or more component waves are at the same point in a medium at the same time, the resultant displacement of the medium is equal to the sum of the amplitudes of the component waves.

- Interference occurs when two or more waves meet at the same point in a medium.

- Interference may be either constructive or destructive.

- A standing wave with stationary nodes and antinodes is produced when two periodic waves with the same shape, amplitude, and wavelength travel in opposite directions in the same linear medium. Adjacent nodes are spaced half a wavelength ($\frac{1}{2}\lambda$) apart, as are adjacent antinodes.

- Standing waves can be set up in a linear medium by vibrating one end of the medium at the natural frequency for the medium. The lowest natural frequency is referred to as the fundamental.

- Waves that originate from a point source move outward in circular wavefronts because the speed of the wave in the medium is the same in all directions.

- Two-dimensional waves are reflected from straight barriers so that the angle of reflection is equal to the angle of incidence. Straight waves are reflected from a parabolic barrier so that they converge through a single point.

- When straight waves pass through an opening in a barrier, they diffract around the edges of the barrier and spread out in all directions. Diffraction is greater for smaller openings and larger wavelengths.

- The circular waves moving out from two point sources will produce two-dimensional interference patterns consisting of nodal and antinodal lines. If the two sources are vibrating in phase, there will be an antinodal line along the perpendicular bisector of the lines.

## Knowledge/Understanding

1. Explain in your own words what periodic motion is. What quantities involved in periodic motion are variables? What concepts do we use to describe these variables?

2. How are frequency and period related? What is a hertz?

3. What is resonance, and how is it related to the natural frequency of an object?

4. In your own words, explain what a wave is. How do transverse and longitudinal waves differ?

5. What determines the frequency of a wave?

6. How could you increase the speed of a wave pulse in a large-diameter spring?

7. Both amplitude and wavelength are linear measurements used to describe waves. Explain the difference between these measurements. If you wanted to increase the amplitude of a wave in a large-diameter spring, what would you do? If you wanted to increase the wavelength of a wave in the spring, what would you do?

8. If the frequency of a wave travelling in a rope is doubled, what will happen to the speed of the wave? What will happen to the wavelength of the wave?

9. A 1 cm-high wave crest is travelling toward a 2 cm-high wave crest in the same spring. What will be produced when they meet? What kind of interference is this?

10. A 1-cm high wave crest is travelling toward a 2-cm deep wave trough in the same medium. What will be produced when they meet? What kind of interference is this?

11. What is a standing wave? What conditions are necessary to produce a standing wave? What are nodes? How far apart are adjacent nodes?

12. What happens to straight waves when they pass through an opening in a barrier? What do we call this effect?

## Inquiry

13. Suppose an upright wave pulse travels from a spring where its speed is 20 cm/s into a second spring where its speed is 10 cm/s.

(a) What will happen to its frequency and wavelength in the second spring?

(b) Describe what the two springs will look like 2 s after the incident pulse has reached the boundary between the two springs.

(c) Suppose the wave pulse had gone from the 10 cm/s spring into the 20 cm/s spring. Describe and draw the two springs, 2 s after the incident pulse had reached the boundary between the two springs.

14. A large erect wave pulse is moving to the left on a large spring at the same time that a smaller inverted wave pulse is moving to the right. Draw what the spring would look like
(a) shortly before they meet
(b) shortly after they meet

15. Design an experiment to determine the speed of a wave. You are free to select any equipment you need in your summarized design.

## Communication

16. Sketch a diagram illustrating how a straight wavefront is reflected from a straight barrier. (Be sure to include the rays that indicate the direction in which the wavefront is moving.)

17. Sketch a diagram illustrating how a straight wavefront is reflected from a parabolic barrier.

18. Sketch the kind of interference pattern produced by two in-phase point sources in a two-dimensional medium.

(a) What kind of interference will there be on the perpendicular bisector of the line connecting the two sources? Why is this?

(b) Even though the two sources are in phase and produce crests and troughs at the same time, there are nodal lines in the pattern where destructive interference occurs. Explain why this is the case.

## Making Connections

19. The speed of a wave in a string depends on its tension (the greater the tension, the greater the speed) and its mass per unit length (the greater the mass per unit length, the lower the speed). Multi-stringed musical instruments, such as the

guitar or violin, typically play high frequency notes on strings that are under considerable tension and are relatively thin. They play low frequency notes on strings that are under less tension and relatively thick. Explain why this is the case. (You may want to examine a guitar or violin to help you with this question.)

20. Scientists have no way of observing Earth's centre directly. What data, then, gave them evidence that Earth has a solid core surrounded by a thick layer of molten, liquid rock? The answer lies in the study of earthquakes and the types of waves that they generate. Research earthquake waves and how these waves were used to hypothesize the characteristics of our Earth's core.

## Problems for Understanding

21. A pendulum takes 1.0 s to swing from the rest line to it's highest point. What is the frequency of the pendulum?

22. By what factor will the wavelength change if the period of a wave is doubled?

23. A wave with an amplitude of 50.0 cm travels down a 8.0 m spring in 4.5 s. The person who creates the wave moves her hand through 4 cycles in 1 second. What is the wavelength?

24. A sound wave has a frequency of 60.0 Hz. What is its period? If the speed of sound in air is 343 m/s, what is the wavelength of the sound wave?

25. Water waves in a ripple tank are 2.6 cm long. The straight wave generator used to produce the waves sends out 60 wave crests in 42 s.
    (a) Determine the frequency of the wave.
    (b) Determine the speed of the wave.

26. A rope is 1.0 m long and the speed of a wave in the rope is 3.2 m/s. What is the frequency of the fundamental mode of vibration?

27. A tsunami travelled 3700 km in 5.2 h. If its frequency was $2.9 \times 10^{-4}$ Hz, what was its wavelength?

28. A storm produces waves of length 3.5 m in the centre of a bay. The waves travel a distance of 0.50 km in 2.00 min.
    (a) What is the frequency of the waves?
    (b) What is the period of the waves?

29. A grandfather clock has a long pendulum with an adjustable mass that is responsible for the clock's ability to keep regular time. This pendulum is supposed to have a period of 1.00 s. You discover that the pendulum executes 117 compete vibrations in 2.00 min.
    (a) Calculate the period of the pendulum.
    (b) Calculate the percentage error in the time the clock records.
    (c) How many hours slow will the clock be after a year?
    (d) How might you adjust the pendulum so that its period is exactly 1.00 s?

---

**Numerical Answers to Practice Problems**

1. 0.98 Hz; 1.0 s  2. 7.5 to 11 Hz  3. 27.9 s  4. 0.40 Hz; 2.5 s
5. 7.5 m/s; 0.80 s  6. $1.4 \times 10^9$ Hz  7. $3.1 \times 10^{-4}$ Hz  8. (a) 8.80 Hz
(b) 853 m (c) constant frequency  9. (a) 1.34 m (b) 0.670 m

The howl of a wolf, the cry of a baby, the beat of a drummer — all of these sounds focus your attention and provide information about events in the world around you. More than any other mode of communication, sound evokes strong emotions — fear, concern, happiness, or excitement. Human ingenuity has fashioned sound into complex forms, such as language and music that are characteristic of cultures, nations, and generations. Sound is at the heart of who you are, what groups you bond with, and how you perceive yourself.

Much the same as light, sound is a powerful tool for investigating the world. Bats and dolphins explore their environments and locate prey by emitting high-frequency sound pulses and interpreting the resultant echo. Physicians use ultrasound imaging devices to obtain critical information about a beating heart or a fetus in the womb.

Sound is a complex phenomenon that can enrich your own life and the lives of those around you. However, sound can also be problematic — provoking conflict and even causing physical harm. While the wise use of sound involves much more than science, a study of the physics of sound provides important knowledge for thinking critically and creatively about the multifaceted role that sound plays in society.

# Sounds and Their Sources

## Musical Rulers

How can you make a ruler produce different notes? Position a ruler so that about two thirds of its length projects over the edge of your desk. Firmly clamp the end of the ruler to the desktop with one hand, and then pluck the free end with your other hand. Change the length of the ruler projecting beyond the desk, and observe the effect on the sound produced.

### Analyze and Conclude

1. What vibrates to produce the sound? Describe the vibration.

2. What determines the pitch of the sound produced?

3. How can the sound be made louder or softer?

## Sound from a Tuning Fork

Hypothesize about how a tuning fork produces sound. Strike a tuning fork with a rubber mallet. Record what is vibrating and explain how you can tell. Strike the fork harder and record how the sound changes. Compare the sound produced by two different tuning forks.

### Analyze and Conclude

1. Sketch the way that the tuning fork moves when you strike it with a rubber mallet.

2. How does striking the tuning fork harder affect the sound?

3. What is responsible for the difference in the sounds of the two tuning forks?

## Sound from a Graduated Cylinder

How can you make a 100 mL graduated cylinder produce different notes? Hold the open end of a clean graduated cylinder just below your lower lip and blow strongly across the top. Practise this a few times, until you can produce a sound consistently. Fill the cylinder about one third full of water, and blow again. Produce sounds with different water levels. Record what you hear each time.

Predict how the sound will change as you slowly fill the cylinder with water, while blowing across it. Now test your prediction.

### Analyze and Conclude

1. Describe how the sound changes when you change the water level in the cylinder.

2. How can you make the sound louder?

3. What is vibrating to make the sound?

4. How did the sound change when you were adding water and blowing at the same time? Was your prediction correct?

5. Give a possible explanation for the change in the sound.

- Describe and illustrate how sound is produced.

- Analyze and interpret the properties of sound.

- Identify the relationship between velocity, frequency, and wavelength of sound modelled as longitudinal waves.

**KEY TERMS**

- loudness
- pitch
- quality
- natural frequency
- oscilloscope
- compression
- rarefaction

To the average person, sound is simply part of the everyday world — something that is used for communication and entertainment. To a physicist, however, sound is a more complex entity that, as it turns out, can be explained in terms of mechanical waves. How did physicists make the connection between sound and waves?

## Waves and Sound — Some Interesting Similarities

Physicists who first studied sound discovered that it had many properties in common with mechanical waves. As you saw in Chapter 7, waves diffract around corners. Sound also travels around corners. Similarly, the way that waves are reflected by barriers could nicely account for the echo that is heard a short time after shouting toward a rocky cliff.

Table 8.1 provides a summary of the basic properties that physicists first observed to be shared by mechanical waves and sounds.

**Table 8.1** Early Comparisons of Mechanical Waves and Sounds

| Property | Mechanical waves | Sounds |
|---|---|---|
| transmit energy | yes | yes |
| travel around corners | yes | yes |
| pass through each other | yes | yes |
| reflect off barriers | yes | yes |
| require a medium | yes | ? |
| exhibit constructive and destructive interference | yes | ? |

**PHYSICS FILE**

For Sir Karl Popper (1902-1994), a philosopher of science, scientific theories had to be testable, and the role of experiments was to test theories. In fact, he believed experiments should try to *disprove* theories. Scientific theories are reliable because they have survived many attempts to prove them wrong.

The similar characteristics of mechanical waves and sounds that are listed in Table 8.1 illustrate why researchers would have considered using a wave model to explain the properties and behaviour of sound. Like mechanical waves, sounds transmit energy, travel around corners, pass through each other, and are reflected by barriers. The fact that some wave-like properties had still not been confirmed (for example, the requirement of a medium and the ability to produce interference effects) only served to motivate physicists to conduct further research into the possibility of explaining sound by means of a wave model. And, as you will discover throughout this chapter, it is a model that works very well.

# Testing a Prediction:
# Does Sound Require a Medium?

In order to test the prediction that sound requires a medium through which to travel, a method of producing a vacuum was needed. It is not surprising, then, that it was Otto von Guericke, the inventor of the air pump, who carried out the task. In 1654, von Guericke demonstrated that the sound from a bell inside a jar decreased in intensity as air was removed from the jar. A modern demonstration of this effect is shown in Figure 8.1.

If the vacuum pump is a good one, the sound of the bell will almost be eliminated. It would be interesting to perform this experiment in the nearly perfect vacuum of space where no sound would be expected at all (see Figure 8.2).

**Figure 8.2** What would these astronauts have to do to hear each other speak?

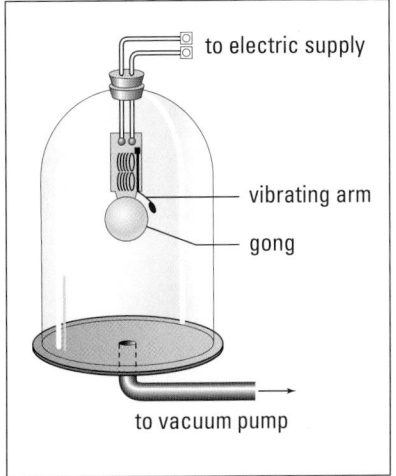

**Figure 8.1** An electric bell is sealed inside a bell jar and a vacuum pump removes the air. When the electric bell is turned on, it produces a loud ringing sound. As the vacuum pump removes the air from the bell jar, the loudness of the ringing decreases.

# Describing Sound with a Wave Theory

Humans can distinguish between sounds in a variety of ways. Sounds vary in **loudness** (perceived intensity). Jet aircraft engines are so loud that airport workers have to wear ear protection when working near them. On the other hand, the breathing of a sleeping baby is so quiet that new parents can become anxious about their child's welfare.

Sounds also vary in **pitch** (perceived frequency). Flutes and piccolos produce very similar sounds. The main difference between the two is a matter of pitch. In general, the sound of the piccolo is higher and that of the flute is lower. Sounds also vary in another important way called **quality**. The sound of a flute or whistle is described as pure, and that of a cello or organ as rich. It is the quality of a sound that enables you to identify it as being made by a piano rather than a trumpet, even when the two instruments play notes with the same loudness and pitch.

A wave model for sound must relate the loudness, pitch, and quality of the sounds that you hear to specific properties of sound

## Computer Link

The difference in sound quality produced by different voices has been the major difficulty in producing word processors that recognize human speech. To use current voice-recognition software, speakers need to "train" the program to recognize the unique quality of their voices. This involves a person correcting the errors that the program makes in typing out words. These corrections are then stored in a particular file for that person that can be accessed when that same person wants to use the software again.

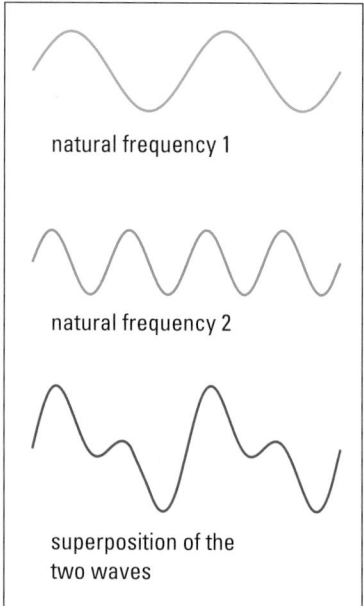

natural frequency 1

natural frequency 2

superposition of the two waves

**Figure 8.3** When only two frequencies are added together, the resultant wave becomes complex. The quality of this sound is richer than a pure fundamental tone.

waves. Everyday experience makes it clear that loudness is related to energy. To produce a louder sound from a bell, you have to hit it with more force. Yelling requires significantly more effort than whispering. Loudness, then, is in some manner connected to the **amplitude** of the sound wave. Pitch, on the other hand, is related to the frequency of the sound wave. You might recall from the "musical ruler" activity on page 365 that the shorter projections of the ruler that produced higher frequencies of vibrations also produced higher-pitched notes. Likewise, the longer projections produced lower frequencies and lower-pitched notes.

Pure sounds are produced by sources vibrating at only one **natural frequency**. Sound quality arises when the source of the sound vibrates at *several* of its natural frequencies at the same time. As shown in Figure 8.3, the superposition of these component waves — even just two of them — produces a complex wave form with a variety of smaller crests and troughs.

The conceptual links between sound perceptions and their corresponding sound wave characteristics are summarized in Figure 8.4.

| Sound perceptions | Sound wave characteristics | |
|---|---|---|
| **Loudness** | **Amplitude** | |
| loud | large | |
| quiet | small | |
| **Pitch** | **Frequency** | |
| high | high | |
| low | low | |
| **Quality** | **Wave form** | |
| pure | simple | |
| rich | complex | |

**Figure 8.4** Characteristics of sounds and sound waves

## "Seeing" Sound Waves

An **oscilloscope** is an electronic instrument that displays the form of electronic signals on a small monitor similar to a television screen. Sounds can be "seen" by using a microphone to convert them into electronic signals, an amplifier to amplify these signals, and an oscilloscope to display their form (see Figure 8.5). With an oscilloscope, you can visualize the difference between the sounds made by a variety of musical instruments, as shown in Figure 8.6.

**ELECTRONIC LEARNING PARTNER**

Go to your Electronic Learning Partner to enhance your learning about vibrations and sound.

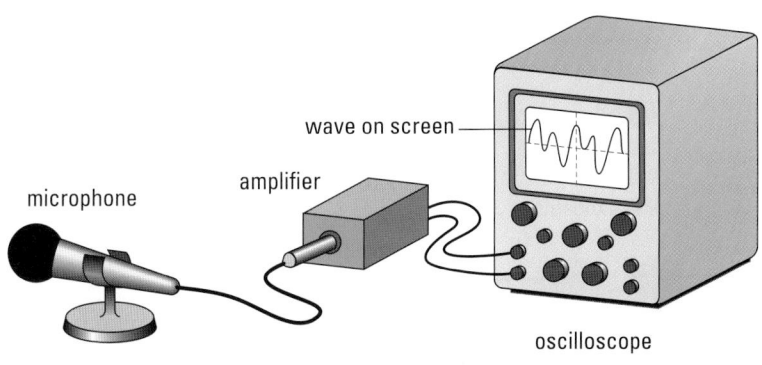

**Figure 8.5** By using an oscilloscope, you can "see" sound waves.

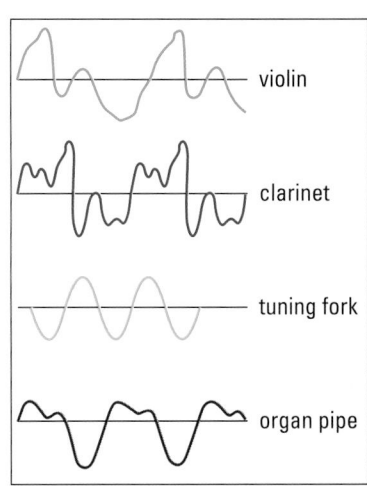

**Figure 8.6** These oscilloscope tracings show why you can distinguish among different instruments.

## Sound Waves Are Longitudinal Waves

If you use a strobe light to make the vibrations of a large speaker cone appear in slow motion, you will see that the cone is moving in and out, toward and away from the listener. When the speaker cone moves out, the air molecules in front of it are pushed together to produce a small volume of higher pressure air called a **compression**. When the speaker cone moves back, it produces an expanded space for the air molecules to spread out in. The result is a volume of lower pressure air called a **rarefaction**. This alternating pattern of compressions and rarefactions spreads outward through the room.

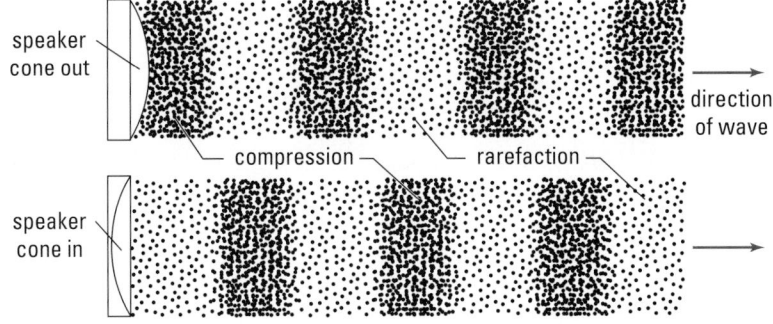

**Figure 8.7** When a loudspeaker cone moves out, it exerts a force on the molecules in the air. The molecules move outward until they collide with more molecules. Individual molecules vibrate back and forth, but the collisions carry the sound energy throughout the room.

Tuning forks produce compressions and rarefactions in a somewhat different manner than speaker cones do. When one prong of a tuning fork is struck with a rubber mallet, both prongs move in and out together. As they move away from each other, they produce compressions on their outward sides (and a rarefaction between them). As they move toward each other, they produce rarefactions on their outward sides (and a compression between them).

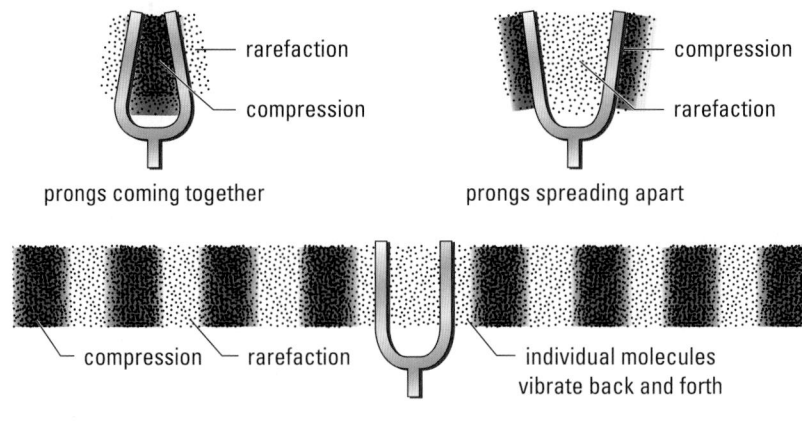

**Figure 8.8** A tuning fork directs sound waves outward in two directions.

As you will recall from Chapter 7, there are two distinct types of waves — transverse waves and longitudinal waves. For transverse waves, the vibrations are perpendicular to the direction of the wave motion; for longitudinal waves, the vibrations are parallel to the direction of the wave motion. The above analysis of the sound produced by speakers and tuning forks demonstrates that sound behaves as a longitudinal wave. As shown in Figure 8.9, the vibrations in a sound wave correspond to the changes in air pressure at a point in space — that is, crests that are produced by compressions and troughs that are produced by rarefactions.

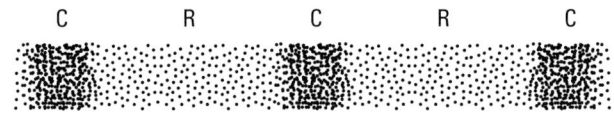

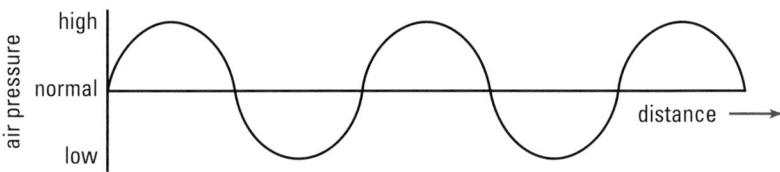

**Figure 8.9** Compressions are volumes of maximum pressure, and rarefactions are volumes of minimum pressure.

1. **K/U** Identify the vibrating source of sound for each of the following:

    (a) guitar          (d) singer

    (b) drum           (e) piano

    (c) clarinet

2. **C** Describe in your own words how the wave theory of sound explains the following phenomena involving sound:

    (a) Sounds can be heard around barriers.

    (b) Sometimes after a sound is produced, an echo is heard.

    (c) Two people can be talking, face to face, at the same time and still hear what the other is saying.

    (d) A large explosion occurs in a fireworks-manufacturing plant, and several seconds later windows shatter in a nearby housing development.

3. **C** Mechanical waves require a medium to travel through. Thus, the wave theory of sound predicts that sound cannot travel through a vacuum. Describe an experiment designed to test this prediction. Explain how the results of this experiment support the wave theory of sound.

4. **I** Use your knowledge of mechanical waves to make two more predictions about the behaviour of sound. Describe an experimental set-up that could be used to test each of your predictions. In each case, what results would you expect to observe?

5. **C** Sounds can be described in terms of their loudness, pitch, and quality.

    (a) Explain how each concept enables you to differentiate sounds.

    (b) How is each of these characteristics of sound represented in a sound wave?

6. **K/U** The molecules in air are relatively far apart (about 10 molecular diameters). Thus, there are basically no forces of attraction among them. Explain why this fact would lead you to conclude that transverse waves cannot travel through air. How, then, can longitudinal sound waves travel through air?

7. **K/U** Sketch a graph of pressure versus position for a sound wave of wavelength 15 cm at some instant in time. Mark appropriate scales for pressure on the vertical axis and for position on the horizontal axis.

---

**UNIT ISSUE PREP**

Is it possible to predict noise pollution hazards within a community by considering the wave nature of sound?

- Identify specific properties of sound waves that should be considered when creating a noise policy document.

- Identify specific activities that would generate pervasive sounds in a community setting.

- Conduct an experiment to investigate factors that affect the speed of sound.

- Analyze the factors that affect sound intensity and its effect in nature.

- Compare sound travelling in different media.

KEY
TERMS

- infrasonic
- bel
- audible
- decibel
- ultrasonic
- sound intensity level

**Biology Link**

Your range of hearing can be affected by age, illness, or injury. Injury can be caused by, among other things, prolonged exposure to loud noise or music. Research what percentage of teenagers have permanently lost some hearing by listening to loud music.

Convinced that sound is a type of mechanical wave, investigators set out to discover more about the properties of sound waves. What is the speed of sound in various media? What is the range of frequencies of sound? What frequencies within the full range can humans hear?

## The Range of Hearing in Humans

Scientific investigation always begins with the senses. Because the senses, including hearing, have limited precision, more precise and reliable measuring instruments for measuring phenomena are continually being developed.

### The Frequency Range of Human Hearing

There are both upper and lower limits to the sound frequencies that humans can hear. A healthy young person can typically hear frequencies in a range from about 20 to 20 000 Hz (20 kHz). You may well have had experience with a dog whistle that seems to produce no sound at all when blown, but still brings your pet dog bounding back. The frequency of the sound produced by these whistles is higher than 20 kHz. While it is outside the audible range for humans, it is obviously not outside the audible range for dogs.

As you will observe in the Quick Lab opposite, individuals have quite different ranges of hearing. Thus, in order to accurately study sound, physicists have developed instruments that measure the frequency and intensity of sound waves with increased reliability and precision. These instruments provide more objective measures of the frequency and intensity of sound waves than our more subjective perceptions of pitch and loudness. Using these instruments, investigators have discovered that different animals can hear sounds over extremely different frequency ranges (see Figure 8.10). Nevertheless, physicists have established a three-part classification of sound, based on the range of human hearing. Sound frequencies lower than 20 Hz are referred to as **infrasonic**, those in the 20 to 20 000 Hz range are **audible**, and those higher than 20 000 Hz are **ultrasonic**. There is no real qualitative difference in the behaviour of these three kinds of sound.

# Determining the Upper and Lower Frequency Limits of Hearing

If you have an audio frequency generator with a loudspeaker available, it is fairly simple to determine the upper and lower frequency limits of your own hearing, and to compare them with those of other students in your class. Start the audio frequency generator at a frequency below 20 Hz and gradually turn it up. Have people in the room raise their hands as soon as they can just hear the sound, so that the lower frequency limit of their hearing can be recorded. Once everyone in the group can hear the sound, the frequency can be turned up more rapidly until it reaches about 12 000 Hz. From 12 000 Hz onward, the frequency should be increased more slowly so that as people lower their hands,

when they can no longer hear the sound, the frequency can be noted.

## Analyze and Conclude

1. How do the lower frequency limits of individuals in your group compare? How do the upper frequency limits compare?

2. Are there any apparent reasons for the differences you found in question 1?

3. Determine your teacher's lower and upper frequency limits. How do they compare with those of the students in your group? Why might this be?

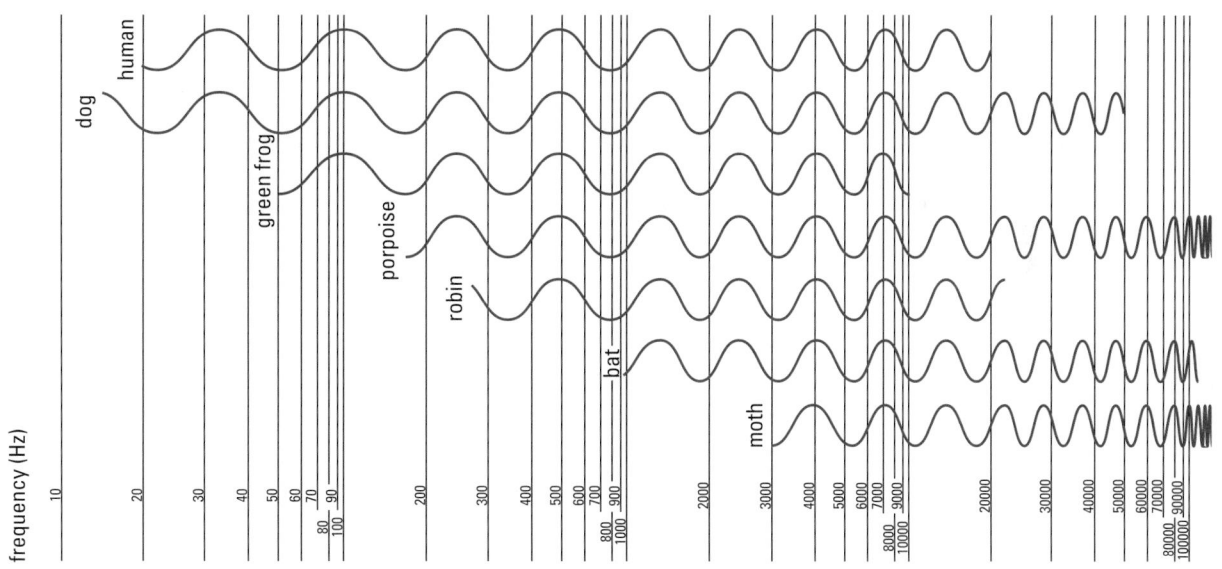

**Figure 8.10** The range of hearing in various animals is extremely broad. In this chart showing the audible frequency ranges for some animals, do you see a relationship between the pitch that different animals hear and some characteristic of their bodies?

**Math Link**

In Figure. 8.10, notice that the distance between the lines corresponding to 10 Hz and 100 Hz is the same as the distance between the lines for 100 Hz and 1000 Hz. In fact, the distance between any two adjacent powers of 10 is always the same. This type of scale is referred to as a *logarithmic* scale. Describe the difficulty you would have drawing this type of graph using a standard, linear scale.

**Biology Link**

Human ears are very sensitive instruments. They respond to sound intensities as low as a few picowatts per square metre. The ear also deals with a wide range of intensities, from a few picowatts per square metre to ten billion ($10^{10}$) picowatts per square metre! How does this compare with microphone technologies?

## Intensity Range of Human Hearing

Loudness is a measure of the response of the ear to sound waves. Individuals, however, may hear the same sound wave as differing in loudness, depending on how sensitive their ears are. In the previous section, you saw that humans are unable to hear sounds whose frequencies are less than 20 Hz or more than 20 000 Hz. In other words, the loudness of these sounds to humans is zero, although other animals can hear them. In order to accurately compare sound waves, investigators use instruments to measure a property of sounds called intensity. Intensity is a measure of the amount of sound energy reaching a unit of area in 1 s (power per unit area). Sound intensity is measured in units of picowatts per square metre ($pW/m^2$). (Note that 1.0 pW is $1.0 \times 10^{-12}$ W.) Intensity is an *objective* property of the sound wave — in fact, it is related to the square of the wave amplitude, and does not depend on the particular characteristics of a person's ears. Loudness, on the other hand, is a *subjective* property of the sound that depends on the human ear, the sensitivity of the ear to the frequency of the sound, and the distance from the source of the sound.

The range from $1.0 \times 10^0$ (one) to $1.0 \times 10^{10}$ (ten billion) picowatts is an extremely wide range to describe and compare. So, scientists decided that it might be easier to describe intensity according to the exponent of 10 instead of the number itself. They defined a new measurement, in terms of the exponent of 10, called **sound intensity level** and named the unit the **bel** (B), after Alexander Graham Bell. In math, you can find the exponent of 10 that corresponds to any specific number by taking the logarithm of the number. For example, the logarithm of 1000 is 3. Thus, the scale in bels is called a logarithmic scale. When the sound intensity increases by a factor of 10 (is 10 times larger), the sound intensity *level* increases by 1 bel. When the sound intensity increases by a factor of 100 (is 100 times larger), the sound intensity *level* increases by 2 bels. Notice the relationship between the number of bels and the number of zeros in the factor by which the sound intensity increases.

1 bel is equal to 10 $pW/m^2$, the lower threshold of human hearing. Detailed hearing tests have shown that the smallest increase in sound intensity level that humans can distinguish is 0.1 bels, or a **decibel**. Therefore, the decibel (dB) has become the most commonly used unit to describe sound intensity levels.

## Telephone Invention an Accident

Alexander Graham Bell (1847–1922)

Few inventors have had as great an impact on our everyday lives as Alexander Graham Bell, creator of the world's first telephone. The invention of the telephone in 1876 changed our social lives and the world of business, and it paved the way for today's information age.

Bell was born in Edinburgh, Scotland, but moved to Brantford, Ontario, with his family as a young adult. Although he eventually became a U.S. citizen and taught the deaf in the United States, he is often considered a "Canadian inventor" because he did much of his scientific work in his summer home on Cape Breton Island in Nova Scotia, and developed his idea for the telephone in Brantford.

From an early age, Bell had a keen interest in speech and sound. It was a passion that came naturally — his mother was deaf and his father was a speech therapist who developed an alphabet to teach deaf people to speak.

Despite his knowledge and his enthusiasm for sound and human communication, Bell did not set out to invent the telephone. He was working on improving the telegraph, the first instrument used to send a message using electricity. The telegraph was already in full use but it could not send multiple messages simultaneously. Bell and his assistant, Thomas Watson, were trying to solve this problem when Bell discovered he could send sound using electrical current.

Bell created the beginnings of the telephone using reeds arranged over a magnet. The reeds vibrated up and down, toward and away from the magnet in response to sound wave variations. The vibrations of the reeds generated a current that could then be carried by wire to a receiver and converted back into sound waves.

On March 10, 1876, Bell accidentally spilled acid while he and Watson were working on the apparatus. He called out, "Mr. Watson. Come here. I want to see you." Watson, working in another room, heard Bell's voice over his receiver. The first transmission of the human voice over a telephone had occurred.

The telephone may be the greatest of Bell's inventions, but it hardly sums up his accomplishments. After inventing the telephone, he continued his experiments in communications and developed the photophone — a device that transmitted a voice signal on a beam of light, and was a forerunner of today's fibre-optic technology.

His interests also extended beyond human communications. The *Silver Dart*, an airplane he played a large part in creating, made aviation history in 1909 as Canada's first successful heavier-than-air flying machine. Yet another of Bell's many inventions was an electric probe that was used in surgery before X rays were discovered.

### Web Link

**www.school.mcgrawhill.ca/resources/**
For more information about Alexander Graham Bell, including diagrams by him and photographs, go to the above Internet site. Follow the links for **Science Resources** and **Physics 11** to find out where to go next.

## Math Link

The logarithm of a number is equal to the exponent to which the base 10 must be raised in order to be equal to the number. For example, the logarithm of 10 is 1 ($10 = 10^1$); the logarithm of 100 is 2 ($100 = 10^2$); and the logarithm of 1000 is 3 ($1000 = 10^3$). It follows that the logarithm of 1 is zero ($1.0 = 10^0$) and that the logarithm of 2 is greater than 0 but less than 1. Actually, the logarithm of 2 is about 0.30103. If you have a calculator with a "log" button on it, try entering the number 2 and then pressing the log button to confirm this.

Even though sound intensity levels are almost always given in decibels, it is easier to see the logarithmic nature of this scale by looking at their values in bels. Table 8.2 provides equivalent values for the sound intensity levels in bels and decibels with the sound intensity in picowatts per square metre.

**Table 8.2** Logarithmic Intensity Levels and Sound Intensity

| Intensity level (dB) | Intensity level (B) | Intensity (pW/m$^2$) |
|---|---|---|
| 0 | 0 | 1 |
| 10 | 1 | 10 |
| 20 | 2 | 100 |
| 30 | 3 | 1 000 |
| 50 | 5 | 100 000 |
| 100 | 10 | 10 000 000 000 |

Sound intensity levels range from barely audible (0 dB), to the threshold of pain (130 dB), to a space rocket booster (160 dB). The sound of the rocket booster would instantly break workers' eardrums if they did not wear the appropriate ear protection.

**20 dB**
students whispering

**50 dB moderate**
class working on assignment

**70 dB noisy**
school hallway

**10 dB**
rustling leaves

**30 dB quiet**
quiet living room (student reading)

**90 dB**
passing subway train

**Figure 8.11** The sound intensity levels of some common sounds

Because the human ear is not equally sensitive to all frequencies, sounds of different frequencies may be perceived as equally loud even though they have quite different sound intensity levels. Figure 8.12 illustrates this disparity by displaying a set of *constant-loudness* curves. For these graphs, the sound frequencies are plotted along the horizontal axis and the sound intensity levels (in decibels) are plotted on the vertical axis. The curves are called constant-loudness curves because they display the sound intensity level required at each frequency for a sound to be perceived as having the same loudness. The curves are labelled with their sound intensity levels at 1000 Hz. This means that the curve labelled 40 represents all sounds perceived as equally loud as a 1000 Hz sound with an intensity level of 40 dB. The lowest curve can be thought of as the threshold-of-hearing curve. It displays the intensity levels at which sounds of different frequencies are barely audible (0 dB). For example, a 60 Hz sound requires an intensity level just under 50 dB in order to be barely audible, while a 1000 Hz sound requires an intensity level of 0 dB. This means that the 60 Hz sound needs to have an intensity $10^5$ or 100 000 times greater to be barely audible.

Note the way that these constant-loudness curves become much "flatter" as the loudness increases. This relative flatness indicates that the ear is almost equally sensitive to all frequencies when the sound is loud. Thus, when the volume on a sound system is turned up, the low, middle, and high frequencies are heard about equally well. However, when the volume is decreased to produce a quieter sound, the low and high frequencies are not heard nearly as well, destroying the balance of the music. Many better sound systems have controls that enable you to individually adjust the amplification of the sound for different frequency ranges.

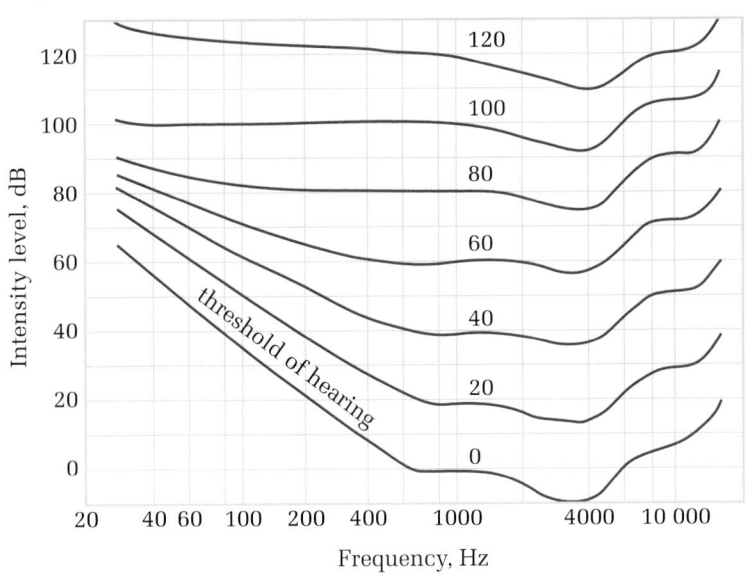

**Figure 8.12** Curves of constant loudness for different frequencies

## The Speed of Sound in Air: A Wave Property

As you learned in Chapter 7, the speed of a wave is determined by the properties of the medium through which it travels. Friction in a medium slowly decreases the amplitude of a wave, but does not affect its speed. If sound is a wave, then it should have a constant speed that is determined by the medium through which it travels. Sound should also travel at different speeds in different media.

Sound waves travel through air by means of moving molecules. Molecular speeds in a gas, however, depend on the temperature of the gas. Therefore, the speed of sound in air should increase with temperature. This is indeed the case. At a temperature of 0°C and a pressure of 101 kPa, the speed of sound in air is 331 m/s, and for each 1°C rise in temperature, the speed of sound increases by 0.59 m/s.

### THE SPEED OF SOUND IN AIR

The speed of sound in air is 331 plus the product of 0.59 and the Celsius temperature.

$$v = 331 + 0.59T_C$$

| Quantity | Symbol | SI unit |
|---|---|---|
| speed of sound | $v$ | $\frac{m}{s}$ (metres per second) |
| temperature of air | $T_C$ | not applicable* (°C is not an SI unit) |

**Unit Analysis**

$$\frac{m}{s} + \frac{\frac{m}{s}}{°C}(°C) = \frac{m}{s}$$

***Note:*** This formula is based on the Celsius temperature scale and cannot be used with the Kelvin scale. How would you modify the equation so that it would apply to the Kelvin scale?

## The Speed of Sound in Solids and Liquids

In general, sound travels fastest in solids, slower in liquids, and slowest in gases. The speed of sound in water is almost five times faster than its speed in air. This difference is great enough to be noticed by the human ear. A swimmer who is 1500 m away from a loud noise (perhaps a cannon being fired) would hear the sound that travelled through the water 1 s after it was produced. The same sound travelling through the air, however, would not be heard until 5 s after it was produced.

Knowledge of the speed of sound in different materials is the basis for a variety of techniques used in exploring for oil and minerals, investigating the interior structure of Earth, and locating objects in the ocean depths.

**Table 8.3** The Speed of Sound in Some Common Materials

| Material | Speed (m/s) |
|---|---|
| Gases (0°C and 101 kPa) | |
| carbon dioxide | 259 |
| oxygen | 316 |
| air | 331 |
| helium | 965 |
| Liquids (20°C) | |
| ethanol | 1162 |
| fresh water | 1482 |
| seawater (depends on depth and salinity) | 1440–1500 |
| Solids | |
| copper | 5010 |
| glass (heat-resistant) | 5640 |
| steel | 5960 |

## MODEL PROBLEMS

### Applying the Speed of Sound Equation

1. Suppose the room temperature of a classroom is 21°C. Calculate the speed of sound in the classroom.

### Frame the Problem

- At a *temperature of 0°C* and a pressure of 101 kPa, the *speed of sound in air* is 331 m/s.

- For each *1°C rise in temperature* the speed of sound increases by 0.59 m/s.

*continued* ▶

*continued from previous page*

## Identify the Goal

The speed of sound, $v$

## Variables and Constants

| Involved in the problem | Known | Unknown |
|---|---|---|
| $T_C$ | $T_C = 21°C$ | $v$ |
| $v$ | | |

| Strategy | Calculations |
|---|---|
| Use the formula for the velocity (speed) of sound in air. | $v = 331 + 0.59T_C$ |
| Substitute in the known values. | $v = 331 \frac{m}{s} + 0.59 \frac{\frac{m}{s}}{°C} (21°C)$ |
| Simplify. | $v = 331 \frac{m}{s} + 12.39 \frac{m}{s}$ $v = 343.39 \frac{m}{s}$ |

The speed of sound in the classroom is $3.4 \times 10^2$ m/s.

---

## Validate

The speed of sound increases with temperature. At 0°C, the speed is about 330 m/s. The temperature is about 20°C above that, and increases approximately 0.5 m/s for each degree of increase in temperature. Thus, the speed should be about 340 m/s.

---

2. The temperature was 4.0°C one morning as Marita hiked through a canyon. She shouted at the canyon wall and 2.8 s later heard an echo. How far away was the canyon wall?

---

## Frame the Problem

- The *speed of sound* can be calculated from the *temperature*, which is given.

- The *distance* the sound travels can be calculated from the *speed* of sound and the *time* for the echo to return.

- The sound travels to the wall and back, so the distance to the wall is half the distance that the sound travels.

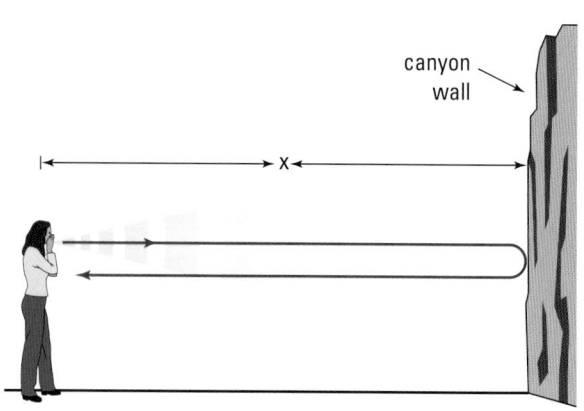

Sound echoing off a canyon wall

---

## Identify the Goal

The distance, $x$, to the wall

## Variables and Constants

| Involved in the problem | | Known | Unknown |
|---|---|---|---|
| $\Delta t$ | $\Delta d$ | $\Delta t = 2.8$ s | $v$ |
| $T_C$ | $x$ | $T_C = 4.0°C$ | $\Delta d$ |
| $v$ | | | $x$ |

## Strategy

Use the formula for the *velocity* (speed) of sound in air.

Substitute in the known values.

Simplify.

Find the *distance* to the wall and back by rearranging the formula for the *velocity* of any entity.

Substitute in the known values.

Simplify.

The distance to the wall, $x$, is half the distance to the wall and back, $\Delta d$.

Substitute in the known values.

Simplify.

The canyon was $4.7 \times 10^2$ m away.

## Calculations

$$v = 331 + 0.59T_C$$

$$v = 331\,\frac{m}{s} + 0.59\,\frac{\frac{m}{s}}{°C}(4.0°C)$$

$$v = 331\,\frac{m}{s} + 2.36\,\frac{m}{s}$$
$$v = 333.36\,\frac{m}{s}$$

$$v = \frac{\Delta d}{\Delta t}$$

$$\Delta d = v(\Delta t)$$

$$\Delta d = 333.36\,\frac{m}{s}\,(2.8\,\cancel{s})$$

$$\Delta d = 933.41\ m$$

$$x = \frac{d}{2}$$

$$x = \frac{933.41\ m}{2}$$
$$x = 466.7\ m$$

## Validate

Sound at 4°C is not travelling much faster than it is at 0°C. Thus, the sound is travelling at approximately 330 m/s. In 2.8 s, this sound will have travelled about $3 \times 330$ m, or 1000 m. However, the sound travelled to the wall and back, so the distance to the wall is only half of 1000 m, or about 500 m.

> **PROBLEM TIP**
>
> In problems such as this one, always remember that the sound travels from the source *to the reflector and back.* Thus, the distance from the source to the reflector is only half of the distance that the sound travelled.

*continued* ▶

*continued from previous page*

## PRACTICE PROBLEMS

1. What is the speed of sound in air at each of the following temperatures?

   (a) −15°C        (b) 15°C

   (c) 25°C        (d) 33°C

2. For each speed of sound listed below, find the corresponding air temperature.

   (a) 352 m/s       (b) 338 m/s

   (c) 334 m/s       (d) 319 m/s

3. A ship's horn blasts through the fog. The sound of the echo from an iceberg is heard on the ship 3.8 s later.

   (a) How far away is the iceberg if the temperature of the air is −12°C?

   (b) How might weather conditions affect the accuracy of this answer?

4. An electronic fish-finder uses sound pulses to locate schools of fish by echolocation.

What would be the time delay between the emission of a sound pulse and the reception of the echo if a school of fish was located at a depth of 35 m in a lake? Assume the temperature of the water is 20°C.

5. You want to estimate the length of a large sports complex. You generate a loud noise at one end of the stadium and hear an echo 1.2 s later. The air temperature is approximately 12°C. How far away is the far wall of the stadium from your position?

6. (a) How long does it take for sound to travel 2.0 km in air at a temperature of 22°C?

   (b) The speed of light is $3.0 \times 10^8$ m/s. How long does it take for light to travel 2.0 km?

   (c) The rumble of thunder is heard 8.0 s after a flash of lightning hits a church steeple. How far away is the church in 22°C air?

---

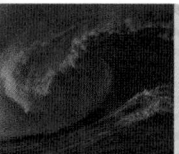

## QUICK LAB

# At the Speed of Sound

### TARGET SKILLS

- **Performing and recording**
- **Analyzing and interpreting**
- **Communicating results**

A rough determination of the speed of sound in air can be performed using echolocation. Stand at least 50 m from a large outside wall of your school. Measure the distance to the wall. Clap two wooden blocks together, and listen for the echo. Estimate the time interval (a fraction of a second) that it takes for the echo to return. Clap the blocks together repeatedly with a steady rhythm, so that the time interval between a clap and its echo is the same as the time between the echo and the next clap. When you do this, the time between your claps is equal to twice the time that it takes the sound to travel to the wall and back. Thus, the period of your clapping is equal to the length of time it takes sound to travel four times the distance to the wall. Maintaining this rhythm, have your partner measure the length of time that it takes for 20 claps and echoes.

### Analyze and Conclude

1. Determine the period of your clapping and the time it takes the sound to travel to the wall. Calculate the speed of sound in air. How accurate is your answer?

2. Assume that your answer to question 1 is correct. Calculate the temperature of the air. Is this a reasonable value? Explain.

3. Use a thermometer to measure the actual temperature of the air. Discuss the discrepancy between your predicted value and your measured value.

1. **K/U**

   (a) What is the range of sound frequencies audible to the human ear?

   (b) What is the term for sound frequencies that are higher than those humans can hear? Name four animals that can hear these frequencies.

   (c) What is the term for sound frequencies that are lower than those humans can hear? Name two animals that can hear these frequencies.

   (d) How does the human range of hearing compare with that of other animals?

2. **C** For a physicist, sound includes pressure waves of all frequencies, even though many of them are inaudible to humans. Explain how this definition is more general and less subjective than a definition that includes only those frequencies that are audible to humans.

3. **C** Explain the difference between *loudness* and *sound intensity*.

4. **K/U** The sound intensity level measured in bels or decibels is expressed on a logarithmic scale.

   (a) If a sound has an intensity of 10 pW/m$^2$, what is the sound intensity level in bels? In decibels?

   (b) How much greater than the intensity of a 1 dB sound is the intensity of a
   - 2 dB sound?
   - 3 dB sound?
   - 4 dB sound?

5. **I** Estimate the decibel reading on a sound meter for

   (a) a quiet room

   (b) a school cafeteria at lunchtime

   (c) a rock concert

6. **K/U** Each of two students, when talking, produces a 60 dB reading on a sound meter.

   (a) Explain why you would not expect to get a reading of 120 dB when these two students were speaking at the same time.

   (b) What would be a reasonable reading to expect?

7. **C**

   (a) What do the constant-loudness curves on page 377 tell you about how the human ear hears different frequencies?

   (b) As the loudness increases, what does the increased "flatness" of these curves tell you?

8. **C** Explain why you would expect that an increase in air temperature would increase the speed of sound in air.

9. **I** When jelly sets, it changes from a liquid to a solid. What would you expect to happen to the speed of sound in the material as the jelly sets? Explain your reasoning.

10. **K/U** When a batter hits a home run during a baseball game, it is possible to notice a time delay between when the bat is seen to hit the ball and when the bat hitting the ball is heard. Explain why this is the case.

---

**UNIT ISSUE PREP**

Investigate how sound intensity changes as the distance from the source changes.

- Numerical simulations or actual data will help clarify the type of noise pollution policy required for an area.

## SECTION EXPECTATIONS

- Explain and illustrate the principle of superposition.

- Identify examples of constructive and destructive interference.

- Investigate interference and compare predicted results with experimental results.

- Describe the characteristics common to all forms of potential energy.

## KEY TERMS

- beats
- beat frequency

## TRY THIS...

Working with a partner, try to describe the location of the quiet regions around a tuning fork. Strike a tuning fork and then hold it vertically beside the ear of your partner. Slowly rotate the fork, asking your partner to indicate when the sound is soft and then loud. After you have identified the locations of the quiet regions that your partner observed, reverse roles so that your partner can locate the quiet regions that you observed. Repeat the activity until you can agree on the location of the quiet regions.

# There Must Be Interference!

The property that seems most characteristic of waves is their ability to interfere. Look back at Table 8.1 on page 366. If sound is a wave, then it must also experience constructive and destructive interference. In Chapter 7, you saw how two waves could combine to produce a resultant wave of either an increased or decreased amplitude. In looking for interference in sound, then, you will look for increases and decreases in loudness and intensity level.

# Quiet Regions Near a Tuning Fork

The arms of a tuning fork vibrate back and forth symmetrically so that the air between them is first squeezed (producing a compression) and then pulled apart (producing a rarefaction). At the same time as a compression is produced between the arms, the arms are pulling apart the air outside the arms (producing rarefactions). As a result, we can think of sound waves radiating outward from one source at a point between the arms and from two other sources outside the arms. The source between the arms, however, is in the opposite phase to the sources outside the arms. Where do you expect to find regions of destructive interference around a tuning fork?

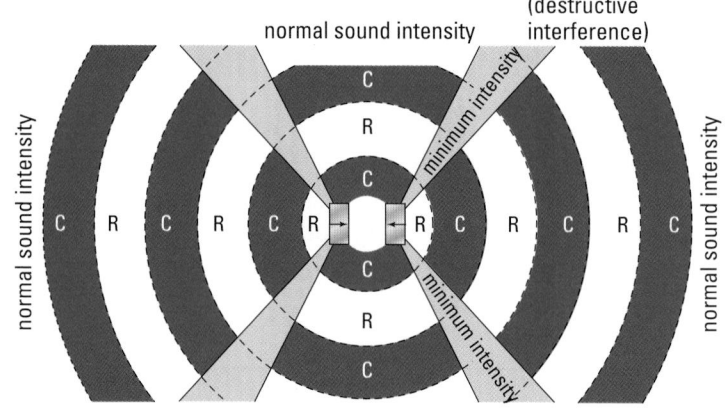

**Figure 8.13** Destructive interference occurs in the regions where compressions and rarefactions overlap.

# Beats: A More Complex Interference Effect

So far in this section, you have observed interference effects as differences in loudness at *different positions* relative to two or more sound sources. Interference can also be observed as a difference in loudness at different *times*. In this case, what you hear is a kind of wavering of the sound intensity as it becomes alternately louder and quieter. This phenomenon is known as **beats** and is the consequence of two sources of similar (but not identical) frequency producing sound waves at the same time. One beat is a complete cycle from loud to quiet to loud, and the **beat frequency** is the number of cycles of loud-quiet-loud produced per second. The beat frequency depends on the difference in frequency of the two sounds. A larger difference in frequency produces a greater beat frequency, and a smaller difference in frequency produces a smaller beat frequency.

**PROBEWARE**

Visit the **Science Resources** section of the following web site if your school has probeware equipment: *www.school.mcgrawhill.ca/ resources/* and follow the **Physics 11** links for interesting activities on the interference of sound.

## QUICK LAB
## Locating Interference Between Two Loudspeakers

**TARGET SKILLS**
- **Performing and recording**
- **Predicting**
- **Communicating results**

Two point wave sources vibrating in phase produce a characteristic interference pattern of nodal and antinodal lines. To create sound interference, you can use two loudspeakers placed side by side, driven by the same audio frequency generator, as two in-phase sound sources. Since destructive interference is produced when one sound wave travels half a wavelength farther than the other (the crest of wave A meets the trough of wave B), you should be able to locate a point of destructive interference. A frequency of 340 Hz will produce a wavelength of approximately 1 m, so you are looking for a point that is 0.5 m farther from one loudspeaker than the other. By walking across the expected pattern, you should be able to hear a variation of loudness as you walk through nodal and antinodal lines. Set up speakers as shown in the illustration and locate the nodal and antinodal line. Walk back and forth in front of the speakers until you are convinced that you have heard interference in sound.

(**Note:** Unwanted reflections of sound waves from hard surfaces in the room can complicate the sound pattern produced and make it difficult to hear clear points of destructive interference. Try to keep your loudspeakers away from the walls of the room or any other potential sound reflectors.)

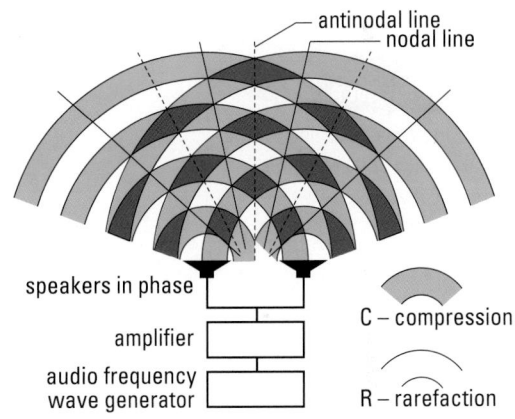

Sound interference produced by two loudspeakers vibrating in phase

### Analyze and Conclude

1. What do you expect to hear as you cross the centre line (perpendicular bisector) of a line between the two loudspeakers?

2. How much farther along your path do you expect the first quiet point to be?

3. How far apart do you expect two adjacent quiet points to be?

4. How could you change your walking path to make these quiet points farther apart and, hence, easier to identify?

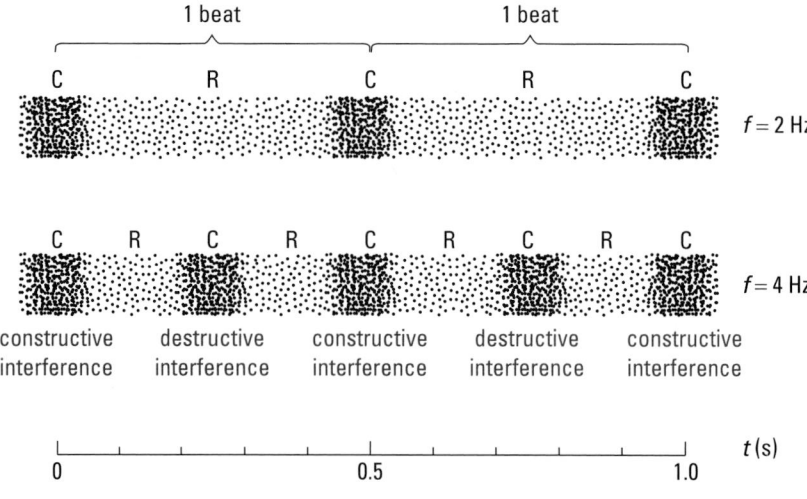

**Figure 8.14** Beats are produced by alternating instances of constructive and destructive interference over time.

1 beat       1 beat

C   R   C   R   C     $f = 2$ Hz

C R C R C R C R C    $f = 4$ Hz

constructive interference    destructive interference    constructive interference    destructive interference    constructive interference

$t$ (s)

0       0.5       1.0

Using the principle of superposition, the resultant pressure wave produced by two component sound waves of similar frequency can be constructed, as shown above in Figure 8.14. This resultant wave clearly illustrates the regular variations in loudness characteristic of beats. A second method of visualizing beats is shown in Figure 8.15. Here, air pressure is plotted against time.

**TRY THIS...**

Sketch several pairs of component waves and their resultant wave to convince yourself that the beat frequency is equal to the difference in frequency of the two component waves.

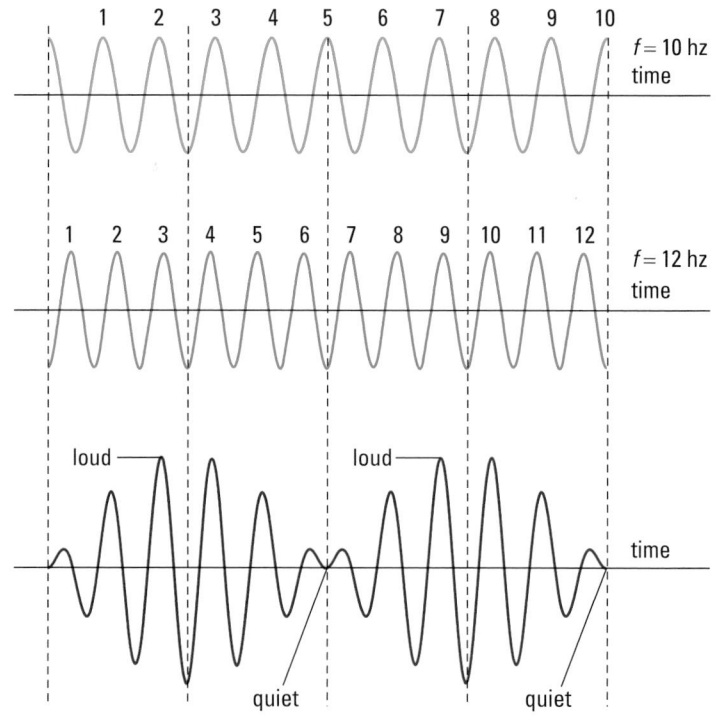

**Figure 8.15** The superposition of two similar sound waves produces a resultant sound wave with an intensity that alternates between loud and quiet.

## BEAT FREQUENCY

The beat frequency is the absolute value of the difference of the frequencies of the two component waves.

$$f_{beat} = |f_2 - f_1|$$

| Quantity | Symbol | SI unit |
|---|---|---|
| beat frequency | $f_{beat}$ | Hz (hertz) |
| frequency of one component wave | $f_1$ | Hz (hertz) |
| frequency of other component wave | $f_2$ | Hz (hertz) |

When musicians playing in an ensemble are tuning their instruments, they listen for beats produced by the note they are sounding interfering with a reference note sounded by one musician. They then adjust their own instruments to reduce the beat frequency to zero. Which instrument in an orchestra plays the reference note, and why?

## MODEL PROBLEM

### Finding the Unknown Frequency

A tuning fork of unknown frequency is sounded at the same time as one of frequency 440 Hz, resulting in the production of beats. Over 15 s, 46 beats are produced. What are the possible frequencies of the unknown-frequency tuning fork?

### Frame the Problem

- Two tuning forks of different *frequency* are sounding at the same time.

- This results in the production of *beats*.

- The absolute value of the difference between the two *frequencies* is equal to the beat *frequency*.

### Identify the Goal

The possible frequencies, $f_2$, of the unknown-frequency tuning fork

### Variables and Constants

| Involved in the problem | | Known | Unknown |
|---|---|---|---|
| $f_1$ | $N$ | $f_1 = 440$ Hz | $f_2$ |
| $f_2$ | $\Delta t$ | $N = 46$ | $f_{beat}$ |
| $f_{beat}$ | | $\Delta t = 15$ s | |

*continued* ▶

continued from previous page

## Strategy

You can find the beat *frequency* because you know the number of beats and the time interval. Substitute these values into the equation.

The absolute value of the difference between the two frequencies is equal to the beat frequency. Substitute the known values into the equation.

Simplify. Notice that when you remove the absolute value sign, you cannot know whether the value is positive or negative. Therefore, use both possibilities.

Using the positive value

Using the negative value

The frequency of the second tuning fork is either 443 Hz or 437 Hz.

## Calculations

$$f_{beat} = N/\Delta t$$

$$f_{beat} = 46 \text{ beats}/15 \text{ s}$$

$$f_{beat} = 3.1 \text{ Hz}$$

$$f_{beat} = |f_2 - f_1|$$

$$3.1 \text{ Hz} = |f_2 - 440 \text{ Hz}|$$

$$(f_2 - 440 \text{ Hz}) = \pm 3.1 \text{ Hz}$$

$$f_2 - 440 \text{ Hz} = 3.1 \text{ Hz}$$
$$f_2 = 440 \text{ Hz} + 3.1 \text{ Hz}$$
$$f_2 = 443.1 \text{ Hz}$$

$$f_2 - 440 \text{ Hz} = -3.1 \text{ Hz}$$
$$f_2 = 440 \text{ Hz} - 3.1 \text{ Hz}$$
$$f_2 = 436.9 \text{ Hz}$$

## Validate

Having 46 beats in 15 s gives a beat frequency of about 3 Hz, but the beat frequency is equal to the absolute value of the difference between the two tuning forks. Thus, the unknown frequency is either 3 Hz above 440 Hz or 3 Hz below 440 Hz. It is either 443 Hz or 437 Hz.

> **PROBLEM TIP**
>
> The beat frequency provides you with only the *difference* in frequency between two sounds. If the frequency of one of the sounds is known, then the other sound could have one of two possible frequencies — one higher and one lower than the known frequency. You would need some additional information to determine which of these two possible frequencies is the right one.

### PRACTICE PROBLEMS

7. Two tuning forks of frequencies 512 Hz and 518 Hz are sounded at the same time.

   (a) Describe the resultant sound.

   (b) Calculate the beat frequency.

8. A 440 Hz tuning fork is sounded at the same time as a 337 Hz tuning fork. How many beats will be heard in 3.0 s?

9. A trumpet player sounds a note on her trumpet at the same time as middle C is played on a piano. She hears 10 beats over 2.0 s. If the piano's middle C has been tuned to

   256 Hz, what are the possible frequencies of the note she is sounding?

10. A string on an out-of-tune piano is struck at the same time as a 440 Hz tuning fork is sounded. The piano tuner hears 12 beats in 4.0 s. He then slightly increases the tension in the string in order to increase the pitch of the note. Now he hears 14 beats in 4.0 s.

   (a) What was the original frequency of the string on the out-of-tune piano?

   (b) Is the piano more or less in tune after he tightens the string? Explain.

# Determining the Effect of a Load on the Frequency of a Tuning Fork

**TARGET SKILLS**
- Performing and recording
- Analyzing and interpreting
- Communicating results

Strike two identical tuning forks and listen to the sound that they produce. Then, using a small quantity of modelling clay, add equal loads to both arms of one of the forks, equal distances from the ends. Again, strike the two forks and listen to the sound produced. Repeat with an increased load, then a load at a different distance from the ends.

## Analyze and Conclude

1. How do you know that the two tuning forks were emitting sounds of the same frequency?

2. How do you know that the loaded tuning fork had a different frequency? Was it higher or lower?

3. Determine the frequency of the loaded tuning fork.

4. How was the frequency of the loaded tuning fork affected by (a) an increase in the load? (b) an increase in the distance of the load from the end of the fork?

## Apply and Extend

The A above middle C on a piano is supposed to be tuned to 440 Hz. Use a 440 Hz tuning fork and your knowledge of beats to determine how far a piano is out of tune. Provide a quantitative measure for how far out of tune the piano is.

---

## 8.3 Section Review

1. **K/U**
   (a) What would you listen for when searching for evidence of interference effects in sound?

   (b) What would you expect to hear as a result of destructive interference? As a result of constructive interference?

2. **I** Two loudspeakers are placed side by side about 0.5 m apart, and driven in phase with the same audio frequency signal.

   (a) Describe the sound that you expect to hear along the perpendicular bisector of the line that connects the two speakers. Why do you expect to hear this along the perpendicular bisector?

   (b) If a frequency of 340 Hz is used, it will produce sound waves with a wavelength of about 1 m. Where would you expect destructive interference to occur? Use a diagram to help you explain.

   (c) If the two speakers were driven in opposite phase to each other, what would you expect to hear along the right bisector? Why?

3. **C**
   (a) Explain how a tuning fork vibrates.

   (b) Explain how this mode of vibration can be seen as resulting from more than one sound source.

   (c) Where would you expect to find quiet regions near the tuning fork? Why?

4. **C** What are "beats"? What conditions are needed to produce them? What is the beat frequency?

5. **C** Outline the arguments and the evidence that you might use to convince someone that the properties of sound can be explained very effectively by using a wave model.

## SECTION EXPECTATIONS

- Identify the conditions required for resonance.

- Analyze, through experimentation, the components of resonance.

- Describe how knowledge of the properties of sound have been applied to musical instruments.

## KEY TERMS

- noise
- music
- harmonics
- fundamental frequency
- sound spectrum
- closed air column
- resonance length
- displacement node
- displacement antinode
- open air column

# Music, Noise, and Resonance in Air Columns

If you strike two stones together, you produce a sound that is immediately recognizable, but which has no specific pitch. A **noise**, such as this, is a mixture of many sound frequencies with no recognizable relationship to each other. **Music**, on the other hand, is a mixture dominated by sound frequencies known as **harmonics** that are whole-number multiples of the lowest frequency or **fundamental frequency**. By plotting the intensity of the various sound frequencies that make up a sound, you can see graphically the difference between music and noise. The **sound spectrum** of music consists of a number of discrete frequencies, while the sound spectrum for noise shows a continuous or nearly continuous range of frequencies.

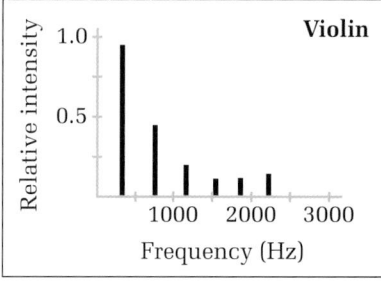
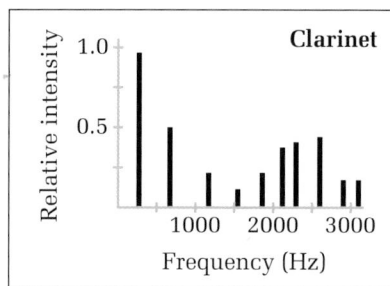
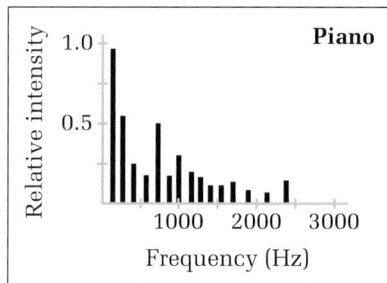
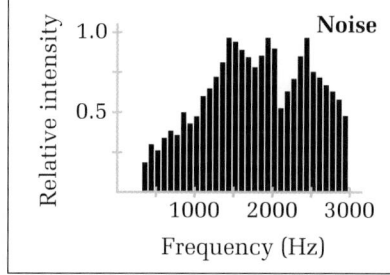

**Figure 8.16** Sound spectra of notes played on musical instruments and of a noise

Musical instruments can produce sounds that are harmonious because the vibrations that are set up are essentially in one dimension only. In a guitar or violin, the vibrations are set up in a string. In a wind instrument, like a trombone or clarinet, vibrations are set up in a narrow column of air. Many musical sounds, including the human voice, are produced by resonance in air columns. Resonance in a linear air column occurs when a standing wave is established. The air column (the same as a spring) can sustain a standing wave only for frequencies of vibration that are whole-number multiples of a fundamental. Thus, the sound waves that are emitted from the air column are all whole-number multiples (or integral multiples) of a fundamental, and the total effect is perceived as musical.

## Resonance Lengths of a Closed Air Column

An air column that is closed at one end and open at the other is called a **closed air column**. If a tuning fork is held over the open end and the length of the column is increased, the loudness of the sound will increase very sharply for specific lengths of the tube, called **resonance lengths**. If a different tuning fork is used, there will still be distinct resonance lengths, but they will be different than those you found with the first fork.

## Resonance in a Closed Air Column

Resonance occurs in an air column when the length of the air column meets the criteria for supporting a standing wave. The

**TRY THIS...**

Borrow a trumpet, trombone, or other brass instrument. Take the mouthpiece out of the instrument and blow through it. The sound that your vibrating lips make is noise. Re-insert the mouthpiece into the instrument, and then blow again. (You might need a few practice tries to produce a note.) The air column of the instrument will resonate with only one of the frequencies your lips are producing. It is this frequency that you hear.

---

**QUICK LAB**

## Resonance Lengths of a Closed Air Column

**TARGET SKILLS**
- **Performing and recording**
- **Identifying variables**

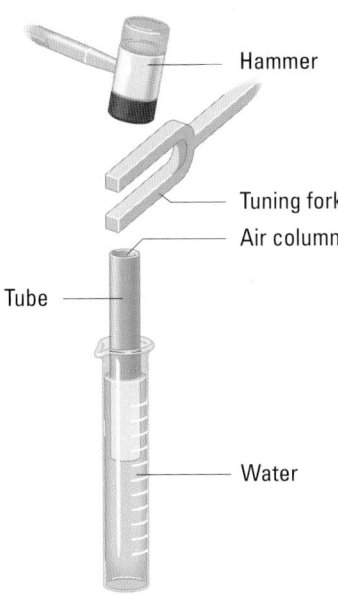

Hammer

Tuning fork

Air column

Tube

Water

Place a 50 cm long piece of plastic pipe inside a large graduated cylinder almost completely filled with water that is at room temperature. Sound a 512 Hz tuning fork and hold it over the open end of the plastic pipe. Raise the pipe slowly out of the water while keeping the tuning fork positioned over the open end. Measure the lengths of the air column for which resonance occurs. Repeat the procedure using a 1024 Hz tuning fork.

### Analyze and Conclude

1. Use a thermometer to measure the room temperature. Calculate the speed of sound in air, and from that, the wavelength of the sound produced by the 512 Hz tuning fork.

2. By how much is one resonance length longer than the previous one? (If you were able to determine three or more resonance lengths, was this increase in length constant?) What fraction of a wavelength is this increase in resonance length?

3. Repeat questions 1 and 2 for the 1024 Hz tuning fork.

tuning fork produces a sound wave, which travels down the air column and is reflected off the closed end. This reflected wave interferes with the wave from the tuning fork, producing a standing wave.

It is easier to see how resonance occurs if sound waves are represented in the same manner as standing waves on a rope or spring, as presented in Chapter 7. In the case of sound waves, however, the amplitude of the wave does not represent the actual position of the molecules in the air as it does for the rope. The amplitude of the sound wave represents the extent of the longitudinal displacement of the molecules in the air at that specific point along the air column.

In Figure 8.17, the arrows on the right hand side show the distance that the air molecules move (displacement) at different points along the tube. At the closed end of a tube, the air molecules are prevented from moving entirely. Thus, when the standing wave is set up, there has to be a **displacement node** at this closed end. (There will also be a pressure antinode at the same point because the closed end will experience the greatest variation in pressure over time.) At the open end of the air column, however, the air molecules are free to move back and forth relatively easily. Thus, when a standing wave is set up in the air column, there must be a **displacement antinode** at this open end.

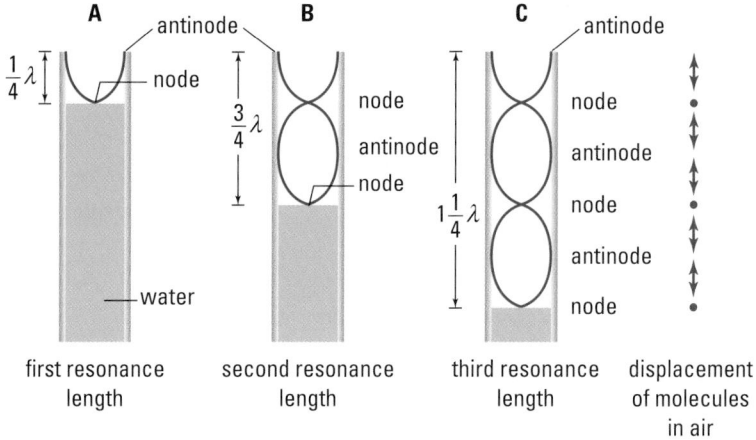

**Figure 8.17** First three resonance lengths of a closed air column and their displacement standing wave patterns

In Figure 8.17 (A), you can see that the first resonance length is equal to one quarter of a wavelength, or $\frac{\lambda}{4}$. In (B), the second resonance length is equal to three quarters of a wavelength, or $\frac{3\lambda}{4}$. Finally, in (C), the third resonance length is equal to one and a quarter wavelengths, or $\frac{5\lambda}{4}$. Each of the subsequent resonance lengths is half a wavelength longer than the previous one. Thus, for the first three resonance lengths, the pattern is $\lambda = 4L_1$, $\lambda = \frac{4L_2}{3}$, $\lambda = \frac{4L_3}{5}$, $\lambda = \frac{4L_4}{7}$. To find the resonance lengths, solve for $L_n$ in the equations, where $n$ is a positive integer.

**PHYSICS FILE**

Theories attempt to explain the complexities of real-world phenomena. In deriving the resonance lengths for a closed air column, it was assumed that the behaviour of the air in the column could be treated as being one-dimensional (like the standing waves in a spring). This is a very good assumption for the air inside the air tube, but is not as good at the open end, particularly if the diameter of the tube is large. As a result, the antinode at the open end of the tube actually lies a short distance inside the end. The nodes inside the tube, however, are spaced one half wavelength apart, regardless of the diameter of the tube.

## RESONANCE LENGTHS OF A CLOSED AIR COLUMN

The resonance lengths of a closed air column are odd integer multiples of the first resonance length, $\frac{1}{4}\lambda$.

$$L_n = (2n - 1)\frac{\lambda}{4}$$

where $n$ is a positive integer.

### TRY THIS...

Convince yourself that the equation in the box will give you the expressions for the resonance wavelengths by first solving for each $L$ in the expressions at the bottom of page 392, and then by substituting the values one through four into the equation.

## Resonance Lengths of an Open Air Column

An air column that is open at both ends is called an **open air column**. At these open ends, the air molecules are free to move easily, so that when a standing wave is set up, there is a displacement antinode at each end, as shown in Figure 8.18.

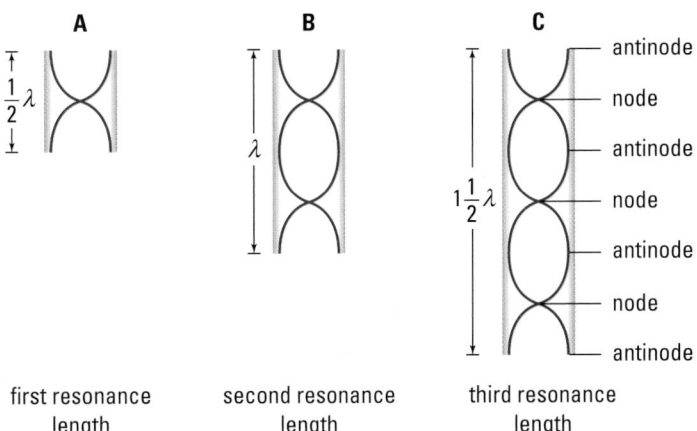

| A | B | C |
|---|---|---|
| first resonance length | second resonance length | third resonance length |

**Figure 8.18** First three resonance lengths of an open air column and their displacement standing wave patterns

As you can see in Figure 8.18, the first three resonance lengths are $\frac{\lambda}{2}$, $\lambda$, $\frac{3\lambda}{2}$, with each of the subsequent resonance lengths half a wavelength longer. Thus, for the first three resonance lengths, the pattern is $\lambda = 2L_1$, $\lambda = L_2$, $\lambda = \frac{3L_3}{2}$.

## RESONANCE LENGTHS OF AN OPEN AIR COLUMN

The resonance lengths of an open air column are integral multiples of the first resonance length, $\frac{1}{2}\lambda$.

$$L_n = \frac{n\lambda}{2}$$

### Music Link

Wind instrument players play different notes by varying the lengths of the air columns in their instruments. In a trombone, this is most evident as the slide is pushed in and out. In trumpets, tubas, and other brass instruments, different valves are used to open or close auxiliary lengths of tubing. In clarinets, saxophones, and other woodwinds, key pads are lifted off holes in the side of the instrument. In what other ways do these instruments vary?

## Resonance Lengths of a Closed Air Column

A vibrating tuning fork is held near the mouth of a narrow plastic pipe partially submerged in water. The pipe is raised, and the first loud sound is heard when the air column is 9.0 cm long. The temperature in the room is 20°C.

(a) Calculate the wavelength of the sound produced by the tuning fork.

(b) Calculate the length of the air column for the second and third resonances.

(c) Estimate the frequency of the tuning fork.

### Frame the Problem

- Since one end of the pipe is submerged in water, it is a *closed air column*.

- The shortest *resonance length* is one quarter of the wavelength of the sound.

- The resonance lengths of a closed air column are $\frac{\lambda}{4}, \frac{3\lambda}{4}, \frac{5\lambda}{4}$.

- The *wave equation* applies to all kinds of waves.

### Identify the Goal

(a) The wavelength, $\lambda$, of the sound

(b) The length of the air column for the second and third resonances, $L_2$ and $L_3$

(c) The frequency, $f$, of the tuning fork

### Variables and Constants

| Involved in the problem | | Known | Unknown | |
|---|---|---|---|---|
| $L_1$ | $v$ | $L_1 = 9.0$ cm | $L_2$ | $v$ |
| $L_2$ | $f$ | $T = 20°C$ | $L_3$ | $f$ |
| $L_3$ | $T$ | | $\lambda$ | |
| $\lambda$ | | | | |

### Strategy

You can find the wavelength of the sound because you know the first resonance length. Substitute this value into the equation.

### Calculations

$$L_1 = \frac{\lambda}{4}$$

$$\lambda = 4L_1$$

$$\lambda = 4\,(9.0 \text{ cm})$$

$$\lambda = 36 \text{ cm or } 0.36 \text{ m}$$

(a) The wavelength of the sound is 0.36 m.

## Strategy

You can find the second and third resonance lengths because you know the wavelength and the sequence of formulas for resonance lengths.

Substitute the wavelength into the equations for second and third resonance lengths.

## Calculations

$$L_2 = \frac{3\lambda}{4}$$

$$L_2 = \frac{3}{4}(36 \text{ cm})$$

$$L_2 = 27 \text{ cm or } 0.27 \text{ m}$$

$$L_3 = \frac{5\lambda}{3}$$

$$L_3 = \frac{5}{4} (36 \text{ cm})$$

$$L_3 = 45 \text{ cm or } 0.45 \text{ m}$$

(b) The second and third resonance lengths are 0.27 m and 0.45 m, respectively.

You can find the frequency of the tuning fork if you know the speed of sound in the room and the wavelength of the sound.

$$v = f\lambda$$

You can find the speed of sound if you know the temperature in the room.

$$v = 331 + 0.59T_\text{C}$$

Substitute the value of the temperature in the formula for the speed of sound.

$$v = 331 \text{ m/s} + 0.59 \frac{\frac{\text{m}}{\text{s}}}{\text{°C}} (20\text{°C})$$

$$v = 331 \text{ m/s} + 11.8 \text{ m/s}$$

$$v = 342.8 \text{ m/s}$$

**Substitute first**

**Solve for *f* first**

Substitute the values of the wavelength (in metres) and the speed of sound in the formula for the frequency.

$$v = f\lambda$$

$$342.8 \text{ m/s} = f(0.36 \text{ m})$$

$$f = \frac{342.8 \frac{\text{m}}{\text{s}}}{0.36 \text{ m}}$$

$$f = 952.2 \text{ s}^{-1}$$

$$v = f\lambda$$

$$f = \frac{v}{\lambda}$$

$$f = \frac{342.8 \frac{\text{m}}{\text{s}}}{0.36 \text{ m}}$$

$$f = 952.2 \text{ s}^{-1}$$

(c) The frequency of the tuning fork is $9.5 \times 10^2$ Hz.

---

## Validate

If the first resonance length is 9.0 cm, the second is three times the first (27 cm), and the third is five times the first (45 cm).

At a temperature of 20°C, the speed of sound is about 340 m/s. The wavelength of the sound is $4 \times 9 = 36$ cm, which is about $\frac{1}{3}$ m. The frequency is then approximately 340 m/s divided by $\frac{1}{3}$ m, or 1000 Hz.

*continued* ▶

## PRACTICE PROBLEMS

**11.** A narrow plastic pipe is almost completely submerged in a graduated cylinder full of water, and a tuning fork is held over its open end. The pipe is slowly raised from the water. An increase in loudness of the sound is heard when the pipe has been raised 17 cm and again when it has been raised 51 cm.

  **(a)** Determine the wavelength of the sound produced by the tuning fork.

  **(b)** If the pipe continues to be raised, how far from the top of the pipe will the water level be when the next increase in loudness is heard?

**12.** The first resonance length of an air column, resonating to a fixed frequency, is 32 cm.

  **(a)** Determine the second and third resonance lengths, if the column is closed at one end.

  **(b)** Determine the second and third resonance lengths, if the column is open at both ends.

**13.** The third resonance length of a closed air column, resonating to a tuning fork, is 95 cm. Determine the first and second resonance lengths.

**14.** The second resonance length of an air column, open at both ends and resonating to a fixed frequency, is 64 cm. Determine the first and third resonance lengths.

**15.** A particular organ pipe, open at both ends, needs to resonate in its fundamental mode with a frequency of 128 Hz. The organ has been designed to be played at a temperature of 22°C.

  **(a)** How long does the organ pipe need to be?

  **(b)** If this pipe is closed at one end by a stopper, at what fundamental frequency will it resonate?

# Resonance Frequencies for Fixed-Length Air Columns

How can a bugler play a melody when there are no valves, keys, or slides? The length of the air column is fixed, so the fundamental frequency is fixed. When trumpeters and trombonists blow into the mouthpiece, the vibration of their lips creates a range of frequencies. The length of the air column determines the fundamental frequency and overtones, which it "amplifies" by setting up a standing wave. The resonating air column is, in fact, what makes the sound musical.

The pitch of a wind instrument can be changed not only by increasing or decreasing the length of its air column, but also by increasing or decreasing the tension in the player's lips (the embouchure). This change in the player's embouchure produces a different range of frequencies as the player blows through the mouthpiece. High tension of the lips creates only frequencies that are higher than the fundamental. The result is that the pitch is

**Figure 8.19** Bugles have no valves, keys, or slides, so how can the bugler play a tune?

perceived as the lowest frequency of the overtone frequencies that set up standing waves in the air column. Although buglers have no means of varying the length of the air columns, they can still play a melody by varying their embouchures. They can access a range of different notes when playing "Reveille," "Last Post," or "Taps."

## Open Air Columns

Most wind instruments, such as the bugle and the flute, behave the same as air columns that are open at both ends. The major exceptions are reed instruments such as the clarinet, oboe, and bassoon. They behave like closed air columns. For an open air column of given length $L$, the fundamental and overtone frequencies can be calculated in a straightforward way. The wave equation tells you that the frequency is the quotient of the speed of the wave and the wavelength, $f = \frac{v}{\lambda}$. Thus, you can find the relationship between the frequency and the length of the air column by substituting the expression for wavelength in terms of air column length, $L$, as shown in Figure 8.20.

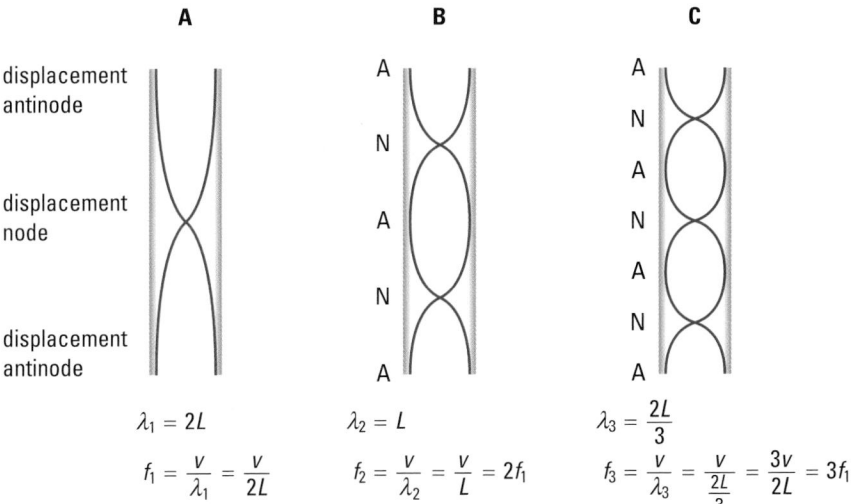

displacement antinode

displacement node

displacement antinode

**A**

$\lambda_1 = 2L$

$f_1 = \frac{v}{\lambda_1} = \frac{v}{2L}$

**B**

$\lambda_2 = L$

$f_2 = \frac{v}{\lambda_2} = \frac{v}{L} = 2f_1$

**C**

$\lambda_3 = \frac{2L}{3}$

$f_3 = \frac{v}{\lambda_3} = \frac{v}{\frac{2L}{3}} = \frac{3v}{2L} = 3f_1$

**Figure 8.20** **(A)** Fundamental mode or first harmonic; **(B)** first overtone or second harmonic; **(C)** second overtone or third harmonic

In general, the harmonics of an open air column form the series $f_1$, $2f_1$, $3f_1$, $4f_1$, $5f_1$, and so on. That is, all integer multiples of the fundamental frequency are produced when an air column is open at both ends. Thus, the effect of these frequencies sounding together is music rather than noise. This is why all of these frequencies are referred to as harmonics — the fundamental frequency as the first harmonic, the first overtone as the second harmonic, and so on.

By doubling the frequency of a note, a wind player jumps an octave in pitch. For brass players, this is easily accomplished by tightening up their embouchure to move from the first to the second harmonic of an open air column. For a clarinet player, however, this will not work because of the odd integer harmonic structure of a closed air column. Use the equation for resonance frequencies in closed air columns to show why this is true.

Clarinet players cannot jump an octave by changing their embouchure.

---

**RESONANCE FREQUENCIES OF A FIXED-LENGTH OPEN AIR COLUMN**

The resonance frequencies of a fixed-length open air column are integral multiples of the first resonance frequency, $f_1$.

$$f_n = nf_1$$

where $f_1 = \dfrac{v}{2L}$

---

## Closed Air Columns

For the clarinet, which behaves like a closed air column, the overtones produced are somewhat different.

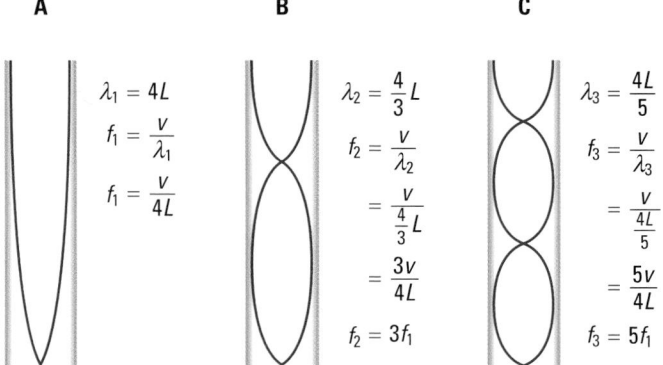

**A**

$\lambda_1 = 4L$

$f_1 = \dfrac{v}{\lambda_1}$

$f_1 = \dfrac{v}{4L}$

**B**

$\lambda_2 = \dfrac{4}{3}L$

$f_2 = \dfrac{v}{\lambda_2}$

$= \dfrac{v}{\frac{4}{3}L}$

$= \dfrac{3v}{4L}$

$f_2 = 3f_1$

**C**

$\lambda_3 = \dfrac{4L}{5}$

$f_3 = \dfrac{v}{\lambda_3}$

$= \dfrac{v}{\frac{4L}{5}}$

$= \dfrac{5v}{4L}$

$f_3 = 5f_1$

**Figure 8.21** **(A)** Fundamental mode or first harmonic; **(B)** first overtone or second harmonic; **(C)** second overtone or third harmonic

In general, the harmonics of a closed air column form the series $f_1$, $3f_1$, $5f_1$, $7f_1$, $9f_1$, and so on. That is, only odd integral multiples of the fundamental frequencies are produced when the air column is closed at one end.

---

**RESONANCE FREQUENCIES OF A FIXED-LENGTH CLOSED AIR COLUMN**

The resonance frequencies of a fixed-length closed air column are odd integer multiples of the first resonance frequency, $f_1$.

$$f_n = (2n - 1)f_1$$

where $f_1 = \dfrac{v}{4L}$

---

# INVESTIGATION 8-A

## Determining the Speed of Sound in Air

### Problem

How can resonance in air columns be used to obtain a more accurate value for the speed of sound in air?

### Prediction

Make a prediction about the relative precision of measurements of the speed of sound. Will measuring the speed of sound using air columns have greater precision than the method you used in the Quick Lab on page 382? On what do you base your prediction?

### Equipment

- 512 Hz and 1024 Hz tuning forks
- rubber mallet
- hollow plastic pipe (do *not* use a glass pipe)
- 1000 mL graduated cylinder
- water at room temperature
- thermometer
- metre stick

**CAUTION** Use only plastic pipes with tuning forks. If glass pipes are used, there is a danger of flying glass if the vibrating tuning fork touches the pipe.

### Procedure

1. After placing the plastic pipe in the graduated cylinder, fill the cylinder with water, as near to the top as possible. Using the thermometer, measure the temperature of the air in the room.

2. Sound the 512 Hz tuning fork, and hold it over the open end of the plastic pipe. Raise the pipe slowly out of the water while keeping the tuning fork positioned over the end of the plastic pipe. Locate the positions for which the sound increases dramatically. (Ignore positions of slightly increased sound of different frequencies from your tuning fork.) Have your partner carefully measure the length of the air column for each of these resonance points.

3. Arrange your data in four columns, with the headings: Resonance length, Change in resonance length, Wavelength, Speed of sound ($v = 512 \times \lambda$).

   Frequency = 512 Hz
   Temperature = _____

4. Repeat the procedure with the 1024 Hz tuning fork. Arrange your data and calculations in a similar table.

### Analyze and Conclude

1. Using the data you obtained with the 512 Hz tuning fork, calculate an average value for the speed of sound in air. Estimate the precision of this value.

2. Using the data you obtained with the 1024 Hz tuning fork, calculate an average value for the speed of sound in air. Estimate the precision of this value.

3. Using your measurement of the temperature of the air, calculate a value for the speed of sound in air, using $v = 331 + 0.59T_C$.

4. Earlier in this chapter, you used an echolocation procedure to determine the speed of sound in air. Compare the two values you obtained in this experiment with each other and with your earlier result. Compare the precisions of the three results. Which result do you think is the most accurate? Why?

5. How do your experimental results for the speed of sound in air compare with the calculated value? Which do you trust the most? Why?

## Harmonics in a Fixed-Length Air Column

1. An air column, open at both ends, has a first harmonic of 330 Hz.

   (a) What are the frequencies of the second and third harmonics?

   (b) If the speed of sound in air is 344 m/s, what is the length of the air column?

### Frame the Problem

- The air column is *open at both ends*, so the *harmonics* are integral multiples of $f_1$.

- The frequency of the first harmonic, or $f_1$, is 330 Hz.

- The frequency of the first harmonic is equal to the speed of sound in air divided by twice the length of the air column.

### Identify the Goal

(a) The frequencies of the second and third harmonics, $f_2$ and $f_3$

(b) The length, $L$, of the air column

### Variables and Constants

| Involved in the problem | | Known | Unknown |
|---|---|---|---|
| $f_1$ | $v$ | $f_1 = 330$ Hz | $f_2$ |
| $f_2$ | $L$ | $v = 344$ m/s | $f_3$ |
| $f_3$ | | | $L$ |

### Strategy

You can find the frequencies of the second and third harmonics because you know the frequency of the first harmonic.

Substitute this value into the equations.

### Calculations

$f_n = nf_1$

$f_2 = 2f_1$

$f_2 = 2(330 \text{ Hz})$

$f_2 = 660 \text{ hz}$

$f_3 = 3f_1$

$f_3 = 3(330 \text{ Hz})$

$f_3 = 990 \text{ Hz}$

(a) The frequencies of the second and third harmonics are 660 Hz and 990 Hz, respectively.

## Strategy

You can find the length of the air column because you know the speed of sound in air and the frequency of the first harmonic.

Substitute these values into the equation.

## Calculations

**Substitute first**

$$f_1 = \frac{v}{2L}$$

$$330 \text{ Hz} = \frac{344\frac{\text{m}}{\text{s}}}{2L}$$

$$330 \text{ s}^{-1}(2L) = \frac{344\frac{\text{m}}{\text{s}}}{2L}(2L)$$

$$\frac{660 \text{ s}^{-1}\,L}{660 \text{ s}^{-1}} = \frac{344\frac{\text{m}}{\text{s}}}{660 \text{ s}^{-1}}$$

$$L = 0.5212 \text{ m}$$

**Solve for L first**

$$f_1 = \frac{v}{2L}$$

$$f_1 L = \frac{v}{2L}(L)$$

$$\frac{f_1 L}{f_1} = \frac{v}{2f_1}$$

$$L = \frac{344\frac{\text{m}}{\text{s}}}{2\,(330 \text{ s}^{-1})}$$

$$L = 0.5212 \text{ m}$$

**(b)** The length of the air column is 0.52 m.

## Validate

For an open air column, the harmonics are all integral multiples of the first harmonic. Thus, the frequency of the second harmonic is $2 \times 330 = 660$ Hz, and the frequency of the third harmonic is $3 \times 330 = 990$ Hz.

For the first harmonic, the resonance length is $\frac{1}{2}\lambda$ and the wavelength can also be calculated from $\lambda = \frac{v}{f}$. The length is equal to $\frac{1}{2} \times \frac{344}{330}$ or approximately $\frac{1}{2} \times 1$ or 0.5 m. These values are very close to the calculated values.

---

**2.** An air column, closed at one end, has a first harmonic of 330 Hz. If the speed of sound in air is 344 m/s, what is the length of the air column?

---

## Frame the Problem

- The air column is *closed at one end*, so the *harmonics* are $f_1, 3f_1, 5f_1, 7f_1, 9f_1, ....$

- The frequency of the first harmonic, or $f_1$, is 330 Hz.

- The frequency of the first harmonic is equal to the speed of sound in air divided by four times the length of the air column.

## Identify the Goal

The length, $L$, of the air column

## Variables and Constants

| Involved in the problem | Known | Unknown |
|---|---|---|
| $f_1$ | $f_1 = 330$ Hz | $L$ |
| $v$ | $v = 344$ m/s | |
| $L$ | | |

*continued* ▶

## Strategy

You can find the length of the air column because you know the speed of sound in air and the frequency of the first harmonic.

Substitute these values into the equation.

The length of the air column is 0.26 m.

## Calculations

**Substitute first**

$$f_1 = \frac{v}{4L}$$

$$330 \text{ Hz} = \frac{344\frac{\text{m}}{\text{s}}}{4L}$$

$$330 \text{ Hz}\,(L) = \frac{344\frac{\text{m}}{\text{s}}}{4\cancel{L}}\,(\cancel{L})$$

$$\frac{\cancel{330 \text{ s}^{-1}}\,L}{\cancel{330 \text{ s}^{-1}}} = \frac{86\frac{\text{m}}{\text{s}}}{330 \text{ s}^{-1}}$$

$$L = 0.2606 \text{ m}$$

**Solve for *L* first**

$$f_1 = \frac{v}{4L}$$

$$f_1 L = \frac{v}{4\cancel{L}}\,(\cancel{L})$$

$$\frac{f_1 L}{f_1} = \frac{v}{4f_1}$$

$$L = \frac{344\frac{\text{m}}{\text{s}}}{4\,(330 \text{ s}^{-1})}$$

$$L = 0.2606 \text{ m}$$

## Validate

For the first harmonic of a closed air column, the resonance length is $\frac{1}{4}\lambda$ and the wavelength can be calculated from $\lambda = \frac{v}{f}$. Thus, the length is $\frac{1}{4} \times \frac{344}{330}$, or approximately $\frac{1}{4} \times 1$, or 0.25 m.

### PRACTICE PROBLEMS

**16.** An air column, open at both ends, resonates with a fundamental frequency of 256 Hz. Determine the frequencies of its first and second overtones (second and third harmonics).

**17.** A bugle is essentially a 2.65 m pipe that is open at both ends.

(a) Determine the lowest frequency note that can be played on a bugle.

(b) Determine the next two higher frequencies that will produce resonance.

**18.** A trombone is playing F (87.3 Hz) as its first harmonic. A trombone functions as an air column that is open at both ends.

(a) Determine the second harmonic.

(b) If the speed of sound is 344 m/s, what is the length of the tubing being used? Why would this note be difficult to play?

## 8.4 Section Review

**1.** **C** Explain the difference between noise and music.

**2.** **K/U** Why are the resonance lengths of open and closed air columns different?

**3.** **I** A fixed-length air column has only one fundamental frequency. How is it possible to play a melody on an instrument such as a bugle, which has a fixed-length air column?

**4.** **C** Explain the statement: "The first overtone is the second harmonic."

**ELECTRONIC LEARNING PARTNER**

Go to the Electronic Learning Partner to quiz yourself on these concepts.

## REFLECTING ON CHAPTER 8

- A wave model of sound predicts that sound requires a medium and that sound waves will exhibit interference effects with each other.
- Sound does not travel through a vacuum.
- For a wave theory of sound:
  - loudness corresponds to amplitude
  - pitch corresponds to frequency
  - quality corresponds to harmonic structure
- Sound waves are longitudinal waves.
- Human beings can hear sounds in a frequency range from 20 Hz to 20 000 Hz.
- Sound frequencies are classified as infrasonic (less than 20 Hz), audible (20 to 20 000 Hz), and ultrasonic (greater than 20 000 Hz).
- Loudness is what the ear perceives. Sound intensity is the energy delivered by a sound wave in 1 s over an area of $1m^2$. Intensity is measured in picowatts per square metre ($pW/m^2$).
- The intensity level of a sound wave is a logarithmic scale of sound intensity measured in decibels. A 20 dB sound has a sound intensity 10 times greater than a 10 dB sound. A 30 dB sound has an intensity 100 times greater than a 10 dB sound ($10\ dB = 1\ B = 10pW/m^2$).
- The speed of sound in air at standard atmospheric pressure (101 kPa) is constant for a given temperature, and given by the equation $v = 331 + 0.59T_C$.
- In general, the speed of sound is slowest in gases, faster in liquids, and fastest in solids.

- When two sounds of similar frequency are sounded at the same time, alternately loud and quiet sounds are produced. This wavering effect is called beats.
  The beat frequency is given by the equation
  $$f_{beat} = |f_2 - f_1|$$
- Music is a sound made up of whole number multiples of a lowest or fundamental frequency. Noise is a sound made up of a multitude of sound frequencies with no recognizable relationship to each other.
- A closed air column (closed at one end) has resonance lengths
  $$L_n = \frac{(2n - 1)\lambda}{4}$$
- An open air column (open at both ends) has resonance lengths
  $$L_n = \frac{n\lambda}{2}$$
- A closed air column of fixed length has resonance frequencies
  $$f_n = (2n - 1)f_1$$
  where $f_1 = \frac{v}{4L}$.
- An open air column of fixed length has resonance frequencies
  $$f_n = nf_1$$
  where $f_1 = \frac{v}{2L}$.

### Knowledge/Understanding

1. What properties of sound suggest a wave theory for sound? How does the wave theory explain each?
2. What properties of sound does a wave theory predict should exist that otherwise might not have been considered?
3. Explain ways in which the simple wave theory of sound is testable.
4. Describe the bell-jar experiment and its obser-

vations. How do these observations help increase our confidence in the wave theory of sound?

5. Sketch pairs of sound waves that illustrate the following contrasts in sound.
   (a) pitch (low versus high)
   (b) loudness (quiet versus loud)
   (c) quality (pure versus rich)
6. Why, in physics, is sound not defined in terms of the pressure waves that humans can hear?

7. **(a)** What is the standard audible range of frequencies for the human ear?

   **(b)** Explain the meaning of the terms "infrasonic" and "ultrasonic."

   **(c)** How does the human audible frequency range compare to the range for dogs? Bats? Porpoises?

8. **(a)** A sound intensity level of 1 bel is equal to an intensity of $10pW/m^2$. What are the intensities for intensity levels of 2 bels? For levels of 3 bels?

   **(b)** What term is used to describe a scale such as the sound intensity scale?

9. Estimate a reasonable sound intensity level in decibels for

   **(a)** students whispering

   **(b)** a passing train

   **(c)** a jet engine

10. As the temperature of air increases, what happens to a sound's (a) speed? (b) frequency? (c) wavelength?

11. Does sound travel faster in a gas or in a solid? Explain why you think that this is the case.

12. Some animals use short pulses of high-frequency sound to locate objects. If the echo is received a time interval $\Delta t$ after the pulse of sound is transmitted, how could you use the $\Delta t$ and the speed of sound to calculate the distance between the animal and the object?

13. Describe the location of the quiet regions near a tuning fork. Explain how these are created.

14. Describe the phenomenon of beats and the conditions necessary to produce them.

15. A trumpet player is tuning the instrument to concert A (440 Hz) being played by the oboe. As the trumpet player moves the tuning slide in, beat frequency increases. Is the trumpet getting closer to, or farther away from, the correct pitch? Explain.

16. How is music different from noise?

17. When a trombone player pushes the slide of the instrument out, the pitch of the sound being produced decreases. Explain why this happens.

18. When a stream of water is used to fill a graduated cylinder, a sound of steadily rising pitch is heard. Explain why this is the case.

19. When a stream of air is directed over the end of a 40 cm long piece of plastic pipe, open at both ends, a sound is produced.

   **(a)** Explain why this occurs.

   **(b)** If the bottom end of the pipe is covered, what will happen to the pitch of the sound? Explain.

20. How do the harmonics of a fixed air column closed at one end compare with the harmonics of a fixed air column open at both ends?

## Inquiry

21. Design an experiment to test the assumption that sound waves are influenced by the medium through which they travel. Be sure to identify the variables that you would control and those you would test.

22. A student constructs a home-made flute. She collects the data below, which represent different effective lengths and different harmonics, at an air temperature of 22°C. Copy the table and fill in the third column with an appropriate fraction. Should her instrument be considered an open or closed resonator?

| Length at which resonance occurs (m) | Frequency of sound | Resonance length as fraction of wavelength |
|---|---|---|
| 0.39 | 440 Hz | |
| 0.67 | 512 Hz | |
| 0.88 | 584 Hz | |

## Communication

23. Typically, thunder is heard a short time after lightning is observed. Describe how this time interval can be used to calculate the distance that the lightning is away. Explain why this strategy yields a reasonable answer. Would you be able to tell if a storm is approaching using this method?

## 24.
What are the first three resonance lengths of a closed air column? Sketch displacement standing wave patterns for each of these.

## 25.
What are the first three resonance lengths of an open air column? Sketch displacement standing wave patterns for each of these.

## Making Connections

**26.** Human hearing receptors are more sensitive to frequencies above 400 Hz than those below. Consider how this variation affects (a) hearing speech on a telephone, (b) noise concerns in a community, and (c) the intensity level at which various types of music sound most pleasant.

**27.** Thoughtfully consider and discuss how you believe science affected the production of musical instruments, and conversely, how music affected scientific inquiry into sound.

## Problems for Understanding

**28.** Calculate the speed of sound in air for each temperature.
- **(a)** −40.0°C
- **(b)** 5.0°C
- **(c)** 21.0°C
- **(d)** 35.0°C

**29.** Calculate the temperature of the air if the speed of sound is
- **(a)** 355 m/s
- **(b)** 344 m/s
- **(c)** 333 m/s
- **(d)** 318 m/s

**30.** A hunter wanted to know the air temperature. The echo from a nearby cliff returned 1.5 s after he fired his rifle. If the cliff is 250 m away, calculate the air temperature.

**31.** On a crisp fall day, a cottager looks across the lake and sees a neighbour chopping wood. As he watches, he notices that there is a time delay of 2.1 s between the time the axe hits the log and when he hears the sound of its impact. If the air temperature is 8.0°C, how far is he from his neighbour?

**32.** A guitar player is tuning his guitar to A (440 Hz) on the piano. He hears 14 beats in 4.0 s when he tries to play A on his guitar.
- **(a)** What two frequencies might he be playing?
- **(b)** Explain how he could determine which of the two he was playing.

**33.** Two tuning forks of frequencies 441 Hz and 444 Hz are sounded at the same time.
- **(a)** Describe what is heard.
- **(b)** Calculate the beat frequency, and explain what it is.

**34.** When a violin plays concert A, it produces a sound spectrum with a very intense line at 440 Hz, and less intense lines at 880 Hz, 1320 Hz, 1760 Hz, 2200 Hz, and 2640 Hz.
- **(a)** Explain why this sound would be described as music.
- **(b)** Sketch a possible sound spectrum (intensity against frequency) for noise.

**35.** A narrow plastic pipe is placed inside a large graduated cylinder almost filled with water. An 880 Hz tuning fork is held over the open end as the pipe is slowly raised out of the water.
- **(a)** Describe what will be heard as the pipe is raised.
- **(b)** Assuming the air temperature to be 22°C, calculate the first four resonance lengths.

**36.** A slightly smaller-diameter plastic pipe is inserted inside a second plastic pipe to produce an air column, open at both ends, whose length can be varied from 35 cm to 65 cm. A small loudspeaker, connected to an audio frequency generator, is held over one of the open ends. As the length of the air column is increased, resonance is heard first when the air column is 38 cm long and again when it is 57 cm long.
- **(a)** Calculate the wavelength of the sound produced by the audio frequency generator.
- **(b)** Assuming the air temperature to be 18°C, calculate the frequency setting of the audio frequency generator.

---

**Numerical Answers to Practice Problems**

**1. (a)** $3.2 \times 10^2$ m/s **(b)** $3.4 \times 10^2$ m/s **(c)** $3.5 \times 10^2$ m/s
**(d)** $3.2 \times 10^2$ m/s **2. (a)** 35.6°C **(b)** 11.9 °C **(c)** 5.1 °C **(d)** −20.3 °C
**3. (a)** $6.2 \times 10^2$ m **4.** 0.005 s **5.** $2.0 \times 10^2$ m **6. (a)** 5.8 s
**(b)** $6.7 \times 10^{-6}$ m **(c)** 2.8 km **7. (b)** 6.00 Hz **8.** 9.0 beats
**9.** 251 Hz or 261 Hz **10. (a)** 443 Hz **11. (a)** 68 cm **(b)** 85 cm
**12. (a)** 96 cm, 160 cm **(b)** 64 cm, 96 cm **13.** 19 cm, 57 cm
**14.** 32 cm **15. (a)** 1.34 m **(b)** 64 Hz **16.** 512 Hz, 768 Hz
**17. (a)** 64.9 Hz **(b)** 130 Hz, 195 Hz **18. (a)** 175 Hz **(b)** 1.97 m

A blue whale's 180 dB rumble is the loudest animal sound ever recorded. Whale sounds also appear to be part of a highly evolved communication system. Some whales are thought to communicate over hundreds and maybe thousands of kilometers. This is possible, in part, because sound waves travel five times faster in water than in air. In addition, the temperature characteristics of ocean water — decrease in temperature with depth — create a unique sound phenomenon. Layers of ocean water at specific temperatures "focus" the sound energy in channels that allow the sound to travel tremendous distances with little dissipation.

Artificial sounds in the oceans may be partially responsible for some recent threats to whale populations. In June of 2000, 16 whales and dolphins were found dead or dying on beaches in the Bahamas. Scientists found that the whales had torn and bleeding eardrums consistent with a "distant explosion or intense acoustic event." Some people attributed the beachings to ongoing U.S. Navy sonar experiments involving sound intensities of over 200 dB. More and more, the oceans are becoming awash in noise pollution, from the constant hum of cargo ships to remote sensing systems like sonar. The background noise in much of the ocean has reached the 85 dB level. The increased noise pollution of the underwater environment is prompting exciting research dealing with the mysteries of sound in the depths of our oceans.

This one example alone shows you the importance of understanding sound energy and its applications in the living world. In this chapter you will learn about many practical applications of sound as well as concerns relating to sound.

# Analyzing Sound

TARGET SKILLS

- Performing and recording
- Analyzing and interpreting
- Communicating results

## Musical Pipes

You can make musical instruments out of PVC™ pipe by cutting it into specific lengths. Make several pipes and determine the pitch of each by comparing the sound to a piano or other tuned instrument. Label the pipes with their musical notes. Form small groups and work out some tunes.

Try this:

**F F C C D D C**
**B♭ B♭ A A G G F**

### Analyze and Conclude

1. How is the resonant frequency of each pipe related to its length?
2. Sketch the standing waves in a pipe. Label nodes and antinodes.

## Seeing Speech

Cut or melt a hole in the bottom of a plastic cup. Cut out a piece of balloon or cellophane that is larger than the top of the cup. Stretch the piece of balloon over the mouth of the cup and secure it with an elastic band. Observe the stretched balloon while your lab partner speaks into the hole in the bottom of the cup. When you speak into the hole, touch the balloon lightly to see how sound "feels."

### Analyze and Conclude

1. Describe and explain your observations of the balloon.

2. Would this device work if a balloon is attached to both ends of the cup? Explain.

3. Would the device in question 2 work if the space between the balloons is a vacuum? Explain.

## Motion and Sound

Strike a tuning fork and then rest it gently against a suspended pith ball. Observe the effect the tuning fork has on the pith ball. Hold your palm tightly against your ear and press the base of a struck tuning fork on your elbow.

### Analyze and Conclude

1. Describe and explain the motion of the pith ball.

2. Describe and explain the results of holding a sounded tuning fork against your elbow. What general statements can you make about sound waves based on this activity?

- Describe the perception mechanisms that allow the ear to distinguish between different frequency sounds.

- Analyze and interpret energy transformations in the human ear.

KEY TERMS

- outer ear
- middle ear
- inner ear
- place theory of hearing
- temporal theory of hearing

- conductive hearing loss
- sensorineural hearing loss
- vocal cords
- articulators
- resonators

The human body is one of the finest examples of physics at work that you will ever discover. In your study of hearing, what you might expect to be a simple process, you will encounter compressional waves in air, resonance in pipes, mechanical vibrations, compressional waves in fluids, resonance in a membrane, and electrical signals.

## The Human Ear

You have already learned that the human ear can detect sound waves with frequencies between 20 Hz and 20 000 Hz. As well, you learned that the ear can detect sound intensities as low as a few picowatts per square metre and may withstand sound intensities as high as ten billion picowatts per square metre. The next important question to ask is, "How does the ear accomplish these tremendous feats?" In this section, you will look at the structure and function the ear as an instrument of sound detection and interpretation.

Scientists divide the ear into three parts, the outer, middle, and inner ear. The **outer ear** captures and guides sound waves in air. The **middle ear** converts sound waves into mechanical vibrations of three tiny bones. The **inner ear** converts these vibrations into sound waves in a fluid. The fluid motion stimulates specialized cells in the inner ear, which communicate with neurons (nerve cells) that send coded messages to the brain. Parts of the inner ear also sense changes in motion and the direction of the gravitational force. These functions, however, do not involve waves or sound so you will not examine the structures.

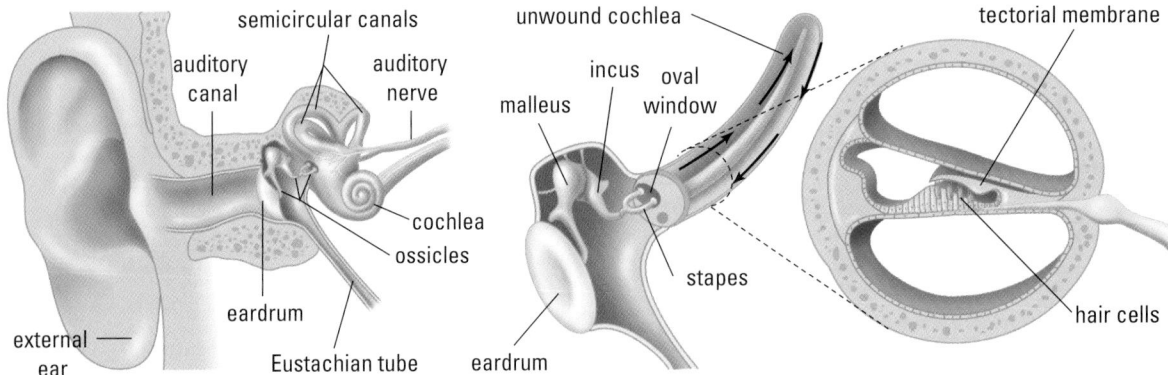

**Figure 9.1** The outer, middle, and inner ear.

# The Outer Ear

The external ear and the auditory canal, shown in Figure 9.1, constitute the outer ear. The external ear captures sound waves and funnels them down the auditory canal. As you might expect from its shape, the auditory canal functions as a closed-end, resonating air column. Its resonant frequencies vary from approximately 2.5 kHz to 4.5 kHz, corresponding to the range of frequencies to which the human ear is most sensitive.

# The Middle Ear

The middle ear begins with the eardrum or tympanic membrane, a cone-shaped membrane that stretches across the end of the auditory canal. Sound waves arriving at the eardrum carry slight changes in pressure, causing the eardrum to vibrate at the frequency of the sound waves (Figure 9.2). Since the eardrum must respond to very small variations in pressure, it is critical that, in the absence of sound, the pressure is the same on the two sides of the delicate membrane. This "resting" condition is maintained by the eustachian tube by connecting the air filled cavity of the middle ear to the outside air through the throat and mouth. If the outside pressure changes significantly and puts pressure on the eardrum, the eustachian tube might open suddenly, causing the familiar "popping" of the ear.

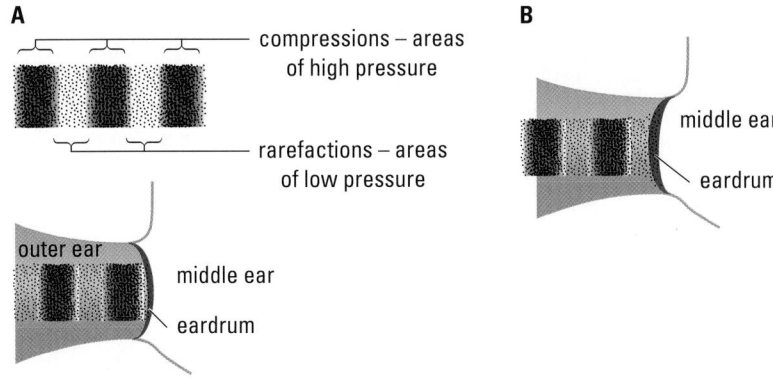

**Figure 9.2** **(A)** Compressions cause the eardrum to bend inward. **(B)** Rarefactions cause the eardrum to bend outward because their pressure is lower than that of the middle ear.

Three small bones, called ossicles, form a chain that transmits and amplifies the vibrations of the eardrum across the middle ear to the oval window of the inner ear. The first bone, the malleus, attaches to the eardrum from one edge to the centre, causing the eardrum to assume its cone shape. The malleus vibrates with the eardrum and transmits the vibrations on to the incus, which then pushes and pulls the stapes. The stapes passes the amplified

**PHYSICS FILE**

Biophysicist John van Opstal, who works at the University of Nijmegen in the Netherlands, conducted a six-week-long study into the function of the external ear. He and his students wore plastic inserts in their ears continuously for six weeks. The inserts effectively changed the shape of their sound-capturing external ears. Our brains compare the relative loudness and arrival time of sounds reaching each ear in order to locate the sound source in the horizontal plane. Dr. Opstal found that vertical source location was related to the shape of our external ear's nooks and crannies and was learned by our brain over time. Students wearing the plastic inserts were unable to determine if a sound was originating above or below them. During the study, each person began to be able to detect vertical location, until finally they had all learned to do so with their "new" ears. Interestingly, when they removed the plastic inserts, they were still able to detect vertical changes in a source's position, demonstrating that our brains are able to store more than one acoustic imprint.

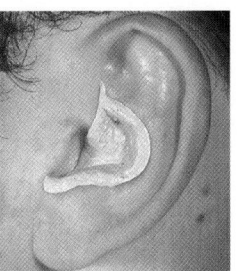

motion to the oval window, part of the inner ear. You may have heard of the ossicles by their common names, hammer, anvil, and stirrup, which describe the shapes of these tiny bones.

Muscles in the middle ear respond to loud sounds by putting tension on the stapes, thus preventing it from causing excessively large vibrations of the oval window. Researchers have theorized that in some young children the stapes actually loses contact with the oval window during periods of excessive noise, thus protecting the child's hearing.

## The Inner Ear

The inner ear is a small bony structure containing membranous, fluid-filled tubes and cavities. The cochlea is the snail-shaped segment of the inner ear that functions in hearing. Figure 9.3 shows the cochlea as though it were unwound. In the figure you can see the continuous channel of fluid that vibrates back and forth with the vibrations of the oval window. At the far end of the bony channel is the round window with its flexible membrane that relieves the pressure of the moving fluid.

The cross section of the channels, also shown in Figure 9.3, shows yet another, smaller channel. The sensory cells, called hair cells, sit on the basilar membrane of this channel. Vibrations of the fluid in the outer channel initiate vibrations of the basilar membrane. The tiny protrusions on the hair cells are imbedded in a gelatinous structure called the tectoral membrane. When the basilar membrane vibrates up and down, the tectorial membrane pulls and pushes the "hairs" from side to side. The hair cells are not themselves neurons (nerve cells) but they respond to the motion of the "hairs" by releasing chemicals that stimulate neurons. The neurons then carry electrical impulses to the brain where the impulses are interpreted as sound.

**Figure 9.3** If you could unwind the cochlea, you would see two outer channels that connect at the apex (tip) of the cochlea. Vibration of the stapes starts compressional waves in the fluid that initiate vibrational waves in the flexible basilar membrane.

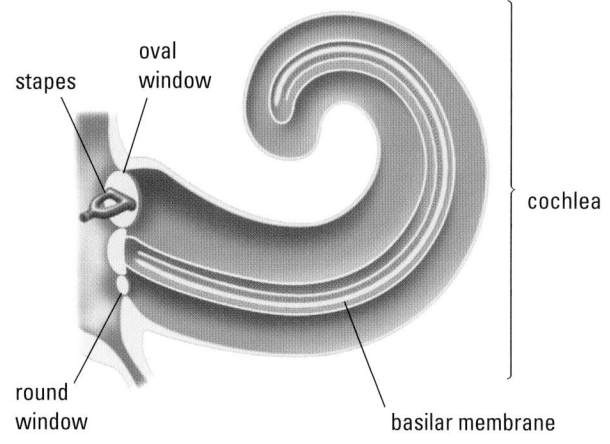

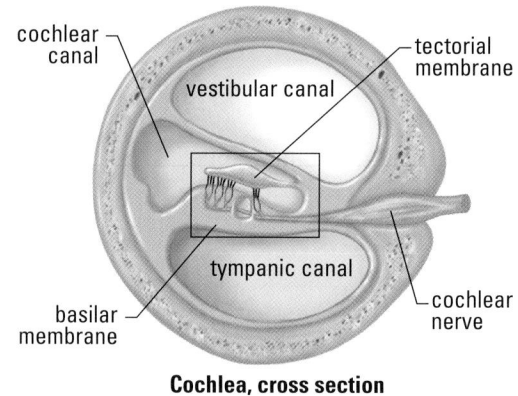

**Cochlea, cross section**

All hair cells are fundamentally the same. How, then, does the ear distinguish so precisely between such a wide range of frequencies of sound, and send information to the brain that can be interpreted as pitch? Scientists have been studying this problem for hundreds of years and have provided several models that can be divided into two general theories, the place theory and the temporal theory.

The term "place" in the **place theory of hearing** refers to the position along the basilar membrane that resonates in response to a given frequency of vibration. The shape of the basilar membrane changes gradually along the cochlea. Near the oval window, it is narrow and stiff but becomes wider and more flexible as it approaches the apex or tip of the winding cochlea. The narrow, less flexible end of the basilar membrane resonates at high frequencies much like a taut string. As the frequency decreases, portions of the basilar membrane further from the oval window resonate. Figure 9.4 shows the positions along the cochlea that resonate at various frequencies of the compressional waves in the fluid. You can see an example of larger vibrations in one region of the membrane than in the rest of the membrane.

The **temporal theory of hearing** proposes that the brain receives information about sound waves in the form of impulses arriving at the brain with the same frequency as the sound wave. A single neuron is not capable of firing at any but the very lowest frequencies of sound waves that the ear detects. Thus the theory proposes that a group of nerve fibres works as a unit to send impulses to the brain with the same frequency as the incoming sound wave.

The place theory is firmly established by experimental observations but the temporal theory cannot be ruled out. It quite possibly plays a role in coding messages for the brain to interpret. As researchers learn more about the details of the coding and decoding of sound messages to the brain, scientists and physicians will be better able to develop technologies to assist the hearing-impaired.

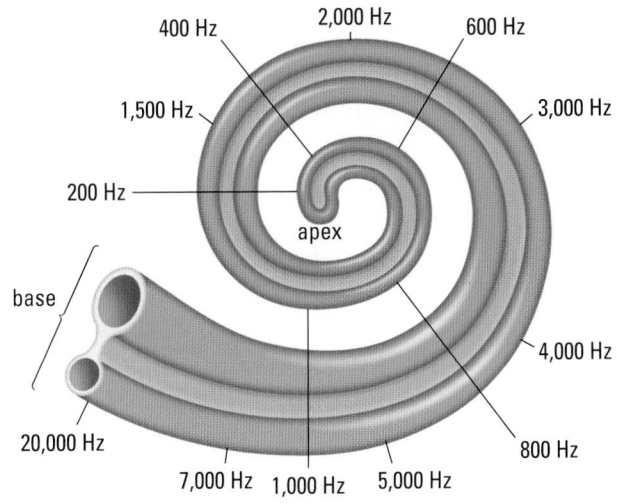

**Figure 9.4** When a wave travels down the basilar membrane, the part of the membrane with a natural frequency similar to the travelling wave vibrates at a much greater amplitude than the rest of the membrane.

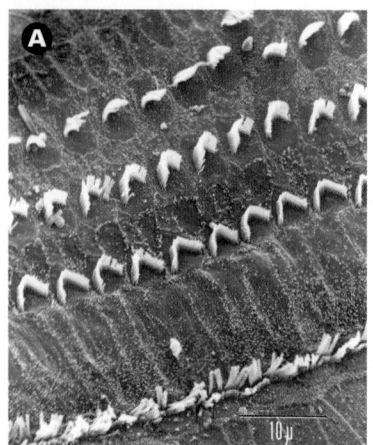

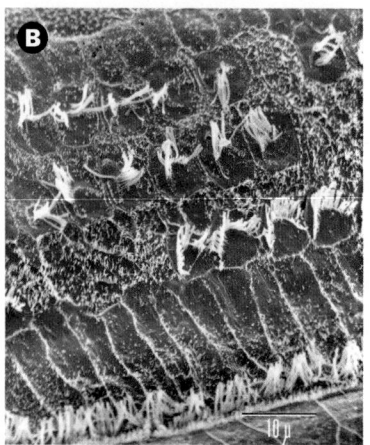

**Figure 9.5** The hair cells in **(A)** are from a healthy inner ear of a guinea pig. The hair cells in **(B)** are from a guinea pig that had been exposed to 120 dB sound, very similar to sound at a loud concert.

# Hearing Impairment

Hearing loss is divided into two categories based on the structures in the ear that are responsible for the impairment. **Conductive hearing loss**, as the name suggests, results from improper conduction of sound waves through the ear passages. The simplest form of conductive hearing loss is the accumulation of earwax in the auditory canal. Ear wax, a sticky substance secreted in the ear canal, serves several purposes such as catching dirt particles, preventing germs from reaching the eardrum, and, due to its unpleasant odour, repelling insects. If, however, an excess of earwax accumulates, it will reduce the amount of sound energy reaching the eardrum. The problem is easily corrected by removal of the wax by a physician.

Middle ear infections also cause conductive hearing loss. Infections often cause an accumulation of liquid in the normally air-filled cavity around the tiny ossicles. The amount of energy required for the bones to vibrate is much greater in a liquid than in air, because liquid friction is much greater than air friction. Hearing is usually restored when the infection subsides. Persistent infections, however, can cause permanent damage. For example, frequent or long-term infections might stimulate the growth of a type of skin over the eardrum that makes it harder and less flexible, making it difficult to hear anything but loud sounds. Other types of growths may also form as a result of long term infections.

The second category of impairment, **sensorineural hearing loss**, includes damage to the sensory cells (hair cells) or to the nerves that conduct the electrical impulses to the brain. The two most common causes of sensorineural hearing loss are aging and long-term exposure to loud sounds. Roughly one third of the population over age 65 experiences some loss of hearing, especially in the high frequency range, the frequencies detected by the part of the basilar membrane nearest the oval window. Most people in this category retain their sensitivity to mid-range and low frequencies and thus can still hear normal speech. The loss of high frequency sound often makes it difficult to distinguish certain words, as though the speaker was mumbling.

Prolonged exposure to sounds over 85 dB can permanently damage the sensory hair cells. Figure 9.5 shows the damage to guinea pig hair cells after the animal was exposed to sounds equivalent to a loud concert. Table 9.1 is a guide to the length of time the ear might endure a sound level before permanent damage is likely to occur.

**Table 9.1** General Exposure Guidelines

| Exposure time (h) | Sound level (dB) |
| --- | --- |
| 8 | 90 |
| 4 | 95 |
| 2 | 100 |
| 1 | 105 |
| 1/2 | 110 |
| <1/4 | 115 |

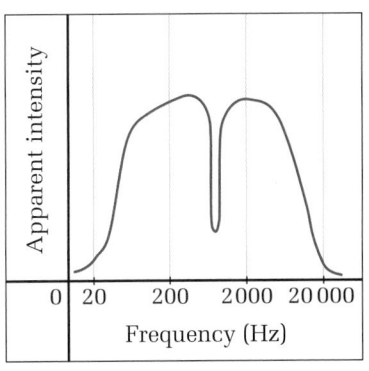

**Figure 9.6** In a test, a person was exposed to sounds at the same intensity level through the entire range of frequencies. The apparent intensity of sound that the person heard shows that he or she had lost the ability to hear a very narrow range of frequencies.

### Think It Through

- Notice, in Table 9.1, that the exposure time is cut in half when the sound goes up by five decibels. Based on your knowledge of the decibel scale, explain why a small increase on the decibel scale results in a much greater danger of suffering permanent damage to the sensory cells in the ear.

Exposure to sounds of a specific, small range of frequecies over a long period of time, can permanently damage just those sensory cells that respond to that particular frequency. For example, a person might operate a piece of machinery that emits a sound with a frequency around 1000 Hz. The individual would eventually lose the ability to hear those frequencies, as illustrated in Figure 9.6.

## Sound Technologies

Hearing aid technology has seen dramatic advances in recent decades. Digital hearing aids, such as the one in Figure 9.7, that the wearer can adjust by remote control are now available. The wearer can turn them up or down, tune out background noise, and even set the amplification for specific frequencies. Such hearing aids are often the best solution to partial hearing loss caused by damage or degeneration of sensory cells.

Total hearing loss can sometimes be treated with cochlear implants. The "implant" consists of an array of tiny wire electrodes that are threaded directly into the cochlea. The electrodes are connected to a receiver-stimulator that is implanted in the bone just behind the ear. An external microphone receives sound waves and a processor converts the sound into electrical signals. A transmitter, attached just behind the ear, sends signals through the skin to the receiver that then stimulates the auditory nerve in a combination of positions along the cochlea. The implant recipient does not hear sounds in the same way that a hearing person does.

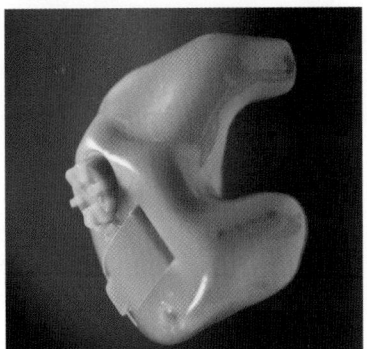

**Figure 9.7** A hearing aid is composed of three distinct components: a microphone, an amplifier, and a speaker. Improvements in technology allow for higher quality components of smaller and smaller sizes. The amplifier often contains a programmable computer chip.

Nevertheless, with practice, they are able to recognize words and phrases — in some cases, for the first time in their lives. As the technology improves, cochlear implants may dramatically increase the quality of life for individuals with complete hearing loss.

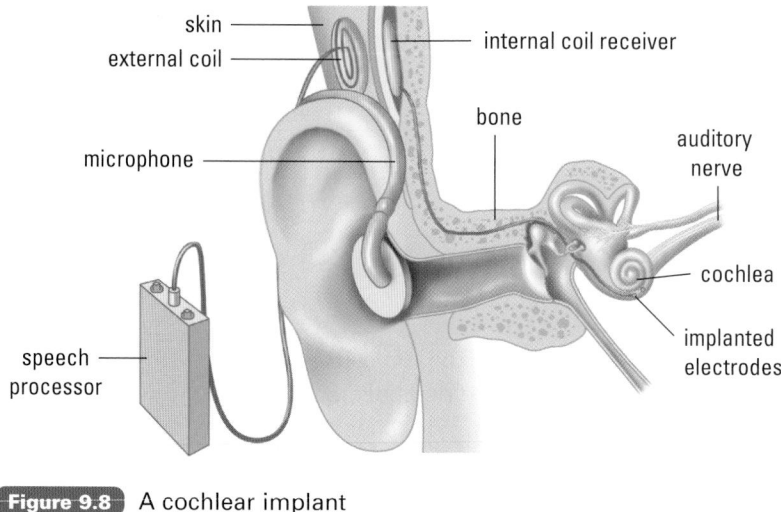

**Figure 9.8** A cochlear implant

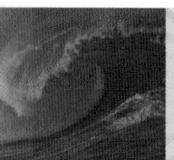

## QUICK LAB

# Test Your Hearing

**TARGET SKILLS**
- **Analyzing and interpreting**
- **Communicating results**

Read and answer each of the following questions:

- Do you have trouble hearing over the telephone?

- Do you have trouble following conversation when two or more people are talking at the same time?

- Do you have to strain to understand conversations?

- Do others complain that you turn the TV volume too loud?

- Do you miss hearing common sounds like the phone or doorbell ringing?

- Do you get confused about the direction from which a sound comes?

- Do you misunderstand some words in a sentence and need to ask people to repeat themselves?

- Do you especially have trouble understanding the speech of small children or people with high voices?

- Do you have a history of working in noisy environments?

- Do you experience ringing or buzzing sounds in your ears?

### Analyze and Conclude

If you answered yes to any of the questions, you may have hearing loss.

Explain the reason why F, S, P, and TH sounds are commonly the first sounds that people who develop hearing loss find difficult to hear.

# The Human Voice

When you blow on PVC pipe, your lips vibrate and create sounds having a wide range of frequencies. The length of the bottle or pipe cause certain frequencies to resonate and those are the sounds that you hear. Your voice works in much the same way. Your **vocal cords** create the initial vibrations and your throat and oral cavity act somewhat like a resonating closed-end pipe, with your vocal cords at the closed end and your mouth at the open end. In fact, your vocal cords are never completely closed, but their small size makes a pipe closed at one end an acceptable model of your vocal tract.

Your vocal cords are two thin folds of muscle and elastic tissue that relax when you are not speaking. To speak, your muscles stretch your vocal cords. Your diaphragm (a wall of muscle underneath your lungs that acts like a bellows) creates air pressure behind your vocal cords. A steady pressure from your diaphragm causes your vocal cords to vibrate. Classical singers know how to concentrate their energy in the diaphragm, so the air passes effortlessly over the vocal cords. Rock singers usually over-tense their vocal cords, which creates a dramatic sound but can cause vocal injuries such as nodules (small growths on the vocal cords that must be surgically removed). Greater tension in the vocal cords and higher air pressure create vibrations at higher frequencies.

**Biology Link**

Healthy vocal cords are smooth and moist. Laryngitis is an inflammation and swelling of the vocal cords. What causes your vocal cords to become inflamed? How can you prevent this from occurring? How should you treat inflamed vocal cords?

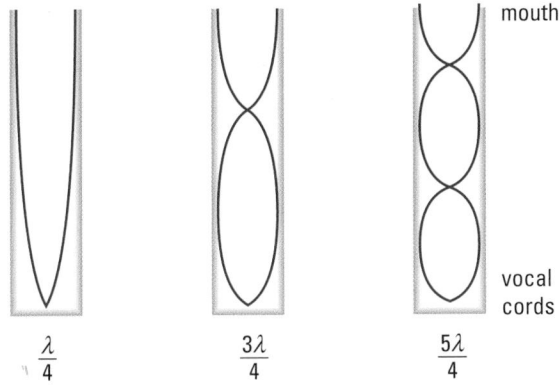

$$\frac{\lambda}{4} \qquad \frac{3\lambda}{4} \qquad \frac{5\lambda}{4}$$

**Figure 9.9** Sound is manipulated by changing the resonance of your vocal tract. The vocal tract behaves somewhat like a closed-end resonator.

The typical adult vocal tract functions as a tube approximately 17 cm to 18 cm long. As shown in Figure 9.9, resonant frequencies occur at $\frac{\lambda}{4}$, $\frac{3\lambda}{4}$, and $\frac{5\lambda}{4}$. For a 17 cm tube, these frequencies are 500 Hz, 1500 Hz, and 2500 Hz. Most sounds in normal human speech occur between 300 Hz and 3000 Hz. Thus the first three harmonics of the vocal tract are the most important.

## TRY THIS...

Try making the sounds in Table 9.2 and see if you can "feel" how the parts of your oral cavity are functioning in the production of these sounds. Compare the following combinations of sounds and decide how they are alike and how they differ. What is the fundamental difference between the pairs of sounds. (a) *k* and *g*, (b) *sh* and *ch*, (c) *s* and *z*, (d) *t* and *d*.

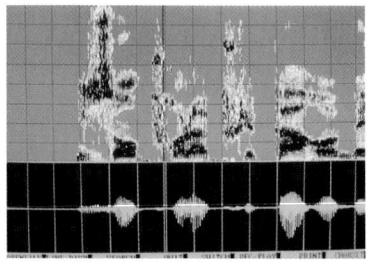

**Figure 9.10** This is an actual voice print. Compare it with the simulated voice prints in Figure 9.11. Do you think you could determine the sound a person was making from their voice print alone?

Human speech is far more complex than simply producing one frequency followed by another. Your brain has learned how to direct your vocal tract to create the extremely large variety of sounds required by language. One technique is to modify sounds by changing the shape and position of your **articulators**, which are your lips, tongue, and teeth. Table 9.2 shows which articulators you use to make some very common consonant sounds. To produce different vowel sounds, you also make use of your **resonators**: the elements of your vocal tract, which include your mouth (oral and laryngial pharynx) and your nasal cavities.

**Table 9.2** Sounds and Related Articulators and Resonators

| Sound | Articulator/Resonator |
|---|---|
| p, b, m, w | lips |
| t, d, th | tongue and teeth |
| s, z, sh, ch | tip of tongue |
| n | tongue |
| k, g | middle or back of tongue |

Although everyone must create sounds in much the same way, no two voices sound identical. Every person's vocal tract is unique and creates a specific combination of frequencies that can be identified by a "voice print" like the one shown in Figure 9.10. You create particular sounds by combining a specific set of frequencies. A few examples are shown in Figure 9.11.

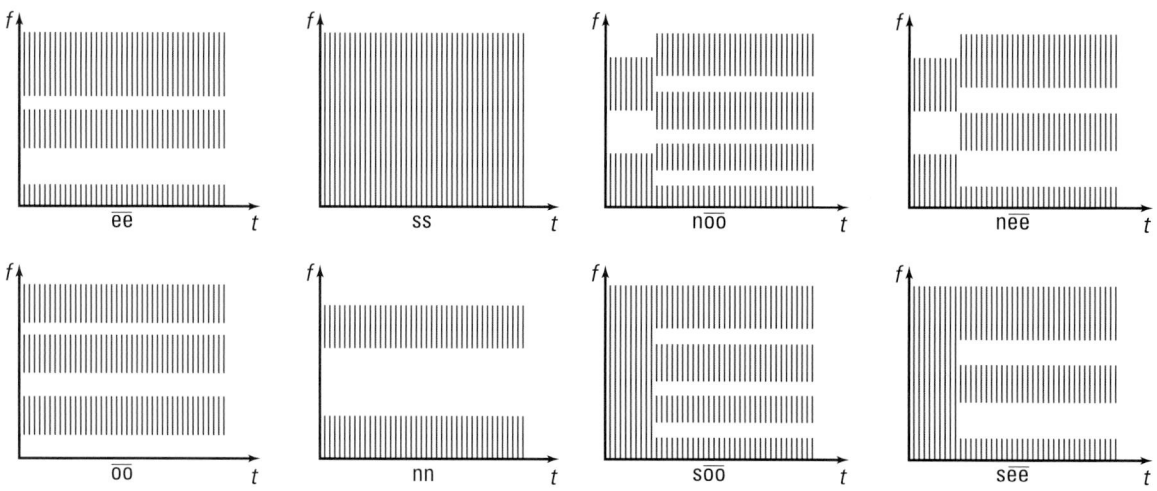

**Figure 9.11** Idealized voice prints on frequency versus time axes. Try making each of the sounds represented here. Notice which part of your vocal tract you adjust and how you modify it to make each sound.

1. **K/U** State the range of hearing of an average human being.

2. **K/U** Trace the energy path from a sound wave reaching your ear to the signal in your brain.

3. **C** Describe how the ossicles function to affect the way we hear very soft sounds and very loud sounds.

4. **C** Describe the process that converts mechanical motion into electrical signals in the cochlea.

5. **K/U** Summarize (a) the place theory and (b) the temporal theory of hearing.

6. **K/U** What range of frequencies can an average human generate?

7. **C** A friend tells you that a dog whistle does not actually make a sound, but rather that the dog is reacting to your breathing change when you blow into the whistle. Describe how you would explain the physics behind a dog whistle to your friend.

8. **C** Several animal species rely on their hearing as a means of survival. Often, these animals have very large, visible external ears. Is there an advantage for animals to have large external ears? Explain.

9. **C** Explain why it is difficult to talk while breathing in.

10. **K/U** List and describe the function of human vocal articulators and resonators.

11. **MC** Describe each of the two main categories of hearing loss.

12. **MC** How does the temporary hearing loss associated with a middle ear infection illustrate conservation of energy?

13. **MC** How will the type of hearing loss associated with long-term exposure to a specific frequency sound over 85 dB most likely affect someone as they age?

14. **C** As people age, some hearing loss is inevitable. Explain how a digital hearing aid is able to improve hearing by doing more than simply amplifying the sound.

15. **C** Several friends talking on a front porch hear the screeching of tires, and all turn immediately to face the same direction. Explain how is it possible that each person was able to know in which direction to look.

### UNIT ISSUE PREP

Your hearing system is both incredibly dynamic, able to sense varied intensities and frequencies, and very sensitive and susceptible to damage.

- Identify specific mechanisms within your ear that are most susceptible to damage.
- Investigate types of sound energy exposure that can result in damage.

KEY
TERMS

- Doppler effect

- sonic boom

- Mach number

- echolocation

- sonar

- ultrasound

**ELECTRONIC LEARNING PARTNER**

Learn more about the Doppler effect by going to your Electronic Learning Partner.

Submarines navigate through the murky depths of the oceans by sending and receiving sound waves. Whales and bats do the same, but with considerably more precision. For people, the ability to develop sound technologies as tools for navigation requires an understanding of the effect of motion on sound waves. If the source of the sound is moving, how does that affect the perception of the observer? Does motion of the observer influence the perception of sound?

## The Doppler Effect

Have you ever noticed how the sound of an emergency vehicle's siren seems to increase in pitch as it approaches? Then, just as the vehicle passes, the pitch suddenly appears to drop. The phenomenon responsible for the apparent change in pitch is called the **Doppler effect**. The siren of an emergency vehicle generates exactly the same sound at all times. The apparent changes in pitch as the vehicle approaches, then recedes, results from the motion of the vehicle as the source of the sound.

When the source of a sound is stationary, the sound will radiate outward in the shape of a three-dimensional sphere. This effect is shown in two dimensions in Figure 9.12 A. When the source of a sound is moving relative to the observer, the wave fronts appear as shown in Figure 9.12 B. During the time between successive compressions, which initiate a wave front, the source has moved toward the observer. Therefore, each new compression is nearer to the front of the previously created compression. This motion reduces the effective wavelength of the sound wave and the frequency appears greater. The same line of reasoning explains why the pitch seems to drop as the source of the sound moves away from an observer.

**Figure 9.12** The Doppler effect

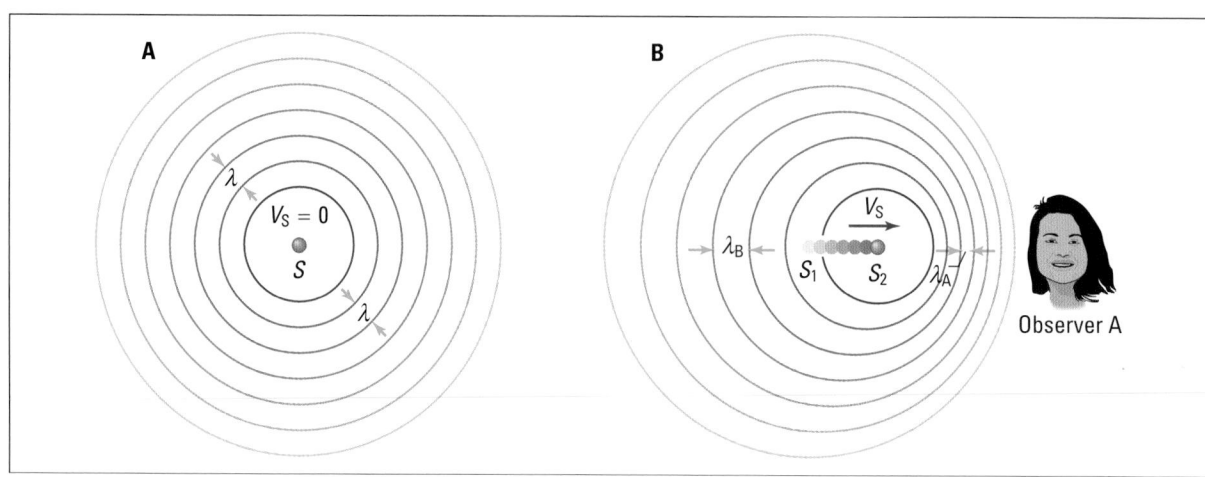

## Think It Through

- Draw a diagram similar to Figure 9.12 B, that demonstrates the Doppler effect as (a) a source moves away from your stationary position and (b) you move toward a stationary source.

- A friend puts a battery and a siren from a toy into a Nerf™ ball. She connects the battery and tosses you the ball with the siren wailing. Describe what you will hear and what she will hear as the ball moves through the air.

- Why does the wavelength of a sound generated from a moving source *decrease* as the speed that the source moves toward you increases? Why (in terms of the wave equation) does the frequency *increase*?

- Draw a diagram that illustrates sound waves surrounding a source that is moving at the speed of sound.

**TRY THIS...**

Tie a tuning fork securely to a string, strike it, and then let it swing like a pendulum. How does the pitch change as the tuning fork swings toward you and away from you? Explain why. Try it again but stand in front of the swinging tuning fork (so that you are facing it as if it was a grandfather clock). Does the pitch change? Why or why not?

## Sonic Booms

An extreme case of the Doppler effect occurs when an object travels at or beyond the speed of sound as shown in Figure 9.13. Figure 9.14 demonstrates compression waves generated by a source moving at various speeds.

If you have ever put your hand out of the window of a moving car, you will know that the pressure of the air on your hand is substantial. Your hand is colliding with air molecules and generating a longitudinal wave that moves out from your hand. Moving your hand through water provides a more visual example of the same phenomenon, except with transverse waves.

Imagine that you are able to keep holding your hand outside of the car as it accelerates toward the speed of sound. The pressure on your hand would become immense, but nothing compared to the pressure that you would feel as you reach and exceed the

**PROBEWARE**

The **Science Resources** section on the following web site: **www.school.mcgrawhill.ca/ resources/** has an excellent laboratory activity on the Doppler effect using probeware equipment. Navigate to the investigation by following the **Physics 11** links.

**Figure 9.13** The shock wave as a jet breaks the sound barrier is visible because the increased pressure causes water vapour to condense.

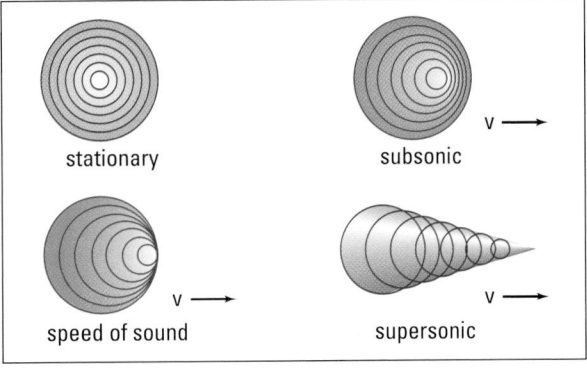

**Figure 9.14** Sound waves propagating outward from a source moving at various speeds

**Figure 9.15** The large bow wave is the result of the boat moving faster than the speed of the waves in water.

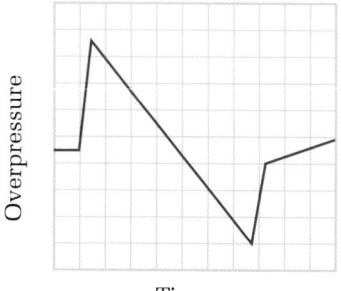

**Figure 9.16** Pressure, measured on the ground, versus time as an F-18 fighter jet flies overhead at an altitude of 9000 m and a speed of Mach 1.20. The "N-wave" change in pressure is characteristic of sonic booms.

speed of sound. If you are travelling slower than the speed of sound, although the pressure is great because of the increased number of air molecule collisions that you are experiencing, the wave fronts are still moving away from your hand. At the speed of sound, you are moving with the same speed as the wave fronts. Now the compressive wave fronts that you generate with your hand cannot move away from your hand. Each successive wave front combines with the ones made before it, creating a massive compression. This area is called the overpressure.

The photo of the large bow wave in front of the boat moving through the water in Figure 9.15 demonstrates how the wave fronts build up when the source is travelling faster than the waves can move in the medium. Likewise, behind your hand a massive rarefaction, or area of extremely low pressure, exists. This rapid change in air pressure, from very high to very low, is called a shock wave and is heard as a **sonic boom**. Figure 9.16 shows what a typical sonic boom pressure profile looks like.

Examine the wavefronts for supersonic motion in Figure 9.14. Notice that many wavefronts converge along a V-shaped path behind the source. Of course, in three dimensions, it forms a cone. This superposition of compressional wavefronts creates extremely large pressure changes as the cone moves. You hear the results as a sonic boom. Figure 9.17 shows you how sonic boom trails behind the source that is moving faster than the speed of sound. Supersonic jets usually fly at very high altitudes because although the sonic boom cone then covers more area on the ground, it is significantly weaker. The sonic boom cone sweeps over a path on the ground that is approximately 1 km wide for every 250 m of altitude. Therefore, a jet flying faster than the speed of sound at an altitude of 10 000 m will create a sonic boom across a 40 km wide path of land beneath the jet. The higher a jet flies, the wider the path of the sonic boom, but the lower the energy level on the ground.

A certain jet flying at an altitude of 9000 m generates a 128 dB sonic boom on the ground. The same jet, flying at an altitude of 30 m would generate a 263 dB sonic boom! The pressure change from typical atmospheric pressure to the maximum pressure of the compressional wave of a sonic boom occurs over a time interval of 100 ms for a fighter jet and in approximately 500 ms for the

shock wave ⎯

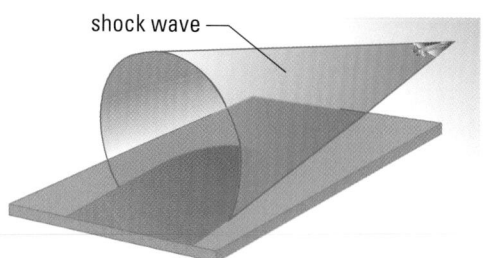

**Figure 9.17** The sonic boom trails behind the jet, growing larger but weaker as it expands.

space shuttle or the *Concorde*. Such a sudden, drastic pressure change can force windows to flex beyond their elastic limit, and shatter.

Since the speed of sound in air varies with temperature and pressure, it is not possible to classify specific speeds as subsonic, sonic, and supersonic. Austrian physicist Ernst Mach devised a method to describe these classes of speeds as ratios of the speed of the jet (or other object) to the speed of sound in air that has the temperature and pressure of the air in which the jet is flying. The ratio is now known as the **Mach number**. A mach number of less than one indicates that an object is moving slower than the speed of sound. Mach one means that it is flying at precisely the speed of sound and a Mach number greater than one indicates that the object is moving faster than the speed of sound.

---

## MACH NUMBER

The Mach number of a moving object is the ratio of its speed to the speed of sound in air at conditions identical to those in which the object is moving.

$$\text{Mach number} = \frac{v_{\text{object}}}{v_{\text{sound}}}$$

| Quantity | Symbol | SI Unit |
|---|---|---|
| speed of object | $v_{\text{object}}$ | m/s (metres/second) |
| speed of sound | $v_{\text{sound}}$ | m/s (metres/second) |
| Mach number | | pure number (no units) |

---

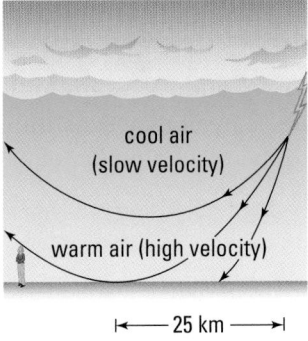

cool air
(slow velocity)

warm air (high velocity)

|←——— 25 km ———→|

---

### MODEL PROBLEM

## Calculating Mach Number

**Calculate the Mach number of a Canadian Forces jet flying through 15.0°C air at standard atmospheric pressure with a velocity of $2.20 \times 10^3$ km/h near Cold Lake, Alberta.**

---

### Frame the Problem

- *Mach number* is a *ratio* of the *speed of the jet* compared to the *speed of sound*.

- Both speed values must have the same units, so that they will cancel, leaving a Mach number without units.

- The *speed of sound* in air depends on the temperature.

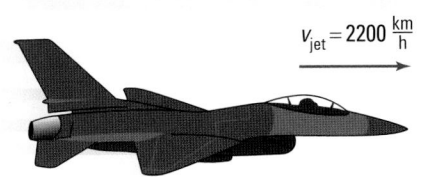

$v_{\text{jet}} = 2200 \frac{\text{km}}{\text{h}}$

$T_{\text{air}} = 15.0°C$

*continued* ▶

*continued from previous page*

## Identify the Goal

The speed of the plane in terms of the Mach number

## Variables and Constants

| Involved in the problem | Known | Unknown |
|---|---|---|
| $v_{jet}$ | $v_{jet} = 2.20 \times 10^3$ km/h | $v_{sound}$ |
| $v_{sound}$ | $T = 15.0°C$ | Mach number |
| $T$ | | |
| Mach number | | |

## Strategy

Convert $v_{jet}$ to SI units.

Calculate $v_{sound}$ using approximation given by equation (from Chapter 8, Section 2):

Both $v_{jet}$ and $v_{sound}$ are in the same units, so you may substitute them into the Mach number equation.

## Calculations

$$2.20 \times 10^3 \frac{\text{km}}{\text{h}} \times \frac{1\text{ h}}{3600\text{ s}} \times \frac{1000\text{ m}}{1\text{ km}} = 6.11 \times 10^2 \frac{\text{m}}{\text{s}}$$

$$v_{sound} = 331 + 0.59\,T$$

$$v_{sound} = 331\,\frac{\text{m}}{\text{s}} + 0.59\,\frac{\frac{\text{m}}{\text{s}}}{°C}(15.0\,°C)$$

$$v_{sound} = 339.85\,\frac{\text{m}}{\text{s}}$$

$$\text{Mach number} = \frac{v_{jet}}{v_{sound}}$$

$$\text{Mach number} = \frac{611.11\,\text{m/s}}{339.85\,\text{m/s}}$$

$$\text{Mach number} = 1.80$$

The jet is flying at Mach 1.80.

## Validate

A jet flying faster than the speed of sound will have a Mach number greater than 1.0, which is the case here.

## PRACTICE PROBLEMS

1. Calculate the Mach number of a bullet travelling at 385 m/s in air at standard conditions.

2. How fast is a jet flying if it is travelling at Mach 0.500 through 25.0° C air at standard atmospheric pressure?

3. What is the minimum speed with which the tip of a whip must travel in air at standard pressure and temperature to make a cracking sound?

# Echolocation

Sound can be used to measure distance. The slight time difference between a single sound reaching our two ears allows our brain to pinpoint the source. Some animals have a far more refined auditory system. Toothed whales or *odontocetes* and most species of bats are able to generate and interpret ultrasonic pulses that reflect off obstacles and prey. This process, called **echolocation,** allows animals to pinpoint not only the object's exact location, but also (using Doppler shifts) its speed and direction.

Dolphins, part of the toothed whale family, have very specialized vocal and auditory systems. They make whistle-type vocal sounds using their larynx, which does not contain vocal cords. The echolocation and navigation clicks are produced in their nasal sac region. Dolphins are able to generate sounds ranging in frequency from 250 Hz to 150 kHz. The lower range frequencies, from 250 Hz to 50 kHz, are thought to be used primarily for communication between dolphins, while the higher frequencies are used for echolocation.

Echolocation clicks produced by dolphins, each lasting from 50 ms to 128 ms, are grouped into "trains." A large, flexible, gelatinous outcropping on the front of the dolphin's skull, called a melon, focuses the clicks. The melon is filled mostly with fatty tissue and is easily shaped by the muscles to act as an acoustical lens for the high-pitched clicks. Research has shown that the high-pitched sounds do not travel as far in water as lower frequency noises, resulting in effective echolocation ranges from 5 to 200 m away for targets of 5 cm to 15 cm in length. Dolphins produce a wide range of sounds, varying in frequency, volume, wavelength, and pattern. They are able to identify size, shape, speed, distance, direction, and even some internal structure of the objects in the water. For example, a dolphin can detect flatfish lying beneath a layer of sand on the seabed.

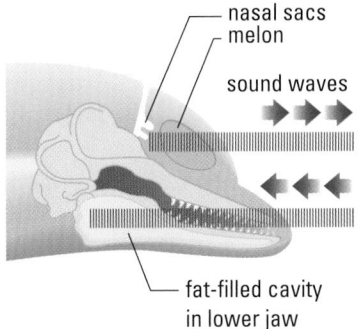

**Figure 9.18** The echolocation clicks are generated in the nasal sacs, focussed by the melon, and received by the fat-filled region in the lower jaw.

**Figure 9.19** Notice the slightly contorted shape of the melon as this dolphin focusses echolocation clicks. Tests have demonstrated that dolphins are able to distinguish between different-shaped objects contained within a closed box. Essentially, they are able to accomplish naturally what we do with ultrasounds to examine expectant mothers.

**Figure 9.20** Most species of bats are able to "see" using ultrasonic sounds, enabling them to catch their prey in the darkness of night.

## Determining Distance

A boat approaches a large cliff in a dense fog. The captain sounds his foghorn and hears an echo 2.4 s later. He assumes the speed of sound to be 343 m/s. Calculate the distance the boat is from the cliff.

### Frame the Problem

- Sketch the situation.

- The *sound* must *travel* to the cliff and then *back again*.

- The sound travels *twice the distance* the boat is from the cliff.

- Sound travels at a *constant speed*.

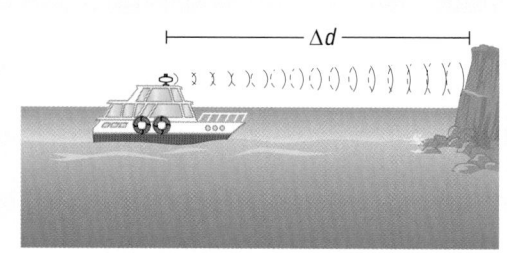

### Identify the Goal

The distance $\Delta d$, of the boat from the cliff

### Variables and Constants

| Involved in the problem | | Known | Unknown |
|---|---|---|---|
| $\Delta d$ | $\Delta t_{total}$ | $v_{sound} = 343$ m/s | $\Delta d$ |
| $v_{sound}$ | $\Delta t_{1/2}$ | $\Delta t_{total} = 2.4$ s | $\Delta t_{1/2}$ |

| Strategy | Calculations |
|---|---|

**Strategy**

Divide the total time in half to find the time required for the sound to reach the cliff.

**Calculations**

$$\Delta t_{1/2} = \frac{\Delta t_{total}}{2}$$

$$= \frac{2.4 \text{ s}}{2}$$

$$= 1.2 \text{ s}$$

Calculate the distance to the cliff using the speed of sound and the time for half the round trip.

$$\Delta d = v_{sound}\Delta t_{1/2}$$

$$\Delta d = \left(343 \ \frac{\text{m}}{\text{s}}\right)(1.2 \text{ s})$$

$$\Delta d = 412 \text{ m}$$

The boat is $4.1 \times 10^2$ m from the cliff.

### Validate

Using the round-trip time and then dividing the final distance in half could also complete this question. The trick to echo questions is remembering that the sound makes a round trip from the source to the reflecting barrier and back.

4. While vacationing in northern Ontario, an ingenious physics student decides to calculate the distance from her cottage to the cliff at the far edge of the lake. She finds that the time between her dog barking and the echo returning is 3.5 s. The temperature of the air is 26°C. How far away is the cliff?

5. The human ear is just able to distinguish between two sounds about 0.10 s apart. How far from a large wall must someone stand in order to just hear an echo? (Assume the speed of sound is 343 m/s.)

# CAREERS IN PHYSICS

## Sounds from the Seabed

The survey ship glides across the inky-blue ocean. Sensors on its hull sweep back and forth, sending sound waves to the seabed and receiving the sound waves that bounce back. This is multi-beam sonar (SOund Navigation And Ranging). The exact amount of sound energy bouncing back is determined by what is on the seabed. Harder materials send back more energy than softer ones. The sensors on the hull, called transducers, convert the sound waves they receive into electrical signals. These are processed by computer to produce three-dimensional, photographic-quality maps of what is on and beneath the seabed, perhaps an offshore oil pipeline, debris from a downed aircraft, or a sea cage used in fish farming. Viewers can even "fly through" the data to see the seabed in real time.

Scientists in the University of New Brunswick's Geodesy and Geomatics Engineering Department (GGE) are among those who collect, process, and interpret such data. Dr. Susan E. Nichols and her colleagues in the GGE Department apply their expertise to projects involving ocean governance. For example, countries sometimes disagree over offshore boundaries. Fish farmers sometimes conflict with people in traditional fisheries. Companies interested in developing offshore oil or gas need to know where to explore and where to lay pipelines. Federal, provincial, and municipal governments want to ensure that

petroleum development will not conflict with property rights or harm environmentally sensitive areas. In such cases, three-dimensional maps like those from the GGE are a valuable basis for discussing and resolving conflicts.

Besides ocean mapping, other uses for sonar include finding and sizing schools of fish, detecting submarines or icebergs, and determining whether there are valuable gravel, gold, or other mineral deposits in the ocean floor.

### Going Further

1. Research sonar in encyclopedias, Internet sites, and/or other sources. Prepare a labelled diagram showing how it works.

2. Find out what is involved in geodesy and geomatics. What are some career possibilities in these fields?

Exploring salmon and lobster farming sites in the Bay of Fundy. Professional geomatics engineer Dr. Susan E. Nichols, is second from the right. With her are (left to right) researchers Michael Sutherland, Rosa Tatasciore, and Sue Hanham.

# Locating a Low-Pitched Sound

Test your ability to locate a sound. Two funnels, some rubber tubing, and a T-connector are all you need. Cut one tube to 2.0 m and the other to 3.0 m long. Attach the tubes using the T-connector. Connect a funnel to the end of each tube. Hold the funnels to each ear and have a partner gently tap the open end of the T-connector. From which side of your head does the sound seem to be originating? Switch the funnels to the opposite ears and repeat the experiment.

### Analyze and Conclude

1. Did the sound seem to be originating from the funnel attached to the longer or shorter tube?

2. Explain how this demonstration provides insight into how the human hearing system locates the horizontal position of low-pitched sounds.

**Figure 9.21** Enhanced technology allows very detailed, colour images to be constructed from sonar data.

## SONAR

Sonar was developed during World War I (1914–1918) to detect German submarines. The technology has progressed dramatically and so have the applications. If you enjoy fishing, you may have used a fish-finder or depth meter that uses sonar. Figure 9.22 demonstrates the basic principle. In **sonar**, sound pulses are sent out and the reflected signal received. A computer is able to measure the time between the outgoing and incoming signals, which is then used to calculate an object's depth. The computer is also able to determine the relative size of an object based on the intensity of the reflected signal.

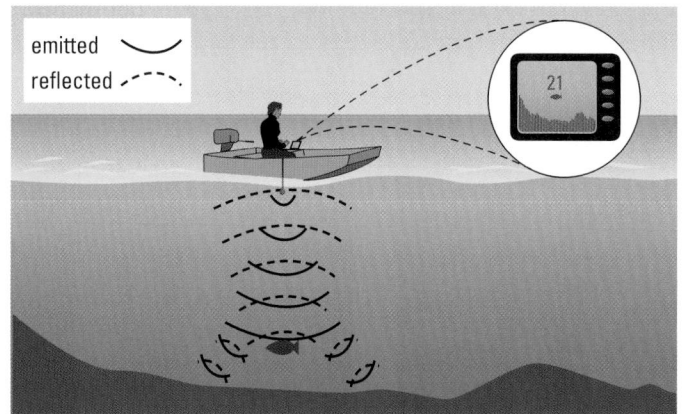

**Figure 9.22** Ultrasonic pulses are sent toward the lake bottom. They reflect off fish and off the bottom of the lake. A computer measures the time between the emitted and the received signal and calculates the object's depth. The intensity of the signals provides information about the size of the object.

Uses for sound technologies have grown as computer technology developed speeds required to handle more and more data in real time. Figure 9.23 shows some of the varied applications of **ultrasound** (frequencies above 200 kHz), audible sounds, and infrasound in use today.

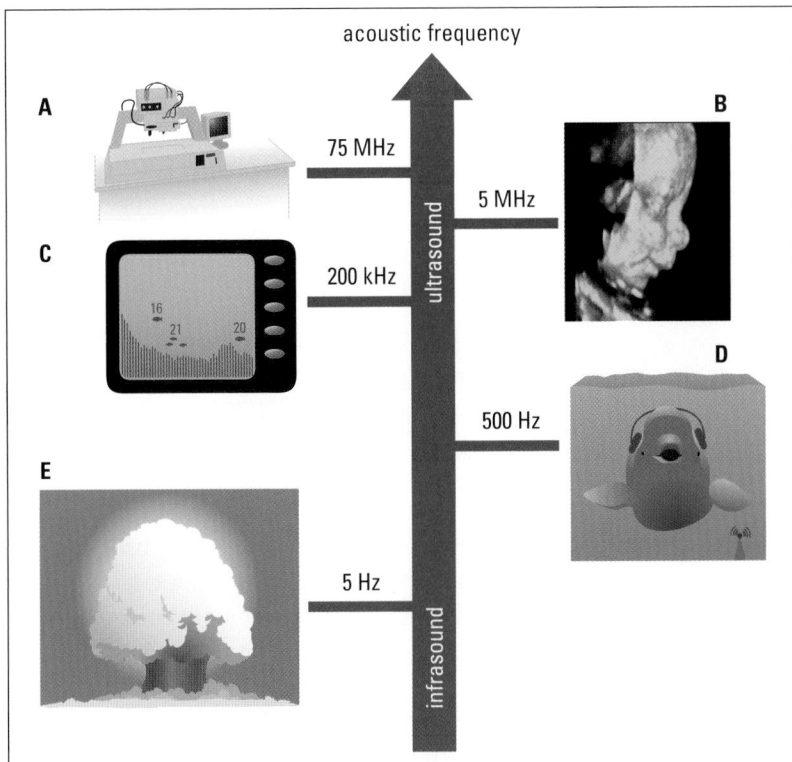

**Figure 9.23** **(A)** An acoustic microscope can detect defects in certain materials. **(B)** Doctors use three-dimensional ultrasound images during pregnancy. **(C)** Fish-finders locate fish. **(D)** Long-range military sonar may affect marine life. **(E)** Infrasonic vibration detectors warn of volcanoes, earthquakes, avalanches, or nuclear explosions.

## QUICK LAB

## "Sonar" in the Classroom

**TARGET SKILLS**
- Identifying variables
- Performing and recording

In a location where your partner cannot see, set up a "rocky ocean floor," with high ridges and deep valleys. Use boxes, books, basketballs, and anything else that will provide an interesting profile. Once your "sea floor" is complete, your partner will attempt to build a map of the underwater terrain using a motion sensor and computer interface. Work together to determine the best way to conduct the experiment, but remember that your partner needs to construct a terrain map without ever actually looking at the model set-up. Repeat the procedure with your partner building the model and you using the probeware to make the map.

**Analyze and Conclude**

1. How is the computer able to measure the distance from the "surface" to the "sea bottom"?

2. Describe the factors that you needed to control to get the best image possible.

**Apply and Extend**

3. How could you improve the mapping procedure to differentiate between large fish and slender, tall peaks rising from the sea floor or perhaps to make a two-dimensional map? If time permits, test your ideas.

1. **C** Explain the cause of the Doppler effect.

2. **K/U** Sketch a frequency versus position graph to describe the pitch of a train whistle as it approaches and then recedes from your position.

3. **MC** You are in a car travelling toward the base of a large cliff in an otherwise completely open area. As you approach the massive wall of rock, you sound the horn.

   **(a)** Will there be a change in frequency of the reflected sound?

   **(b)** Is the change due to a moving source or receiver? Explain.

4. **C** Explain why a sonic boom is capable of breaking windows.

5. **K/U** Which of the following photographs illustrate a situation or item that could produce a sonic boom?

6. **K/U** Explain why jets that fly at supersonic speeds should maintain very high altitudes?

7. **(a)** **K/U** What frequencies does a dolphin use to echolocate its food?

   **(b)** **MC** Give two reasons why the frequency from part (a) is a good choice for tracking fast moving prey.

8. **C** Explain how the human auditory system determines the direction from which a sound originated. Use diagrams to support your answer.

9. **K/U** How does sonar detect objects?

10. **K/U** What is ultrasound and how is it used?

11. **MC** Ultrasound can be produced by applying an alternating voltage across opposite faces of piezoelectric crystals such as quartz. The applied voltage causes the crystal to expand and contract extremely quickly with the same frequency as the alternating current. Certain frequencies will cause the crystals to resonate, producing very strong ultrasonic waves. Identify as many applications of this effect as you can.

Music has been an important part of human culture throughout history and even in prehistoric times. Some exquisite musical instruments date back thousands of years. Clearly, the technology for making fine instruments existed long before scientists discovered the principles of sound waves and resonance. Understanding the nature of the sounds that blend together to make music has not changed the importance of the music itself to culture and society. The knowledge has, however, made it possible to mimic the sounds of instruments electronically and also to record and reproduce music. Try to imagine the reaction of some great, historic musician such as Mozart if you played a CD for him for the first time.

Modern recording techniques are only the most recent innovations in the performance and enjoyment of music. Every instrument in the classical orchestra has evolved over the centuries, in reaction to new material technologies and the demands of composers such as Bach, Beethoven, and Mahler. More recently, rock instruments have benefited from the invention of multitrack recording, the wow pedal, and the electronic drum machine. Nevertheless, it is still possible to categorize all these instruments according to the way in which they produce their sound.

**SECTION EXPECTATIONS**

- Explain and illustrate the principle resonance in musical instruments.
- Define standing waves and harmonic structure.
- Investigate how notes are produced.

**KEY TERMS**

- brass instruments
- woodwind instruments
- stringed instruments
- consonance
- dissonance
- timbre
- harmonic structure
- percussion instruments

## Wind Instruments

Today there are two main categories of wind instruments — brass and woodwind. The fundamental difference between the categories is the source of the vibrations that produce initial sound. The musicians' lips provide the vibration for **brass instruments** such as the trumpet and trombone. Trumpet players tighten their lips then force air through them, creating vibrations having a wide range of frequencies. A reed that vibrates when air flows by it supplies the initial sounds for **woodwinds** such as the clarinet and saxophone.

**Figure 9.24** An instrument's mouthpiece produces sound, but the instrument's resonant cavity produces music.

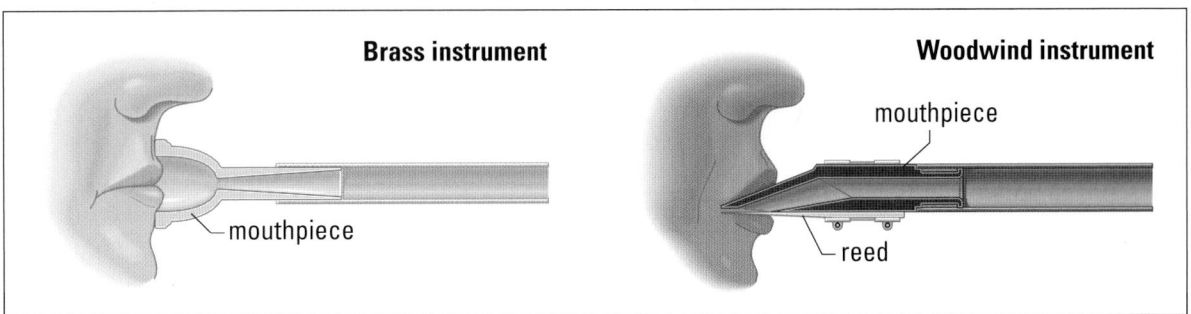

**Brass instrument**

mouthpiece

**Woodwind instrument**

mouthpiece

reed

**Figure 9.25** Each instrument produces a characteristic sound.

**Figure 9.26** A seashell acts as a resonator for selected background noise frequencies, producing a sound like distant ocean waves.

**PROBEWARE**

Visit the **Science Resources** section of the following web site if your school has probeware equipment: **www.school.mcgrawhill.ca/ resources/** and follow the **Physics 11** links for an interesting activity on the quality of sound.

The sounds from the mouthpieces alone are far from musical. Only when they are attached to the instruments are pleasant tones heard. Different notes are obtained in brass instruments by varying both the frequency of the initial sound waves produced by the lips and the length of the resonating column. Woodwind instruments change the effective length of the resonating column by covering and uncovering small finger holes along the instrument.

The mouthpiece also determines the type of resonances available. Brass instruments and flutes act as open-end resonating columns. Other woodwinds, like the oboe, clarinet, and basson, function as closed-end resonators, because they have reeds in the mouthpiece that behave like human vocal cords.

The length and shape of an instrument determine the range of musical notes that it can produce. Early instruments were created through trial and error. Musicians experimented with the length of the resonating column, the position of the finger holes, and the internal shape (cylindrical or conical) of an instrument in an attempt to create pleasant sounds. Table 9.3 provides a list of some instruments and their approximate musical ranges. The sequence of notes in a single octave is C D E F G A B C; subscripts represent higher or lower octaves. Composers must be aware of these ranges because it would be silly to write, say $D_2$, to be played on the violin since that note is below the instrument's range.

**Table 9.3** Selected Instruments and Their Musical Ranges

| Instrument | | Lowest note | Frequency (Hz) | Highest note | Frequency (Hz) |
|---|---|---|---|---|---|
| **woodwind** | soprano recorder in C | $C_5$ | 523.2 | $D_7$ | 2349.3 |
| | flute | $C_4$ | 261.6 | $D_7$ | 2349.3 |
| | soprano clarinet in A | $C_3$ | 130.8 | $A_6$ | 1760.0 |
| | baritone saxophone in $E^\flat$ | $D_2$ | 73.4 | $A_4$ | 440.0 |
| | oboe | $B_5$ | 987.8 | $G_6$ | 1568.0 |
| **brass** | C trumpet | $F_3^\sharp$ | 174.6 | $C_6$ | 1046.5 |
| | French horn in F | $B_1$ | 61.7 | $F_5$ | 698.5 |
| | trombone in $B^\flat$ | $E_2$ | 82.4 | $D_5$ | 587.3 |
| | tuba in $B^\flat$ | $B_0^\flat$ | 30.9 | $B_3^\flat$ | 246.9 |
| **string** | violin | $G_3$ | 196.0 | $E_7$ | 2637.0 |
| | double bass | $E_1$ | 41.2 | $B_3$ | 246.9 |
| | harp | $C_1$ | 32.7 | $G_7$ | 3136.0 |
| | guitar | $E_2$ | 82.4 | $A_5$ | 880.0 |
| **keyboard** | piano | $A_0$ | 27.5 | $C_8$ | 4186.0 |

## QUICK LAB  Singing Straws

**TARGET SKILLS**
- **Performing and recording**
- **Analyzing and interpreting**

Press one end of each of two straws flat and then cut them as shown in the diagram. Set one straw aside. Carefully cut small holes into the top of one straw. (Alternatively, a heated nail melts nice clean holes into the straw.) Blow into the straw end that has been cut. Cover and uncover the holes and notice how the pitch changes. Begin with all of the holes covered and then uncover them one at a time starting from the far end and moving toward your mouth.

Using the second straw, carefully cut small pieces off the far end of the straw as you blow through it. Observe the change in pitch.

A  B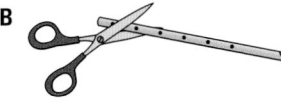

### Analyze and Conclude

1. Did covering and uncovering the holes in the straw affect the pitch of the sound produced?

2. Describe a relationship between the pitch and the distance from the flattened end to the first uncovered hole.

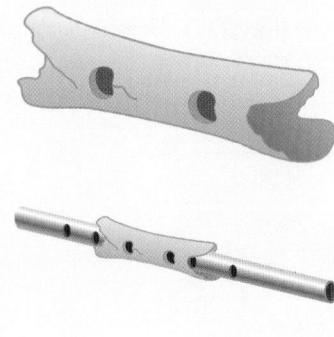

## The Neanderthal Flute

In 1995, in a cave occupied by Neanderthals between 40 000 and 80 000 years ago, paleontologist Dr. Ivan Turk discovered a broken section of the leg bone of a bear. Close examination revealed two holes drilled into the sides and two more at each of the broken ends. Dr. Turk realized that he had found part of a flute, the oldest known musical instrument.

Specialists in radiocarbon dating at Simon Fraser University in British Columbia determined that the flute, and pieces of charcoal found in the soil layers around it, were approximately 45 000 years old. This discovery was exciting because it suggested that Neanderthals, distant cousins of humans thought to have become extinct some 20 000 years ago, had a culture that included musical entertainment.

A short time later, the flute was examined by Dr. Bob Fink, a Saskatchewan musicologist. Dr. Fink carefully measured the spacing of the holes. Comparing the distances to the separation of his fingers, he proposed the length and structure of the original instrument. Then, making use of the physics of open air columns, Dr. Fink determined what notes the flute would have played. His findings created quite a stir: Fink's analysis suggested that the flute would have played a standard major or Do-Re-Mi scale familiar to any piano student or viewer of *The Sound of Music*. The major scale forms the basis of most music in the Western world.

Fink's claim is very interesting and quite controversial. The major scale is only one of many possible musical scales. Much Chinese music is based on a five-note or pentatonic scale. Gregorian chants composed in the Middle Ages are based on six-note scales called hexachords. Since the origin of the major scale is the strongest harmonics of the fundamental note, Fink argued that it is a natural scale. The fact that the Neanderthals used that scale, he said, showed its universality and their sophistication.

The discussion about the validity of Dr. Fink's claim continues. It is amazing how a couple of holes in an old bone can lead to so much good scientific and artistic discussion, and tell us so much about our ancestors and ourselves.

### Analysis

1. What does a musicologist do?

2. On what factors did Dr. Fink base his claim that Neanderthals used our familiar musical scale?

3. Find examples of speculation (as opposed to solid evidence) in the article.

# Stringed Instruments

Strumming stretched elastic bands will produce sound, but only someone very near to you will hear the music you are creating. All **stringed instruments** include strings that can either be plucked, strummed, or bowed, and some form of a resonator, or sound box. An acoustic guitar has a uniquely shaped body filled with air and a large central hole. A violin has a characteristic shape with two openings into the air-filled centre. A harp and a harpsichord have their strings tightly bound to the large frame or soundboard. You can demonstrate the process of forcing the vibration of another object to amplify sound by pressing the base of a vibrating tuning fork against your desk. The desktop will be made to oscillate with the same frequency as the tuning fork and, because the surface area of the desk is large, the amount of sound generated will be amplified.

Figure 9.27 illustrates the standing wave pattern that is set up in a vibrating guitar string. Pressing the string down onto the frets effectively reduces the length of the string and raises the note. The vibrations of the string alone can be heard only by those very close to the instrument. The vibrating string causes the air in the cavity of an acoustic guitar to resonate, providing the energy necessary to project the sound. An electric guitar uses a device called a pickup to transform the string's vibration energy into electric energy fed to an amplifier.

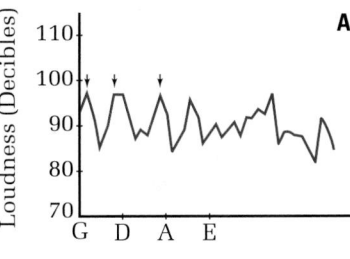

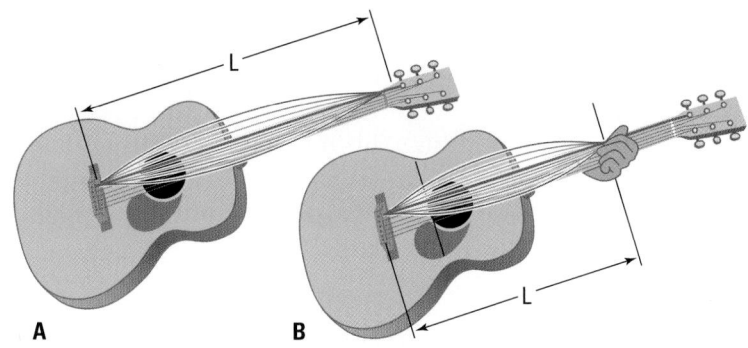

**Figure 9.27** The frequency of the longer wavelength is lower than that of the shortened string. Only the fundamental frequency is shown.

A well-designed and well-built violin will have a low and high resonance due to the wood and a third resonance due to the air in the cavity. All three resonant cavities should be of approximately equal loudness. Figure 9.28 illustrates two loudness curves (sound created by bowing equally up the scale), one from a good violin and one from a poor violin. Violins are amazing instruments, and each has its very own characteristic sound due to specific harmonics that result from both the material used and its design. In the late seventeenth century, an Italian craftsman named Stradivarius built beautiful-sounding violins, of which 650 are still played today.

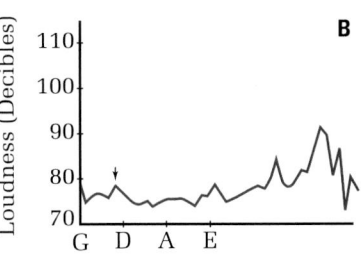

**Figure 9.28** These are loudness curves for **(A)** a good violin and **(B)** a poor violin. Notice low and high resonance due to the wood and the middle resonance due to the air inside the cavity.

Interestingly, a violin actually sounds better the more it is played. This is probably because the wood making up the body of the instrument gains flexibility in specific areas due to repeated harmonic oscillations in the two-dimensional surfaces.

## The Piano

The only musical constellation in the sky is Lyra, The Lyre, easily recognized high overhead in summer and fall by its bright star, Vega. The lyre is a plucked instrument, like a harp, and was popular in Greece more than 2500 years ago. The piano is a distant cousin to the ancient lyre. Instead of using fingers to pluck strings, the piano uses a keyboard-operated mechanism to "hammer" strings. Over the last 200 years, the piano has evolved into one of the most popular musical instruments, thanks to its range in both pitch and loudness and to the beauty of the music it can produce. Its popularity is due to the advent of cast-iron frames. Each of the 230 metal strings is under tremendous tension, resulting in a very strong combined compressive force, up to $2.6 \times 10^5$ N, that wood frames could not withstand.

When a pianist presses down on the keys, felt-covered wooden hammers strike metal strings, causing them to vibrate. (Technically, this makes the piano a percussion instrument.) Vibrating strings themselves would be too soft to hear, so they are connected to the soundboard, which generates the large amplitude sound waves. High tension in the strings ensures that the transfer of energy from the strings to the soundboard is very efficient. Notes of higher pitch have triplets of strings, middle notes are strung in pairs, and low notes have only one string. This distribution of strings helps to equalize the loudness of all notes played. Finally, the large amplitude sound waves reflect off the open lid of a grand piano, or off a nearby wall for an upright piano.

### Language Link

The name piano is shortened from pianoforte, a word which reflected the instrument's ability to hammer the strings to play both soft and loud, or, in Italian, "piano" and "forte." What is a "fortepiano"?

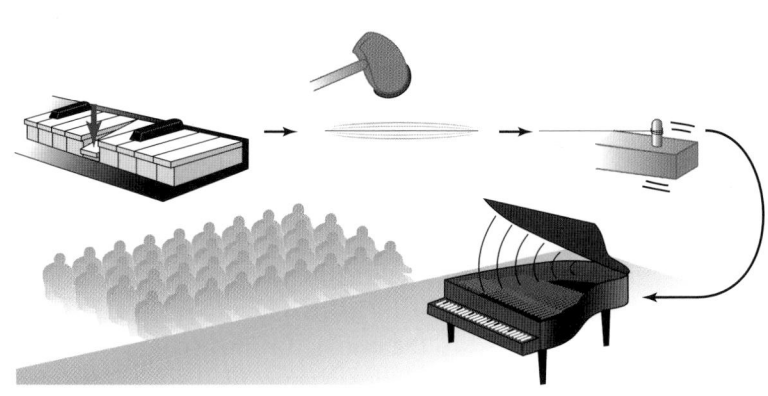

**Figure 9.30** Mechanical energy is transformed into sound energy.

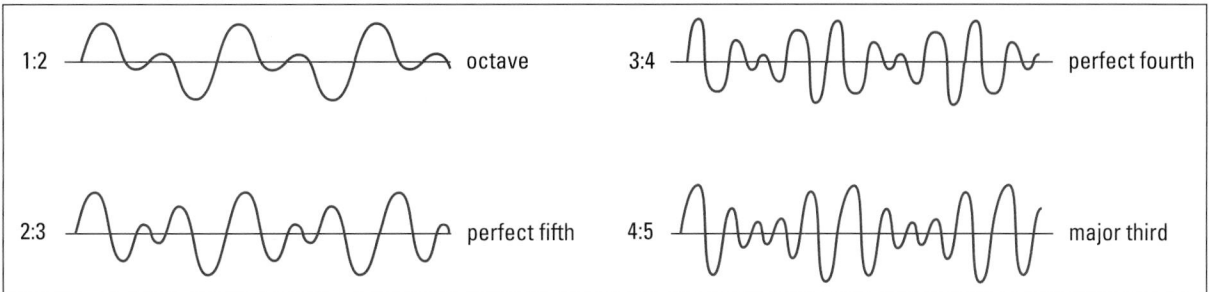

| 1:2 | octave | 3:4 | perfect fourth |
| 2:3 | perfect fifth | 4:5 | major third |

## Consonance and Dissonance

What sounds good to you may or may not sound good to someone else. Surprisingly, though, almost all humans are generally pleased by certain sounds and displeased by others. When more than one pitch is played at a time, the resulting sound is called a chord. Different cultures find different chords to be more pleasing, and therefore often have very distinctive-sounding music.

The ancient Greek mathematician Pythagoras, famous for the Pythagorean theorem, conducted extensive research into the mathematics of pleasing chords. He used identical strings under identical tensions, varying only the length. He found that pleasing sounds were generated when the strings' lengths were whole number fractions of the original length, such as 1/2, 2/3, and 3/4. As you have learned, the length differences translate directly into frequency or pitch differences. Musicians have discovered that several pleasant-sounding musical intervals exist in those ratios as shown in Figure 9.31. Musical notes that sound pleasant together are said to be in **consonance**. Combinations that sound unpleasant are said to be in **dissonance**.

Musical instruments each produce a distinctive sound. Even when different instruments are playing the same note, they produce sounds with different qualities that are created by each instrument's specific harmonic structure. The difference in sound is described as a difference in musical **timbre**. Figure 9.32 shows three sound intensity versus frequency graphs illustrating the relative strengths of different harmonics for each instrument. This pattern of intensities is the **harmonic structure**.

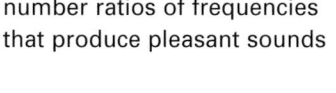

**Figure 9.31** Common whole number ratios of frequencies that produce pleasant sounds

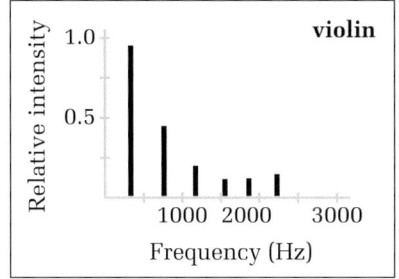
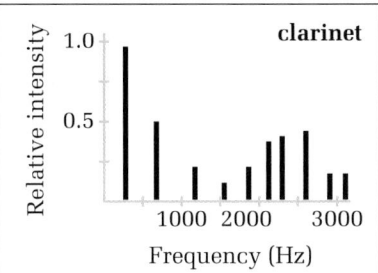
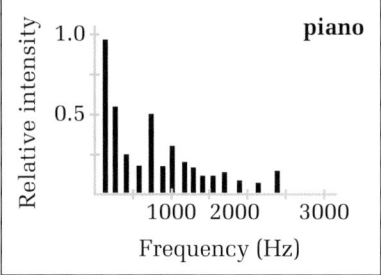

**Figure 9.32** A violin, clarinet, and piano produce characteristic sound spectra.

# Consonance versus Dissonance

**TARGET SKILLS**
- Performing and recording
- Analyzing and interpreting

The simplest Pythagorean ratio that produces consonance is 1/2. That is, reducing the string length by a factor of 1/2 causes the pitch to go up an octave. Using a piano or an electronic keyboard, play two notes exactly one octave apart simultaneously. Repeat the procedure for several different notes. What you are hearing is said to be in *consonance*. Now return to your original position on the keyboard and begin to vary only one of the previous notes. Move up one note at a time for three or four notes. Repeat the procedure moving down one note at a time.

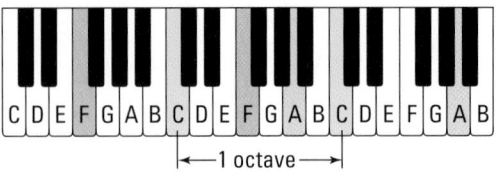

## Analyze and Conclude

1. Describe the note combinations that sounded pleasant to your ear.

2. Determine the exact frequency of the note pairs and then find their ratios.

**ELECTRONIC LEARNING PARTNER**

View musical vibrations by going to your Electronic Learning Partner.

## Percussion Instruments

Pounding on a drum, chiming a bell, or making music on a xylophone creates specific sounds based on each instrument's shape and the way in which it is struck. A large bass drum produces a substantially lower note than a much smaller snare drum. A large church bell rings out much lower tones than a small hand-held bell, and a xylophone generates a range of notes based on the length of the bar you choose to strike. **Percussion instruments** (instruments that sound when struck) have specific fundamental frequencies and harmonics associated with their size and the way they are played, as do wind and stringed instruments.

The membrane of a drum oscillates when struck. Unlike a string that settles into a one-dimensional mode of oscillation, a two-dimensional surface like the surface of a drum oscillates in more complicated two-dimensional modes. Figure 9.33 illustrates three possible modes of vibration for the surface of a drum. Striking a drum near the edge or near its centre will produce different modes of vibration and therefore different sounds.

**Figure 9.33** Three modes of vibration of a two-dimensional surface. Several much more complicated modes also exist.

| | | | |
|---|---|---|---|
| radial nodal lines (m) | 0 | 0 | 1 |
| nodal circles (n) | 1 | 2 | 1 |
| | A | B | C |

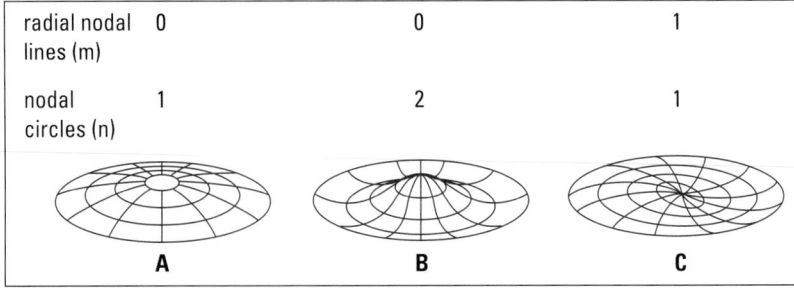

The bars of a xylophone behave in the same way as open-ended resonant air columns. The bars are fastened to the frame at nodal points of the fundamental frequency of vibration for each bar (see Figure 9.34). The bars also resonate at various harmonic frequencies. Xylophones vary widely in quality, as you can easily hear by the purity of the tones they produce.

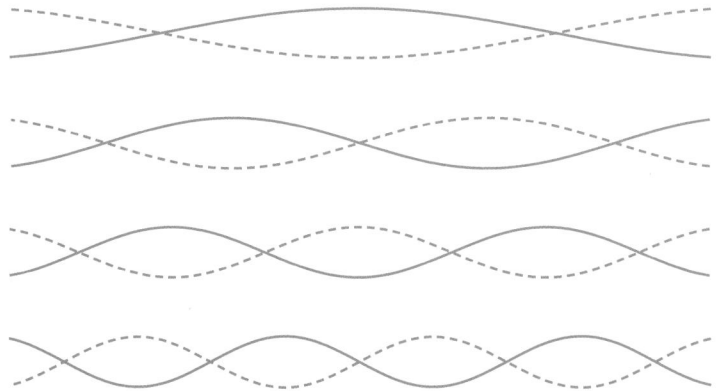

**Figure 9.34** Bars on a xylophone vibrate in a similar fashion to open-ended resonant columns.

# 9.3    Section Review

1. **K/U** How does the length of a brass instrument affect its sound?

2. **K/U** What is the purpose of the body of a guitar or a violin?

3. **MC** You and a friend are learning to play the recorder. Your instrument is less than half the length of your friend's. Whose instrument will resonate with a lower pitch? Why?

4. **C** While playing your recorder, you cover all of the finger holes, and then uncover just one near the middle. Describe what will happen to the sound as you uncover the single hole and explain why.

5. **C** Describe how a xylophone is able to produce a range of pitches.

6. **K/U** Figure 9.32 illustrates the harmonic structure of three instruments. Draw the harmonic structure of an ideal tuning fork with a fundamental frequency of 512 Hz.

7. **K/U** Calculate the fundamental frequency of a metal bar held in position by supports 8.0 cm apart. The speed of sound in the metal is 5050 m/s.

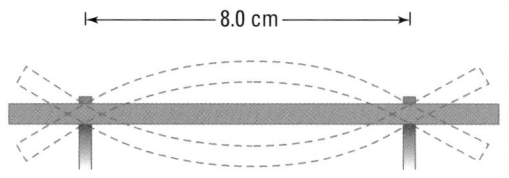

The first fundamental is superimposed

8. **MC** For some of the composers named on page 429, or other historic composers or modern musicians of your choosing, describe innovations in instrument design that the musician either brought about or popularized.

- Describe how knowledge of the properties of waves is applied to the design of buildings.
- Analyze the acoustical properties of technological equipment and natural materials.

KEY
TERMS

- acoustics
- reverberation time
- absorption coefficient
- liveness
- intimacy
- fullness
- clarity
- warmth
- brilliance
- texture
- ensemble
- blend
- anechoic chamber
- acoustical shadow

Why do speeches and oral presentations sound clear and easy to understand in some large rooms while you can hardly hear a speaker in other rooms of the same size? How do theatres designed for orchestras differ from movie theatres? Why is it difficult to understand the words from a television set in certain rooms, even when the volume is turned up? Answers to all of these questions all relate to acoustic qualities of the rooms. How do sound waves interact with the walls, floor, ceiling, and the contents of the room, including people?

## Acoustics of a Room

**Acoustics** is the science or nature of sound quality. Thus, to design a room with appropriate acoustics, you need a scientific understanding of echoes, frequency variations and reflection and absorption of sound waves. Echoes are one of the most important factors affecting the acoustics of a room. Too many echoes cause speech to sound muddled while too few echoes make a room sound eerily dead. The term **reverberation time** describes the relative sound intensity of echoes in a room in terms of the time required for echoes to become inaudible. Quantitatively, reverberation time is the time for a sound to drop by 60 dB from its maximum intensity.

You can engineer a room's reverberation time to a desired length by carefully choosing the texture of the wall coverings, the ceiling texture and even the properties of the furniture. A theatre will have a substantially different reverberation time when empty compared to when it is full of people, because reflected sound is the main factor contributing to reverberation time. Materials that absorb sound will limit the amount of reflected energy in a room and thereby reduce the reverberation time. Likewise, surfaces that reflect a great deal of sound will increase the reverberation time of a room. Table 9.4 lists **absorption coefficients** of several different materials for specific frequencies. The higher the coefficient, the more sound is absorbed and the less sound is reflected. Perfectly reflective material would have an absorption coefficient of 0.00.

Multi-purpose rooms in many schools and public gathering places create problems for an acoustical engineer. The appropriate room acoustics for a presentation such as a speech differ from the acoustics that are desirable for a musical performance. In fact, the appropriate acoustics for a string ensemble are very different than those for a rock band. The engineer must make serious compromises in the design of a multi-purpose room.

**Table 9.4** Absorption Coefficients for Different Materials

| Material | 125 Hz | 500 Hz | 4000 Hz |
|---|---|---|---|
| concrete, bricks | 0.01 | 0.02 | 0.03 |
| glass | 0.19 | 0.06 | 0.02 |
| drywall | 0.20 | 0.10 | 0.02 |
| plywood | 0.45 | 0.13 | 0.09 |
| carpet | 0.10 | 0.30 | 0.60 |
| curtains | 0.05 | 0.25 | 0.45 |
| acoustical board | 0.25 | 0.80 | 0.90 |

## QUICK LAB — Reverberation Time

**TARGET SKILLS**
- **Predicting**
- **Performing and recording**
- **Analyzing and interpreting**

How do the rooms in your school measure up in regard to appropriate acoustics? In this lab, you will select a variety of rooms and measure the reverberation times using a "loudness" or decibel meter and a stopwatch.

Make a list of the rooms you will test such as a classroom, the cafeteria, gymnasium, and music room. Before making any measurements, predict the relative reverberation times. List the rooms in order from longest to shortest reverberation time according to your predictions.

Select a sound source such as a trumpet. Keep the sound intensity level and the frequency of your source the same for all trials. A frequency near 250 Hz ($C_4$) would be appropriate. Since reverberation times will be quite short, possibly

on the order of 1.0 s, they will be difficult to measure accurately. Therefore, carry out several trials for each room and use the average value of the results for your final comparison of the different rooms. Position your sound source in the place where speakers or performers would normally be located. Take your sound measurements in the centre of the room.

### Analyze and Conclude

1. Organize your results from longest to shortest reverberation times. How accurate were your predictions?

2. Attempt to find similarities or differences between the rooms as the reverberation times increase down your list (for example, do the rooms get larger as reverberation times increase?).

3. Compare the surface of the walls and the ceiling between the rooms with the shortest and longest reverberation times. Suggest how the surfaces might affect the reverberation times of each room.

This graph illustrates typical reverberation times for performance-specific rooms versus the volume of the rooms. Rooms designed for speaking (Figure 9.36) require much shorter reverberation times than rooms constructed for musical performances (Figure 9.37).

**Figure 9.35** Notice that the volume scale is logarithmic, beginning in a small room with dimensions 3 m × 3 m × 3 m and moving to a very large auditorium 30 m × 30 m × 30 m.

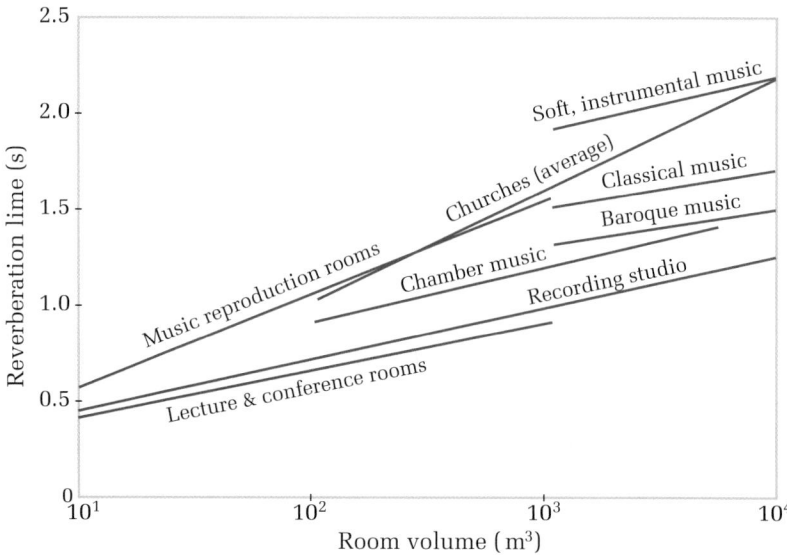

The entertainment industry has coined several terms to describe the acoustical properties of a room. Each term may sound very qualitative but, in fact, can be attributed to some very specific and measurable properties of sound. Physicists have quantified the terms that describe how well a room reacts to different sound situations, as shown in Table 9.5 on the next page.

**Figure 9.36** Speaking requires a room that is not too live, and has good clarity and brilliance.

**Figure 9.37** Singing with the quiet accompaniment of an acoustic guitar sounds best in a room that is very live and has good intimacy and fullness.

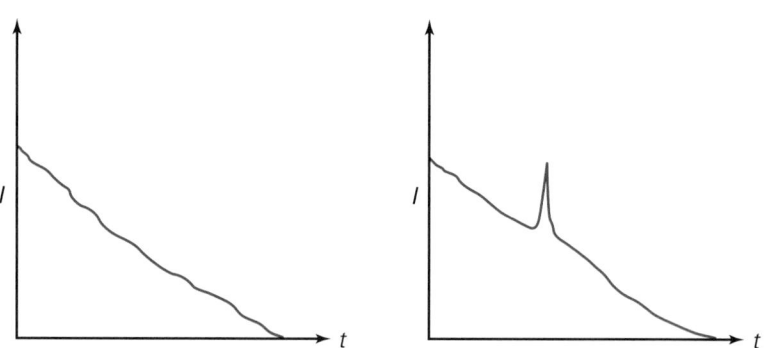

**Figure 9.38** The first graph shows a room with good texture; the sound intensity drops off consistently. The second graph illustrates a room that has poor texture; a large late echo peaks the sound intensity, distorting what the audience hears.

**Table 9.5** Acoustical Properties of Buildings

| Term | Description | Adjusting techniques | Performance type |
|------|-------------|----------------------|------------------|
| liveness | Liveness is a direct measure of reverberation time. The longer the reverberation time, the more "live" a room is. | Increase reverberation time by using more reflective materials on walls and ceiling. | soft, slow tempo instrumental music |
| intimacy | A room is more intimate if the listener feels very close to the performing group. This is accomplished if the first reflected sound reaches the listener 20 ms or less after the direct sound. | Placing a reflective canopy above the performers improves intimacy by reflecting sound to the audience. | all music and all speaking |
| fullness | The closer the reflected sound intensity is to the direct sound intensity, the more fullness a room has. | Fullness is generally increased by increasing reverberation time. | soft, slow tempo instrumental music, pipe organs |
| clarity | Clarity is the acoustical opposite of fullness. Reflected sound intensity should be very low for good clarity. Clarity is measured by repeating consonants. The maximum allowable loss of understanding is taken to be 15%. | Reducing reflected sound intensity is usually accomplished by reducing reverberation time. | speaking |
| warmth | Reverberation time for low frequencies (< 500 Hz) should be up to 1.5 times longer than frequencies above 500 Hz. | Select wall and ceiling materials to enhance low frequency reverberation time. | all music |
| brilliance | A brilliant room has the reverberation time of all frequencies nearly equal. If high-frequency reverberation time is too long, a constant high-pitched sound may occur. | Wall and ceiling materials can be selected to maintain consistent reverberation time for all frequencies. | speaking |
| texture | Good texture is achieved when at least 5 separate echoes reach the listener within 60 ms of the direct sound. Also, the sum of the reflected sound intensity should decrease uniformly. | The shape of the room is most related to texture. Sound focussing or the creation of large late groups of reflections produces poor texture. | all music |
| ensemble | Good ensemble means that each member of the performing group is able to hear what every other member is doing. No strong reflections should exist past the shortest notes being played by the group. | Like texture, the shape of the performing area is crucial to achieving good ensemble. Sounds must be softly reflected to all participants. | all music |
| blend | Blend is ensemble, but for the listening audience. Good blend is desired in every location of the room, but is much more easily achieved in the middle of any auditorium. | Good blend is achieved by mixing the sound before it is dispersed to the audience, using reflecting surfaces around the performing group. | all music |

## Creating Appropriate Acoustics

A singer is preparing to give a performance in your school's auditorium and complains that the room is too lively and lacks acoustical warmth. Describe the problem and a possible solution.

### Frame the Problem

The singer is identifying two acoustical properties of the room.

- Liveness: Longer reverberation times make a room more lively.

- Acoustical warmth is a room's tendency to have longer reverberation times for lower frequencies (< 500 Hz).

- Materials exist that preferentially reflect lower frequencies (see Table 9.5 on the previous page).

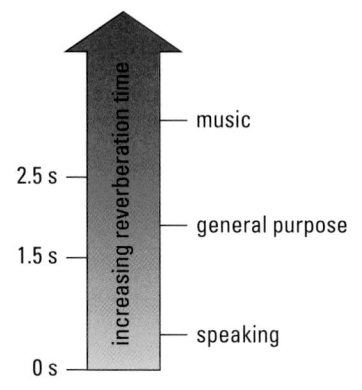

### Identify the Goal

To decrease the room's liveness (shorten reverberation time)
To increase the room's acoustical warmth (longer reverberation time for lower frequencies)

### Variables and Constants

| Involved in the problem | Known | Unknown |
|---|---|---|
| materials covering walls | room acoustics: | low frequency reflecting |
| | (a) too lively | materials |
| room acoustics | (b) lacks warmth | materials covering walls |
| low frequency reflecting materials | | |

### Strategy

Identify the definition of acoustical liveness.
Identify the definition of acoustical warmth.

Identify materials to

**(a)** decrease reverberation time (high absorption coefficient)

**(b)** decrease reverberation time of higher frequencies more than lower frequencies (high absorption coefficient for higher frequencies)

Refer to Table 9.5 for absorption coefficients.

### Calculations

Acoustical board has an absorption coefficient of 0.25 for 125 Hz and 0.90 for 4000 Hz.

The acoustical board will reduce the room's overall reverberation time, making the room less acoustically lively.

The acoustical board will absorb more of the high frequencies and less of the lower ones, making the room increase in acoustical warmth.

## Validate

Acoustical board may be available in the music room of the school, making the proper adjustments to the auditorium possible.

PRACTICE PROBLEMS

6. Why does a room designed for speaking require a relatively short reverberation time?

7. A physics student moves about in an auditorium as an orchestra plays. He notices that a region exists near the back where the bass sounds drown out the treble. What is a possible cause of the poor acoustics and how might it be corrected?

8. A classroom has one entire side wall made of windows. During oral presentations, the room lacks clarity. Should the curtains be open or drawn closed during the speeches? Explain.

9. Examine the Physics File below. Why does a person seated in the audience have an absorption coefficient greater than 1?

## Sound Focussing

For optimal presentations, domes, curved walls, and other shapes that resemble ellipses or parabolas should be avoided. Curved surfaces focus sound to specific areas of a room, rather than spreading it evenly throughout the audience. Focussing sound using a microphone with a parabolic reflector demonstrates the principle that causes problems with curved surface construction. Concave surfaces focus sound and convex surfaces diffuse sound.

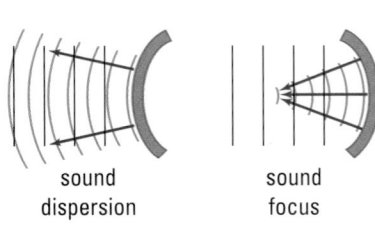

sound            sound
dispersion       focus

**Figure 9.39** Convex surfaces (left) diffuse sound while concave surfaces focus sound.

**Figure 9.40** A parabolic reflector around a microphone focusses sounds and allows distant noises to be clearly recorded.

**PHYSICS FILE**

An empty auditorium has a substantially different reverberation time than when filled with an audience. An adult person absorbs a great deal of the direct and reflected sound during a performance. Consider the absorption coefficients at 1000 Hz:

Wooden seat              0.28
Upholstered seat         3.2
Adult person             4.6
Adult person in an
upholstered seat         5.0

Both the person and the seat have absorption coefficients greater than one because sound is not only absorbed but also blocked from progressing through the room. It is analogous to computing a brick wall's absorption coefficient, not from the reflected sound, but rather the transmitted sound that manages to penetrate the wall.

# Acoustical Diffraction

Obtain a very high-pitched and a very low-pitched tuning fork. Find a very quiet location away from walls or other sound-reflecting surfaces. Gently sound the high-pitched fork and hold it at arm's length out from your partner's right ear. Practise striking the tuning fork to give it just enough energy to be barely audible. Repeat this procedure, but this time, have your partner cover his or her right ear. Can your partner hear the sound? Repeat the entire procedure with the low-pitched tuning fork. Repeat the procedure and have your partner hold the tuning fork about 1 m from your right ear.

## Analyze and Conclude

1. Were you able to hear the high-pitched tuning fork when your right ear was covered?

2. Were you able to hear the low-pitched tuning fork when your right ear was covered?

3. Explain how this experiment tests the diffraction of sound. Make a general statement about how the amount of diffraction of sound depends on frequency.

**Figure 9.41** An anechoic chamber that is literally "without echo" yields sound power reduction that follows the inverse square law ($1/r^2$).

## Echoes

Outdoor band shells are often in the shape of a large parabola with the performers positioned at the focal point. In the open expanse of the outdoors, this design projects the sound out to the audience very effectively. This design causes a drastic echo problem however when used indoors. An audience will hear the direct sound followed by a large echo caused by the sound reflecting off the rear wall. Interior design requires that the sound be mixed at the source by using small reflective boards and then evenly dispersed to the audience. Large flat surfaces of an auditorium should be covered with a sound-absorbing material to reduce the amount of reflected sound, thereby minimizing echoes.

Anechoic chambers are used in the study of acoustics. The term anechoic means "without echo" and is pronounced "ann-e-KO-ic." An **anechoic chamber** has sound absorbing material on the floor, walls, and ceiling. Speaking in a room that has absolutely no echoes generates a feeling that your words are being sucked out of you, only to disappear.

## Shadows

In Chapter 8, you discovered that sound waves, like all waves, can interact to cause interference. Constructive interference, when compressions overlap with compressions and rarefactions with rarefactions, produces an increase in the loudness of a sound experienced at that point. Destructive interference occurs when compressions overlap with rarefactions causing a reduction in the loudness. If destructive interference occurs in an auditorium, the region where it is occurring is called an **acoustical shadow**. An audience member in an acoustical shadow may have trouble hearing a performance.

Shadows often occur in places where large objects, such as balconies, protrude into the hall. The sound diffracts around the objects. As you have learned, the amount of diffraction depends on the frequency of the sound. A large amount of diffraction causes the performance to be distorted. Limiting the intrusion of balconies into the main performance hall reduces shadows.

## Resonance

Resonance becomes an issue only in small rooms, where the length of the room may be only a few wavelengths of some lower frequencies. A bathroom shower is a good example of a room exhibiting resonance. The material, glass or tile, is highly reflective and the shower's sides are symmetrical. The reflecting surfaces come in pairs, allowing standing waves to set up in every dimension of the room (see Figure 9.42). Auditoriums and theatres avoid the use of parallel walls to help reduce the resonance of certain frequencies.

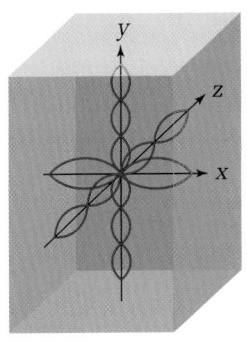

**Figure 9.42** Each pair of parallel surfaces provides a resonant chamber for a standing wave to form. Combine the three resonant lengths associated with a standard rectangular shower stall and the very reflective surfaces and it is no wonder your voice in the shower does not sound half bad.

---

## QUICK LAB

## Singing in the Shower

**TARGET SKILLS**

- **Predicting**
- **Performing and recording**
- **Analyzing and interpreting**

Measure the dimensions of an enclosed shower stall. Measure the height, width, and depth of the shower. Calculate the fundamental resonant frequency for each dimension. Determine what notes should resonate in your shower. Be sure to include one or two harmonics for each fundamental. Test your predictions.

### Analyze and Conclude

1. Why does your voice seem to sound "better" when you sing in the shower?

2. Calculate the wavelength range of the musical spectrum, approximately 20 Hz to 15 000 Hz.

## Acoustic Design in Buildings

The performer on stage is speaking, but you cannot quite make out the words. This problem, so annoying for theatregoers, is not always the actor's fault. Some elegant theatres have bad acoustics.

Acoustics is the physics of sound. When designing theatres and concert halls, architects consult acoustic engineers, in the hope that music and speech will sound perfect in their buildings. Sometimes experimentation and changes are required after the hall is constructed. Roy Thomson Hall, home of the Toronto Symphony Orchestra, has large circular panels hanging from the ceiling. These can be raised, lowered, or tilted in an attempt to reflect sounds equally to all regions of the theatre. An extensive renovation in 2002 should further improve the hall's acoustics.

For residential, office, and industrial buildings, the architect's concerns are different. Noise reduction becomes important. Acoustic engineers have to consider the sources, frequencies, and intensities of the sounds produced in various parts of the building. Then, they recommend materials to use to block the sound. For example, various materials transmit sound differently, depending on the pitch. To test a material for its sound-blocking ability,

a wall of the material is placed between a speaker and a microphone. The intensity is plotted on a graph against the frequency. From the graph, the acoustic engineer can suggest suitable materials for the specific building being designed.

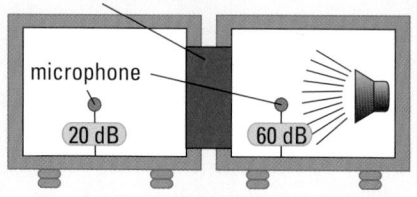
wall material sample

microphone

20 dB          60 dB

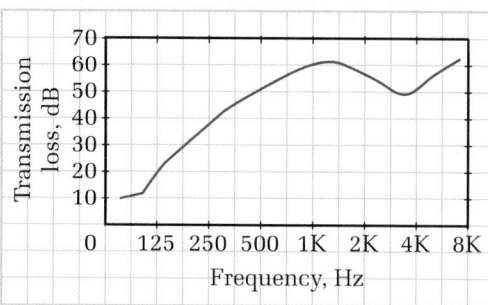

Sometimes the best-designed concert halls have disappointing sound qualities. Even the most well-planned buildings also can sometimes have regions where noise pollution is high. Fixing such problems involves knowledge of physics and experimentation skills. The application of acoustics to building design is as much an art as a science.

### Analysis

1. Why might "dead spaces" occur in concert halls where some notes produced by a singer cannot be heard?

2. Despite the use of appropriate materials in the walls, ceiling, and floor, sometimes even low intensity sounds at a certain pitch can produce annoying buzzes in rooms, such as the rattling of windows. Why might this happen?

# INVESTIGATION 9-A

## Design and Test Your Own Theatre

**TARGET SKILLS**
- **Initiating and planning**
- **Performing and recording**
- **Modelling concepts**

Models are often used to perform critical tests before a full-scale, expensive structure is built. In this activity, you are to use the acoustical design knowledge you have gained to design a movie theatre. Produce a scale drawing of your theatre design that conforms to the specifications listed below. Remember to choose a scale that will fit into a wave tank.

### Theatre Specifications

- seating capacity: 75
- seat dimensions (including personal space): 0.60 m × 1.20 m to 0.80 m × 1.60 m
- aisle minimum width: 3.0 m
- minimum distance from screen: 5.0 m
- maximum distance from screen: 20 m
- screen dimensions: 10.0 m × 8.0 m

Sound will be generated at the front of the theatre only.

### Building your Model

Construct a scale model of your theatre's *perimeter* in a wave tank. Use waterproof materials arranged so that the interior of the model matches the design of your theatre. The photograph shows a simple example.

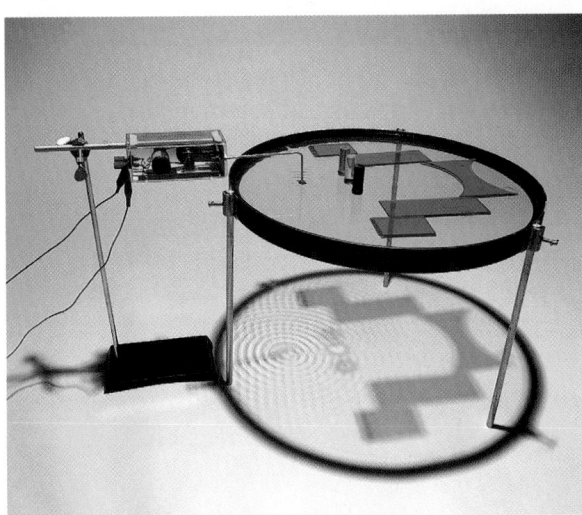

### Testing your Model

Record the scale used to construct the model.

Test your model using (a) a straight wave source and (b) a point source.

Look for acoustical problems such as

- areas of sound focussing
- large reflections
- large late reflections
- sound shadows
- resonance (interference)
- reverberation time between low and high frequencies (acoustical warmth)

(Refer to Table 9.5, Acoustical Properties of Buildings, for possible properties to test.)

Mark problem areas on your scale model drawing. Attempt to improve the acoustical properties of your theatre by employing strategies discussed in Table 9.5, Acoustical Properties of Buildings. Test your modified design.

### Analyze and Conclude

1. Describe acoustical problems you discovered in your initial design.

2. Assuming the model and wave tank test was valid, describe the acoustical properties of your theatre using the terms of the music industry (Table 9.5) before and after design revisions.

3. How valid was the wave tank test? Explain.

### Apply and Extend

4. Is it possible to equate the speed of the water waves in the tank to the speed of sound in air? How?

5. Use your results from question 4 to determine the frequencies of sound that were actually tested for during the wave tank testing.

## External Noise

The final consideration architects of auditoriums must take into account is the amount of external noise. Designing the perfect room can be rendered useless if, during each performance, a rumbling train roars by and drowns out the famous singers' voices. If highway or rail traffic noise exists outside of a theatre, engineers must insulate the building to prevent the unwanted noises from entering the room.

## 9.4 Section Review

1. **C** Describe the physical characteristics a physicist attributes to:

   **(a)** acoustical warmth

   **(b)** acoustical clarity

   **(c)** acoustical texture

2. **C** Describe the main acoustical difference between an auditorium designed for speaking and one designed for pipe organ music.

3. **K/U** You wish to renovate your bedroom so that you can enjoy music the way you like it—lots of bass and very little treble or high frequencies. What material(s) would be best for you to put on your walls?

4. **K/U** What is an anechoic chamber?

5. **MC** Could you use some very powerful speakers to mask the noise of a helicopter's rotors? If so, how?

6. **C** Explain why singing in the shower sounds better than singing in a large room.

7. **C** Consider the graphs in Figure 9.38 showing how the intensity of sound within a room fades. Describe the acoustical problem that is demonstrated by the graph.

8. **C** Based on your wavelength calculation, would it be wise for a movie theatre to have a length of 17.0 m? Explain.

### UNIT ISSUE PREP

Acoustical design principles are applied to the design and layout of communities.

- Identify materials and design features that would help control sound pollution.
- Determine which acoustical characteristics generate noise pollution complaints by members of a community.

## REFLECTING ON CHAPTER 9

- The outer ear collects sound energy and directs it toward the eardrum.
- The eardrum, in the middle ear, vibrates in response to the compressional waves in the air. The eardrum initiates the mechanical vibrations of the three ossicles.
- The stapes, the third ossicle, vibrates the oval window on the surface of the inner ear, starting compressional waves in the liquid in the cochlea. Travelling waves move down the basilar membrane and create resonance in specific locations.
- Hair cells on the basilar membrane stimulate neurons in the auditory nerve, which then carries signals to the brain.
- Programmable hearing aids can benefit people with partial deafness. Cochlear implants, which stimulate auditory neurons directly, can benefit profoundly hearing impaired persons.
- Air forced through the vocal cords cause them to vibrate. The sounds resonate in the vocal tract including the throat, mouth, and nasal cavity. Articulators create specific sounds from resonating air columns.

- The Doppler effect is the apparent change in the frequency of a sound due to relative motion of the source of sound and the observer.
- When an object is moving faster than the speed of sound in the air through which it is travelling, each new compressional wavefront is ahead of the previous one. The overlapping of wavefronts along a cone creates extremely large compressions that are heard as a "sonic boom."
- Mach number $= \dfrac{\text{speed of object}}{\text{speed of sound}}$
- Most bats and whales use echolocation to detect prey or solid objects.
- Technologies such as sonar and ultrasound are based on the principle of echolocation.
- Musical instruments create sound with a wide spectrum of frequencies. The resonant properties of the instruments determine which frequencies are amplified.
- The reverberation time of a room describes the time required for the intensity of echoes to drop by 60 dB. The purpose of the room dictates the appropriate reverberation time.

### Knowledge/Understanding

1. Does the human ear have any mechanisms to protect against extremely loud sounds?
2. Draw an energy path diagram detailing the path of a sound from when it is generated by a falling tree right up to when the electrical signal reaches the brain.
3. Name and explain the function of the three smallest bones in the body.
4. What is the frequency range of normal human speech?
5. A friend comes to school and is barely able to speak, claiming to have laryngitis. How does laryngitis limit the ability to generate sound?
6. What physical characteristic(s) makes human voices unique?

7. Define and provide an example of conductive and sensorineural hearing loss.
8. Complete the following chart.

| Sound | Articulator/resonator |
|---|---|
| p, b, m, w | |
| | tip of tongue |
| k, g | |

9. Sketch a diagram that traces the energy path from the beginning of an incident sound wave all the way to the electrical signal reaching the brain of a patient with a cochlear implant.
10. Sketch a diagram to model the sound waves emanating from an object that is travelling away from you at a speed that is slower than the speed of sound.

11. Dolphins are able to use ultrasonic waves to locate and track prey.
    (a) How do they generate and receive the sounds?
    (b) What frequencies do the dolphins use primarily for echolocation?
    (c) What are the advantages of using the frequencies from part (b)?

12. Draw an energy path diagram that traces the energy imparted to a string on an acoustic guitar by someone strumming it right up to the eventual sound that is heard by the audience.

13. Define reverberation time and explain how it affects a room's acoustical properties.

## Communication

14. Describe the features of the human ear that aid in the amplification of sound.

15. The auditory canal of most humans has resonant frequencies that vary between 2.5 kHz to 4.5 kHz. Explain the significance of this resonance in terms of hearing sensitivity and communication.

16. Human speech is an incredibly complex process that enables us to communicate effectively. Explain how are we able to produce so many complex sounds.

17. (a) Briefly describe the place theory of hearing.
    (b) Describe the temporal theory of hearing.

18. While waiting at a railway crossing, you hear the approaching train sound its horn. Describe the change in frequency of the sound you hear as it approaches and then departs from your position at high speed.

19. High frequency and low frequency sounds behave differently. High frequency sounds diffract (bend) less around obstacles than lower frequency sounds. Describe why this characteristic makes it beneficial to put high frequency sirens on emergency vehicles and low frequency fog horns on boats.

20. Sketch and label a pressure versus time graph to describe how a sonic boom is able to break a window.

21. Explain the relationship between the absorption coefficient of wall and ceiling material and reverberation time.

22. Describe the acoustical property of a room with good
    (a) liveness
    (b) fullness
    (c) brilliance
    (d) blend

23. Describe the acoustical property of a room that has a reverberation time for low frequencies (< 500 Hz) that is 1.5 times longer than for frequencies above 500 Hz.

24. Describe how each of the following acoustical problems could be reduced.
    (a) poor texture
    (b) sound focussing
    (c) excessive liveness

25. Describe the acoustical properties of a room that is well designed for speaking performances.

26. Describe the acoustical effects that are represented by this photograph of a water droplet.

## Making Connections

27. While driving in a car, your favourite music is replaced by a newsperson giving a winter storm report dealing with school closures. You find that it is difficult to hear what is being said, even when the volume is increased. How should you adjust the treble and bass settings to help you hear?

28. Canadian Forces jets flying from Cold Lake, Alberta are restricted from flying at supersonic speeds over certain areas. Discuss the possible effects on wildlife subjected to regular low-level supersonic air traffic. Explain why high-altitude supersonic flights affect a greater area, but have a diminished impact.

29. Current oceanographic research involves deploying high frequency sound beacons that may travel hundreds of kilometres throughout the oceans. They are used to measure the temperature of the oceans or for military surveillance. Discuss whether the benefits of these applications are worth the possible damage caused to marine life.

30. Mice hear sounds that are well above the range of human hearing. Devices that emit continuous, high intensity ultrasonic sounds will drive away mice without being heard by human ears. Do you believe this type of device is safe? Is the use of such a device ethical, not only in terms of the mice, but also in terms of the unknown effects it may have on other nearby wildlife or on neighbours' pets?

## Problems for Understanding

31. A neighbour explains that the old well behind his house is 500 m deep. You decide to see for yourself. You drop a stone from rest and measure the time interval until you hear the splash of the stone striking the water. You find it to be 6.0 s. You assume the speed of sound in air to be 343 m/s. How deep is the well?

32. A car horn produces a 4200 Hz sound. You hear the sound as 4700 Hz. Is the vehicle approaching or leaving your position?

33. Calculate the wavelength of the sound from the car horn in question 32 when (a) stationary and when (b) perceived by you as 4700 Hz.

34. The siren of an emergency vehicle produces a 5500 Hz sound. If the speed of sound is 340 m/s in air, calculate the wavelength of the sound.

35. A jet is travelling at Mach 2.4 in air, with a speed of sound of 320 m/s. How fast is the jet flying in km/h?

36. A bullet leaves the barrel of a gun at 458 m/s and goes into air that has an ambient temperature of 26.5°C. Determine the Mach number of the bullet.

37. A sonar depth finder is capable of receiving a reflected signal up to 1.75 s after sending it. Calculate the maximum depth that the device could accurately measure in fresh water.

38. A vacant city lot sits next to a large building. You hope to convince City Council to turn the lot into a soccer field. First you need to determine how far it is from the edge of the lot to the building. You bark out a loud call and the echo returns in 0.75 s. Calculate the distance to the building if the air temperature is 31°C.

39. Intimacy is achieved when the first reflected sound arrives less than 20 ms after the direct sound. Calculate the maximum extra path length that the reflected sound may travel in order to arrive in 20 ms. (Assume air temperature of 20°C.)

40. Calculate three fundamental resonant frequencies for a small room with dimensions of 2.0 m × 3.0 m × 1.5 m. (Assume standard atmospheric pressure at 0°C.)

41. Some friends drop a water balloon out of a window 12.0 m above the ground. As it falls, one of the pranksters cries out to warn the person below, 1.5 s after releasing the balloon. If the air temperature is 28.0°C and if the person below is able to move infinitely fast upon hearing the warning, will the person avoid the balloon?

The environment is alive with sounds.

## Background

Sounds that are continuous, or loud, or both, are called noise pollution. Air and road traffic, construction, and loud music, are common forms. The effects of noise pollution on hearing loss are well documented. Studies have found that people living in remote regions, far from the noises of an industrialized environment, have much better hearing in old age than people that have lived in urban settings. These studies suggest that although hearing loss may be a natural result of aging, external factors can also have a dramatic impact. Studies also suggest that noise pollution can have an effect on our emotional well-being. As a result, many towns and cities attempt to protect citizens by passing legislation aimed at regulating noise levels.

**Hearing Loss with Age at 3000 Hz**

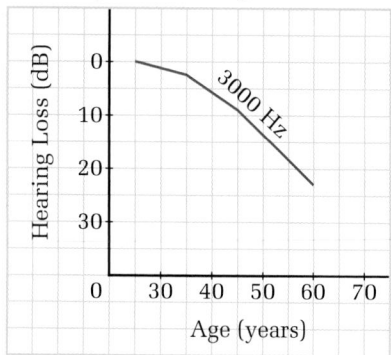

In recent court proceedings brought by a town in British Columbia against a local company, noise pollution bylaws were put to the test. Because of the company's proximity to a residential area, bylaws prohibited the company staff from making noise outside the hours of 6:00 a.m. to 8:00 p.m. on weekdays. Apparently, however, company personnel were routinely receiving shipments at 5:00 a.m., which prompted a group of angry residents to file a complaint. The judge in the case threw out the complaint, claiming the town's bylaws did not sufficiently define unacceptable noise. The judge further suggested that noise deemed annoying by one person could easily go unnoticed by another.

This case raises important questions about the planning and implementation of noise pollution bylaws. What constitutes acceptable versus unacceptable levels of noise? Who should have a say? What steps must municipal planners and lawmakers take to ensure that such bylaws do not unfairly discriminate against businesses, or prevent people from going about everyday tasks, while at the same time ensuring that residents of local communities have periods of quiet time?

Bearing these issues in mind, your task will be to apply your knowledge of waves and sound to draft a noise pollution policy for a specific community.

Many regions have by-laws governing when construction noise is allowed.

## Plan and Present

1. As a class, establish clear guidelines for the finished product. Discuss specifics such as overall length, required sections, proper sourcing, and timelines.

2. In groups of three or four, brainstorm potential communities for which a noise pollution policy could be useful. Communities to consider include a hospital, an apartment building, a retirement residence, or your school. Each group should share its choice of community with the class.

3. Once your group has selected a community, begin identifying the stakeholders (any person or group that will be affected by the policy).

4. Develop your proposal by:
   - conducting experiments to determine current noise levels
   - interviewing stakeholders
   - researching existing policies
   - researching appropriate noise levels as prescribed by unbiased scientific studies
   - determining reasonable penalties for bylaw infractions

you find current and relevant information?

- assess your communication skills: how effectively were you able to share your ideas with the other members of your group?

- assess your ability to make connections to the world outside the classroom: did you address social and economic considerations in your policy document?

## Evaluate

1. What components were common to each group's policy document, regardless of the community?

2. List three items that you found most interesting after analyzing all of the documents. Explain your selections.

3. Describe two challenges and two successes associated with your group's attempt to draft a useful and enforceable document.

4. Describe how you feel about noise pollution. Respond to the comment, "Societal noise regulation has often been left to individual municipalities because of the perception that noise pollution is simply an irritation."

Logging vehicles are often restricted from operating during specific times of the day. In some areas seasonal operating restrictions also exist.

Urban planners often set aside specific geographic regions for industrial, entertainment, and housing developments to help reduce problems one may cause another.

## Knowledge and Understanding

### True/False

In your notebook, indicate whether each statement is true or false. Correct each false statement.

1. The time required to complete one cycle is called the frequency.

2. When a particle oscillates, its maximum displacement from its rest position is called its amplitude.

3. Pushing a friend on a swing is an example of mechanical resonance.

4. The medium through which a wave is passing experiences a net movement in the direction of travel of the wave.

5. The speed of a wave depends on the amount of energy used to create it.

6. Water waves are an example of transverse waves.

7. When two waves interact to produce destructive interference, there is a net loss of energy.

8. Sound travels around corners in a phenomenon known as refraction.

9. Constructive and destructive interference cannot occur in longitudinal waves.

10. Infrasonic sound waves have frequencies below what humans can hear.

11. A sound intensity of 0 dB indicates that a sound wave has no amplitude.

12. When the source of a sound moves toward a stationary observer, the apparent pitch of the sound seems to be lower than the actual pitch of the source.

### Multiple Choice

In your notebook, write the letter of the best answer for each of the following questions.

13. The period of a pendulum oscillating in periodic motion is
    (a) the time to complete one cycle.
    (b) the position the pendulum assumes when allowed to hang freely.
    (c) one complete repeat of the pattern.
    (d) the number of cycles completed in a specific time interval.

(e) the amplitude of the pendulum divided by its velocity.

14. The main reason the Tacoma Narrows bridge collapsed was
    (a) extremely high winds
    (b) the steel used was faulty, and could not support the weight of the traffic
    (c) certain wind speeds caused the bridge to vibrate at its natural frequency
    (d) the bridge was not made massive enough
    (e) all of the above

15. In the diagram below, which of the features labelled (a) – (e) is the amplitude?

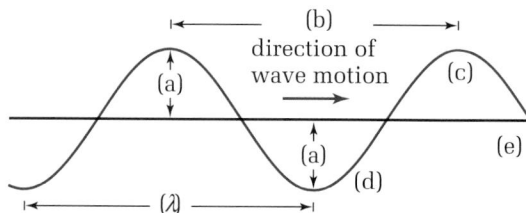

16. In the diagram for question 19, which of the features labelled (a)–(e) is the wavelength?

17. In the diagram for question 19, which of the features labelled (a)–(e) is the rest position?

18. Identify the false statement. When a wave travels from one medium to another,
    (a) the frequency of the wave changes
    (b) the amplitude of the wave changes
    (c) the speed of the wave changes
    (d) some of the energy is transmitted and some is reflected
    (e) the period of the wave remains constant

19. The frequency range of human hearing is approximately
    (a) 2 Hz to 2000 Hz
    (b) 20 Hz to 20 000 Hz
    (c) 200 Hz to 200 000 Hz
    (d) 2 Hz to 2 kHz
    (e) 2 kHz to 2 MHz

20. Two sound sources that vary in frequency of only a few hertz, when sounded together will produce
    (a) resonance
    (b) acoustical liveness

(c) a closed air column

(d) beats

(e) nodal lines

21. Where are incident sounds amplified in the human ear?

    (a) ear drum, ossicles, cochlea

    (b) external ear, ear canal, oval window

    (c) ear canal, ear drum, ossicles

    (d) ear drum, Eustachian tube, auditory nerve

    (e) external ear, ear canal, ossicles

22. Pleasant-sounding combinations of notes are said to

    (a) be in dissonance with each other

    (b) be in consonance with each other

    (c) interfere destructively

    (d) interfere constructively

    (e) produce harmonic structure

23. An auditorium is designed so that at least five reflected sounds reach the listener within 60 ms of the direct sound. This auditorium would be said to have good

    (a) blend        (b) warmth

    (c) texture    (d) ensemble

    (e) clarity

## Short Answer

In your notebook, write a sentence or a short paragraph to answer each of the following questions.

24. Compare the detected wavelength and frequency of a source of sound as it moves toward a stationary listener with the detected wavelength and frequency as the source moves away from the stationary listener.

25. If the fundamental mode of vibration for a standing wave on a vibrating string is 200 Hz, could a standing wave of 250 Hz exist on the same string? Explain why or why not.

26. How does a tuning fork produce sound? Why is the sound produced at a specific frequency?

27. What evidence exists to support the assertion that sound requires a medium to travel through?

28. How would the intensity of a sound be affected if you moved

    (a) twice as far away from the source?

    (b) seven times farther away from the source?

    (c) one third as far away from the source?

29. How does the distance between the first and second resonant lengths compare for an air column that is closed at one end and an air column that is open at both ends?

30. How is a harmonic mode of vibration related to the fundamental mode of vibration for a string?

31. Why is sound intensity, measured in dB, presented using a logarithmic scale?

32. List and describe the three main functional areas of the human ear.

33. (a) How does the human hearing system amplify sounds?

    (b) How does the human hearing system protect against very loud sounds?

34. Summarize the place theory and the temporal theory of hearing. Ensure that you identify their strengths and weaknesses.

35. Describe how a sonic boom is an extreme case of the Doppler effect.

36. Define the term echolocation and provide two examples of animals that use it.

37. How can a single drum be made to produce different sounds?

38. (a) Define reverberation time.

    (b) What types of performances are better suited for rooms with long reverberation times? Why?

## Inquiry

39. Using your understanding of the diffraction of waves in general, describe how you could carry out an experiment to show that sound waves do or do not diffract.

40. Two physics students conducted an experiment to test how the speed of sound in air varied with changes in temperature. Use the graph of their data (below) to develop a mathematical relationship that could be used to predict speed of sound in air at specific temperatures.

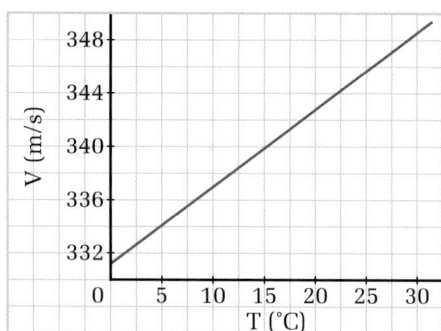

41. Describe how you could use a meter stick, a pair of binoculars, a classmate with a drum, and a large pendulum hanging from the branch of a tree in an open valley to assist you in the approximate calculation of the speed of sound.

42. Your school plans to conduct a "coffee house" to allow students to share their musical talents. Two locations are suggested, the cafeteria and the gymnasium. The cafeteria is a large L-shaped room with sound-absorbing ceiling tiles and wall coverings. The gymnasium is a large rectangular box with concrete walls and a flat steel ceiling. Design a series of tests and plan a report format to provide the "coffee house" committee with a detailed acoustical analysis of each room, with highlighted strengths and weaknesses of each.

## Communication

43. Extremely high water waves at sea are sometimes known as rogue waves. They may occur following a storm and seem to come out of nowhere. Explain, using your knowledge of wave interference why these waves occur and have these special characteristics.

44. A police car is equipped with a siren type horn located under its hood that rotates very quickly. Explain the reason for the rotation of this horn as it emits a sound of constant pitch and loudness.

45. Dolphins and whales often make use of submarine "sound channels" to communicate over very large distances. Find out and describe how these sound channels operate and are able to carry sound so far without losing much of its intensity.

## Making Connections

46. Explain why an empty auditorium has a substantially different reverberation time when it is filled with an audience. Does the audience increase or decrease the reverberation time?

47. Draw an energy path diagram illustrating how incident sound energy is eventually transformed into electrical impulses by the human hearing system.

48. Many countries use the sky above the wilderness area around Goose Bay, Labrador to train fighter pilots in the skills of low level flying using supersonic military aircraft. Write a research paper on the environmental impact of low level flying of supersonic aircraft over wilderness areas such as in the case of the Labrador wilderness.

## Problems for Understanding

Show complete solutions for all problems that involve equations and numbers.

49. Calculate the speed of water waves hitting the shore if adjacent crests are $3.0 \times 10^1$ m apart and a wave hits the beach every $1.0 \times 10^1$ s.

50. A pendulum takes 1.50 s to swing from the rest line to its highest point. What is the frequency of the pendulum?

51. A wave with an amplitude of 50.0 cm travels down a 12.0 m spring in 3.00 s. The student who creates the wave moves his hand through 5 cycles in 1 s. What is the wavelength?

52. A klystron tube in a microwave oven generates radiation of wavelength 4.20 cm. What is the frequency of the microwave radiation? (Microwaves travel at the same speed as light.)

53. The international tuning note (A above middle C) has a frequency of 440 Hz. If the speed of sound in air is 320 m/s, what is the wavelength of the note in air?

54. A sound wave reflects from the end of an air column with a distance between any two consecutive nodes of 54.0 cm. If the air temperature is 10.0°C, what is the frequency of the vibration?

55. A violin string vibrates with a frequency of 990 Hz at the second octave. If the speed of

sound is 343 m/s, what is the wavelength of the same string vibrating at its fundamental frequency?

56. A tuning fork with a frequency of 324 Hz is held over a tube whose length can be changed by raising and lowering a column of water in the tube. The surface of the water, initially very near to the top of the tube, is gradually lowered. If the speed of sound in air is 336 m/s, how far from the top of the tube is the surface of the water when the first point of constructive interference is detected?

57. When two tuning forks vibrate simultaneously the sound grows louder and softer, with 100 intensity peaks every 80.00 seconds. If one tuning fork is known to have a true frequency of 384.0 Hz, what are the possible frequencies of the other tuning fork?

58. While scuba diving the Pacific Ocean you bump into your partner and your oxygen tanks make a loud clank. How far are you from the nearest underwater reflecting surface if the sound returns to you in 3.00 s?
(Assume speed of sound in salt water to be 1440 m/s)

59. (a) What is the Mach number of an airplane travelling at $2.0 \times 10^3$ km/h through 8°C air?
(b) Will the Mach number increase or decrease as the temperature falls?

60. Calculate the range of wavelengths of sound travelling through air at 20°C that are audible to the human ear.

61. In archery class you shoot an arrow with a constant speed of 22.0 m/s at a target that is $5.0 \times 10^1$ m away. How long after you release the arrow will you hear it hit the target?

62. Professional cliff diving experts often leap from heights of $5.0 \times 10^1$ m into deep water. How much time would pass between the moment the professional jumped to the moment spectators would hear a splash if they were watching from a boat located 35 m from where the diver strikes the water?

63. A bat uses echolocation to find insects. Waves can only be used to detect objects that are one

wavelength long or greater. If the bat's echolocation is to be able to detect an insect that is 0.450 cm long on a night when the temperature is 15.0 °C, what must be the frequency that it uses to locate the insect? (Hint: Some bats use frequencies as high as 150 kHz.)

64. The sonar of a submarine uses a sonic "ping" with a frequency of 698 Hz. The echo returns from a distant submarine 5.60 km away after 8.00 s. What is the wavelength of the sound the first submarine is using to echolocate the other submarine?

65. People love to sing in the shower, since the dimensions of the shower cabinet are usually such that it causes resonance, making their voices sound better than normal. If a shower cabinet measures 1.0 m by 1.0 m square and 2.0 m high, would a male or a female singer most likely appreciate the effect of the resonance? Explain.

66. A student had no thermometer, so in order to measure the temperature, she resourcefully used the following procedure. She accurately measured a distance of $1.00 \times 10^2$ m from a wall, and then struck two stones together so that each new strike coincided with the echo from the previous strike. The student found that the time to make $1.50 \times 10^2$ strikes was 92 seconds. To the nearest degree, what was the air temperature?

**COURSE CHALLENGE**

**Space-Based Power**

Plan for your end-of-course project by considering the following:

- Are you able to incorporate analysis of waves and wave properties into your project?

- Begin to consider time and equipment requirements that may arise as you design project-related investigations.

- Examine the information that you have gathered to this point. Produce a detailed plan, including a timeline, to guide you as you continue working on your project portfolio.

# Light and
# Geometric Optics

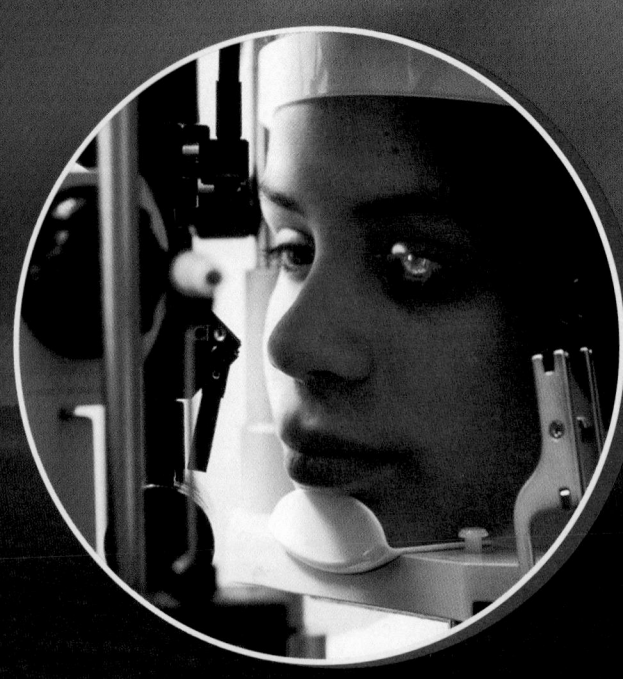

## OVERALL EXPECTATIONS

**DEMONSTRATE** an understanding of the properties of light and its transmission

**INVESTIGATE** and predict the behavior of light using ray diagrams and algebraic equations

**EVALUATE** the contributions to entertainment, communications, and health made by optical devices

Sometimes, when the conditions are just right, you can see sunbeams piercing the clouds during a sunset. Aside from their beauty, these sunbeams convey information about the nature of light. For example, you can see that light travels in straight lines. What property of light is revealed when tiny waves on the water sparkle brightly? When and why can you see images reflected from the surface of still water? Sometimes, on a hot summer day, you may look at a road surface off in the distance and think you see shimmering water, but there is no water there at all. How is this mirage effect created?

Early observers and scientists asked and found answers to questions similar to those above. In the process, they learned so much about light that they were able to develop a variety of technologies that enhance everyday life. For example, the woman in the small photograph may no longer have to wear corrective lenses. Her doctor is using specialized lenses to examine her eyes. He will decide whether he can use laser light to change the shape of the corneas of her eyes. In this unit you will develop an understanding of the properties of light that make these incredible technologies such as laser eye surgery possible.

### UNIT PROJECT PREP

Refer to pages 596–597. In this unit project, you will have the opportunity to build a reflecting telescope or a periscope.

- How will the light ray model help you in the design of your optical instrument?
- What is the nature of the images produced by concave and convex lenses?

Have you ever been in a house of mirrors at an amusement park? Finding your way out of a mirror maze can sometimes be quite challenging. The placement of the mirrors and the lighting effects create so many different images that some people become quite disoriented as they try to find the exit. You can also create some amazing optical illusions by using different combinations of curved mirrors instead of plane mirrors. However, mirrors serve many functions other than having fun.

Interior decorators use mirrors to make rooms appear larger or to create dramatic effects. You use mirrors at home to check your own appearance. Mirrors in cars help you avoid collisions. Also, curved reflectors help focus car headlights to provide maximum visibility without blinding oncoming drivers. How many ways do you use mirrors in a single day?

# Mirrors and Images

## Reflection from Plane and Curved Mirrors

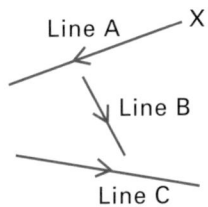

Draw and label three lines at random on a blank page, like the ones shown here. Mark arrows on each line to indicate a direction. Place a ray box at point X on line A, and shine a single light ray along the line. Position a plane mirror at the other end of line A, such that the reflected light ray touches the beginning of line B. Mark the position of the mirror, and draw in the light ray from A to B. Repeat for lines B and C, so that the final reflected light ray touches point X.

## Analyze and Conclude

1. Find a pattern in the angles of the incident and reflected rays.

2. Repeat the procedure, using the concave and convex mirror shapes. Does the same relationship exist for the curved mirrors as for the plane mirror? Explain your answer.

3. What difficulties did you encounter measuring the angles?

4. Suggest a practical solution to the problem.

## Your Mirror Image

Hold a plane mirror so close to your face that you cannot see your image. Very slowly move it away from your face. Observe the size and orientation (upright or inverted) of your image as you move the mirror further from your face. Repeat the procedure with concave and convex mirrors.

### Analyze and Conclude

1. At what relative distances and with which mirrors was your image (a) upright, (b) inverted, (c) larger than your face, and (d) smaller than your face?

2. For each mirror, summarize the changes that took place as you moved the mirror further from your face.

## Reflecting Images

Place a concave mirror behind a light source such as a small bulb or a candle. Move a paper screen around through many positions until you can see an image of the light source on the screen. You may have to move the light source also. Once you have found an image, note the distances of the mirror and screen from the light source. Also note the size and orientation of the image compared to the light source. Attempt to carry out the procedure with plane and convex mirrors.

### Analyze and Conclude

1. Under which conditions could you project an image of the light source onto the screen?

2. Describe the conditions under which it was not possible to form an image on the screen.

3. Make a general statement about the conditions needed to project an image on a screen using a mirror.

- Describe the wave theory and the wavefront and light ray models of light, and use them to explain how we see objects.

- Illustrate and predict the reflection of light, according to the law of reflection.

**KEY**
**TERMS**

- wave theory of light
- wavefront model
- linear propagation of light
- ray model
- light ray
- law of reflection
- regular reflection
- diffuse reflection

---

**PHYSICS FILE**

Isaac Newton believed that light consisted of extremely rapidly moving *particles* and many other scientists of his time followed his teachings. Later, Christian Huygens demonstrated that light travelled with *wave*-like properties such as constructive and destructive interference. Eventually, Max Plank, Albert Einstein, and others demonstrated that, when light interacts with matter, it behaves like particles or packets of energy now called photons. Today, scientists accept the "wave-particle" duality of the nature of light. Light travels like a wave and interacts with matter like a particle. Nevertheless, physicists still seek a more complete model of light.

More than 80% of the information humans receive about their surroundings is carried by the light that enters their eyes. You can read this book and see your classmates because light reflected from all these objects and people reaches your eyes. The information that you will learn from this book about light and images will come to you in the form of light reflected from the pages of this book and form images on the retinas of your eyes.

## The Wave Theory of Light

When possible, scientists try to develop a theory that can explain as many aspects as possible of the phenomena they are investigating. As scientists gathered data and information about the propagation of light, they discovered many wave-like properties. Light can be reflected, refracted, and diffracted in essentially the same way as water waves and other forms of mechanical waves. However, no material objects are in motion when light energy travels from one location to another so light cannot be a mechanical wave. In fact, light can travel through a vacuum. Nevertheless, the similarities outweigh the differences and scientists now accept the **wave theory of light**, in which light energy travels in the form of electromagnetic waves.

You can apply all of the principles of waves that you studied in Chapter 7 to the transmission of light, except that light does not need a medium in which to travel. You cannot see the wavefronts of light but you can envision them as very similar to water wavefronts. However, you need to try to picture them in three dimensions as shown in Figure 10.1. Light wavefronts from a point source (very small source) move out as spherical surfaces. If you use a parabolic reflector like the one you used to experiment on water waves, you can guide the waves so that the wavefronts are flat, planar surfaces. The **wavefront model** of light allows you to explain and predict many of the properties of light propagation.

The observation in the unit opener on page 459 that light travels in straight lines is such a fundamental property that it has been accorded the status of a principle. The principle of **linear propagation of light** leads to the very useful **ray model** of light. A **light ray** is an imaginary arrow that points in the direction of the propagation of light. Like mechanical waves, the direction in which light energy travels is perpendicular to the wavefronts. Thus light rays are always perpendicular to wavefronts. The wavefront model can explain the properties of light, but wavefronts are

**A**

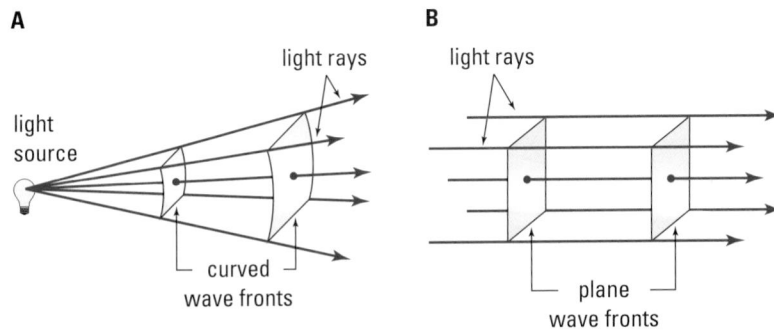

light source

light rays

curved
wave fronts

**B**

light rays

plane
wave fronts

**Figure 10.1** **(A)** When the light waves are close to the light source, the wavefronts appear as parts of a sphere, and the light rays gradually diverge (spread out) from one another. **(B)** When the light waves are at a great distance from the light source, the wavefronts can be considered parallel to one another. The light rays are also parallel, indicating the direction in which the light is travelling.

more difficult to sketch and interpret than light rays. Therefore, you will be using light rays to understand, describe, and predict the properties of light through out this unit.

## Reflection

In Chapter 7, you briefly considered plane water waves reflecting from a straight barrier. The law of reflection, introduced in connection with water waves, applies to all types of waves but it takes on greater importance as you learn about the formation of images by light rays. Examine Figure 10.2 to review the terms and symbols scientists use in discussing reflection. You always draw a normal (or perpendicular) line from the reflecting surface from the point at which a light ray strikes a reflecting surface. The angle between the incident ray and the normal line is the angle of incidence, $\theta_i$. The angle between the reflected ray and the normal line is the angle of reflection, $\theta_r$. The **law of reflection** states that the angle of reflection is always equal to the angle of incidence. This law applies to all surfaces.

---

### THE LAW OF REFLECTION

The angle of incidence, $\theta_i$, equals the angle of reflection, $\theta_r$, and the incident light ray, the reflected light ray, and the normal to the surface all lie in the same plane.

---

**Language Link**

The word "ray" is derived from the Latin word *radius*. Usually, light fans out radially from a point source of light. What are some other examples of radial phenomena?

**PHYSICS FILE**

In Unit 3, Mechanical Waves, it was established that all waves travelled through a medium. In the seventeenth century, Huygens, a Dutch physicist, developed the first significant theory proposing that light travelled as a wave. Scientists supporting this theory believed that light (electromagnetic) waves travelling through space from the Sun must also travel through a medium. By the mid-nineteenth century the wave theory was well established, and scientists "invented" a medium, called "ether," that was considered to permeate the entire universe. The electromagnetic waves were called "ether waves." In 1887, a series of brilliant experiments by Michelson and Morley proved conclusively that ether did not exist, and it was established that electromagnetic waves like light could, in fact, travel through a vacuum without the need of a medium of any kind.

**Language Link**

The word "normal" is derived from the Latin word *norma*. What does *norma* mean? Why is this appropriate for the use of "normal" in mathematics and physics?

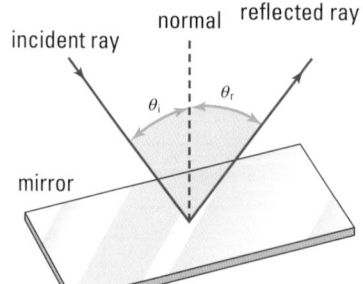

**Figure 10.2** The angle of incidence, $\theta_i$, equals the angle of reflection, $\theta_r$. Both of these angles are measured with respect to the normal, a line drawn at right angles to the mirror surface at the point of incidence.

There is a legend that the ancient Greek mathematician and inventor, Archimedes, used mirrors to set fire to Roman ships around 212 B.C.E. In 1973, the Greek historian, I. Sakkas, set out to test whether Archimedes could have done so. He lined up 70 soldiers with flat copper shields and directed them to reflect sunlight to a rowboat anchored 50 m from shore. The boat soon caught fire.

This doesn't prove the story was true, but it does show that Archimedes could have used such a method.

## Regular and Diffuse Reflection

Why is it that you can see yourself in a mirror, but you cannot see yourself in a piece of paper, such as this page in your textbook, when you hold it in front of your face? The quick answer is that the mirror surface is much smoother than the surface of the paper. Try looking at other shiny surfaces, such as the cover of your textbook. You might be able to see a faint reflection of your face on the book cover.

How smooth does a surface have to be to behave like a mirror? If you magnify the surface of a mirror several hundred times, the surface still appears flat and smooth. However, the magnified surface of a page in a book is quite irregular, as shown in Figure 10.3.

To understand what happens to the light when it strikes these two kinds of surfaces, use the light ray model to see the path of the light. When a set of parallel light rays hit the mirror surface in Figure 10.4, the reflected light rays are also parallel to one another. This type of reflection is called **regular** (or specular) **reflection**. Each of the parallel light rays that hit the irregular surface in Figure 10.4 is reflected in a different direction. This type of reflection is called **diffuse reflection**. However, the law of reflection still applies for every light ray reflecting from the irregular surface. The surfaces of many common objects, such as most types of cloth, paper, brick and stone surfaces, and pieces of wallpaper, reflect light in this way.

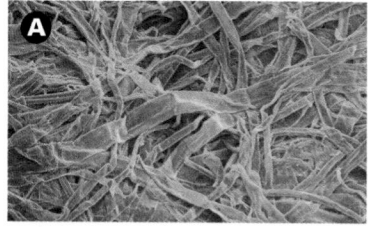

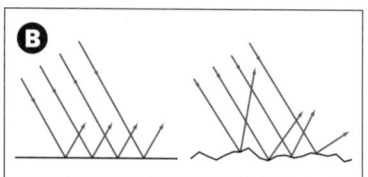

**Figure 10.3** (A) The page of a book viewed under a microscope. (B) Which of the ray diagrams illustrates reflection off paper?

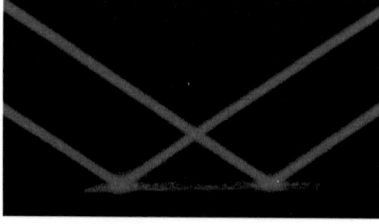 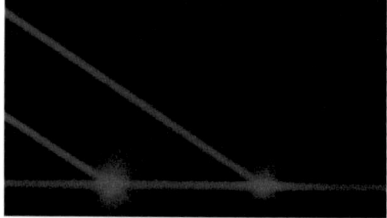

**Figure 10.4** A laser beam is a group of parallel light waves that are close enough together to form a fine beam of light. Notice that the laser beams reflected from the mirror are parallel, but those that strike the rough surface appear as fuzzy, round dots because the incoming parallel light rays in the beam are reflected in many different directions.

## Concept Organizer

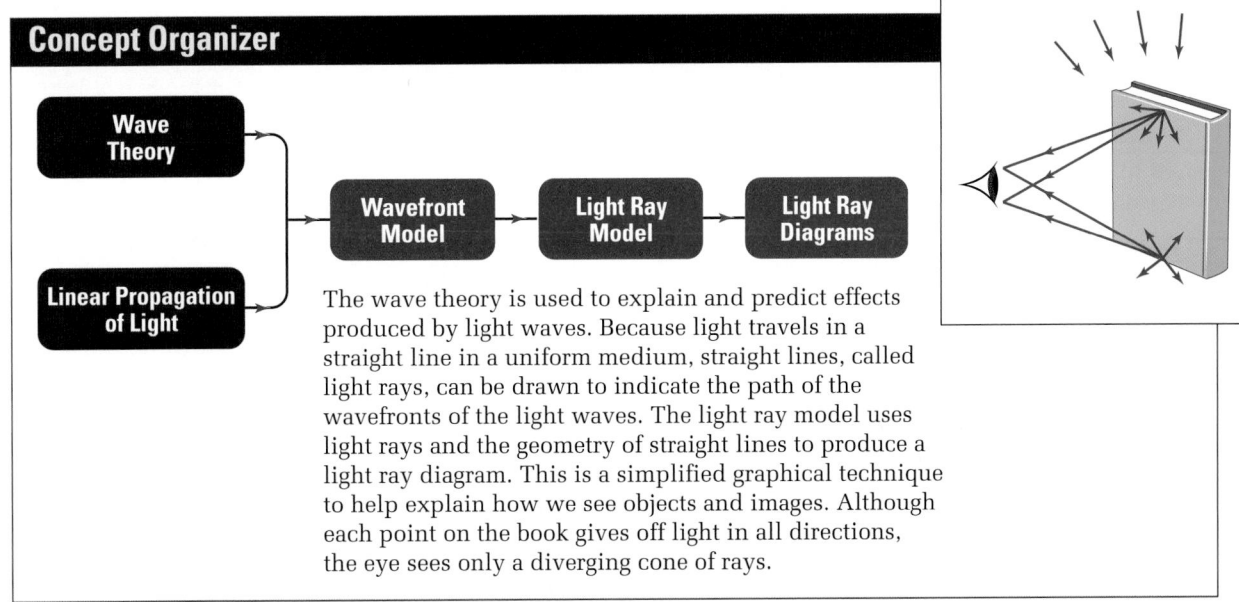

Wave Theory → | Linear Propagation of Light → | Wavefront Model → Light Ray Model → Light Ray Diagrams

The wave theory is used to explain and predict effects produced by light waves. Because light travels in a straight line in a uniform medium, straight lines, called light rays, can be drawn to indicate the path of the wavefronts of the light waves. The light ray model uses light rays and the geometry of straight lines to produce a light ray diagram. This is a simplified graphical technique to help explain how we see objects and images. Although each point on the book gives off light in all directions, the eye sees only a diverging cone of rays.

**Figure 10.5** Using light ray diagrams to model how you see an object.

## QUICK LAB

# Predicting Reflections

**TARGET SKILLS**
- Predicting
- Performing and recording
- Analyzing and interpreting

Obtain a white piece of paper, a pencil, a ruler, a plane mirror, and a ray box. Draw a line across the upper left corner at approximately a 60° angle with the side of the page to represent a mirror surface. Near the middle of the line, mark a point and label it A. On the right side of the paper, mark a point and label it B. Imagine a light ray coming from B and hitting the "mirror surface" at A. Predict the direction of the reflected ray and mark it on the paper. Repeat the procedure for two more pairs of points $A_2$, $B_2$ and $A_3$, $B_3$.

Place a mirror directly on top of the "mirror" line on the paper. With the ray box, find and mark the direction of the reflected rays for each of the "rays" that you marked on the paper. Compare the directions of the actual ray with your predictions.

### Analyze and Conclude

1. Explain the reasoning that you used when you were making your predictions about the directions of the reflected rays.

2. Comment on the accuracy of your predictions.

3. Formulate a method that would allow you to make accurate predictions. Then test your method.

1. **K/U** List four effects that can be observed that demonstrate light travels in straight lines.

2. **C**
   (a) State the law of reflection for light.
   (b) Draw a fully labelled diagram to show the reflection of light from a plane mirror surface.

3. **K/U** Why does it become more difficult to read a page in a glossy magazine than a page in this book when reading in bright sunlight?

4. **C**
   (a) Explain the difference between the wavefront and light ray models of light.
   (b) Which of the two models is most useful for explaining effects produced by optical devices? Why?

5. **K/U** A light ray strikes a mirror at an angle of 72° to the normal.
   (a) What is the angle of reflection?
   (b) What is the angle between the incident ray and the reflected ray?

6. **K/U** A ray of light strikes a plane mirror at an angle of 26° to the mirror surface. What is the angle between the incident ray and the reflected ray?

7. **I** Design and carry out an experiment, using one or more ray boxes and several reflecting surfaces, to show the law of reflection still applies when light undergoes diffuse reflection.

8. **K/U** Light is shining onto a plane mirror at an angle of incidence of 48°. If the plane mirror is tilted such that the angle of incidence is increased by 17°, what will be the total change in the angle of reflection from the original reflected light?

9. **C** In the room illustrated in the figure, the room is dark and has perfectly black walls. The air in the room is clear, without dust or smoke. If you stay at the position indicated by the eye, can you see the mirror on the opposite wall when a collimated beam of light enters the room in the direction indicated by the line segment? Explain your answer.

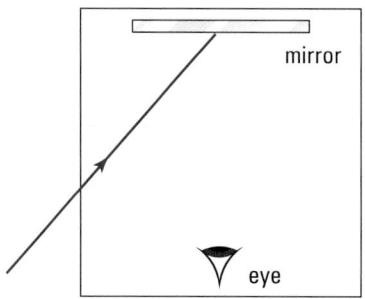

10. **MC** The windows of some large buildings are coated with a very thin of layer of gold. Why was this material used? From an economic point of view, is the use of gold justifiable for this purpose?

---

### COURSE CHALLENGE

**We're Losing Power**

Signal attenuation refers to energy loss. This energy loss can be result of absorption or reflection. Design an experiment to test the attenuation properties of various substances to microwaves radiation. (Hint: Start with an ice cube and a microwave oven.) Have your teacher verify that your procedures are safe before conducting any experiments.

Learn more from the Science Resources section of the following web site: *www.school.mcgrawhill.ca/resources/* and find the *Physics 11 Course Challenge*

The dentist in Figure 10.6 is using a mirror to see inner side of the patient's teeth. Dentists can determine precisely where an object is by using mirrors. When mirrors are built into instruments such as periscopes, cameras, telescopes, and other devices, they must be placed in precisely the correct position before the instrument is ever used. The engineer that designs the instrument must have a method to determine exactly how light rays will be reflected, in order to design the instrument. In the remainder of this chapter, you will learn how to use the law of reflection and the geometry of light rays to predict the location and characteristics of images formed by mirrors.

**SECTION EXPECTATIONS**

- Classify images formed by reflection according to their characteristics.

- Use the light ray model to predict image characteristics for a plane mirror, and test these predictions experimentally.

**KEY TERMS**

- object
- image
- real image
- virtual image
- plane mirror
- geometric optics

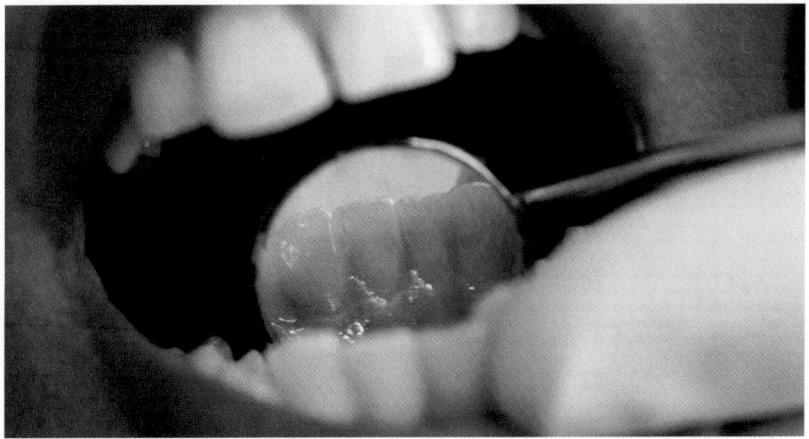

**Figure 10.6** A dentist is using a mirror to see the inner side of the patient's teeth.

## Objects and Images

As you have read many times in this text, everyday terms used in physics often have specific meanings. In the study of geometric optics, two of the most important terms are object and image. The **object** is the tangible item that you see in the absence of any optical devices. You can see an object because light rays are spreading out or diverging from every point on the object. In most cases, light from a luminous source is shining on the object and the object reflects the light by diffuse reflection. Your eyes collect the diverging rays and focus them on the retinas of your eyes. When an optical device redirects the diverging light rays and makes the rays appear to be coming from a point that is not really on the object, the device has formed an **image**.

**History Link**

Polished metal mirrors used in Egypt around 2000 B.C.E. were used until the sixteenth century when the first glass mirrors were made in Venice. Early glass mirrors were sheets of glass attached to very thin sheets of tin coated with mercury. In 1857, Foucault developed a process for coating glass surfaces with silver. How have mirrors developed since then?

When you were observing images in different mirrors in the Multi-Lab on page 461, you saw a variety of sizes, shapes, and orientations of images. You can completely describe any image by defining four characteristics. The magnification is the ratio of the image size to the object size. If the magnification is greater than one, the image is larger than the object. When it is equal to one, the object and image are the same size. If the magnification is less than one, the image is smaller than the object. The attitude of an image indicates whether the image is oriented the same way as the object (upright) or upside down (inverted) with respect to the object. The image position is the distance between the image and the optical device — mirror or lens.

The fourth characteristic, the type of image, indicates whether the image is real or virtual. An image is **real** if light rays are actually converging at a point then continuing on beyond that point and diverging. In other words, if you placed a screen at the image position, the image would appear on the screen. The student in Figure 10.7 B is using his hand as a screen to demonstrate the meaning of a real image.

If an image is not real, it is **virtual**. If you placed a screen at the position of a virtual image, nothing would appear on the screen. There are no light rays actually converging on the image position, as Figure 10.7 A reveals. Light rays only appear as though they are diverging from the image location. This will become more clear as you practice drawing ray diagrams. The four image characteristics are summarized in Table 10.1.

 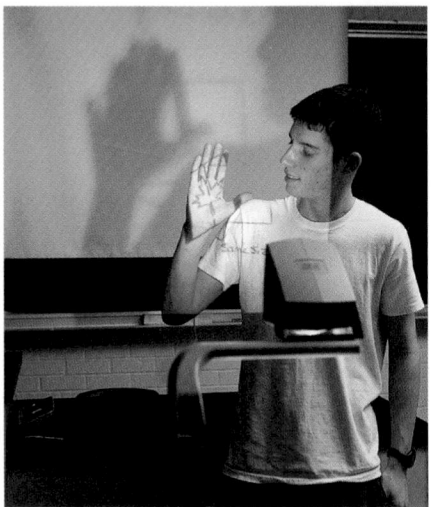

**Figure 10.7** **(A)** No image is formed on the screen behind the plane mirror because a virtual image is formed.

**(B)** When the student places his hand in front of an overhead projector screen, a real image is formed on his hand.

**Table 10.1** Image Classification Properties

| Property or characteristic | Possible values | Comments |
|---|---|---|
| magnification | larger same size smaller | |
| attitude | upright | The image has the same vertical orientation as the object. |
| | inverted | The image is upside down in comparison to the object. |
| type | virtual | If a screen is placed at the image position, no image will appear on the screen. |
| | real | If you placed a screen at the image position, an image would appear on the screen. |
| position | measured from optical device | |

## Images and the Light Ray Model

You can determine all of the characteristics of an image by using the light ray model. Since you are very familiar with **plane mirrors** (mirrors with a flat surface), they will provide a good learning tool. By simply looking at an image in a plane mirror, you can qualitatively describe the image. Look at the Canadian flag in Figure 10.8. As you can see, the image is upright. It also appears to be the same size as the object and the same distance behind the mirror as the object is in front of the mirror. Although the image appears to be behind the mirror, it is obvious that there are no light rays coming from behind the mirror and passing through it. Therefore, the image must be virtual.

**Figure 10.8** The image of the flag is the same size, is upright, and is the same distance behind the mirror as the flag is in front of it. However, if you hold a screen behind the mirror where the image appears to be, no light from the flag is able to reach the screen. The image formed by any plane mirror is a virtual image.

The light ray model provides a method for predicting the four properties of an image without using a mirror. Study Figure 10.9 to see how light rays and the law of reflection allow you to predict the image characteristics quite precisely. Once again, the Canadian flag is the object. The straight line represents the mirror. As you now know, light rays diverge from every point on the flag. You can choose any point or combination of points and draw several rays emanating outward. In the figure, rays are diverging from Point $O$. Those rays that strike the mirror will reflect according to the law of reflection. You can see that the rays continue to diverge after they reflect from a plane mirror. The point $O'$ from which those diverging rays *appear* to be diverging, is the location of the image of point $O$ on the flag. To find the image location, you simply continue the light rays back behind the mirror. The point where two or more extended rays meet is the location of the image. You should always use dashed lines to extend the rays to indicate that these are not actually light rays but instead, an imaginary path. When used in this way, dashed lines will always indicate that the image is virtual. You can repeat the process from as many points on the image as you choose in order to visualize the image as a whole. For example, if you sketched rays from point $A$ on the flag, you would find that they appear to be diverging from point $A'$ behind the mirror. For practice using the ray model, complete the following investigation.

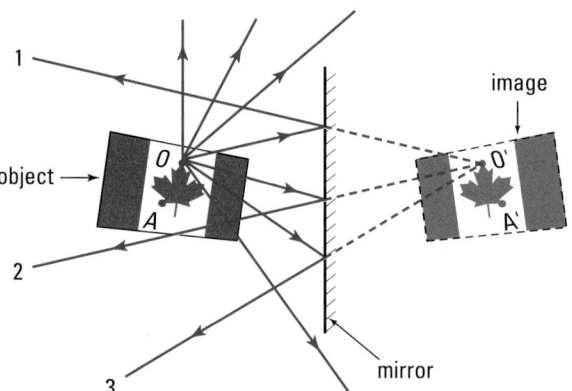

**Figure 10.9** The image of point $O$ can be seen in the mirror from any location in front of the mirror where the reflected light rays from point $O$ are travelling.

Notice that point $A$ is on the left side of the flag but $A'$ is on the right side of the image of the flag. This apparent right-left reversal is a characteristic of all mirror images.

# INVESTIGATION 10-A

## Predicting Image Characteristics Using the Light Ray Model

### Problem

How can you use light ray diagrams to predict the characteristics of an image formed by a plane mirror?

### Equipment

- plane mirror
- mirror supports
- ray box with single ray

### Procedure

1. Draw a diagram similar to the one shown here. Place the "object" about 10 cm in front of the centre of the mirror.

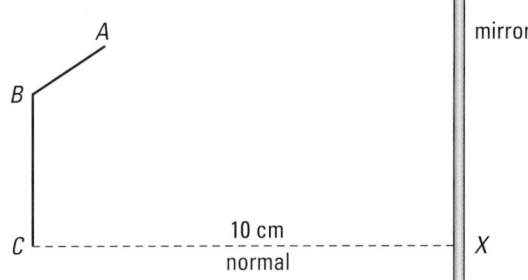

2. Draw a normal line from point $C$ to the mirror. This is your reference line.

3. Support the mirror vertically. Using the ray box, shine a single light ray so that it appears to travel from point $A$ toward the mirror surface along a normal line to the mirror.

4. Mark a small dot just in front of the mirror to indicate the path of the light ray on the paper.

5. Move the ray box, and shine a second light ray toward the mirror so that it appears to travel from point $A$ to point $X$ where the reference line meets the mirror. Mark dots on the paper to indicate the path of both the incident and reflected rays.

6. Move the ray box, and shine a third light ray from point $A$ toward the mirror so that it reflects from the mirror surface at any point above or below the other two rays. Mark the path of the incident and reflected rays.

7. Remove the ray box and mirror from the paper. Draw lines to represent the three incident and reflected rays. Draw arrows on the lines to represent the direction in which the light rays are travelling.

8. Draw dotted lines that extend the reflected ray behind the mirror line until they cross one another.

9. Repeat steps 3 to 8 for the point marked "$B$" on the object in the diagram.

10. Modify the procedure in steps 3 to 8 as required, and draw the path of three light rays coming from point $C$.

11. Draw a dotted line to show the position of the image behind the mirror.

### Analyze and Conclude

1. Describe the four characteristics of your image.

2. What is the minimum number of light rays that need to be drawn to establish the location of a point on an image? Why?

## Drawing Light Ray Diagrams

You could use any light ray from the object that strikes a mirror to help locate an image. However, certain rays are much easier to use. In fact, two specific rays will allow you to locate an image formed by any type of mirror or lens. Often, you will want to use a third ray, just as a test to ensure that you drew the first two rays correctly. Table 10.2 outlines the procedure for using the first two rays for plane mirrors.

**Table 10.2** Light Ray Diagrams for Plane Mirrors

| Description | Comments | Illustration |
|---|---|---|
| **Reference line**<br><br>Draw a normal line from the base of the object to the mirror. | Since the line is perpendicular to the mirror, an incident ray would reflect directly backward. The ray extended behind the mirror would form a continuous, straight line. Thus the base of the image will lie on this line. | |
| **Light ray #1**<br><br>Draw a ray from the top of the object to the mirror parallel to the reference line. | Since this line is parallel to the reference line, it is normal to the mirror. The reflected ray will go back along the same line. | |
| **Light ray #2**<br><br>Draw a line from the top of the object to the point where the reference line meets the mirror. Draw the reflected ray according to the law of reflection. | Instead of measuring angles to draw the reflected ray, you can construct congruent triangles. Mark a faint point directly below the object that is as far below the line as the top of the object is above the line ($T$). Draw the reflected ray from the mirror to this point. Since the triangles are congruent, the angle of reflection must be equal to the angle of incidence. | |
| **Extended rays**<br><br>Extend both rays behind the mirror, using dashed lines, until they intersect. | The point at which the extensions of the rays meet is the position of the top of the image. | |

# Geometry and the Properties of Images

Qualitatively, observation of images in mirrors and light ray diagrams give a strong indication that an image in a plane mirror is the same size as the object. However, physicists always prefer to demonstrate such relationships quantitatively. One reason that the light ray model is so useful is that you can apply the mathematics of geometry and develop quantitative relationships. For this reason, this branch of physics is known as **geometric optics**.

Figure 10.10 shows a ray diagram of an object (an arrow) with the two rays described in Table 10.2 and the image. Notice that the distance from the object to the mirror is labelled $d_o$ and the distance from the mirror to the image is labelled $d_i$. Consider the following aspects of the figure.

- $\angle\beta = \angle\theta$ because they are opposite angles formed by intersecting lines.

- $\angle\alpha_1 = \angle\alpha_2$ because they are complementary to equal angles.

- The angles at point $C$ are both right angles.

- Triangle $ABC$ is congruent to triangle $DBC$ because they have two equal angles with a common side between the equal angles. (angle-side-angle)

- $d_i = d_o$ because they are equivalent sides of congruent triangles.

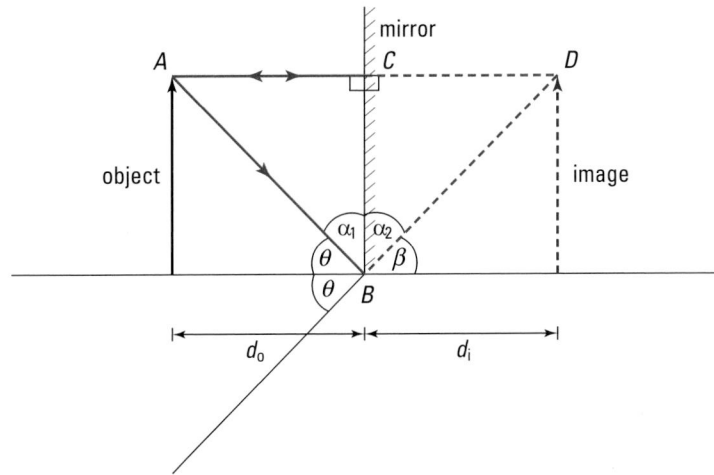

**Figure 10.10** The triangles formed by the incident and extended light rays are similar triangles. By using geometry, it is possible to show that the image distance $d_i$ equals the object distance $d_o$.

## Think It Through

- Make a ray diagram identical to Figure 10.10. Add the labels, $h_o$ for the height of the object and $h_i$ for the height of the image. Use geometry to verify that the heights of the image and object are the same.

*Career Link*

"Chips" Klein is an inventor and successful businesswoman. She created and produced a special kind of make-up mirror that consists of three mirrors mounted in an arc. The mirrors are being sold throughout the world. Klein is one of the founding members of the Women Inventors Project, an organization that provides information, expertise, and encouragement to women inventors throughout Canada. Find out more about the Women Inventors Project and some of the inventions associated with it.

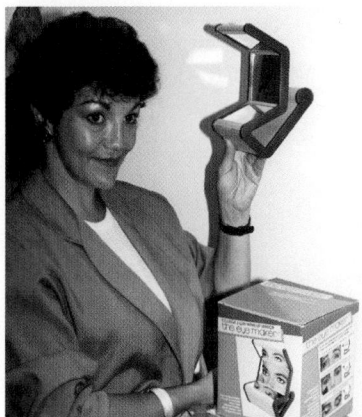

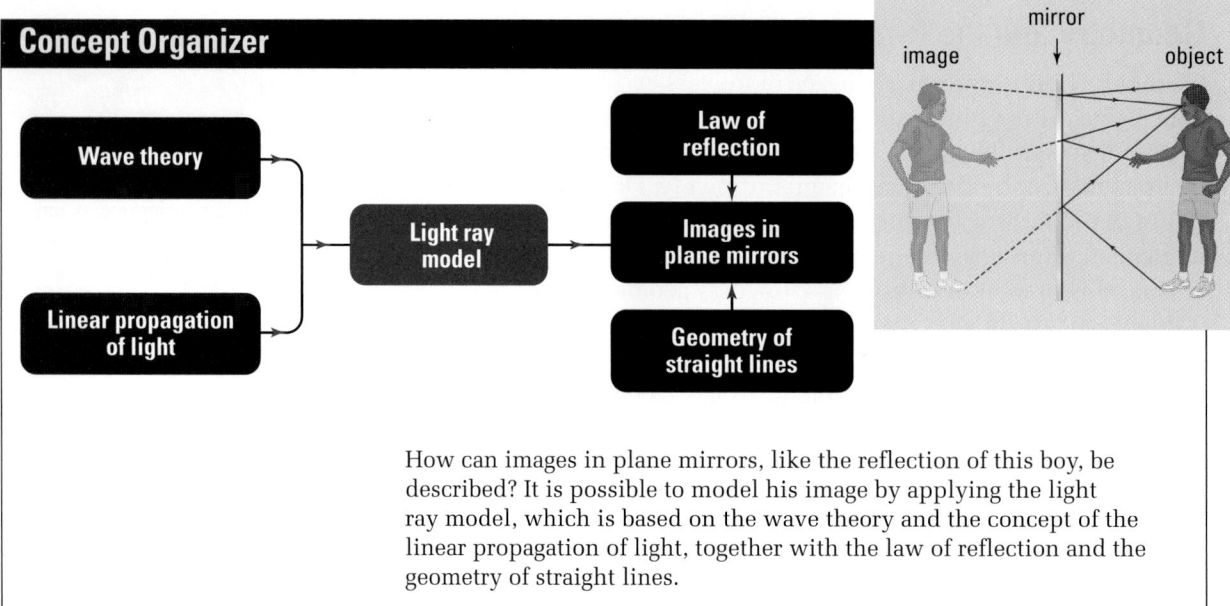

## Concept Organizer

Wave theory

Linear propagation of light

Light ray model

Law of reflection

Images in plane mirrors

Geometry of straight lines

mirror

image    object

How can images in plane mirrors, like the reflection of this boy, be described? It is possible to model his image by applying the light ray model, which is based on the wave theory and the concept of the linear propagation of light, together with the law of reflection and the geometry of straight lines.

**Figure 10.11** Understanding images in plane mirrors

In summary, the characteristics of images in plane mirrors are:

- the image is the same distance from the mirror as is the object
- the image is the same size as the object
- the image is always upright
- the image is always virtual

## Applying Plane Mirrors: MEM Chips

Micro-electro-mechanical (MEM) computer chips are silicon chips the size of a thumbnail that have tiny movable plane mirrors mounted on them. When a small electric current is applied to the computer chip, the mirror changes its angle. MEM chips are used in "optical switches." Optical switches are electronic devices that control beams of light travelling inside optical fibres about as thick as a single human hair. The light coming out of one optical fibre can be reflected by the mirror into a nearby optical fibre. A single light beam can carry huge amounts of telecommunications information such as TV channels and telephone messages all at the same time. Bundles of these efficient optical fibres are now connected throughout most parts of North America, and by submarine cable to Europe and the rest of the world.

1. **K/U** Distinguish between an object and an image.

2. **K/U** Distinguish between converging light rays and diverging light rays.

3. **K/U** Why is it necessary to draw only two light rays from any given point on an object to determine the position of the image when light is reflected from a (plane) mirror?

4. **C** Copy the following diagrams into your notebook and draw a light ray diagram for each example to show how the image is formed in the plane mirror. Show all construction lines.

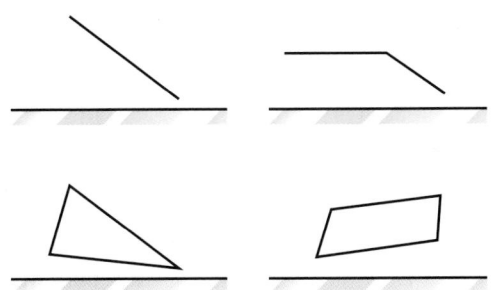

5. **K/U** A person is facing a mirror, observing a light bulb. At what position does the person "see" the image of the light bulb?

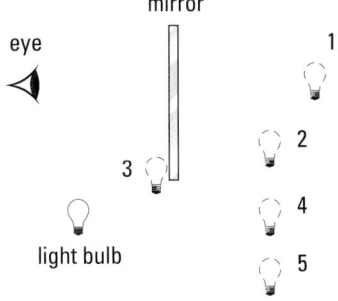

6. **C** Explain how it is possible to observe a given point on an object or an image from many different locations.

7. **K/U** State the characteristics of the image formed by a plane mirror. What characteristics of the image remain the same, and what ones change as the object position is changed? Explain your answer.

8. **C** A small, round bead, 1 mm in diameter, is placed 5.0 cm in front of a plane mirror. Draw an eye 7.0 cm to one side of the bead and 6.0 cm in front of the mirror. Draw a light ray diagram to locate the position of the image and to show how the light travels from the object to the eye to form the image.

9. **MC** Predict the shortest length a mirror could be that would enable all members of a family ranging in height from 2.1 m to 0.85 m to see full-length images of themselves. Assume that everyone's eyes are 10 cm below the top of their head, and that they are standing 1.5 m in front of the mirror. Does the distance the person stands in front of the mirror make any difference to the length of mirror required? Explain your answer, and draw a light ray diagram to illustrate the path of the light for each person.

10. **I** Design an experiment to illustrate the difference between real and virtual images.

11. **MC** Why are full-length mirrors in clothing stores sometimes tiltable?

**UNIT PROJECT PREP**

Plane mirrors are useful in the construction of optical devices.

- How will reflections from plane mirrors affect your final image?

- Describe, with the aid of light ray diagrams, the images formed in concave and convex mirrors by objects at different distances from the mirror.

- Develop and use equations to calculate the heights and distances of images formed by concave and convex mirrors.

- Use the light ray model to predict image characteristics for a concave mirror, and test these predictions experimentally.

KEY
TERMS

- spherical mirror
- concave mirror
- convex mirror
- centre of curvature
- radius of curvature
- vertex
- principal axis

- focal point
- focal length
- spherical aberration
- parabolic reflector
- mirror equation
- magnification equation

What do a shiny, metal soupspoon and the reflector in a car headlight have in common? Both are examples of curved mirrors. Some mirrors, such as those used to apply make-up or to shave, are designed to form images. Other mirrors are used for focussing light energy in a particular direction, such as those used in flash-lights or in halogen yard lights. Using modern manufacturing techniques, it is possible to produce very complex mirror shapes to suit any required need.

## Concave Mirrors

Next time that you are in a parking lot, look at the reflectors in the headlights of the different makes of car. Almost every model of car has a different-shaped headlight reflector. Many car manufacturers design the headlights to be a streamlined part of the car body, and so the curved mirrors in the headlights also have to be a special shape. However, to understand what happens to the light as it is reflected from curved mirrors, you must first analyze how light reflects from a relatively simple shape — a **spherical mirror**.

As Figure 10.12 shows, a spherical mirror has the shape of a section sliced from the surface of a sphere. If the hollowed inside surface is made into the mirror surface, it is called a **concave mirror** (think of the hollow entrance of a cave). If the outside of the sphere is made into a mirror surface, it is called a **convex mirror** (think of how your eyeball bulges outwards).

Consider the concave mirror surface shown in Figure 10.13. The **centre of curvature** (C), is located at the centre of an imaginary sphere with the same curvature as the mirror. The **radius of curvature** (R) is any straight line drawn from the centre of curvature to the curved surface. The geometric centre of the actual curved mirror surface is called the **vertex** (V). The straight line passing through both the vertex, V, and the centre of curvature, C, is called the **principal axis** (PA). One way to visualize the principal

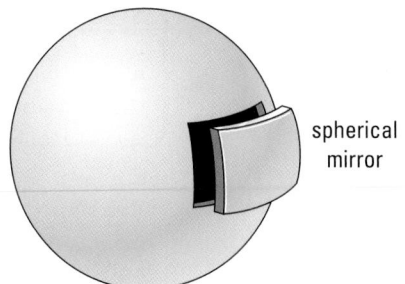

spherical mirror

**Figure 10.12** A spherical mirror has the shape of a segment of the surface of a sphere.

axis passing through the vertex of the curved mirror is to think of the handle of an open umbrella, and the way it appears to come from the centre of the umbrella covering, as seen in Figure 10.14. Figure 10.14 also illustrates an important point. All normal lines run along radii of curvature, thus they all meet at the centre of curvature. Each incident ray, its normal line, and reflected ray lie in a plane. Therefore, all reflected rays generated by incident rays that are parallel to the principal axis pass through the principal axis.

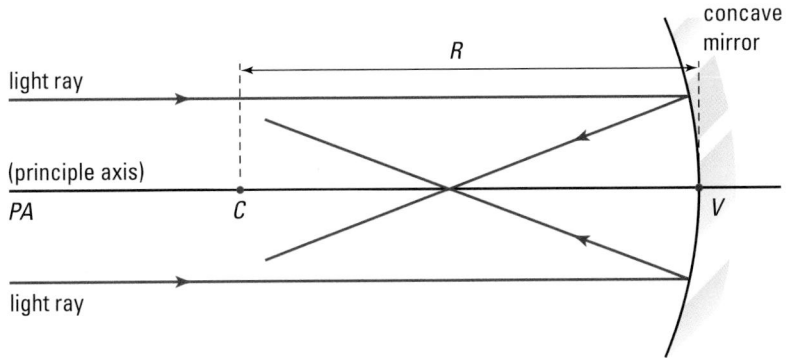

**Figure 10.13** $C$ on the principal axis of the spherical, concave mirror surface in this diagram is called the "centre of curvature." Light rays travelling parallel to the principal axis reflect through the same point.

**Biology Link**

A number of animals, including cats, have a mirror-like reflecting layer at the back of their eyes. When light is shone into one of their eyes, some of the light is reflected back towards the light source. What is this reflecting layer for?

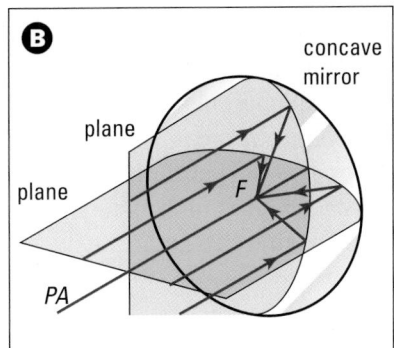

**Figure 10.14** **(A)** The handle of an umbrella passes through the fabric in a similar way to the principal axis passing through the vertex of the mirror. **(B)** Incident rays parallel to the principal axis create reflected rays that pass through the principal axis.

Consider what happens when light rays that are parallel to the principal axis, strike a concave mirror. Follow the path of the ray in Figure 10.15 on page 478. The light ray strikes the mirror at point $P$. The line segment $CP$ is the radius of the mirror and, therefore, is the normal to the spherical surface of the mirror at $P$. The light ray reflects from the mirror such that the angle of reflection $\theta_r$ equals the angle of incidence $\theta_i$. Furthermore, the angle $PCF$ is also $\theta_i$, because the radial line segment $CP$ crosses two parallel lines. Since two of its angles are equal, the (coloured) triangle $CPF$ is an isosceles triangle; thus, sides $CF$ and $FP$ are equal.

When the incoming parallel light ray lies close to the principal axis, the angle of incidence, $\theta_i$, is small, and the distance $FP$ becomes similar in length to distance $FV$. Because $\theta_i$ is small,

$CF = PF = FV$. Therefore, $FV = \frac{1}{2}CV$ and so point $F$ lies halfway between the centre of curvature and the vertex of the mirror. Point $F$ is called the **focal point**. The distance from the focal point to the vertex is called the **focal length** and is symbolized, $f$. The focal length, $f$, is one-half of the radius of curvature, $R$, for the spherical mirror. Point $F$ is called the focal point because all the incident light rays that are parallel and close to the principal axis of the mirror, reflect from the mirror and pass through that one point.

**Focal length** of a concave mirror: $f = \frac{1}{2}R$

**Figure 10.15** When the parallel light ray is close to the principal axis, the length segment *PF* is very close to the length *FV*. This shows that all light rays travelling close to the principal axis reflect through the same point, *F*.

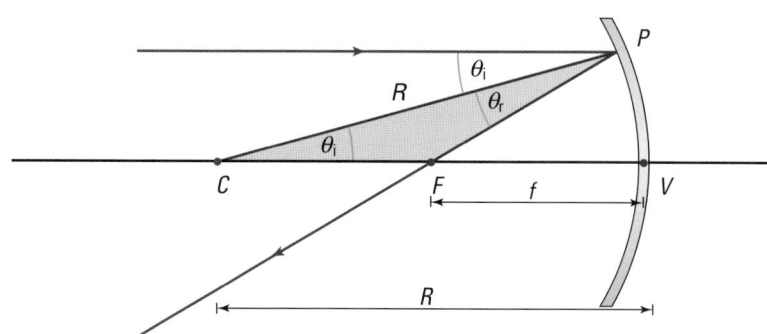

• *Think It Through*

• What are the focal length and radius of curvature of a *plane* mirror? To answer this, imagine a plane mirror as a section of a spherical mirror as in Figure 10.12. For a small sphere, any section is obviously curved. However, considering larger and larger spheres, you can see that the curvature of any subsection appears to "get flatter." A weakly curved, or flatter, surface has a larger radius of curvature than a strongly curved surface. Following this logic, answer the original question.

**COURSE CHALLENGE**

**From the Equator to the Poles**

Investigate the efficiency of photovoltaic cells from a local electronics shop. Is it possible to incorporate a parabolic reflector into a photovoltaic system to improve power output?

Learn more from the **Science Resources** section of the following web site: **www.school.mcgrawhill. ca/resources/** and find the *Physics 11 Course Challenge*

## Spherical Aberration

The formula above only applies to light rays that are close to the principal axis. As you can see in Figure 10.16 (A), parallel incident light rays that are not close to the principal axis reflect across it between the focal point and the vertex of the mirror. If you use a large spherical concave mirror as a make-up mirror, you may see a partially blurred image of your face, because all the light rays that are not close to the principal axis do not focus as they should. To avoid this **spherical aberration** and reflect all of the parallel incident light to a single focal point, the mirror's surface must have a parabolic shape, as shown in Figure 10.16 (B). A spherical mirror can be used to form reasonable images if the diameter of the mirror is small compared to its focal length.

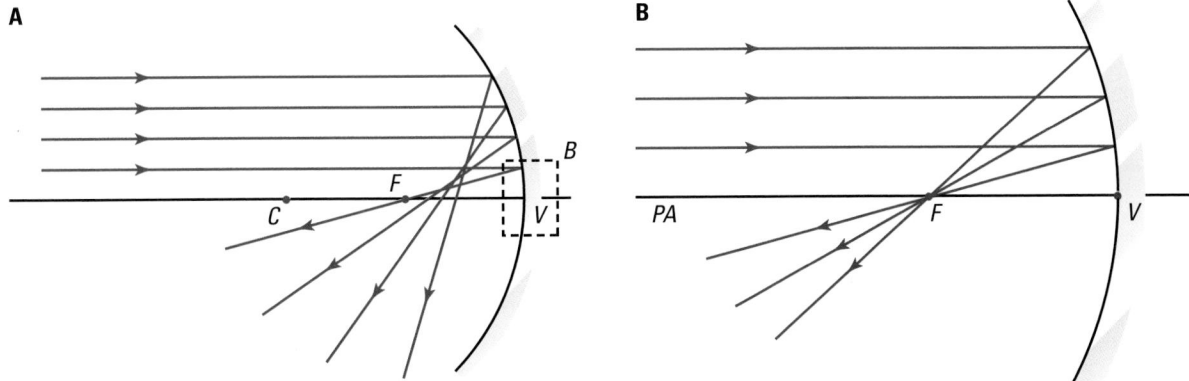

**A**

**B**

**Figure 10.16** **(A)** Incident light rays that are parallel but not close to the principal axis of a spherical concave mirror do not reflect through the focal point. **(B)** All incident light rays that are parallel to the principal axis of a parabolic reflector pass through its focal point.

When concave parabolic mirrors are used to reflect light energy, rather than to produce images, they provide an interesting example of the reversibility of the path of light. When light from a distant source such as the Sun is reflected by a **parabolic reflector** (mirror), all the light energy is focussed at the focal point of the mirror. Concave mirrors are used in parabolic reflectors to focus sunlight for solar-thermal energy facilities, such as those located in the Mojave Desert, near Barstow, California, and in the Pyrénées of southern France. The television satellite dish is another example of a parabolic reflector. In this case, electromagnetic signals are being received from orbiting transmitters. If the path of the light is reversed, and a light source is placed at the focal point of the mirror, light from the bulb is reflected from the mirror in a parallel beam, such as is formed by a searchlight.

Concave mirrors that are not parabolic also have many practical uses. Most car headlights are not perfectly parabolic because they are designed to reflect a beam of light that spreads out a little to illuminate the complete width of the road (see Figure 10.17). Solar ovens used for cooking are designed to reflect the sunlight onto a small area of the bottom of the pot (cooking utensil). Why would it not be practical to use a parabolic mirror for this purpose?

**Figure 10.17** The reflectors used in solar cookers and headlights are deliberately designed so that they do not have a perfectly parabolic shape.

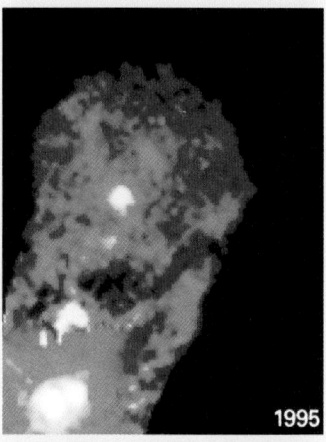

1995

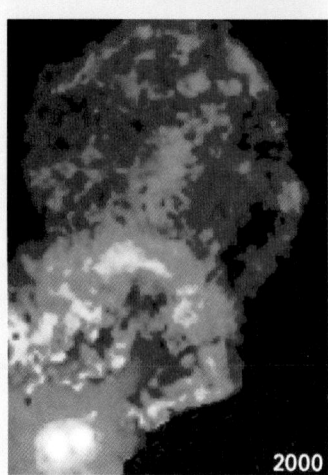

2000

## The Hubble Space Telescope

Astronomers need clear skies for their telescopes. That's why most of the large telescopes are on mountaintops — higher than most of the clouds, pollution, and hot-air currents that could degrade the images. For many years, astronomers realized that a perfect location for telescope would be in space, above the atmosphere. Astronomers' dreams were realized when, on April 24, 1990, the space shuttle *Discovery* carried the Hubble Space Telescope into orbit. Unfortunately, unexpected problems soon arose. As the telescope went from −150°C temperatures in the dark to +200°C in the sunlight, Hubble's solar arrays expanded and vibrated. Worse still, the primary mirror produced spherical aberration.

During subsequent space shuttle visits, astronauts corrected the problems, actually improving the telescope beyond the original specifications. Today, the Hubble Space Telescope produces breathtaking new images of the planets, stars, galaxies, and nebulas.

Light from distant objects enters the telescope through a door that can be closed to protect the inside from direct sunlight or space debris. The light reflects off the 2.4 m primary concave mirror to the secondary mirror. This 25 cm convex mirror reflects the light back through a hole in the primary mirror to the focal plane. Here, special mirrors correct the blurring and send the light to various optical instruments.

A Near Infrared Camera and Multi-Object Spectrometer use infrared radiation to produce images that highlight temperature differences. Visible light from nearby planets and distant galaxies is sent to the Wide Field Planetary Camera or to a smaller, high-resolution camera. The Space Telescope Imaging Spectrograph separates ultraviolet light into different frequencies in order to identify the chemical makeup of celestial objects. Special Faint Object Cameras can produce images in the visible and ultraviolet ranges.

The Hubble Space Telescope has recorded explosions of stars, found new planets around nearby stars, produced images of Pluto and its moon, photographed volcanoes on Jupiter's moon Io, and shown amazingly intricate webs of dust and gas in space almost to the edge of the universe.

### Analyze and Conclude

1. What is the great advantage of the Hubble Space Telescope?

2. What kinds of lenses or mirrors are used on the telescope?

3. What kinds of radiation does the Hubble Space Telescope detect? What can astronomers learn from each kind?

# Properties of Concave and Convex Mirrors

**TARGET SKILLS**

- Predicting
- Performing and recording
- Analyzing and interpreting

Place a concave mirror on edge in a piece of modelling clay, and point the mirror toward the Sun, tilting it slightly so that the light is reflected slightly upward. Place an opaque white screen in the path of the reflected light. Move the screen back and forth in the reflected beam, and locate the position where the light is focussed to the smallest possible spot of light. Repeat the procedure with the convex mirror.

**CAUTION** Take care not to reflect the sunlight into your eyes.

## Analyze and Conclude

1. What is special about the position of the bright spot of light?

2. What does the bright spot of light actually represent?

3. Determine the focal length of the concave mirror.

4. Comment on the observations made when using the convex mirror. Suggest reasons for these observations.

## The Geometry of Spherical Mirrors

Draw a line representing the principal axis on a blank piece of paper. Position a concave mirror shape at right angles to the principal axis line, with the vertex of the mirror touching the line. With a ray box shine several light rays parallel to the principal axis toward the mirror. Draw the path of the incident and reflected rays on the page.

Predict what will happen if you shine a set of light rays parallel to the principal axis of a convex mirror that is positioned the same way as the concave mirror. Check your prediction. For each light ray, mark the path of the light ray and draw it on the paper.

## Analyze and Conclude

1. What is the significance of each of the two points on the principal axis where the light rays cross?

2. What mathematical relationship exists between these two points?

3. Does the concave mirror surface demonstrate spherical aberration?

4. Determine the position of the focal point of the convex mirror, and measure the focal length.

# Images in Concave Mirrors

When you move a plane mirror away from your face, the only image characteristic that changes is the distance of the image from the mirror. The other three characteristics remain the same. The images formed by curved mirrors are much more varied, particularly in the case of the concave mirror. If you have a make-up or shaving mirror at home, try propping it up so that you can view your image in the mirror as you back away from it across the room. As you first move away from the mirror, the image of your face is upright and gets larger, then it disappears. As you continue moving away, your image re-appears but it is now upside down. As you continue to move away, the image continues to decrease in size. How can a single mirror shape form so many different

 **Web Link**

**www.school.mcgrawhill.ca/ resources/**
For more pictures and information about the Hubble Space Telescope, go to the above Internet site. Follow the links for **Science Resources** and **Physics 11** to find out where to go next.

kinds of images? You can answer this question by constructing ray diagrams for several different distances from a concave mirror. The directions for drawing ray diagrams for concave mirrors are given in Table 10.3. Note that a third, "test" ray has been added to the list.

**Table 10.3** Light Ray Diagrams for Concave Mirrors

| Description | Comments | Illustration |
|---|---|---|
| **Reference line**<br><br>Draw a principal axis from the centre of curvature to the vertex of the mirror. Place the base of the object on this line. | Since the line is perpendicular to the mirror, an incident ray would reflect directly backward. The ray extended behind the mirror would form a continuous, straight line. Thus the base of the image will lie somewhere along this line. | |
| **Light ray #1**<br><br>Draw a ray from the top of the object to the mirror parallel to the principal axis.<br><br>Draw the reflected ray through the focal point. | Since this incident ray is parallel to the principal axis, it will reflect through the focal point. | |
| **Light ray #2**<br><br>Draw a ray from the top of the object to the vertex.<br><br>Draw the reflected ray according to the law of reflection. | Instead of measuring angles to draw the reflected ray, you can construct congruent triangles. Mark a faint point directly below the object that is as far below the line as the top of the object is above the line. Draw the reflected ray from the mirror to this point. Since the triangles are congruent, the angle of reflection must be equal to the angle of incidence. | |
| **Light ray #3**<br><br>Draw a ray from the top of the object through the focal point and on to the mirror.<br><br>Draw the reflected ray parallel to the principal axis. | Any incident light ray passing through the focal point will reflect back parallel to the principal axis. | |
| **Extended rays**<br><br>If the reflected rays are converging, extend them until they all cross. If the reflected rays are diverging, extend them back behind the mirror with dashed lines. | The point at which the rays meet is the position of the top of the image. If the reflected rays meet, the image is real. If the rays must be extended backward, behind the mirror, the image is virtual. | |

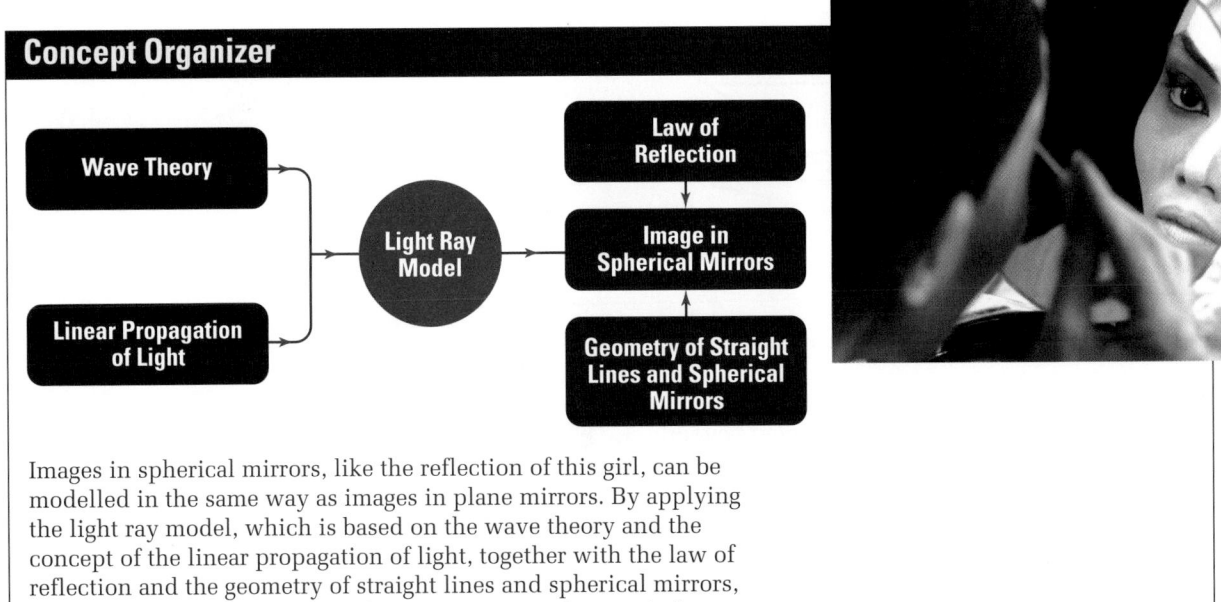
**Figure 10.18** Understanding images in spherical mirrors

You can now draw ray diagrams for concave mirrors with the object in three, representative regions. The results will provide examples of all possible classes of images that can be formed by concave mirrors. The regions are (1) beyond the centre of curvature, (2) between the centre of curvature and the focal point, and (3) between the focal point and the vertex.

The ray diagram in Figure 10.19 on page 484 has an object beyond the centre of curvature. As you can see, the reflected rays converge in front of the mirror and below the principal axis, making the image inverted. If you placed a screen at the image position, the reflected rays would focus on the screen and form an image. Therefore, the image is real. In addition, you can see that the image is smaller than the object.

Because your eyes can only focus diverging rays, you must be behind the image in order to see it. After the rays converge at the image position, they pass through each other and begin to diverge. These are the rays that you see.

In Figure 10.20 on page 484, the light ray diagram shows that when the object is placed between the centre of curvature, $C$, and the focal point, $F$, the image is still real and inverted, but is now larger than the object.

### • Think It Through

- At what position on the principal axis will the size of the image be the same as the size of the object?

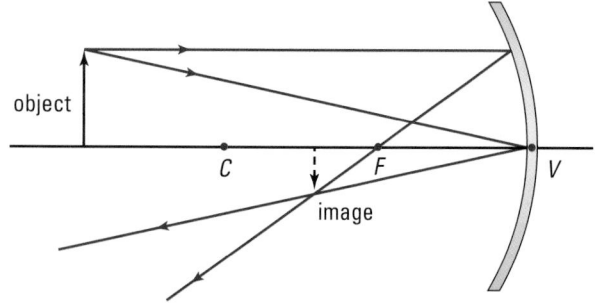

**Figure 10.19** When the object is located beyond the centre of curvature, the image is always inverted, real, and smaller than the object, and is located between the focal point and the centre of curvature.

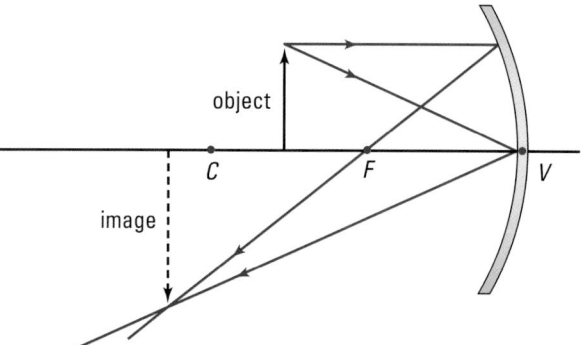

**Figure 10.20** When the object is located between the centre of curvature and the focal point, the image is always inverted, real, and larger than the object, and is located beyond the centre of curvature.

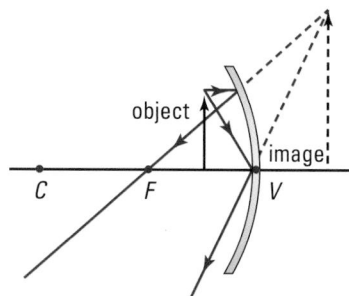

**Figure 10.21** When the object is located between the focal point and the vertex of the concave mirror, the image is always upright, virtual, and larger than the object. The image is located behind the mirror.

When the object is placed at the focal point, light rays are reflected parallel to one another and, thus, no image is formed because they never cross in front of the mirror or behind it. When the image of your face in a make-up mirror disappears, it is because your face is positioned at the focal point of the mirror. When the object is positioned between the focal point and the vertex of the mirror, the image properties change completely. As shown in Figure 10.21, the light rays reflecting from the concave mirror are now diverging. When *reflected light rays diverge*, no image will be formed in front of the mirror, and thus the image will be virtual. To locate the position of the image, the reflected light rays must be extended behind the mirror where they converge to a point. The virtual image is upright and larger than the object. A concave mirror can be used as a make-up or shaving mirror when the object, your face, is placed between the focal point and the vertex of the mirror.

## Applications of Concave Mirror Images

In movies such as *Top Gun*, some of the most exciting scenes occur when it seems that you are in the pilot's seat looking out through the cockpit window of the fighter plane. Because pilots operating these high performance aircraft do not have time to look down at their instruments, a special information display system called a "head-up display" (HUD) is used. The information given by the critical instruments in the cockpit is displayed directly on the window that the pilot looks through. This method of displaying information is now used in some cars and trucks (see Figure 10.22). The driver sees the speed of the car displayed on the windshield, and does not need to look down at the speedometer.

**Figure 10.22** HUD display in a jet

A concave mirror is a key part of the HUD technology. As shown in Figure 10.23, the digital readout of the speedometer is placed between the focal point and the vertex of the concave mirror. The mirror produces a virtual, enlarged, and upright image of the digital readout that reflects up onto the windshield of the car. A special layer of material called a "combiner" is imbedded in between the layers of glass in the windshield in front of the driver. The light from the concave mirror is reflected into the driver's eyes by the combiner material. The combiner material has special optical properties. All the colours in the light coming through the windshield to the driver's eyes pass through the combiner, except the colour given off by the digital speedometer display. The combiner also acts as a plane mirror to the colour of the light from the digital display, thus making the digital readout always visible to the driver. The display is usually located at the bottom of the windshield to provide the minimum distraction to the driver.

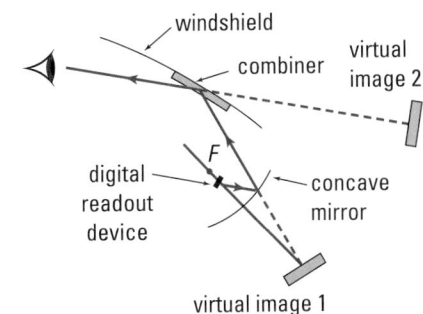

**Figure 10.23** The HUD display is formed by combining a concave mirror and a plane mirror (the combiner) that is mounted inside the glass of the windshield.

## CAREERS IN PHYSICS

### "Good Seeing"

The first time Dr. Neil de Grasse Tyson looked at the Moon through binoculars, he knew he wanted to be a scientist. Today, he is an astrophysicist doing research into dwarf galaxies and the "bulge" at the centre of the Milky Way. At a recent high school reunion, his former classmates voted him the one with the "coolest job."

Astrophysicists such as Dr. Tyson study the physical properties and behaviour of celestial bodies. They make observations using optical telescopes that generate visible images of stars and other celestial bodies by means of concave mirrors. Using computers, they record the images and then examine them in detail.

To find the kinds of images they seek, astrophysicists often need to travel. One of Canada's best optical observation sites is the Dominion Astrophysical Observatory (DAO) near Victoria, B.C. The Victoria area generally has clear nights and stable weather, which make for what astrophysicists call "good seeing." A more recently developed facility is the Canada-France-Hawaii Telescope (CFHT) on the extinct Hawaiian volcano Mauna Kea. Mauna Kea, with its dark skies and super-sharp star images, is the northern hemisphere's best site for optical observations.

Mauna Kea is also one site for a new optical-infrared telescope project. Called the Gemini project, it has another site on Cerro Pachon, a mountain in central Chile. Together, the twin Gemini telescopes give astrophysicists total coverage of both the northern and southern hemispheres' skies. That's "good seeing!"

### Going Further

Research one of these: the Dominion Astrophysical Observatory (DAO), the Canada-France-Hawaii Telescope (CFHT), or the Gemini Project.

# Images in Convex Mirrors

Do you remember reading the warning printed on the convex mirror on the passenger side of a car? This warning gives you a hint about the image formed by the mirror. What does the warning say, and why is it necessary? Next time you sit in the driver's seat of a car, look at an object in the rearview mirror inside the car, and then look at the same object through the passenger-side outside mirror. The image in the passenger-side mirror is smaller than the one in the plane mirror inside the car.

You can determine the properties of an image formed in a convex mirror by using the same light rays as those you used for the concave mirror. The relationship between the focal point of the convex mirror and its centre of curvature is the same as that of the concave mirror, except that for the convex mirror both points are behind the mirrored surface, rather than in front of it. The convex mirror has a virtual focal point. The three light rays used to determine the properties of the image are described in Table 10.4.

The properties of images formed by convex mirrors are very similar to those formed by plane mirrors. Look at the reflected light rays in the last illustration in Table 10.4. The reflected light rays will always diverge for all positions of the object in front of the convex mirror, so the images are always virtual. As well, the images will always be upright. The differences occur in the size and position of the images formed by the mirror. When the object is close to the mirror, the image size is slightly smaller than that of the object. As the object moves away from the mirror, the image becomes smaller and its position moves toward the virtual focal point of the mirror. The image will always be formed within the shaded triangle on the diagram.

Convex mirrors are generally used because they provide a very wide field of view, compared to a plane mirror. In many stores where they are used for security purposes, it is possible to see almost the entire store from one location. Convex mirrors are also used in hospital corridors and in factories for safety purposes to help prevent people and sometimes vehicles from colliding.

**Figure 10.24** Convex mirrors are used when you need to be able to view a large area with just one mirror. The smaller, distorted images formed by convex mirrors can sometimes cause safety problems when used for car rearview mirrors.

**Table 10.4** Light Ray Diagrams for Convex Mirrors

| Description | Comments | Illustration |
|---|---|---|
| **Reference line**<br><br>Draw a principal axis from the front of the mirror through the vertex to the centre of curvature. Place the base of the object on this line. | Since the line is perpendicular to the mirror, an incident ray would reflect directly backward. The ray extended behind the mirror would form a continuous, straight line. Thus the base of the image will lie somewhere along this line. | |
| **Light ray #1**<br><br>Draw a ray from the top of the object to the mirror parallel to the principal axis.<br><br>Draw the reflected ray as though it were coming from the focal point. | Any incident ray that is parallel to the principal axis, will reflect along a line from the focal point, through the point where the parallel ray meets the mirror. | |
| **Light ray #2**<br><br>Draw a ray from the top of the object to the vertex.<br><br>Draw the reflected ray according to the law of reflection. | Instead of measuring angles to draw the reflected ray, you can construct congruent triangles. Mark a faint point directly below the object that is as far below the line as the top of the object is above the line. Draw the reflected ray from the mirror to this point. Since the triangles are congruent, the angle of reflection must be equal to the angle of incidence. | |
| **Light ray #3**<br><br>Draw a ray from the top of the object toward the focal point behind the mirror. Stop the ray at the mirror.<br><br>Draw the reflected ray parallel to the principal axis. | Any incident light ray directed toward the virtual focal point will reflect back parallel to the principal axis. | |
| **Extended rays**<br><br>Extend the rays behind the mirror until they cross. | The point at which the rays meet is the position of the top of the image. Since the reflected rays meet behind the mirror, the image is virtual. | |

# The Mirror Equations

Drawing light ray diagrams provides a strategy for both explaining and predicting the properties of images formed in plane, concave, and convex mirrors. To determine the magnification of the image, and its position with precision, it is necessary to draw very accurate scaled diagrams.

For the best precision, you can calculate the image distance and magnification. When studying the ray diagrams for plane mirrors, you used the geometry of congruent triangles to show that the image distance from the mirror and the height of the image was the same as that of the object. By using a similar strategy, you can use similar triangles to derive two equations, known as the *mirror equation* and the *magnification equation*, to provide a complete quantitative description of the images formed in curved mirrors. The following quantities are included in these two equations:

$f$ = the focal length of the mirror

$h_o$ = the height of the object

$h_i$ = the height of the image

$d_o$ = the distance of the object from the mirror

$d_i$ = the distance of the image from the mirror

$m$ = the magnification of the image

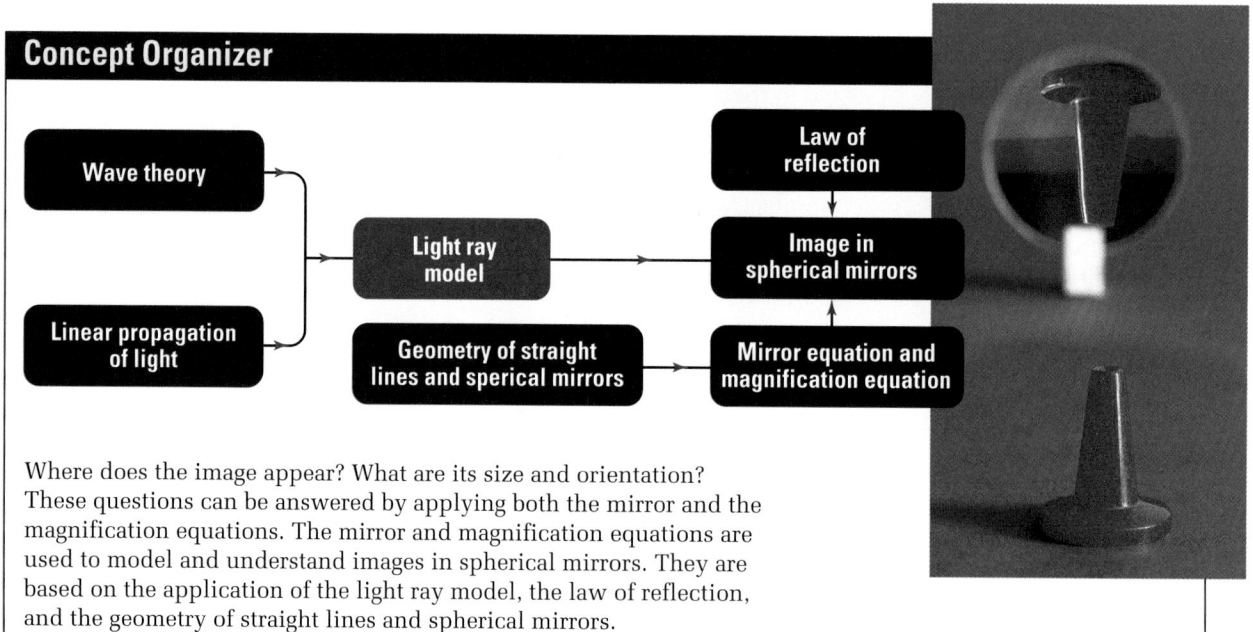

**Concept Organizer**

Where does the image appear? What are its size and orientation? These questions can be answered by applying both the mirror and the magnification equations. The mirror and magnification equations are used to model and understand images in spherical mirrors. They are based on the application of the light ray model, the law of reflection, and the geometry of straight lines and spherical mirrors.

**Figure 10.25** Understanding the properties of images in spherical mirrors

# CONVENTION FOR THE MIRROR AND MAGNIFICATION EQUATIONS

To include all of the possible properties of both images and objects, the following sign convention has been established for both concave and convex spherical mirrors.

## OBJECT DISTANCE

$d_o$ is positive for objects in front of the mirror (real objects)

## IMAGE DISTANCE

$d_i$ is positive for objects in front of the mirror (real images)

$d_i$ is negative for objects behind the mirror (virtual images)

## IMAGE ATTITUDE

$h_i$ is positive for images that are upright, compared to the object

$h_i$ is negative for images that are inverted, compared to the object

## FOCAL LENGTH

$f$ is positive for concave mirrors

$f$ is negative for convex mirrors

Examine the ray diagram in Figure 10.26. The shaded triangles are formed by a light ray travelling from the top of the object to the vertex of the mirror and the reflected ray travelling back at an angle that obeys the law of reflection. The angle of incidence is equal to the angle of reflection and the angles at $C$ and at $A$ are right angles. Since the triangles have two angles that are equal to each other, the third angles must be equal and the triangles are similar.

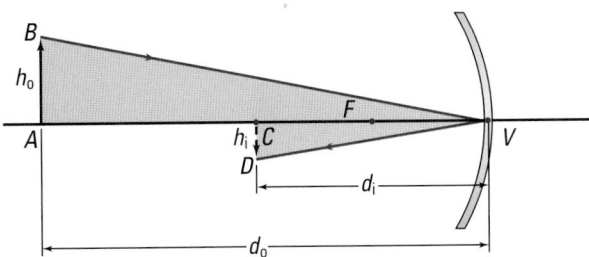

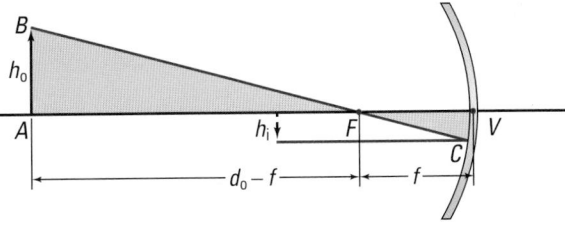

**Figure 10.26** The coloured triangles *ABV* and *CDV* are similar triangles. Remember that the mirror equation and the magnification equation are only valid if the parallel light rays are close to the principal axis of the mirror.

**Figure 10.27** The coloured triangles *ABF* and *VCF* are similar triangles.

Figure 10.27 shows different rays for the same object and image. Each of the shaded triangles has a right angle. Also, the small angles in each triangle are equal because they are opposite angles between two intersecting lines. Since the triangles have two equal angles, the third angles must also be equal and the triangles are similar.

The following procedure uses the rules of similar triangles to and geometry to develop the mirror equation.

- Ratios of equivalent sides of similar triangles (Figure 10.26) are equal. The image height is negative because it is inverted.

$$\frac{h_o}{-h_i} = \frac{d_o}{d_i}$$

- Using the same rule of similar triangles and the triangles in Figure 10.27, the relationship to the right holds true.

$$\frac{h_o}{-h_i} = \frac{d_o - f}{f}$$

- Since both of the equations have expressions equal to $\frac{h_o}{-h_i}$, you can set those expressions equal to each other.

$$\frac{d_o}{d_i} = \frac{d_o - f}{f}$$

- To isolate $f$, separate the right side as shown.

$$\frac{d_o}{d_i} = \frac{d_o}{f} - \frac{f}{f}$$

$$\frac{d_o}{d_i} = \frac{d_o}{f} - 1$$

- Add 1 to both sides of the equation.

$$\frac{d_o}{d_i} + 1 = \frac{d_o}{f}$$

- Create a common denominator on the left side.

$$\frac{d_o}{d_i} + \frac{d_i}{d_i} = \frac{d_o}{f}$$

$$\frac{d_o + d_i}{d_i} = \frac{d_o}{f}$$

- Divide both sides by $d_o$.

$$\frac{d_o + d_i}{d_o d_i} = \frac{\cancel{d_o}}{f \, \cancel{d_o}}$$

$$\frac{1}{f} = \frac{d_o + d_i}{d_o d_i}$$

- Separate the right side into two terms and simplify.

$$\frac{1}{f} = \frac{\cancel{d_o}}{\cancel{d_o} \, d_i} + \frac{\cancel{d_i}}{d_o \, \cancel{d_i}}$$

$$\frac{1}{f} = \frac{1}{d_i} + \frac{1}{d_o}$$

The final equation above is known as the **mirror equation**. It relates the distances of the object and the image to a concave mirror, in terms of the mirror's focal length.

## MIRROR EQUATION

The reciprocal of the focal length is the sum of the reciprocals of the object distance and the image distance.

$$\frac{1}{f} = \frac{1}{d_i} + \frac{1}{d_o}$$

| Quantity | Symbol | SI unit |
|---|---|---|
| focal length | $f$ | m (metres) |
| object distance | $d_o$ | m (metres) |
| image distance | $d_i$ | m (metres) |

### Think It Through

- Use the mirror equation to show that when the object distance is very large for concave mirrors, the image will be formed at the focal point. Hint: Let $d_o \rightarrow \infty$ so that $\frac{1}{d_o} \rightarrow 0$ and solve for $d_i$.

The magnification of the image in a mirror is the ratio of the image height to the object height: $m = \frac{h_i}{h_o}$. If the image is larger than the object, the magnitude of $m$ is greater than one, and if the image is smaller than the object, then $m$ is less than one. Rearranging the first equation in the derivation of the mirror equation, you find:

**Magnification Equation**  $\quad m = \dfrac{\text{image height}}{\text{object height}} = \dfrac{h_i}{h_o} = \dfrac{-d_i}{d_o}$

When determining the magnification, it should be noted that a positive value indicates that the image is upright, and a negative value indicates that the image is inverted, compared to the object.

### Think It Through

- Use the mirror and magnification equations to show that the image distance for a plane mirror is negative, and thus a virtual image is formed.

- Show that the magnification for a plane mirror is $m = +1$. Hint: begin by recalling that the focal length for a plane mirror is infinite, so that $\frac{1}{f} = \frac{1}{\infty} = 0$, and solve for $d_i$.

## Images Formed by Concave and Convex Mirrors

1. A concave mirror has a radius of curvature of 24.0 cm.
   An object 2.5 cm tall is placed 40.0 cm in front of the mirror.

   (a) At what distance from the mirror will the image be formed?

   (b) What is the height of the image?

### Frame the Problem

- Visualize the problem by sketching the light ray diagram.

- Because the object distance is greater than the radius of curvature, you would expect the image to be inverted, smaller than the object, and real.

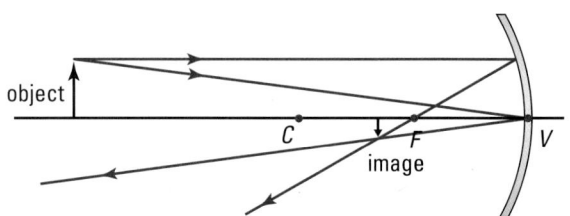

### Identify the Goal

The distance, $d_i$, from the mirror that an image is formed

The height, $h_i$, of the image

### Variables and Constants

| Involved in the problem | | Known | Unknown |
|---|---|---|---|
| $d_o$ | $f$ | $d_o = 40.0$ cm | $d_i$ |
| $d_i$ | $h_o$ | $R = 24.0$ cm | $h_i$ |
| $R$ | $h_i$ | $h_o = 2.5$ cm | $f$ |

### Strategy

Calculate the focal length, $f$, from the radius of curvature, $R$.

Use the mirror equation to find the image distance, $d_i$.

Substitute values for the focal length, $f$, and the object distance, $d_o$.

Rearrange.

### Calculations

$$f = \frac{R}{2}$$

$$f = \frac{24.0 \text{ cm}}{2}$$

$$f = 12.0 \text{ cm}$$

$$\frac{1}{f} = \frac{1}{d_o} + \frac{1}{d_i}$$

$$\frac{1}{12.0 \text{ cm}} = \frac{1}{40.0 \text{ cm}} + \frac{1}{d_i}$$

$$\frac{1}{12.0 \text{ cm}} - \frac{1}{40.0 \text{ cm}} = \frac{1}{d_i}$$

Find a common denominator.

$$\frac{10}{120 \text{ cm}} - \frac{3}{120 \text{ cm}} = \frac{1}{d_i}$$

$$\frac{7}{120 \text{ cm}} = \frac{1}{d_i}$$

Invert.

$$\frac{120 \text{ cm}}{7} = d_i$$

$$d_i = 17.14 \text{ cm}$$

(a) The image is formed 17 cm in front of the mirror. Because the image distance is positive, the image is real.

**PROBLEM TIP**

Remember to take the reciprocal of $1/d_i$ to find $d_i$.

Use the magnification equation to find $h_i$.

$$\frac{h_i}{h_o} = \frac{-d_i}{d_o}$$

Substitute in known values.

$$\frac{h_i}{2.5 \text{ cm}} = \frac{-17.1 \text{ cm}}{40 \text{ cm}}$$

Simplify.

$$\frac{h_i}{2.5 \text{ cm}}(2.5 \text{ cm}) = \frac{-17.1}{40}(2.5 \text{ cm})$$

Round to one decimal place.

$$h_i = -1.07 \text{ cm}$$

(b) The height of the image is −1.1 cm.

## Validate

The units are in centimetres.

The height of the image is a negative value because the image is inverted (see the previous ray diagram).

The height of the image is less than the height of the object.

The magnitudes of both the image position and the height seem reasonable when compared to the ray diagram.

2. **An object 3.5 cm tall is located 7.0 cm in front of a concave make-up mirror that has a focal length of 10.0 cm. Where is the image located, and what is the height of the image?**

## Frame the problem

- Visualize the problem by sketching the light ray diagram.

- Because the object distance is less than the focal length, you would expect the image to be upright, larger than the object, and virtual.

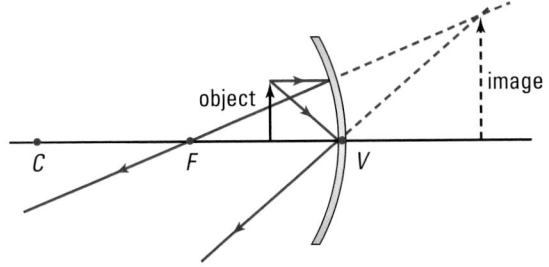

*continued* ▶

## Identify the Goal

The location, $d_i$, and height, $h_i$, of the image in a concave mirror

## Variables and Constants

| Involved in the problem | | Known | Unknown |
|---|---|---|---|
| $d_o$ | $h_i$ | $d_o = 7.0$ cm | $d_i$ |
| $d_i$ | $f$ | $h_o = 3.5$ cm | $h_i$ |
| $h_o$ | | $f = 10.0$ cm | |

| Strategy | Calculations |
|---|---|
| Use the mirror equation to find the image location, $d_i$. | $\dfrac{1}{f} = \dfrac{1}{d_o} + \dfrac{1}{d_i}$ |
| Substitute in the known values. | $\dfrac{1}{10.0 \text{ cm}} = \dfrac{1}{7.0 \text{ cm}} + \dfrac{1}{d_i}$ |
| Rearrange. | $\dfrac{1}{10.0 \text{ cm}} - \dfrac{1}{7.0 \text{ cm}} = \dfrac{1}{d_i}$ |
| Find a common denominator. | $\dfrac{7}{70 \text{ cm}} - \dfrac{10}{70 \text{ cm}} = \dfrac{1}{d_i}$ |
| | $\dfrac{-3}{70 \text{ cm}} = \dfrac{1}{d_i}$ |
| Invert. | $\dfrac{-70 \text{ cm}}{3} = d_i$ |
| | $d_i = -23.3$ cm |
| Use the magnification equation to find $h_i$. | $\dfrac{h_i}{h_o} = \dfrac{-d_i}{d_o}$ |
| Substitute in known values. | $\dfrac{h_i}{3.5 \text{ cm}} = \dfrac{-(-23.3) \text{ cm}}{7.0 \text{ cm}}$ |
| | $\dfrac{h_i}{3.5 \text{ cm}}(3.5 \text{ cm}) = \dfrac{23.3}{7.0}(3.5 \text{ cm})$ |
| Simplify. | $h_i = +11.7$ cm |

The image is formed 23 cm behind the mirror.

The height of the image is 12 cm.

## Validate

The units are in centimetres.

The height of the image is a positive value because the image is upright (see the previous ray diagram).

The image location is a negative value because the image is virtual (behind the concave mirror).

The image height is larger than the object height.

The magnitudes of both the image position and the height seem reasonable when compared to the ray diagram.

3. A convex supermarket surveillance mirror has a radius of curvature of 80.0 cm. A 1.7 m tall customer is standing 4.5 m in front of the mirror.

(a) What is the location of the customer's image in the mirror?

(b) What is the height of the customer's image?

## Frame the problem

- Visualize the problem by sketching the light ray diagram.

- From your knowledge of convex mirrors, you expect the image to be behind the mirror and smaller than the object.

- The focal length is negative for a convex mirror.

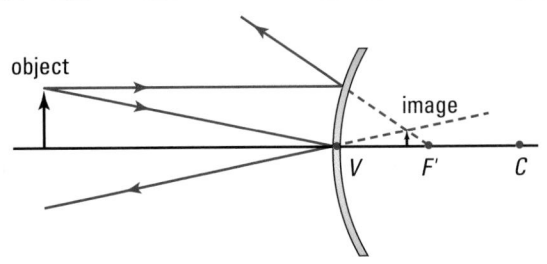

## Identify the Goal

The location, $d_i$, and height, $h_i$, of an image in a convex mirror

## Variables and Constants

| Involved in the problem | | Known | Unknown |
|---|---|---|---|
| $d_o$ | $h_i$ | $d_o = 4.5$ m | $d_i$ |
| $d_i$ | $R$ | $h_o = 1.7$ m | $h_i$ |
| $h_o$ | $f$ | $R = 80.0$ cm $= 0.8$ m | $f$ |

## Strategy

Calculate the focal length, $f$, from the radius of curvature, $R$.

Use the convention that the focal length is negative for convex mirrors.

Use the mirror equation to find the image distance, $d_i$.

Substitute in known values.

Rearrange.

Find a common denominator.

## Calculations

$$f = \frac{R}{2}$$

$$f = \frac{0.8 \text{ m}}{2}$$

$$f = 0.4 \text{ m}$$

$$f = -0.4 \text{ m}$$

$$\frac{1}{f} = \frac{1}{d_o} + \frac{1}{d_i}$$

$$\frac{1}{-0.4 \text{ m}} = \frac{1}{4.5 \text{ m}} + \frac{1}{d_i}$$

$$\frac{1}{-0.4 \text{ m}} - \frac{1}{4.5 \text{ m}} = \frac{1}{d_i}$$

$$\frac{-4.5}{1.8 \text{ m}} - \frac{0.4}{1.8 \text{ m}} = \frac{1}{d_i}$$

$$\frac{-4.9}{1.8 \text{ m}} = \frac{1}{d_i}$$

*continued* ▶

continued from previous page

Invert.

$$d_i = \frac{-1.8 \text{ m}}{4.9}$$

$$d_i = -0.367 \text{ m}$$

(a) The image is formed 0.37 m behind the mirror.

Use the magnification equation to find $h_i$.

$$\frac{h_i}{h_o} = \frac{-d_i}{d_o}$$

Substitute in known values.

$$\frac{h_i}{1.7 \text{ m}} = \frac{-(-0.367) \text{ m}}{4.5 \text{ m}}$$

$$\frac{h_i}{1.7 \text{ m}} (1.7 \text{ m}) = \frac{0.367}{4.5} (1.7 \text{ m})$$

Simplify.

$$h_i = 0.1386 \text{ m}$$

(b) The height of the image is 0.14 m.

## Validate

The units are in metres.

The height of the image is a positive value because the image is upright (see the ray diagram above).

The height of the image is smaller than the height of the object.

The image location is a negative value because the image is virtual (behind the convex mirror).

The magnitude of both the image position and the height seem reasonable when compared to the ray diagram.

## PRACTICE PROBLEMS

1. An object 6.0 mm tall is 10.0 cm in front of a concave mirror that has a 6.0 cm focal length. Find the image distance and its height by means of

   (a) a ray diagram drawn to scale.

   (b) the mirror and magnification equations.

2. What is the radius of curvature of a concave mirror that magnifies an object placed 30.0 cm from the mirror by a factor of +3.0?

3. A convex security mirror in a warehouse has a radius of curvature of 1.0 m. A 2.2 m tall forklift is 6.0 m from the mirror. What is the location and size of the image?

4. A dancer is applying make-up in a concave mirror. Her face is 35 cm in front of the mirror. The image is 72 cm behind the mirror. Using the mirror equation, find the focal length of the mirror. What is the magnification of the image?

# INVESTIGATION 10-B

## Image Formation in Concave Mirrors

**TARGET SKILLS**
- Predicting
- Hypothesizing
- Performing and recording
- Analyzing and interpreting

One of the fundamental purposes of the scientific theories, models, and equations related to the study of light is that they can be used to make predictions, as well as to provide explanations.

### Problem

Can the light ray model and the mirror and magnification equations predict the properties of the images formed by concave mirrors?

### Equipment

- concave mirror
- optical bench (with attachments)
- metre stick
- light source (candle)
- screen

### Procedure

1. Read the procedure and prepare a suitable table to record the predictions and observations in the investigation.

2. In a darkened room, point the mirror at a distant light source or at an object that can be seen through the window. Move the screen back and forth until the object is clearly focussed on the screen. Measure the distance from the mirror to the screen. Determine the focal length, $f$, of the mirror.

3. Using the focal length, determine the object distance from the mirror for object positions of 2.5$f$, 2.0$f$, 1.5$f$, and 0.5$f$.

4. Draw a scale light ray diagram for each of the four object positions.

5. Interpret and record the four properties of the image in the table.

6. Using the mirror and magnification equations, predict the four characteristics of the image for each object position and record them in the table.

7. Place the mirror at the zero end of the optical bench.

8. Place the object (light source) at the 2.5$f$ position from the mirror. Move the screen back and forth until you can see a clearly focussed image on the screen. It might be necessary to tilt the mirror slightly upward to prevent the source light from being blocked by the screen. Record the image properties in the table.

9. Repeat step 8 for the other object positions in step 5. It will not be possible to record the image distance for virtual images.

### Analyze and Conclude

1. Compare your predictions with your experimental results for each object position, and account for any discrepancies.

2. Which method for predicting the image properties is most useful? Explain why.

3. What image properties cannot be obtained easily by experiment? Why?

4. In terms of the focal length, $f$, for what part of the total range of object positions is it most difficult to use each of the three methods for obtaining the image properties?

1. **K/U** State what happens when a set of parallel light rays is reflected from (a) a plane mirror, (b) a concave mirror, and (c) a convex mirror. Draw a light ray diagram for each of the three mirror shapes to show how the light is reflected.

2. **K/U** In what position must an object be placed in front of a concave mirror to form (a) enlarged, virtual images, and (b) enlarged, real images? Draw light ray diagrams to show how each of these images is formed.

3. **K/U** Why are convex mirrors often used instead of plane mirrors in stores?

4. **K/U** What shape of curved mirror should be used to avoid problems caused by spherical aberration? Draw a light ray diagram to show why spherical aberration occurs.

5. **C** Explain the basic principle that determines where the image will be formed by light rays when they are reflected from any shape of mirror.

6. **C** Carry out research to find out how "see-through" mirrors work. Explain their operation, with the aid of light ray diagrams, and list applications for plane and curved mirrors of this type.

7. **K/U**
   (a) A concave mirror has a focal length of 24 cm. What is the radius of curvature of the mirror?
   (b) A convex mirror has a radius of curvature of 2.4 m. What is the focal length of the mirror?

8. **C** Draw light ray diagrams to show the characteristics of the image formed by a concave mirror for an object placed at (a) 0.5 $f$, (b) 1.75 $f$, and (c) 3.0 $f$. State the characteristics of the image in each case.

9. **I** Draw a light ray diagram to show that an object placed at the focal point of a concave mirror will not form an image.

10. **I** Draw light ray diagrams to determine the image characteristics for an object 1.5 cm high, when it is placed in front of a concave mirror with a focal length of 8.0 cm at a distance of (a) 4.0 cm, (b) 12.0 cm, and (c) 20.0 cm.

11. **I** Draw scaled light ray diagrams to determine the image characteristics for an object 2.0 cm high, when it is placed in front of a concave mirror with a focal length of 24.0 cm at a distance of (a) 10.0 cm, (b) 40.0 cm, and (c) 60.0 cm.

12. **I** Determine the image characteristics for the object in question 11 using the mirror equations.

13. **I** Design and carry out an investigation to measure the focal length of a convex mirror mounted on the door of a car. Check to see if the radius of curvature of the mirror is within the legal requirements for such mirrors. (Refer to the Canadian Motor Vehicle Safety Standards).

14. **MC** Research how parabolic reflectors are used to focus sunlight and create solar furnaces in solar-thermal electrical energy facilities. How much energy is produced compared to other energy generation methods? Describe some advantages and disadvantages of using solar furnaces to generate energy.

**UNIT PROJECT PREP**

The mirror and magnification equations can be used to determine the properties of curved mirror images.

- Construct a schematic diagram for your optical device based on the equations.

- The wave theory of light is based on the concept that light is a series of electromagnetic waves. Two scientific models, the wavefront model and the light ray model, help explain light phenomena in terms of the wave theory.

- Regular reflection occurs when light reflects from a smooth surface, such as a mirror. Diffuse reflection occurs when light reflects from a rough surface.

- Each point on an object is a source of light rays that diverge in all directions. The rays coming into your eye from any single point on the object form a cone. When all these cones of divergent rays, from every point on the object, come into your eye, you see the object.

- A likeness of an object, such as that seen in a plane mirror, is called an image. A real image is an image that can be projected onto a screen. Actual light rays that converge after reflection from a mirror surface form it.

- The image formed in a plane mirror is the same size as the object, and is virtual and upright. The image is located as far behind the mirror as the object is in front of the mirror.

The light ray model can be used to determine the properties of the image formed by a plane mirror. The use of the model is based on the geometry related to straight lines.

- A spherical mirror has the shape of a section from the surface of a sphere, and can be used to form concave and convex mirror surfaces. The principal axis of a spherical mirror is the straight line drawn through the centre of curvature and the vertex of the mirror.

- When parallel light rays reflect from a concave mirror, they pass through the focal point. Parallel light rays reflecting from a convex mirror appear to diverge from the virtual focal point behind the mirror surface.

- Concave mirrors form real, inverted images if the object is farther from the mirror than the focal point, and virtual, upright images if the object is between the focal point and the mirror.

- The mirror and magnification equations can be used to determine the location, size, and orientation of an image, and whether the image is real or virtual.

## Knowledge/Understanding

1. Why are parallel light rays used to show the difference between regular and diffuse reflection?

2. When driving a car on a rainy night, does a wet part of the road appear lighter or darker than a dry portion of the road in the car's headlights? Explain your answer in terms of the reflection of the light from the surface of the road.

3. List some situations where the letters on signs are reversed. Explain why this is done.

4. How could you set up mirrors to be able to read the image of printing seen in a plane mirror? Explain your answer.

5. As the object is moved along the principal axis toward the vertex of a concave mirror, the type of image changes. Where does the change in

type of image occur? Explain why, in terms of the path of the light rays.

6. Explain the difference between a real and a virtual image.

7. Why can you always tell by looking at a light ray diagram whether an image is real or virtual? Why are the lines drawn in this way?

8. Explain why concave mirrors can produce real or virtual images, but convex mirrors can produce only virtual images.

## Communication

9. Describe how you can determine the type of image formed in a plane mirror (a) by experiment, and (b) by using the light ray diagram.

10. Copy the diagrams in the following illustration into your notebook, and draw a light ray diagram for each example to show how the image is formed in the plane mirror. Show all construction lines.

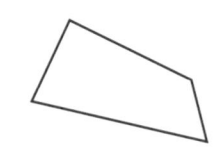

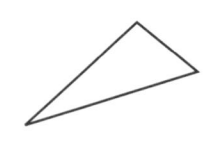

11. Copy the diagrams in the following illustration into your notebook and draw a light ray diagram for each example to show how the image is formed in the plane mirror. Show all construction lines.

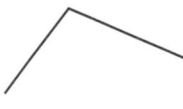

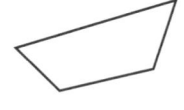

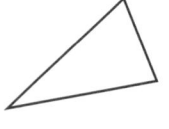

12. Prepare a series of overhead projector slides to use in a presentation to the class to explain changes in the characteristics of the images formed by (a) a concave mirror, and (b) a convex mirror as the object is moved from the surface of the mirror to infinity.

13. Concave mirrors are used in searchlights and solar ovens. Draw two diagrams to show the path of light in each case, and explain why these two applications demonstrate the principle of the reversibility of light.

14. Using a diagram, explain why a diverging mirror provides a wide field of view.

## Inquiry

15. Sometimes campers use a small mirror to light paper to start a campfire. What shape of mirror would they use? What is happening to the sunlight when they do this? Draw a light ray diagram to illustrate your answer.

16. Draw a fully labelled light ray diagram to show the position of the image formed by a concave mirror for each example listed in the table below. The height of the object is 1.5 cm. State the four properties for each image below the ray diagram.

| Focal length of mirror (cm) | Distance of object from mirror (cm) |
|---|---|
| 5.0 | 15.0 |
| 12.0 | 3.0 |
| 9.0 | 16.0 |

17. Draw a fully labelled light ray diagram to show the position of the image formed by a convex mirror for each example listed in the table below. The height of the object is 1.5 cm. State the image characteristics for each image below the appropriate ray diagram.

| Radius of curvature of mirror (cm) | Distance of object from mirror (cm) |
|---|---|
| 16.0 | 5.0 |
| 10.0 | 15.0 |

## Making Connections

18. Describe the similarities and differences of the properties of the images formed in a plane mirror and a convex mirror. Why are convex mirrors used for security purposes in stores, rather than plane mirrors?

19. Satellite dishes are now commonly used for receiving television and radio signals. What shape are they? What method is used to align them? What is the position of the satellite receiver (the device which detects the signal)?

## Problems for Understanding

20. A student wishes to take a photograph of his own image in a plane mirror. At what distance should the camera lens be focussed if it is positioned 2.3 m in front of the mirror?

21. A ray of light strikes a mirror at an angle of 55° to the normal.
    (a) What is the angle of reflection?
    (b) What is the angle between the incident ray and the reflected ray?

22. A ray of light strikes a plane mirror at an angle of 57° to the mirror surface. What is the angle between the incident ray and the reflected ray?

23. What is the angle of incidence if the angle between the reflected ray and the mirror surface is 34°?

24. Light is shining on to a plane mirror at an angle of incidence of 27°. If the plane mirror is tilted such that the angle of incidence is reduced by 8°, what will be the total change in the angle of reflection from the original reflected light?

25. A patient is sitting in an optometrist's chair, facing a mirror that is 2.25 m from her eyes. If the eye chart she is looking at is hanging on a wall behind her head, 1.75 m behind her eyes, how far from her eyes does the chart appear to be? Why would charts used for this purpose have to be specially made?

26. If you move directly along a normal line toward a plane mirror at a speed of 3.5 m/s, what is the speed of the image relative to you?

27. What is the speed of the image, relative to you, if you walk away from the mirror surface at 3.5 m/s at an angle of 30° to the mirror surface?

28. Look at a kaleidoscope and determine how many images of the group of brightly coloured objects are formed by the two mirrors inside. Using the equation $N = \frac{360°}{\theta} - 1$ (where N is the number of images and $\theta$ is the angle between the mirrors), calculate the angle between the mirrors.

29. A concave mirror has a focal length of 26 cm. What is the radius of curvature of the mirror?

30. The radius of curvature of a convex mirror is 60 cm. What is the focal length of the mirror?

31. The light from a star reflects from a concave mirror with a radius of curvature of 1.70 m. Determine how far the image of the star is from the surface of the mirror.

32. Use the mirror equation to find the image location and the height of an object placed at the centre of curvature of a concave mirror. Also find the magnification. Hint: What is the relation between the focal length and the object distance, $d_o$, for this situation?

33. Using the mirror equation and the magnification equation, find the four properties of the image formed in a concave mirror with a focal length of 50.0 cm, if the object is 1.5 m from the mirror and is 2.5 cm high.

34. Using the mirror equation and the magnification equation, find the four properties of the image formed in a concave mirror with a focal length of 35 cm, if the object is 50.0 cm from the mirror and is 3.0 cm high.

35. Using the mirror equation and the magnification equation, find the four properties of the image formed in a concave mirror with a focal length of 1.2 m, if the object is 0.80 m from the mirror and is 2.0 cm high.

36. Using the mirror equation and the magnification equation, find the four properties of the image formed in a convex mirror with a focal length of 90 cm, if the object is 2.5 m from the mirror and is 0.4 m high.

37. Using the mirror equation and the magnification equation, find the four properties of the image formed in a convex mirror with a radius of curvature of 1.5 m, if the object is 80.0 cm from the mirror and is 25 cm high.

38. A concave mirror has a radius of curvature of 50.0 cm. At what position should an object be placed to produce an upright virtual image that is 3.0 times as large as the object?

---

**Numerical Answers to Practice Problems**

1. (b) $d_i = 15$ cm; $h_i = -0.90$ cm  2. $9.0 \times 10^1$ cm
3. $d_i = -0.46$ m; $h_i = 0.17$ m  4. $f = 66$ cm

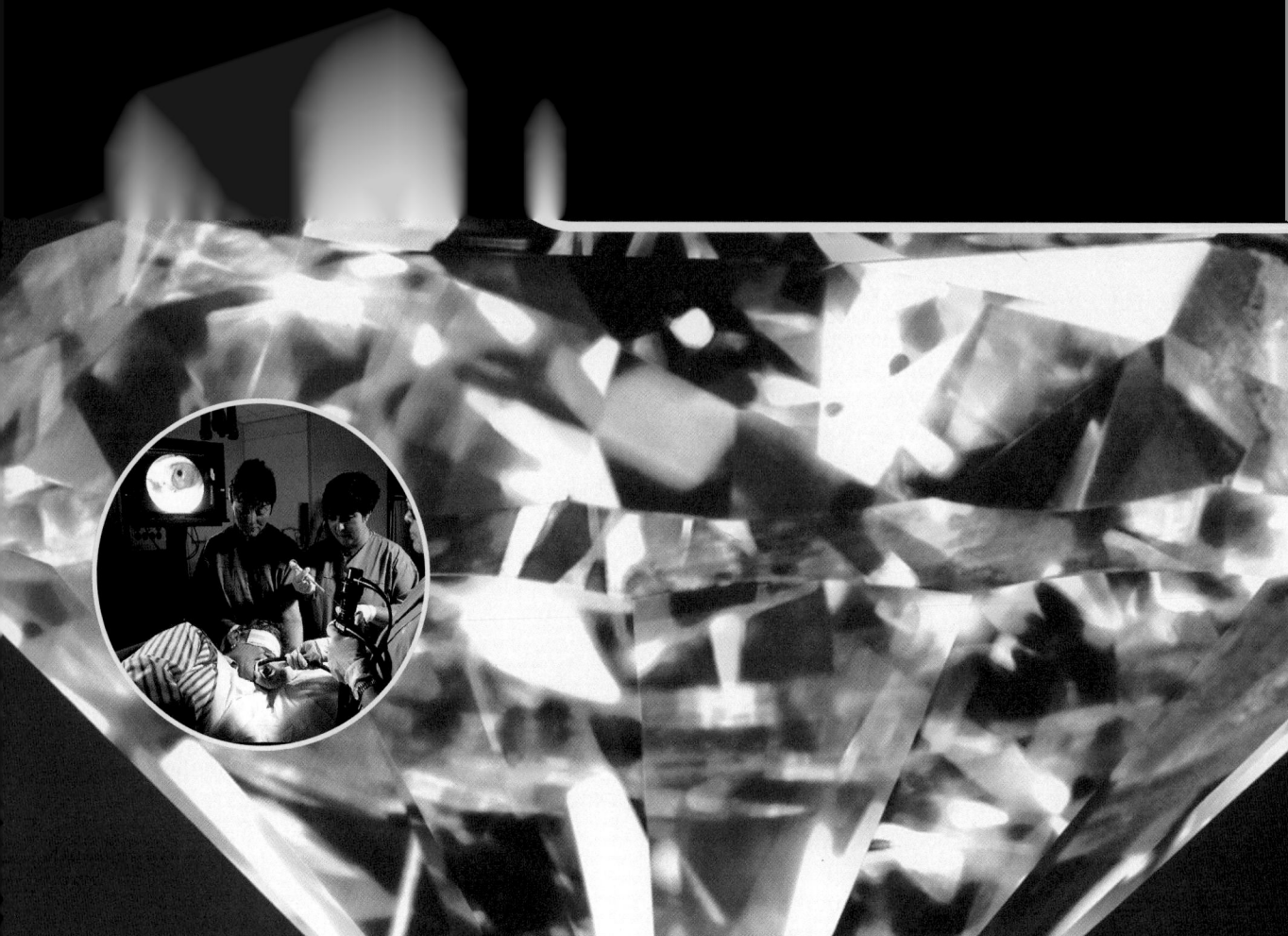

**W**hen you pass a jewellery store, look at the displays of diamond necklaces, earrings, and engagement rings in the window. Notice the light sparkling from the jewels, often glistening with different colours. Sometimes, jewellers mount their displays on rotating platforms to enhance the glittering of the colours. How does light interact with diamonds to create colour from an apparently colourless material? What other materials produce a similar effect? Could you get this effect by using glass instead of diamonds?

How could the glittering of diamonds be related to the surgical procedure shown in the small photograph? Some of the same properties of light that make diamonds glitter help the surgeon to see inside a patient's stomach. The device, called an "endoscope," contains optical fibres that carry light into the patient's stomach, allowing the doctor to observe the tissues. If light travels in straight lines, as you learned in Chapter 10, something unusual must be happening to the light inside the endoscope that causes the light to follow the curved path of the tube. In this chapter, you will learn about refraction of light, which is responsible for this and for many other interesting phenomena.

**MULTI LAB**

# Light Travelling Through Matter

**TARGET SKILLS**

- Hypothesizing
- Analyzing and interpreting
- Communicating results

## Beaker Optics

Pour about 300 mL of water into a 500 mL beaker. Hold the beaker with the markings facing away from you. Observe the markings from above the water, then through the water. Finally, look up through the bottom of the beaker.

### Analyze and Conclude

Describe the changes in the appearance of the markings as you looked down from the top of the beaker, then directly through the beaker, and, finally, up through the bottom of the beaker.

## Apparent Depth

Fill a sink or a plastic bowl or bucket approximately three quarters full of water. Hold a ruler about half-submerged in the water, and slowly change the angle of the ruler relative to the water. Tilt the ruler to the side, and then tilt it away from you. Move your head up and down so that the level of your eyes is at first well above the water and then very close to the water. In each case, note any changes in the appearance of the ruler. Move your head and the ruler until the submerged end of the ruler disappears.

### Analyze and Conclude

1. Describe any patterns in the changes of the appearance of the ruler.

2. Formulate a hypothesis that accounts for the fact that the ruler seems to disappear at certain angles. As you study this chapter, look for explanations of the disappearance of the ruler and compare those with your hypothesis.

## Shimmering

Adjust a Bunsen or alcohol burner to burn with the hottest flame. From a safe distance, position your eyes level with the flame. Look past the flame as close to its edges as possible.

### Analyze and Conclude

Describe how the flame affected the appearance of objects that were beyond and just to the side of the flame. What phenomenon do you think is responsible for these effects?

## SECTION EXPECTATIONS

- Define and describe refraction and index of refraction.

- Predict, in quantitative terms, the effect of a medium, on the velocity of light.

## KEY TERMS

- index of refraction (refractive index)

- optically dense

When you look through a window, objects usually appear just the same as they do when looking only through air. Why do the windowpanes in Figure 11.1 cause so much distortion? Why does the curved side of a beaker cause distortion? Windowpanes and beakers are transparent, so if light penetrates these materials, why do some of them cause distortions, while others do not? Can you think of any situations in which air alone can cause distortion?

**Figure 11.1** Imagine looking at the world through these windowpanes. Why was this kind of glass once used in windows?

## Refraction and Light Waves

When you were looking at the ruler that was half-submerged in water, was the ruler really bent? You are probably thinking, "Of course not!" It was actually the light rays that were bent, not the ruler. The change in direction of the light as it passes at an angle from one material into another is a result of refraction.

**Refraction** is the changing of the speed of a wave when it travels from one medium into another. Light travels as a wave, and therefore experiences refraction. When light travels from one medium into another at an angle, the difference in the speed causes a change in the direction of the light (Figure 11.2). The shimmering effect you see above a barbecue grill, the distorted view of objects viewed through a glass bottle, and even rainbows are all results of the refraction of light.

Light travels through a vacuum at the extremely high speed of $3.00 \times 10^8$ m/s (300 000 km/s). The speed of light in a vacuum is such an important fundamental physical constant in the study of physics that it has been assigned its own symbol, $c$.

**Figure 11.2** Light refracts, or bends, toward the normal when it passes from air into water or glass.

**History Link**

Until the seventeenth century, people generally accepted that light appeared instantaneously. In fact, the first few scientists who attempted to measure the speed of light were ridiculed by the entire scientific community for the values that they suggested. Finally, in the late 1800s, U.S. physicists A.A. Michelson (1852–1931) and E.W. Morley (1838–1923) performed the critical experiments that established the speed of light. This determination played a major role in the development of both classical and modern physics. What is the connection between the Michelson-Morley experiments and Einstein's formulation of the theory of relativity?

## Index of Refraction

The values for the speed of light in different media are extremely large, unwieldy numbers. Physicists developed a more useful constant called the **index of refraction** (or **refractive index**). The index of refraction is a ratio of the speed of light in a vacuum to the speed of light in a specific medium. Since the term represents a ratio of two values having the same units, the units cancel, leaving the constant unitless.

---

### INDEX OF REFRACTION

The index of refraction of a material is the ratio of the speed of light in a vacuum to the speed of light in that material.

$$n = \frac{c}{v}$$

| Quantity | Symbol | SI unit |
|---|---|---|
| index of refraction | $n$ | none |
| speed of light in a vacuum | $c$ | $\frac{m}{s}$ (metres per second) |
| speed of light in a specific medium | $v$ | $\frac{m}{s}$ (metres per second) |

**Unit Analysis**

$$\frac{\text{metres per second}}{\text{metres per second}} = \frac{\frac{\cancel{m}}{\cancel{s}}}{\frac{\cancel{m}}{\cancel{s}}} = \text{no unit}$$

---

**Language Link**

The symbol for the speed of light, $c$, represents the Latin word *celeritas*, which means "speed." What term that you used frequently in your study of forces and motion is also derived from this Latin word for speed?

## Light Pulse Breaks Ultimate Speed Limit

Red alert! The nebula your starship is headed toward contains deadly radiation. A faster-than-light signal has arrived from the future, warning you not to enter. You change course, narrowly escaping danger.

Scenarios like this are common in science fiction. They are based on speculation that if faster-than-light travel were possible, time travel into the past would be possible. Among other things, you could send information back in time and change the course of your life. There's just one problem — if you did this, the version of the future from which you sent the information might no longer exist!

This paradox and Einstein's famous theory of relativity both imply that superluminal (faster-than-light) speeds are impossible. Neither mass nor information should be able to travel faster than the speed of light in a vacuum ($3 \times 10^8$ m/s). Yet, in laser experiments conducted by Dr. Lijun Wang and his colleagues at the NEC Research Institute in New Jersey, the crest of a light pulse did exactly that. The crest actually exited from the back end of a cesium gas cell before it entered the front end!

Light pulses, which have no mass, can be coaxed into travelling at superluminal speeds. Physicists first predicted that this was possible in 1970, and the work of Wang's team provides the first unambiguous evidence of this fact.

A light pulse is a packet of light waves of various frequencies. Where the crests of these light waves align, their amplitudes add together to produce the pulse.

In Wang's experiment, a light pulse travels through a cell of specially prepared cesium gas that accelerates higher frequency light waves and slows lower frequency light waves. This is the opposite of what a glass prism does to light. The altered light waves within the pulse interfere with each another and rephase, shifting the pulse ahead in time by 62 ns, as shown in the diagram. The shifted crest of the pulse exits from the cesium gas cell well before the original crest enters the cell.

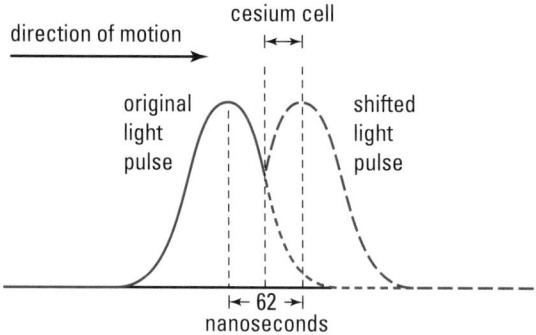

This light pulse leaves the cesium gas cell *before* it enters.

Wang insists that information cannot be transmitted superluminally by this technique. However, not all physicists agree. There is now considerable debate over what counts as "information."

In the meantime, the technique might have practical applications. Today's technologies, such as computers and the Internet, transfer information at velocities well below the speed of light. Wang feels his research might boost transfer velocities to this ultimate speed limit.

### Web Link

**www.school.mcgrawhill.ca/resources/**
For more historical detail on the various methods used to measure the speed of light, go to the above Internet site. Follow the links for **Science Resources** and **Physics 11** to find out where to go next. You will also find links to faster-than-light physics, including the theoretical basis for a "space warp."

**Table 11.1** Index of Refraction of Various Substances*

| Substance | Index of Refraction ($n$) |
|---|---|
| vacuum | 1.00000 |
| gases at 0°C, 1.013 × 10⁵ Pa | |
| hydrogen | 1.00014 |
| oxygen | 1.00027 |
| air | 1.00029 |
| carbon dioxide | 1.00045 |
| liquids at 20°C | |
| water | 1.333 |
| ethyl alcohol | 1.362 |
| glycerin | 1.470 |
| carbon disulfide | 1.632 |

| Substance | Index of Refraction ($n$) |
|---|---|
| solids at 20°C | |
| ice (at 0°C) | 1.31 |
| quartz (fused) | 1.46 |
| optical fibre (cladding) | 1.47 |
| optical fibre (core) | 1.50 |
| Plexiglas™ or Lucite™ | 1.51 |
| glass (crown) | 1.52 |
| sodium chloride | 1.54 |
| glass (crystal) | 1.54 |
| ruby | 1.54 |
| glass (flint) | 1.65 |
| zircon | 1.92 |
| diamond | 2.42 |

* Measured using yellow light, with a wavelength
  of 589 nm in a vacuum.

The index of refraction of a material can be determined by methods that do not involve measuring the speed of light in that material. Once the value is measured, however, you can use it to calculate the speed of light in that material. The index of refraction has been determined for a large number of materials, some of which are listed in Table 11.1.

Notice that the values for the index of refraction are always greater than one, because the speed of light in a vacuum is always greater than the speed of light in any material. Notice, also, that media in which the speed of light is low have large indices of refraction: When you divide a constant by a small number, the ratio is large. For example, the speed of light in water ($n = 1.33$) is $2.26 \times 10^8$ m/s, whereas the speed of light in zircon ($n = 1.92$) is $1.56 \times 10^8$ m/s.

The term **optically dense** refers to a refractive medium in which the speed of light is low in comparison to its speed in another medium. For example, based on the previous values given for zircon and water, you would say that zircon is more optically dense than water. The speed of light in air and other gases is so close to that of the speed of light in a vacuum that the index of refraction of these materials is considered to have a value of one, or unity, for most practical purposes. You will notice in Table 11.1 that the index of refraction in gases differs from that in a vacuum only in the fifth significant digit.

## Index of Refraction and Speed of Light

The speed of light in a solid is $1.969 \times 10^8$ m/s.

(a) What is the index of refraction of the solid?

(b) Identify the material, using Table 11.1.

## Frame the Problem

- The *speed of light in the medium* is known.

- The *index of refraction* is unknown.

- The speed of light in a medium is related to the index of refraction of that medium.

## Identify the Goal

(a) The index of refraction, $n$, of the medium

(b) The identity of the medium

## Variables and Constants

| Involved in the problem | Known | Implied | Unknown |
|---|---|---|---|
| $v$ | $v = 1.969 \times 10^8 \ \frac{m}{s}$ | $c = 3.00 \times 10^8 \ \frac{m}{s}$ | $n$ |
| $n$ | | | medium |
| $c$ | | | |
| medium | | | |

## Strategy

Use the equation that defines the index of refraction.

All of the values are known, so substitute them into the equation.

## Calculations

$$n = \frac{c}{v}$$

$$n = \frac{3.00 \times 10^8 \ \frac{\cancel{m}}{\cancel{s}}}{1.969 \times 10^8 \ \frac{\cancel{m}}{\cancel{s}}}$$

$$n = 1.5236$$

(a) The index of refraction of the material is 1.52.

(b) Based on the information in Table 11.1, the material is probably crown glass. However, more information would be needed to confirm its identity, since many materials do not appear in the table.

## Validate

All of the units cancelled, leaving the answer unitless.
This is in agreement with the nature of the index of refraction.

The magnitude is between 1.00 and 2.00, which is quite realistic.

1. What is the index of refraction of a solid in which the speed of light is $1.943 \times 10^8$ m/s?

2. Determine the speed of light in diamond.

3. What is the speed of light in glycerin?

4. Determine the time taken for light to travel a distance of 3500 km along the core of an optical fibre.

5. Determine the change in the speed of light as it passes from ice into water.

The index of refraction, and therefore the speed of light, for all wavelengths (colours) of light is the same when light is travelling in a vacuum. For any other given material, the index of refraction varies slightly, however, depending on the wavelength of the light passing through it. Therefore, different wavelengths of light travel at different speeds in the same medium. To avoid confusion, physicists have chosen to report the values for the speed of light of a specific colour. They have chosen the yellow light emitted by the sodium atom, which has a wavelength of 589 nm in a vacuum.

**Figure 11.3** Streetlights are usually filled with sodium vapour. The yellow light it produces is preferred, because it penetrates fog and mist better than the blue-green light produced by mercury vapour.

**PHYSICS FILE**

Objects such as a stove burner emit visible light when the temperature is high enough to excite the electrons within an atom. Many substances emit many different wavelengths (colours) of light, which are distributed throughout the visible part of the electromagnetic spectrum. Some elements, such as sodium and hydrogen, release only a few wavelengths of visible light. In fact, sodium releases only two wavelengths of light and they are both in the yellow part of the visible spectrum. The wavelength for yellow light, 589 nm, is the standard wavelength used to determine the index of refraction for all materials.

## 11.1    Section Review

1. **K/U** What are the units for the index of refraction?

2. **K/U** As the speed of light in a medium increases, what happens to the index of refraction of the medium?

3. **K/U** Explain the term "optically dense."

4. **K/U** What is the benefit of choosing light of a specific colour to report values for the speed of light?

5. **C** Describe, in your own words, how the index of refraction is related to the speed of light, and suggest a reason why gases have refractive indices close to 1.

For thousands of years, people have known that various types of crystals can change the direction of light when it passes from air into the crystals. Scientists also knew long ago that the amount of refraction depended on the both the material that made up crystal and the angle at which the incident light crossed the boundary between the air and the crystal. Numerous scientists tried to identify a mathematical relationship that could describe refraction, but it was not until 1621 that Willebrord Snell (1580–1626), a Dutch mathematician, discovered the relationship experimentally.

## Snell's Observations

In Chapter 10, you learned the meaning of the terms "angle of incidence," "angle of reflection," and "normal line." In Figure 11.4, in which a light ray is travelling from air into water, you see a new, but related, angle — the **angle of refraction**, $\theta_R$. The angle of refraction is defined as the angle that the refracted ray makes with the normal line.

Snell discovered that when light travels across a boundary from one medium into another, the ratio of the sine of the angle of incidence to the sine of the angle of refraction is a constant value for all angles of incidence greater than zero. When the angle of incidence is zero, the angle of refraction is also zero and the direction of the light is unchanged. Snell's observations can be described mathematically by the following expression.

$$\frac{\sin \theta_i}{\sin \theta_R} = \text{a constant}$$

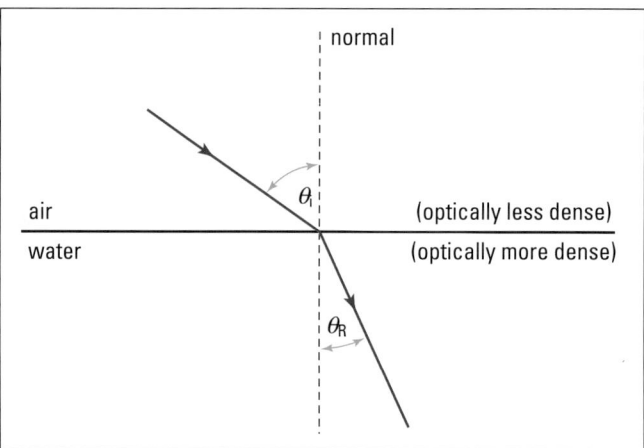

**Figure 11.4** Light refracts (bends) toward the normal when passing from air into water.

When the incident medium is a vacuum (or air), the constant in Snell's relationship is actually the same value as the index of refraction, $n$, of the refracting medium. For example, the constant for light travelling from air into water is 1.33. This relationship provides a means of determining the speed of light in any specific medium. For a given angle of incidence, simply measure the angle of refraction when light travels from air into the medium, and determine $n$. Use the calculated value of $n$ to determine the velocity, $v$, for that medium. In addition, since many indices of refraction are known, you can identify an unknown material by measuring the angle of refraction in that medium, calculating $n$, and then comparing it to values of $n$ of known materials.

**History Link**

Astronomer and mathematician Ptolemy of Alexandria (Claudius Ptolemaeus, approximately 85–168 B.C.E.) made accurate measurements and kept data tables of angles of incidence and refraction of several materials. At a different time in his career, he also developed a set of trigonometric ratios, including sine ratios. He never identified the link between these ratios and the angles of incidence and refraction, however. Which scientific theory is Ptolemy most famous for?

## MODEL PROBLEM

### Index of Refraction

**Light travels from air into an unknown liquid at an angle of incidence of 65.0°. The angle of refraction is 42.0°. Determine the index of refraction of the unknown liquid.**

### Frame the Problem

- Make a sketch of the problem.

- The *angles of incidence and refraction* are known.

- The index of refraction relates the angles of incidence and refraction when the incident medium is *air*.

- *Air* is the incident medium.

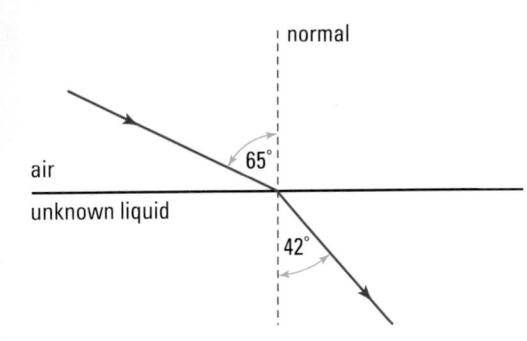

### Identify the Goal

The index of refraction, $n$, of the unknown liquid

### Variables and Constants

| Involved in the problem | Known | Unknown |
|---|---|---|
| $\theta_i$ | air | $n$ |
| $\theta_R$ | $\theta_i$ | |
| $n$ | $\theta_R$ | |
| incident medium: air | | |

*continued* ▶

*continued from previous page*

## Strategy

Use Snell's constant

Since the incident medium is air, the constant is the index of refraction, $n$, of the liquid.

Substitute in the known values.

## Calculations

$$\frac{\sin \theta_i}{\sin \theta_R} = \text{a constant}$$

$$n = \frac{\sin \theta_i}{\sin \theta_R}$$

$$n = \frac{\sin 65.0°}{\sin 42.0°}$$

$$n = \frac{0.9063}{0.6691}$$

$$n = 1.354$$

The index of refraction of the unknown liquid is 1.35.

## Validate

The absence of units is in agreement with the unitless nature of the index of refraction.

The value is between one and two, which is very reasonable for an index of refraction.

### PRACTICE PROBLEMS

6. Light travels from air into a material at an angle of incidence of 59°. If the angle of refraction is 41°, what is the index of refraction of the material? Identify the material by referring to Table 11.1, Index of Refraction of Various Substances.

7. A beam of light travels from air into a zircon crystal at an angle of 72.0°. What is the angle of refraction in the zircon?

8. What is the angle of incidence of light travelling from air into ethyl alcohol when the angle of refraction is 35°?

Snell's constant ratio is observed for any two media, even when neither medium is air. However, the value of the constant is not the same as the index of refraction of either medium. For example, when light is travelling from water into crown glass, the constant is found to be 1.143. This is a different value than that shown in Table 11.1 for either material. An additional observation shows that the constant might be less than one. In the case of light travelling from quartz into water, the ratio of the sine of the angle of incidence to the sine of the angle of refraction is 0.914. Nevertheless, there is still a relationship between Snell's constants and the indices of refraction of any two media. Many observations and comparisons revealed that Snell's constant is, in fact, the ratio of the indices of refraction of the two media:

$$\frac{\sin \theta_i}{\sin \theta_R} = \frac{n_R}{n_i}$$

For example, Snell's constant for light travelling from water into crown glass (1.143) is the ratio of the index of refraction of crown glass (1.523) to that of water (1.333). This relationship, known as **Snell's law**, is usually rearranged and expressed mathematically as shown in the box below.

---

**SNELL'S LAW**

The product of the index of refraction of the incident medium and the sine of the angle of incidence is the same as the product of the index of refraction of the refracting medium and the sine of the angle of refraction.

$$n_i \sin \theta_i = n_R \sin \theta_R$$

| Quantity | Symbol | SI unit |
|---|---|---|
| index of refraction of the incident medium | $n_i$ | unitless |
| angle of incidence | $\theta_i$ | none (degree is not a unit) |
| index of refraction of the refracting medium | $n_R$ | unitless |
| angle of refraction | $\theta_R$ | none (degree is not a unit) |

---

In addition to the mathematical relationship, a geometric relationship exists between the incident ray and refracted ray. *When light travels across a boundary between two different materials, the refracted ray, the incident ray, and the normal line at the point of incidence all lie in the same plane.*

Inspection of Snell's law reveals an important general result of the relationship. When light travels from an optically less dense medium into an optically more dense medium, the refracted ray bends toward the normal. Conversely, when light travels from an optically more dense medium into an optically less dense medium, the refracted ray bends away from the normal.

### Think It Through

- Use Snell's law to verify the statements in the last paragraph about the directions of bending of the refracted ray and the optical density of the incident and refracting media.

# INVESTIGATION 11-A

## Verifying Snell's Law

When a scientist collects data and finds a mathematical relationship from the data alone, the relationship is said to be based on empirical evidence. The mathematical relationship is not derived from more fundamental principles. Often, the empirical relationship leads a scientist to more experiments that reveal more fundamental scientific principles.

Snell's law was originally based on empirical evidence. It was more than 300 years before scientists determined the speed of light and were then able to explain refraction on the basis of the ratio of the speed of light in a vacuum and in a medium. In this investigation, you will verify Snell's law empirically and then use it to make predictions.

### Problem

Verify Snell's Law.

### Equipment

- ray box
- semicircular block of a material
- protractor
- sheet of paper (or polar coordinate graph paper)

### Procedure

1. Make a data table with the following headings: Angle of Incidence ($\theta_i$), Angle of Refraction ($\theta_R$), sin $\theta_i$, and sin $\theta_R$.

2. Draw a horizontal line across the sheet of paper so that the paper is divided into half.

3. Down the centre of the page, from top to bottom, draw a dashed line at right angles to the horizontal line. This line is the normal line for the air-medium interface.

4. Using a protractor, draw lines on the top half of the page to represent the path of the incident light in air for all of the angles shown in the diagram at right.

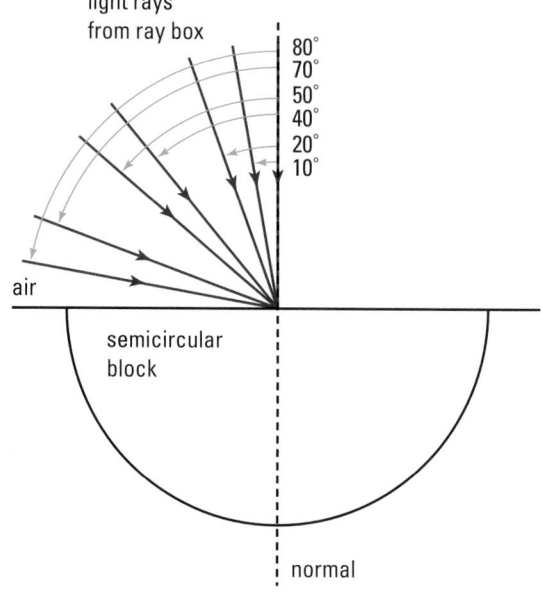

light rays from ray box

80°
70°
50°
40°
20°
10°

air

semicircular block

normal

5. Place the semicircular block of material on the page so that the straight side is positioned on the horizontal line. Position the centre of the flat side of the block where the normal line crosses the horizontal line (refer to the diagram).

6. Shine a single light ray from the ray box along the 10° line that you drew on the paper. The ray should strike the flat side of the block of material at the centre and pass through the block. Mark the paper at the point where the light ray exits the semicircular side of the block. Label the incident light ray and the corresponding refracted light ray at the point where it comes out through the semicircular side.

7. Repeat step 6 for each of the angles of incidence shown in the diagram.

8. Remove the semicircular block and draw lines from the centre point where the rays entered the block to the points where they exited. These lines are the paths of the refracted rays.

9. Measure each angle of refraction and enter the data in your table.

10. Determine the values of the sines of the angles of incidence and refraction, and enter the data in your table.

## Analyze and Conclude

1. What happens to the light when it passes from air into the refracting medium?

2. As the angle of incidence increases, does the angle of refraction increase more rapidly or less rapidly?

3. Predict what will happen if the angle of incidence continues to increase. Is there a maximum value for the angle of incidence?

4. Plot a graph of $\theta_i$ versus $\theta_R$.

5. Plot a graph of $\sin \theta_i$ versus $\sin \theta_R$.

6. Compare the two graphs. How might such a comparison help scientists to develop empirical relationships?

7. Determine the slope of the graph of $\sin \theta_i$ versus $\sin \theta_R$. What does the slope of this line represent?

8. Compare the accepted value of the index of refraction of the material used in the investigation with the experimental value you just obtained.

9. Comment on the experimental results in terms of verification of Snell's law. Suggest reasons for any discrepancies.

10. Using your experimental value of the index of refraction of the material, predict the angles of refraction for angles of incidence of 30° and 60°.

11. Test your predictions.

## Apply and Extend

12. You have just learned about the principle of reversibility of light. Imagine doing the experiment in reverse. Your refracted rays become incident rays and your incident rays become refracted rays. Which angles (incidence or refraction) will approach 90° more rapidly?

13. Predict what will happen as the incident angle approaches 90°.

14. If possible, develop a method for testing your prediction and carry out the test.

## Finding the Angle of Refraction

Light travels from air into a ruby crystal at an angle of incidence of 45°. Determine the angle of refraction of the light in the ruby.

## Frame the Problem

- Sketch and label a ray diagram.

- Light travels from *air*, an *optically less dense* medium, into ruby, an *optically more dense* medium.

- The refracted ray should *bend toward the normal* line.

- You can use *Snell's law* to determine the extent of the bending of the refracted ray.

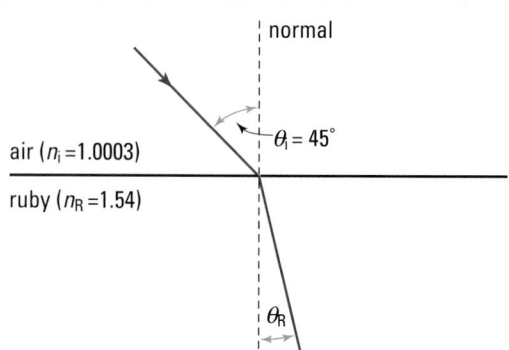

## Identify the Goal

The angle of refraction, $\theta_R$, in ruby

## Variables and Constants

| Involved in the problem | Known | Implied | Unknown |
|---|---|---|---|
| $\theta_i$   $n_i$ | $\theta_i = 45°$ | $n_i = 1.00$ | $\theta_R$ |
| $\theta_R$   $n_R$ | | $n_R = 1.54$ | |

## Strategy

Use Snell's law to solve the problem.

## Calculations

$n_i \sin \theta_i = n_R \sin \theta_R$

**Substitute first**

$1.00 \ \sin \ 45° = 1.54 \ \sin \theta_R$

$$\frac{1.00(0.7071)}{1.54} = \frac{\cancel{1.54} \sin \theta_R}{\cancel{1.54}}$$

$\sin \theta_R = 0.4592$

$\theta_R = \sin^{-1}(0.4592)$

$\theta_R = 27.33°$

**Solve for $\theta_R$**

$$\frac{n_i \sin \theta_i}{n_R} = \frac{\cancel{n_R} \sin \theta_R}{\cancel{n_R}}$$

$$\sin \theta_R = \frac{n_i \sin \theta_i}{n_R}$$

$$\theta_R = \sin^{-1}\left(\frac{n_i \sin \theta_i}{n_R}\right)$$

$$\theta_R = \sin^{-1}\left(\frac{1.00 \sin 45°}{1.54}\right)$$

$$\theta_R = \sin^{-1}\left(\frac{0.7071}{1.54}\right)$$

$$\theta_R = \sin^{-1}(0.4592)$$

$$\theta_R = 27.33°$$

The angle of refraction in ruby is 27°.

## Validate

The angle of refraction is less than the angle of incidence, which is to be expected when light travels from an optically less dense medium (air) into an optically more dense medium (ruby). The magnitude of the angle of refraction is realistic.

**PRACTICE PROBLEMS**

**9.** A beam of light passes from air into ethyl alcohol at an angle of incidence of 60.0°. What is the angle of refraction?

**10.** A beam of light passes from ethyl alcohol into air. The angle of refraction is 44.5°. Determine the angle of incidence.

## Refraction and the Wave Model for Light

Have you been wondering how a change in the speed of a wave can cause it to change its direction? The ray model is excellent for visualizing the angles and helping you to set up the calculations, but it is not as helpful as the wave model for visualizing the *reason* for the change in direction.

Although you cannot see the shape of light waves, recall your study of water waves and envision light waves in the same way. What happens to the shape of the waves when they pass from an optically less dense medium into a more dense medium, where the velocity is slower? Recall from your study of mechanical waves that the frequency of waves does not change. From the wave equation, $v = \lambda f$, you can see that if the velocity decreases and the frequency remains the same, then the wavelength must decrease. Figure 11.5 shows the behaviour of wavefronts of light waves as they pass from air into water at an angle of incidence of 0°. In this case, there is no change in direction, but the wavefronts are closer together because the wavelength is shorter.

Figure 11.6 (A) on the next page shows light wavefronts approaching the interface between two media (air and water) at an angle. You can take the same approach as shown in Figure 11.5 to see why the change in direction occurs. In Figure 11.6 (A), the left edge of each wavefront reaches the interface and begins to slow down, while the rest of the wavefront continues at the faster speed. The dashed lines show where the wavefronts would have been if the speed had not decreased. In a sense, the wavefronts themselves are bending, because one end is travelling more slowly than the other. The direction of the wavefronts (which is, of course, the direction of a light ray) bends toward the normal.

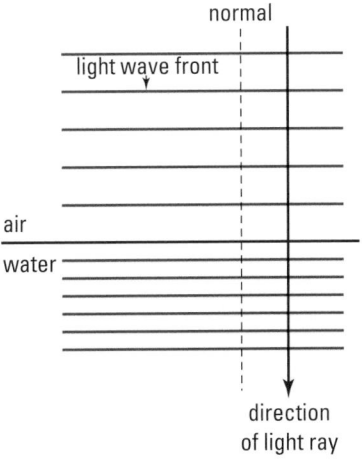

**Figure 11.5** Plane light waves, represented by a ray, are travelling from air into water.

**COURSE CHALLENGE**

**Twinkling Stars Point to a Problem**

Design an experiment to test whether microwave frequencies are more or less likely than light to be refracted by variations in the atmosphere.

Learn more from the **Science Resources** section of the following web site: **www.school.mcgrawhill.ca/ resources** and find the *Physics 11 Course Challenge*.

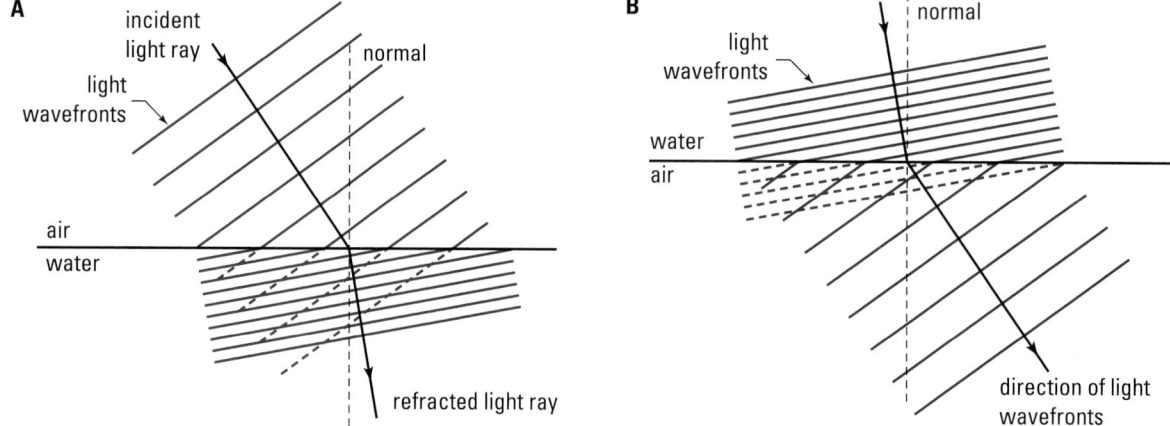

**Figure 11.6** **(A)** The left edges of the wavefronts are bent backwards because the waves travel slower in the water than they do in air. **(B)** The left edges of the wavefronts bend forward, because the waves travel faster in air than in water.

The opposite situation is illustrated in Figure 11.6 (B), where light is travelling from water into air. The edge of the wavefront that reaches the air first speeds up, leaving the other side behind. Again, the wavefronts bend, causing the light ray to bend away from the normal.

## Reversibility of Light Rays

How do parts (A) and (B) of Figure 11.6 differ? How are they the same? In Figure 11.6 (A), the ray is travelling from air into water, and the refracted ray bends toward the normal. In part (B), the ray is travelling in the opposite direction, from water into air. The refracted ray bends away from the normal. If you rotated Figure 11.6 (B) and then superimposed the two figures, you would see that the rays that are travelling in opposite directions have exactly the same shape. This situation is always true. If a new ray of light is directed backward along the path of a refracted ray, it will follow the same path after passing across the boundary between the two media. This outcome is called the **principle of reversibility of light**.

### Think It Through

- Use Snell's law to verify the principle of reversibility. Explain your reasoning in detail.

1. **K/U** Why is the value for Snell's constant for light travelling from water into glass a different value than that shown in Table 11.1 for either material?

2. **K/U** Explain, in terms of the wave model, why light changes direction when it crosses the interface between two different media.

3. **C** Explain Snell's law without the use of an equation.

4. **K/U** Is the relationship between the incident angle and the refracted angle linear or non-linear?

5. **C** Draw ray diagrams to illustrate the principle of the reversibility of light.

6. **C** Describe why stars twinkle, even when the night sky is cold and clear.

7. **C** Consider the diagram showing how wave fronts change direction at the interface between to different optical media.

   (a) Design an analogy to explain this bending behaviour of light as it crosses a boundary between media of different optical densities.

   (b) Recall the reversibility of lights rays as described on the previous page. Analyze your analogy from part (a) to ensure that it also predicts the behaviour of wave fronts moving in the opposite direction.

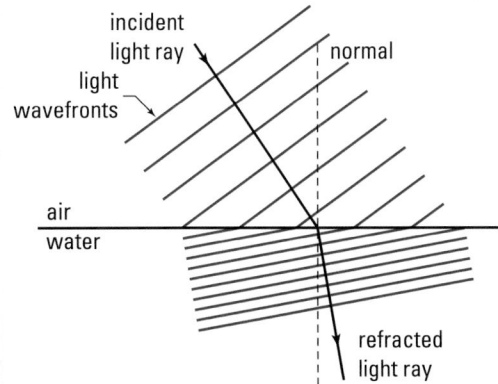

8. **K/U** As the angle of incidence increases, does the angle of refraction increase more or less rapidly?

9. **K/U** Is there a maximum value for the incident angle?

10. **I** Investigation 11-A provided opportunity to verify Snell's law. As part of the procedure, you were required to plot a graph of $\sin \theta_i$ and $\sin \theta_R$.

   (a) Describe the type of relationship that was represented by the graph?

   (b) What does the slope of the line represent?

---

**UNIT PROJECT PREP**

Snell's law provides a mathematical model to predict the behaviour of light as it passes from one optical medium to another.

- How will an understanding of Snell's law assist you with your project design?

- You may want to individualize your project by selecting materials with extraordinarily high refractive indices. How would these materials affect your design?

- Apply the ray model of light to explain optical effects that occur as natural phenomena such as a rainbow.

- Define and describe partial reflection and refraction.

KEY
TERMS

- apparent depth

- angle of deviation

- lateral displacement

- partial reflection and refraction of light

- mirage

 **Math Link**

You can use the formula for Snell's law ($n_i \sin \theta_i = n_R \sin \theta_R$) to develop the relationship shown below for calculating the apparent depth of an object under water, when the object is viewed from directly above. In this example, $n_R$ is the index of refraction of the air, and $n_i$ is the index of refraction of the water.

$$d_{apparent} = d_{actual} \left( \frac{n_R}{n_i} \right)$$

So, for $n_R < n_i$, the apparent depth, $d_{apparent}$, is less than the actual depth, $d_{actual}$.

Attempt to develop the proof for the relationship and write an explanation, using diagrams. Hint: For small angles, $\sin \theta \approx \tan \theta$.

Without realizing it, you make decisions nearly every waking moment, using information that is transmitted to your eyes by light. When you take a step, you assume that you know where the floor is located. When you reach for an object, your brain tells your hand how far and in what direction to reach. All of these decisions are based on the assumption that light always travels in straight lines. Usually your observations are correct, because the light from the objects you observe is travelling only through air. You start having problems, however, when the light has travelled through more than one medium and refraction changes the path of the light.

## Apparent Depth

How can you make people appear to be shorter without physically altering their bodies in any way? At first, this task would seem to be impossible, but by using an optical effect caused by refraction, it is, in fact, quite easy to accomplish. Think about your results from "Apparent Depth" in the Multi-Lab at the beginning of the chapter. Based on these observations, suggest how you could make people appear to be shorter.

When you look at objects that are under water, they appear to be closer to the surface than they actually are. This effect, known as **apparent depth**, is due to the refraction of the light that is coming up from the object through the water and out into the air toward your eyes. The light ray diagram in Figure 11.7 shows what happens to the light travelling from an object on the bottom of a swimming pool up to the eyes of a person standing at the pool's edge.

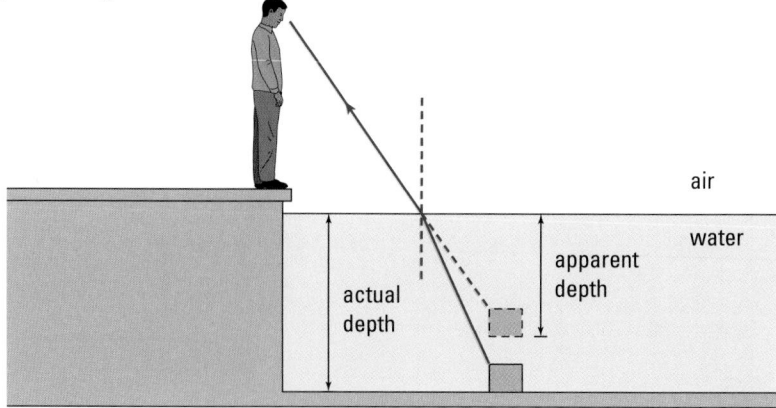

**Figure 11.7** The object on the bottom of the pool appears to be closer to the surface than it is, because the light is refracted away from the normal as it enters the air from the water.

Water is usually about 30 percent deeper than it appears to be. A person who thinks that the water in a stream is deep enough to come only halfway up his thighs might be unpleasantly surprised to find himself waist-deep in water. This illusion also makes it difficult for people fishing among shallow reefs and shoals to judge when the water is deep enough to operate the outboard motors on their boats.

**Biology Link**

What problem would the phenomenon of apparent depth pose for diving birds such as ospreys and seagulls?

## MODEL PROBLEM

### Apparent Depth of a Lake

**A woman in a motorboat sees a wristwatch on the bottom of the lake, just beside the boat. The boat's depth finder indicates that the water is 6.55 m deep. What is the apparent depth of the water?**

### Frame the Problem

- Visualize the problem by drawing a light ray diagram.

- Light rays from the watch travel a distance, $d_{actual}$, through the water and refract on entering the air. Extending the refracted rays straight back gives the image position.

- The indices of refraction of air and water are listed in Table 11.1, Index of Refraction of Various Substances. You can use the equation for the apparent depth to solve the problem.

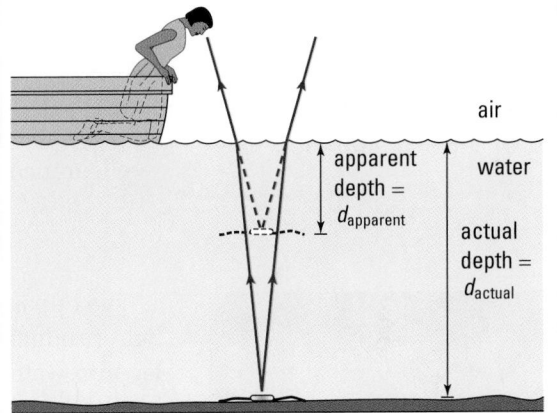

### Identify the Goal

The apparent depth, $d_{apparent}$, of the wristwatch below the water surface

### Variables and Constants

| Involved in the problem | Known | Implied | Unknown |
|---|---|---|---|
| $n_{air}$ | $d_{actual} = 6.55$ m | $n_{air} = 1.000$ | $d_{apparent}$ |
| $n_{water}$ | | $n_{water} = 1.333$ | |
| $d_{actual}$ | | | |
| $d_{apparent}$ | | | |

*continued* ▶

*continued from previous page*

## Strategy

You can use the formula for the apparent depth, $d_{apparent}$.

You have the indices of refraction and the actual depth, so all of the required variables are known.

Substitute values into the formula.

Evaluate.

The apparent depth of the wristwatch is 4.91 m.

## Calculations

$$d_{apparent} = d_{actual}\left(\frac{n_R}{n_i}\right)$$

$d_{apparent} = 6.55 \text{ m } (1.000/1.333)$

$d_{apparent} = 4.9137 \text{ m}$

## Validate

The units are in metres.
The apparent depth is less than the actual depth.

---

### PRACTICE PROBLEMS

**11.** A girl is holding a clear, thin plastic bag of water that contains a goldfish. If she looks directly at the goldfish when it is 15 cm away from the side through which she is looking at it, how far away from the plastic will the fish appear to be?

**12.** A worker is looking down at a sample of radioactive waste that is encased in a rectangular glass block. The sample appears to be 0.55 m below the top surface of the glass. What is the actual depth of the sample in the glass block?

**Figure 11.8** Why does it appear that the girl in this photograph has two left arms?

The apparent change in position caused by the refraction of light produces some unusual optical illusions. If you dip your leg into water and someone looks at it, the leg will look as though it is broken and bent upward. This is another example of the apparent depth effect. The human eye-brain system assumes that light travels in a straight line and, therefore, that the light is travelling directly from the leg toward the observer's eyes.

## Deviation

When light from an incandescent light bulb that has a clear glass envelope shines on a wall or ceiling, the glass has no effect on the light — all of the walls are illuminated evenly. If light from the same bulb shines through crystal chandeliers or crystal glassware, however, beautiful patterns of light appear on the walls and ceiling. You might have seen this effect in displays of chandeliers in stores or when the sun shines through a piece of crystal glassware. The patterns occur because the light is entering, passing through, and then leaving a piece of glass in which the two sides of the glass are not parallel to each another.

In the prism shown in Figure 11.9, the incident light ray travels through the air and enters the left side of the glass. The light bends toward the normal in the glass, because glass has a higher index of refraction (optical density) than the air. When the light leaves the glass and emerges into the air on the other side of the prism, the light is refracted away from the normal. Notice that the direction of the light leaving the glass is different from that of the light entering it. The change in direction of the light as it passes through the glass is known as its deviation. The amount of change is called the **angle of deviation**. You can determine the angle of deviation for any shape of prism by applying Snell's law at each air-glass interface.

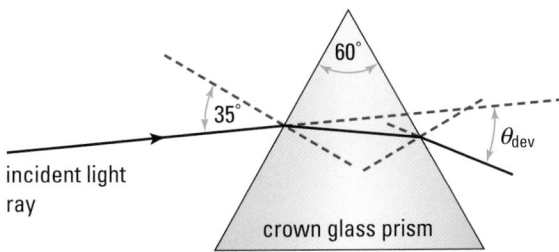

**Figure 11.9** The light in a prism is refracted at both air-glass interfaces. The change in direction of the emergent light ray compared to the original incident light ray is called the angle of deviation.

**TRY THIS...**

Fill a glass beaker with water. Place a 30 cm ruler vertically into the water so that the flat side of the ruler is touching one side of the beaker. The millimetre markings should be facing in toward the water. With your face near the top of the ruler, look straight down into the water. Compare the size of a 1 cm space on the ruler, both above the water and below it. Then, move your head to one side, away from the ruler, to increase the angle of incidence at which you observe the submerged part of the ruler, and make the same comparison. Estimate the minimum and maximum changes in the apparent size of the submerged ruler.

**MODEL PROBLEM**

## Calculating The Angle of Deviation

**Light enters the side of an equilateral, crown glass prism at an angle of incidence of 35.0°. Determine the angle of deviation for the light after it has passed through the prism.**

### Frame the Problem

- Draw a ray diagram of the light passing through the prism.

- Light enters the prism and is *refracted toward the normal*.

- The light *passes through* the prism and exits on the other side.

- As the light re-enters the air, it is *refracted away from the normal*.

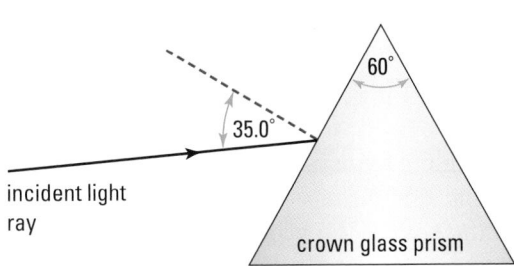

*continued* ▶

- The *angle of deviation* is the angle formed by the extension of the incident ray and the extension of the emergent ray.

- *Snell's law* applies individually to light travelling across each air-glass interface.

- The *rules of geometry* provide the method for finding the angle of incidence for the second interface.

## Identify the Goal

The angle of deviation of the light ray as it passes through the prism

## Variables and Constants

| Involved in the problem | | Known | Implied | Unknown |
|---|---|---|---|---|
| $\theta_1$ | $n_1$ | $\theta_1 = 35.0°$ | $n_1 = 1.000$ | $\theta_2$ |
| $\theta_2$ | $n_2$ | $\theta_{prism} = 60.0°$ | $n_2 = 1.523$ | $\theta_3$ |
| $\theta_3$ | $n_3$ | | $n_3 = 1.523$ | $\theta_4$ |
| $\theta_4$ | $n_4$ | | $n_4 = 1.000$ | $\theta_{dev}$ |
| $\theta_{dev}$ | | | | |
| $\theta_{prism}$ | | | | |

## Strategy

Determine the angle of refraction in the glass at the first air-glass interface, using Snell's law.

All of the values except the angle of refraction are known, so substitute and solve.

## Calculations

$$n_1 \sin \theta_1 = n_2 \sin \theta_2$$

$$n_1 \sin \theta_1 = n_2 \sin \theta_2$$

$$\frac{n_1 \sin \theta_1}{n_2} = \frac{\cancel{n_2} \sin \theta_2}{\cancel{n_2}}$$

$$\sin \theta_2 = \frac{n_1 \sin \theta_1}{n_2}$$

$$\sin \theta_2 = \frac{1.000 \sin 35°}{1.523}$$

$$\sin \theta_2 = \frac{0.57357}{1.523}$$

$$\sin \theta_2 = 0.3766$$

$$\theta_2 = 22.12°$$

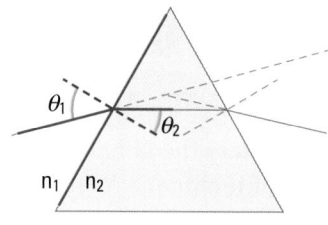

Find angle $a$ by using the rule that a normal line forms a 90° angle with the side of the prism.

$$\theta_2 + a = 90°$$

$$a = 90° - 22.12°$$

$$a = 67.88°$$

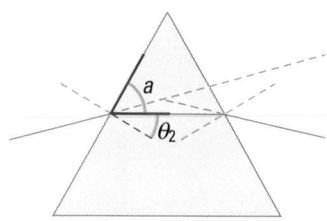

Find angle $b$ from the rule that the sum of the angles in any triangle is 180°.

$$60° + 67.88° + b = 180°$$
$$b = 180° - 127.88°$$
$$b = 52.12°$$

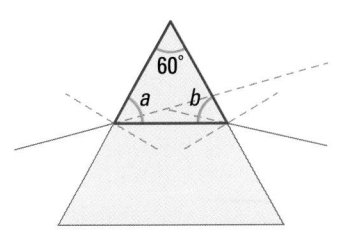

Find $\theta_3$ from the rule that a normal line makes a 90° angle with the side of the prism.

$$b + \theta_3 = 90°$$
$$\theta_3 = 90° - b$$
$$\theta_3 = 90° - 52.12°$$
$$\theta_3 = 37.88°$$

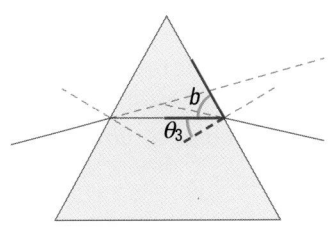

Find the angle of refraction at the second air-glass interface, using Snell's law.

$$n_3 \sin \theta_3 = n_4 \sin \theta_4$$
$$\frac{n_3 \sin \theta_3}{n_4} = \frac{\cancel{n_4} \sin \theta_4}{\cancel{n_4}}$$
$$\sin \theta_4 = \frac{n_3 \sin \theta_3}{n_4}$$
$$\sin \theta_4 = \frac{1.523 \sin 37.88°}{1.000}$$
$$\sin \theta_4 = \frac{0.9351}{1.000}$$
$$\sin \theta_4 = 0.9351$$
$$\theta_4 = 69.24°$$

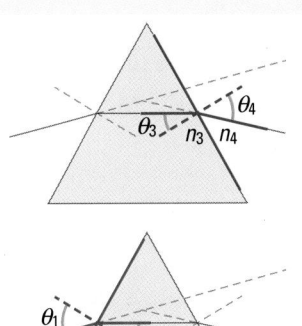

Now, focus on the small triangle formed by the ray inside the prism and the extensions of the incident and emergent rays. If you can find the value of angle $f$, you can find the angle of deviation, $\theta_{\text{dev}}$, because they are supplementary angles. Supplementary angles add to 180°. If you find angles $c$ and $e$, you can use them to find $f$.

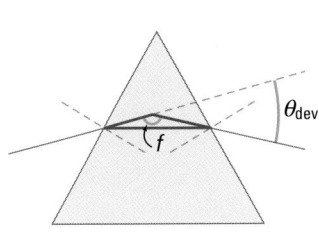

*continued* ▶

Find angle $c$ by using the rule of geometry that when two straight lines intersect, the opposite angles are equal. The incident ray and its extension intersect with the normal.

$$\theta_1 = c + \theta_2$$
$$35° = c + 22.12°$$
$$35° - 22.12° = c + 22.12° - 22.12°$$
$$c = 12.88°$$

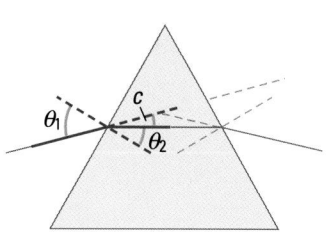

Find angle $e$, using the same techniques that you used above. The emergent ray and its extension intersect with the normal on the second interface.

$$\theta_4 = e + \theta_3$$
$$69.24° = e + 37.88°$$
$$69.24° - 37.88° = e + 37.88° - 37.88°$$
$$e = 31.36°$$

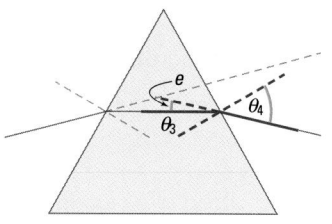

Find angle $f$ from the rule of geometry that the sum of the angles of a triangle is 180°.

$$c + e + f = 180°$$
$$12.88° + 31.36° + f = 180°$$
$$f = 180° - 12.88° - 31.36°$$
$$f = 135.76°$$

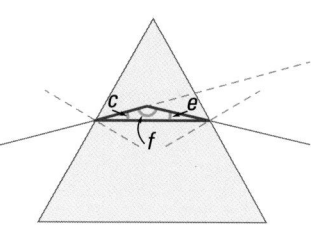

Angles $f$ and $\theta_{\text{dev}}$ are supplementary; therefore, they add to 180°.

$$\theta_{\text{dev}} + f = 180°$$
$$\theta_{\text{dev}} + 135.76° = 180°$$
$$\theta_{\text{dev}} + 135.76° - 135.76° = 180° - 135.76°$$
$$\theta_{\text{dev}} = 44.24°$$

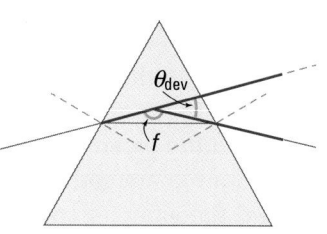

The angle of deviation is 44°.

---

## Validate

From the diagram, the answer appears to be reasonable. You could also draw a scale diagram and determine that the answers were nearly the same.

**13.** Light enters the side of a Plexiglas™ prism, which has an apex angle of 30.0°, at an angle of incidence of 45.0°. Determine the angle of deviation for the light after it has passed through the prism.

**14.** Light leaves the second interface of a crystal glass prism, which has an apex angle of 60.0°, at an angle of refraction of 45°.

(a) Determine the angle of incidence for the light as it first entered the prism.

(b) Determine the angle of deviation for the light after it has passed through the prism.

Look at the object positioned behind the aquarium in Figure 11.11. Notice how the part of the object that you are viewing through water appears to be shifted to the side relative to the top of the object, which you can see through air only. This apparent shift to the side is a special case of deviation that occurs whenever light penetrates a medium for which the two sides are parallel. In these cases, the angle of deviation is always zero.

**TRY THIS...**

Obtain a triangular-shaped prism, a protractor, and a light source from your teacher. Choose an angle of incidence and use Snell's law to predict the angle of deviation for the prism. Verify your prediction experimentally. Then, make a prediction about how the angle of incidence affects the angle of deviation. Test your prediction.

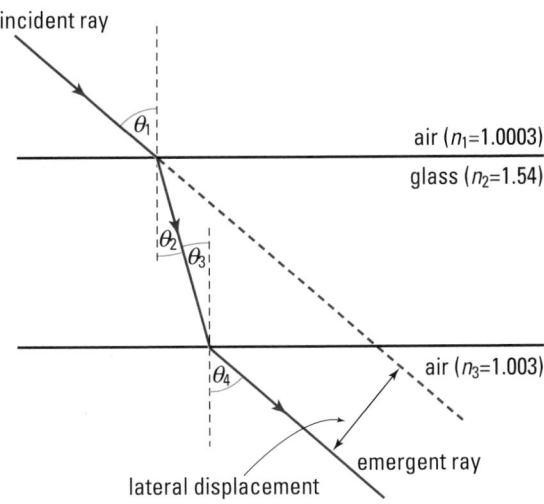

**Figure 11.10** The light does not change direction, but is shifted to the side, so that it emerges travelling parallel to the original path of the light.

**Figure 11.11** The top and bottom parts of the object appear to be separated.

The path of the light ray is not a straight line, however. Inside the refracting medium, the ray is shifted to the side, but the ray that emerges from the medium is parallel to the incident ray; Figure 11.10 illustrates how this occurs. The incident ray enters the medium and is bent toward the normal. When the ray exits from the other side of the medium, the ray is refracted again and, this time, it is bent away from the normal. This shifting of light to the side is called **lateral displacement**. The amount of lateral displacement depends on the angle of incidence, the thickness of the material, and the index of refraction of the material.

Lateral displacement occurs every time you look at an angle through a window at an object such as a flagpole, so the flagpole is not positioned exactly where it appears to be. The light coming from the flagpole is refracted twice as it passes through the glass, and is therefore shifted to the side.

## Think It Through

- Why do you usually not notice lateral displacement? Suggest a way in which you could prove to someone that lateral displacement is occurring when looking at objects such as flagpoles through a window.

- Does the position at which the incident light enters the side of a prism affect the path of the light? Explain your answer.

- Does the position at which the incident light enters the side of a rectangular block affect the path of the light? Explain your answer.

The discussion above describes lateral displacement qualitatively. To prove that the emerging ray is, in fact, exactly parallel to the incident ray, you can use Snell's law, along with the laws of geometry.

- Write Snell's law for the first air-glass interface, using the notations from Figure 11.10.

$$n_{air} \sin \theta_1 = n_{glass} \sin \theta_2$$

- Write Snell's law for the second glass-air interface.

$$n_{glass} \sin \theta_3 = n_{air} \sin \theta_4$$

- Recall from the rules of geometry that when a straight line (ray) cuts two parallel lines (normal lines), the internal angles are equal.

$$\theta_2 = \theta_3$$

- Substitute $\theta_2$ for $\theta_3$ into Snell's law for the second interface.

$$n_{glass} \sin \theta_2 = n_{air} \sin \theta_4$$

- Notice that the first and third statements of Snell's law contain identical terms. Therefore,

$$n_{air} \sin \theta_1 = n_{air} \sin \theta_4$$

- Divide both sides of the equation by $n_{air}$.

$$\frac{n_{air} \sin \theta_1}{n_{air}} = \frac{n_{air} \sin \theta_4}{n_{air}}$$

$$\sin \theta_1 = \sin \theta_4$$

$$\theta_1 = \theta_4$$

- Based on the rules of geometry, since the lines forming one side of the equal angles are parallel (normal lines), the second sides of those angles must be parallel.

The emergent ray is parallel to the incident ray.

## Atmospheric Effects

Since the index of refraction of air is so close to that of a vacuum, you might conclude that you could never detect any refractive effects of air. When you consider that the atmosphere is hundreds of kilometres thick, though, it should not surprise you that the refraction of light from the Sun and stars might produce a noticeable effect.

When the Sun is setting or rising and is very close to the horizon, the bottom half of the Sun appears to be flattened. This illusion occurs because the light from the bottom portion of the Sun is refracted more than the light from the upper portion, as shown in the diagram in Figure 11.12.

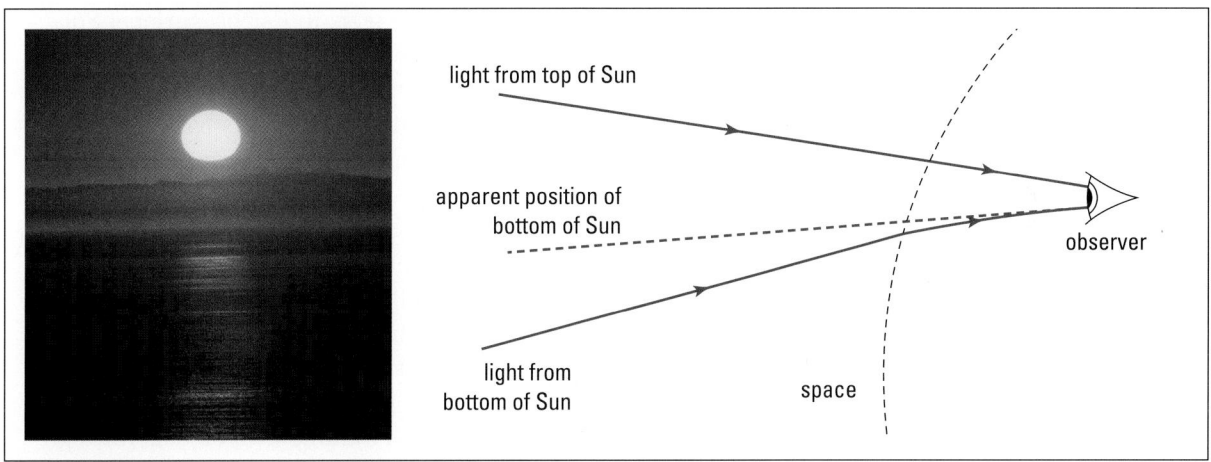

**Figure 11.12** Refraction of light from the Sun in the upper atmosphere causes the bottom of the Sun to appear flattened as it rises and sets.

The bending of the light from the Sun just as it is setting produces yet another optical illusion — you can still see the Sun above the horizon after it has actually set. Due to the change in density and temperature, and therefore a gradual change in the refractive index of the atmosphere, the light from the Sun follows a curved path as it passes through the atmosphere, as shown in Figure 11.13.

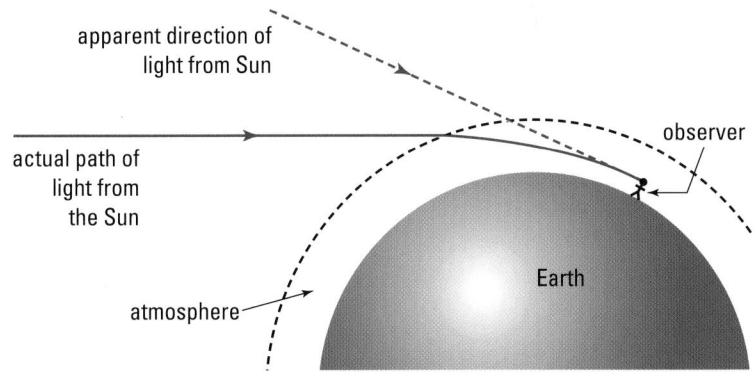

**Figure 11.13** The changing optical density of Earth's atmosphere causes sunlight to follow a curved path toward the observer. The Sun still appears to be positioned above the horizon, even though it has already set.

Refraction of Light • MHR **529**

Both effects — the flattened look of the Sun at sunrise and sunset and the fact that you can still see the Sun above the horizon after it has set — are produced in a similar way to the apparent depth effect. The change in the index of refraction of Earth's atmosphere is produced by variations in both the composition and temperature of the air. The index of refraction of a material varies slightly with temperature. In Table 11.1, the refractive index of each of the gases was specified for a temperature of 0°C.

The twinkling of stars at night is another effect of the refraction of light as it travels through Earth's atmosphere. The starlight passes through many different layers of air that are continuously moving and that vary in temperature and density. These variations cause tiny changes in the refractive index of the air and, therefore, in the path of the star's light as it travels to Earth's surface. For a fraction of a second, the direction of the light from the distant source is changed sufficiently that the star's location appears to shift slightly in the sky, and then it quickly shifts back again, close to its original position. The effect occurs so quickly and so often that the star appears to twinkle. The orbiting Hubble Space Telescope is located beyond Earth's atmosphere and provides astronomers with opportunities to view celestial phenomena without the distortions produced by atmospheric refraction.

The variation of the refractive index of air with temperature causes the shimmering effect you see above barbecues and the surface of hot roads. Because the heat that creates these effects only occurs in the daytime, you see a number of objects through the heated air instead of a single one. Therefore, you see a constantly changing (shimmering), distorted image of all of the objects, rather than a twinkling effect.

## Partial Refraction

Sometimes, when a car goes by, you are momentarily blinded by a flash of light, due to sunlight reflecting from a car window. At night, when you are inside a lighted room looking out through a window into the darkness, you can see an image of yourself and the room's contents clearly reflected in the glass. In both of these situations, you know that light is passing through the window, as well as being reflected. When light strikes any interface between two media, some of it reflects back, while some light refracts through the second medium.

As illustrated in Figure 11.14, when light travelling through the air (the first medium) strikes the flat surface of a transparent plastic block (the second medium), the incident light separates into two distinct light rays. Most of the light passes (is refracted) into the plastic block. A small amount of the light is reflected back into the air from the surface of the plastic block, according to the law of reflection. This phenomenon is called the **partial reflection and refraction of light**.

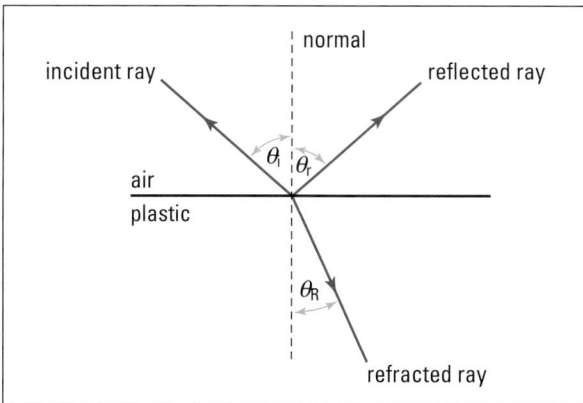

**Figure 11.14** Light is simultaneously reflected and refracted when it passes from one medium into another.

The amount of the incident light that is refracted, as opposed to reflected, depends on the angle of incidence and on the indices of refraction of the two transparent materials. When light travelling from air into glass shines directly down onto the glass surface, almost all of the light is refracted and very little is reflected. When the angle of incidence approaches 90°, however, only a small portion of the light is refracted into the glass and most of it is reflected.

You can observe how the amount of reflected light changes with the angle of incidence by standing at the edge of a shallow pond on a sunny day and looking across the pond's surface. If you look at the surface just in front of you, you can see down into the water quite clearly. Very little sunlight reflects from the water surface toward you. Most of the sunlight refracts down into the water, then reflects off the bottom, and travels back up to your eyes. However, if you look across the surface farther from shore, the amount of sunlight reflecting from the water surface is much greater. Much less of the sunlight is entering the water at this larger angle of incidence. Figure 11.15 illustrates this.

When you are driving at night, the glaring reflection in your rearview mirror of the headlights of vehicles behind you can be a very real safety hazard. Most rearview mirrors in cars have a small lever mechanism that allows the driver to flip the mirror to another preset position, to reduce the glare to safer levels.

As you can see in Figure 11.16 (A), on the following page, the structure of the rearview mirror is quite different from that of normal plane mirrors. The mirror consists of a glass prism (wedge) that has a mirrored back surface. During the day, the mirror is positioned so that the light is refracted as it enters the glass wedge. The light is then reflected from the mirror surface at the back, travels back through the glass, and is refracted out into the air toward the driver's eyes. Glare is usually not a problem in daylight, even if a following vehicle has its headlights switched on.

**Figure 11.15** As the angle of incidence increases, more light is reflected from the surface than is refracted into the medium.

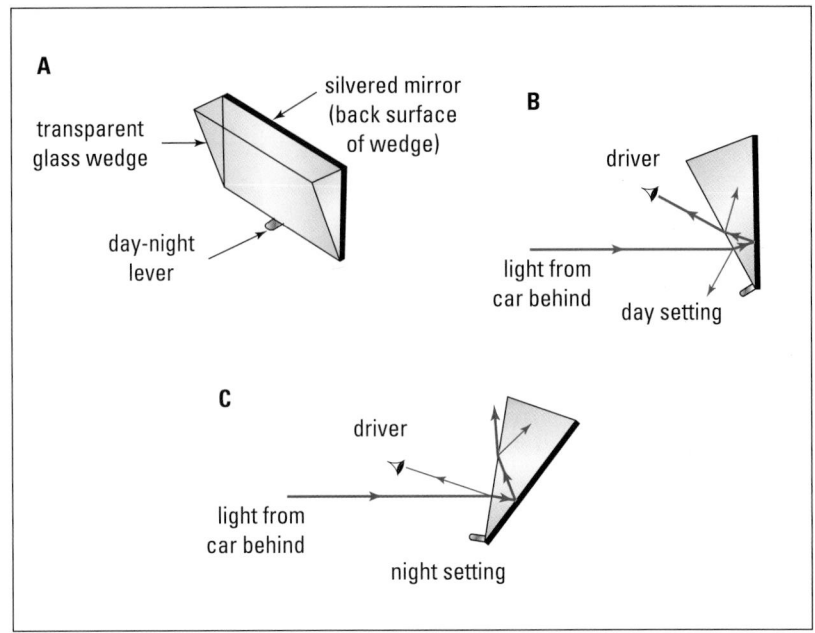

**Figure 11.16** A rearview mirror has a mechanism to reduce the glare of headlights. The mirror can be adjusted so that the light is reflected from the front surface of the glass, rather than being reflected by the mirror surface on the back of the glass.

Figure 11.16 (B) shows that a small amount of partial reflection also occurs at the air-glass interface each time the light is refracted. In both cases, however, the reflected light is not directed toward the driver's eyes. At night, when the glare reflected from the mirror is too great, the driver uses the lever to flip the mirror to the preset position shown in Figure 11.16 (C). When the mirror is in the new position, the bright light that is reflected from the mirror surface travels back through the glass wedge, emerges into the air, and shines up toward the roof of the car, thus avoiding the driver's eyes. The shape of the glass wedge is such that its front surface is now in just the right position to act as a mirror surface, reflecting the light toward the driver's eyes. Because the incident light is so intense, the partial reflection that occurs at the front surface is sufficient to provide the driver with enough light to observe the traffic behind.

**Language Link**

From which French verb is the word "mirage" derived, and what does it mean?

## Mirages

A **mirage** is a virtual image that occurs naturally when particular atmospheric conditions cause a much greater amount of refraction of light than usual. A mirage might be observed when some layers of Earth's atmosphere are warmer than others. On a hot, sunny day, the layers of air above a road surface vary in temperature and form a mirage, as shown in Figure 11.17 (A). Light coming from the sky is refracted to such a great extent as it passes through the different layers of air that some of the light follows a curved path and travels up toward the observer's eye. This refraction effect makes objects that are just above the horizon appear as if they are formed as images in a mirror-like road surface, as shown in Figure 11.17 (B).

Sometimes the layers of air are reversed, with the hot air being above a layer of cold air that is next to the ground. When this happens, light from a distant object undergoes refraction in the hot air, and the image appears to be higher in the air than it actually is. These kinds of mirages sometimes make distant mountains seem higher and closer than they actually are.

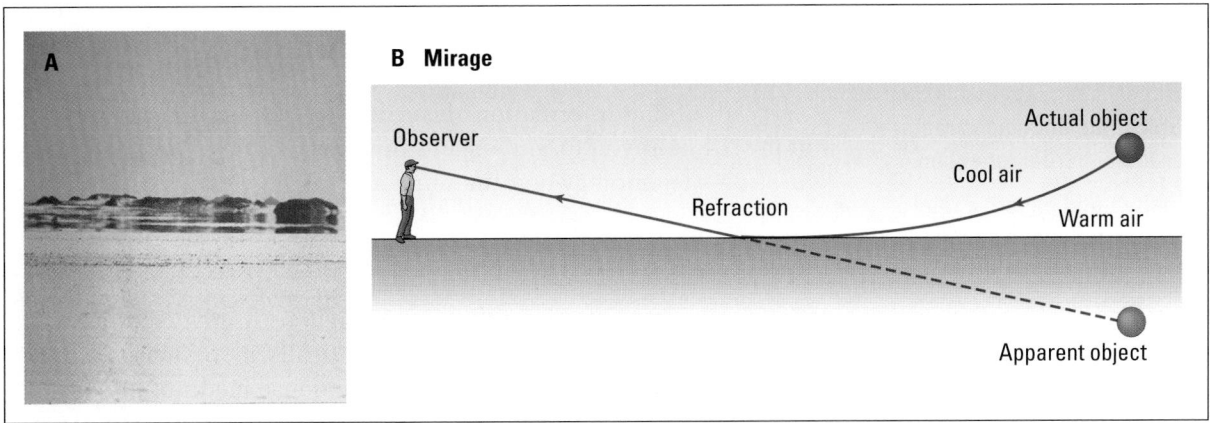

**Figure 11.17** Mirages are caused by refraction of light by air when extreme temperature changes occur near the ground. Cool air above very warm air that has been heated by the ground creates ideal conditions for the formation of a mirage.

## 11.3 Section Review

1. **K/U** Why is there more glare from car headlights on a rainy night than on a night when the roads are dry?

2. **K/U** Why is it easier to see objects on the floor of a lake or river when looking directly down, rather than when looking at an angle?

3. **K/U** List several optical distortions or illusions that you have observed. Explain how they occur.

4. **MC** Planets do not twinkle, so they can easily be distinguished from stars. Explain why this is the case, based on the effects of refraction.

5. **C** Explain, using diagrams, why you can still see the Sun even after it has set.

**UNIT PROJECT PREP**

Refraction effects are a key element of many optical instruments.

- How do refraction effects influence where you place your lenses in your optical instrument?

- Consider lens properties such as size, thickness and index of refraction when designing your optical instrument.

SECTION
EXPECTATIONS

• Define and describe total internal reflection and critical angle.

• Explain the conditions required for total internal reflection.

• Analyze and describe total internal reflection situations using the light ray model.

KEY
TERMS

• critical angle

• total internal reflection

• retroreflector

The next time you are talking on the telephone or watching a television program, look down at the hairs on your arm. A single glass fibre, thinner than one of those hairs, is now able to transmit tens of thousands of telephone calls and several dozen television programs simultaneously. How can such a thin piece of glass fibre carry all of that information? Learning about total internal reflection will help you to understand this remarkable — and extremely useful — technology and the impact it has on your daily life.

## Conditions for Total Internal Reflection

So far, you have studied refraction of light as it travels from an optically less dense medium into an optically more dense medium, and as it travels from an optically more dense medium into an optically less dense medium. Have you noticed anything unusual that occurs in the second case, but not in the first? Complete the Quick Lab on the opposite page and focus on any differences that you observe between light travelling from the optically less dense into the more dense medium, compared to light travelling in the opposite direction.

When light travels from an optically more dense material into an optically less dense one, the refracted light bends away from the normal. As you can see in Figure 11.18, for all angles of refraction up to and including 90°, the light is simultaneously reflected and refracted, as expected. The amount of reflected light gradually increases and the amount of refracted light gradually decreases as the angle of incidence increases. Notice that the angle of refraction increases more rapidly than the angle of incidence.

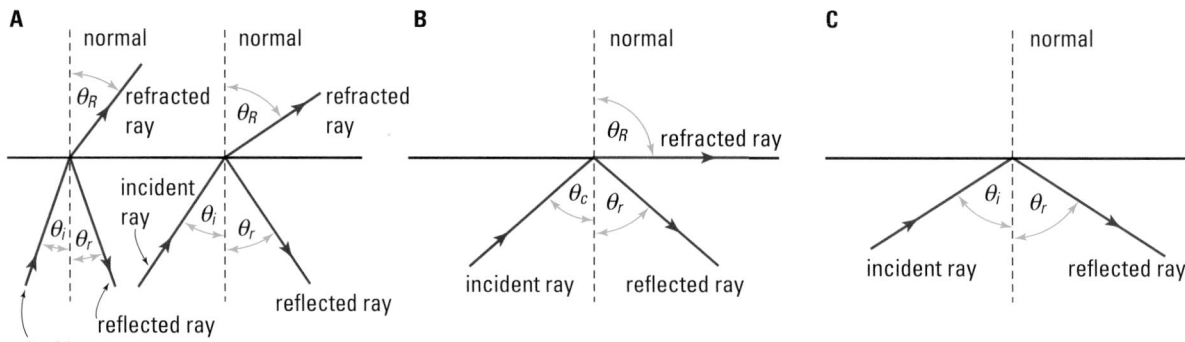

**Figure 11.18** The incident light in the Plexiglass™ is both partially refracted and partially reflected for all angles of incidence **(A)** up to the critical angle, $\theta_c$, for the Plexiglas™ and **(B)** at the critical angle. **(C)** As soon as the angle of incidence exceeds the critical angle, total internal reflection occurs.

# Total Internal Reflection

In this lab, you will observe the same object, a pencil, from three different perspectives. Position a 1.0 L beaker so that you can observe it from the top, side, and bottom. Fill the beaker with water. Observe the motion of a pencil as it is slowly submerged into the water at a very small angle with the horizontal, as shown in the first part of the diagram.

In each of the following cases, gradually move the pencil under the surface of the water, after referring to the indicated part of the diagram. Observe the appearance of the pencil:

- from almost directly above the water, as shown in (B)

- from the side, so that you are looking toward the surface of the water as shown in (C)

- from the side, looking at the surface of the water from below, as shown in (D)

- from below, looking up through the bottom of the beaker

## Analyze and Conclude

1. Describe in detail the appearance of the pencil as you saw it from each perspective.

2. Describe the most significant difference that you observed between one perspective and the others.

3. Explain how and why the appearance of the pencil differed so greatly from the different perspectives. What properties of light and its interaction with various materials were responsible for these differences?

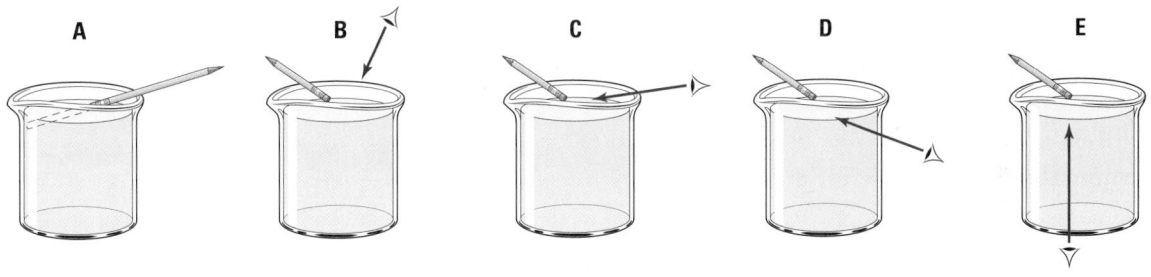

A          B          C          D          E

The angle of refraction can be increased only until it reaches its maximum possible value of 90°, as seen in Figure 11.18 (A). The maximum angle of refraction in any medium is 90°, because it is not physically possible to exceed this value without having the light remain in the original medium. In that case, the light would no longer be refracted, but would be reflected.

The angle of incidence, for which the angle of refraction is exactly 90°, is called the **critical angle**, $\theta_c$ as illustrated in Figure 11.18 (B). In fact, something very interesting occurs when the angle of incidence is increased beyond the critical angle — all of the incident light is completely reflected back into the optically more dense Plexiglas™, as shown in Figure 11.18 (C). None of the

**ELECTRONIC LEARNING PARTNER**

Go to your Electronic Learning Partner to find out more about total internal reflection.

Pour about 400 mL of water into a 500 mL beaker and place an empty test tube in the water at an angle. Observe the test tube from above and below the water surface, and describe any differences. Now, remove the test tube from the water in the beaker and pour water into the test tube until it is about one third full. Replace the test tube in the water in the beaker and repeat your observations. Explain the effect produced by adding the water to the test tube.

light is refracted into the optically less dense air. The interface between the two media behaves as if it was a perfect mirror surface. This phenomenon is called **total internal reflection**.

Even the most efficient (metallic) mirrored surfaces absorb from 8 percent to 11 percent of the incident light. By contrast, when total internal reflection occurs in a glass prism, *none* of the light is absorbed during reflection and only about 2 percent of the light is absorbed by the glass itself.

---

**TOTAL INTERNAL REFLECTION**

The two conditions required for total internal reflection to occur are as follows.

- The light must travel from an optically more dense medium into an optically less dense medium.

- The angle of incidence must exceed the critical angle, $\theta_c$, associated with the material.

---

You can determine the critical angle, $\theta_c$, for light travelling from diamond into air by using Snell's law, as demonstrated in the following problem.

**MODEL PROBLEM**

## Finding $\theta_c$

**Determine the critical angle for diamond.**

### Frame the Problem

- Make a sketch of the problem. Label all of the media, angles, and rays.

- Light is travelling *from* an optically *more dense* material *into* an optically *less dense* material

- The *critical angle* of incidence corresponds to an angle of refraction equal to 90°.

- The needed *indices of refraction* are listed in Table 11.1.

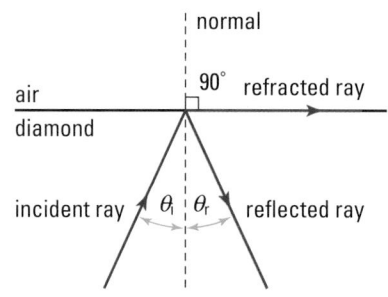

A light ray striking a diamond-air interface, with an angle of incidence equal to $\theta_c$

### Identify the Goal

Calculate the critical angle, $\theta_c$, for diamond.

## Variables and Constants

**Involved in the problem**

$n_{air}$

$n_{diamond}$

$\theta_{air}$

$\theta_{c/diamond}$

**Implied**

$n_{air} = 1.0003$

$n_{diamond} = 2.42$

$\theta_{air} = 90°$

**Unknown**

$\theta_{c/diamond}$

## Strategy

The critical angle of incidence, $\theta_{c/diamond}$, occurs when the angle of refraction is exactly 90°.

Use Snell's law.

Find the required indices of refraction by using Table 11.1.

Substitute values into the formula, using the fact that $\theta_i = \theta_{c/diamond}$ when $\theta_R = 90°$.

Evaluate.

## Calculations

$n_i \sin \theta_i = n_R \sin \theta_R$

$2.42 \ \sin \theta_{c/diamond} = 1.0003 \sin \ 90°$

$\sin \theta_{c/diamond} = \dfrac{1.0003 \times 1.000}{2.42}$

$\sin \theta_{c/diamond} = 0.4133$

$\theta_{c/diamond} = \sin^{-1} 0.4133$

$\theta_{c/diamond} = 24.415°$

The critical angle of incidence for diamond is 24.4°.

## Validate

The angle of incidence is less than the angle of refraction.

---

### PRACTICE PROBLEMS

15. Determine the critical angle for ethyl alcohol.

16. The critical angle for a new kind of plastic in air is 40°. What is the critical angle for this plastic if it is immersed in water?

17. Optical fibres, made of a core layer surrounded by cladding, trap transmitted light by ensuring that the light always strikes the core-cladding interface at an angle greater than the critical angle. Calculate the critical angle between the core-cladding interface.

18. While swimming in a friend's pool, you allow yourself to slowly sink to the bottom exactly 3.0 m from the edge of the pool. As you sink you fix your gaze on the edge. Calculate how deep your eyes must be below the surface for total internal reflection to occur.

### PROBLEM TIP

When using Snell's law to determine the critical angle for a transparent material, the value for the sine of the angle of refraction will always be unity (one), because the maximum angle of refraction is 90°.

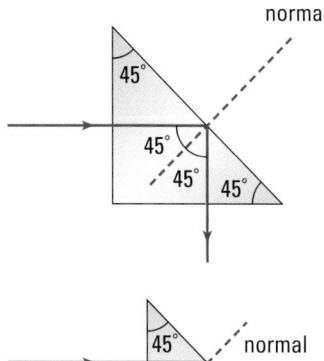

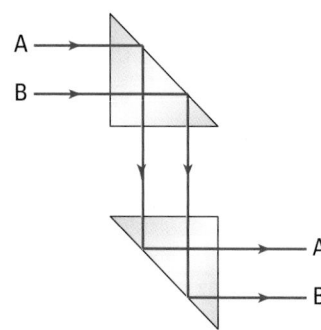

**Figure 11.19** The direction of the light can be changed by either 90° or 180° as the light undergoes total internal reflection inside a single 45°–45°–90° prism.

## Applications of Total Internal Reflection

What do driveway reflectors, binoculars, and optical fibres all have in common? Each technology is based on total internal reflection of light. You can understand the operation of these devices by studying total internal reflection in a simple glass prism, such as the one illustrated in Figure 11.19. Notice that the cross section of the prism is an isosceles right triangle. The critical angle for most types of glass and plastic is less than 45°. When light enters the prism perpendicular to the glass surface on one of the shorter sides of the prism, it will strike the longer side at an angle of 45°, as shown in Figure 11.19 (A). Because the angle of incidence at the glass-air interface is greater than the critical angle for glass ($\theta_c$ is in the range of 40° to 43°), the light undergoes total internal reflection and passes out into the air perpendicular to the third glass surface.

If the light enters the glass prism on the longer side, as shown in Figure 11.19 (B), the light strikes both glass-air interfaces and is totally internally reflected twice inside the prism. The light finally leaves the prism through the same side that it entered, having been reflected back parallel to its original path.

Optical devices such as periscopes, binoculars, and cameras use transparent glass or plastic prisms as reflectors to change the direction of light by either 90° or 180°, as shown in Figure 11.20. In some cameras, a single pentaprism, similar to that shown in Figure 11.20 (C), reflects the light into the viewing eyepiece. As discussed earlier, prisms are particularly efficient at reflecting light — even more efficient than high-quality mirrors. The extra amount of light reflected by prisms provides a brighter image than a mirror, which can be very important when using periscopes, binoculars, cameras, and other optical devices in dim light.

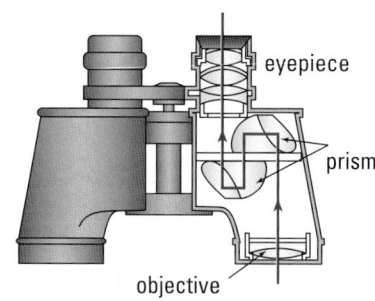

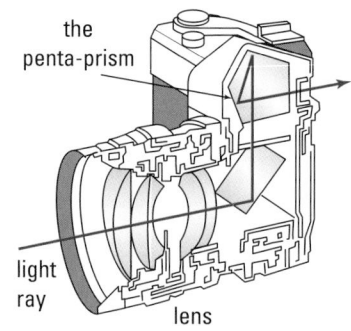

**Figure 11.20** **(A)** In a periscope, total internal reflection occurs once in each prism to laterally displace the light the desired amount. Notice that the image remains upright.

**(B)** In a pair of binoculars, total internal reflection occurs twice in each prism, making it possible both to reduce the length of the binoculars and to achieve lateral displacement of the image.

**(C)** The principle of total internal reflection is also incorporated into the design of some cameras.

One of the most widely used applications of total internal reflection is found in a familiar device known as a **retroreflector**. Anyone travelling in any type of road vehicle, or even on a bicycle, uses retroreflectors as a safety precaution. A retroreflector is an optical device that reflects light directly backward, parallel to its original path. Car and bicycle reflectors and markers used to identify hazards on highways and roads are all designed to reflect light by means of total internal reflection.

As the light ray diagram in Figure 11.21 shows, light travelling from a car headlight toward a bicycle reflector passes through the reflector's glass and strikes the many sets of flat, angled surfaces at the back of the device. The light undergoes total internal reflection at the 45° surfaces, before passing out through the front of the reflector and back toward the light source. If a single plane mirror was used, the light would be reflected uselessly off to the side of the road.

Yet another application of total internal reflection can be found in gemstones, such as diamonds and emeralds, and various crystal ornaments, such as those seen on elaborate chandeliers — objects that appear to sparkle as you move past them. This sparkling effect is produced by the total internal reflection of light as it enters the many flat surfaces (facets) of a gem or crystal and is then reflected internally and back out through other surfaces toward your eyes. In addition to being valuable, diamond is very popular because its large index of refraction (and, therefore, small critical angle) allows a gem-cutter to cut more facets for total internal reflection than any other gemstone. Each additional facet creates more opportunities for light to reflect and create more "sparkles."

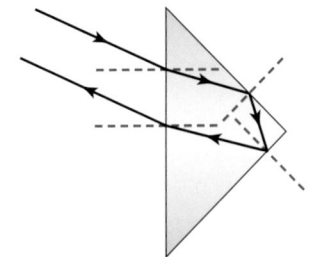

**Figure 11.21** To show the basic principles of retro-reflectors, light is shown being refracted and then reflected from just two reflecting surfaces. In actual retroreflectors, there are three reflecting surfaces, each positioned at 90° to one another. Why are three reflecting surfaces required?

## Optical Fibres

Perhaps one of the most innovative and exciting ways in which total internal reflection is integrated with technology is in the transmission of information carried by light energy in optical fibres. An optical fibre is a very fine strand of a special kind of glass. When light shines into one end of an optical fibre, total internal reflection causes the energy to be confined within the fibre (Figure 11.22). The light travels along the inside of the length of the fibre, carrying information in the form of pulses. Even if the optical fibre is literally tied in knots, the light will still travel through the fibre until it reaches the other end.

A typical optical fibre is about the thickness of a human hair. As illustrated in Figure 11.23, on the next page, the fibre consists of a glass core, roughly 50 $\mu$m in diameter, surrounded by a thin layer known as "optical cladding," which is made of another type of glass. The cladding increases the total outside diameter to about 120 $\mu$m. The glass core has a slightly higher index of refraction ($n = 1.5$) than the optical cladding layer ($n = 1.47$), as is required for total internal reflection. By surrounding the entire length of the

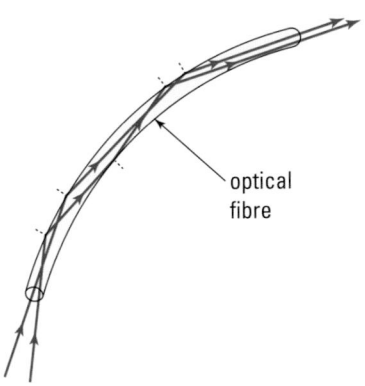

**Figure 11.22** The light travelling inside the optical fibre will continue to undergo total internal reflection along the interior wall of the fibre until it reaches the other end.

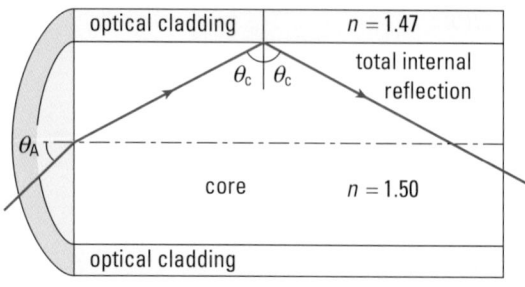

**Figure 11.23** Only light that has an angle of incidence of more than 78.5° will be totally internally reflected down the entire length of the optical fibre.

core, the outer layer of glass cladding ensures that the critical angle of incidence of the core remains constant throughout the entire length of the fibre.

Because the difference in refractive indices of the two materials that make up an optical fibre is very small, the critical angle of the glass core can be as high as 78.5°. Consequently, only light entering the optical fibre at angles of incidence greater than 78.5° can be transmitted along it. What happens to any light that enters the optical fibre at angles up to and including the critical angle of incidence for the core? The angle, $\theta_A$, that is shown in the Figure 11.23 is called the "acceptance angle" for the fibre, and in the case of this fibre, the $\theta_A$ is 11.5° (90° − 78.5°). Any light entering the fibre at an angle less than the acceptance angle will meet the cladding at an angle of incidence greater than the critical angle, $\theta_c$, and will experience total internal reflection. The light enters the fibre and then travels along it, reflecting from side to side in a zigzag path until it reaches the other end.

Groups of optical fibres are often bundled together to produce what are called "optical fibre cables," which are usually covered with a protective plastic covering (Figure 11.24). Because these cables have a small diameter (some less than a millimetre) and are flexible, they can fit into spaces too small for metallic wires. They can also be used safely in corrosive and explosive environments.

The optical fibres are made from a very special form of ultra-pure fused silica (sand). An optical fibre cannot be made of ordinary glass, because internal impurities reduce by a factor of 100 000 the intensity of any light entering it, after the light travels a distance of only 5 m. Currently, the very pure form of glass used in optical fibres is so transparent that light can travel up to 50 km along the fibre. To appreciate the purity of this glass, note that normal sunlight penetrates to a depth of only about 100 m in ordinary seawater before it is completely absorbed, and at depths below this, there is total darkness.

Clearly, sending information along an optical fibre requires a unique light source. The unique properties of laser light allow audio, video, and text based data to be transmitted simultaneously over long distances. When the signals are received they can be separated and processed individually. The light produced by a laser is called "coherent light," because in laser light, the light waves emitted by the source leave perfectly in phase with each another and are all of the same wavelength. All other types of light sources give off light waves randomly, so the light waves are not in phase.

The development in 1970 of the ultra-pure glass required to transmit light signals many kilometres made it possible to use optical fibres as a practical telecommunications medium.

**Figure 11.24** Total internal reflection between the optical fibre core and cladding allows the transmission of information in the form of light pulses.

Developments in laser light sources and receivers signals meant that, by 1980, it was possible to establish a worldwide installation of optical fibre communication systems. Some of the advantages of using optical fibres rather than metal conductors are the ability to significantly increase the amount of transmitted data, reductions in both weight and size, lower operating costs, and increased security in transmitting sensitive data.

## CANADIANS IN PHYSICS

### Photons, Lasers, and Superatoms

Physicist Milena Imamovic-Tomasovic

Born in Sibenik, Croatia, Milena Imamovic-Tomasovic was fascinated with science in elementary school. In high school, she read Kurt Mendelssohn's book *The Quest for Absolute Zero*, and decided to study low-temperature physics. She was particularly intrigued by the work of Indian physicist, Satyendra Nath Bose.

In 1924, Bose wrote to Albert Einstein, explaining his theory of the statistics of photons (the particles that make up light). Einstein extended Bose's theory to other particles and atoms. He and Bose predicted that, at a low enough temperature, a gas would condense into a "superatom," a completely new state of matter.

This dramatic transformation, called "Bose-Einstein condensation" was not actually observed until 1995, while Imamovic-Tomasovic was studying physics at the University of Belgrade. That year, researchers in Boulder, Colorado, used laser cooling, and other techniques, to cool a gas to less than a millionth of a degree above absolute zero (–273°C) — and produce the new form of matter.

Bose condensates are now being created every day in more than 30 laboratories around the world. Imamovic-Tomasovic, at the University of Toronto, is one of a number of physicists studying their unique properties.

"Bose-Einstein condensation is still new," Imamovic-Tomasovic explains. "It's hard to tell what practical application one might expect in the future. Bose condensate is very much like laser light. What makes laser light different from ordinary light is that all of its photons are exactly the same (the same energy and the same phase). The same is true for atoms in the Bose condensate; all are exactly the same (the same energy state). Therefore, one could hope to build an atom laser that would be used in very sensitive measurement instruments. In fact, experiments demonstrating the first atom lasers have already been done."

Asked what advice she would give to high school students, Imamovic-Tomasovic recalls how she followed her heart into low-temperature physics: Look closely and deliberately at possible career paths. Then do what your heart tells you. If you work on something that you really like, almost nothing is too hard for you.

### Web Link

***www.school.mcgrawhill.ca/resources***
For more information on Bose condensates and atom lasers, go to the above Internet site. Follow the links for **Science Resources** and **Physics 11** to find out where to go next.

## Optical Fibres and Medicine

In the field of medicine, the use of optical fibres has significantly
changed many types of surgical and diagnostic procedures. In
arthroscopic surgery, the endoscope is used to carry out many
different types of internal surgical operations. The endoscope
consists of a flexible tubular sheath only millimetres in diameter,
containing two separate optical fibre cables, minute surgical tools,
fine tubes to carry cleansing water down to and away from the site
of the operation, and a microscope for viewing the interior area of
a person's body. One optical cable shines light down to illuminate
the interior of the body, and the other cable carries the reflected
light back so that the surgeon can see the affected area. The surgeon makes a small incision in the skin and carefully slides the
long tube of the endoscope into the area requiring surgery.

The reflected images can also be fed to a video camera and displayed on a television screen. Sometimes, the patient is conscious
during the operation. The minimal damage caused to the tissues
near the area of the operation allow the patient to recover much
more rapidly than with conventional surgery.

One type of endoscope, the bronchoscope, is inserted through
the throat or the nose and used to view the bronchial tubes or
lungs. A similar instrument can be used to view the upper parts of
the digestive system. Tissue samples can be obtained and surgical
procedures performed using all of these devices.

## QUICK LAB — Determining the Critical Angle

### TARGET SKILLS

- Initiating and planning
- Predicting
- Performing and recording

So far in this unit, the light ray model has been
a very useful visual scientific model for analyzing and predicting the behaviour of light. A
mathematical formula is another very powerful
and commonly used type of scientific model
that can be used for the same purposes. For
example, Snell's law can be used to predict the
critical angle of incidence for any material.

- Given the index of refraction for a known
  material, use Snell's law to predict the critical
  angle of incidence for the material, and then
  design an investigation to experimentally
  verify the prediction.

- Prepare an appropriate list of required
  equipment and materials. The transparent
  material to be used in the investigation will
  be specified by your teacher.

- Design a procedure for testing your prediction
  and present it to your teacher for approval.
  After it has been approved, carry out the
  procedure and record and analyze all
  required data.

### Analyze and Conclude

Compare the theoretical predictions and the
experimental (empirical) evidence, and account
for any discrepancies.

## Lasers in Medicine

In many action films, lasers have been agents of destruction, used to burn through heavy metal, aim powerful weapons, or act as "trip wires" to detect intruders. Lasers have, however, many peaceful, life-enhancing uses. They read the music engraved on your compact disk, help bricklayers align a wall, and identify the brand and price of items at stores' checkout counters.

The pinpoint accuracy of a laser makes it a useful tool for surgeons. Instead of knotting a thread around a tiny artery that is leaking blood into the abdomen, a surgeon can "zap" the end of the vessel with some laser light. The very local heating cauterizes the artery, effectively sealing the end.

With a laser, an eye surgeon can reshape the eye to correct a problem that would otherwise require glasses or contact lenses. For example, by taking a tiny slice off of the front of the cornea with a high energy $CO_2$ laser, the eye surgeon can correct a patient's near-sightedness.

Lasers are used in a wide variety of cosmetic surgical procedures, from wrinkle, freckle, and tattoo removal, to repairing drooping eyelids, to reducing snoring.

Lasers also have a place in major operations. Some patients with severe angina, a condition in which portions of the heart are scarred and not working properly, are too ill for open-heart surgery. In a new procedure called "transmyocardial revascularization" (TMR), a $CO_2$ laser is used to burrow tiny channels in the heart muscle. Small blood vessels grow through these channels, restoring blood flow and oxygen to the muscle. Although experimental and currently being used only for patients with few other options, TMR does not require that the surgeon open the chest wall, so patients can go home in just a few days.

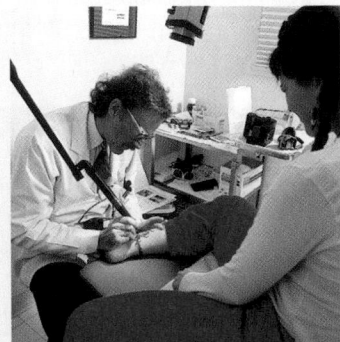

Tattoo removal using a laser can be quite painful.

Tattoo removal before and after images.

### Analyze

What property of a laser makes it useful for surgical procedures?

## 11.4 Section Review

1. **K/U** List the two conditions required for total internal reflection to occur.

2. **K/U** Why are prisms, rather than mirrors, often used in optical devices to reflect light?

3. **K/U** How does the speed of light relate to the critical angle for a given material?

4. **MC** Why would it be desirable to have such a large critical angle for optical fibres? What problems could be caused if the critical angle was reduced?

EXPECTATIONS

• Demonstrate and illustrate, using light ray diagrams, how light behaves at the interface of various media.

• Define and explain dispersion.

KEY TERMS

• dispersion

• visible spectrum

• recombination

The beautiful, sparkling colours produced by ice crystals on a twig in winter, the vibrant colours of a rainbow, and the brilliant flashes of colour you see when light passes through glass chandeliers and diamonds — these are all examples of a phenomenon known as **dispersion**. Dispersion is the separation of visible light into its range of colours. Although this effect has been observed for thousands of years, it was Sir Isaac Newton who, in 1666, initiated the first systematic study of dispersion.

**Figure 11.25** Ice crystals formed on blades of grass disperse sunlight separating it into a collage of colour.

Newton's research was prompted by problems that he encountered with the lenses he was using to build a refracting telescope, a device invented by Hans Lippershey (1570–1619), a Dutch lensmaker. The lenses used in early telescopes suffered from a problem now known as "chromatic (colour) aberration." The glass at the outer edges of the lenses refracted the light so much that rainbow-coloured fringes appeared around the perimeter of all objects observed through a telescope.

## The Spectrum

Newton found that sunlight was separated into a range of colours (called the **visible spectrum**) when it passed through a glass prism, as shown in Figure 11.26. Some skeptics believed that, instead of being a property of the light, the colours were somehow produced by the prism. So Newton added a lens and a second prism to show that the coloured light could be put back together into white light again through a process referred to as **recombination** — something that should not happen if the prism was the source of the colours. Figure 11.27 shows Newton's experiment.

Sunlight has a continuous visible spectrum that, for convenience, has been grouped into seven colours: red, orange, yellow, green, blue, indigo, and violet. More careful analysis shows that there is an infinite number of different colours that blend together continuously to form the entire visible spectrum.

As Figure 11.26 shows, red light is refracted the least and violet light the most. This difference in the degree of refraction of the colours indicates that the index of refraction of the material varies with the colour of the light. For all commonly used transparent materials, the index of refraction is smaller for red light than it is for violet light. Since the index of refraction is related to the speed of light in a medium, the smaller index of refraction of red light indicates that red light travels through common media faster than violet light.

Each type of light source, such as incandescent, fluorescent, and neon lights, produces a characteristic spectrum of light. When the light from an incandescent lamp passes through a prism, a continuous band of colour is produced, with no one colour being brighter than any other. A normal fluorescent source produces a continuous spectrum of colour, but four discrete wavelengths, or colours, predominate. A neon source produces only a few discrete colours of the spectrum. The laser light is unique because it consists of only one wavelength out of the entire visible spectrum of light.

## Dispersion in Nature

Water droplets and ice crystals in the atmosphere produce some beautiful natural displays of colour. Rainbows form when sunlight passes through raindrops, which disperse the light into the different spectral colours. The various wavelengths in light are refracted by differing amounts when they enter the top of the water droplet.

**Figure 11.26** A prism separates white light into its spectrum of colours.

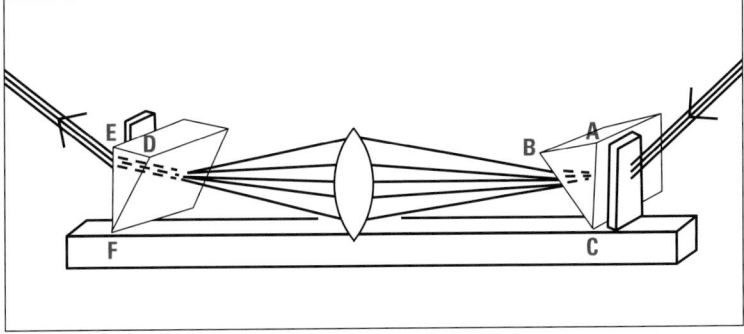

**Figure 11.27** This sketch illustrates Newton's demonstration with prisms and lenses. Prism ABC separates white light into the spectral colours. The dispersed light then passes through the lens, and prism DEF recombines the colours back into white light.

The colours then undergo total internal reflection at the back of the droplet and are refracted again as they leave the droplet near the bottom (see Figure 11.28). The different colours exit the droplets at slightly different angles. The red light leaving the droplets makes an angle of 42° in relation to the sunlight entering them. The violet light leaving the droplets makes an angle of 40° in relation to the incident sunlight, and so on.

When you look at a rainbow, you see only the red light from the droplets higher in the sky. The other colours leaving those droplets pass above your eye. You see only the violet light from droplets lower in the sky, because the other colours from these droplets pass below your eye. All of the other different colours of light in the spectrum leave the millions of droplets between the droplets that produce the red and violet light, allowing you to see the complete spectrum in the rainbow.

As shown in Figure 11.29, conditions sometimes to produce a second rainbow that is higher in the sky and fainter than the first (primary) one. This secondary rainbow is formed by light that undergoes total internal reflection *twice* inside each raindrop.

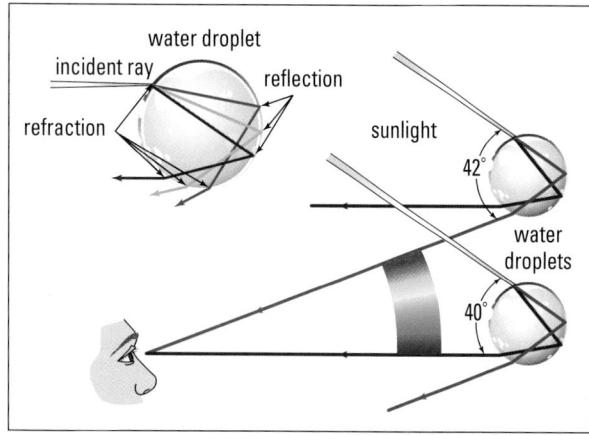

**Figure 11.28** Sunlight undergoes both refraction and total internal reflection to produce a rainbow.

**Figure 11.29** What do you notice about the colour distribution in this double rainbow?

## 11.5 Section Review

1. **C** Describe how Newton proved that sunlight is a combination of all of the spectral colours?

2. **K/U** Which of the spectral colours travels slowest in a glass prism? Explain.

3. **K/U** How could the shape of a rainbow be affected by the altitude of your location on the side of a mountain?

4. **K/U** Which colours in a rainbow do you see at the lowest elevation in the sky? Explain.

- Refraction of light, as of other waves, is the change in velocity when light passes from one medium into another. Refracted light may also change direction.

- The index of refraction of a medium is the ratio of the speed of light in a vacuum to the speed of light in the medium: $n = \frac{c}{v}$. A medium with a high index of refraction is optically dense.

- The angle of refraction is the angle of a light ray exiting from a refractive boundary. For any two given media, the ratio of the sine of the angle of incidence to the sine of the angle of refraction is a constant: $\frac{\sin \theta_i}{\sin \theta_R}$ = constant.

- Snell's law relates angles of incidence and refraction to indices of refraction: $n_i \sin \theta_i = n_R \sin \theta_R$. The incident ray, the refracted ray, and the normal all lie in the same plane.

- Refraction can be explained in terms of the wave model for light and the wave equation.

- The principle of reversibility of light states that, if a new ray of light is directed backwards along the path of a refracted ray, it will follow the same path after crossing the boundary between the media.

- Refraction causes effects such as the smaller apparent depth of objects submerged in water, deviation in prisms with sides that are not parallel, lateral displacement in prisms with parallel sides, atmospheric refraction in sunsets and mirages, and partial refraction and reflection.

- Total internal reflection occurs when light in an optically more dense medium strikes the boundary with an optically less dense medium at an angle of incidence greater than the critical angle for the medium.

- Applications of total internal reflection include prisms for reflection in periscopes, binoculars and cameras; retroreflectors; and optical fibres, in telecommunication, computing, and medicine.

- Dispersion is the separation of visible light into its component colours resulting from refraction. Water droplets disperse sunlight to create primary and secondary rainbows.

## Knowledge/Understanding

1. Define the term index of refraction.

2. An observer looking down at a mug at an angle is not able to see a coin resting on the bottom of the mug. As the mug is filled with water, the coin becomes visible. Explain why.

3. Light travels from medium Y to medium X. The angle of refraction is larger than the angle of incidence. In which medium does the light travel at a lower speed? Explain your logic.

4. When light passes from Plexiglass™ into ice at an angle, which will be smaller, the angle of refraction or the angle of incidence? Explain why.

5. Light travels from medium C to medium D. The angle of incidence is larger than the angle of refraction. Which medium has the lower index of refraction? Explain your logic.

6. How does the size of the critical angle change as the index of refraction decreases?

7. Which pair of media, air and water or air and glass, have the smaller critical angle? Explain why.

8. When white light passes through a prism and is dispersed into the spectral colours, which colour is refracted the least? Explain why in terms of the index of refraction and the speed of light in the medium.

## Inquiry

9. **(a)** You are given a rectangular, transparent block of an unknown solid material. Design an experiment to determine the index of refraction of the material and the speed of light in the material.

**(b)** You are given a sample of an unknown liquid material. Design an experiment to determine the index of refraction of the material.

10. Design an investigation to measure the index of refraction of a sample of a salt solution with a known concentration. Devise and describe a procedure that could be used to determine the concentration of a salt solution (or any other kind of solution) with an unknown concentration.

11. Design an investigation, using scaled light ray diagrams, to measure the change in the apparent depth of an object submerged in water, at a series of angles of incidence, increasing by 10°, from 0° to 70°. Plot a graph of the ratio of apparent depth to actual depth versus angle of incidence, and interpret the results obtained. Create an activity to demonstrate the effects you have identified.

12. Design an investigation to determine the maximum angle of deviation for a given triangular prism. Identify any practical limitations that restrict the maximum angle of deviation. How does the maximum angle of deviation depend on the angles of the prism?

13. For a given rectangular, transparent glass block, predict the maximum amount of lateral displacement possible, and check your prediction by experiment. Explain your prediction.

14. Try shining a flashlight at a car or bicycle reflector at different angles. Over what range of angles does the reflector direct the light back to you?

## Communicating

15. Use a variety of information technology resources to research the use of optical fibres in such areas as the transmission of telecommunications signals. Prepare a written report, with photographs or diagrams to show typical uses.

16. Carry out research on optical illusions caused by refraction effects. Summarize and present the general categories of illusions, using diagrams and photographs where appropriate, and explain why each type of illusion is produced.

17. **(a)** Draw an enlarged copy of the ray diagram below, such that side A is four times the size shown. Using Snell's law and the law of reflection, draw light rays to show the path of the light through the prism.

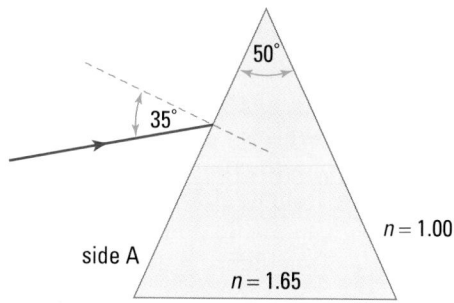

**(b)** Draw an enlarged copy of the ray diagram below, such that side A is four times the size shown. Using Snell's law and the law of reflection, draw light rays to show the path of the light through the prism.

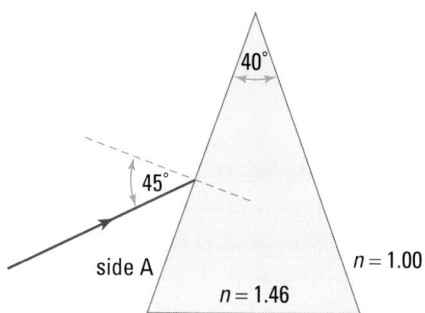

## Making Connections

18. Evaluate the impact of the development of optical fibres in our modern society. In which area of application will optical fibres have the greatest influence in the near future? Summarize your findings in a report.

19. Imagine you are swimming underwater and wearing goggles. As you look up at the air beyond the water surface, what effect can you expect to see? Draw a diagram to illustrate your answer. What implications does this underwater view of the airspace above the water have for fish? (Hint: Think about the critical angle of water.)

## Problems for Understanding

20. Light passes from water into a block of unknown material. If the angle of incidence in the water is 70.0° and the angle of refraction is 40.0°, what is the index of refraction of the unknown material?

21. A ray of light passes from Lucite™ into water at an angle of incidence of 20.0°. What is the angle of refraction in the water?

22. A light ray enters the longest side of a 45°–45° – 90° crown glass retroreflector. Assume that the light ray enters the longest side of the retroreflector at a point one quarter of the length of the side from one corner, at an angle of incidence of 30°. Use Snell's law and the law of reflection to determine the complete path of the light ray until it leaves the retroreflector. Draw an accurate diagram to show the complete path of the light ray. The length of each of the two shorter sides of the retroreflector is 10 cm.

23. Determine the time it takes for light to travel 54 cm through glycerin in an aquarium.

24. (a) What is the index of refraction in a medium if the angle of incidence in air is 57°, and the angle of refraction is 44°?
    (b) What is the angle of refraction if the angle of incidence in air is 27°, and the index of refraction of the medium is 2.42?
    (c) What is the angle of incidence in air if the angle of refraction is 28°, and the index of refraction of the medium is 1.33?

25. Light passes from crystal glass into ethyl alcohol. If the angle of refraction is 25°, determine the angle of incidence.

26. Red light travels from air into liquid at an angle of incidence of 39.0° and an angle of refraction of 17.0°. Calculate the wavelength of the red light in the liquid if its wavelength in air is 750 nm.

27. Make a careful copy of Figure 11.6A on page 518, and mark the angles of incidence and refraction. At each end of the boundary, add a line perpendicular to the wavefront to make a right-angled triangle. Using these two triangles, mathematically derive Snell's law.

28. A diver is standing on the end of a diving board, looking down into 2.5 m of water. How far does the bottom of the pool appear to be from the water's surface?

29. The critical angle for a special glass in air is 44°. What is the critical angle if the glass is immersed in water?

30. When astronauts first landed on the Moon, they set up a circular array of retroreflectors on the lunar surface. Scientists on Earth were then able to shoot a laser beam at the array and receive a reflection of the original laser light. By measuring the time interval for the signal's round trip, scientists were able to measure the distance between Earth and the Moon with great accuracy.
    (a) If the time interval could be measured to the nearest $3 \times 10^{-10}$ s, how accurate would the distance measurement be? (Hint: Remember that the signal makes a round trip.)
    (b) Suggest two reasons why a laser beam was used for the measurement.

---

**Numerical Answers to Practice Problems**

**1.** 1.54 **2.** $1.24 \times 10^8$ m/s **3.** $2.04 \times 10^8$ m/s **4.** $1.8 \times 10^{-2}$ s
**5.** It slows by 1.5 %. **6.** 1.31, ice **7.** 29.7° **8.** 51° **9.** 39.5°
**10.** 31.0° **11.** 11 cm **12.** 0.84 m **13.** 18.1° **14.** (a) 56° (b) 41°
**15.** 47.2° **16.** 58.9° **17.** 78.5° **18.** 2.6 m

At night, astronomers probe the universe using devices such as the 500 cm telescope at the Hale Observatory on Mount Palomar in California. Built in 1948, until 1976 it was the largest optical telescope ever built. It is able to photograph galaxies as far as five billion light years away. Its monolithic (one-piece) primary mirror is a massive 14 tonnes and is near the upper limit in size for monolithic mirrors. At the other end of the scale, medical and biological researchers use powerful optical microscopes that can magnify images up to 1000 times. Whether they are designed to observe macroscopic or microscopic objects, lenses are a basic element of almost all optical instruments.

In Chapter 10, you learned how mirrors reflect light to create real and virtual images. In this chapter, you will learn how lenses refract light to form images. You will investigate the construction of optical devices such as telescopes, microscopes and cameras. In preparation for an in-depth study of these instruments, you will learn how to apply ray diagrams and the thin lens equations to predict the properties of images formed by lenses.

## Comparing Kepler and Galileo Telescopes

**TARGET SKILLS**

- Hypothesizing
- Performing and recording
- Analyzing and interpreting

Simple telescopes use two lenses. One lens is called the eyepiece and the second lens is the objective lens. The Galileo telescope (developed by Galileo Galilei in 1608) uses a convex objective lens and a concave eyepiece. An astronomical or Kepler telescope (invented by Johannes Kepler in 1611) uses a convex objective lens and a convex eyepiece.

**CAUTION** Never look at the Sun through any optical instrument. You could permanently damage your eyes. Handle all lenses with care.

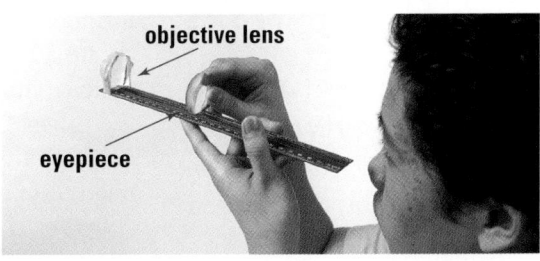

objective lens

eyepiece

### Problem

How do the different eyepieces used in Galileo and Kepler telescopes affect the images created?

### Equipment and Materials

- 2 convex lenses ($f \approx 20$ cm and $f \approx 5$ cm)
- concave lens ($f \approx 5$ cm)
- 30 cm ruler
- masking tape or modelling clay

### Procedure

1. To simulate an astronomical telescope, attach the convex lens with the longer focal length at one end of the ruler with masking tape or a bit of modelling clay. Be sure that the lens is adjusted so that when you look down along the ruler from the other end, you are looking directly through the lens. This will be your objective lens.

2. Hold the ruler with one hand so that you can look through the lens from the other end of the ruler, as shown in the photograph. Hold the second convex lens in your other hand and place it between your eye and the objective lens. This lens acts as your eyepiece. Adjust the eyepiece until you can see a clear image of an object on the other side of the room. Record the distance between the lenses at the point where the image is sharpest. Record the distance between the lenses. Record the appearance of the image.

3. Walk closer to the object and repeat the process. Record the distance between the lenses the point where the image is the sharpest. Record the appearance of the image.

4. To simulate a Galileo telescope, use the same objective lens but use the concave lens as the eyepiece. Carry out the same procedure with the concave eyepiece that you completed with the convex eyepiece. Be sure to stand the same two distances from the same object that you used to take data with the astronomical telescope. Record the same type of information.

### Analyze and Conclude

Answer the following questions for each telescope.

1. How does the distance to the object affect the distance between the lenses?

2. Is the image upright or inverted?

3. Is the image larger or smaller than the object? Estimate the difference in size. (Hint: Look at the image through the telescope with one eye and directly at the object with the other eye.)

4. Is the image closer to or farther away from you than the object?

### Apply and Extend

5. With objects at the same distance, for which telescope were the lenses closer together?

6. Which telescope gave the largest image for a given object at a given distance? Which was the easiest to adjust to get a clear image?

## SECTION EXPECTATIONS

SECTION
EXPECTATIONS

- Explain, using light ray diagrams, the characteristics of images formed by convex lenses.

- Predict, using ray diagrams and the lens equation, the characteristics and positions of images formed by convex lenses.

- Conduct experiments to compare theoretical predictions and empirical evidence.

## KEY TERMS

- convex lens
- principal axis
- vertical axis
- secondary axis
- principal focus or focal point
- focal length
- secondary focus or focal point

- biconvex or double convex
- concavo-convex or convex meniscus
- plano-convex
- lens-maker's equation
- thin-lens equations
- magnification equation
- mirror/lens equation

Compare the two images shown in Figure 12.1, which were formed by the same convex lens. Obviously, a convex lens can create an upright image or an inverted image much the same as a concave mirror does. Although it is not obvious from Figure 12.1, the upright image is virtual and the inverted image is real. To find out how the same lens can create such contrasting images, you need to examine convex lenses in detail.

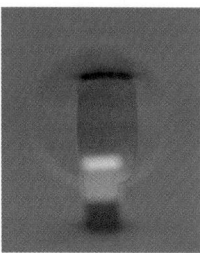 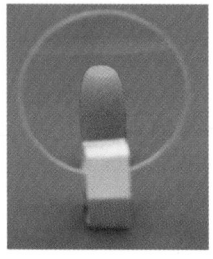

**Figure 12.1** These two images were formed by the same convex lens.

**Convex lenses** are also called converging lenses because any set of parallel rays that strike the lens, will *converge* on a single point on the opposite side of the lens. The concepts you learned about refraction and prisms in Chapter 11 will help you understand how a convex lens functions. In Chapter 11, you traced the path of light through a prism and measured the angle of deviation. If the refracting angle at the apex of a prism increases, the angle of deviation increases as well. Thus, by using proper arrangements of prisms, beams of light can be deviated so that they pass through a common area as shown in Figure 12.2. Such arrangements simulate crude convex lenses. As you can see in Figure 12.2 B, by adding more segments of prisms, you can more closely simulate a convex lens. When the prisms are combined into one solid object with smooth, rounded surfaces, you have a convex lens that causes the light rays to converge to a distinct point, as shown in Figure 12.2 C.

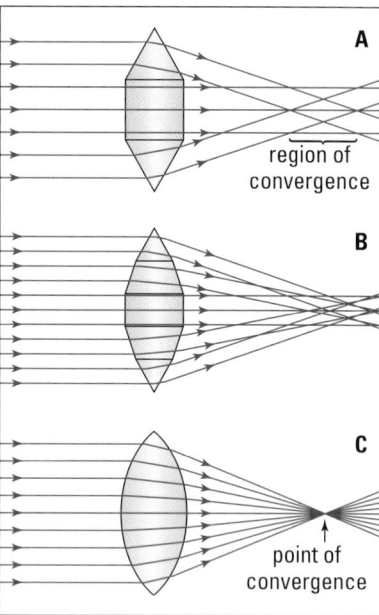

**Figure 12.2** Sets of prisms can be used to create a "lens" that crudely focusses a set of parallel incident rays.

# Properties of Convex Lenses

When you examine the simple diagram of a convex lens in Figure 12.3, you can see many similarities between mirrors and lenses. The term, convex, means curved outward for both mirrors and lenses. The surfaces of simple mirrors and lenses are spherical and described by radii of curvature. As well, any line passing through the geometric centre of the lens is called an "axis." The axis that passes through the centres of curvature of the lens is the **principal axis**. The upright axis that is perpendicular to the principal axis is the **vertical axis**. All other axes are called **secondary axes**.

**TRY THIS...**

Hold up each of the convex lenses from Investigation 12-A, one at a time, close to your eyes, and look through them at an object that is about 1 m away from you. Gradually move the lens farther from your eye until you can see a clear image of the object. Write a brief description of the image that you observed for each lens, including answers to the following questions.

• Is the image upright or inverted?

• Is the image larger or smaller than the object?

• How far from your eye do you have to hold the lens before you can see a clear (in focus) image?

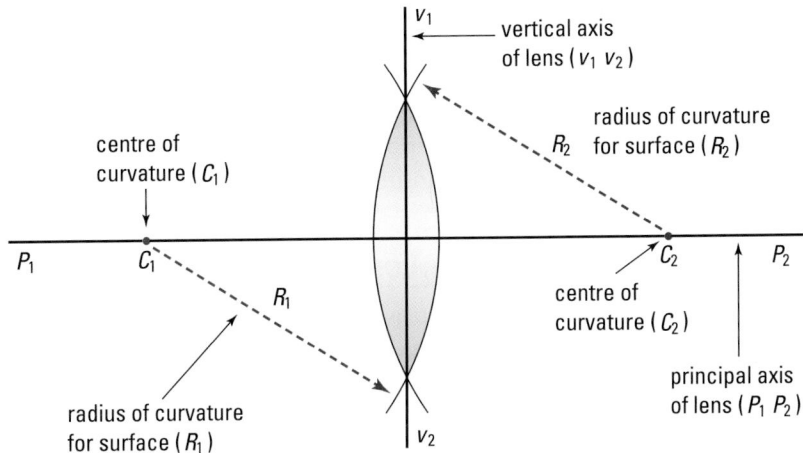

**Figure 12.3** The simplest way to draw a representation of a convex lens is to draw two intersecting circular arcs.

The similarities between mirrors and lenses end with the method for determining the focal point and focal length. Mirrors have only one spherical surface and one radius of curvature and the focal length was half the length of the radius of curvature. Lenses, however, have two spherical surfaces and two radii of curvature. As well, light passes through a refracting medium. All of these properties contribute to the focal length. Nevertheless, the focal length is quite easily found experimentally. You have already seen that parallel rays that pass through a convex lens converge to a point on the other side of the lens. As shown in Figure 12.4, the point of convergence of the rays that enter parallel to the principal axis is called the **principal focus** or **focal point**. Since the rays of light can enter the lens from either side, there is a principal focus or focal point on the principal axis on each side of the lens. The distance from the vertical axis to either of the focal points is called the **focal length** of the lens.

**Figure 12.4** A convex lens refracts rays that are parallel to its principal axis to converge on the focal point on the opposite side of the lens. Rays can enter the lens from either side; thus, each lens has two principal foci that lie on the principal axis, equidistant from the vertical axis of the lens.

**Language Link**

When lenses were first discovered, they were thought to resemble the seeds of the lentil plant, known in Latin as *lens culinaris*. Similarly, the word "lenticular" comes from the Latin word *lenticularis*, meaning "like a lentil". What is the exact meaning of lenticular?

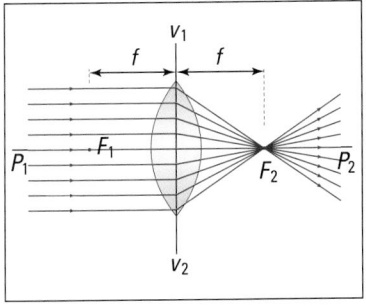

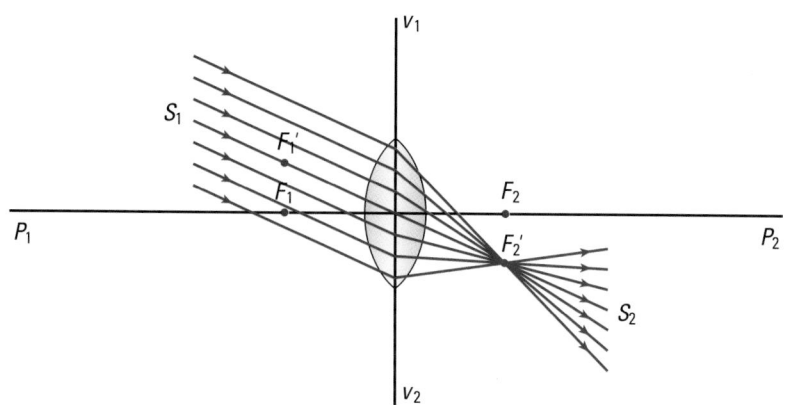

**Figure 12.5** The line $S_1$-$S_2$ is a secondary axis with secondary focal points $F_1'$ and $F_2'$. Rays parallel to a secondary axis are refracted to pass through the secondary focal points on that axis.

When rays enter a convex lens parallel to a secondary axis, they converge to a point called a **secondary focal point**, as shown in Figure 12.5. All secondary focal points are the same distance from the lens as the principal focal points.

## Practical Lenses and the Lensmakers' Equation

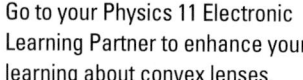

**ELECTRONIC LEARNING PARTNER**

Go to your Physics 11 Electronic Learning Partner to enhance your learning about convex lenses.

If you examine any pair of eyeglasses, you will discover that both surfaces curve in the same direction. One side is convex but the other side is concave. How can you classify such lenses? Early in this section, you read that convex lenses were also called converging lenses. Any lens that is thicker in the centre than around the edges will cause parallel rays to converge and is therefore classified as a convex lens. Figure 12.6 shows three variations of convex lenses. The first lens, shaped like those you have been examining, is called a **biconvex** or **double-convex** lens. The second lens, similar to eyeglass lenses for people who are farsighted, is called a **concavo-convex** or **convex meniscus** lens. The last lens is similar to one that you might find in a projector. This lens, called a **plano-convex** lens, produces a concentrated light beam that makes a projected image very bright.

It is not difficult to experimentally determine the focal length of a lens, but how does a lensmaker know what curvatures are needed to make a lens of a desired focal length? This knowledge goes back to the time of Galileo and Newton. As the interest in telescopes increased, scientists derived an equation that related the radii of curvature and the index of refraction of the material to the focal length. When using the lensmakers' equation given in the box on the next page, it is necessary to know how to distinguish the difference between convex and concave sides of a lens. By convention, the radius of curvature of a convex surface is positive and of a concave surface is negative. If the lens is plano-convex, the flat side has "infinite radius" and $\frac{1}{R_2} = 0$.

The **lensmakers' equation** allows you to calculate the focal length of a lens from the index of refraction of the lens' material and the radii of curvature of the lens' two surfaces.

---

### LENSMAKERS' EQUATION

The reciprocal of the focal length is the product of one less than the index of refraction and the sum of the reciprocals of the radii of curvature.

$$\frac{1}{f} = (n - 1)\left(\frac{1}{R_1} + \frac{1}{R_2}\right)$$

| Quantity | Symbol | SI unit |
|---|---|---|
| focal length | $f$ | m (metre) |
| refractive index | $n$ | no unit |
| first radius of curvature | $R_1$ | m (metre) |
| second radius of curvature | $R_2$ | m (metre) |

**Unit Analysis**

(no unit)((length)$^{-1}$ + (length)$^{-1}$) = m$^{-1}$ + m$^{-1}$ = m$^{-1}$

---

A    B    C

**Figure 12.6** Three types of converging lenses: **(A)** double convex; **(B)** convex meniscus or concavo-convex; **(C)** plano-convex.

---

## MODEL PROBLEM

### The Focal Length of a Convex Lens

A block of crown glass ($n = 1.50$) is ground into a biconvex lens. One surface has a radius of curvature of 15.0 cm, while the opposing surface has a radius of curvature of 20.0 cm. What is the focal length of the lens?

---

### Frame the Problem

- The medium is *crown glass* and the *index of refraction* is given.

- The two surfaces of the lens have different *radii* of *curvature*.

- The *lensmakers' equation* applies in this situation.

---

*continued* ▶

*continued from previous page*

## Identify the Goal

The focal length, $f$, of the lens

## Variables and Constants

| Involved in the problem | | Known | Unknown |
|---|---|---|---|
| $R_1$ | $n$ | $n = 1.50$ | $f$ |
| $R_2$ | $f$ | $R_1 = 15.0$ cm | |
| | | $R_2 = 20.0$ cm | |

### Strategy

Identify the equation that relates the known to the unknown variables.

Since the unknown variable is already isolated in the equation, replace the variables on the right-hand side of the equation with values and solve.

The lens has a focal length of 17.1 cm.

### Calculations

$$\frac{1}{f} = (n - 1)\left(\frac{1}{R_1} + \frac{1}{R_2}\right)$$

$$\frac{1}{f} = (1.50 - 1)\left(\frac{1}{15.0 \text{ cm}} + \frac{1}{20.0 \text{ cm}}\right)$$

$$\frac{1}{f} = (0.50)\left(\frac{4}{60 \text{ cm}} + \frac{3}{60 \text{ cm}}\right)$$

$$\frac{1}{f} = \left(\frac{1}{2}\right)\left(\frac{7}{60 \text{ cm}}\right)$$

$$\frac{1}{f} = \frac{7}{120 \text{ cm}}$$

$$f = 17.1 \text{ cm}$$

### Validate

The units of the focal length are centimetres. The focal length is positive, as expected for a convex lens.

---

### PRACTICE PROBLEMS

1. What is the focal length of a biconvex lens made of quartz ($n = 1.46$) if the radii of curvature of the opposing surfaces are 12.0 cm and 15.0 cm?

2. You grind a biconvex lens with opposing surfaces whose radii of curvature are 8.0 cm each. You thought the material was a block of crown glass ($n = 1.50$) but it turns out that it was flint glass ($n = 1.65$). What is the difference in the focal lengths of the lens you thought you were making and the one you did make?

3. A lens is ground using the same radius of curvature ($R = +13.0$ cm) for each of the opposing surfaces. What index of refraction would result in a lens with a focal length of 10.0 cm? Use the table of indices of refraction in Chapter 11 to find out what this material is (see Table 11.1, page 507).

4. Find the focal length of a plano-convex lens made of fused quartz ($n = 1.46$) if the radius of curvature of the spherical surface is 25.0 cm.

# Ray Diagrams for Convex Lenses

You can now determine the focal length of a convex lens experimentally and through calculations. The next, and most important, step is using the focal length to predict the nature of an image formed by the lens. Consider any point on an object that is distant from the lens in relation to the focal point. (In optics, this usually means any distance that is greater than two focal lengths from the vertical axis of the lens.) A convex lens causes all of the rays diverging from that point to converge on a single point on the opposite side of the lens (see Figure 12.7). Most importantly, after the rays converge on that point, they continue on, but now the rays are diverging. All of the rays diverging from the object that were incident on the lens are now diverging from a new point on the opposite side of the lens. In order to see the image you must put your eye *in the path* of these diverging rays. The image you see is projected in front of the lens at the point of divergence. If you place your eye in front of the image in the cone where the rays are still converging, you cannot see the image. You can see images only if the rays are diverging. If you always keep this in mind, it will make locating and understanding the properties of images a straightforward task.

Obviously, you do not want to have to draw all of the rays shown in Figure 12.7 when you draw a ray diagram. Since all rays from the object are refracted to one point, any two of the rays are sufficient to locate that point. There are three rays that are very easy to construct when drawing a ray diagram. These rays are described and illustrated in Table 12.1 on the next page. Although two rays are sufficient for locating an image, it is always a good idea to use the third ray as a check to ensure that you drew the first two correctly.

Before starting to draw ray diagrams, there is one important feature to consider. As you know, light rays refract at *both* surfaces of the lens. For ease in drawing, however, physicists approximate the two changes in direction of the ray by one change and place it on the vertical axis. As a result, ray diagrams are *approximations* of the path of the light. Nevertheless, the diagrams are very accurate when the lens is thin relative to the focal length and distances between the lens and the object and image.

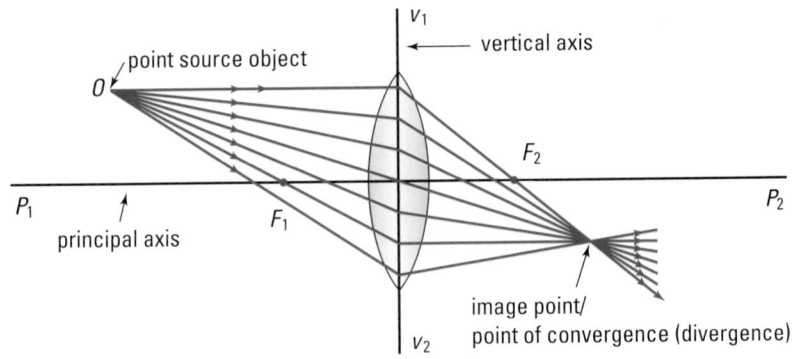

**Figure 12.7** All rays diverging from a single point on one side of the lens are refracted so that they converge on, then diverge from, a point on the opposite side of the lens. The image can only be seen by using the diverging rays.

**Table 12.1** Light Ray Diagrams for Convex Lenses

| Description | Comments | Illustration |
|---|---|---|
| **Reference line** Draw the principal axis through the centre of the lens. Mark and label the focal points on this axis on both sides of the lens. Place the base of the object on this reference line. | Since this line is perpendicular to the lens, any ray travelling along the line will pass through the lens without changing direction. Therefore, the base of the image will lie on this line. | |
| **Light ray #1** Draw a ray from the top of the object to the lens parallel to the principal axis. Draw the refracted ray through the focal point on the far side of the lens. | By the definition of the focal point, any ray entering the lens parallel to the principal axis will converge on the focal point on the far side of the lens. | |
| **Light ray #2** Draw a ray from the top of the object through the centre of the lens and extend it straight through to the far side of the lens. | At the very centre of the lens, the sides are parallel. Recall studying lateral displacement in Chapter 11. When light passes through a refracting medium with parallel sides, the emergent ray is parallel to the incident ray but displaced to the side. In the case of thin lenses, the displacement is so small that it can be ignored. | |
| **Light ray #3** Draw a ray from the top of the object through the focal point and on to the lens. Draw the refracted ray on the far side of the lens parallel to the principal axis. | Two principles determine the path of this ray. As in the case of ray #1, any ray entering the lens parallel to the principal axis will converge on the focal point on the other side of the lens. Then apply the principle of reversibility of light. The result is that any ray entering a lens from a focal point will leave parallel to the principal axis. | |
| **Identify the Image** If the refracted rays are converging, extend them until they all cross. If the refracted rays are diverging, extend them backward with dashed lines until they intersect. | The point at which the rays meet is the position of the top of the image. If the refracted rays converge, the image is real. If the rays must be extended backward to converge, the image is virtual. | |

Each point on the object is a source of a set of diverging rays that carries the image of that point. In the ray diagrams in this text, the object is always drawn with one end (its bottom) on the principal axis. You only need to locate the top of the image because the base of the image will lie on the principal axis.

## Images Formed by Convex Lenses

The distance between the object and the convex lens determines the nature of the image formed by the lens. The series of ray diagrams below covers all of the general types of images that convex lenses can produce.

### Case 1: $d_o > 2f$

When the object distance is greater than twice the focal length,

- the image lies on the side of the lens opposite the object,
- the image distance is less than twice the focal length,
- the image is real,
- the image is inverted, and
- the image is smaller than the object.

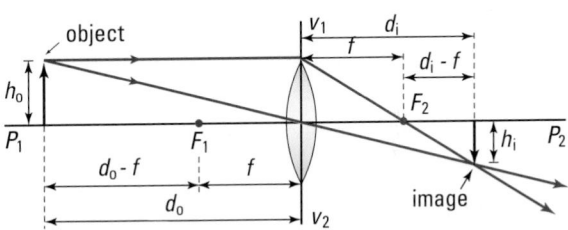

### Case 2: $d_o = 2f$

When the object distance is equal to twice the focal length,

- the image lies on the side of the lens opposite the object,
- the image distance is equal to twice the focal length,
- the image is real,
- the image is inverted, and
- the image is the same size as the object.

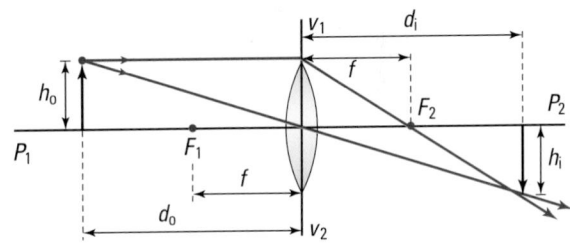

### Case 3: $2f < d_o < f$

When the object distance is less than twice the focal length but greater than the focal length,

- the image lies on the side of the lens opposite the object,
- the image distance is greater than twice the focal length,
- the image is real,
- the image is inverted, and
- the image is larger than the object.

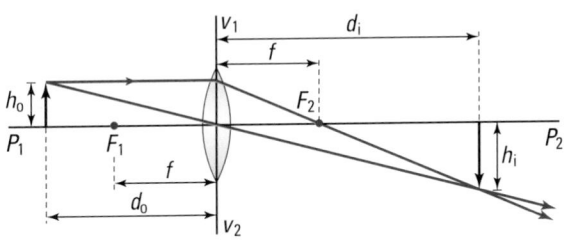

**Case 4: $d_o = f$**

When the object distance is equal to the focal length,

- no image exists because the refracted rays are parallel. You could say that the image lies at infinity.

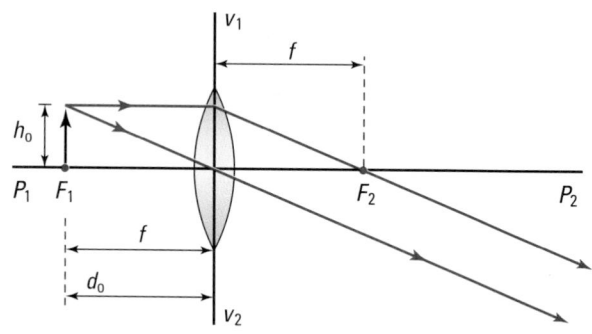

**Case 5: $f > d_o > O$**

When the object distance is less than the focal length but greater than zero,

- the image lies on the same side of the lens as the object,
- the image distance is greater than the object distance,
- the image is virtual,
- the image is upright, and
- the image is larger than the object.

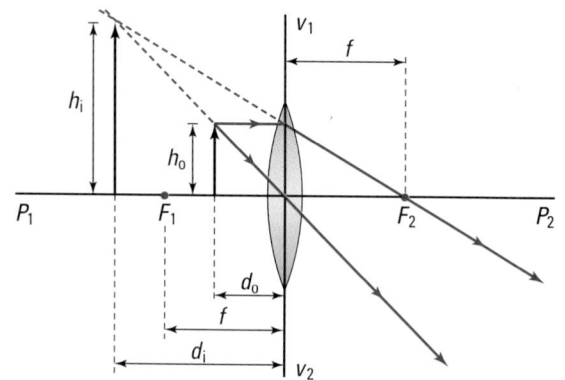

### Think It Through

- For cases 1 through 3 above, explain why the image is real.
- Use the concept of the "point of divergence" to explain why no image exists in case 4.
- For case 5, explain why the image is virtual.
- Use the rules of geometry to prove that the size of the object and image in case 2 are the same.

If you follow the sequence presented in the previous cases, it is very easy to see the pattern that exists. Start with the object at a very great distance from the lens. A very tiny, real, inverted image is formed just outside the focal point on the *opposite* side of the lens. As the object moves closer and closer to the focal point of the lens, the real image grows and moves away from the focal point (and the lens).

The point of equilibrium occurs when the object reaches a point that is two focal lengths from the lens. At that instant, the real image is also two focal lengths from the lens and the same size as the object (Case 2). As the object moves toward the focal point, the real image moves farther away from the lens and continues to grow larger (Case 3). Just before the object reaches the focal point, the real image would almost be at an infinite distance and infinitely large. When the object reaches the focal point, there is no image (Case 4). The instant the object moves inside the focal point, a virtual, upright image appears on the *same* side of the lens as the object. The image is larger and farther away than the object (Case 5). As the object moves closer to the lens, the virtual image moves toward the lens as well.

No matter whether a real or virtual image is formed, if the image is farther from the lens than the object, then the image is larger than the object, and vice versa. The relationship between an object's position and the nature of the image is summarized in Table 12.2.

**Table 12.2** Relationship between an Object's Position and the Nature of the Image

| Position of object ($d_o$) | Nature of image |
|---|---|
| $d_o > 2f$ | The image is real, inverted, smaller than the object, and closer to the lens than the object. |
| $d_o = 2f$ | The image is real, inverted, and equal in size and distance from the lens as the object. |
| $f < d_o < 2f$ | The image is real, inverted, larger than the object, and farther from the lens than the object. |
| $d_o = f$ | There is no image formed. This is the boundary between the formation of real and virtual images. |
| $0 < d_o < f$ | The image is virtual, upright, larger than the object, and farther away from the lens than the object. |

**Web Link**

*www.school.mcgrawhill.ca/ resources/*
For more animated examples of ray diagrams for mirrors and lenses, go to the above Internet site. Follow the links for **Science Resources** and **Physics 11** to find out where to go next.

## Thin-Lens Equations

You can use ray diagrams to develop equations for calculating the relative sizes and positions of the object and images. These are called the **thin-lens equations**, since they are very accurate only if the thickness of the lens is small compared to its diameter.

You probably recall from Chapter 10, that sign conventions for the symbols on the ray diagrams were critical for the development of the mirror equation. The same is true for the lens equations. Table 12.3 defines all of the sign conventions needed for the thin lens equations.

**Table 12.3** Sign Conventions for Convex Lenses

| Quantity | Sign property |
|---|---|
| $h_o$ | The height of a real object is always positive. |
| $d_o$ | The distance from the vertical axis to the object is positive for real objects. |
| $f$ | The focal length is positive for convex lenses and negative for concave lenses. |
| $h_i$ | The height of the image is *positive* if the image is *upright* relative to the object; negative if the image is *inverted*. |
| $d_i$ | The distance from the vertical axis to the image is *positive* if the image is on the *opposite side of the lens* from the object and *negative* if the image is on the *same side* of *the lens* as the object. |

Magnification for lenses is defined in the same way as it is for mirrors.

$$M = \frac{h_i}{h_o}$$

To find a second way to calculate magnification, examine Figure 12.8. The shaded triangles to the left and right of the lens are similar, since the corresponding angles for the triangles are equal. Thus, the ratios of corresponding sides must be equal.

Hence,

$$\frac{-h_i}{h_o} = \frac{d_i}{d_o}$$

Therefore,

$$M = \frac{h_i}{h_o} = -\frac{d_i}{d_o}$$

This is the **magnification equation.** The distance from the vertical axis to the object, $d_o$, the height of the object, $h_o$, and the distance from the vertical axis to the image, $d_i$, are all positive. The image is inverted, therefore, its height, $h_i$, should be represented by a negative number.

**Figure 12.8** The thin-lens equations are approximations based on ray diagrams.

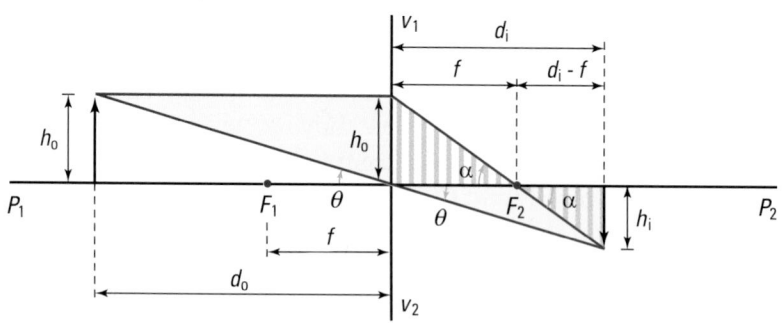

To develop the other thin lens equation, again examine Figure 12.8. The right triangles with striped shading to the right of the lens have equal corresponding angles. By geometry, they are similar and the ratios of corresponding sides for these triangles must be equal. Thus, we can write $\dfrac{-h_i}{h_o} = \dfrac{d_i - f}{f}$.

- Compare the two equations that you obtained from similar triangles in Figure 12.8.

$$\frac{-h_i}{h_o} = \frac{d_i}{d_o} \text{ and } \frac{-h_i}{h_o} = \frac{d_i - f}{f}$$

- Since the left sides of the equations are the same, you can set the right sides equal to each other.

$$\frac{d_i}{d_o} = \frac{d_i - f}{f}$$

- Multiply both sides by $d_o f$.

$$\left(\frac{d_i}{d_o}\right) d_o f = \left(\frac{d_i - f}{f}\right) d_o f$$

- Simplify.

$$d_o (d_i - f) = d_i f$$

- Multiply through by $d_o$.

$$d_o d_i - d_o f = d_i f$$

- Rearrange to place all terms with $f$ on the right side of the equation.

$$d_o d_i = d_o f + d_i f$$

- Factor out $f$.

$$d_o d_i = (d_o + d_i) f$$

- Divide both sides of the equation by $d_i d_o f$.

$$\frac{d_o d_i}{d_o d_i f} = \frac{(d_o + d_i) f}{d_o d_i f}$$

- Simplify and write the right side as the sum of two fractions.

$$\frac{1}{f} = \frac{d_o}{d_o d_i} + \frac{d_i}{d_o d_i}$$

- Simplify.

$$\frac{1}{f} = \frac{1}{d_i} + \frac{1}{d_o}$$

You will probably recognize this equation as the mirror equation. Because the lens and mirror equations are the same, the equation is often called the **mirror/lens equation**.

---

**THIN-LENS EQUATIONS**

**Magnification equation:** The ratio of the image and object heights is the negative of the ratio of the image and object distances.

$$M = \frac{h_i}{h_o} = \frac{-d_i}{d_o}$$

**Mirror/lens equation:** The reciprocal of the focal length is the sum of the reciprocals of the image and object distances.

$$\frac{1}{f} = \frac{1}{d_o} + \frac{1}{d_i}$$

| Quantity | Symbol | SI unit |
|---|---|---|
| image height | $h_i$ | m (metre) |
| object height | $h_o$ | m (metre) |
| image distance | $d_i$ | m (metre) |
| object distance | $d_o$ | m (metre) |
| focal length | $f$ | m (metre) |

---

## Images Formed by a Convex Lens

An object 8.5 cm tall is placed 28 cm from a convex lens that has a focal length of 12 cm. Find the size and location of the image. Describe the image.

### Frame the Problem

- According to Table 12.2, we expect the image to be real, inverted, smaller than the object, and closer to the lens than the object.

- Visualize the problem by sketching the light ray diagram.

- The mirror/lens equation and the magnification equation can be applied.

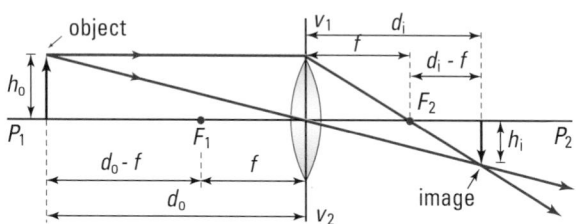

### Identify the Goals

(a) The size, $h_i$, and location, $d_i$, of the image

(b) A description of the properties of the image: real/virtual, upright/inverted, larger/smaller than the object.

### Variables and Constants

| Involved in the problem | Known | Unknown |
|---|---|---|
| $d_o$ | $d_o = 28$ cm | $d_i$ |
| $d_i$ | $h_o = 8.5$ cm | $h_i$ |
| $h_o$ | $f = 12$ cm | |
| $h_i$ | | |
| $f$ | | |

### Strategy

Use the mirror/lens equation, to find the image distance.

Rearrange.

### Calculations

$$\frac{1}{f} = \frac{1}{d_o} + \frac{1}{d_i}$$

$$\frac{1}{d_i} = \frac{1}{f} - \frac{1}{d_o}$$

## Strategy

Substitute the known values into the equation.

Solve.

Now you can use $d_i$ to find $h_i$.

Begin with the magnification equation.

Rearrange.

Substitute known values.

Solve.

**(a)** The image is 6.4 cm tall and 21 cm from the lens on the opposite side from the object.

**(b)** The image is real, inverted, and smaller than the object.

## Calculations

$$\frac{1}{d_i} = \frac{1}{12 \text{ cm}} - \frac{1}{28 \text{ cm}}$$

$$\frac{1}{d_i} = \frac{7}{84 \text{ cm}} - \frac{4}{84 \text{ cm}}$$

$$\frac{1}{d_i} = \frac{3}{84 \text{ cm}}$$

$$\frac{1}{d_i} = \frac{1}{21 \text{ cm}}$$

$$d_i = 21 \text{ cm}$$

$$\frac{h_i}{h_o} = -\frac{d_i}{d_o}$$

$$h_i = -\frac{h_o d_i}{d_o}$$

$$= -\frac{(8.5 \text{ cm})(21 \text{ cm})}{28 \text{ cm}}$$

$$= -6.375 \text{ cm}$$

## Validate

The units are in centimetres.

The height of the image is a negative value, because the image is inverted.

The height of the image is less than the object, and the image is closer to the lens than the object.

---

### PRACTICE PROBLEMS

**5.** A convex lens has a focal length of 12.0 cm. An object 6.30 cm tall is placed 54.0 cm from the vertical axis of the lens. Find the position and size of the image. Describe the image.

**6.** An object 7.50 cm tall is placed 1.50 m from a convex lens that has a focal length of 90.0 cm. Find the position and size of the image. Describe the image.

**7.** Calculate the size and position of the image if an object that is 4.20 cm tall is placed 84.0 cm from a convex lens that has a focal length of 120.0 cm. Describe the image.

**8.** A real image 96.0 cm tall is formed 144 cm from a convex lens. If the object is 36.0 cm from the lens,

   **(a)** what is the focal length of the lens?

   **(b)** what is the size of the object?

**9.** A virtual image 35.0 cm tall is formed 49.0 cm from a convex lens. If the object is 25.0 cm tall, find (a) the position of the object and (b) the focal length of the lens.

**10.** Calculate the position and size of the image if an object is placed at the focal point of a convex lens with a focal length of 20 cm.

# INVESTIGATION 12-B

## Converging Lens Images

**TARGET SKILLS**

- Performing and recording
- Modelling concepts
- Analyzing and interpreting
- Communicating results

## Problem

Confirm the thin-lens equations (the mirror/lens and magnifications equations) for convex lenses. Use a screen to project real images (see page 563).

## Equipment and Materials

- 2 convex lenses (each with a different focal length)
- ticker tape
- modelling clay
- small light source
- power supply
- square of white cardboard (approximately 10 cm by 10 cm)
- pencil or pen
- vernier callipers

## Procedure

1. Attach a long piece of ticker tape to the surface of the lab table with masking tape.

2. Near the middle of the tape, draw a line across the tape, to represent the location of the vertical axis of the lens. Accurately place the vertical axis of the lens, supported by a small piece of modelling clay, over the mark. Be sure to position the lens so that its principal axis is nearly horizontal.

3. Draw a line across the ticker tape near one end of the ticker tape. Place the light source (object) as accurately as possible so that its filament is above the line. (**Note:** If you place the bulb so that the filament is across the tape, rather than parallel, you will get a sharper image to project on the screen. See the diagram of the apparatus set-up.)

4. On the opposite side of the lens, move the screen slowly away from the lens until a clear, sharp image is formed on the screen. (**Note:** Be sure to keep the screen perpendicular to the principal axis of the lens.)

5. Place a line across the ticker tape to mark the position of the image. (**Note:** Place a trial number (1) beside the lines for the object and the image so that later on you can be sure which value for $d_o$ corresponds with which value of $d_i$.)

6. Draw a line about 10 cm closer to the lens from the position of the object; place the light source (object) on that line and repeat steps 4 and 5.

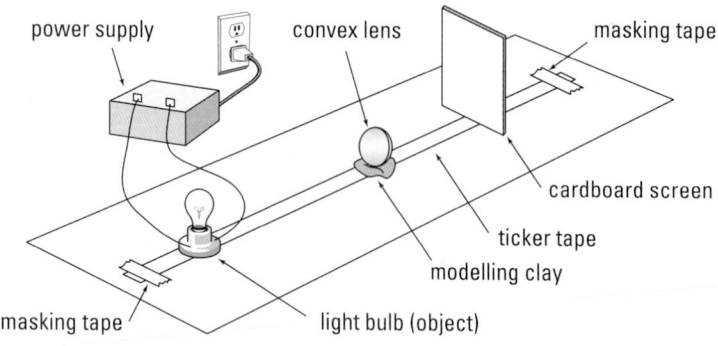

7. Find images for the position of the object as it is moved closer and closer to the lens until images can no longer be located. Remember to mark lines for the corresponding object and image positions with a trial number.

8. Move the object and the lens off the ticker tape. Measure the distance from the lens to the positions of the object and image for each trial.

9. Record your data in a table as shown below.

| Trial | Object distance: $d_o$ (cm) | Image distance: $d_i$ (cm) | Focal length: $f$ (cm) | Percent error |
|-------|-----------|-----------|-----------|---------|
|       |           |           |           |         |
|       |           |           |           |         |

10. When the object is at a very large distance from the lens, the rays from the object to the lens are very nearly parallel. Thus, the image is very close to the focal point. Move the object as far from the lens as it is possible for you to do and still locate the image. Mark the image position on the tape. This value of $d_o$ should be very close to the focal length.

11. Change the ticker tape and repeat the experiment for the other convex lens.

## Analyze and Conclude

Complete the following for each set of data.

1. For each trial, calculate the focal length from $d_o$ and $d_i$.

2. Calculate the average focal length for the trials.

3. Compare the measure of the focal length you found in step 10 to the average previously calculated.

4. For each trial, calculate the percent deviation for the focal length from the average focal length.

5. Calculate the average percent deviation for the trials.

6. What are the possible sources of deviation? If the percent deviation for some trials was large, give reasons why this might have occurred.

7. Do the results of your investigation support or reject the mirror/lens equation?

8. Of the lenses used, which one gave the smallest average percent deviation? Suggest reasons why this might occur.

## Apply and Extend

9. Do a set of trials as above, but measure the size of the object and image for the trials. The length of the filament for the light bulb can be reasonably accurately measured if you place a set of vernier callipers in front of the bulb and close them until the gap appears to be the same size as the filament.

10. Record your data in a table, as shown.

| Trial | $h_o$ (cm) | $d_o$ (cm) | $h_i$ (cm) | $d_i$ (cm) | $h_i / h_o$ | $d_i / d_o$ |
|-------|-----------|-----------|-----------|-----------|-----------|-----------|
|       |           |           |           |           |           |           |
|       |           |           |           |           |           |           |

11. Make a graph of $h_i/h_o$ versus $d_i/d_o$.

12. Is the graph a straight line? What is its slope?

13. Which trials were the most prone to errors in measurements?

## Projecting Images

It would not be very convenient if the only way you could view images formed at the point of divergence was to put your eye in the path of the diverging rays, as mentioned previously. However, if you place a screen at the exact point at which the rays converge, rather than continuing to travel away from the lens, the rays will be diffusely reflected back into the room (see Figure 12.9). In this way, everyone in the room will be able to see the image "on the screen." Clearly this phenomenon is possible only because real images are the result of "real" points of divergence. Thus, virtual images cannot be projected onto screens since imaginary points of divergence produce them. In the laboratory, real images are almost always studied by projecting them onto a screen. Notice that, if the screen is not exactly at the point of divergence, a small circle of light will exist rather than a point. In this case, the image is "out of focus."

**Figure 12.9** Rays that converge on a point on a screen will then, by diffuse reflection, diverge to all points in front of the screen. A sharply focussed image can be seen on the screen.

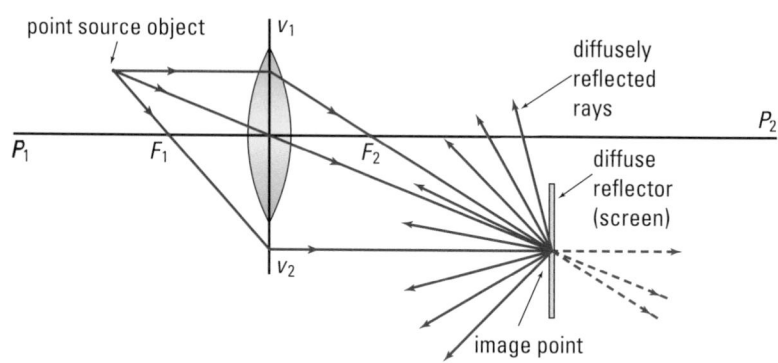

---

## 12.1    Section Review

1. **K/U** What is a convex lens?

2. **K/U** Draw an example of the following lenses:

   a) double convex

   b) convex meniscus

   c) plano-convex

3. **C** Draw and label a convex lens. Include the following in your diagram: axis, principal axis, vertical axis, focal point, focal length, secondary axis, secondary focal point.

4. **K/U** How can you determine the focal length of a convex lens?

5. **C** Describe how it is possible for you to see an object.

6. **MC** Draw ray diagrams and explain similarities between convex lenses and concave mirrors in the creation of (a) real images and (b) virtual images.

7. **C** Explain how a convex lens forms an image.

8. **C** Describe what is meant by the term "out of focus."

9. **K/U** Why does a convex lens invert an image?

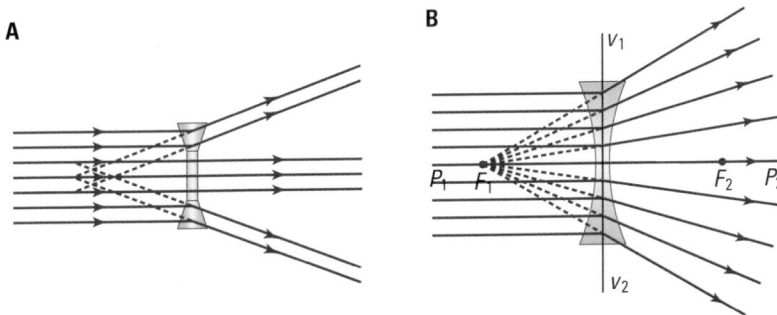

**A**

**B**

$V_1$

$P_1$ $F_1$

$F_2$ $P_2$

$V_2$

**Figure 12.10** **(A)** Prisms can be combined to create a crude concave lens. **(B)** The focal length of concave lenses is negative.

The most common reason for wearing eyeglasses is near-sightedness. This condition is easily corrected with concave lenses. Any lens that causes parallel rays to diverge after passing through the lens, can be classified as a **concave lens**. You can sim-ulate a crude concave lens with prisms in much the same way that you can simulate a convex lens. As you can see in Figure 12.10, rays entering the lens parallel to the principal axis, exit as though they were diverging from a single point on the opposite side of the lens. This point from which the rays *appear* to be coming is the focal point of the concave lens. Because the focal point for the diverging rays is on the side of the lens *from which the light is coming* and because it could be considered as a *virtual* focal point, the focal length must be negative.

Just as convex lenses have three basic forms, so do concave lenses. Figure 12.12 illustrates the shape of the double concave, **convexo-concave or concave meniscus**, and **plano-concave lenses**. Being thinner at the middle, they all act as diverging lenses.

## Images from Concave Lenses

The ray diagrams used to locate images are drawn much like those for convex lenses — with one major difference. Rays parallel to the principal axis of a convex lens are refracted to pass through the focal point on the opposite side of the lens. For a concave lens, the rays parallel to the prin-cipal axis are refracted so that they appear to be diverging from the focal point on the side of the lens where they began.

**SECTION**
**EXPECTATIONS**

- Describe the characteristics of images formed by concave lenses using light ray diagrams.

- Describe the effects of diverging lenses on light.

- Investigate the characteristics of images formed by concave lenses.

**KEY**
**TERMS**

- concave lens

- convexo-concave or convex meniscus lens

- plano-concave lens

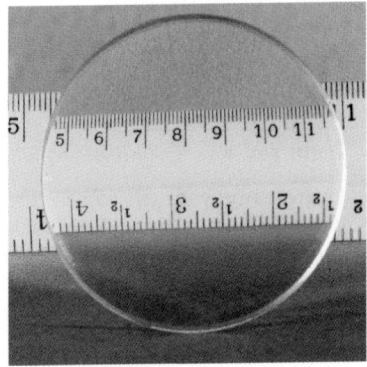

**Figure 12.11** A concave lens.

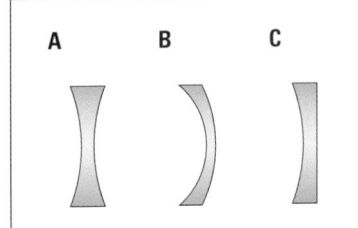

**A** **B** **C**

**Figure 12.12** Diverging lenses come in three shapes: **(A)** double concave, **(B)** concave meniscus, and **(C)** plano-concave.

**Table 12.4** Light Ray Diagrams for Concave Lenses

| Description | Comments | Illustration |
|---|---|---|
| **Reference line**<br><br>Draw the principal axis through the centre of the lens. Mark and label the focal points on this axis on both sides of the lens. Place the base of the object on this reference line. | Since this line is perpendicular to the lens, any ray travelling along the line will pass through the lens without changing direction. Therefore, the base of the image will lie on this line. | |
| **Light ray #1**<br><br>Draw a ray from the top of the object to the lens parallel to the principal axis.<br><br>Draw the refracted ray as though it were coming from the focal point on the same side of the lens as the object. | By the definition of the focal point, any ray entering the lens parallel to the principal axis will diverge as though it were coming from the focal point on the same side of the lens as the object. | |
| **Light ray #2**<br><br>Draw a ray from the top of the object through the centre of the lens and extend it straight through to the far side of the lens. | At the very centre of the lens, the sides are parallel. When light passes through a refracting medium with parallel sides, the emergent ray is parallel to the incident ray but displaced to the side. In the case of thin lenses, the displacement is so small that it can be ignored. | |
| **Light ray #3**<br><br>Draw a ray from the top of the object directly toward the focal point on the opposite side of the lens.<br><br>Draw the refracted ray on the far side of the lens parallel to the principal axis. | As in the case of ray #1, any ray entering the lens parallel to the principal axis will diverge as though it were coming from the focal point on the object side of the lens. Then apply the principle of reversibility of light. The result is that any ray entering a lens toward the opposite focal point will leave parallel to the principal axis. | |
| **Identify the Image**<br><br>Refracted rays always diverge. Therefore, extend the rays backward with dashed lines until they intersect. | The point at which the extended rays meet is the position of the top of the image. The image is always virtual. | |

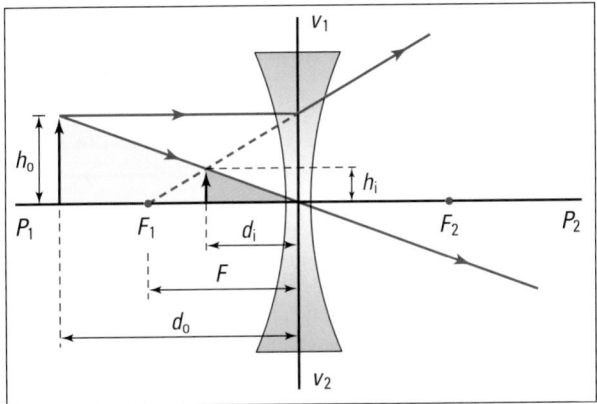 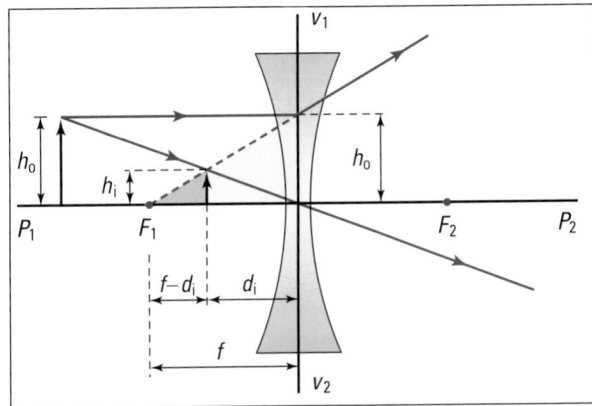

You can use Figure 12.13 to prove that the thin-lens equations developed for convex lenses also apply to concave lenses. The triangles in Figure 12.13 (A) are similar. Comparing the ratios of their corresponding sides and applying the sign conversions results in the magnification equation, $\dfrac{h_i}{h_o} = \dfrac{-d_i}{d_o}$. Using the shaded triangles in Figure 12.13 (B) you can show that

$$\frac{h_i}{h_o} = \frac{-d_i - (-f)}{f} = \frac{-d_i + (f)}{f} = -\frac{(d_i - f)}{f}$$

**Figure 12.13** The thin-lens equations can be shown to apply to concave lenses by using the property of similar triangles.

Since these are exactly the same relationships you used in your derivation of the lens equation for convex lenses, they will reduce to the same lens equation.

$$\frac{1}{f} = \frac{1}{d_o} + \frac{1}{d_i}$$

Again, these equations are valid only if the lens thickness is small compared with its diameter.

## QUICK LAB

# Concave Lens Images

**TARGET SKILLS**
- Modelling concepts
- Communicating results

Draw a series of ray diagrams to examine the difference in the image that results when the object moves from a point about 2.5 $f$ from the lens to a point very close to the lens. (You could use the same set of object positions that were used in Cases 1 through 5 in Section 12.1.)

### Analyze and Conclude

1. Are there any object positions for which a concave lens forms a real image? Explain.

2. What is the greatest distance from the lens that the image can appear?

3. Explain why the ray diagrams predict that the largest image that can be formed is the same size as the object.

4. Write a summary of what happens to the image as the object moves closer and closer to the lens.

# INVESTIGATION 12-C

## Focus Length of A Concave Lens

**TARGET SKILLS**
- **Predicting**
- **Performing and recording**
- **Analyzing and interpreting**

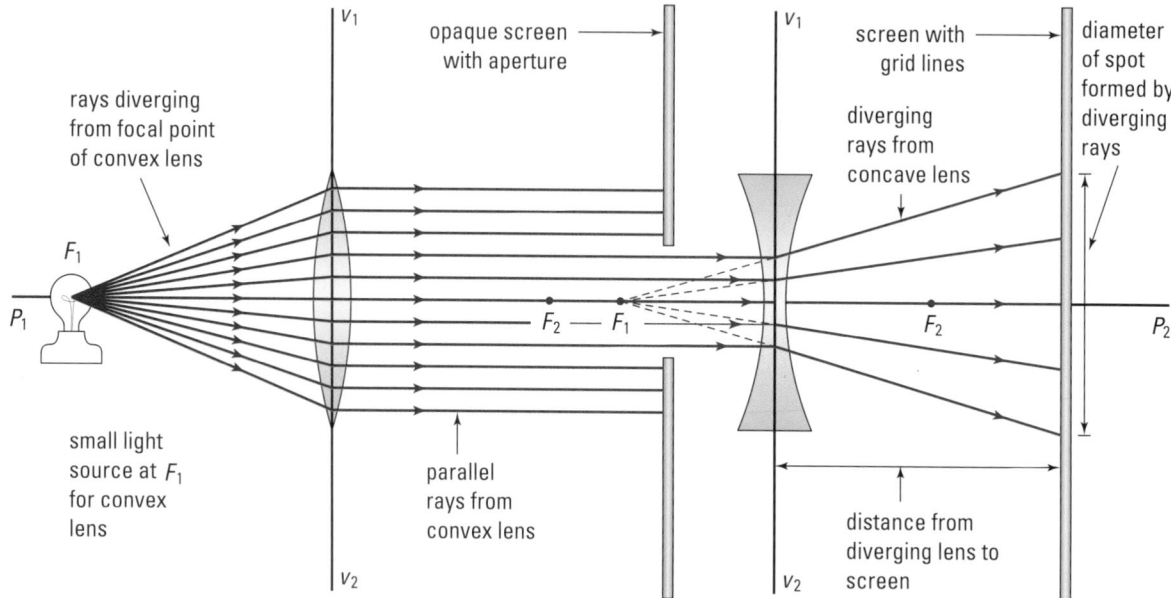

Rays parallel to the principal axis of a concave lens are refracted so that they diverge from the focal point of the lens. By measuring the diameter of a set of spots formed by the diverging beam at different distances from the lens, you should be able to predict the point from which they are diverging. The source of the beam of parallel rays is a light source placed at the focal point of a convex lens, as shown in the diagram.

## Problem

What is the focal length of a concave lens?

## Prediction

As you are making the measurements, make a prediction of what the focal length of the lens will be.

## Equipment

- convex lens (with a known focal length, used to create a parallel light beam)
- concave lens (with unknown focal length)
- 2 retort stands and test tube clamps to support the lenses

- cardboard screen (Hint: Covering the screen with a piece of graph paper makes taking measurements much easier.)
- cardboard screen with a circular hole about 1 cm in diameter
- retort stands to support the screens
- small electric light source
- power supply
- box or stand to support the lamp
- masking tape
- metre stick

## Procedure

1. Set up the lamp at the focal point of the convex lens so that the beam is refracted parallel to the principal axis. A spot about the size of the lens should be projected onto the screen, as shown in the diagram below. Place the cardboard screen with the hole in it into the path of the beam to limit the size of the beam. Care must be taken to be sure that the axis of the convex lens and the concave lens are collinear.

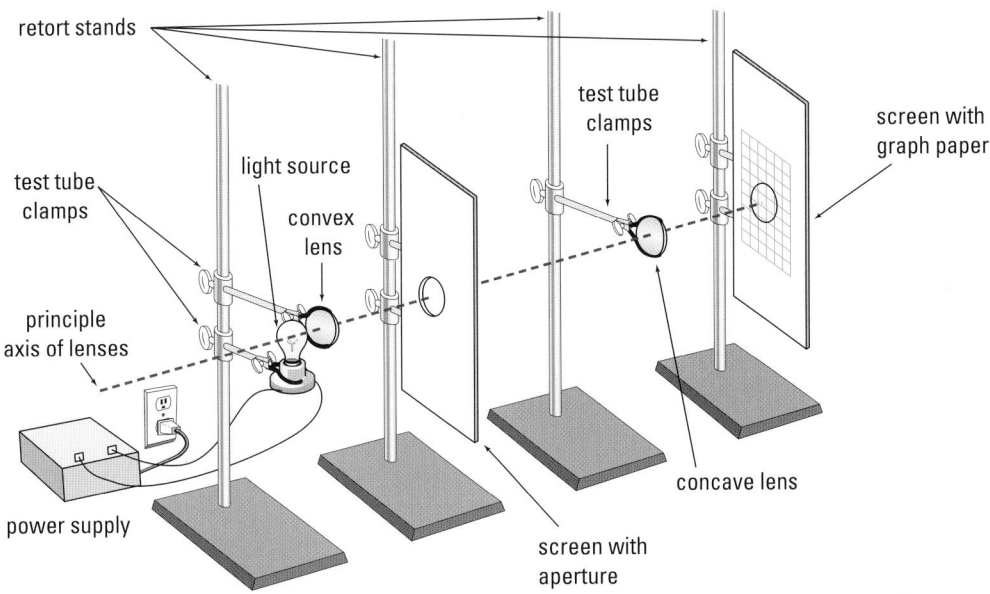

retort stands

test tube
clamps

light source

convex
lens

test tube
clamps

screen with
graph paper

principle
axis of lenses

concave lens

power supply

screen with
aperture

2. Place the concave lens in the path of the
beam, perpendicular to its path, and project
the spot onto the graph paper screen.

3. Move the screen to a distance of 10 cm from
the lens and measure the diameter of the
spot on the screen.

4. Record your data in a table similar to the
one shown below.

| Trial | Distance to screen (cm) | Diameter of the light spot (cm) |
|---|---|---|
| | | |
| | | |

5. Move the screen 10 cm farther away from
the lens and repeat your readings.

6. Repeat steps 4 and 5 until you have
completed at least five trials.

**Analyze and Conclude**

1. Graph the results with the diameter as the
dependent variable and the distance from
the lens as the independent variable.

2. Extend your graph back to predict the
position at which the diameter would be
zero. This is the predicted point of divergence and thus the focal point of the lens.
(**Note:** If the graph is a straight line, you
could find its equation, then calculate the
value for distance from the screen that
would give a diameter of zero.)

3. What is the focal length of the concave lens?

**Apply and Extend**

Either a light ray box, or, if your instructor will
permit it, a laser can be used as the source of
the parallel beam rather than the lens aperture
arrangement.

CAUTION You must be very careful when
handling a laser. Never let the laser shine into
your own or a classmate's eyes or look at reflected
laser light; severe retinal damage can occur.

## Images in Concave Lenses

An object that is 48.0 cm tall is located 60.0 cm from a concave lens with a focal length of 40.0 cm. Calculate the size and position of the image. Describe the image.

### Frame the Problem

- Because this is a concave lens, we expect the image to appear on the same side of the lens as the object, and to be upright.

- Visualize the problem by drawing the light ray diagram.

- Use mirror/lens and magnification equations.

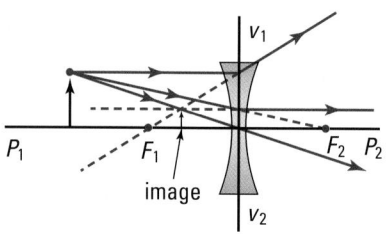

### Identify the Goals

(a) The size, $h_i$, and the position, $d_i$, of the image

(b) A description of the image

### Variables and Constants

| Involved in the problem | | Known | | Unknown |
|---|---|---|---|---|
| $d_o$ | $h_o$ | $d_o = 600$ cm | $d_i$ | |
| $d_i$ | $h_i$ | $h_o = 48.0$ cm | $h_i$ | |
| | $f$ | $f = -40.0$ cm | | |

### Strategy

Use the mirror/lens equation to find $d_i$.

Rearrange.

Substitute known values.

Solve.

Now you can use $d_i$ to find $h_i$.
Begin with the magnification equation.

### Calculations

$$\frac{1}{f} = \frac{1}{d_o} + \frac{1}{d_i}$$

$$\frac{1}{d_i} = \frac{1}{f} - \frac{1}{d_o}$$

$$\frac{1}{d_i} = \frac{1}{-40.0 \text{ cm}} - \frac{1}{60.0 \text{ cm}}$$

$$\frac{1}{d_i} = -\frac{3}{120 \text{ cm}} - \frac{2}{120 \text{ cm}}$$

$$\frac{1}{d_i} = -\frac{5}{120 \text{ cm}}$$

$$\frac{1}{d_i} = -\frac{1}{24.0 \text{ cm}}$$

$$d_i = -24.0 \text{ cm}$$

$$\frac{h_i}{h_o} = -\frac{d_i}{d_o}$$

Rearrange.

$$h_i = -\frac{h_o \cdot d_i}{d_o}$$

Substitute known values.

$$= -\frac{(48.0 \text{ cm})(-24.0 \text{ cm})}{60.0 \text{ cm}}$$

Solve.

$$= 19.2 \text{ cm}$$

(a) The image is 19.2 cm tall and 24.0 cm from the lens on the same side as the object.

(b) The image is virtual, upright, and smaller than the object.

## Validate

The units are in centimetres.

The negative value for the position of the image is expected since the image from a concave lens is always found on the same side of the lens as the object.

You expect that the image height will be positive since a concave lens produces only virtual, upright images.

## PRACTICE PROBLEMS

11. Calculate the size and position of the image formed when an object 84.0 cm tall is placed 120 cm from a concave lens with a focal length of 80.0 cm. Describe the image.

12. A concave lens is used to form the image of an object that is 35.0 cm tall and located 25.0 cm from the lens. If the image is 28.0 cm tall, what is the focal length of the lens?

13. A concave lens with a focal length of 180 cm is used to form an image that is 75.0 cm tall from an object that is 45.0 cm from the lens. What is the height of the object?

## 12.2    Section Review

1. **K/U** What is a concave lens?

2. **K/U** Draw an example of the following lenses:

    (a) double concave,

    (b) convexo-concave, and

    (c) plano-concave.

3. **MC** Explain, with the aid of ray diagrams, how concave lenses and convex mirrors are similar.

4. **C** Explain why the image from a concave lens is upright.

5. **K/U** Where is an image located if it is the same size as the object?

## SECTION
### EXPECTATIONS

- Explain why different lens types are used in optical instruments.

- Analyze, describe, and explain optical effects that are produced by technological devices.

- Evaluate the effectiveness of devices or procedures related to human perception of light.

**Figure 12.14**   The Orion nebula

When Galileo turned his telescope on the skies, he changed forever how humans view the universe. Besides offering visual proof that Earth was not the centre of all planetary motion, Galileo mapped the surface of the Moon. Today, using telescopes such as the Hubble Space Telescope, scientists have mapped the universe far beyond what Galileo even imagined might exist. Photographs such as that of the Orion nebula are now available to anyone with an Internet connection.

## Magnification and Magnifying Glasses

The simplest and perhaps most commonly used optical apparatus is the simple magnifying glass. As you have learned, **magnification** ($M$) refers to the ratio of the size of the image to the size of the object. Mathematically, this is

$$M = \frac{h_i}{h_o}$$

and, therefore, can also be found using

$$M = -\frac{d_i}{d_o}$$

   For example, if a convex lens creates a real image that is three times as big as the object, then the magnification will equal −3. The negative sign indicates that the real image is inverted.

   When a convex lens is used as a magnifier, the lens is moved so that the object is inside the focal point. If you draw ray diagrams for an object at various positions between the convex lens and the focal point, it will become obvious that the closer to the focal point you locate the object, the greater the size of the image.

On average, the closest distance that people can clearly focus on an object is about 25 cm from their eye. You can find the best focussing distance for each of your eyes by bringing this page closer and closer to your eyes (one at a time) until the page begins to become unfocussed. Note the distance at which the print on the page begins to blur. When estimating the magnifying power of a convex lens, assume that the image should be 25 cm from your eye, which, like Sherlock Holmes' eye, is very close to the lens for best vision. Now, to get the greatest magnification the object should be very close to, but inside, the focal point. This means the distance to the object is almost the same as the focal length ($d_o \approx f$). Thus, the magnifying glass can produce effective magnifications of

$$M \approx -\frac{d_i}{d_o}$$

where $d_i = -25$ cm and $d_o \approx f$. Therefore, the best magnification that you can expect from a convex lens being used to produce a virtual image is

$$M \approx -\frac{-25 \text{ cm}}{f}$$
$$\approx \frac{25 \text{ cm}}{f}$$

Hence, for a convex lens with a focal length of 10 cm, the best you can expect for its magnifying power is

$$M \approx \frac{25 \text{ cm}}{10 \text{ cm}}$$
$$\approx 2.5$$

The image is a virtual, upright image; thus, the magnification is positive. Obviously, the shorter the focal length of the convex lens, the more powerful it becomes as a magnifying glass.

## MODEL PROBLEM

### Magnification with a Convex Lens

Find the magnification when an object 45.0 cm tall is placed 48.0 cm from a convex lens with a focal length of 64.0 cm.

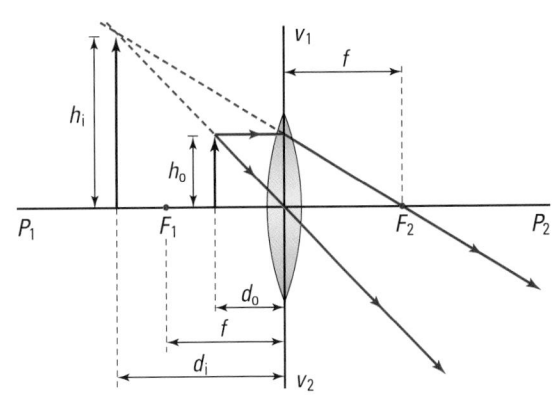

### Frame the Problem

- The object is located within the focal length of the lens, so we expect the image to be larger than the object and farther from the lens than the object.

- Visualize the problem by sketching the ray diagram.

*continued* ▶

*continued from previous page*

## Identify the Goal

The magnification, $M$, of an object by a convex lens

## Variables and Constants

| Involved in the problem | | Known | Unknown |
|---|---|---|---|
| $d_o$ | $h_i$ | $d_o$ = 48.0 cm | $d_i$ |
| $d_i$ | $f$ | $h_o$ = 45.0 cm | $h_i$ |
| $h_o$ | $M$ | $f$ = 64.0 cm | $M$ |

## Strategy

Use the mirror/lens equation to find the image location, $d_i$.

Isolate the unknown, $d_i$.

Substitute known values.

Solve.

Now, use the magnification equation, to find $M$.

## Calculations

$$\frac{1}{f} = \frac{1}{d_o} + \frac{1}{d_i}$$

$$\frac{1}{d_i} = \frac{1}{f} - \frac{1}{d_o}$$

$$\frac{1}{d_i} = \frac{1}{64.0 \text{ cm}} - \frac{1}{48.0 \text{ cm}}$$

$$\frac{1}{d_i} = \frac{3}{192 \text{ cm}} - \frac{4}{192 \text{ cm}}$$

$$\frac{1}{d_i} = \frac{-1}{192 \text{ cm}}$$

$$d_i = -192.0 \text{ cm}$$

As expected, the image is farther from the lens than the object and on the same side of the lens.

$$M = -\frac{d_i}{d_o}$$

$$= -\frac{(-192.0 \text{ cm})}{48.0 \text{ cm}}$$

$$= +4.0$$

The image is a virtual (upright) image that is 4.0 times larger than the object.

## Validate

The units for the image distance are in centimetres. There are no units for the magnification.

The value of the magnification seems reasonable.

**14.** Suppose you want to magnify a photograph by a factor of 2.0 when the photograph is 5.0 cm from the lens. What focal length is needed for the lens?

**15.** What is the magnification of a lens with a focal length of 10.0 cm when the object is 9.5 cm from the lens? Where is the image located?

**16.** The fine print in a contract that you want to read is only 1.0 mm tall. You have a magnifier with a focal length of 6.0 cm. Where must you hold the magnifier so that the image appears to be 1.0 cm tall? Where is the image located?

# Refracting Telescopes

Refracting telescopes generally use two lenses to magnify and focus on a distant object. The **objective lens** converges the almost-parallel rays from the object, and the **eyepiece lens** produces a magnified, virtual image. Modern refracting telescopes are **Kepler** or **astronomical telescopes**, in which the eyepiece and the objective lens are both convex. The object viewed in astronomical observations is obviously very far away; thus, the rays from that object are almost parallel. The refracted rays converge on a point just a bit farther from the objective lens than its second focal point, which lies inside the telescope (See Figure 12.15). The image formed by the objective lens is a very tiny, real (inverted) image. This image is viewed as the object for the eyepiece. The eyepiece is a convex lens used as a magnifying glass. As seen in the previous section, the best magnification is achieved when the object being magnified is just inside the first focal point of the eyepiece. This type of magnifier produces an upright image of the object. Since the object for the eyepiece was an inverted image from the objective, the final image is therefore inverted compared

**History Link**

Johannes Kepler (1571–1630) was famous as the creator of Kepler's laws of planetary motion. His modifications, in 1611, to Galileo's telescope design enabled him to achieve magnifying powers of almost 1000 times, greatly enhancing its field of view. With his telescope, he observed that the tail of a comet always pointed away from the Sun. He suggested that there must be solar radiation pressure to cause this. It was more than three hundred years before he was proven to be correct.

**Figure 12.15** The lens arrangement for an astronomical refracting (Kepler) telescope

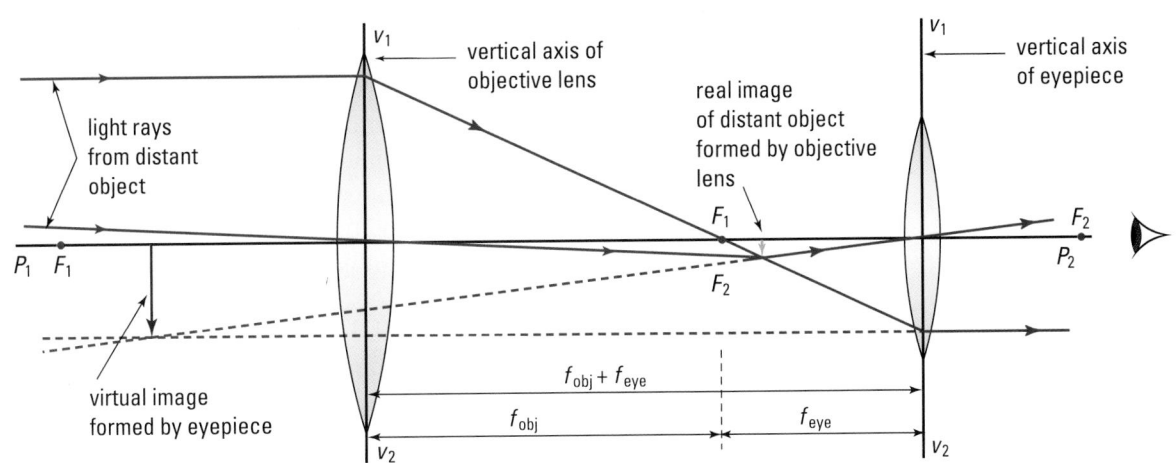

**Figure 12.16** High-powered refracting telescopes have very long tubes.

to the original object. For astronomy, it is irrelevant whether the image of the star is upright or inverted.

The telescope is constructed so that the second focal point of the objective lens and the first focal point of the eyepiece are at the same point inside the telescope. Thus, the image from the objective lens that is just outside its focal point will become the object for the eyepiece that is just inside its focal point. Hence, the length of the telescope tube should be $f_{obj} + f_{eye}$, where $f_{obj}$ and $f_{eye}$ are the focal lengths of the objective lens and the eyepiece, respectively. The magnification of the telescope is found by

$$M = -\frac{f_{obj}}{f_{eye}}$$

The negative value tells us that the image is inverted. To produce a high-powered telescope, the focal length of the eyepiece must be small compared with the focal length of the objective. In theory, astronomical (Kepler) telescopes have no limit to their possible magnification. Obviously, that is not the situation in reality. Other factors come into play.

Telescopes with very high powers need to have an objective lens with a long focal length. The magnification is obviously increased by using an eyepiece with a short focal length. But if you make the focal length of the eyepiece too small, there is little room for error in the position of the image from the objective. Even minor changes in the length of the telescope tube make locating the image impossible.

The longer you make the focal length of the objective lens, the longer you must make the length of the telescope. As a result, most astronomical telescopes must be mounted on a tripod or other base to hold them still, as in Figure 12.16. If you want to use an astronomical telescope design to make binoculars, two problems arise. The first is that they become very long and unwieldy. The second problem is that the image is inverted (not a problem in astronomy but a definite drawback for bird watching).

Fortunately, both problems are solved by the same design adaptation. The characteristic "shifted" tube shape of high quality binoculars comes from using prismatic mirrors to make the light path double back on itself to shorten the tube.

By arranging the prismatic mirrors as shown in Figure 12.17, not only is the length of the tube shortened, but also the image is inverted while being reflected. This re-inverts the image so that it is seen as being upright.

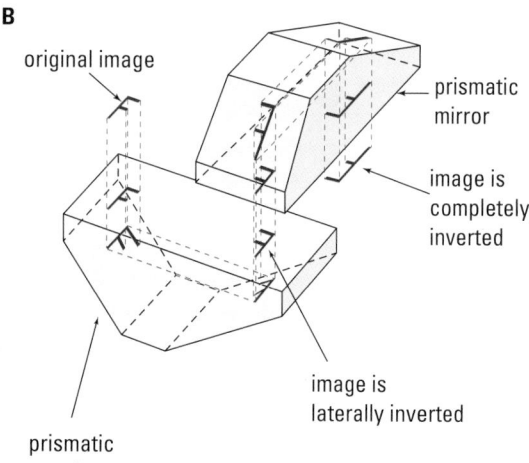

**A**

**B**

**Figure 12.17** **(A)** Binoculars use prismatic mirrors to shorten the tube length and **(B)** re-invert the image to make it appear upright.

• Return to your notes on Investigation 12-A, at the beginning of this chapter, and review your findings on the Kepler telescope. How accurately did you answer the questions at that time?

## Design Limitations for Refracting Telescopes

Regardless of magnification, if you want to see very distant and very dim objects, you need to increase the light-gathering ability of the telescope. This means that you need to make the surface area of the objective lens larger, so that more light falls on it. As the diameter of the lens increases, the lens becomes thicker and the problems of lens construction increase. If you are viewing very tiny images of distant stars, even a very slight loss in image quality can render the image useless.

The first problem in the production of large lenses is the process of construction itself. Manufacturing a large piece of flawless glass is very difficult. Even small internal flaws will disturb the image quality. At large magnifications, even small flaws can produce a significant decrease in image quality. It is, therefore, very expensive to make glass of a sufficiently high quality for a large lens. Once the high-quality glass base for the lens is formed, the surfaces must then be ground to an extremely high degree of accuracy. Computers have increased our ability to do this but it is still a very challenging technological process. That is why very large astronomical telescopes use concave mirrors rather than convex lenses for the objective.

Second, the thin-lens equations no longer apply to such thick lenses. The image is still formed but as you move from the centre of the lens toward its edges, the focal length of the lens changes causing the image to become blurry (see Figure 12.18). This is called **spherical aberration**.

**Technology Link**

Since the mid-1970s, optical engineers and astronomers have been designing huge reflecting optical telescopes with composite primary mirrors that are upwards of 50 m across — bigger than the infield of a baseball diamond. Once built, they will greatly improve the images available in optical astronomy. (For comparison, the largest optical telescope at present is the Keck telescope, located in Hawaii. It is 10 m across and comprised of 36 segments of 2 m each.) Presently, many different telescopes are used to observe the universe using non-visible "light" such as radio waves. Because the mirrors for radio-telescopes do not use specular reflectors, these can generally be made much larger than the mirrors for telescopes designed to reflect visible light. The world's largest radio telescope, at Arecibo in Puerto Rico, is 305 m across. Arrays of radio telescopes are now being used to increase resolution and sensitivity. Find out more about the sensitivity of radio telescopes. What is a jansky?

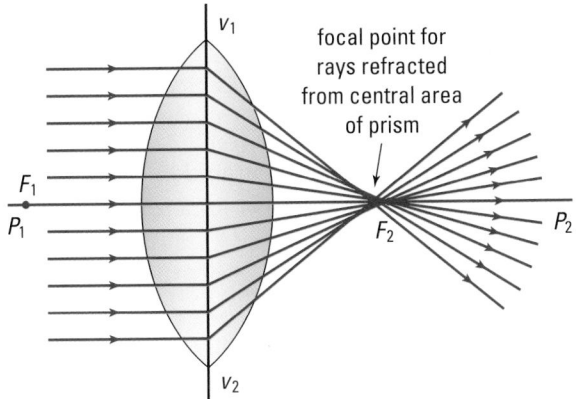
**Figure 12.18** Spherical aberration causes deterioration in the quality of an image.

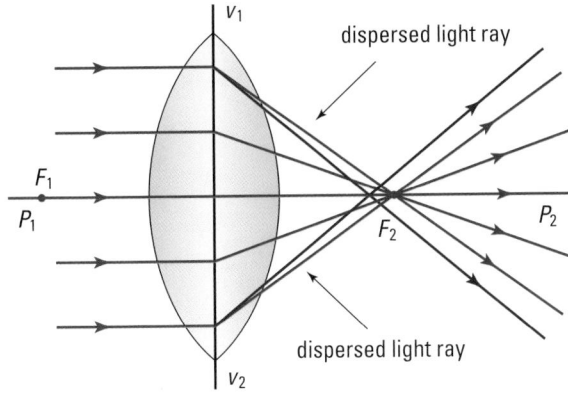
**Figure 12.19** Chromatic aberration is the result of dispersion of the light by the edges of the lens.

Third, the edges of the lens become very fat prisms and dispersion of the light from the object into its spectrum disturbs image formation. Because the angle of deviation for violet light is greater than that for red light, the edge of the lens, acting as a prism, refracts the violet light closer to the lens than the red light. Thus, the focal length for violet light is shorter than the focal length for red light. This produces an effect called **chromatic aberration**, as illustrated in Figure 12.19 on the previous page. The result is images that are tinged with a halo of colour.

Infrared photographs can be used to increase the contrast and reduce atmospheic distortion.

Camera lenses might have an infrared focussing scale to allow you to take pictures using infrared light.

## PHYSICS & TECHNOLOGY

### Focussing on Infrared

If you are a photographer, you can photograph objects using infrared light, otherwise known as radiant heat. Because the index of refraction for infrared light is even smaller than the index of refraction for red light, for any given lens its focal length is even longer than the focal length of red light.

So that you can be sure your infrared pictures are in focus, the lenses on many professional cameras have an infrared focussing scale marked on them. First, the camera is focussed for visible light; then, the setting for accurate focus is turned to the infrared focussing mark. This adjusts the focal distance from visible light to infrared light.

The leaves of plants are good reflectors of the Sun's infrared rays, while the ground and water are very good absorbers of infrared light. The result is that landscapes show an almost eerie increase in contrast when photographed using infrared light.

A further effect results because it is the blue end of the spectrum that is most strongly scattered by the atmosphere, while infrared light at the opposite end of the spectrum undergoes very little scattering. Hence, photographs made with infrared light can often be used to "see" through haze and give a clear picture when a photograph made with visible light would not.

### Going Further

1. Find and explain several examples of how infrared photographs are used to gain information that is difficult or impossible to obtain using visible light.

2. Explain clearly why the sky is blue and why the Sun appears red at sunset.

# Galileo Telescopes

The Galileo telescope is popular today in the form of opera glasses. Some binoculars (usually inexpensive) that have straight tubes are forms of Galileo telescopes. Galileo telescopes are, however, more limited in their magnification compared to Kepler telescopes.

The advantage offered by **Galileo telescopes** is that the image is upright, compared to the original object, without the use of prismatic mirrors. This is the result of a very clever application of a concave eyepiece. Galileo placed a concave lens in the path of the converging rays from the convex objective lens. He made the length of the telescope tube so that the second focal point of the objective lens coincided with the second focal point of the eyepiece *outside* the telescope tube. Thus, the length of a Galileo telescope tube was $f_{obj} - f_{eye}$. In other words, the rays had not yet reached the point at which they would form an image.

If these rays had not passed through the eyepiece, they would have converged just beyond the second focal point of the eyepiece. They appeared to diverge from a point in front of the concave eyepiece inside the telescope. Because the rays converging from the objective lens had not yet reached the point of convergence they had not crossed and the image was still upright. The image from the objective lens could be considered to be a virtual (or an imaginary) object for the concave lens (see Figure 12.21). For more detail, see Section 12.4, which discusses virtual objects.

**Figure 12.20** Opera glasses are usually a pair of Galileo telescopes.

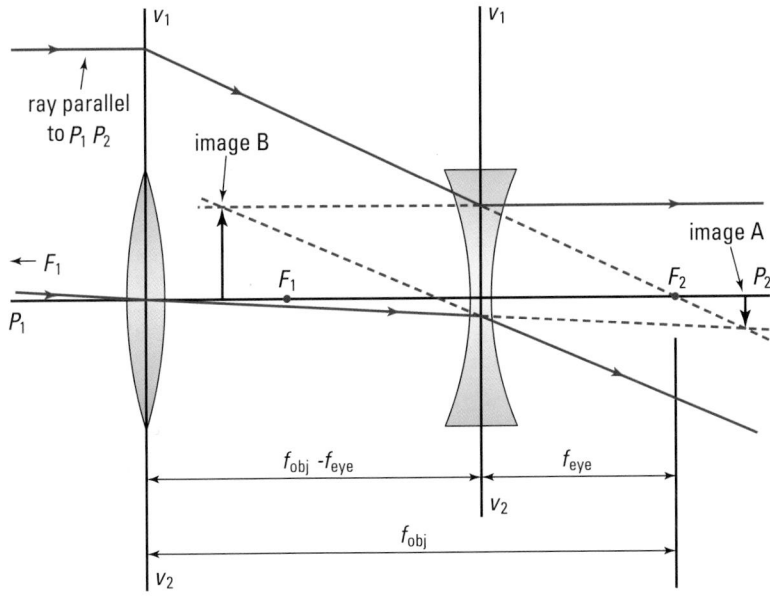

**Figure 12.21** In this diagram, the rays from the objective lens are converging to a point beyond the concave eyepiece. The eyepiece causes these rays to diverge from a point inside the telescope, the same point where you see the image.

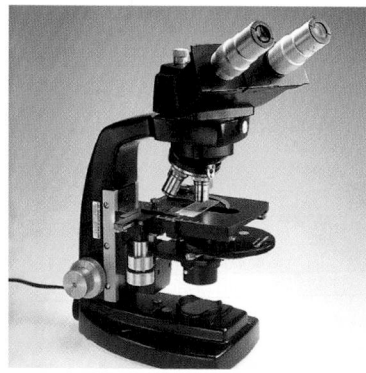
### • *Think It Through*

- Check your notes from Investigation 12-A at the start of this
  chapter and see how you explained the Galileo telescope at that
  time. How accurately have you explained how the telescope
  formed its image? Did you find that the distance between the
  lenses for the Galileo telescope was less than for the Kepler
  (astronomical) telescope?

## Microscopes

The structures of an astronomical telescope and a microscope are
very similar. They both use a convex eyepiece to observe the
real image formed by a convex objective. While the telescope has
an objective lens with a long focal length to view large objects at
great distances, the microscope has an objective lens with a very
short focal length to view small objects at very small distances.

   Unlike a telescope, the objective lens for a microscope has the
shorter focal length and the eyepiece has the longer focal length.
But, like the telescope, the real image from the objective lens is
projected inside the focal point of the eyepiece. The eyepiece is
being used as a standard magnifying glass, which leaves the image
inverted. In order to make the image from the objective lens as
large as possible, you put the object as close as possible to, but
still outside the focal point of, the objective lens, as shown in
Figure 12.23 (see also the convex lens ray diagrams on
pages 559–560). The length of the microscope tube then is the sum
of the length of the image distance for the objective lens plus the
focal length of the eyepiece.

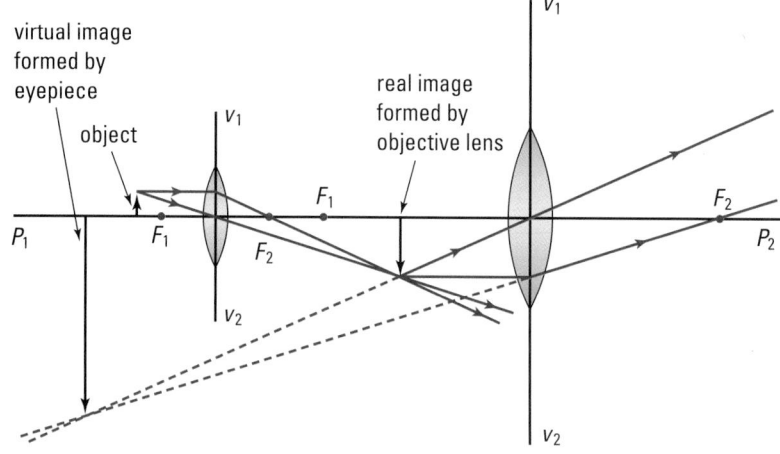

**Figure 12.22** The lens
arrangement of a microscope is
similar to that of an astronomical telescope. In a binocular
microscope, the light from the
objective lens is split in two by
prismatic mirrors and directed
to the eyepiece lenses.

**Figure 12.23** A ray diagram for a microscope is illustrated here. For
maximum magnification, the real image from the objective lens of a
microscope is located just inside the focal point of the eyepiece.

# The Eye

The human eye is a marvel of biological design. It is able to focus on objects at different distances, record images, and detect subtle changes in colour and brightness. The focussing occurs at the front of the eye and everything else occurs at the back, or in the brain.

Figure 12.24 shows how the cornea refracts incoming light and projects an image onto the retina. In fact, the cornea contributes approximately seventy percent of the total converging power of the eye. The purpose of the lens, located behind the cornea, is to make focal length adjustments as the distance to the object changes. As the distance from an object to the eye decreases, the distance from the lens to the image increases. The image, originally on the retina, shifts to a position behind the retina. To move the image back onto the retina, the ciliary muscle contracts, allowing the lens, by virtue of its elasticity, to become more convex, shortening its focal length. By shortening its focal length, the lens moves the image so that it falls exactly on the retina. This phenomenon is known as **accommodation**. When an object is far from the eye, the ciliary muscle relaxes allowing ligaments to increase tension on the lens and cause it to flatten.

The retina is spread across the back of the eye. It contains millions of specialized receptor cells of two types: rod and cone cells. Rod cells are highly sensitive and enable you to see in dim light. Cone cells are comparatively less sensitive — they do not work in dim light — but they are able to distinguish fine details and colours. Light from an object you are looking directly at, falls on a small region of your retina called the fovea. The fovea contains only cone cells making the image very clear and distinct. The rest of the retina has a combination of rod and cone cells. Both the rods and cones respond to light from the image and trigger nerve cells to send impulses along the optic nerve to the visual cortex of the brain where "seeing" is actually done.

**PHYSICS FILE**

Looking straight at an object is not always the best way to see it, especially if the object is faint. When you look at an object directly, you focus it on the centre of your retina, where there are few rods. If you look slightly away, or "avert" your vision, you use more rods and can then see more faint detail of the object. The next time you are looking at the night sky through a telescope or binoculars, remember to use "averted vision"—you'll see more stars!

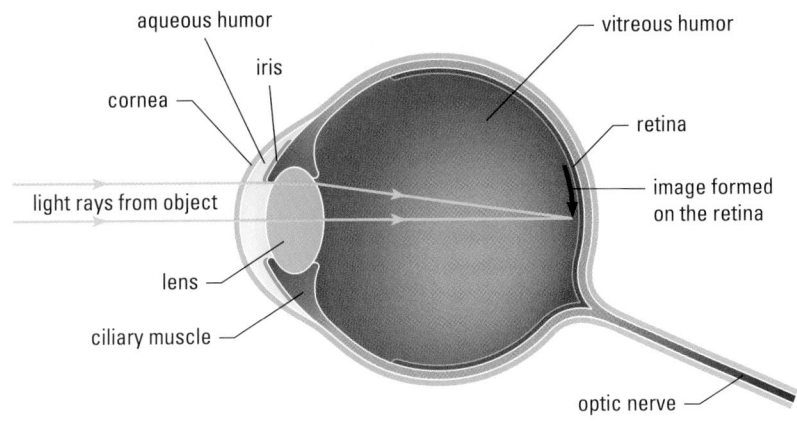

**Figure 12.24** The eye focusses an image on its retina.

**Language Link**

The term "myopia," commonly known as near-sightedness, comes from Greek for "shut-vision." To improve their vision, people suffering from myopia will squint or partially close their eyes. The term "hyperopia," commonly known as far-sightedness, comes from the Greek for "vision beyond." Many words in English are combined to form words from other languages. In Greek, the word *presbus* means "old man." Can you create a word that means "old man's vision"? This term will be explained later in this chapter.

## Myopia (Near-sightedness)

As people grow older, the shape of their eye sockets sometimes changes, which affects the shape of their eyeballs. If an eyeball is forced out of a round shape, the lens might not be able to accommodate enough to focus the image exactly on the retina.

In teenagers, their growth spurt often results in the eyeball's shape becoming slightly elongated. The lens cannot project the image far enough back in the eyeball to fall on the retina. This condition is called "**myopia**" or "near-sightedness." Using a diverging lens in front of the eye causes the rays from the object to diverge slightly before they strike the cornea. If the amount of increased divergence is correct, the image will be formed a bit farther behind the lens on the retina. The diverging lens is usually in the form of a concave meniscus (see Figure 12.25). Recently, medical scientists have learned how to use lasers to reshape the cornea to correct for myopia.

**Figure 12.25** A concave meniscus is used to correct myopia.

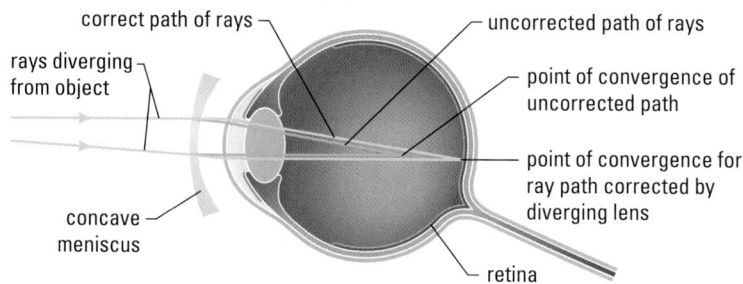

correct path of rays

rays diverging from object

concave meniscus

uncorrected path of rays

point of convergence of uncorrected path

point of convergence for ray path corrected by diverging lens

retina

## Hyperopia (Far-sightedness)

When the eyeball becomes too short for the lens to focus the image to the retina, the condition is known as "**hyperopia**" or "far-sightedness." This condition is corrected by the use of a converging lens in the form of a convex meniscus. The effect of the corrective lens is to assist the eye to shorten the distance from the lens to the image, as shown in Figure 12.26. Although laser surgery has been developed to correct hyperopia, it is much more complicated, and thus not as popular, as the laser surgery used to correct myopia.

**Figure 12.26** A convex meniscus lens is used to correct hyperopia.

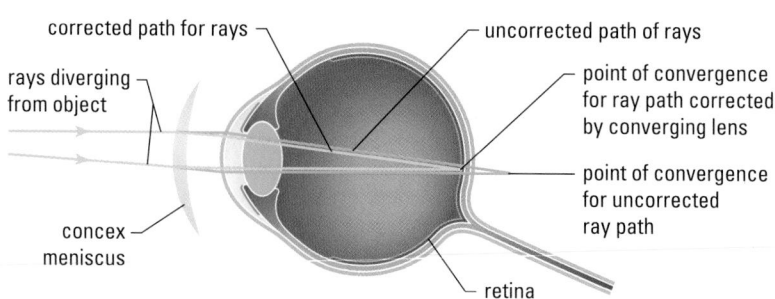

corrected path for rays

rays diverging from object

concex meniscus

uncorrected path of rays

point of convergence for ray path corrected by converging lens

point of convergence for uncorrected ray path

retina

### Have your eyes ever been checked?

When you look at a cherry pie, grizzly bear, or anything else, the rays of light from the image ideally focus right on your retinas, the light-sensitive tissues lining the backs and sides of your eyes. However, if you are nearsighted, images focus in front of your retinas because your eyes are longer than normal. If you are farsighted, images focus behind your retinas because your eyes are shorter than normal. Either way, blurry vision can result. An optometrist or ophthalmologist can check your eyes to detect visual problems, if any, and write a prescription for corrective lenses, if needed.

Dispensing opticians interpret such prescriptions. They take facial measurements and measure the distance between the pupils of your eyes. Based on such measurements and your lifestyle, they recommend eyeglass frames and lens types. If you are considering contact lenses, they inform you of the choices involved.

Once you and the dispensing optician decide what to order, the eyeglasses or contact lenses are made, usually in a laboratory supervised by an optician. Based on the prescription, laboratory technicians make concave spherical lenses for near-sightedness. They make convex spherical lenses for far-sightedness. For astigmatism, a condition which is caused by incorrectly shaped cornea, they make cylindrical lenses. For cases where the pupils do not quite line up, they use prisms to help the eyes work together better. For more than one problem, the laboratory combines lens forms. When your eyeglasses or contact lenses are ready, the dispensing optician checks that they are made correctly. He or she adjusts eyeglass frames so they fit comfortably. In the case of contact lenses, the dispensing optician shows you how to insert, remove, and care for them.

Becoming an optometrist (doctor of optometry) requires about four years of university. Becoming an ophthalmologist takes eight to ten years since an ophthalmologist becomes a Doctor of Medicine (M.D.) first and then specializes in eye disorders. Becoming an optician takes one to three years of community college, depending on where you live. For all eye care professionals, useful high school courses include physics, mathematics, biology, chemistry, and English.

### Going Further

1. Ask people wearing eyeglasses or contact lenses about their experiences with eye care professionals.

2. Find out what types of eye care are paid for by your province's or territory's public health system. If you have not had your eyes checked in the last year, consider doing so.

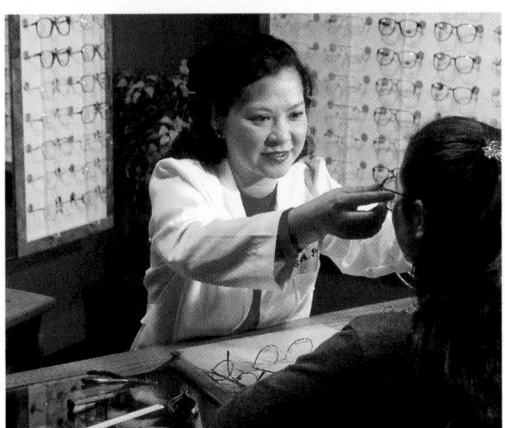

## Presbyopia

Rather than resulting from a change in the shape of the eyeball, the inability to see objects that are close to you can also occur because the lens loses its flexibility. This condition is known as **presbyopia**. Because the rays from a far object are diverging less than rays from a near object, the lens can still focus rays from distant objects on the retina. Normally, to cause the more sharply diverging rays from a near object to be focussed on the retina, the lens has to become quite round to shorten its focal length. As you age, the lens often loses its flexibility and cannot become round enough to create clear images of near objects.

Presbyopia can occur in conjunction with myopia or hyperopia. If one already wears glasses to correct for myopia, as presbyopia occurs, the lens prescription becomes one for bifocals, or even trifocals, to accommodate both conditions. People who have undergone laser surgery to correct for myopia often still have to wear reading glasses because of presbyopia.

## Astigmatism

Another vision problem is associated with a change in shape of the eyeball. If the cornea becomes "out of round" laterally or vertically, then the focus of either horizontal or vertical lines might be affected. The eye focusses in one plane slightly better than the other planes. This is defined as **astigmatism** and often occurs in conjunction with myopia or hyperopia. In this case, adjustments can be made to the lenses of spectacles to correct this problem. If any lines seem darker than others when you look at Figure 12.27, you might have an astigmatism.

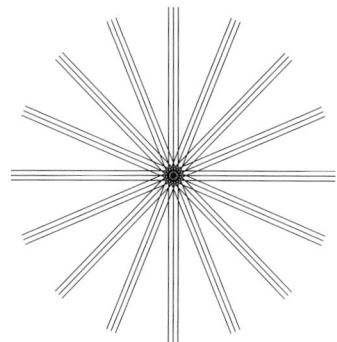

**Figure 12.27** If one set of bars seems darker than the rest, it indicates that your cornea is astigmatic.

---

## 12.3 Section Review

1. **K/U** Where should the object be placed to obtain the greatest magnification from a convex lens?

2. **K/U** Define the magnification of a refracting telescope.

3. **K/U** What are some problems associated with large lens construction?

4. **C** Discuss the differences and similarities between a telescope and a microscope.

5. **K/U** List the differences and similarities between a Galileo and a Kepler telescope.

6. **K/U** Why are Kepler telescopes longer than Galileo telescopes?

7. **C** Explain accommodation as it applies to vision.

8. **K/U** What are the roles of the rod and the cone cells in the eye?

9. **K/U** Define
   (a) myopia,
   (b) hyperopia,
   (c) presbyopia, and
   (d) astigmatism.

In the Kepler telescope and the microscope, the real image, formed by the objective lens, is used just as a real object by the eyepiece. Even virtual images can be used as objects for other lenses, since the rays from them are diverging. However, in the Galileo telescope, the concave eyepiece intercepts the converging rays from the objective before they reach the point at which the real image would be formed (in other words, the point of divergence). The concave lens (remember this is a diverging lens) then refracts these converging rays so that they appear to diverge from a point inside the telescope. That point of divergence is where the observer sees the image.

Imagine a set of rays that converge on a concave lens, as shown in Figure 12.28. If they had not passed through the lens, the rays would have come to a point of convergence and the real image would have been formed at $P$, just beyond the focal point of the lens ($F_2$).

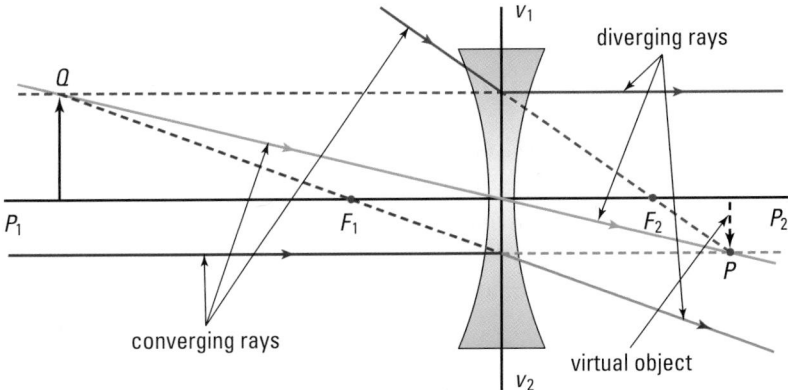

**Figure 12.28** When a set of converging rays are incident on a lens, the image they would have formed when they reached the point of convergence acts as a virtual object for the lens. For the object, $d_o$ is negative and the thin-lens equations can be used to calculate the size and position of the image.

The paths of the three rays normally used to locate images are shown. First, the ray through the centre of the lens (in green) goes straight through without bending. Second, the ray that would have passed through $F_2$ (in red) is refracted parallel to the principal axis. Third, the ray parallel to the principal axis (in blue) refracts so that it appears to have passed through the first focal point ($F_1$). In each case, the solid line indicates the actual path of the ray, the dotted line indicates the path of the rays had they not been refracted, and the segmented line represents the path of the rays projected back to the imaginary point of divergence at $Q$.

If the rays had not passed through the lens, they would have converged on and then diverged from point $P$, creating a real image there. Instead, the rays are refracted so that they appear to be diverging from $Q$; thus, a virtual image will be visible there.

Either concave or convex lenses can create an image when they intercept the converging rays from a convex lens. Depending on the type of lens and where the point of convergence (and thus the location of the unformed real image) would be located in relation to the focal point of the lens, the images might be real or virtual. On the surface, it might seem that the calculations for the position and size of the image might be very complex. But if you think of the not-yet-formed real image as a **virtual object**, it makes calculations using the thin-lens equations quite straightforward. When lenses refract diverging rays to form images, the point of divergence defines the position of the object. The distance from the lens to the object or point of divergence ($d_o$) is considered positive. For virtual objects, the distance from the lens to the virtual object is considered to be negative. All other values are interpreted as for other conditions, as shown in Table 12.2.

### Using an Image as an Object

A set of converging rays is incident on a concave lens with a focal length of 25.0 cm. If they had not passed through the concave lens, the rays would have formed an inverted real image 10.0 cm high at a point 50.0 cm beyond the concave lens. Calculate the location and size of the actual image.

### Frame the Problem

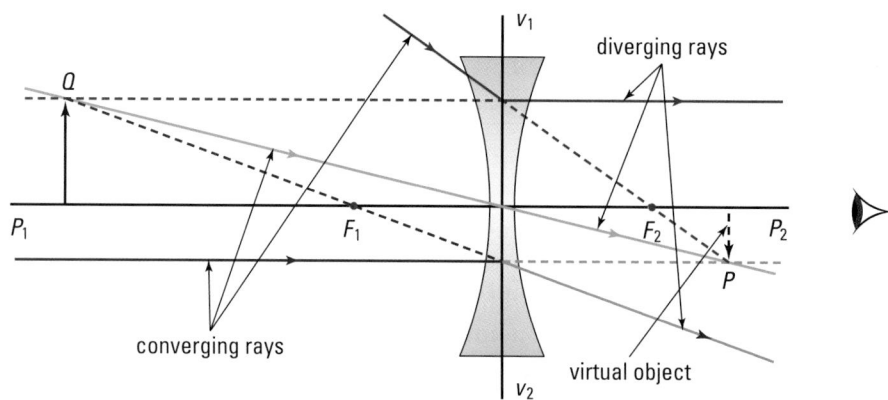

- Visualize the problem by sketching the light ray diagram.

- The *virtual object* would have been a *real image*, so its *size* is *negative*.

- The lens is *concave*; therefore, *focal length* is *negative*.

## Identify the Goal

(a) The location, $d_i$, of the image

(b) The size, $h_i$, of the image

## Variables and Constants

| Involved in the problem | | Known | Unknown |
|---|---|---|---|
| $f$ | $h_o$ | $f = -25.0$ cm | $d_i$ |
| $d_o$ | $h_i$ | $d_o = -50.0$ cm | $h_i$ |
| $d_i$ | | $h_o = -10.0$ cm | |

## Strategy

## Calculations

Use the mirror/lens equation to find $d_i$.

$$\frac{1}{f} = \frac{1}{d_i} + \frac{1}{d_o}$$

Rearrange.

$$\frac{1}{d_i} = \frac{1}{f} - \frac{1}{d_o}$$

Substitute known values.

$$= \frac{1}{-25.0 \text{ cm}} - \frac{1}{-50.0 \text{ cm}}$$

$$= \frac{-2}{50.0 \text{ cm}} + \frac{1}{50.0 \text{ cm}}$$

Solve.

$$= \frac{-1}{50.0 \text{ cm}}$$

$$d_i = -50.0 \text{ cm}$$

Now you can use the magnification equation to find $h_i$.

$$\frac{h_i}{h_o} = -\frac{d_i}{d_o}$$

Rearrange.

$$h_i = -h_o \frac{d_i}{d_o}$$

Substitute known values.

$$= -(-10.0 \text{ cm}) \left( \frac{-50.0 \text{ cm}}{-50.0 \text{ cm}} \right)$$

Solve.

$$= +10.0 \text{ cm}$$

(a) The image is located −50.0 cm from the lens.

(b) The image is +10.0 cm tall.

*continued* ▶

*continued from previous page*

## Validate

The units are in centimetres.

The image is on the same side of the lens as the incident light; therefore, the image distance is negative.

The image is on the opposite side of the lens as the virtual object.

The image is upright and virtual.

### PRACTICE PROBLEMS

17. Make an accurately measured scale diagram using the following values. Use a concave lens with a focal length of 3.0 cm. Assuming the light is coming from the right, place point P at a distance of 4.0 cm to the left of the vertical axis and 1 cm below the principal axis. Accurately draw the ray diagram as shown in Figure 12.28. Measure the size and location of the image. Use the thin-lens equations to calculate the size and location of the image. Remember that the focal length ($f$) is −3.0 cm. The distance to the virtual object ($d_o$) is −4.0 cm since it is on the opposite side of the lens from which the light came. Use the thin-lens equations to calculate the distance to and the size of the image and see if it agrees with the measurements in your diagram.

18. Use the same parameters as in problem 17 but place the virtual object 2.0 cm from the lens.

19. Use the same parameters as in problem 17, but place the virtual object 3.0 cm from the lens.

20. Repeat problems 17 through 19 using a convex lens with at 3.0 cm focal length ($f$ = +3.0 cm). Be careful to make sure you obey the rules of refraction for convex lenses.

## 12.4    Section Review

1. **K/U** What is a virtual object?

2. **C** Explain how a virtual object is produced.

3. **K/U** How are virtual objects used in optical instruments such as the Galileo telescope?

4. **C** Use ray diagrams to sketch how Kepler and Galileo telescopes form an image of the Moon.

### UNIT PROJECT PREP

Lenses and the principles of refraction have been used to improve eyesight, view microscopic objects, and peer deep into space.

■ What advantage, if any, does the application of virtual images provide to the design of your optical instrument?

■ Sketch ray diagrams to predict the path of light using the principle of virtual images.

## REFLECTING ON CHAPTER 12

- When refracted rays cross at the point of divergence, the image of each point is inverted. If the rays actually pass through the point of divergence, the image is real. If the rays diverge from a point through which they never actually pass, the image is virtual.

- A convex or converging lens is thicker in the middle than at the edges. Parallel rays that strike the lens will converge on the focal point on the opposite side of the lens.

- A concave or diverging lens is thinner in the middle than at the edges. Parallel rays that strike the lens diverge from one another and appear to come from the focal point on the same side of the lens as the incident light.

- The lens-maker's equation is used to predict the focal length of a lens from the index of refraction of the lens material and the radii of curvature of the two surfaces of the lens:

$$\frac{1}{f} = (n - 1)\left(\frac{1}{R_1} + \frac{1}{R_2}\right)$$

- The magnification equation relates the image size to the object size, and the image distance to the object distance,

$$M = \frac{h_i}{h_o} = -\frac{d_i}{d_o}$$

  Negative magnifications mean the image is inverted.

- The mirror/lens equation relates the focal length to the object and image distance:

$$\frac{1}{f} = \frac{1}{d_o} + \frac{1}{d_i}$$

- A refracting telescope uses an objective lens with a long focal length to view large objects at great distances.

- Keplerian telescopes use convex lenses as an objective and as an eyepiece.

- Galileo telescopes use a convex lens as an objective and a concave lens as an eyepiece.

- Problems with large lenses include spherical aberration and chromatic aberration.

- A microscope has an objective lens with a very short focal length to view small objects at very small distances. Both the objective and eyepiece are convex and a magnified, inverted, virtual image is produced.

- In the eye, the cornea focusses light onto the retina. The focus can be adjusted by changing the tension on the lens, located behind the cornea. Loss of flexibility of the lens or a change in shape of the eyeball affects the focussing of the eye and can be corrected with lenses.

- A virtual object is created when a concave or convex lens is placed *ahead of* the point of convergence.

### Knowledge/Understanding

1. What is the difference between the angle of deviation and the angle of dispersion?
2. Define the term "radius of curvature."
3. Name the three basic shapes for convex lenses.
4. What is the difference between a convex meniscus and a concave meniscus? Illustrate your answer by a diagram.
5. Why can real images be projected onto a screen?
6. How does the shape of the eyeball differ for myopia and hyperopia?

7. Is the image that forms on the retina of the eye a real or a virtual image?
8. What is the difference between the eyepieces of a Galileo and a Kepler telescope?
9. What causes astigmatism?
10. If the magnification for an image is equal to −3.5, what is the nature of the image?
11. What is meant by the term "thin-lens equations"?
12. Why do very large astronomical telescopes use a mirror rather than a lens for the objective optic?

13. Suppose a convex lens produced an image and then was replaced by a larger lens with the same focal length. How would the second image compare to the first image?

## Communication

14. Explain why a concave lens cannot create a real image from a real object or real image.

15. Explain how the cornea and the lens of the eye create the image that falls on the retina.

16. The image formed on the retina of the eye, like all real images, is inverted. Discuss why the world does not look inverted to you.

17. Reducing the aperture through which light can enter a lens reduces the amount of spherical aberration. Explain why.

18. Discuss the concept of accommodation.

19. Explain why the tube lengths of telescopes and microscopes are set so that the image formed by the objective lens falls just inside the focal point of the eyepiece.

20. Explain how you would design a microscope that can magnify 1000 times.

## Inquiry

21. Eyeglass lens technology continues to improve the quality, durability, and comfort related to wearing glasses. Imagine that you are an entrepreneur, preparing to open an eyeglass manufacturing company. You hope to market your lenses based on new and exciting design advances. Design and conduct several simple experiments to test physical properties of eyeglass lenses (e.g. weight, lateral displacement, etc.). Compare glasses from several years ago with more modern models. Complete your investigation by generating a marketing brochure to promote your new and improved lens design.

22. During an investigation the following data was collected. Complete the table and describe the image characteristics for each scenario.

| $d_o$ (cm) | $d_i$ (cm) | $f$ (cm) |
|------------|------------|----------|
| 60         |            | 30       |
| infinity   | 15         |          |
| 20         |            | 40       |
| 30         | 40         |          |

## Making Connections

23. To put a telescope such as the Hubble Space Telescope into space is a very costly venture. Obviously scientists think that it is worth the expense or they would not have done it. Do a research project to find out the advantages and disadvantages of having the telescope in orbit around Earth rather than on the surface.

24. Investigate laser eye surgery to find out what the pros and cons are for using it to correct myopia.

25. In theory, microscopes should be able to produce unlimited magnification. Investigate microscope design to find out why this is not the case.

26. The camera is often compared to an eye. While there are many features that are very similar, there are many significant differences as well. With the aid of diagrams, compare and contrast the camera and the eye.

## Problems for Understanding

27. Suppose a convex lens is ground so that one surface has a radius of curvature of 30.0 cm and the opposing surface has a radius of curvature of 40.0 cm. The focal length is 34.2 cm. What is the index of refraction of the lens? What kind of glass is the lens?

28. What is the focal length of a plano-convex lens made of flint glass ($n=1.65$) if the radius of curvature of the spherical surface is 15.0 cm?

29. An object 17 cm tall is placed 56 cm from a convex lens that has a focal length of 24 cm. Find the size and location of the image. Describe the image.

30. An object 10.0 cm tall is located 20.0 cm from a concave lens with a focal length of 20.0 cm. What is the size and location of the image? Describe the image and draw a ray diagram of the situation.

31. If a single lens forms an image with a magnification of +4.5, describe the lens and the position of the object.

32. Draw an accurate ray diagram to find the size and location of the image formed by a double convex lens with a focal length of 8.0 cm. The object, 3.0 cm tall, is placed 14 cm from the vertical axis of the lens. Check your answers by using the thin-lens equations.

33. Repeat problem 32 using a concave lens rather than a convex lens.

34. Use an accurate ray diagram to establish the size and position of the image formed by a convex lens with a focal length of 10 cm. Place an object that is 2.0 cm tall 5.0 cm from the lens. Check your answers by using the thin-lens equations.

35. A telescope with a magnification of 30.0 has an 18 mm eyepiece. How long is the tube of the telescope?

36. A bug is on a microscope slide 3.4 mm from the objective lens of a microscope. If the focal lengths of the objective lens and eyepiece are 3.2 mm and 1.6 cm, respectively, what is the magnification?

37. A microscope is made in a tube 25.0 cm long. It magnifies $5.00 \times 10^2$ times, 50.0 times by the objective lens and 10.0 times by the eyepiece. What are the focal lengths of the two lenses?

38. Determine the focal length of the lens for the situation when an object is placed 12.0 cm in front of it and the magnification is (a) -6.00, and (b) +6.00.

39. What is the ratio of the focal length to the distance from the lens to the image if the image is (a) a real image 0.25 times as big as the object and (b) a virtual image 0.60 times as big as the object?

40. A projector takes an image from a slide that is 35.0 mm tall and projects it onto a screen that is 15.0 m away. If the image on the screen is 2.80 m tall, what is the focal length of the lens and how far from the lens was the slide?

---

**Numerical Answers to Practice Problems**

1. 14.5 cm 2. 1.85 cm 3. 1.65; it is probably flint glass.
4. 54.3 cm 5. $d_i = 15.4$ cm; $h_i = -1.80$ cm. The image is real, inverted, and smaller than the object. It is on the opposite side of the lens from the object. 6. $d_i = 225$ cm; $h_i = -11.2$ cm. The image is real, inverted, and larger than the object. It is on the opposite side of the lens from the object. 7. $d_i = -2.80$ m; $h_i = 14.0$ cm. The image is virtual, upright, and 3.33 times larger than the object. It is on the same side of the lens as the object. 8. $f = 28.8$ cm; $h_o = 24.0$ cm 9. $d_o = 35.0$ cm; $f = 122$ cm 10. $\frac{1}{d_i} = 0$; $d_i$ is undefined, as is $h_i$; no image is formed. 11. $d_i = -48$ cm; $h_i = 34$ cm; It is a 34 cm tall virtual image that is upright, on the same side as the object, and 48 cm from the lens. 12. $-1.00$ m 13. 94 cm 14. 0.10 m 15. $2.0 \times 10^1$; $d_i = -1.9 \times 10^2$ cm, the image is farther from the lens than the object and on the same side of the lens 16. $d_o = 5.4$ cm; $d_i = -54$ cm 17. $d_i = -0.12$ m; $h_i = 3.0$ cm 18. $d_i = 6.0$ cm; $h_i = -3.0$ cm 19. $d_i$ and $h_i$ are undefined 20(17). $d_i = 1.7$ cm; $h_i = -0.43$ cm 20(18). $d_i = 1.2$ cm; $h_i = -0.60$ cm 20(19). $d_i = 1.5$ cm; $h_i = -0.50$ cm

**ELECTRONIC LEARNING PARTNER**

Review your knowledge of light and geometric optics with a quiz available on your Electronic Learning Partner.

## Background

In this unit, several optical instruments that use two lenses were discussed. Other important optical instruments, such as reflecting telescopes, periscopes, and projectors, were omitted. Use your knowledge of lenses and mirrors to create a reflecting telescope or a periscope.

The reflecting telescope is similar to that of the Keplerian telescope except that a concave mirror is used to replace the objective lens, as shown in the following art. This creates viewing problems. Check reference books or the Internet (search: "reflecting+telescopes") for design hints.

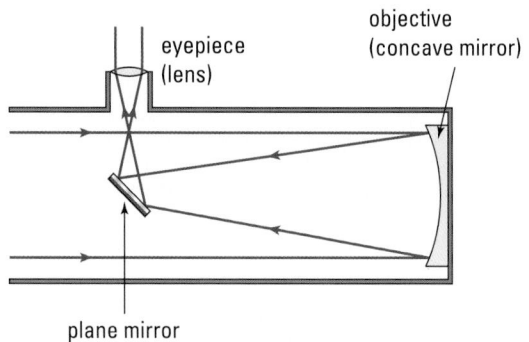

A reflecting telescope

The periscope is merely a refracting (Keplerian) telescope that uses two mirrors to look around corners. Check reference books or the Internet (search: "refracting+telescopes") for design hints.

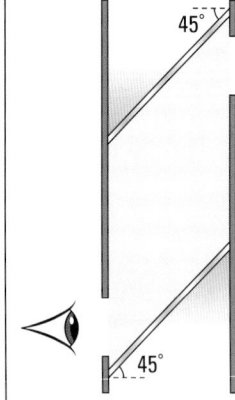

A simple model of a periscope.

## Challenge

In this project, you are to research, design, and build a periscope or a reflecting telescope using three to five optical devices capable of focussing light (in other words, convex or concave mirrors and lenses). Both instruments will be based, with refinements, on the design of a refracting (Keplerian) telescope. Since both instruments can be used to view terrestrial objects, the viewed image should be upright and enlarged. Try for a magnification of about seven times. It should be able to focus to create a sharp image of objects from a distance of about 10 m to infinity.

## Materials

Aside from lenses and/or curved mirrors, choose materials for building your instrument that are easily available and commonplace. Convex lenses are commercially available in the form of magnifying glasses, and concave mirrors can be found as make-up mirrors. Both convex and concave lenses and mirrors may be obtained from scientific supply companies.

Plane mirrors, prismatic or otherwise, may be required to bend the path of the light. If used, they are not to be considered among the three to five optical components since they bend but do not focus light. Concave and convex mirrors, on the other hand, both bend and focus light.

Be creative in constructing the supporting structure or body of the instrument. For example, plastic (either ABS or PVC) plumbing pipe could be used to build the body of your instrument. Light plywood also makes a good building material. A hot-glue gun can be used to attach together pieces of either material, but the adhesives specially created for ABS or PVC pipe work best with those materials. Both materials may be cut and drilled with normal woodworking tools such as handsaws, power saws, coping saws, and drills.

## Safety Precautions

- When using hand or power tools, be sure to wear eye protection.
- Hot-glue guns take several minutes to cool after they are disconnected. If the hot glue gets on your hands, it can cause burns.
- Make sure you clean your hands after using either hot-glue, PVC, or ABS glue.

## Design Criteria

**A.** Work in groups of three or four.

**B.** Obtain your mirrors and lenses. If the focal lengths are not previously known, do measurements to determine them.

**C.** Prepare a written presentation, including

- a title page with your names
- a design blueprint (see Action Plan)
- a ray diagram
- a log describing your work
- presentation of the completed instrument

**D.** Be prepared to have the other students use your telescope or periscope. Students will compare your designs with their own.

## Action Plan

**1.** Brainstorm possible design ideas for your optical instrument. Develop ideas for including more than two optical components. Where would you use them? How would they affect magnification? How do you create the focussing mechanism? Try to create design innovations.

**2.** Prepare a design brief that explains

- what you are building and why you have chosen to build it
- the intended use and users of the instrument

tasks it was designed for. Are the images clear, upright, and enlarged?

- **assess the instrument by letting others operate it. Is it easy to use?**
- **rate the innovation of your design on a scale of 1 to 5 (1 being very innovative, 5 being not very innovative at all.) In other words, did you get all your design ideas from a manual or did you create some or all of them yourself?**

**3.** Prepare a full sized blueprint of the instrument you are designing.

**4.** Create a full sized ray diagram, showing the path of light through the instrument and how the image is formed.

**5.** Prepare a list of all the materials you used.

**6.** Keep a log in which you detail the planning and building processes, including any successes and difficulties encountered. Describe how the problems you encountered were solved.

**7.** Create a construction manual for someone who might want to use your design to build your instrument.

## Evaluate

**1.** Is the magnification as good as you expected? Estimate your actual magnification.

**2.** Can it be focussed easily for objects at different distances? What are the closest and farthest distances at which an object can be seen clearly?

**3.** Are the design blueprints and instructions for building the instrument clear? How could you make them clearer and/or simpler?

**4.** How could you improve your design?

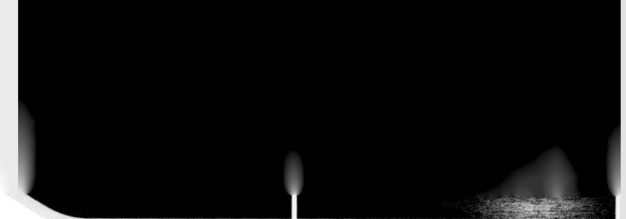

## Knowledge and Understanding

### True/False

In your notebook, indicate whether each statement is true or false. Correct each false statement.

1. Images in plane mirrors are formed at the surface of the mirror.

2. For a plane mirror, the angle of incidence equals the angle of reflection. This does not apply to curved mirrors.

3. The size of the image formed by a plane mirror depends on the distance the object is from the mirror.

4. The image formed by a concave mirror is always a real image.

5. The focal point of a curved mirror is always in front of the mirror.

6. The lower the index of refraction, the faster the speed of light in the substance.

7. For any given angle of incidence greater than zero, the lateral displacement produced by a rectangular prism decreases as the index of refraction increases.

8. The shimmering effect noticed above a hot fire is due to total internal reflection between the layers of air at different temperatures above the flames.

9. The critical angle for a substance decreases as the index of refraction increases.

10. A convex lens that is curved on one side and flat on the other only has one focal point, and that is on the curved side of the lens.

11. A convex lens can form both upright and inverted images.

12. A concave lens can form both real and virtual images.

13. The focal point of a lens is the point at which the image is formed.

14. Chromatic aberration results because the focal length for red light is less than the focal length for violet light.

15. The cornea, rather than the lens, does most of the refraction in the eye.

## Multiple Choice

In your notebook, write the letter of the best answer for each of the following questions.

16. The path of a light ray travelling from crown glass into air is shown in the diagram below. The angle of refraction is
    (a) AOX               (d) BOC
    (b) XOB               (e) YOC
    (c) DOY

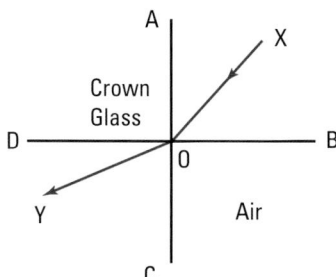

17. The index of refraction of light in water is 1.33. The speed of light in a vacuum is $3.00 \times 10^8$ m/s. The speed of light in water is
    (a) $2.25 \times 10^8$ m/s       (d) $4.33 \times 10^8$ m/s
    (b) $0.443 \times 10^8$ m/s      (e) $1.67 \times 10^8$ m/s
    (c) $3.99 \times 10^8$ m/s

18. The focal length of a double convex lens is 6 cm. If an object is placed 10 cm from the optical centre of the lens, the characteristics of the image formed are
    (a) real, inverted, and larger
    (b) real, inverted, and smaller
    (c) real, upright, and larger
    (d) virtual, upright, and larger
    (e) virtual, inverted, and smaller

19. To rectify the defect of farsightedness in the human eye, the eyeglasses must use
    (a) plano-convex lenses
    (b) diverging lenses
    (c) converging lenses
    (d) double convex lenses
    (e) bifocal lenses

20. A very small light source is placed at the focal point of a convex lens. Which of the following best describes the pattern of light after it passes through the lens?

(a) The rays of light are diverging.

(b) The rays of light are perpendicular.

(c) The rays of light are parallel.

(d) The rays of light are converging.

(e) The rays of light are scattered.

## Short Answer

In your notebook, write a sentence or a short paragraph to answer each of the following questions.

21. Define the terms: light ray, wavefront, normal, regular reflection, and diffuse reflection.

22. Describe an experiment you could do to demonstrate the difference between a real and virtual image.

23. A convex mirror can produce an image that is larger than the object. Comment on this statement, and justify your answer. A diagram may be used.

24. What is the speed of light in fused quartz?

25. When light passes from glycerine into crown glass at an angle, which will be smaller, the angle of incidence or the angle of refraction? Explain why.

26. How does the apparent depth of an object change as the index of refraction of the substance in which the object is immersed increases?

27. List three examples where a shimmering effect can be observed above an object or surface. Select one example, and explain why this effect occurs.

28. Why is it impossible for you ever to stand at the end of a rainbow, even if you are producing the rainbow by using a hose to spray a stream of fine water droplets into the air?

29. A concave lens is used to project a real image onto a screen. Explain how the image would be affected if an opaque piece of cardboard was inserted between the object and the lens so that it gradually covered more and more of the lens.

30. Explain the role of the following parts of the eye: the cornea, the lens, the ciliary muscles, the iris, and the retina.

## Inquiry

31. Design an experiment, using one or more ray boxes and several reflecting surfaces, to show the law of reflection still applies when light undergoes diffuse reflection.

32. Analyze and describe the structure of a kaleidoscope. Complete a drawing and a set of instructions to show how to make one using cardboard and masking tape.

33. Design an investigation to measure the focal length of a convex mirror mounted on the door of a car. Check to see if the radius of curvature of the mirror is within the legal requirements for such mirrors. (Refer to the Canadian Motor Vehicle Safety Standards).

34. You are given a rectangular, transparent block of an unknown solid material. Design an experiment to determine the critical angle of incidence for the material.

35. Using one or more light ray boxes, design and construct three combinations of mirror, prisms, and lenses to demonstrate the principle of the reversibility of light.

36. Research and analyze data related to the structure of different types of optical fibres. Over what range of angles of incidence is it possible for the light beams to enter each type of optical fibre? Explain why these restrictions are imposed on the angles of incidence.

37. Given a concave lens of unknown focal length,
    (a) design an experiment to determine its focal length a) using natural sunlight, and b) using a light ray box.
    (b) Compare the focal lengths obtained by both methods, and account for any differences.

## Communication

38. Copy the diagrams below into your notebook and draw a light ray diagram for each example to show how the image is formed in the plane mirror. Show all construction lines.

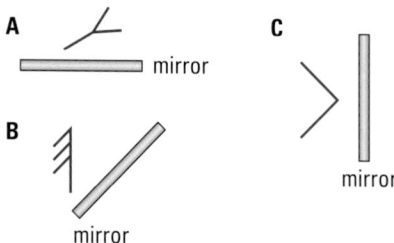

A

mirror

B

mirror

C

mirror

**39.** At the left-hand side of the page, draw a semi-circle of a spherical concave mirror shape that has a radius of curvature of 6 cm, with the principal axis drawn across the width of the page. Directly below it construct and draw a parabolic concave mirror shape, also 12 cm in diameter, with the principal axis drawn across the width of the page. By drawing appropriate light ray diagrams, show why spherical aberration occurs, and explain how it is possible to use spherical mirrors to minimize this defect.

**40.** Design and draw a concept map to show the relationships between the properties and principles of light, the two scientific models, the laws of reflection, plane and curved mirrors, and any associated technological applications.

**41.** Draw a labelled diagram to show the reflection of three rays of light that are travelling parallel to the principal axis of
(a) a concave mirror surface
(b) a convex mirror surface

**42.** Develop a report on the use of optical fibres in the medical field.

**43.** Compare and contrast disposable cameras and conventional cameras. Summarize your findings.

**44.** Interview an optometrist and develop a report that summarizes the relative merits of eye glasses versus contact lenses.

## Making Connections

**45.** The windows of some large buildings are coated with a very thin of layer of gold. Why was this material used? From an economic point of view, is the use of gold justifiable for this purpose?

**46.** "People should not be viewed by means of a "see-through" mirror, unless they have been informed of its use." Debate this issue in terms of a person's right to privacy.

**47.** Research the various factors that are considered by jewelers as they determine the best positions and angles to cut the facets on diamonds and other precious stones.

**48.** Identify the different methods of correcting defects in the human eye related to the cornea and the crystalline lens inside the eye. Propose a set of criteria that could be used to determine which of the different methods should be used for each of the defects, and justify your choice of criteria. Present your findings in an appropriate format.

## Problems for Understanding

Show complete solutions for all problems that involve equations and numbers.

**49.** Draw light ray diagrams to show the characteristics of the image formed by a concave mirror for an object placed at (a) 0.5 f, (b) 1.75 f, and (c) 3.0 f. State the characteristics of the image in each case.

**50.** A ray of light striking a plane mirror makes an angle of 42° with the mirror surface. What is the angle of reflection to the normal?

**51.** Light is shining on to a plane mirror at an angle of incidence of 12° to the mirror surface. If the plane mirror is tilted such that the angle of incidence is halved, what will be the total change in the angle of reflection from the original reflected light?

**52.** A student walks towards a plane mirror at a speed of 1.2 m/s.
(a) Determine the speed of the image relative to the student when the direction is directly towards the mirror.
(b) Determine the speed of the image relative to the student when the direction is at an angle of 60.0° with respect to the normal to the mirror.

**53.** A concave mirror has a focal length of 35 cm. What is the radius of curvature of the mirror?

54. A convex mirror has a radius of curvature of 1.8 m. What is its focal length?

55. Draw scaled light ray diagrams to determine the image characteristics for an object 2.00 cm high when it is placed in front of a concave mirror with a focal length of 18.0 cm at a distance of
    (a) 12.0 cm           (c) 40.0 cm
    (b) 24.0 cm
    In each of parts (a)–(c), determine the image characteristics using the mirror equations.

56. Draw scaled light ray diagrams to determine the image characteristics for an object 2.00 cm high when it is placed in front of a convex mirror with a focal length of 20.0 cm at a distance of
    (a) 12.0 cm           (c) 50.0 cm
    (b) 30.0 cm
    In each of parts (a)–(c), determine the image characteristics using the mirror equations.

57. A small vase is placed 40.0 cm in front of a concave mirror of focal length 15.0 cm. Determine the characteristics of the image, and its magnification.

58. A can on a shelf in a convenience store is 4.50 m away from the surface of a convex mirror on a nearby wall. The mirror has a focal length of 50.0 cm. Use the mirror equations to determine the characteristics of the image, and its magnification.

59. Determine the speed of light in a solid that has an index of refraction of 1.87.

60. Determine the time it takes for light to travel 35 cm through the water in an aquarium.

61. What is the index of refraction a medium in air if the angle of incidence is 68°, and the angle of refraction is 42°?

62. What is the angle of refraction if the angle of incidence is 55°, and the index of refraction of the medium is 1.92?

63. Light passes from hydrogen into sodium chloride at an angle of incidence of 57°. What is the angle of refraction?

64. Yellow light travels from air into a liquid at an angle of incidence of 35.0° and an angle of refraction of 20.0°. Calculate the wavelength of the yellow light in the liquid if its wavelength in air is 580 nm.

65. The speed of light in a clear plastic is $1.9 \times 10^8$ m/s. A beam of light strikes the plastic at an angle of 24°. At what angle is the beam refracted?

66. A diver is standing on the end of a diving board, looking down into 5.5 m of water. How far does the bottom of the pool appear to be from the water's surface?

67. A block of glass has a critical angle of 46.0°. What is its index of refraction?

68. The critical angle for a special glass in air is 40°. What is the critical angle if the glass is immersed in water?

69. Light is travelling from a diamond into the block of plexiglass in which it is embedded. Determine the critical angle of the light in the diamond.

70. The index of refraction of crown glass is 1.51 for red light, and 1.53 for violet light.
    (a) What is the speed of red light in crown glass?
    (b) What is the speed of violet light in crown glass?

71. Calculate the magnification if an object 12.0 cm tall is placed 25.0 cm from a convex lens with a focal length of 15.0 cm. Describe the image that is formed.

72. The image formed by a concave lens is 24.0 cm tall and 42.0 cm from the lens. Find the size and location of the object if focal points are 60.0 cm from the lens.

**COURSE CHALLENGE**

**Space-Based Power**

Continue to plan for your end-of-course project by considering the following:

- How are the properties and uses of mirrors and lenses related to your project?

- How can you incorporate newly learned skills such as drawing schematic or ray diagrams into your project?

- Analyze the information contained within your research portfolio to identify knowledge or skills gaps that should be filled during the last unit of the course.

# Electricity and Magnetism

## OVERALL EXPECTATIONS

**DEMONSTRATE** an understanding of the principles and laws related to electricity and magnetism.

**INVESTIGATE** properties of magnetic fields.

**DESCRIBE** technologies developed on the basis of the scientific understanding of magnetic fields.

The lightning bolts in the photograph are radiating from a Tesla coil, invented in 1891 by Nikola Tesla (1856–1943), a Croatian-born U.S. inventor. In 1899, Tesla successfully generated lightning bolts more than 40 m in length — the largest ever artificially created. Unfortunately, Tesla's work set the town's electric generators on fire, and his experiments were cut short.

As a university student, Tesla earned the scorn of his teachers and classmates by arguing that the direct current (DC) electrical systems proposed by Thomas Edison's electric company were too inefficient. Tesla proposed that alternating current (AC) systems were the only practical way to transmit electric energy. As you proceed through this unit, you will learn why Tesla was correct. At the time, no one had ever built an AC dynamo (generator) that ran efficiently. With the backing of George Westinghouse, Tesla invented and built an efficient and inexpensive AC dynamo. As a result, Westinghouse was able to undercut Edison's bid to harness Niagara Falls for power, by about 50%. Tesla built the first power station at Niagara Falls and then transmitted electric energy 35 km to Buffalo. The statue of Tesla at Niagara (inset) is a testament to his genius as an inventor and scientist. In this unit, you will explore this and more discoveries that have led to our understanding of magnetism, electricity, and electromagnetism.

### UNIT PROJECT PREP

Look ahead to pages XX-XX. You will design and build a simple electric motor. Begin to think about questions like:

- What is the electromotive force?
- What design modifications will make your motor more efficient?

The massive conductors of a power transmission line carry huge currents at high potentials from power stations to your community. The hair-like lines on the printed circuit board in the photograph are the conductors that carry the microcurrents inside your computer and portable music system. Regardless of size, the conductors serve one purpose—to move electric energy, with minimal loss, to a device that will transform it into a desired form of energy.

To ensure that all communities have the electric energy they need, physicists and electrical engineers must know exactly how the properties of the conductor affect the current inside it. To design the circuit board for a computer, an electronic engineer must know how resistance can be used to control the amount of current to the branches of the circuit, so that each computer component gets the correct amount of current at the correct voltage.

In this chapter, you will develop an understanding of how and why electric circuits behave as they do. You will revisit the relationships among potential difference, current, and resistance. You will learn how to diagram, connect, and analyze complex electric circuits to learn how much energy is available to each of the circuits' loads.

# INVESTIGATION 13-A

## Potential Differences along Current-Carrying Conductors

### TARGET SKILLS

- **Manipulating and recording**
- **Analyzing and interpreting**
- **Communicating results**

Have you ever wondered why birds can sit on bare, high voltage power lines without being electrocuted? After you have completed this lab, you should be able to explain why birds seem unaffected by the current in the wire.

Potential difference (or voltage) is a measure of the electrical effort that is being exerted on a system. If a voltage exists between two points on a conductor then a current will flow between those points.

### Problem

On a current-carrying conductor, how does the potential difference between two points on the conductor vary with length between the points?

### Equipment

- power supply
- metre stick
- Nichrome™ wire (22 gauge)
- insulated conductors with alligator clips
- voltmeter
- thumbtacks

### Procedure

1. Connect the apparatus as shown in the diagram.

2. Attach the leads from the voltmeter to the wire near the opposite ends of the metre stick. Have your teacher confirm that the voltmeter is connected correctly.

3. Increase the voltage from the power supply until the voltmeter reads about 2.5 V.

Record the length of wire between the clips and the voltmeter reading in a data table.

**CAUTION** As long as you touch the alligator clips with only one hand you can move them along the wire with no danger of shock.

4. Without adjusting the power supply, move one alligator clip along the wire about 15 cm closer to the other and record the length between the clips and the voltmeter reading. Repeat until the length between the clips is zero.

5. Plot a graph of the line of best fit for voltage versus length between clips.

### Analyze and Conclude

1. What does the graph tell you about the relationship between voltage and length between clips?

2. In light of your findings, explain why birds are not electrocuted when they sit on bare power lines.

### Apply and Extend

3. How is the voltage affected if you move both clips along the wire keeping them a constant distance (say, 15 cm) apart?

4. Does increasing the voltage affect the nature of the result? Try the experiment with the voltage set at 4.0 V.

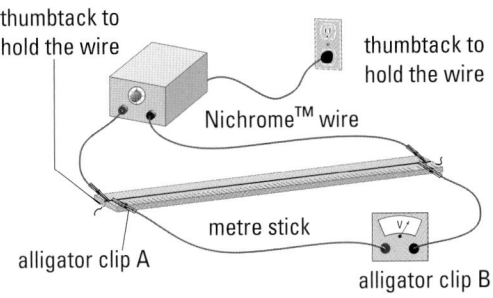

thumbtack to hold the wire

thumbtack to hold the wire

Nichrome™ wire

metre stick

alligator clip A

alligator clip B

SECTION
EXPECTATIONS

• Define and describe electric potential difference.

• Analyze in quantitative terms, problems involving electric potential difference and electric charge.

KEY
TERMS

• conductor
• insulator
• electrostatics
• voltaic cell
• battery
• electrode
• electrolyte

• anode
• cathode
• gravitational potential difference
• potential difference

When you comb your hair with a plastic comb, the comb becomes electrically charged and will attract bits of paper. If your comb is made of metal, however, the bits of paper are unaffected by the comb when it is held close to them. Why do metal combs *not* become charged?

**Figure 13.1** Combing your hair with a plastic comb results in an electrostatic charge on the comb

## Conductors and Insulators

Stephen Gray (1696–1736), an English scientist, made the first recorded explanation of electric conduction in 1729. He classified materials as **conductors** and **insulators**, depending on their ability to allow charges to flow. Although it was a novel idea for him, you probably take for granted that, in general, conductors are metals and insulators are non-metals. Gray also identified Earth as a conductor and gave us the term "ground" to mean "provide a path for charge to escape." Even though Earth is not generally thought of as a metal, it is still a very good conductor, due to its size and the ions dissolved in the moisture in the soil.

## The Voltaic Pile

Gray's discovery marked the first step of the journey from **electrostatics** (the study of charges at rest) to the control of electric current. The second, more crucial, step occurred in 1800, when Italian physicist Alessandro Volta (1745–1827) invented the electrochemical cell. Volta discovered that if he placed a layer of salt-water-soaked paper between disks of two different metals, such as silver and zinc, an electric charge appeared on each of the metal disks. When he made a pile of these cells (for example,

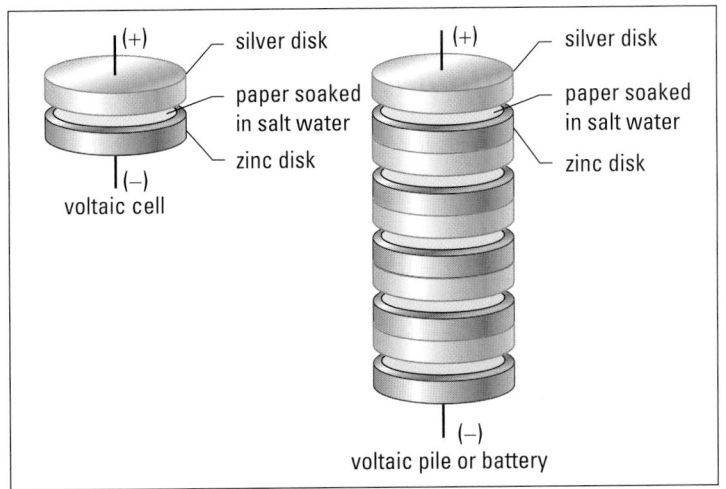

**Figure 13.2** The voltaic pile, or battery, supplied scientists with a source of continuous charge flow. For the first time, scientists were able to experiment with steady currents, rather than with the brief bursts of intense charge flow provided by electrostatic generators.

**Figure 13.3** Alessandro Volta was a professor at the University of Pavia when he invented the electrochemical (voltaic) cell. For his invention, Napoleon made him a Count of the French Empire. The unit of potential difference, the volt, was named in his honour.

silver/paper/zinc/silver/paper/zinc), the electric strength increased. One pair of such disks became known as a **voltaic cell**; the stack of disks became known as a voltaic pile or **battery** (see Figure 13.2). The metal plates in the cell are the **electrodes**, while the solution between them is the **electrolyte**.

Using an electroscope, Volta determined that the charges on the silver disk of his cell were positive, and the charges on the zinc disk were negative. Since the electron had not yet been discovered, physicists had no way of knowing what positive and negative charges actually were, or which type of charge was moving when they connected conductors to the poles of a voltaic pile. They agreed on the convention that positive charges were moving in electric conductors. Consequently, the positive pole of the battery must be considered to be at a higher electric potential energy than the negative pole. The positive pole would be repelling the positive charges and pushing them "downhill" toward the negative pole. Hence, the positive pole of the battery became known as the **anode** (Greek for "upper path") and the negative pole became the **cathode** (Greek for "lower path").

## Potential Difference

If you imagine a model in which a positive charge moving through a circuit is going downhill, then the battery is analogous to a ski lift taking the charge back to the top of the hill. When you ride a ski lift from the bottom to the top of a hill, the lift uses energy from its motor's fuel and transforms that energy into gravitational potential energy of your body. You probably have gained a different amount of gravitational potential energy from other skiers during your ride. However, you all gained exactly the same amount of gravitational potential energy per kilogram of your body mass.

By defining the **gravitational potential difference** as the difference in gravitational potential energy per unit mass, $\Delta E_p/m$, you can develop a term that no longer depends on an object's (skier's) mass. Gravitational potential difference depends only on the height of the hill ($h$) and the acceleration due to gravity ($g$).

**Figure 13.4** A skier of mass $m$ riding up a ski lift to the top of a hill gains a potential energy of $\Delta E_g$.

### Think It Through

- Think about the skier mentioned in the caption of Figure 13.4. Write the equation for gravitational potential energy. Use the equation to show that the gravitational potential difference, $\Delta E_g/m$, depends on the height ($h$) of the hill and the gravitational acceleration ($g$), not the mass of the skier.

- How would the gravitational potential difference change under the following circumstances?

  (a) The skier went three times as far up the hill.

  (b) The skier's mass doubled.

  (c) The skier skied only halfway down the hill.

---

The skiers on our electric hill are similar to positive charges. The chemical action inside a voltaic cell takes positive charges from the cathode (bottom of the electric hill) to the anode (top of the electric hill), giving them electric potential energy. There is no special term or symbol for gravitational potential difference, but there is a special term for electric potential difference. The difference in electrical potential energy ($\Delta E_Q$) per unit charge ($Q$) is defined as the **potential difference** ($V$), sometimes called the "voltage" of the cell, battery, or power supply.

## DEFINITION OF ELECTRIC POTENTIAL DIFFERENCE

The electric potential difference between any two points in a circuit is the quotient of the change in the electric potential energy of charges passing between those points and the quantity of the charge.

$$V = \frac{\Delta E_Q}{Q}$$

| Quantity | Symbol | SI unit |
|---|---|---|
| electric potential difference | $V$ | V (volt) |
| change in electrical potential energy | $\Delta E_Q$ | J (joule) |
| quantity of charge | $Q$ | C (coulomb) |

**Unit Analysis**

$$\frac{(\text{electric potential energy})}{(\text{quantity of charge})} = \frac{J}{C} = V$$

**Note:** One joule per coulomb is equivalent to one volt.

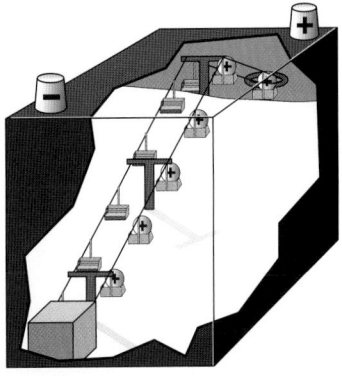

**Figure 13.5** The chemical action of the cell gives electric potential energy to the charges deposited on the anode and cathode. This creates a potential difference between the anode and cathode.

One volt (1 V) of potential difference across the poles of a cell is created when the chemical action inside the cell does one joule (1 J) of work on each coulomb of charge (1 C) that it lifts internally from the cathode to the anode.

### MODEL PROBLEM

### Energy and Potential Difference

**A battery has a potential difference of 18.0 V. How much work is done when a charge of 64.0 C moves from the anode to the cathode?**

### Frame the Problem

- Since the battery has a *potential difference* between the poles, chemical reactions in the battery did *work* to *separate* positive and negative *charges*.

- The *work done* transforms chemical potential energy into *electric potential energy*. Thus,

the electric potential energy is equal to the amount of *work* done.

- The expression that defines *potential difference* applies to this problem.

### Identify the Goal

The amount of work, $W$, done to separate charges

*continued* ▶

*continued from previous page*

## Variables and Constants

| Involved in the problem | Known | Unknown |
|---|---|---|
| $W$ | $V = 18.0$ V | $W$ |
| $\Delta E_Q$ | $Q = 64.0$ C | $\Delta E_Q$ |
| $V$ | | |
| $Q$ | | |

## Strategy

Use the expression for potential difference.

Solve for $\Delta E_Q$.

Since $1\frac{J}{C}$ is equivalent to 1 V, then

$$V = \frac{J}{C}$$

$$VC = \frac{J}{\cancel{C}}\cancel{C}$$

$$VC = J$$

The work done is the same as the potential energy.

## Calculations

$$V = \frac{\Delta E_Q}{Q}$$

**Substitute first**

$$18.0 \text{ V} = \frac{\Delta E_Q}{64.0 \text{ C}}$$

$$(120 \text{ V})(1440 \text{ C}) = \frac{\Delta E_Q}{\cancel{1440\text{C}}}\cancel{1440\text{ C}}$$

$$\Delta E_Q = 1150 \text{ VC}$$

$$\Delta E_Q = 1150 \text{ J}$$

$$W = \Delta E_Q$$
$$\Delta E_Q = 1150 \text{ J}$$
$$W = 1150 \text{ J}$$

**Solve for $\Delta E_Q$ first**

$$VQ = \frac{\Delta E_Q}{\cancel{Q}}\cancel{Q}$$

$$\Delta E_Q = (18.0 \text{ V})(64.0 \text{ C})$$

$$\Delta E_Q = 1150 \text{ VC}$$

$$\Delta E_Q = 1150 \text{ J}$$

If a charge of 64.0 C is transferred by a potential difference of 18.0 V, then $1.15 \times 10^3$ J of work are done.

## Validate

The units cancel to give joules, which is the correct unit for work.

## PRACTICE PROBLEMS

**1.** What is the potential difference of a battery if it does $7.50 \times 10^{-2}$ J of work when it moves $3.75 \times 10^{-3}$ C of charge onto the anode?

**2.** A 9.00 V battery causes a charge of $4.20 \times 10^{-2}$ C to move through a circuit. Calculate the work done on the charge.

**3.** A 12 V battery does 0.75 J of work on a quantity of charge it moved through a circuit. Calculate the amount of charge that was moved.

**TARGET SKILLS**

- **Communicating results**
- **Conducting research**

Sara Goodchild is a science editor, and also a published author of science articles. She graduated from university with a degree in chemistry, and has combined her interests in both writing and science in her present career.

Robotics, global warming, the space station, genetically modified organisms, and cloning — these are just a few of the hot topics being reported in the media. The demand for science writers is increasing as radio, television, magazines, newspapers, encyclopedias, the Internet and even texts such as this one publish more and more reports on science. Science reporters must be able to distinguish between good and bad science and then present their findings in a clear well-written manner so that their point of view can be understood by the public.

Right now is a great time to be a science writer. Many science topics are becoming increasingly controversial. The ability to research stories and present an accurate balanced report will be extremely difficult and ever more important.

If you have an interest in science and a talent for writing, you may have a career in science journalism.

**Going Further**

1. Volunteer with a scientific organization like the Royal Astronomical Society to gain experience in writing and editing.

2. Attend meetings of and/or join professional organizations such as the Science Writers' Association of Canada, Periodical Writers' Association of Canada or the Editors' Association of Canada.

3. Submit reports for your local paper on science events such as the Science Fair.

## 13.1    Section Review

1. **MC** Why was very little known about current electricity and potential difference before the time of Alessandro Volta?

2. **C** Explain the difference between electric potential energy and electric potential difference.

3. **K/U** Which of the following changes would increase the gravitational potential energy of every skier at the top of a chair lift compared to the bottom of the lift?

   (a) Increase the number of runs to accommodate more skiers.

   (b) Extend the top of the lift to a location 20 m higher up the mountain.

   (c) Install a new high-speed quad lift to carry more skiers to the top of the lift at a higher rate.

   Explain your reasoning.

4. **C** Develop another analogy, different from the ski lift, that would help a classmate understand the concept of electric potential difference.

- Define and describe electric current.

- Describe two conventions used to denote the direction of movement of electric charge.

- Use a circuit diagram to model and quantitatively predict the movement of elementary charge.

- current
- electron flow
- elementary charge
- open circuit
- closed circuit
- loads
- power supply
- circuit elements
- ammeter
- voltmeter
- series
- parallel

Volta's invention of the battery provided other scientists with a source of constant electric current for the first time. As a result, many other discoveries relating to current electricity followed quickly. Less than 25 years after Volta published his findings, scientists such as Ohm, Oersted, and Ampère published the results of their experiments, opening the door to the age of electricity.

## Electric Current

To develop an understanding of the flow of electric charge, you can compare it to the flow of water. If you were asked to describe the flow of water over Niagara Falls, you might give your answer in litres per second or cubic metres per second. In an electric conductor, **current** ($I$) is described as a quantity of charge ($Q$) passing a given point during an interval of time ($\Delta t$).

---

**ELECTRIC CURRENT**

Electric current is the quotient of the quantity of charge that moves past a point and the time interval during which the charge is moving.

$$I = \frac{Q}{\Delta t}$$

| Quantity | Symbol | SI unit |
|---|---|---|
| current | $I$ | A (ampere) |
| amount of charge | $Q$ | C (coulomb) |
| time interval | $\Delta t$ | s (second) |

**Unit Analysis**

$$\frac{\text{coulomb}}{\text{second}} = \frac{\text{C}}{\text{s}} = \text{A}$$

**Note:** One coulomb per second is equivalent to one ampere.

---

**Figure 13.6** In Niagara Falls, Ontario, the rate of water flow over the Canadian (Horseshoe) Falls is approximately $2.25 \times 10^6$ L/s.

## Electric Current and Charge

The electrical system in your home operates at a potential difference of 120.0 volts. A toaster draws 9.60 A for a period of 2.50 min to toast two slices of bread.

(a) Find the amount of charge that passed through the toaster.

(b) Find the amount of energy the toaster converted into heat (and light) while it toasted the bread.

### Frame the Problem

- Power lines transport *electric energy* to your home and provide a constant *potential difference*.

- When the toaster is connected to the power source and turned on, the *potential difference* drives a *current* through the toaster elements.

- As *charges* pass through the element, *electric energy* is converted into *heat*.

- The amount of energy that was converted into heat is the same as the *change* in the *potential energy* of the *charges* as they pass through the toaster.

### Identify the Goal

The amount of charge, $Q$, that passes through the toaster elements in a given time

The amount of energy, $\Delta E_Q$, converted into heat (and light)

### Variables and Constants

| Involved in the problem | | Known | Unknown |
|---|---|---|---|
| $V$ | $Q$ | $V = 120.0$ V | $Q$ |
| $I$ | $\Delta E_Q$ | $I = 9.60$ A | $\Delta E_Q$ |
| $\Delta t$ | | $\Delta t = 2.50$ min | |

### Strategy

Use the definition for current to find the amount of charge.

Convert time to SI units.

### Calculations

$$I = \frac{Q}{\Delta t}$$

$$2.5 \, \cancel{\text{min}} \frac{60 \text{ s}}{\cancel{\text{min}}} = 150 \text{ s}$$

**Substitute first**

$$9.60 \text{ A} = \frac{Q}{150 \text{ s}}$$

$$(9.60 \text{ A})(150 \text{ s}) = \frac{Q}{\cancel{150 \text{ s}}} \cancel{150 \text{ s}}$$

1 A · s is equivalent to 1 C.

$$Q = 1440 \text{ A} \cdot \text{s}$$

$$Q = 1440 \text{ C}$$

**Solve for $Q$ first**

$$(I)(\Delta t) = \frac{Q}{\cancel{\Delta t}} \cancel{\Delta t}$$

$$Q = (9.60 \text{ A})(150 \text{ s})$$

$$Q = 1440 \text{ A} \cdot \text{s}$$

$$Q = 1440 \text{ C}$$

*continued* ▶

*continued from previous page*

**(a)** In 2.5 min, $1.44 \times 10^3$ C of charge pass through the toaster.

## Strategy

Find the change in potential energy of the charges by using the definition of potential difference.

A V · C is equivalent to a J.

## Calculations

$$V = \frac{\Delta E_Q}{Q}$$

**Substitute first**

$$120 \text{ V} = \frac{\Delta E_Q}{1440 \text{ C}}$$

$$(120 \text{ V})(1440 \text{ C}) = \frac{\Delta E_Q}{1440 \text{ C}} \cdot 1440 \text{ C}$$

$$\Delta E_Q = 1.73 \times 10^5 \text{ V} \cdot \text{C}$$

$$\Delta E_Q = 1.73 \times 10^5 \text{ J}$$

**Solve for $\Delta E_q$ first**

$$VQ = \frac{\Delta E_Q}{Q} \cdot Q$$

$$\Delta E_Q = (120 \text{ V})(1440 \text{ C})$$

$$\Delta E_Q = 1.73 \times 10^5 \text{ V} \cdot \text{C}$$

$$\Delta E_Q = 1.73 \times 10^5 \text{ J}$$

The toaster converted $1.73 \times 10^5$ J of electric energy into heat and light while it toasted the bread.

## Validate

The units combined to give joules, which is correct for energy. Also, appliances that generate heat, such as a toaster, typically draw a larger current and consume more energy than devices that generate light, such as a light bulb.

### PRACTICE PROBLEMS

**4.** A battery sends a 2.25 A current through a circuit for 1.50 min. If a total of $8.10 \times 10^2$ J of work was done by the current, what was the potential difference of the battery?

**5.** How long would it take a 17 V battery, sending a 5.0 A current through a circuit, to do 680 J of work?

**6.** How much work is done by a 25.0 V battery when it drives a 4.70 A current through a circuit for 36.0 s?

**7.** If a 160 V battery did $9.6 \times 10^5$ J of work in 2 min, what was the current?

**8.** A light draws a current of 0.48 A. How long must it be left on for charge of 36 C to pass through it?

**9.** An electric circuit draws 20 A. If the electric potential drop over the entire circuit is 120 V, calculate the total charge passing through the circuit in 1 h.

**10.** A celluar phone battery is recharged in 0.25 h after receiving $2.5 \times 10^3$ C of charge. Calculate the amount of electric current that the battery draws during recharging?

**11.** A physics student wishes to determine the amount of electric energy consumed in one day at his school as a result of classroom and hallway lighting. A quick survey revealed that there were approximately 200 40W fluorescent lights operating under a potential difference of 240 V for 16 hours each day.

# Current versus Electron Flow

Although physicists began to study and use electric current around 1800, it was not until 1876 that an experiment at Harvard University showed that negative charges were moving in current-carrying conductors. It was another 25 years before J.J. Thomson (1856–1940) discovered the electron, and experiments demonstrated that the moving negative charges were electrons. By this time, the concept of a positive current was entrenched in scientific theory and literature. Fortunately, as long as you use a constant frame of reference, circuit analysis does not depend on knowing whether it is actually positive or negative charges that are moving. All measurable effects, such as the amount of energy transformed, are the same whether positive charges move one way or negative charges move the other way. Today, the term current ($I$) means the flow of positive charge (from anode to cathode) in a circuit. The flow of negative charge (from cathode to anode) is called **electron flow**. Since a wealth of theory was developed using positive current, the convention for analyzing circuits is still to use positive or conventional current.

Not all charges that move do so inside metals. In other media, either negative or positive (or both) charges can move. The aurora borealis lights up the sky when high-energy electrons from the sun collide with gas molecules in the air and are captured by Earth's magnetic field (see Figure 13.7).

In the process of electroplating with an aqueous salt solution such as silver cyanide (Figure 13.8), the positive silver ions ($Ag^+$) are attracted to the negative electrode, and the negative cyanide ions ($CN^-$) are attracted to the positive electrode.

**Figure 13.7** The aurora borealis

## COURSE CHALLENGE

### Free Energy?

Investigate the efficiency of photovoltaic cells using a small electric toy and photovoltaic cells from a local electronics shop. Learn more from the **Science Resources** section of the following web site: *www.school.mcgrawhill.ca/ resources/*

**Figure 13.8** A less expensive metal can be silver-plated to produce an attractive and corrosion-resistant surface. The object to be plated is connected to a circuit as the cathode. It is suspended in a solution containing silver cyanide. The silver ions are attracted to the cathode, where they combine with electrons and become solid silver atoms that remain permanently attached to the surface of the cathode.

**PHYSICS FILE**

During the last 20 years of the 1800s, physicists discovered that light had the ability to cause certain metals to emit negative charges. By 1905, Albert Einstein had created an hypothesis for the cause of, and formulated a law for, the photoelectric effect. In 1916, Robert Andrews Millikan carried out very careful and precise experiments in which he confirmed Einstein's predictions. In 1921, Einstein was awarded the Nobel Prize for "services to Theoretical Physics and the discovery of the law of the photoelectric effect."

# Current and the Elementary Charge

Robert Andrews Millikan (1868–1953), a U.S. physicist, won the Nobel Prize in Physics in 1923 for his discovery of the elementary charge and for his research on the photoelectric effect. In 1917, his "oil-drop experiment" revealed that the static charge on a microscopic oil drop was always a whole-number (integral) multiple of a minute electric charge that was fixed in size. He concluded that the minute charge was the smallest size in which electric charge could be found. He designated this minute amount of charge the **elementary charge** ($e$). His measurements revealed that the size of one elementary charge is $e = 1.60 \times 10^{-19}$ C. (The most precise measurement to date is $e = 1.602\ 177\ 33 \times 10^{-19}$ C.) Today, one elementary charge is known to be the magnitude of the charge on a proton (+1 $e$) or an electron (−1 $e$).

When J.J. Thomson (see Figure 13.9) discovered the electron in 1897, he was able to measure only the ratio of the charge to the mass. Many scientists were sceptical about Thomson's proposed charge-carrying particle. They still thought that electric charge might be a fluid that could be divided into infinitely small pieces. However, when Millikan performed his oil-drop experiment in 1917, he established that when charge moved, it moved only as integral (whole-number) multiples of the elementary charge ($e$), just as water must be moved by at least one molecule at a time. He confirmed Thomson's hypothesis. Scientists now know that every quantity of charge can be expressed as an integral number of elementary charges.

**Figure 13.9** J.J. Thomson devised ingenious experiments showing that the mysterious "rays" that caused phosphorus to glow, were in fact, tiny identical particles — he had discovered the electron. Most televisions still use this technology.

---

**ELEMENTARY CHARGE**

The amount of charge is the product of the number of elementary charges (electrons or protons) and the magnitude of the elementary charge.

$$Q = Ne$$

| Quantity | Symbol | SI unit |
|---|---|---|
| amount of charge | $Q$ | C (coulomb) |
| number of elementary charges | $N$ | integer (pure number, no unit) |
| elementary charge | $e$ | C (coulomb) |

---

## Charge and Electrons

**A light bulb draws a current of 0.60 A. If the bulb is left on for 8.0 min, how many electrons (elementary charges) pass through the bulb?**

### Frame the Problem

- When a *current* exists in a light bulb, *electrons* are passing through it.

- If you know the amount of charge that passes through the light bulb, you can use the magnitude of the *elementary charge* to find the number of electrons.

### Identify the Goal

The number ($N$) of electrons passing through the bulb

### Variables and Constants

| Involved in the problem | Known | Implied | Unknown |
|---|---|---|---|
| $I$ | $I = 0.60$ A | $e = 1.60 \times 10^{-19}$ C | $Q$ |
| $\Delta t$ | $\Delta t = 8.0$ min | | $N$ |
| $N$ | | | |
| $e$ | | | |
| $Q$ | | | |

### Strategy

Use the definition of current to find the amount of charge passing through the light bulb in 8.0 min.

First, convert time to SI units.

1 A · s is equivalent to 1 C.

### Calculations

$$I = \frac{Q}{\Delta t}$$

$$8.0 \ \cancel{\text{min}} \ \frac{60 \text{ s}}{\cancel{\text{min}}} = 480 \text{ s}$$

**Substitute first**

$$0.60 \text{ A} = \frac{Q}{480 \text{ s}}$$

$$(0.60 \text{ A})(480 \text{ s}) = \frac{Q}{\cancel{480 \text{ s}}} \cancel{480 \text{ s}}$$

$$Q = 288 \text{ A} \cdot \text{s}$$

$$Q = 288 \text{ C}$$

**Solve for Q first**

$$(I)(\Delta t) = \frac{Q}{\cancel{\Delta t}} \cancel{\Delta t}$$

$$Q = (0.60 \text{ A})(480 \text{ s})$$

$$Q = 288 \text{ A} \cdot \text{s}$$

$$Q = 288 \text{ C}$$

*continued* ▶

## Strategy

Use the relationship between amount of charge and the elementary charge to find the number of electrons.

## Calculations

$$Q = Ne$$

**Substitute first**

$$288 \text{ C} = N \, 1.60 \times 10^{-19} \text{ C}$$

$$\frac{288 \text{ C}}{1.60 \times 10^{-19} \text{ C}} = \frac{N \, 1.60 \times 10^{-19} \text{ C}}{1.60 \times 10^{-19} \text{ C}}$$

$$N = 1.80 \times 10^{21}$$

**Solve for *N* first**

$$\frac{Q}{e} = \frac{N \, e}{e}$$

$$N = \frac{288 \text{ C}}{1.60 \times 10^{-19} \text{ C}}$$

$$N = 1.80 \times 10^{21}$$

In the 8.0 min that the light bulb was on, $1.8 \times 10^{21}$ electrons (elementary charges) passed through it.

## Validate

In the first part, the units combine to give coulombs, which is correct for charge. In the second part, the units cancel to give a pure number. This is correct, because there are no units for number of electrons. The answer is extremely large, which you would expect because the number of electrons in one coulomb is exceedingly large: $N = \frac{1 \text{ C}}{1.60 \times 10^{-19} \text{ C}}$ or $6.26 \times 10^{18}$ electrons.

### PRACTICE PROBLEMS

**8.** Calculate the current if $2.85 \times 10^{20}$ elementary charges pass a point in a circuit in 5.70 min.

**9.** A 16.0 V battery does $5.40 \times 10^4$ J of work in 360.0 s.

   **(a)** Calculate the current through the battery.

   **(b)** Calculate the number of elementary charges that pass through the battery.

**10.** Calculate the number of elementary charges that pass a point in a circuit when a current of 3.50 A flows for 24.0 s.

**11.** In transferring $2.5 \times 10^{20}$ elementary charges in 12 s, a battery does 68 J of work.

   **(a)** Calculate the current through the battery.

   **(b)** Calculate the potential difference of the battery.

# Electric Circuits

Suppose a power supply (battery) is connected to a load such as a light bulb. A switch allows you to open and close the circuit. An **open circuit** means there is a break (perhaps an open switch) somewhere in the circuit that prevents current from flowing. A **closed circuit** means that all connections are complete. A closed, or continuous, path exists, allowing current to move around the circuit. You could represent the above circuit by using realistic drawings of the apparatus involved, as shown in Figure 13.10. That technique would be very cumbersome, however. It is much more efficient to represent and analyze electric circuits by using the electric-circuit symbols shown in Figure 13.11. The circuit shown in Figure 13.10 is redrawn in Figure 13.12, using these symbols.

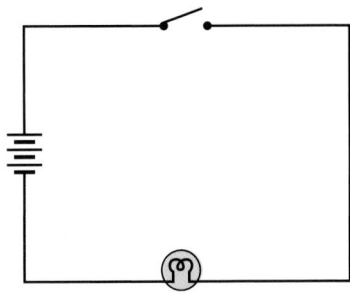

**Figure 13.12** This diagram of the same circuit is easier to draw and to analyze.

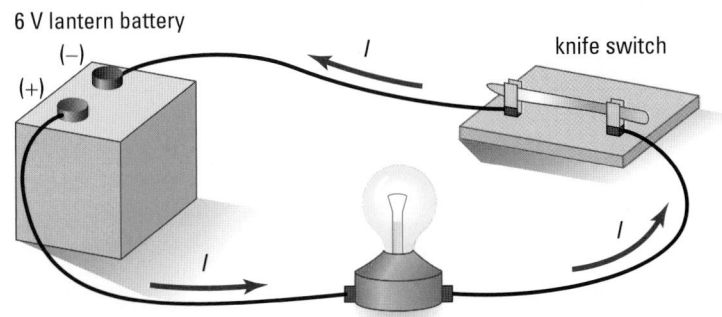

**Figure 13.10** A realistic sketch of even a simple circuit is cumbersome.

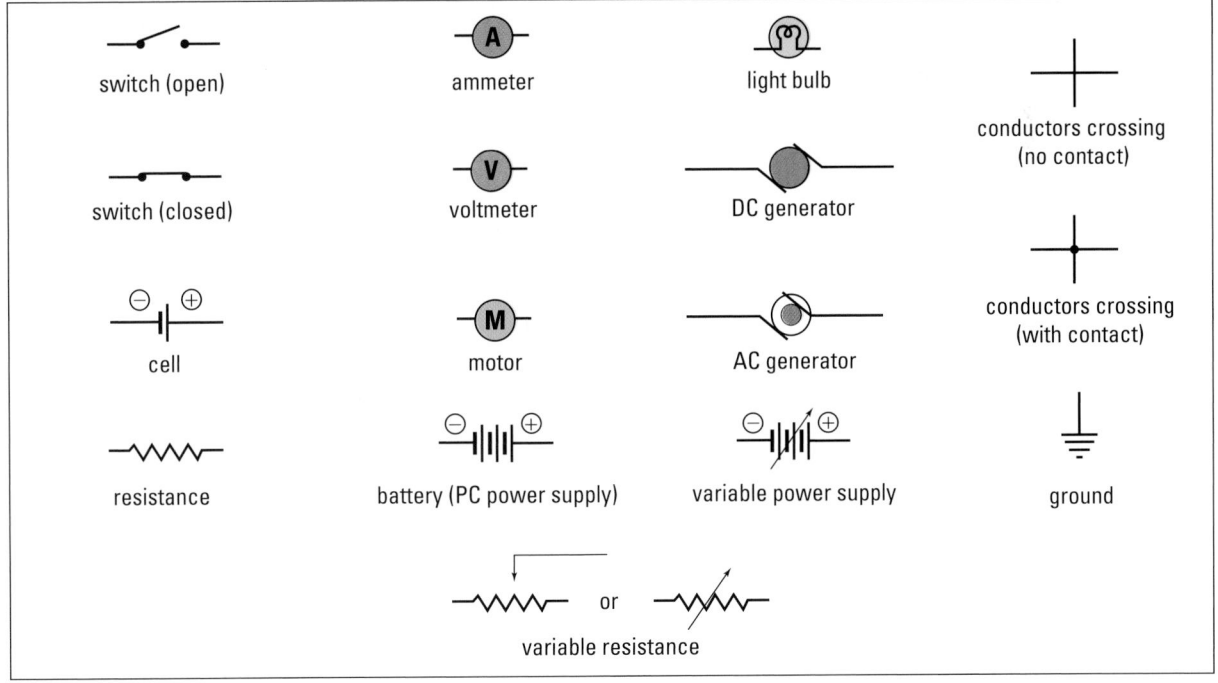

**Figure 13.11** Symbols for elements of an electric circuit

- In the circuit symbol for a battery, the longer line represents the positive pole of the battery and the shorter line is the negative pole. In one of the circuits shown here, the arrows represent conventional current. In another, the arrows represent electron flow. One circuit is drawn incorrectly. Neither conventional current nor electron flow could take the directions indicated by the arrows. Analyze the circuits and determine which illustrates conventional current, electron flow, and neither. Explain your reasoning.

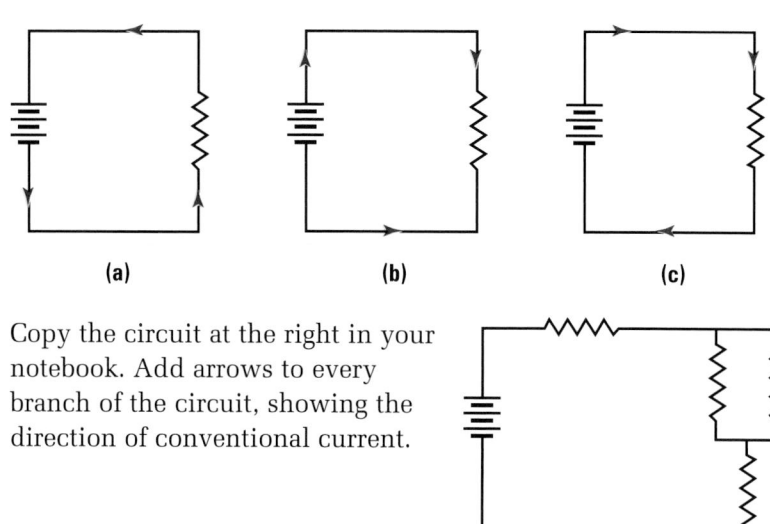

(a)          (b)          (c)

- Copy the circuit at the right in your notebook. Add arrows to every branch of the circuit, showing the direction of conventional current.

## Ammeters and Voltmeters

To find out what is happening inside the parts of a circuit, scientists use an assortment of devices, such as ammeters, voltmeters, galvanometers, and ohmmeters. A simple circuit is composed of **loads** (for example, light bulbs, resistances, motors) and a **power supply** (cell, battery, or an AC or DC generator). These **circuit elements** (loads and power source) may be connected in series or in parallel to each other. A switch is often included but serves only to open or close the circuit. When meters are used to measure current of potential difference, they are connected in a way that will not interfere with the circuit operation. An **ammeter** measures the electric current to or from a circuit element. A **voltmeter** measures the electric potential difference across a circuit element.

Since ammeters measure the current through a circuit element, they must be inserted into the line before or after the circuit element so that all of the current passing through the circuit element also goes through the ammeter. This is called a **series** connection since the current moves through the circuit element and the ammeter one after the other. On the other hand, a voltmeter measures the potential difference from one side of a circuit element to the other. To function properly, voltmeters must be connected to the opposite sides of the circuit element across which you want to know the potential difference. This is a called a **parallel** connection, since the voltmeter presents a path that runs beside the circuit element. Notice that the ammeter is actually part of the circuit. If you disconnect either pole of the ammeter, the circuit is opened. The voltmeter, on the other hand, makes contact with the circuit at two points to measure the potential difference between those points.If you disconnect either pole of the voltmeter, the circuit is still perfectly functional. Figure 13.13 shows the same circuit as in Figure 13.10, with the addition of an ammeter and a voltmeter showing the proper connection.

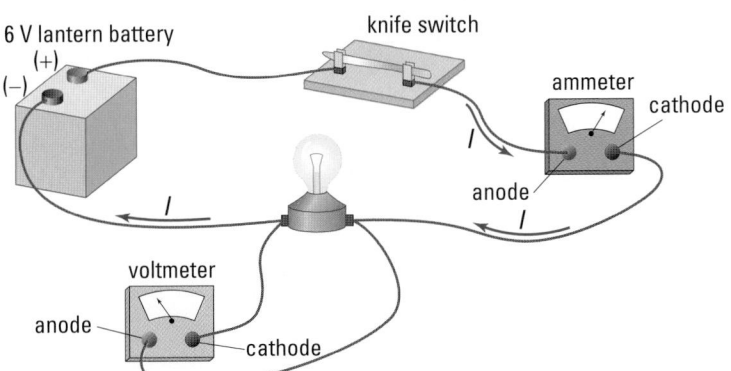

**Figure 13.13** Notice the labels indicating the anodes and cathodes of the meters relative to the anode and cathode of the power supply and the direction of the current. How would you connect a voltmeter to measure the potential difference of the battery?

---

## 13.2    Section Review

1. **MC** Why do physicists define current, (*I*), in a circuit to be opposite to the direction in which electrons flow?

2. **K/U** Give an example of a current in which positive charges move.

3. **C** When Millikan measured the amount of charge on oil drops, he obtained data similar to the following. Explain how he used such data to determine (a) that elementary

charges existed, and (b) the size of the elementary charge.

Data: $6.4 \times 10^{-19}$ C, $1.28 \times 10^{-18}$ C, $1.92 \times 10^{-18}$ C, $8.0 \times 10^{-19}$ C, $1.6 \times 10^{-18}$ C, $6.4 \times 10^{-19}$ C, $1.12 \times 10^{-18}$ C.

4. **C** Explain how a voltmeter must be connected in a circuit in order to measure the potential difference across a light bulb.

- Describe the concepts and units related to electricity and magnetism in terms of electron flow.

- Design and conduct an experiment to investigate major variables relating electric potential, current, and resistance.

KEY
TERMS

- resistivity

- unit of resistance (ohm)

- non-linear or non-ohmic resistance

Everything that moves experiences frictional forces that resist that motion. The energy used to overcome frictional resistance within mechanical and electrical systems, such as automobile engines and electric motors, costs billions of dollars each year. Like every other system, the frictional effects in conductors offer resistance to the current passing through them. Similar to mechanical systems, friction in electric conductors produces thermal energy that is radiated away from the conductor. This energy is no longer available to do work. Electrical engineers' knowledge of the factors that affect resistance enable them to design systems that reduce energy loss to a minimum.

**Figure 13.14** Overcoming resistance is an important aspect of electric-circuit design.

## Factors Affecting the Resistance of a Conductor

You can compare a metal conductor carrying electric current to a water pipe carrying water current. For a water pipe of a fixed diameter, the longer the pipe, the greater the drag it exerts on the water passing through it. Similarly, for an electric conductor with a fixed diameter, resistance increases proportionately with the *length* ($R \propto L$). Thus, a 2 m length of a particular conductor has twice as much resistance as a 1 m length of the same conductor.

For two conductors of equal length, changing the cross-sectional area changes the resistance. Again, for a water pipe, the bigger the cross-sectional area of the pipe, the lower the drag on the water moving inside it. It is the cross-sectional area that provides the space in which the current travels. Therefore, doubling the cross-sectional area doubles the space for the current to move and halves the resistance. For electric conductors, resistance varies inversely as the *cross-sectional area* ($R \propto 1/A$). For very long extension cords, the resistance due to the increased length of the cord can cause a significant energy loss. To lower the resistance, long conductors are made of thicker wire, which increases cross-sectional area and thus reduces resistance. Conversely, light bulb filaments must have a large resistance so that the energy will be transformed into light and thermal energy. To increase resistance, filaments are made very short and very thin.

### Think It Through

- Consider the different electric cords typically found around the home, such as the cords on an iron, lamp, television set, small space heater, and toaster; a standard extension cord; and the cords for plugging in a vacuum cleaner or a car's block heater. Think about the length and thickness of each cord. Explain why each cord has its own specific size. For example, why is a toaster cord shorter and thicker than a lamp cord?

- Often, electric power generating stations are many kilometres from the communities that they serve. The conductors that carry electric energy over many kilometres have a certain amount of electric resistance for every metre of line. You can calculate the amount of power lost to the resistance and the consequent heating of the lines by using the equation $P = I^2R$. Explain how power companies keep their losses of power to a minimum.

If you combine the relationship of the resistance of a conductor to its length and cross-sectional area, the result is $R \propto \frac{L}{A}$. Any proportionality can be written as an equality if a proportionality constant is included. In the case of resistance, the symbol used for this proportionality constant is the Greek letter *rho* ($\rho$). The equation for the resistance of a conductor can now be written: $R = \rho\frac{L}{A}$. The value of the proportionality constant ($\rho$) is called the **resistivity** and is a property of the material from which the conductor is made.

Diameters/Resistances of Some Gauges of Copper Wire

| Gauge | Diameter (mm) | Resistance ($\times 10^{-3}\Omega$/m) |
|---|---|---|
| 0 | 9.35 | 0.31 |
| 10 | 2.59 | 2.20 |
| 14 | 1.63 | 8.54 |
| 18 | 1.02 | 21.90 |
| 22 | 0.64 | 51.70 |

## RESISTANCE OF A CONDUCTOR

The resistance of a conductor is the product of the resistivity and the length divided by the cross-sectional area.

$$R = \rho \frac{L}{A}$$

| Quantity | Symbol | SI unit |
|---|---|---|
| resistance | $R$ | $\Omega$ (ohm) |
| resistivity | $\rho$ | $\Omega \cdot$ m (ohm metres) |
| length of conductor | $L$ | m (metres) |
| cross-sectional area | $A$ | $m^2$ (square metres) |

**Unit Analysis**

(ohm metres) $\dfrac{\text{metres}}{\text{square metres}} = \Omega \cdot \text{m} \frac{\text{m}}{\text{m}^2} = \Omega$

The resistance of a conductor with a particular length and cross-sectional area depends on the *material* from which it is made. At room temperature, copper is one of the best conducting (lowest resistance) metals. Table 13.1 includes resistivity values for carbon and germanium, which are semiconductors, and for glass, which is an insulator. Insulators are sometimes thought of as conductors with extremely high resistances. By examining Table 13.1, you can see that glass has about $10^{18}$ to $10^{22}$ times the resistance of copper.

**Table 13.1** Resistivity of Some Conductor Materials

| Material | *Resistivity, $\rho$ ($\Omega \cdot$ m) |
|---|---|
| silver | $1.6 \times 10^{-8}$ |
| copper | $1.7 \times 10^{-8}$ |
| aluminum | $2.7 \times 10^{-8}$ |
| tungsten | $5.6 \times 10^{-8}$ |
| Nichrome™ | $100 \times 10^{-8}$ |
| carbon | $3500 \times 10^{-8}$ |
| germanium | $0.46$ |
| glass | $10^{10}$ to $10^{14}$ |

*Values given for a temperature of 20°C

Finally, the *temperature* of the conductor affects the resistance. The electrons that move inside a metallic conductor are the electrons from the outermost orbit of the atoms of the metal. Thus, they are the electrons that are most loosely held by the atoms of the metal. These outermost electrons of good conductors can move quite freely within the metal, behaving much like the molecules of a gas. As you heat the metal, these electrons begin to move more randomly at higher speeds inside the metal. As a result, it is more difficult to organize them into a current. Near 20°C, copper increases its resistance by about 0.39% for each degree of temperature increase. Conversely, lowering the temperature reduces the resistance.

## Using Resistivity

**Calculate the resistance of a 15 m length of copper wire, at 20°C, that has a diameter of 0.050 cm.**

## Frame the Problem

- The *electric resistance* of a conductor depends on its *length*, *cross-sectional area*, the *resistivity* of the conducting material, and the *temperature*.

- These variables are related by the equation for resistance of a conductor.

- The *resistivity* of copper at 20°C is listed in Table 13.1.

## Identify the Goal

Resistance, $R$, of the copper conductor

## Variables and Constants

| Involved in the problem | | Known | Implied | Unknown |
|---|---|---|---|---|
| $R$ | $A$ | $d = 0.050$ cm | $\rho = 1.7 \times 10^{-8}\ \Omega \cdot m$ | $R$ |
| $L$ | $\rho$ | $L = 15$ m | | $A$ |
| $d$ (diameter) | | | | |

## Strategy

Use the equation relating resistance to resistivity and dimensions of the conductor.

Convert diameter to SI units. (All others are in SI units.)

Find the cross-sectional area from the diameter.

## Calculations

$$R = \rho \frac{L}{A}$$

$$0.050\,\cancel{cm}\ \frac{m}{100\,\cancel{cm}} = 5.0 \times 10^{-4}\ m$$

$$A = \pi r^2$$

$$r = \frac{d}{2}$$

$$r = \frac{5.0 \times 10^{-4}\ m}{2} = 2.5 \times 10^{-4}\ m$$

$$A = \pi(2.5 \times 10^{-4}\ m)^2$$

$$A = 1.96 \times 10^{-7}\ m^2$$

*continued* ▶

*continued from previous page*

## Strategy

The values are all known, so substitute into the equation for resistance.

## Calculations

$$R = (1.7 \times 10^{-8} \, \Omega \cdot m) \frac{15 \, m}{1.96 \times 10^{-7} \, m^2}$$

$$R = 1.3 \, \frac{\Omega \cdot \cancel{m} \cdot \cancel{m}}{\cancel{m^2}}$$

$$R = 1.3 \, \Omega$$

The conductor has a resistance of 1.3 $\Omega$.

## Validate

The units cancel to give ohms, which is correct for resistance. At first glance, a resistance of 1.3 $\Omega$ seems large for a 15 m length of copper given that copper is a very good conductor. However, the wire is very fine, only 0.5 mm in diameter, giving a cross-sectional area of only $1.96 \times 10^{-7}$ m$^2$. This small area accounts for the resistance.

### PRACTICE PROBLEMS

Use the data provided in Table 13.1 and the table in the Physics File on page 624 to solve the following problems.

**12.** What is the resistance of 250 m of aluminum wire that has a diameter of 2.0 mm?

**13.** What is the length of 18 gauge Nichrome™ wire that has a resistance of 5.00 $\Omega$?

**14.** The resistance of a 100 W ($1.00 \times 10^2$ W) light bulb is 144 $\Omega$. If its tungsten filament is 2.0 cm long, what is its radius? The bulb is not turned on.

**15.** An extension cord is made of 14 gauge aluminum wire. Calculate the resistance of this cord if its length is 35 m.

**16.** A square carbon rod is 24 m long. If its resistance is 140 $\Omega$, what is its width?

**TARGET SKILLS**

- Hypothesizing
- Performing and recording
- Analyzing and interpreting
- Communicating results

If you change one property in a circuit, such as the potential difference or the resistance, you would expect that another characteristic would change in response. Does it? If so, how does it change?

## Problems

- How is current affected by a change in the potential difference if the resistance remains constant?

- How is current affected by a change in the resistance if the potential difference remains constant?

## Hypotheses

Formulate hypotheses for the relationships between potential difference and current in a circuit, and between resistance and current in a circuit.

## Equipment

- variable DC power supply
- multi-range ammeter
- multi-range voltmeter
- Nichrome⋺ wire (1 m, approximately 22 gauge)
- metre stick
- insulated connecting leads with alligator clips
- thumbtacks

**CAUTION** If the current in the Nichrome⋺ wire is large, the wire will become very hot. When your circuit has more than 1 A of current, you could be burned if you touch the wire.

**CAUTION** When wiring your circuit, be sure to connect the conductor to the anode last. Making this connection last will ensure that you do not accidentally create a live circuit while you are making other connections.

**CAUTION** DC ammeters and DC voltmeters must be connected properly to avoid damaging them and to ensure that they make proper measurements for the desired parts of the circuit. To avoid damaging the meters, make sure that they are connected in the correct direction. Like DC power supplies, DC meters have an anode (red or positive post) and a cathode (black or negative post). The meter must always be connected so that the current is moving from the red post to the black post (downhill) as it passes through the meter.

## Procedure
### Part 1
#### Connecting the Circuit

1. Study the figure to see how all connections will be made. Then, follow the order of making connections in the following steps.

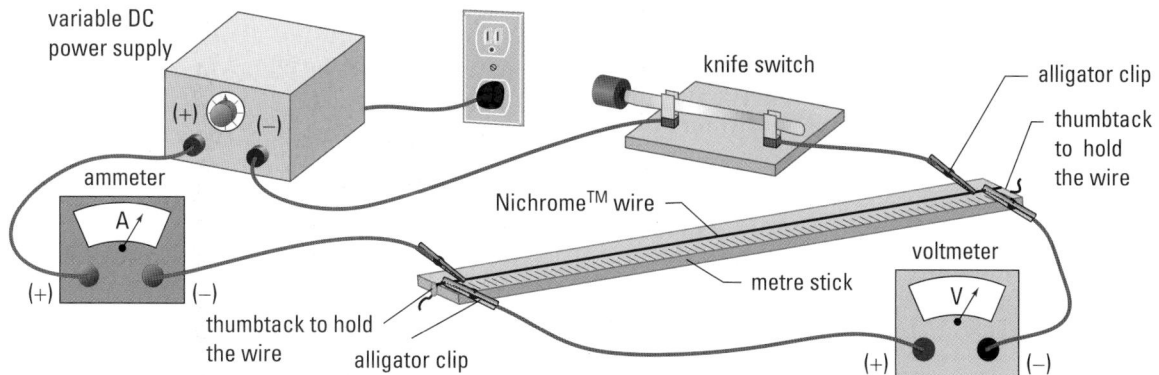

*continued* ▶

*continued from previous page*

2. Connect the cathode (black post) of the power supply to one end of the load. (In the first part of the investigation, the load will be about 50 cm of the Nichromeɘ wire.)

3. Connect the other end of the load to the black post of the ammeter.

4. Connect the red post of the ammeter to the anode of the power supply. Do *not* connect the voltmeter until these connections are complete and have been checked.

5. Before turning on the power supply, be sure that the knob is turned completely down (counterclockwise). Turn on the power supply and increase the potential difference very slightly. Check the ammeter to see if the current is in the correct direction. If everything is correct, turn off the power supply or disconnect the wire at the anode.

6. Connect the voltmeter across the load (power supply). Be sure that the red post of the voltmeter is connected to the end of the load that is closest to the anode (red post) of the power supply. Reconnect the power supply and increase the potential difference of the power supply slightly. Check that the needle of the voltmeter is moving in the correct direction.

## Part 2
### Current versus Potential Difference

7. Choose about 50 cm of the Nichromeɘ wire as the load. Complete the circuit as described above, checking to see that the ammeter and the voltmeter are connected properly. If the voltmeter and the ammeter are multi-range meters, be sure to start on the least-sensitive range (the highest possible voltage or current readings) and move to more-sensitive ranges as conditions permit. If you are unsure of how to read the meter, consult your teacher.

8. Make a data table with the column headings: Trial, Potential Difference ($V$), Current ($I$).

9. With all connections in place, set the power supply at a low value. Read and record the current through, and the potential difference across, the load.

10. Keep the resistance constant; that is, do not move the alligator clips on the Nichromeɘ wire. Increase the potential difference of the power supply slightly and, again, read the current and potential difference.

11. Repeat step 10 several times, until you have five or six readings.

## Part 3
### Current versus Resistance

**CAUTION** Remember that a large current will cause excessive heating.

12. Use the same circuit connections as in Part 1.

13. Choose a length of Nichromeɘ wire (about 12 to 15 cm) as your standard resistance. The resistance of this length of wire will be considered one unit of resistance. (**Note:** Since you have created this standard resistance, you get to name the unit.)

14. Make a data table with the column headings: Trial, Resistance (in the units you gave it), Current. Leave two more columns for data to be used when you make your analysis.

15. Start with one unit of your standard resistance as a load. Increase the potential difference of the power supply until the ammeter registers a current of about 0.5 A. Record this potential difference. In your table, record the current and resistance as Trial 1.

**16.** Set the resistance of the load to be one unit larger than the previous trial ($R \propto L$). Check the potential difference across the load and reset it to the original value. Read and record the new values of current and resistance.

**17.** Repeat step 16 until you have four or five separate trials.

## Analyze and Conclude

**1.** Graph your data for current ($I$) versus potential difference ($V$). Plot $I$ as a function of $V$.

**2.** What does the shape of the line on your graph indicate about the relationship between potential difference and current when resistance is constant? How do these results compare with your hypothesis?

**3.** Write a summary statement describing your conclusion about the relationship of current to potential difference.

**4.** Graph your data for current ($I$) versus resistance ($R$). Plot $I$ as a function of $R$.

**5.** What does the shape of the graph suggest is the probable relationship between the variables, $R$ and $I$?

**6.** Based on your interpretation of the graph, decide how to mathematically modify the variable $R$ so you could plot the adjusted value as a function of $I$ and obtain a straight line. (Hint: Mathematical modifications could be such things as squaring, taking the square root, inverting, or multiplying by a constant or by the other variable.)

**7.** Calculate values for the adjusted independent variable. In one of the empty columns, write your calculated values.

**8.** Graph the current as a function of the newly created variable.

**9.** Was your decision in question 6 correct? If not, continue this process until you have found a modified form of $R$ that gives a straight line.

**10.** Was your original hypothesis about the relationship between current and resistance correct? If not, what is the correct relationship?

**11.** Write a summary statement about the relationship between current and resistance when the potential difference is held constant.

## Apply and Extend

**12.** Combine these two relationships into one. Write your results in the form of a proportionality. (**Note:** This is the basis of what is known as Ohm's law.)

**13.** Convert your relationship into an equation by including a constant ($k$) of proportionality.

**14.** Calculate the value of your constant by using the data from one of the trials in the second part of the investigation. Rewrite your relationship with your value for $k$ included. This is Ohm's law for your standard resistance.

**15.** The value you find for $k$ for your apparatus depends on the length of Nichromeᴐ wire that you arbitrarily chose as your standard resistance. Calculate the length of wire segment that you would have to use as your standard resistance if you wanted the value for $k$ to equal one. By definition, the length of Nichromeᴐ wire that would produce a proportionality constant equal to one ($k = 1$) has a resistance of one ohm.

# Ohm's Law

In 1826, German physicist Georg Simon Ohm conducted the original experiments in resistance in electric circuits, using many lengths and thicknesses of wire. He studied the current passing through the wire when a known potential difference was applied across it. From his data, Ohm developed the mathematical relationship that now bears his name, Ohm's law.

---

### OHM'S LAW

The potential difference across a load equals the product of the current through the load and the resistance of the load.

$$V = IR$$

| Quantity | Symbol | SI unit |
|---|---|---|
| potential difference | $V$ | V (volt) |
| current | $I$ | A (ampere) |
| resistance | $R$ | $\Omega$ (ohm) |

**Unit Analysis**

(potential difference) = (current)(resistance) = A $\cdot$ $\Omega$ = V

**Note:** One ampere times one ohm is equivalent to one volt.

---

The **unit of resistance**, the **ohm**, is defined in accordance with Ohm's law. One ohm is defined as the amount of electric resistance that will allow one ampere of current to move through the resistor when a potential difference of one volt is applied across the resistor.

$$\left(1\ \Omega = \frac{1\ \text{V}}{1\ \text{A}}\right)$$

Initially, Ohm's law seemed to be the answer to the problem of defining how a load would affect current. Unfortunately, its usefulness is limited to metal conductors at stable temperatures. For the majority of loads, such as motors, electronic capacitors, and semiconductors, the resistance changes with a change in the potential difference. Even a light bulb does not obey Ohm's law, because the heating of the filament causes its resistance to increase. When a load does not obey Ohm's law, the graph of $I$ versus $V$ is not a straight line. Devices and materials that do not obey Ohm's law are said to be **non-linear** or **non-ohmic**. Metallic conductors are, however, a sufficiently important and large class of materials that demonstrate the law is still extremely useful.

### Think It Through

- Consider the graph of potential difference versus current such as shown here. What is the significance of the slope of the line? (Hint: The equation for any straight line is $y = mx + b$, where $m$ is the slope of the line and $b$ is the $y$-intercept.)

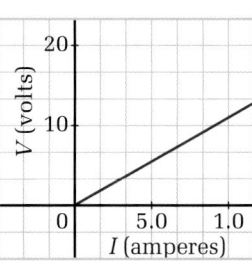

---

## MODEL PROBLEM

### Applying Ohm's Law

**What is the resistance of a load if a battery with a 9.0 V potential difference causes a current of 0.45 A to pass through the load?**

---

### Frame the Problem

- A battery creates a *potential difference* that provides energy to cause a *current* to flow in a circuit.

- A *load* resists the flow of current.

- You can find the *resistance* of the load by using Ohm's law.

---

### Identify the Goal

The resistance, $R$, of the load

### Variables and Constants

| Involved in the problem | Known | Unknown |
|---|---|---|
| $V$ | $V = 9.0$ V | $R$ |
| $I$ | $I = 0.45$ A | |
| $R$ | | |

### Strategy

Apply Ohm's law.

### Calculations

$V = IR$

**Substitute first**

$9.0 \text{ V} = (0.45 \text{ A}) R$

$\dfrac{9.0 \text{ V}}{0.45 \text{ A}} = \dfrac{(\cancel{0.45 \text{ A}}) R}{\cancel{0.45 \text{ A}}}$

$R = 20 \ \Omega$

**Solve for $R$ first**

$\dfrac{V}{I} = \dfrac{\cancel{I} R}{\cancel{I}}$

$\dfrac{9.0 \text{ V}}{0.45 \text{ A}} = R$

$R = 20 \ \Omega$

If a 9.0 V potential difference across a resistance results in a current of 0.45 A, the resistance is $2.0 \times 10^1 \ \Omega$.

---

*continued* ▶

*continued from previous page*

## Validate

The data fit Ohm's law. The number 9 divided by approximately $\frac{1}{2}$ is the same as $9 \times 2 = 18$. The final answer, 20, is close to 18.

### PRACTICE PROBLEMS

17. The heating element of an electric kettle draws 7.5 A when connected to a 120 V power supply. What is the resistance of the element?

18. A toaster is designed to operate on a 120 V $(1.20 \times 10^2$ V) system. If the resistance of the toaster element is 9.60 Ω, what current does it draw?

19. A small, decorative light bulb has a resistance of 36 Ω. If the bulb draws 140 mA, what is its operating potential difference? (**Note:** The prefix "m" before a unit always means "milli-" or one one-thousandth. 1 mA is $1 \times 10^{-3}$ A.)

20. The light bulb in the tail-light of an automobile with a 12 V electrical system has a resistance of 5.8 Ω. The bulb is left on for 8.0 min.

(a) What quantity of charge passes through the bulb?

(b) What was the current in the tail-light?

21. An iron transforms $3.35 \times 10^5$ J of electric energy to thermal energy in the 4.50 min it takes to press a pair of slacks. If the iron operates at 120 V $(1.20 \times 10^2$ V), what is its resistance?

22. In Europe, some countries use 240 V $(2.40 \times 10^2$ V) power supplies. How long will it take an electric kettle that has a resistance of 60.0 Ω to produce $4.32 \times 10^5$ J of thermal energy?

## 13.3    Section Review

1. ① By what factor would the resistance of two copper wires differ if the second wire:

(a) was double the length of the first?

(b) was triple the cross-sectional area of the first?

(c) had a radius that was half the radius of the first?

(d) was half as long and twice as thick (twice the diameter) as the first?

(e) was three times as long and a third the cross-sectional area of the first?

2. ① What happens to the resistance of a conductor when the temperature of the conductor increases?

3. ① Design an experiment that you would carry out to determine whether a particular circuit element was ohmic or non-ohmic. Explain how you would interpret the results.

### UNIT PROJECT PREP

Your electric motor will require current to operate.
- What factors affect the current flow within a conductor?
- How can you increase or decrease the current flow within a motor?

**SECTION EXPECTATIONS**

- Demonstrate understanding of the physical quantities of electricity.

- Synthesize information to solve electric energy problems.

**KEY TERMS**

- series
- parallel
- equivalent resistance
- internal resistance
- electromotive force
- terminal voltage

**Figure 13.15** Skiers arriving at the top of a ski lift have several different runs down the hill available to them. The route down the hill is a complex circuit of series and parallel runs, taking skiers back to the bottom of the lift.

An extremely simple electric device, such as a flashlight, might have a circuit with one power source (a battery), one load (a light bulb), and one switch. In nearly every practical circuit, however, the power source supplies energy to many different loads. In these practical circuits, the loads may be connected in **series** (Figure 13.16) or in **parallel** (Figure 13.17). The techniques used to analyze these complex circuits are very similar to those you used to analyze simple circuits.

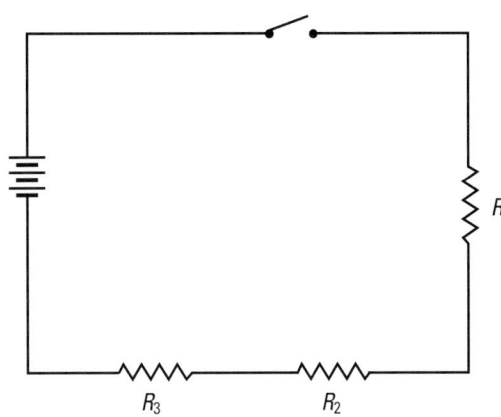

**Figure 13.16** A circuit with resistances in series has only one closed path.

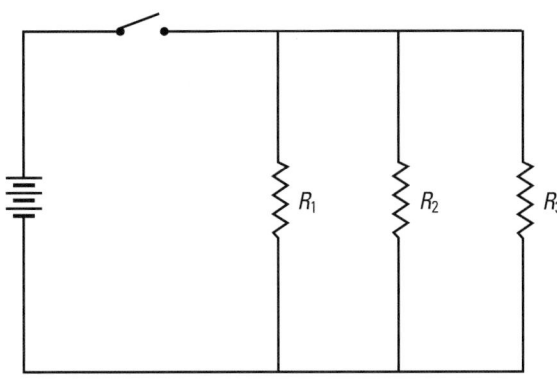

**Figure 13.17** A circuit with resistances in parallel has several closed paths.

## Series Circuits

A ski hill consisting of three downhill runs, one after the other, with level paths connecting the runs, is an excellent analogy for a series circuit. Since there is only one route down, the number of skiers going down each run would have to be the same. The total height of the three runs would have to equal the total height of the hill, as illustrated in Figure 13.18.

**Figure 13.18** As skiers go around a ski circuit, the lift raises them to the top of the hill. Since the ski runs are in series, all of the skiers must ski down each of the runs, so that the number of skiers completing each run is the same. Each of the runs takes the skiers down a portion of the total height given to them by the ski lift. The combined height of the three runs must equal the height of the hill ($h_L = h_1 + h_2 + h_3$).

---

**PHYSICS FILE**

Just as skiers' gravitational potential energy drops as they ski down a hill, the electric potential energy of charges moving in a circuit drops when they pass through a load. Physicists often call the potential difference across a load a "potential drop." When the charges are given more potential energy in a battery or power source, they experience a "potential gain."

---

A series circuit consists of loads (resistances) connected in series, as was shown in Figure 13.16. The current that leaves the battery has only one path to follow. Just as the skiers in the previous analogy must all ski down each run in sequence, all of the current that leaves the battery must pass through each of the loads. An ammeter could be connected at any point in the circuit and each reading would be the same.

Also, just as the total height of the hill must be shared over the three runs, the potential difference of the battery must be shared over all three loads. Thus, a portion of the electric potential of the battery must be used to push the current through each load. If each load had a voltmeter connected across it, the total of the potential differences across the individual loads must equal the potential difference across the battery.

To find the **equivalent resistance** of a series circuit with $N$ resistors, as illustrated in Figure 13.19 on the following page, analyze the properties of the circuit and apply Ohm's law.

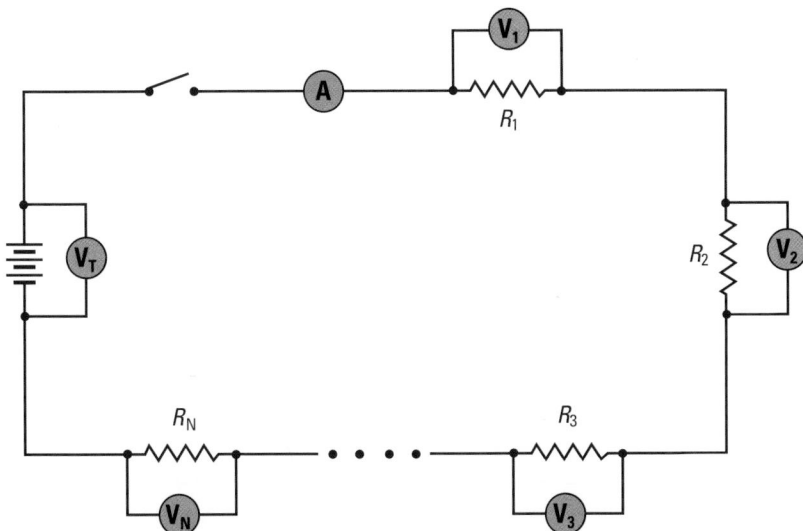

**Figure 13.19** A circuit might consist of any number of loads. If this circuit had eight loads, $R_N$ would represent $R_8$, and eight loads would be connected in series.

- Write, in mathematical form, the statement, "The total of the potential differences across the individual loads must equal the potential difference across the battery."

$$V_S = V_1 + V_2 + V_3 + \cdots + V_N$$

- By Ohm's law, the total potential difference across the battery (source) must be equal to the product of the current through the battery and the equivalent resistance of the circuit.

$$V_S = I_S R_{eq}$$

- For each load, the potential difference across that load must equal the product of the current through the load and its resistance.

$$V_1 = I_1 R_1, \ V_2 = I_2 R_2, \ V_3 = I_3 R_3, \cdots, V_N = I_N R_N$$

- Substitute these expressions into the first expression for potential difference.

$$I_S R_{eq} = I_1 R_1 + I_2 R_2 + I_3 R_3 + \cdots + I_N R_N$$

- Write, in mathematical form, the statement, "All of the current that leaves the battery must pass through each of the loads."

$$I_1 = I_2 = I_3 = \cdots = I_N = I_S$$

- In the expression above, replace the symbol for each separate current with $I_S$.

$$I_S R_{eq} = I_S R_1 + I_S R_2 + I_S R_3 + \cdots + I_S R_N$$

- Factor $I_S$ out of the right side.

$$I_S R_{eq} = I_S(R_1 + R_2 + R_3 + \cdots + R_N)$$

- Divide both sides by $I_S$.

$$R_{eq} = R_1 + R_2 + R_3 + \cdots + R_N$$

## MODEL PROBLEM

### Resistances in Series

Four loads (3.0 $\Omega$, 5.0 $\Omega$, 7.0 $\Omega$, and 9.0 $\Omega$) are connected in series to a 12 V battery. Find

(a) the equivalent resistance of the circuit

(b) the total current in the circuit

(c) the potential difference across the 7.0 $\Omega$ load

### Frame the Problem

- Since all of the resistors are in series, the formula for the equivalent resistance for a series circuit applies to the problem.

- All of the loads are in series; thus, the current is the same at all points in the circuit. The current can be found by using the potential difference across the battery, the equivalent resistance, and Ohm's law.

- Ohm's law applies to each individual circuit element.

### Identify the Goal

The equivalent resistance, $R_S$, for the series circuit, the current, $I$, and the potential difference, $V_3$, across the 7.0 $\Omega$ resistor

### Variables and Constants

| Involved in the problem | | Known | Unknown |
|---|---|---|---|
| $R_1$ | $R_{eq}$ | $R_1 = 3.0\ \Omega$ | $R_{eq}$ |
| $R_2$ | $V_S$ | $R_2 = 5.0\ \Omega$ | $I$ |
| $R_3$ | $I_S$ | $R_3 = 7.0\ \Omega$ | $V_3$ |
| $R_4$ | $V_3$ | $R_4 = 9.0\ \Omega$ | |
| | | $V_S = 12\ \text{V}$ | |

## Strategy

Use the equation for the equivalent resistance of a series circuit.

## Calculations

$$R_{eq} = R_1 + R_2 + R_3 + R_4$$
$$= 3.0\ \Omega + 5.0\ \Omega + 7.0\ \Omega + 9.0\ \Omega$$
$$= 24\ \Omega$$

(a) The equivalent resistance for the four resistors in series is 24 Ω.

Use Ohm's law, in terms of current, and the equivalent resistance to find the current in the circuit.

$$I_S = \frac{V_S}{R_{eq}}$$

$$I_S = \frac{12\ V}{24\ \Omega}$$

$1\frac{V}{\Omega}$ is equivalent to 1A.

$$I_S = 0.50\frac{V}{\Omega}$$

$$I_S = 0.50\ A$$

(b) The current in the circuit is 0.50 1A.

Use Ohm's law, the current, and the resistance of a single resistor to find the potential drop across that resistor.

$$V_3 = I_3 R_3$$

$$I_3 = I_S$$

Since the circuit has only one closed loop, the current is the same everywhere in the circuit, so $I_3 = I_S$.

$$V_3 = (0.50\ A)(7.0\ \Omega)$$

$$V_3 = 3.5\ A \cdot \Omega$$

1 A · Ω is equivalent to 1 V.

$$V_3 = 3.5\ V$$

(c) The potential drop across the 7.0 Ω resistor is 3.5 V.

## Validate

If you now find the potential difference across the other loads, the sum of the potential differences for the four loads should equal 12 V.

$V_1 = (0.50\ A)(3.0\ \Omega)$     $V_2 = (0.50\ A)(5.0\ \Omega)$     $V_4 = (0.50\ A)(9.0\ \Omega)$
$V_1 = 1.5\ V$                          $V_2 = 2.5\ V$                          $V_4 = 4.5\ V$

1.5 V + 2.5 V + 3.5 V + 4.5 V = 12 V

## PRACTICE PROBLEMS

**23.** Three loads, connected in series to a battery, have resistances of 15.0 Ω, 24.0 Ω, and 36.0 Ω. If the current through the first load is 2.2A, calculate

(a) the potential difference across each of the loads

(b) the equivalent resistance for the three loads

(c) the potential difference of the battery

**24.** Two loads, 25.0 Ω and 35.0 Ω, are connected in series. If the potential difference across the 25.0 Ω load is 65.0 V, calculate

(a) the potential difference across the 35.0 Ω load

(b) the potential difference of the battery

*continued* ▶

**25.** Two loads in series are connected to a 75.0 V battery. One of the loads is known to have a resistance of 48.0 Ω. You measure the potential difference across the 48.0 Ω load and find it is 40.0 V. Calculate the resistance of the second load.

**26.** Two loads, $R_1$ and $R_2$, are connected in series to a battery. The potential difference across $R_1$ is 56.0 V. The current measured at $R_2$ is 7.00 A. If $R_2$ is known to be 24.0 Ω, find

**(a)** the resistance of $R_1$

**(b)** the potential difference of the battery

**(c)** the equivalent resistance of the circuit

**27.** A 240 V ($2.40 \times 10^2$ V) power supply is connected to three loads in series. The current in the circuit is measured to be 1.50 A. The resistance of the first load is 42.0 Ω and the potential difference across the second load is 111 V. Calculate the resistance of the third load.

## Resistors in Parallel

The ski hill in Figure 13.20 provides a model for a circuit consisting of a battery and three resistors connected in parallel. The ski lift is, of course, analogous to the battery and the runs represent the resistors. Notice that the hill has three runs beside each other (in parallel) that go all of the way from the top to the bottom of the hill. The height of each of the runs must be equal to the height that the skiers gain by riding the lift up the hill ($h_L = h_1 = h_2 = h_3$).

**Figure 13.20** When skiers are on a hill where there are several ski runs that all are the same height as the hill, the runs are said to be *parallel* to each other.

The gravitational potential difference of the hill is analogous to the electric potential difference across a battery and resistors connected in parallel in a circuit. Similar to the height of the hill, the potential difference across each of the individual loads in a parallel circuit must be the same as the total potential difference across the battery ($V_S$). For example, in Figure 13.21, with $N$ resistors in parallel, the mathematical relationship can be written as follows.

$$V_S = V_1 = V_2 = V_3 = \cdots = V_N$$

In the ski-hill analogy, the skiers themselves represent the current. A skier might select any one of the three runs, but can ski down only one of them. Since the skiers riding up the lift can go down only one of the hills, the sum of the skiers going down all three hills must be equal to the number of skiers leaving the lift.

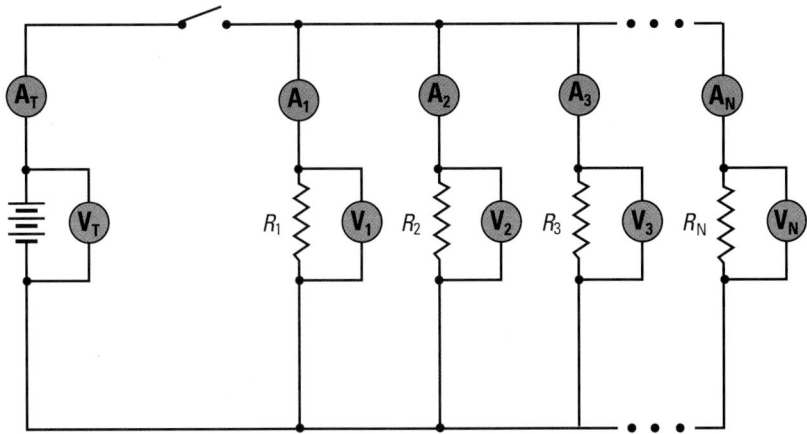

**Figure 13.21** The *N* loads in this circuit are all connected in parallel with each other. The dots indicate where any number of additional loads could be connected in parallel with those present.

When the current leaving the battery ($I_S$) comes to a point in the circuit where the path splits into two or more paths, the current must split so that a portion of it follows each path. After passing through the loads, the currents combine before returning to the battery. The sum of the currents in parallel paths must equal the current entering and leaving the battery.

Knowing the current and potential difference relationships in a parallel circuit, you can use Ohm's law to develop an equation for the equivalent resistance of resistors in a parallel connection.

- Write, in mathematical form, the statement, "The sum of the currents in parallel paths must equal the current through the source."

$$I_S = I_1 + I_2 + I_3 + \cdots + I_N$$

- Write Ohm's law in terms of current.

$$I = \frac{V}{R}$$

- Apply this form of Ohm's law to the current through each individual resistor and for the battery.

$$I_S = \frac{V_S}{R_{eq}} \quad I_1 = \frac{V_1}{R_1} \quad I_2 = \frac{V_2}{R_2}$$
$$I_3 = \frac{V_3}{R_3} \quad I_N = \frac{V_N}{R_N}$$

- Replace the currents in the first equation with the expressions above.

$$\frac{V_S}{R_{eq}} = \frac{V_1}{R_1} + \frac{V_2}{R_2} + \frac{V_3}{R_3} + \cdots + \frac{V_N}{R_N}$$

- Write the relationship for the potential differences across resistors and the battery, when all are connected in parallel.

$$V_S = V_1 = V_2 = V_3 = \cdots = V_N$$

- Replace the potential differences in the equation above with $V_S$.

$$\frac{V_S}{R_S} = \frac{V_S}{R_1} + \frac{V_S}{R_2} + \frac{V_S}{R_3} + \cdots + \frac{V_S}{R_N}$$

- Divide both sides of the equation by $V_S$.

$$\frac{1}{R_{eq}} = \frac{1}{R_1} + \frac{1}{R_2} + \frac{1}{R_3} + \cdots + \frac{1}{R_N}$$

---

### RESISTORS IN PARALLEL

The inverse of the equivalent resistance for resistors connected in parallel is the sum of the inverses of the individual resistances.

$$\frac{1}{R_{eq}} = \frac{1}{R_1} + \frac{1}{R_2} + \frac{1}{R_3} + \cdots + \frac{1}{R_N}$$

| Quantity | Symbol | SI unit |
|---|---|---|
| equivalent resistance | $R_{eq}$ | $\Omega$ (ohm) |
| resistance of the individual loads | $R_1, R_2, R_3, \ldots, R_N$ | $\Omega$ (ohm) |

---

### MODEL PROBLEM

## Resistors in Parallel

A 60 V battery is connected to four loads of 3.0 $\Omega$, 5.0 $\Omega$, 12.0 $\Omega$, and 15.0 $\Omega$ in parallel.

**(a)** Find the equivalent resistance of the four combined loads.

**(b)** Find the total current leaving the battery.

**(c)** Find the current through the 12.0 $\Omega$ load.

### Frame the Problem

- The four loads are connected *in parallel*; therefore, the *potential difference* across each load is the same as the *potential difference* provided by the *battery*.

- The *potential difference* across the battery and the *current* entering and leaving the battery would be *unchanged* if the four loads were replaced with one load having the *equivalent resistance*.

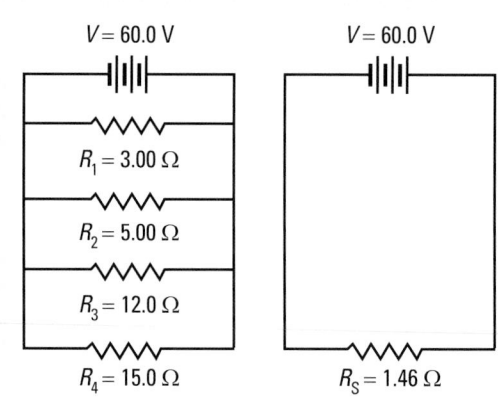

- After the current leaves the battery, it reaches branch points where it separates, and part of the current runs through each load.

- *Ohm's law* applies to each *individual load* or to the *combined load*. However, you must ensure that the current and potential difference are correct for the specific resistance that you use in the calculation.

## Identify the Goal

The equivalent resistance, $R_S$, of the four loads
The current, $I_S$, leaving the battery
The current, $I_3$, through the 12.0 $\Omega$ load

## Variables and Constants

| Involved in the problem | | Known | Unknown |
|---|---|---|---|
| $R_1$ | $R_{eq}$ | $R_1 = 3.00\ \Omega$ | $R_{eq}$ |
| $R_2$ | $V_S$ | $R_2 = 5.00\ \Omega$ | $I_S$ |
| $R_3$ | $I_S$ | $R_3 = 12.0\ \Omega$ | $I_3$ |
| $R_4$ | $I_3$ | $R_4 = 15.0\ \Omega$ | |
| | | $V_S = 60.0\ \text{V}$ | |

## Strategy

Use the equation for resistors in parallel and apply it to the four loads.

Substitute values for resistance and add.

Find a common denominator.
Add.

## Calculations

$$\frac{1}{R_{eq}} = \frac{1}{R_1} + \frac{1}{R_2} + \frac{1}{R_3} + \frac{1}{R_4}$$

$$\frac{1}{R_{eq}} = \frac{1}{3.00\ \Omega} + \frac{1}{5.00\ \Omega} + \frac{1}{12.0\ \Omega} + \frac{1}{15.0\ \Omega}$$

$$\frac{1}{R_{eq}} = \frac{20}{60.0\ \Omega} + \frac{12}{60.0\ \Omega} + \frac{5}{60.0\ \Omega} + \frac{4}{60.0\ \Omega}$$

$$\frac{1}{R_{eq}} = \frac{41}{60.0\ \Omega}$$

## Strategy

Invert both sides of the equation. (If you invert an equality, it remains an equality.)

Divide.

**(a)** The equivalent resistance of the four loads in parallel is 1.46 $\Omega$.

Use Ohm's law, in terms of current, and the equivalent resistance to calculate the current entering and leaving the battery.

## Calculations

$$R_{eq} = \frac{60.0\ \Omega}{41}$$

$$R_{eq} = 1.46\ \Omega$$

$$I_S = \frac{60.0\ \text{V}}{1.46\ \Omega}$$

*continued* ▶

$1 \dfrac{V}{\Omega}$ is equivalent to an A.

$$I_S = 41.0\dfrac{V}{\Omega}$$

$$I_S = 41.0 \text{ A}$$

**(b)** The current entering and leaving the battery is 41.0 A.

Use Ohm's law, in terms of current, to find the current through the 12.0 Ω load.

$$I_3 = \dfrac{60.0 \text{ V}}{12.0 \text{ } \Omega}$$

$$I_3 = 5.00\dfrac{V}{\Omega}$$

$$I_3 = 5.00 \text{ A}$$

**(c)** Of the 41.0 A leaving the battery, 5.00 A are diverted to the 12.0 Ω load.

**Validate**

If you do a similar calculation for the current in each of the loads, the total current through all of the loads should equal 41 A.

$$I_1 = \dfrac{V_1}{R_1} \qquad I_2 = \dfrac{V_2}{R_2} \qquad I_4 = \dfrac{V_4}{R_4}$$

$$I_1 = \dfrac{60.0 \text{ V}}{3.00 \text{ } \Omega} \qquad I_2 = \dfrac{60.0 \text{ V}}{5.00 \text{ } \Omega} \qquad I_4 = \dfrac{60.0 \text{ V}}{15.0 \text{ } \Omega}$$

$$I_1 = 20.0 \text{ A} \qquad I_2 = 12.0 \text{ A} \qquad I_4 = 4.00 \text{ A}$$

$$I_{total} = 20.0 \text{ A} + 12.0 \text{ A} + 5.00 \text{ A} + 4.00 \text{ A}$$
$$I_{total} = 41.0 \text{ A}$$

The answer is validated.

## PRACTICE PROBLEMS

Draw a circuit diagram for each problem below. As an aid, write the known values on the diagram.

**28.** A 9.00 V battery is supplying power to three light bulbs connected in parallel to each other. The resistances, $R_1$, $R_2$, and $R_3$, of the bulbs are 13.5 Ω, 9.00 Ω, and 6.75 Ω, respectively. Find the current through each load and the equivalent resistance of the circuit.

**29.** A light bulb and a heating coil are connected in parallel to a 45.0 V battery. The current from the battery is 9.75 A, of which 7.50 A passes through the heating coil. Find the resistances of the light bulb and the heating coil, and the equivalent resistance for the circuit.

**30.** A circuit contains a 12.0 Ω load in parallel with an unknown load. The current in the 12.0 Ω load is 3.20 A, while the current in the unknown load is 4.80 A. Find the resistance of the unknown load and the equivalent resistance for the two parallel loads.

**31.** A current of 4.80 A leaves a battery and separates into three currents running through three parallel loads. The current to the first load is 2.50 A, the current through the second load is 1.80 A, and the resistance of the third load is 108 Ω. Calculate (a) the equivalent resistance for the circuit, and (b) the resistance of the first and second loads.

# Complex Circuits

Many practical circuits consist of loads in a combination of parallel and series connections. To determine the characteristics of the circuit, you must analyze the circuit and recognize the way that different loads are connected in relation to each other. When a circuit branches, the loads in each branch must be grouped and treated as a single, or equivalent, load before they can be used in a calculation with other loads.

**Figure 13.22** The circuitry shown here, typical of electronic equipment, illustrates the high level of complexity in circuitry of the household devices that we use every day.

# Internal Resistance

The objective of Part 2 of Investigation 13-B (Current, Resistance, and Potential Difference) was to find out how the current varied with resistance. Each time you changed the resistance, you had to reset the potential difference across the load to the desired value. The circuit behaved as if some phantom resistance was affecting the potential difference. To an extent, that is exactly what was happening — the phantom resistance was the internal resistance of the battery or power supply.

**Figure 13.23** A chemical reaction separates electric charge creating a potential difference between the terminals. Internal resistance within the battery reduces the amount of voltage available to an external circuit.

## Resistors in Parallel

**Find the equivalent resistance of the entire circuit shown in the diagram, as well as the current through, and the potential difference across, each load.**

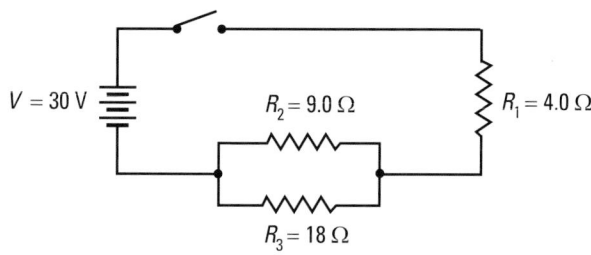

## Frame the Problem

- The circuit consists of a *battery* and *three loads*. The battery generates a specific *potential difference* across the poles.

- The *current* driven by the potential difference of the battery depends on the *effective resistance* of the entire circuit.

- The circuit has two groups of resistors. Resistors $R_2$ and $R_3$ are in parallel with each other. Load $R_1$ is in series with the $R_2$–$R_3$ group.

- Define the $R_2$–$R_3$ group as Group A, and sketch the circuit with the equivalent load, $R_A$.

- Now, define the series group consisting of $R_A$ and $R_1$ as Group B. Sketch the circuit with the equivalent load, $R_B$.

- Load $R_B$ is the equivalent resistance of the entire circuit.

## Identify the Goal

The equivalent resistance, $R_{eq}$, of the circuit
The currents, $I_1$, $I_2$, and $I_3$, through the loads
The potential differences, $V_1$, $V_2$, and $V_3$, across the loads

## Variables and Constants

| Involved in the problem | | | Known | Unknown | |
|---|---|---|---|---|---|
| $R_1$ | $R_{eq}$ | $V_1$ | $R_1 = 4.0\ \Omega$ | $R_{eq}$ | $V_1$ |
| $R_2$ | $I_1$ | $V_2$ | $R_2 = 9.0\ \Omega$ | $I_1$ | $V_2$ |
| $R_3$ | $I_2$ | $V_3$ | $R_3 = 18\ \Omega$ | $I_2$ | $V_3$ |
| $V_S$ | $I_3$ | | $V_S = 30\ V$ | $I_3$ | |

## Strategy · Calculations

| Strategy | Calculations |
|---|---|
| Find the equivalent resistance for the parallel Group A resistors. | $\dfrac{1}{R_A} = \dfrac{1}{R_2} + \dfrac{1}{R_3}$ |
| Find a common denominator. | $\dfrac{1}{R_A} = \dfrac{1}{9.0\ \Omega} + \dfrac{1}{18\ \Omega}$ |
| | $\dfrac{1}{R_A} = \dfrac{2}{18\ \Omega} + \dfrac{1}{18\ \Omega}$ |

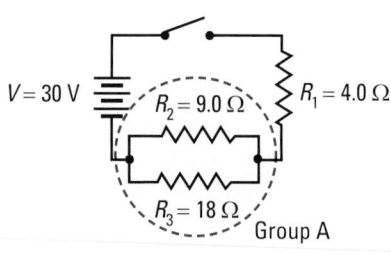

| Strategy | Calculations |
|---|---|
| Simplify. | $\dfrac{1}{R_A} = \dfrac{3}{18\ \Omega}$ |

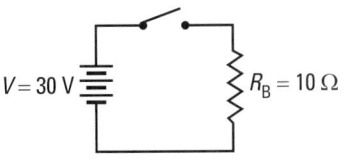

Group B

Invert both sides of the equation.

$$\frac{1}{R_A} = \frac{1}{6.0\ \Omega}$$

$$R_A = 6.0\ \Omega$$

Find the equivalent resistance of the series Group B.

$$R_B = R_A + R_1$$
$$R_B = 4.0\ \Omega + 6.0\ \Omega$$
$$R_B = 10\ \Omega$$

$V = 30\ V$      $R_B = 10\ \Omega$

Since there is now one equivalent resistor in the circuit, the effective resistance of the entire circuit is $R_{eq} = 10\ \Omega$.

Find the current entering and leaving the battery, using the potential difference of the battery and the total effective resistance of the circuit.

$$I_S = \frac{V_S}{R_S}$$

$$I_S = \frac{30\ V}{10\ \Omega}$$

$$I_S = 3.0\ A$$

Since there are no branches in the circuit between the battery and load $R_1$, all of the current leaving the battery passes through $R_1$. Therefore, $I_1 = 3.0$ A.

Knowing the current through $R_1$, and its resistance, you can use Ohm's law to find $V_1$.

$$V_1 = I_1 R_1$$

$$V_1 = (3.0\ A)(4.0\ \Omega)$$

$$V_1 = 12\ V$$

The potential difference across load 1 is 12 V.

The loads $R_1$ and $R_A$ form a complete path from the anode to the cathode of the battery; therefore, the sum of the potential drops across these loads must equal that of the battery.

$$V_S = V_1 + V_A$$

$$30\ V = 12\ V + V_A$$

$$30\ V - 12\ V = V_A$$

$$V_A = 18\ V$$

Since the potential difference across a parallel connection is the same for both pathways, $V_2 = 18$ V and $V_3 = 18$ V.

Knowing the potential difference across $R_2$ and $R_3$, you can find the current through each load by using Ohm's law.

$$I_2 = \frac{V_2}{R_2} \qquad I_3 = \frac{V_2}{R_3}$$

$$I_2 = \frac{18\ V}{9.0\ \Omega} \qquad I_3 = \frac{18\ V}{18\ \Omega}$$

$$I_2 = 2.0\ A \qquad I_3 = 1.0\ A$$

The current through load 2 is 2.0 A, and the current through load 3 is 1.0 A.

*continued* ▶

To summarize, the 30 V battery causes a current of 3.0 A to move through the circuit. All of the current passes through the 4.0 $\Omega$ load, but then splits into two parts, with 2.0 A going through the 9.0 $\Omega$ load and 1.0 A going through the 18 $\Omega$ load. The potential drops across the 4.0 $\Omega$, 9.0 $\Omega$, and 18 $\Omega$ loads are 12 V, 18 V, and 18 V, respectively.

> **PROBLEM TIP**
>
> When working with complex circuits, look for the smallest groups of loads that are connected *only* in parallel or *only* in series. Find the equivalent resistance of the group, then redraw the circuit with one load representing the equivalent resistance of the group. Begin again.

## Validate

The current to Group A, known to be 3.0 A, was split into two parts. The portion of the current in each of the branches of Group A was inversely proportional to the resistance in the branch. The larger 18 $\Omega$ load allowed half the current that the smaller 9.0 $\Omega$ load allowed. The sum of the currents through $R_2$ and $R_3$ is equal to the current through $R_1$, which is expected to be true.

The potential difference across Group A should be the same as that of one load with a resistance, $R_A$, with the total current passing through it. Check this potential difference and compare it with the 18 V found by subtracting 12 V across load 1 from the total of 30 V.

$$V_A = I_S R_A = (3.0 \text{ A})(6.0 \ \Omega) = 18 \text{ V}$$

The values are in agreement. The answers are consistent.

## PRACTICE PROBLEMS

**32.** For the circuit in the diagram shown below, the potential difference of the power supply is 144 V. Calculate

  **(a)** the equivalent resistance for the circuit

  **(b)** the current through $R_1$

  **(c)** the potential difference across $R_3$

**33.** For the circuit shown in the diagram below, the potential difference of the power supply is 25.0 V. Calculate

  **(a)** the equivalent resistance of the circuit

  **(b)** the potential difference across $R_3$

  **(c)** the current through $R_1$

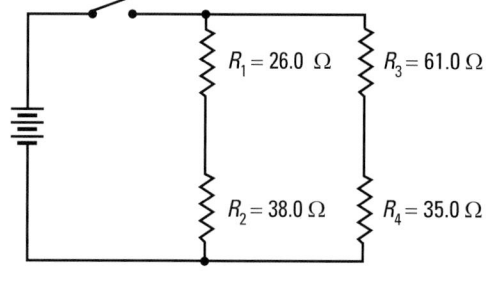

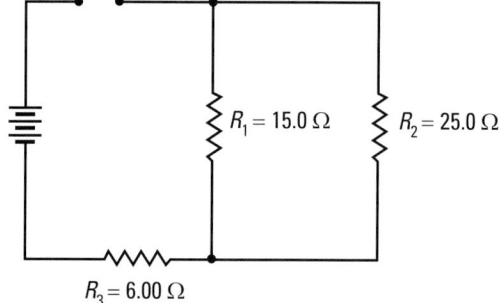

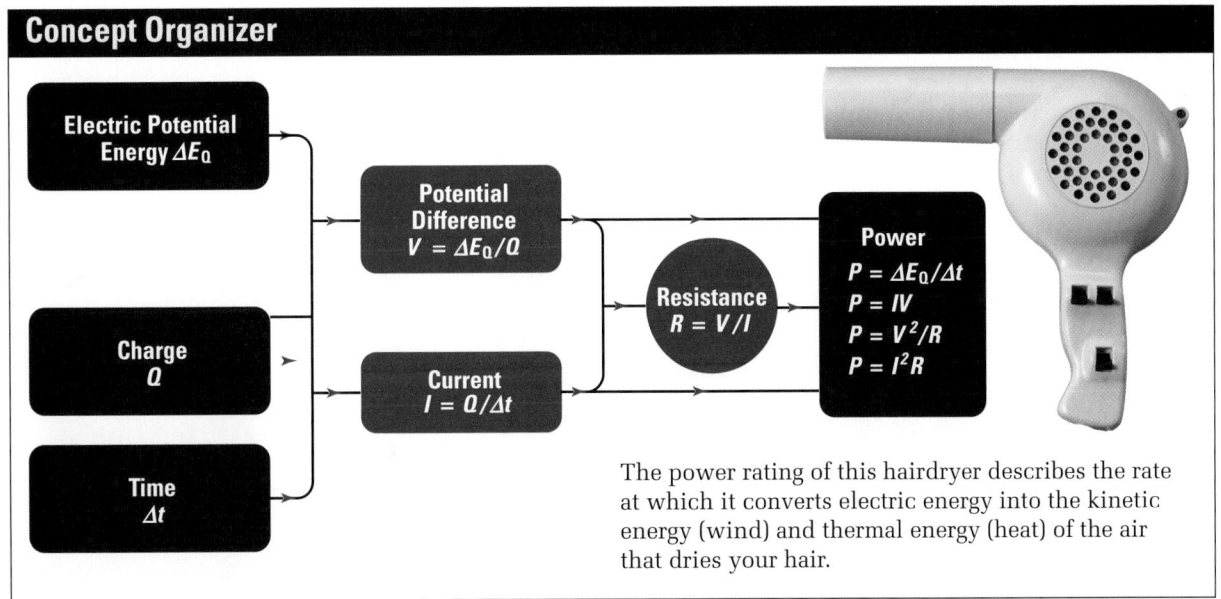

The power rating of this hairdryer describes the rate at which it converts electric energy into the kinetic energy (wind) and thermal energy (heat) of the air that dries your hair.

To develop an understanding of **internal resistance**, consider a gasoline engine being used to power a ski lift. Gasoline engines convert energy from the fuel into mechanical energy, which is then used, in part, to pull the skiers up the hill. No matter how efficient the motor is, however, it must always use some of the energy to overcome the friction inside the motor itself. In fact, as the number of skiers on the lift increases, the amount of energy the motor uses to run itself also increases.

The process involved in an electric circuit is very similar to the motor driving the ski lift. Inside the battery, chemical reactions create a potential difference, called the **electromotive force** (*emf*, represented in equations by $\mathscr{E}$). If there was no internal resistance inside the battery, the potential difference across its anode and cathode (sometimes called the **terminal voltage**) would be exactly equal to the *emf*. However, when a battery is connected to a circuit and current is flowing, some of the *emf* must be used to cause the current to flow through the internal resistance ($r$) of the battery itself. Therefore, the terminal voltage ($V_S$) of the battery is always less than the *emf* by an amount equal to the potential difference across the internal resistance ($V_{int} = I \cdot r$). You can find the terminal voltage by using the equation in the following box.

## TERMINAL VOLTAGE AND *emf*

The terminal voltage (or potential difference across the poles) of a battery is the difference of the *emf* ($\mathscr{E}$) of the battery and the potential drop across the internal resistance of the battery.

$$V_S = \mathscr{E} - V_{int}$$

| Quantity | Symbol | SI unit |
|---|---|---|
| terminal voltage | $V_S$ | V (volt) |
| electromotive force | $\mathscr{E}$ | V (volt) |
| internal potential drop of a battery | $V_{int}$ | V (volt) |

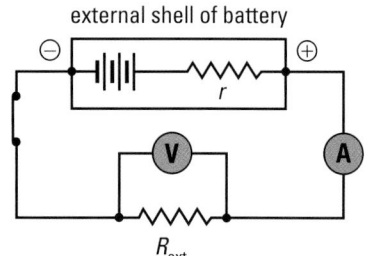

external shell of battery

$R_{ext}$

If no current is passing through a battery, then the potential difference across the internal resistance will be zero ($V_{int} = 0$). As a result, the potential difference across its terminals ($V_S$) will be equal to the *emf* ($\mathscr{E}$).

**Figure 13.24** A battery has an internal resistance ($r$) that is in series with the *emf* ($\mathscr{E}$) of the cell.

### MODEL PROBLEM

## Terminal Voltage versus *emf* of a Battery

**A battery with an *emf* of 9.00 V has an internal resistance of 0.0500 Ω. Calculate the potential difference lost to the internal resistance, and the terminal voltage of the battery, if it is connected to an external resistance of 4.00 Ω.**

### Frame the Problem

- A battery is connected to a closed circuit; thus, a *current* is flowing.

- Due to the *internal resistance* of the battery and the current, a *potential drop* occurs inside the battery.

- The *potential drop* across the internal resistance depends on the amount of *current* flowing in the circuit.

- The *terminal voltage* is lower than the *emf* due to the loss of energy to the internal resistance.

### Identify the Goal

The potential difference ($V_{int}$) lost by current passing through the internal resistance of the battery

The terminal voltage ($V_S$), or potential difference across the poles, of the battery

## Variables and Constants

| Involved in the problem | | Known | Unknown |
|---|---|---|---|
| $\mathcal{E}$ | $V_S$ | $\mathcal{E} = 9.00$ V | $V_S$ |
| $r$ | $V_{int}$ | $r = 0.0500\ \Omega$ | $V_{int}$ |
| $R_{emf}$ | $I_S$ | $R_{ext} = 4.00\ \Omega$ | $I_S$ |
| $R_s$ | | | $\mathcal{E}_s$ |

## Strategy

To find the current flowing in the circuit, you need to know the equivalent resistance of the circuit. Since the internal resistance is in series with the external resistance, use the equation for series circuits.

Use the *emf* ($\mathcal{E}$), and resistance to the *emf* ($R_{emf}$) in Ohm's law to find the current in the circuit.

Find the internal potential drop of the battery by using Ohm's law, the current, and the internal resistance.

Find the terminal voltage from the *emf* and the potential drop due to the internal resistance.

## Calculations

$R_{emf} = r + R_s$

$R_{emf} = 0.0500\ \Omega + 4.00\ \Omega$

$R_{emf} = 4.05\ \Omega$

$I_S = \dfrac{\mathcal{E}}{R_{emf}}$

$I_S = \dfrac{9.00\ \text{V}}{4.05\ \text{V}}$

$I_S = 2.22$ A

$V_{int} = I_S r$

$V_{int} = (2.22\ \text{A})(0.0500\ \Omega)$

$V_{int} = 0.111$ V

$V_S = \mathcal{E} - V_{int}$

$\phantom{V_S} = 9.00\ \text{V} - 0.111\ \text{V}$

$\phantom{V_S} = 8.89$ V

The potential difference across the internal resistance is 0.111 V, causing the *emf* of the battery to be reduced to the terminal voltage of 8.89 V. This is the portion of the *emf* available to the external circuit.

## Validate

In order for a battery to be useful, you would expect that the loss of potential difference due to its internal resistance would be very small, compared to the terminal voltage. In this case, the loss (0.111 V) is just over 1 percent of the terminal voltage (8.898 V). The answer is reasonable.

## PRACTICE PROBLEMS

**36.** A battery has an *emf* of 15.0 V and an internal resistance of 0.0800 $\Omega$.

   **(a)** What is the terminal voltage if the current to the circuit is 2.50 A?

   **(b)** What is the terminal voltage when the current increases to 5.00 A?

**37.** A battery has an internal resistance of 0.120 $\Omega$. The terminal voltage of the battery is 10.6 V when a current of 7.00 A flows from it.

   **(a)** What is its *emf*?

   **(b)** What would be the potential difference of its terminals if the current was 2.20 A?

Electric Energy and Circuits • MHR **649**

# INVESTIGATION 13-C

## Internal Resistance of a Dry Cell

**TARGET SKILLS**
- Performing and recording
- Analyzing and interpreting
- Communicating results

You can usually measure the resistance of a load by connecting a voltmeter across the load and connecting an ammeter in series with the load. However, you cannot attach a voltmeter across the internal resistance of a battery; you can attach a voltmeter only across the poles of a battery. How, then, can you measure the internal resistance of a dry cell?

## Problem

Determine the internal resistance of a dry cell.

## Equipment

- $1\frac{1}{2}$ volt D-cell (or 6 V battery)
- variable resistor
- voltmeter
- ammeter
- conductors with alligator clips

## Procedure

1. Measure and record the electromotive force (*emf*, represented by $E$) of the battery. This is the potential difference of the battery before it is connected to the circuit. Record all data to at least three significant digits.

2. Make a data table with the column headings: Trial, Terminal Voltage ($V_S$), Current ($I_S$), Equivalent Resistance ($R_{eq}$), Resistance to the *emf* ($R_{emf}$), Internal Resistance ($r$).

3. Connect the apparatus as shown in the diagram.

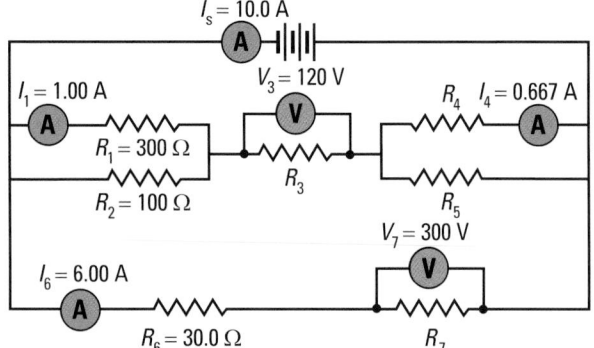

$I_s = 10.0$ A

$V_3 = 120$ V

$I_1 = 1.00$ A

$R_1 = 300\ \Omega$

$R_2 = 100\ \Omega$

$R_4\ \ I_4 = 0.667$ A

$R_3$

$R_5$

$V_7 = 300$ V

$I_6 = 6.00$ A

$R_6 = 30.0\ \Omega$

$R_7$

**CAUTION** Inspect your connections carefully. Refer to the wiring instructions given in Investigation 13-B on page 627 if you need to refresh your memory.

**CAUTION** Open the switch or disconnect the circuit when you are not taking readings.

4. Adjust the variable resistor so that the current is about 1.0 A. Then, record the readings for the voltmeter ($V_S$) and ammeter ($I_S$).

5. Reduce the resistance of the load and record the voltmeter and ammeter readings.

6. Obtain at least four sets of data readings by reducing the resistance after each trial.

7. After taking the last reading from the circuit, remove the cell from the circuit and measure the *emf* of the cell again to confirm that it has not been diminished significantly.

## Analyze and Conclude

1. For each trial, calculate and enter into your data table the value of the equivalent resistance, $R_S$.

$$\left(R_{eq} = \frac{V_S}{I_S}\right)$$

2. For each trial, calculate and enter into your data table the value of the total resistance, $R_T$.

$$\left(R_{emf} = \frac{\mathcal{E}}{I_S}\right)$$

3. For each trial, calculate and record in your data table the value of the resistance offered to the *emf* ($r = R_{emf} - R_{eq}$).

4. Explain why the terminal voltage decreases when the current increases.

5. Is the calculated value for the internal resistance constant for all trials? Find the average and the percent error for your results.

1. **K/U** Classify each of the following statements as characteristics of either a series or a parallel circuit.

   (a) The potential difference across the power supply is the same as the potential difference across each of the circuit elements.

   (b) The current through the power supply is the same as the current through each of the circuit elements.

   (c) The current through the power supply is the same as the sum of the currents through each of the circuit elements.

   (d) The equivalent resistance is the sum of the resistance of all of the resistors.

   (e) The potential difference across the power supply is the same as the sum of the potential differences across all of the circuit elements.

   (f) The current is the same at every point in the circuit.

   (g) The reciprocal of the equivalent resistance is the sum of the reciprocals of each resistance in the circuit.

   (h) The current leaving the branches of the power supply goes through different circuit elements and then combines before returning to the power supply.

2. **K/U** A series circuit has four resistors, A, B, C, and D. Describe, in detail, the steps you would take in order to find the potential difference across resistor B.

3. **K/U** A parallel circuit has three resistors, A, B, and C. Describe, in detail, the steps you would take to find the current through resistor C. Describe a completely different method for finding the same value.

4. **C** A complex circuit has three resistors, A, B, and C. The equivalent resistance of the circuit is greater than the resistance of A, but less than the resistance of either B or C. Sketch a circuit in which this would be possible. Explain why.

5. **K/U** Under what conditions would the equivalent resistance of a circuit be smaller than the resistance of any one of the loads in the circuit? Explain.

6. **I** In Investigation 13-D (Internal Resistance of a Dry Cell), why were you cautioned to open the switch when you were not taking readings?

7. **I** Study the circuit below, in which a battery is connected to four resistors in parallel. Initially, none of the switches is closed. Assume that you close the switches one at a time and record the reading on the voltmeter after closing each switch. Describe what would happen to the voltmeter reading as you close successive switches. Explain why this would happen.

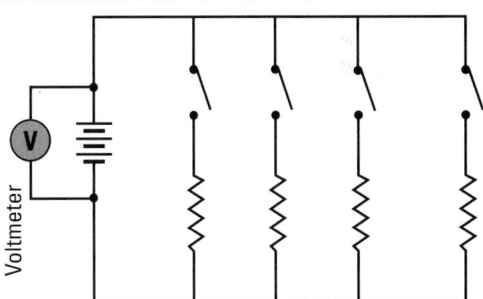

- Explain the concept and related units of electric energy.
- Compare and describe the concepts of electric energy and electric power.

KEY
TERMS

- kilowatt-hour

When a solar storm sent an unexpectedly high flow of electrons into Earth's magnetic field, the energy surge caused an overload of the power grid serving the northeastern United States. An area of more than 200 000 square kilometres was blacked out, including the cities of New York, Buffalo, and Boston. People were trapped in elevators and on subways; hospitals, including operating rooms, were plunged into darkness.

**Figure 13.25** Power outages in large cities, like this one in New York City, can cause major inconveniences.

A major power failure is not required to remind us of how dependent we are on electric energy. Even a brief interruption of the electric energy supply demonstrates how reliant we have become on electricity. A major reason why electricity has become such a dominant energy form is the ease with which it can be transmitted and with which it can be converted into other forms of energy.

## Power Output, Potential Difference, Current, and Resistance

Every appliance is rated for its power output (*P*)—the rate at which it can transform electric energy to a desired form (light, sound, or heat, for example). At the high end of the scale, an electric range might have a power rating greater than 12 000 W. An electric clothes dryer might be rated at 5000 W. At the other end of the scale, an electric shaver might be rated at 15 W and a clock at only 5 W.

In Chapter 5, you learned that power is defined as work done per unit time or energy transformed or transferred per unit time.

$$P = \frac{W}{t} \qquad \text{Power} = \frac{\text{work done}}{\text{time}}$$

$$P = \frac{\Delta E}{t} \qquad \text{Power} = \frac{\text{energy transformed}}{\text{time}}$$

The definition is the same whether you are referring to mechanical energy or electric energy. In a circuit, the loads transform electric potential energy into heat, light, motion, or other forms of energy. In power lines, electric energy is transferred from one location to another.

When working with electric circuits and systems, you typically work with potential difference, current, and resistance. It would be convenient to have relationships among power and these commonly used variables. To develop such relationships, start with the definition of potential difference.

- Definition of electric potential difference $\qquad V = \dfrac{\Delta E_Q}{Q}$

- Solve for electric potential energy. $\qquad VQ = \dfrac{\Delta E_Q}{\cancel{Q}}\cancel{Q}$

$$\Delta E_Q = QV$$

- Recall the definition of power. $\qquad P = \dfrac{\Delta E}{t}$

- Substitute the expression for electric potential energy into the equation for power. $\qquad P = \dfrac{QV}{t}$

- Recall the definition of current. $\qquad I = \dfrac{Q}{t}$

- Substitute $I$ for $\frac{Q}{t}$ in the equation for power. $\qquad P = IV$

**COURSE CHALLENGE**

**How Much Electric Power?**

Apply the electric power formulas, combined with power density information, to create a numerical simulation that tests the feasibility of space-based power to supply energy to a small community. Use data from Chapter 6, page 313, *Solar Energy Transmission from Space*, as well as other sources. Learn more from the **Science Resources** section of the following web site: ***www.school.mcgrawhill.ca/ resources/*** *Physics 11 Course Challenge*

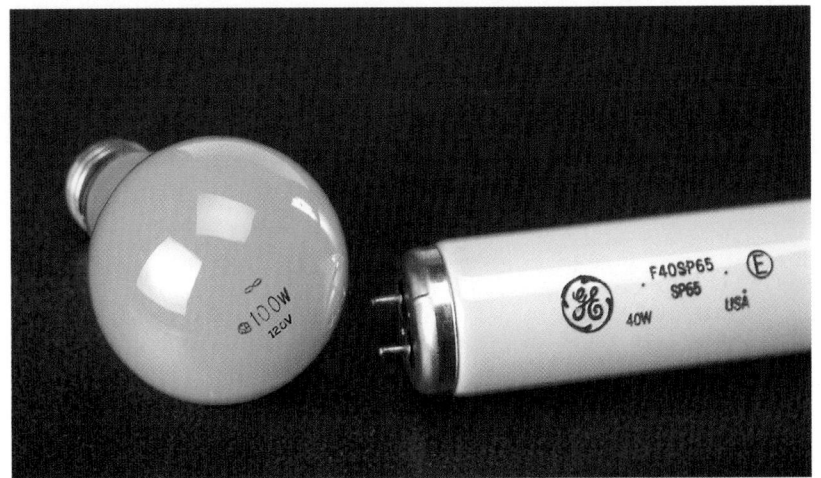

**Figure 13.26** The power rating (sometimes called the "wattage") of a light bulb tells you how fast it will convert electric energy into heat and light. For an incandescent bulb, only about 2 percent of the transformed energy is actually emitted as light; the rest is emitted as heat. A fluorescent bulb, on the other hand, converts about 9.5 percent of its energy into light, making it more than four times as efficient as an incandescent bulb.

## MODEL PROBLEM

### Calculating Electric Power

**What is the power rating of a segment of Nichrome™ wire that draws a current of 2.5 A when connected to a 12 V battery?**

### Frame the Problem

- The *power rating* of an object is the *rate* at which it *converts electric energy* into another form of energy.

- When *current* passes through Nichrome₃ wire, the resistance of the wire causes some energy to be converted into heat.

- You can calculate *power* from the *current* and *potential difference* across the ends of the wire.

## Identify the Goal

The power rating, $P$, of the wire

## Variables and Constants

| Involved in the problem | Known | Unknown |
|---|---|---|
| $P$ | $I = 2.5$ A | $P$ |
| $I$ | $V = 12$ V | |
| $V$ | | |

## Strategy

Use the equation that relates current and potential difference to power.

Substitute the known values.

An A · V is equivalent to a W.

The power rating of the segment of wire is 30 W.

## Calculations

$P = IV$

$P = (2.5$ A$)(12$ V$)$

$P = 30$ A · V

$P = 30$ W

## Validate

The units combine to give watts, which is correct for power.

Refer to the derivation on page 653 to review why current times potential difference is a correct expression for power.

## PRACTICE PROBLEMS

**39.** An electric toaster is rated at 875 W at 120 V.

  **(a)** Calculate the current the toaster draws when it is on.

  **(b)** Calculate the electric resistance of the toaster.

**40.** A light bulb designed for use with a 120 V power supply has a filament with a resistance of 240 Ω.

  **(a)** What is the power output of the bulb when the potential difference is 120 V?

  **(b)** If the bulb is inadvertently connected to an 80.0 V power supply, what would be the power output of the bulb?

  **(c)** If you wanted to construct a bulb to use with an 80.0 V power supply so that it would have the same power output as a 240 Ω bulb connected to a 120 V power supply, what should be the resistance of the bulb's filament?

**41.** A heater has a resistance of 15 Ω.

  **(a)** If the heater is drawing a current of 7.5 A, what is its power output?

  **(b)** If the current to the heater was cut in half, what would happen to the power output?

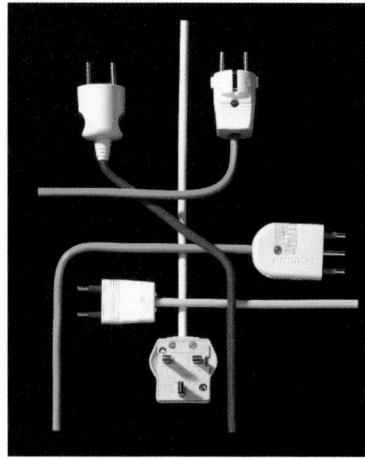
If you read the power rating on any appliance, you will notice that it is always accompanied by the potential difference that produces the specified power output. In Canada and the United States, electric energy is provided at a potential difference of 120 V, while in many other countries (including most of Europe) electric energy is provided at a potential difference of 240 V. Travellers taking appliances such as hair dryers to Europe find that these appliances are often damaged by the higher potential difference of the power supply.

**Figure 13.27** Some small hair dryers designed for travel have switches that allow you to select the voltage rating. The dryer will provide the specified power output and not be damaged, whether you plug it into a 120 V line or a 240 V line.

## MODEL PROBLEM

### Using the Wrong Power Source

**In North America, the standard electric outlet has a potential difference of 120 V. In Europe, it is 240 V. How does the dissimilarity in potential difference affect power output? What would be the power output of a 100 W–120 V light bulb if it was connected to a 240 V system?**

### Frame the Problem

- A light bulb contains a fine filament that has a certain electric *resistance*.

- The specifications (100 W–120 V) on the bulb mean that when it is connected to a 120 V potential difference, it will convert electric energy into light (and thermal energy) at 100 W.

- The resistance will remain approximately the same, even if it is connected to a different

potential difference. Therefore, according to *Ohm's law*, the *current* will be different.

- If both the *current* and *potential difference* have *changed*, the *power* will likely be different.

- The key quantity is the *resistance*.

## Identify the Goal

The power output, $P$, when the light bulb is connected to a 240 V line

## Variables and Constants

| Involved in the problem | | Known | Unknown |
|---|---|---|---|
| $R$ | $V_{240}$ | $V_{120} = 120$ V | $P_{240}$ |
| $V_{120}$ | $P_{240}$ | $P_{120} = 100$ W | $I_{240}$ |
| $P_{120}$ | $I_{240}$ | $V_{240} = 240$ V | $I_{120}$ |
| $I_{120}$ | | | $R$ |

(**Note:** Subscripts refer to the potential difference of the line to which the light bulb is connected.)

## Strategy

Since there is no direct method to calculate the power, several steps will be involved. A tree diagram will help you determine what steps to take.

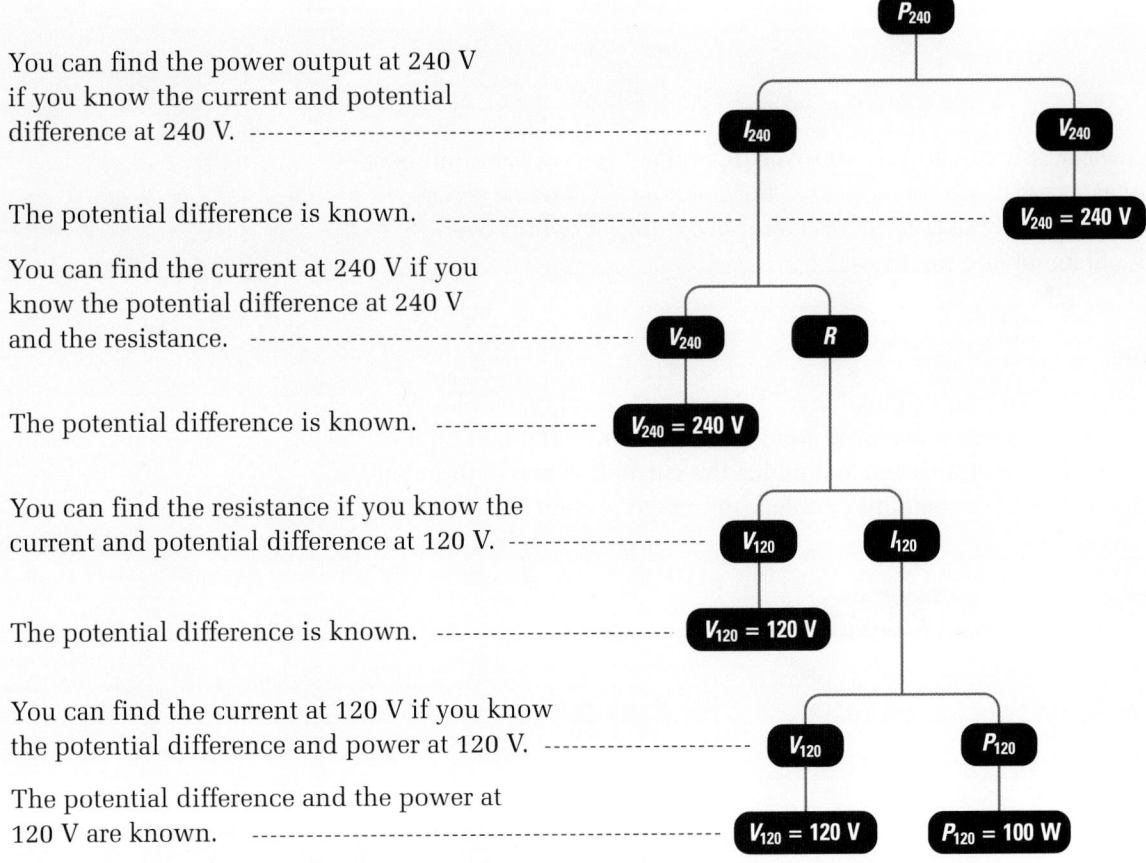

You can find the power output at 240 V if you know the current and potential difference at 240 V. - - - - - - - - - - - - - - - - - - - - - - -

The potential difference is known. - - - - - - - - - - - - - - - - - -

You can find the current at 240 V if you know the potential difference at 240 V and the resistance. - - - - - - - - - - - - - -

The potential difference is known. - - - - - - - - -

You can find the resistance if you know the current and potential difference at 120 V. - - - - - - - - - -

The potential difference is known. - - - - - - - - - - - -

You can find the current at 120 V if you know the potential difference and power at 120 V. - - - - - - - - - - -

The potential difference and the power at 120 V are known. - - - - - - - - - - - - - - - - - - - - - -

The problem is solved. Now do the calculations.

*continued* ▶

*continued from previous page*

| Strategy | Calculations |
|---|---|
| Find the current at 120 V. | $P_{120} = I_{120}V_{120}$ |
| | $I_{120} = \dfrac{P_{120}}{V_{120}}$ |
| | $I_{120} = \dfrac{100 \text{ W}}{120 \text{ V}}$ |
| | $I_{120} = 0.867 \text{ A}$ |
| Find the resistance. | $R = \dfrac{V_{120}}{I_{120}}$ |
| | $R = \dfrac{120 \text{ V}}{0.867 \text{ A}}$ |
| | $R = 144 \text{ } \Omega$ |
| Find the current at 240 V. | $I_{240} = \dfrac{V_{240}}{R}$ |
| | $I_{240} = \dfrac{240 \text{ V}}{144 \text{ } \Omega}$ |
| | $I_{240} = 1.73 \text{ V}$ |
| Find the power output at 240 V. | $P_{240} = I_{240}V_{240}$ |
| | $P_{240} = (1.73 \text{ A})(240 \text{ V})$ |
| | $P_{240} = 400 \text{ W}$ |

The power output at 240 V is 400 W. Notice that the power output is four times as great when you double the potential difference across a given resistance. Such an increase in power output would over-heat the filament and melt it.

## Validate

Although it might seem strange that the power output quadruples when the potential difference doubles, it is logical. Doubling the potential difference for a given resistance doubles the current. When both potential difference and current are doubled, the power output is quadrupled.

## PRACTICE PROBLEMS

42. (a) What is the power output of a 45 $\Omega$ resistance when connected to

   ■ a 180 V power supply?

   ■ a 270 V power supply?

   (b) Find the ratio of the potential differences and the ratio of the power outputs. What is the relationship between the two ratios?

43. A load has a power rating of 160 W when the current in it is 6.0 A. What will be the power output if the current increases to 15 A?

44. (a) What is the power output of a circuit that consists of a 25 $\Omega$ resistance when connected to a 100 V supply?

   (b) If a second 25 $\Omega$ resistance is connected in series with the first, what will be the power output of the circuit? Why has the power output of the circuit decreased?

You probably noticed in the model problem and practice problems on the preceding pages that the resistance of a device is often the key to finding the power output. It would be convenient if relationships were available that would relate power output, resistance, and potential difference ($P$, $R$, and $V$) or power output, resistance, and current ($P$, $R$, and $I$). Since Ohm's law relates resistance, current, and potential difference, you can use it to develop more relationships for power.

- Start with the relationship between potential difference, current, and power.

$$P = IV$$

- To eliminate current and to introduce resistance, write Ohm's law in terms of current.

$$I = \frac{V}{R}$$

- Substitute $\frac{V}{R}$ in place of $I$ in the equation for power.

$$P = \frac{V}{R}V$$

- Simplify.

$$P = \frac{V^2}{R}$$

To find a relationship between $P$, $R$, and $I$, use a similar procedure, but eliminate $V$ from the equation for power.

- Start with the first equation for power.

$$P = IV$$

- Express Ohm's law in terms of potential difference.

$$V = IR$$

- Substitute $IR$ in place of $V$ in the equation for power.

$$P = I(IR)$$

- Simplify.

$$P = I^2R$$

---

**ALTERNATIVE EQUATIONS FOR POWER**

Power is the quotient of the square of the potential difference and the resistance.

$$P = \frac{V^2}{R}$$

Power is the product of the square of the current and the resistance.

$$P = I^2R$$

| Quantity | Symbol | SI unit |
|---|---|---|
| power | $P$ | W (watt) |
| potential difference | $V$ | V (volt) |
| resistance | $R$ | $\Omega$ (ohm) |
| current | $I$ | A (ampere) |

*continued from previous page*

**Unit Analysis**

Recall from the definitions of current and potential difference

$$A = \frac{C}{s} \qquad V = \frac{J}{C}$$

From Ohm's law, $V = A \cdot \Omega$, therefore $\frac{V}{\Omega} = A$.

Use these relationships to analyze the units for the two power formulas.

$$(\text{watt}) = \frac{(\text{volt})^2}{(\text{ohm})} \qquad\qquad (\text{watt}) = (\text{ampere})^2(\text{ohm})$$

$$= \frac{V^2}{\Omega} \qquad\qquad\qquad = (A)^2(\Omega)$$

$$= \left(\frac{V}{\Omega}\right)V \qquad\qquad = (A)(A \cdot \Omega)$$

$$= A \cdot V \qquad\qquad\quad = (A)(V)$$

$$= \left(\frac{\cancel{C}}{s}\right)\left(\frac{J}{\cancel{C}}\right) \qquad = \left(\frac{\cancel{C}}{s}\right)\left(\frac{J}{\cancel{C}}\right)$$

$$= W \qquad\qquad\qquad = \frac{J}{s}$$

$$\qquad\qquad\qquad\qquad\quad = W$$

---

## MODEL PROBLEM

### Resistance and Power

An electric kettle is rated at 1500 W for a 120 V potential difference.

**(a)** What is the resistance of the heating element of the kettle?

**(b)** What will be the power output if the potential difference falls to 108 V?

### Frame the Problem

- An electric kettle will use *energy* at a *rate* of 1500 W if connected to a 120 V line.

- *Power*, *potential difference*, *current*, and *resistance* are all related.

### Identify the Goal

The resistance, $R$, of the heating element of the kettle
The power, $P$, if the potential difference decreases

## Variables and Constants

**Involved in the problem**

| | |
|---|---|
| $P_1$ | $P_2$ |
| $V_1$ | $V_2$ |
| $R$ | |

**Known**

$P_1 = 1500$ W

$V_1 = 120$ V

$V_2 = 108$ V

**Unknown**

$R$

$P_2$

## Strategy

Use the relationship among power, potential difference, and resistance.

## Calculations

$$P_1 = \frac{V_1{}^2}{R}$$

**Substitute first**

$$1.50 \times 10^3 \text{ W} = \frac{(1.20 \times 10^2 \text{ V})^2}{R}$$

$$(1.50 \times 10^3 \text{ W})(R) = \frac{(1.20 \times 10^2 \text{ V})^2}{\cancel{R}}(\cancel{R})$$

$$\frac{(\cancel{1.50 \times 10^3 \text{ W}})(R)}{\cancel{1.50 \times 10^3 \text{ W}}} = \frac{(1.20 \times 10^2 \text{ V})^2}{1.50 \times 10^3 \text{ W}}$$

$$R = 9.60 \frac{\text{V}^2}{\text{W}}$$

$$R = 9.60 \ \Omega$$

**Solve for $R$ first**

$$P_1 R = \frac{V_1{}^2}{\cancel{R}}\cancel{R}$$

$$P_1 R = V_1{}^2$$

$$\frac{\cancel{P_1} R}{\cancel{P_1}} = \frac{V_1{}^2}{P_1}$$

$$R = \frac{V_1{}^2}{P_1}$$

$$R = \frac{(1.2 \times 10^2 \text{ V})^2}{1.50 \times 10^3 \text{ W}}$$

$$R = 9.60 \frac{\text{V}^2}{\text{W}}$$

$$R = 9.60 \ \Omega$$

**Unit Check:** $\dfrac{\text{V}^2}{\text{W}} = \dfrac{\frac{\text{J}^2}{\text{C}^2}}{\frac{\text{J}}{\text{s}}} = \left(\dfrac{\text{J} \cdot \text{J}}{\text{C}^2}\right)\left(\dfrac{\text{s}}{\text{J}}\right) = \left(\dfrac{\text{J}}{\text{C}}\right)\left(\dfrac{\text{s}}{\text{C}}\right) = \text{V}\dfrac{1}{\text{A}} = \Omega$

**(a)** The resistance of the coil in the kettle is 9.60 $\Omega$.

Find the power output when the potential difference is 108 V.

$$P_2 = \frac{V_2^2}{R}$$

$$P_2 = \frac{(108 \text{ V})^2}{9.60 \ \Omega}$$

$$P_2 = 1215\frac{\text{V}^2}{\Omega}$$

$$P_2 = 1.22 \times 10^3 \text{ W}$$

See the box on page 660 for a unit analysis.

**(b)** The power output drops to $1.22 \times 10^3$ W when the potential difference drops to 108 V.

(**Note:** This is a 10% loss of potential difference and a 19% loss in power. The percentage decrease in power is nearly double that of the decrease in potential difference because power is proportional to the square of the potential difference.)

*continued* ▶

*continued from previous page*

## Validate

The units combine properly to give ohms in part (a), and watts in part (b). From past experience, you would expect a proportionately larger drop in the power output than in potential difference.

---

### PRACTICE PROBLEMS

**45.** A filament in a light bulb rated at 192 W, has a resistance of 12.0 Ω. Calculate the potential difference at which the bulb is designed to operate.

**46.** An electric kettle is rated at 960 W when operating at 120 V. What must be the resistance of the heating element in the kettle?

**47.** If a current of 3.50 A is flowing through a resistance of 24.0 Ω, what is the power output?

**48.** A toaster that has a power rating of 900 W $(9.00 \times 10^2 \text{ W})$ draws a current of 7.50 A. If $2.40 \times 10^5$ J of electric energy are consumed while toasting some bread, calculate how much charge passed through the toaster.

**49.** A floodlight filament has an operating resistance of 22.0 Ω. The lamp is designed to operate at 110 V.

**(a)** What is its power rating?

**(b)** How much energy is consumed if you use the lamp for 2.50 hours?

---

## Energy Consumption

The electric meter at your home gives readings in **kilowatt-hours** (kW · h). One kilowatt-hour represents the energy transformed by a power output of 1000 W for one hour. This is equivalent to $3.6 \times 10^6$ J (3.6 MJ). A typical charge by an energy company for consumed energy might be $0.07 per kW · h. That means that for only seven cents, you can buy enough energy to lift 360 kg a vertical distance of more than 1 km.

**Figure 13.28** The electric meter in your home measures how much energy (in kW · h) that you use.

## The Cost of Watching Television

**A family has its television set on for an average of 4.0 h per day. If the television set is rated at 80 W and energy costs $0.070 per kW · h, how much would it cost to operate the set for 30 days?**

## Frame the Problem

- The total amount of *electric energy* a television set uses in a specific amount of *time* depends on its *power output*.

- Power companies charge a specific amount of *dollars* per unit of *energy* used.

## Identify the Goal

The cost, in dollars, for operating a television set for 30 days

## Variables and Constants

| Involved in the problem | Known | Unknown |
|---|---|---|
| $\Delta t$        Cost | $\Delta t = (4.0 \text{ h/day})(30 \text{ days})$ | *Cost* |
| $P$        $\Delta E_e$ | rate = $0.070 kW · h | $\Delta E_e$ |
| rate | $P = 80$ W | |

## Strategy

Find the total time, in hours, that the television set typically is on during one 30-day period.

Find the total amount of energy consumed by using the definition for power.

Find the cost.

The cost of operating a television set for 4.0 h per day is $0.67 for 30 days.

## Calculations

$$\Delta t = \left( 4.0 \frac{\text{h}}{\cancel{\text{day}}} \right)(30 \, \cancel{\text{days}})$$

$$\Delta t = 120 \text{ h}$$

$$P = \frac{\Delta E_e}{\Delta t}$$

**Substitute first**

$$\text{Cost} = \text{rate} \cdot \Delta E_e$$

$$(80 \text{ W})(120 \text{ h}) = \frac{\Delta E_e}{\cancel{120 \text{ h}}} \cancel{120 \text{ h}}$$

$$\Delta E_e = 9600 \text{ W} \cdot \text{h}$$

$$\text{Cost} = \text{rate} \cdot \Delta E_e$$

$$\text{Cost} = \frac{\$0.070}{\cancel{\text{kW} \cdot \text{h}}} \, 9600 \, \cancel{\text{W}} \cdot \cancel{\text{h}} \, \frac{1 \, \cancel{\text{kW}}}{1000 \, \cancel{\text{W}}}$$

$$\text{Cost} = \$0.67$$

**Solve for $\Delta E_e$ first**

$$(P)(\Delta t) = \frac{\Delta E_e}{\cancel{\Delta t}} = \cancel{\Delta t}$$

$$(P)(\Delta t) = \Delta E_e$$

$$\Delta E_e = (80 \text{ W})(120 \text{ h})$$

$$\Delta E_e = 9600 \text{ W} \cdot \text{h}$$

*continued* ▶

*continued from previous page*

## Validate

The units combine and cancel to give dollars. At first, the cost seems to be low. However, a television set does not use as much energy as many other appliances. As well, electric energy is relatively inexpensive.

---

### PRACTICE PROBLEMS

**50.** It takes 25.0 min for your clothes dryer to dry a load of clothes. If the energy company charges 7.20 cents per kW · h, how much does it cost to dry a load of clothes in a 1250 W dryer?

**51.** It takes 12.0 min to dry your hair using a blow dryer that has a resistance of 21.0 Ω and draws a current of 5.50 A. Calculate the cost to dry your hair if electricity costs 8.50 cents per kW · h.

**52.** An electric kettle, which operates on a 120 V supply, has a heating element with a resistance of 10.0 Ω. If it takes 3.2 min to boil a litre of water and the energy charge is 6.5 cents per kW · h, calculate

**(a)** the power rating of the kettle

**(b)** the cost to boil the water

---

## 13.5    Section Review

**1.** **K/U** Write four different equations that could be used to find the power consumed by a circuit element. Explain why it is useful to have so many different equations for power.

**2.** **C** The brightness of a light bulb is directly related to the power it consumes. In the circuits in the diagram on the right, the light bulbs have the same ratings. In which circuit will the light bulbs glow most brightly? Explain your reasoning.

**3.** **I** Using unit analysis, show that a kilowatt-hour is a unit of energy.

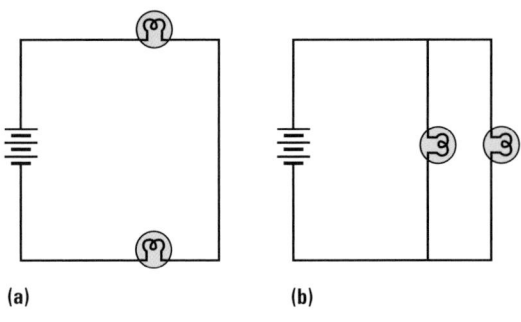

**(a)**          **(b)**

### UNIT PROJECT PREP

Think about these questions in the analysis of your motor.
- How powerful will your electric motor be?
- What factors will affect the power output of your motor?
- How can you determine the power of your motor?

## REFLECTING ON CHAPTER 13

- Materials are classified as conductors, insulators or semi-conductors depending on their ability to pass electric currents.

- The voltaic pile was the first battery. Its invention allowed physicists to carry out the first investigations with current electricity. Potential difference is a measure of the amount of energy that is available per unit charge. One volt is the potential difference that will provide one joule of energy to every coulomb of charge transferred.

- The rate at which charge moves through a circuit is called electric current. The unit of current, the ampere, is the current that transfers charge at the rate of one coulomb per second.

- Current is, by definition, the movement of positive charges. The movement of negative charge is called electron flow.

- Electric charge exists only in whole-number mulitples of the elementary charge, $(1\ e = 1.60 \times 10^{-19}\ C)$

- When a current flows through a load, the resistance of the load converts electric energy to heat, light or other forms of mechanical energy. This results in a potential drop across the load.

- The resistance of a conductor of a particular material varies directly as its length and inversely as its cross-sectional area.

- Ammeters are connected in series with loads; voltmeters are connected in parallel with loads.

- Ohm's law describes the relationship between the potential difference across, the current through and the resistance of a load.

- When loads are connected in series, the equivalent resistance is the sum of their resistances. When loads are connected in parallel, the equivalent resistance is the inverse of the sum of the inverses of their resistances.

- The electromotive force (*emf*) of a battery is the maximum potential difference that the battery can create. When a current flows, some of the *emf* is used to move the current through the internal resistance of the battery. The terminal voltage of the battery is the *emf* less the potential difference across the internal resistance. The terminal voltage is the potential difference available to the circuit to which the battery is attached.

- In a circuit composed of several loads, the equivalent resistance of the circuit is the resistance offered to the terminal voltage of the power supply by all the loads in the circuit.

- The power output (rate of energy transfer) of a circuit is measured in watts. Electric energy is sold in units called kilowatt hours (kW · h).

### Knowledge/Understanding

1. Describe the difference between potential energy and potential difference.

2. What is the difference between electric current and electron flow?

3. Explain why a light bulb should actually be considered to be a non-ohmic resistor.

4. What is the difference between *emf* and terminal voltage?

5. List the four factors that affect the resistance of a conductor.

6. What is the difference between a battery and a cell?

7. What is meant by the resistivity of a conductor?

8. Define what is meant by the term "elementary charge."

9. Materials are classified by their ability to conduct electricity. Describe the three classifications of materials and give an example of a material in each group.

10. Explain the difference between power and energy.

**11.** Why does connecting loads in parallel reduce the equivalent resistance of a circuit?

**12.** A simple circuit consists of a resistance connected to a battery. Explain why connecting two additional resistances in series with the first reduces the power output of the circuit, yet if you connect the two additional resistances in parallel with the first load, the power output increases.

**13.** What is the significance of Millikan's oil-drop experiment?

## Inquiry

**14.** A 100 W light bulb designed to operate at 120 V has a filament with a resistance of 144 Ω. What constraints are on the design of the filament? Why cannot just any 144 Ω resistance be used as a filament for the bulb?

**15.** Make voltaic piles by using five pennies and five nickels separated by paper towels soaked in (a) salt water, (b) vinegar, and (c) dilute sulfuric acid.

- In each case, measure the potential difference of the pile with a voltmeter. Use the voltmeter to measure the potential difference of a single cell. What is the relationship between the values?
- How does the liquid (the electrolyte) on the paper towel affect the potential difference of the pile?

**16.** Research the design and make sketches of the (a) dry cell, (b) lead storage battery, and (c) nickel-cadmium battery. (This information is available on the Internet or in chemistry or physics textbooks.) Label and specify the material or chemical used for the anode, cathode, and electrolyte. What are the advantages and disadvantages of each type of cell or battery?

**17.** The electric circuit to which a television set is connected is protected by a 15 A circuit breaker. If the power rating of the set is 450 W, how many 100 W light bulbs can be operated as well on this circuit without overloading the circuit breaker?

**18.** In the circuit diagram below, values for some of the quantities for each part of the circuit are given. Calculate the missing currents, resis-tances, and the potential differences for each of the loads in the circuit. Find the equivalent resistance for the circuit and the power output for the circuit.

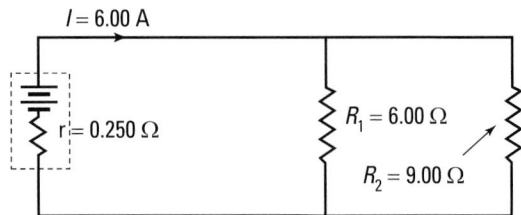

## Communication

**19.** Describe how series and parallel circuits differ in terms of current flow and potential difference.

**20.** On a holiday to Canada, a resident of England decides to bring his electric razor. He has heard that the plugs are different so he buys an adapter that will fit into an electric socket in Canada. When he plugs in and turns on his razor in Canada, it runs very slowly and is very weak. Explain why.

## Making Connections

**21.** When you turn up the heating element of an electric stove from low to high, are you increasing or decreasing the resistance of the circuit? Explain.

**22.** Find the power rating for several of the appliances you have in your home (TV, clothes dryer, electric kettle, iron, hair dryer, and vacuum cleaner, for example). Estimate, for each appliance, how long it is used per month. Make a data table with the column headings: Name of Appliance, Power Rating (kW), Time Used per Month (h), Energy Use per Month (MJ), Cost of Operation (find the charge per kW · h on your electric bill). If you know how, you could use an electronic spreadsheet to organize your calculations.

23. When electricity is transmitted over long distances, very high voltages are used. Consider a hydo-electric plant that has a power output of 25.0 MW. This power output can be achieved by transmitting a small current at a high voltage or a large current at a low voltage.
    (a) What would be the current if it was transmitted at a potential difference of 25.0 MV?
    (b) What would be the current if it was transmitted at a potential difference of 25 kV?
    (c) The transmission lines have a resistance of 0.0100 $\Omega$/km. In each case, how much of the potential difference would be used to push the current through 1000 km of line?

## Problems for Understanding

24. A light bulb is rated at 200 W for a 32.0 V power supply. What is its power output if it is inadvertently connected to a 120 V supply?

25. A 1400 W–120 V toaster requires 3.60 minutes to toast a slice of bread.
    (a) What current does it draw?
    (b) How much charge passes through the toaster in that time?
    (c) How much heat and light would be produced in that time?

26. The heating element of a stove operates at 240 V. How much electric energy is converted to heat if it takes 5.50 minutes to bring a pot of water to a boil? The element draws a current of 6.25 A.

27. How much does it cost to run a 15.0 $\Omega$ load for 12.0 minutes on a 125 V supply if the rate for electric energy is $0.0850/kWh?

28. The equivalent resistance of two loads connected in parallel is 25.0 $\Omega$. If the resistance of one of the loads is 75.0 $\Omega$, what is the resistance of the other load?

29. A load, $R_1$, is connected in series with two loads, $R_2$ and $R_3$, which are connected in parallel with each other. If the potential difference of the power supply is 180 V, find the current through and the potential difference across each of the loads. The loads have resistances of 25.0 $\Omega$, 30.0 $\Omega$ and 6.00 $\Omega$, respectively.

30. A motor draws a current of 4.80 A from a 36.0 V battery. How long would it take the motor to lift a 5.00 kg mass to a height of 35.0 m? Assume 100% efficiency.

31. A 45.0 m extension cord is made using 18 gauge copper wire. It is connected to a 120 V power supply to operate a $1.0 \times 10^2$ W-120 V light bulb.
    (a) What is the resistance of the extension cord? (Remember that there are two wires to carry the current in the cord.)
    (b) What is the resistance of the filament in the light bulb?
    (c) What is the current through the cord to the light bulb?
    (d) What is the actual power output of the light bulb?

32. When a battery is connected to a load with a resistance of 40.0 $\Omega$, the terminal voltage is 24.0 V. When the resistance of the load is reduced to 15.0 $\Omega$, the terminal voltage is 23.5 V. Find the *emf* and the internal resistance of the battery.

---

**Numerical Answers to Practice Problems**

1. 20.0 V  2. 0.378 J  3. $6.5 \times 10^{-2}$ C  4. 40.0 V  5. 8.0 s
6. $4.23 \times 10^3$ J  7. 50 A  8. 57 s  9. $7 \times 10^4$ C  10. 2.8 A
11. 0.73 A  12. 0.133 A  13. (a) 9.38 A (b) $2.11 \times 10^{22}$ elementary charges  14. $5.25 \times 10^{20}$ elementary charges  15. (a) 3.3 A
(b) 1.7 V  16. 2.2 $\Omega$  17. 4.08 m  18. $1.6 \times 10^{-6}$ m  19. 0.45 $\Omega$
20. 2.4 mm  21. 16 $\Omega$  22. 12.5 A  23. 5.0 V  24. (a) $9.9 \times 10^2$ C
(b) $1.2 \times 10^4$ J  25. 11.6 $\Omega$  26. 7.50 min  27. 33 V, 53 V and 79
V respectively (b) 75 $\Omega$ (c) $1.6 \times 10^2$ V  28. (a) 91.0 V (b) 156 V
29. 42.0 $\Omega$  30. (a) 8.00 $\Omega$ (b) 224 V (c) 32.0 $\Omega$  31. 44.0 $\Omega$
32. 0.667 A, 1.00 A and 1.33 A respectively; 3.00 $\Omega$
33. $R_{coil}$ = 6.00 $\Omega$, $R_{bulb}$ = 20.0 $\Omega$, $R_S$ = 4.62 $\Omega$
34. $R_{unknown}$ = 8.00 $\Omega$, $R_S$ = 4.80 $\Omega$  35. (a) 11.2 $\Omega$ (b) 21.6 $\Omega$,
30.0 $\Omega$  36. (a) 38.4 $\Omega$ (b) 2.25 A (c) 91.5 V  37. (a) 15.4 $\Omega$
(b) 9.76 V (c) 1.02 A  38. (a) 14.8 V (b) 14.6 V  39. (a) 11.4 V
(b) 11.2 V  40. (a) 7.3 A (b) 16 $\Omega$  41. (a) $6.0 \times 10^1$ W (b) 27 W
(c) $1.1 \times 10^2$ $\Omega$  42. (a) 840 W (b) The power output drops to $\frac{1}{4}$
its original value, or 210 W  43. $P_a$ = 720 W, $P_b$ = $1.6 \times 10^3$ W
(b) $P_a/P_b$ = 4/9; $V_a/V_b$ = 2/3; $P_a/P_b$ = $(V_a/V_b)^2$  44. $1.0 \times 10^3$ W
45. (a) 400 W (b) 200 W. Increasing resistance decreased the current for the given potential difference.  46. 48.0 V
47. 15 $\Omega$  48. 294 W  49. $2.00 \times 10^3$ C  50. (a) 550 W (b) $5.0 \times 10^6$ J
51. 3.75 cents  52. 1.08 cents  53. (a) $1.4 \times 10^2$ W (b) 0.50 cents

lectromagnets come in all sizes. Large, powerful electromagnets are used in industry to move heavy materials. Tiny electromagnets are used to read and write the data on the disk drive of a microcomputer. While electromagnetism is a common phenomenon today, it was first discovered purely by accident in 1819. Until that date, electricity and magnetism were believed to be completely separate phenomena. Immediately recognized for its importance, the discovery of electromagnetism marked the birth of modern science and technology. Without an understanding of electromagnetism, devices such radios, televisions, computers, tape recorders, VCRs, CD players, lasers, electric motors, and generators, could not have been invented.

In this chapter, you will study the properties of natural magnetism and electromagnetism. You will investigate how electricity and magnetism are related. You will also learn how electromagnetism is used to create electromagnetic devices such as electromagnets, loudspeakers, motors, and meters.

## Invisible Lines

Use a string to create a hanger for a bar magnet. Allow the magnet to hang freely, away from the influence of any other magnets or magnetic materials. Allow enough time for the magnetic to come to rest in the absence of vibrations or air currents.

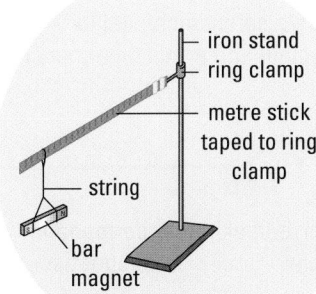

### Analyze and Conclude

1. Note and record the orientation of the long axis of the magnet once the magnet has come to rest.

2. Compare the final direction of your magnet's rest position with the positions of your classmates' magnets.

3. Draw conclusions about the final orientation of the suspended bar magnet.

## Magnets and Materials

In this activity, you will test the properties of magnetic forces between magnets in the presence of various materials using a sensitive magnetic compass. Place a compass on your desktop. Once the needle has come to rest, position the body of the compass so the 0° mark is under the N-pole of the compass needle. Place a bar magnet on a line perpendicular to the axis of the compass needle. Slowly move the bar magnet closer to the compass until it causes the compass needle to deviate about 20° from its original position. Place several different types of materials between the compass and the magnet. Carefully observe any change in the direction of the compass needle. Test the effect of changing the orientation of the material separating the compass and the magnet. Use a second bar magnet to determine which substances are magnetic and which ones are not. Organize your observations in a table. Possible substances to test: copper, zinc, aluminum, iron, lead, plastic, glass, wood

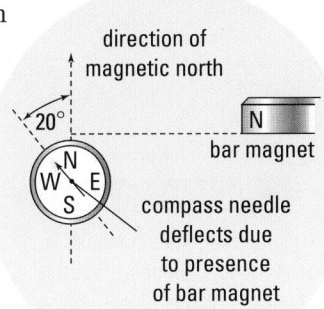

### Analyze and Conclude

1. Did the presence of the material placed between the magnet and the compass affect the amount of deflection of the compass needle?

2. Did the orientation of the material placed between the magnet and the compass affect the amount of deflection of the compass needle?

3. Draw conclusions about the ability of different materials to affect the interaction between magnets (the bar magnet and the compass needle magnet)? How do you think this occurs?

### SECTION EXPECTATIONS

- Define and describe magnetism.

- Use a scientific model to explain and predict magnetism.

### KEY TERMS

- north-seeking pole
- south-seeking pole
- magnetic dipole
- domain
- temporary magnetism
- permanent magnetism
- Curie point
- dipping needle
- magnetic dip
- magnetic declination
- isogonic lines

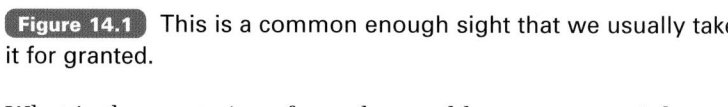

**Figure 14.1** This is a common enough sight that we usually take it for granted.

What is the mysterious force that enables some materials to attract others? The first descriptions of magnetism go back to over 500 years B.C.E. A Greek, named Thales, wrote about a rock that came from the town of Magnesia. Thales (624? – 547 B.C.E.) referred to it as the "magnetes lithos" (Greek for "Magnesian rock"). Eventually, rocks that displayed natural magnetism came to be known as lodestone. Chemical analysis has shown that lodestone gets its magnetic properties from the presence of an oxide of iron, $Fe_3O_4$, known today as magnetite or magnetic iron ore.

## Magnetic Poles

When you were doing the Multi Lab on page 3, you probably realized that a compass needle is simply a small bar magnet which aligns itself on the north-south axis for the same reason as the suspended bar magnet. In the Northern Hemisphere, the presence of the North Star or Polaris, has always been a powerful symbol. Early observers assumed that the compass end was seeking the North Pole, hence the name **north-seeking pole** was given to the end of the magnet that seemed to always end up pointing in that direction. Similarly, the name **south-seeking pole** was given to the other end of the magnet. Gradually, these became known simply as the **North pole (N-pole)** and **South pole (S-pole)** of the magnet.

No matter how often you break a magnet into pieces, each piece is a complete magnet with an N-pole and an S-pole. Scientists assumed that you could do this until each piece had only one atom or molecule. This led scientists to believe that the individual atoms of a material must be magnets. Because magnets always seem to come with an N-pole paired with an S-pole, the bar magnet is often referred to as a **magnetic dipole**. The reason why magnetic poles always come in pairs will become clear when you study electromagnetism.

For early scientists, magnetism was a much easier field to study than static electricity. First, magnets did not discharge when touched and therefore were easier to manipulate. Second, magnetism had a very practical aspect in its direction-finding capabilities. For nations that were interested in exploration, the true nature of magnets was a very important topic.

The distance between magnets affects the strength of their interaction. This has been known since at least C.E. 1300. In 1785, Coulomb devised an experiment, for which he invented the torsion balance, to find the mathematical relationship between the separation and the force. He proved that the force of interaction between magnetic poles, attraction or repulsion, was inversely proportional to the square of the separation between them. Mathematically this is written as:

$$\vec{|F|} \propto \frac{1}{d^2}$$

$\vec{|F|}$ is the magnitude of the magnetic force between the poles and $d$ is the distance between the poles.

This explains why the N-pole of a bar magnet can exert a net force of attraction or repulsion on another bar magnet. When two magnets approach each other, it is almost certain that one pair of poles is going to be closer together. It is the interaction between the two poles nearest each other that will dominate the interaction of the two magnets. Thus, if the poles nearest to each other are unlike poles, then the force you observe will be one of attraction.

## RULES FOR MAGNETIC INTERACTIONS

1. Like poles repel each other.
2. Unlike poles attract each other.
3. The force of attraction varies inversely as the square of the distance between of the poles.

**TRY THIS...**

What happens if you break a magnet in half? Do you get a separate N-pole and S-pole? Take a bar magnet and stroke the side of a hacksaw blade several times in the same direction with one pole of the bar magnet. When you have made about 20 strokes, confirm that the blade has become magnetized by bringing its ends near the poles of a compass. Does a force of attraction between the blade and the poles of the compass prove it is magnetized? Explain.

**Wrap the blade in a piece of cloth to protect your eyes and hands,** and break the blade in half. (They are relatively easy to break.) In turn, hold the ends of each half of the blade near the magnet. Do you have two magnets, each with an N-pole and a S-pole, or are there now separate N- and S-poles?

Wrap one section of the blade in the napkin, and break it in half again. Test each of the new parts of the blade for magnetism.

**History Link**

A compass invented in China about C.E. 1000 consisted of a spoon made from lodestone that was placed on a bronze plate. The spoon represented the constellation Ursa Major (Big Dipper), with the handle end of the spoon representing Ursa Major pointing away from the North Star. Why did they make the base plate of bronze?

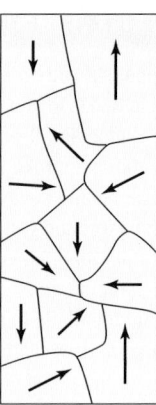

**Figure 14.2** Because the magnetic domains are randomly oriented, the material displays no net magnetic polarity. (The arrows represent the magnetic polarity of the domains.)

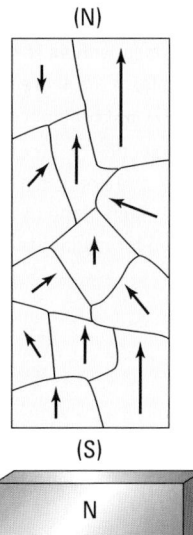

**Figure 14.3** The external magnet has caused the orientation of the domains to shift so that they are in greater alignment with its polarity.

## Magnetic Domains

Imagine you have a box containing thousands of very tiny, very weak magnets. If you shake the box up, the magnets inside the box would tend to orient themselves randomly. A compass needle brought near the box might detect no appreciable net magnetic polarity for the box, even though you know it contains many magnets. Because the magnets inside the box are randomly oriented, more or less, the force of attraction from one magnet or group of magnets will be counteracted by the force of repulsion by other magnets or groups of magnets.

If you opened the box, what would you see? Probably there would be many regions, called **domains**, where a group of tiny magnets were lined up with each other, giving the domain a magnetic polarity (see Figure 14.2). Beside one domain, there might be another domain where the magnets, by chance, were aligned together but in a different orientation to the first domain. Interspersed throughout the box, there might be many domains of different sizes, strengths and orientations. In reality, the tiny magnets which make up the domains are actually the atoms or the molecules of the material.

Now imagine what would happen inside the box if an external magnet is brought near one end of the box. The pole of the external magnet that was near the box would exert a force on the domains and would try to align the domains. Even though the domains are not totally free to move, the force would shift domain orientations into greater alignment. In this way, the direction of the magnetic domains in the box would take on the same polarity in the same direction as the external magnet. The box is now a magnet that is attracted to the external magnet (see Figure 14.3). When the external magnet is removed, the domains would tend to sort themselves randomly again and the polarity of the box would become much weaker and even disappear completely.

In domain theory, the material is affected by the presence of a magnet if the atoms or molecules of the material are magnets. A domain is a group of adjacent atoms whose like poles have "like" orientation within the material. When the domains of a material are randomly oriented, the material shows no permanent magnetism. The presence of an external magnet can induce the domains to become aligned, more or less, with that of the external magnet. Thus, the material becomes a magnet in its own right.

In some magnetic materials, such as iron, the microscopic domains are easily reoriented in the direction parallel to an externally applied field. However, when the external magnet is removed, the domains return to their random orientations and the magnetism disappears. Thus, iron forms a **temporary magnet**. In other materials, such as steel, the internal domains are reoriented only with considerable difficulty. When the external magnet is removed, however random realignment of the domains is also difficult. Thus the material will retain its magnetic properties. These types of materials form **permanent magnets**.

Even though permanent magnets seem to be quite stable, they are fairly fragile. If a magnet is heated, its strength weakens but that strength will generally return when the magnet cools. However, if the magnet is heated above a certain temperature, called the **Curie point**, the magnet will be totally destroyed. Table 14.1 lists the Curie Point for several materials displaying permanent magnetism.

**Table 14.1** Curie Points for Magnetic Materials

| Material | Curie Point (°C) |
|---|---|
| iron | 770 |
| cobalt | 1131 |
| nickel | 358 |
| magnetite | 620 |
| gadolinium | 16 |

## The Magnet Earth

Prior to the sixteenth century, the reason why magnets aligned themselves to point toward the North Star had been a subject of great speculation. Some argued that there must be a huge mass of iron ore at the North Pole. Others suggested that Earth itself must be a magnet. An English scientist named William Gilbert put the argument to rest in 1600. Gilbert fashioned a lodestone into the shape of a sphere. After using a compass to locate the poles of his lodestone sphere, he decided that, since it attracted the N-pole of a compass, the magnetic pole in the Arctic region of Earth had to be a magnetic S-pole. The Antarctic region must contain a magnetic N-pole. A compass needle placed on the surface of Gilbert's lodestone sphere behaved like a compass on the surface of Earth.

Gilbert went one step further to prove his theory. Normally, a compass is held horizontally so that its needle rotates about a vertical axis. Gilbert built a compass that could pivot in the vertical plane about a horizontal axis. When this compass was brought near his lodestone sphere, it dipped toward the surface of the sphere. The closer it came to one pole or other of the sphere, the more it dipped (pointed toward the sphere's surface). At the lodestone sphere's equator, the needle of this compass was parallel to the surface. This type of compass is called a **dipping needle** (Figure 14.4) and measures **magnetic dip**. At the magnetic equator, the amount of dip is zero; at either magnetic pole, it is 90°. A dipping needle can be used to locate one's position with respect to the magnetic equator.

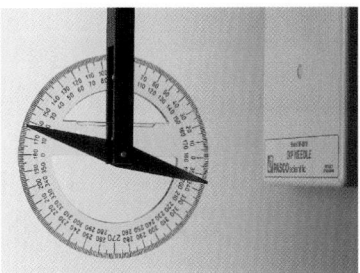

**Figure 14.4** As the dipping needle is moved farther from the magnetic equator and closer to a magnetic pole, the amount of dip increases. At either magnetic pole, the dip is 90°.

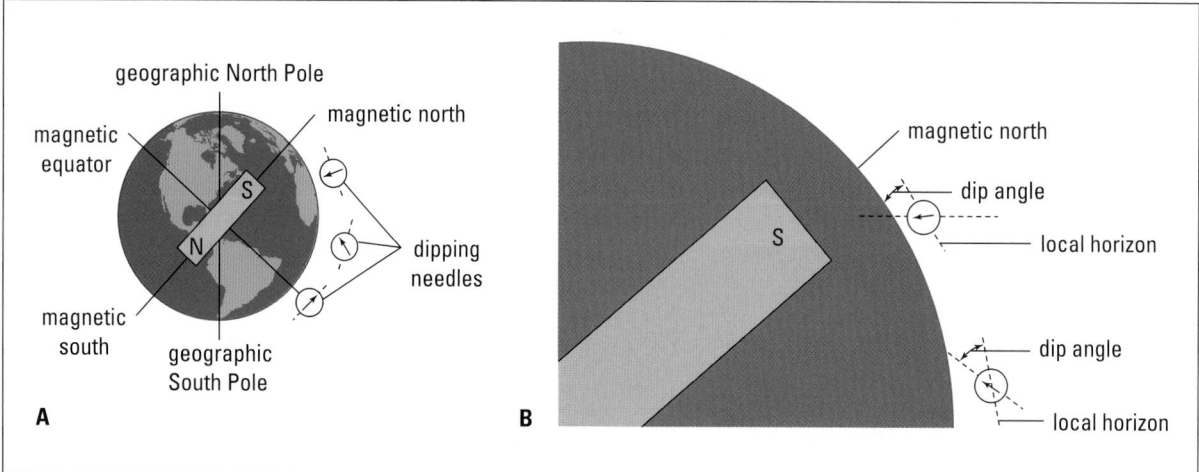

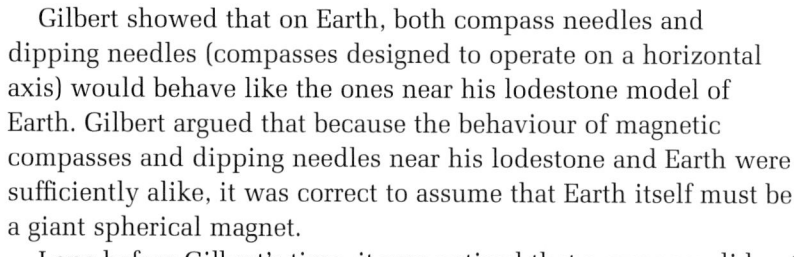

A                                        B

**Figure 14.5** In 1600, Gilbert showed that Earth had the properties of a magnet.

Gilbert showed that on Earth, both compass needles and dipping needles (compasses designed to operate on a horizontal axis) would behave like the ones near his lodestone model of Earth. Gilbert argued that because the behaviour of magnetic compasses and dipping needles near his lodestone and Earth were sufficiently alike, it was correct to assume that Earth itself must be a giant spherical magnet.

Long before Gilbert's time, it was noticed that a compass did not point directly toward true geographic north. In 1580, in London, it had been recorded that a compass needle pointed 11° east of north. This is known as **magnetic declination**. In Gilbert's lifetime, the magnetic declination at London drifted from 11° east of north to 25° west of north.

Clearly, if the magnetic North and South poles were in exactly the same place as the geographic North and South Poles, the compass would always point in a "true" north-south direction and the declination would always be zero. Moreover, the fact that the magnetic declination changes indicates that the magnetic poles must be wandering somewhat. In 1831, James Clark Ross accompanied an expedition to the Canadian Arctic. There, he located the Earth's magnetic North pole on the west coast of the Boothia Peninsula in the Northwest Territories at a latitude of 70° north. When he stood there, the magnetic dip was 89°59'. Figure 14.6 shows its movement from that time to the present.

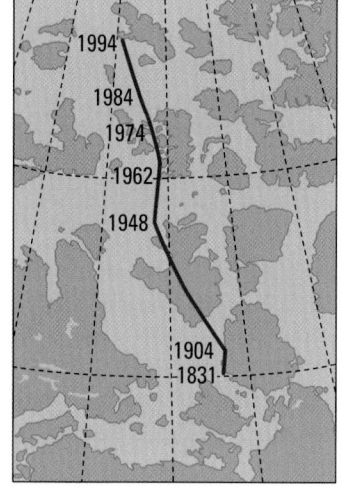

**Figure 14.6** Earth's magnetic poles are not stationary. Since 1831, the magnetic North pole has moved steadily north and west. As a result, magnetic declination constantly changes.

To use a magnetic compass effectively, you must be aware of the magnetic declination at your location on Earth. Detailed maps have been prepared that provide very accurate readings of the magnetic declination everywhere on Earth. The lines on these maps are called lines of constant magnetic declination or **isogonic** lines. Figure 14.7 shows the isogonic lines for a general map of Canada.

In many places on Earth, there are irregularities in the magnetic declination. At these locations, the actual direction of the isogonic lines deviates greatly from the direction that might be expected for that area. For example, north of Kingston, Ontario, the isogonic lines are 90° off the declinations for nearby locations. Anomalous angles of declination are accompanied by anomalous angles of dip.

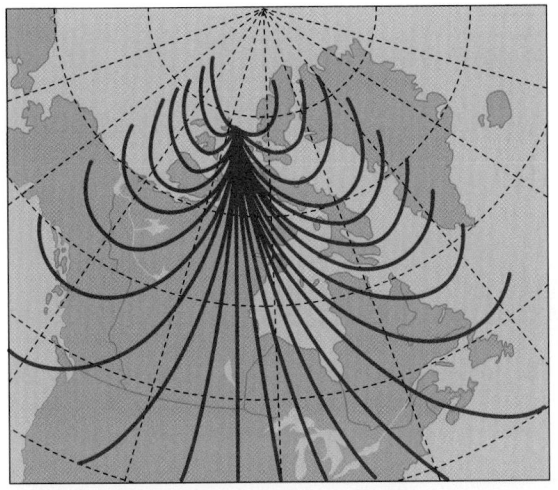

**Figure 14.7** Maps that plot the isogonic lines enable navigation with a magnetic compass. Since magnetic poles are continually shifting, these maps need to be updated regularly. The Geological Survey of Canada (GSC) produces very detailed isogonic maps. In southern Canada, their maps achieve an accuracy of about 0.5°.

# 14.1 Section Review

1. ● Imagine that you are given two iron rods that look identical. When you bring one set of ends together, they attract each other. However, when you reverse the end of one of the rods, they still attract each other. Obviously, one is a magnet and the other is not. Devise a test, using only the interactions of the two rods, to find out which one is the magnet and which one is the plain iron rod.

2. ● It has been found that magnets can be destroyed by several different actions. One of these actions is to heat the magnet. As the temperature of the magnet rises, the magnet gets weaker. The temperature at which the magnet is totally destroyed is called the Curie point. Discovered by Pierre Curie in 1895, the Curie point varies according to the type of material. Use domain theory to explain (a) why heating a magnet weakens its strength and (b) what happens at the Curie point.

3. **K/U** If an iron bar is held in a north-south orientation and then is tapped gently with a hammer, it will become a weak magnet. In light of domain theory, why should this occur?

4. **K/U** If a steel knitting needle is stroked along its length by one pole of a magnet, repeatedly in the same direction, it will gradually become a permanent magnet. How can domain theory be used to explain this?

5. ● Make a hypothesis to explain the cause of the local irregularities in the angles of declination for Earth's magnetic field? Do an Internet search to find information to support your hypothesis. Make sure that you consider the reliability of the source of the information. In your answer, include references to your information sources.

6. ● Explain why it would be necessary to have a detailed map of isogonic lines if you wanted to go wilderness camping.

SECTION
EXPECTATIONS

- Demonstrate an understanding of magnetic fields.

- Describe the properties, including the three-dimensional nature, of magnet fields.

- Conduct experiments to identify the properties of magnetic fields.

KEY
TERMS

- Action-at-a-distance
- magnetic field
- permeability
- magnetic lines of force
- magnetic dipole
- field density
- field strength
- ferromagnetism
- diamagnetism
- paramagnetism

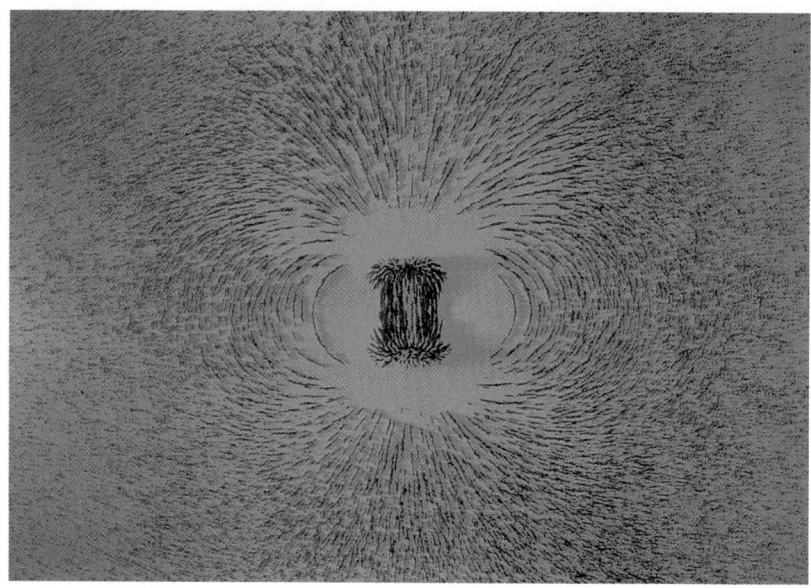

**Figure 14.8** Iron filings reveal a two-dimensional picture of the magnetic lines of force.

One part of Newton's Theory of Gravity, published in 1687, was the concept of the **Action-at-a-distance** theory. Newton argued that somehow the sun acted-at-a-distance across space to pull Earth and the other planets towards it. Newton also proved that the force of gravity varied inversely as the square of the separation between the Sun and the planets. When Coulomb proved that the magnetic force, like the force of gravity, also varied inversely as the square of the separation of the magnets, it was not surprising that the Action-at-a-distance theory was applied readily to magnetism. Action-at-a-distance seemed to serve magnetism very well until Michael Faraday, while studying the patterns in iron filings that had been sprinkled around magnets, created what is known today as "Field theory".

Michael Faraday's concept of electric and magnetic fields revolutionized Physics. Faraday thought that the iron filings had formed the pattern around the magnet because they had been pushed by what he called magnetic lines of force. He observed that the **magnetic lines of force** exited the magnet at one end, passed through space, and re-entered the magnet at the other end.

According to Faraday, the magnetic force experienced by an object was exerted, not by the magnet at a distance, but by the lines of force that were passing through the particular point in space where the object was located. The more lines of force that touched an object, the greater the magnetic force it experienced.

## Charged Up by Electromagnetism

Dr. Catherine Kallin

Superconductors are materials that, at extremely low temperatures (usually below −150°C), conduct electricity with absolutely no resistance. Normal wires always have at least a little resistance, and some of the electricity they carry is lost and becomes thermal energy. Superconducting wires do not do this, no matter how long they are.

Superconductors also have many unusual magnetic properties. They can form perfect barriers to a magnetic field, and can "remember" a magnetic field even when the field is turned off. Many devices that now make use of magnetism — electric motors, power generators, audio devices, and computer memories — will become more powerful, more efficient, and much smaller with advances in superconductivity. In fact, superconductors are already being used in magnetic resonance imaging (MRI) and elsewhere.

To find improved superconductors, we need to understand better how superconductivity works. For example, scientists are continually finding new superconductors that work at warmer and more easily maintained temperatures. Dr. Catherine Kallin, a professor at McMaster University in Hamilton, Ontario, is working at the very frontier of superconductor research, making use of some of the most challenging and difficult theories of magnetism.

"My own research studies 'vortices' in superconductors — small whirlpools of circulating supercurrent, similar to water swirling down a drain or air in a tornado. When a superconductor is placed in a magnetic field, the magnetic field penetrates the superconductor in filaments that lie at the centre of each vortex," she explains.

Dr. Kallin decided on a career in theoretical physics because of her interest in mathematics and a desire to do research that she could talk about concretely: "This element of being able to explain my work to others was very important to me." Dr. Kallin's mathematical interests were encouraged by a Grade 10 math teacher. This teacher suggested that she form a study group with her friends who perhaps "did not share my interest in mathematics." The experience helped improve her friends' grades and kindled in her "a deep love of both discovering new math and teaching it to others."

### Going Further

1. Many companies routinely use superconductivity in their products. Find out the names of some of these companies and contact them to discover how they use superconductivity. How many applications for this interesting phenomenon are already in place?

2. Form your own study group with some of your friends and investigate superconductivity. A good place to start is the Oak Ridge National Laboratory Internet site. Try explaining the effect to another student, a parent, a sibling, or your classmates.

### Web Link

**www.school.mcgrawhill.ca/resources/**
Does Professor Kallin's work make you feel you might be interested in a career in physics? If so, go to the above web site to find out more about careers in physics.

# INVESTIGATION 14-A

## Patterns in Magnetism

The patterns created by sprinkling iron filings around magnets had been observed as far back as the middle of the thirteenth century. In this investigation, you are going to study these patterns from various frames of reference in order to gain a better understanding of the nature of magnetism and magnetic forces.

If you performed the "Try This" experiment on page 671, you know that magnets induce magnetism in materials such as iron. The iron filings that are sprinkled around a magnet are induced to become hundreds of tiny magnets. Because the inertia of the iron filings is very small, they are easily oriented by the magnetic forces exerted on them. In this way, they form patterns that indicate the direction of magnetic forces in the region around a magnet. Because the patterns formed by the iron filings were like the patterns in a field after it has been plowed, these patterns became known as magnetic fields.

### Problem

The object of this investigation is to use iron filings to plot the magnetic field lines for bar magnets.

### Equipment

- bar magnets (2)
- plastic or smooth paper plate (about 20 cm diameter)
- sealable plastic bag (such as a Ziploc bag)
- iron filings

### Part 1: Using one bar magnet
### Procedure

1. Place the plate and some filings in the plastic bag. In this way you will be able to re-use them several times without losing them.

2. Tip the bag and collect the filings on the plate. Gently shake the filings so that they are evenly distributed over the surface of the plate.

3. Stand the bar magnet on its end on the table with the N-pole at the top. Hold the plate over the magnet so that the magnet is perpendicular to the centre of plate. (Someone will need to hold the magnet so that it does not fall over.)

4. Gently tap the plate so that the filings align themselves with the magnetic field of the magnet (see below).

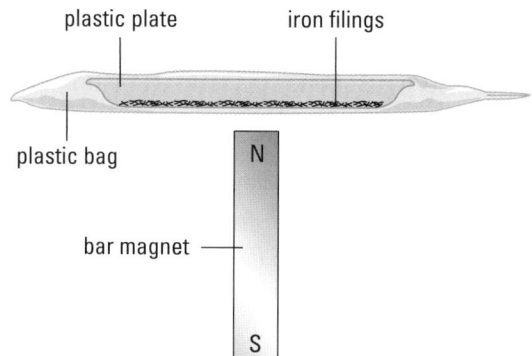

5. Draw an accurate sketch of the pattern formed by the filings. Try to include some sense of the third dimension in your sketches.

6. Move the plate away from the magnet and shake it so that the filings are spread uniformly over the plate again.

7. Reverse the polarity of the magnet on the table (turn the other end up) and repeat steps 3 through 5.

8. Move the plate away from the magnet and shake it again so that the filings are spread uniformly over the plate.

9. Place the magnet flat on the table. Place the plate on the magnet so that the magnet is roughly centred underneath.

10. Gently tap the plate so that the filings align themselves with the magnetic field.

11. Draw an accurate sketch of the pattern formed by the filings.

12. Repeat steps 8 to 11, placing the narrow side of the magnet in contact with the underside of the plate.

## Part 2: Using two bar magnets
### Procedure

1. Spread the iron filings uniformly over the surface of the plate.

2. Place two bar magnets about 3 cm apart, perpendicular to the bottom of the plate. Have the N-pole of one and the S-pole of the other pointing upward.

3. Gently tap the plate so the filings align themselves with the magnetic field of the magnets. Make an accurate sketch of the observed pattern.

4. Spread the iron filings uniformly over the surface of the plate.

5. Place the two bar magnets with their largest surfaces on the table so that the long axes are lined up. The N-pole of one magnet should be about two centimetres away from the S-pole of the other.

6. Place the plate on top of the magnets. Gently tap the plate so that the filings align themselves with the magnetic fields. Make a detailed sketch of the observed pattern.

7. Repeat steps 4 to 6 with the magnets turned so that their narrow edges are on the table.

## Part 3: Two bar magnets, N-poles close
### Procedure

1. Repeat Part 2 of the investigation, placing two N-poles (or two S-poles) near each other, rather than one N-pole and one S-pole.

## Analyze and Conclude

1. Use the following to analyze Part 1 results.
   (a) In the first two sketches, if you didn't know which pole was under the plate, would it be possible to tell from the sketches?
   (b) In the final two sketches, could you tell which pole of the magnet was at which end of the field by looking only at the pattern?
   (c) Examine the shapes of the magnetic fields in your sketches. Can you tell where the field is strongest (where the magnet would exert the largest force on an object)? This becomes a basic premise of Field theory.
   (d) You have seen the field pattern from three different angles. Using all of the information gathered in your sketches, extrapolate them into a three-dimensional picture of the field around the magnet.

2. Use the following to analyze Part 2 results.
   (a) Is there any way to determine which is the N-pole and which is the S-pole of the magnets just by looking at the pattern?
   (b) From your sketches of the three different views of the field pattern, make a three-dimensional drawing of the field.

3. Create a three-dimensional drawing of the field for Part 3, as you did for Part 2.

## Apply and Extend

4. Look at your sketches from Parts 2 and 3 of this investigation. Is there anything about them that would imply that "like poles repel" and "unlike poles attract"? Explain.

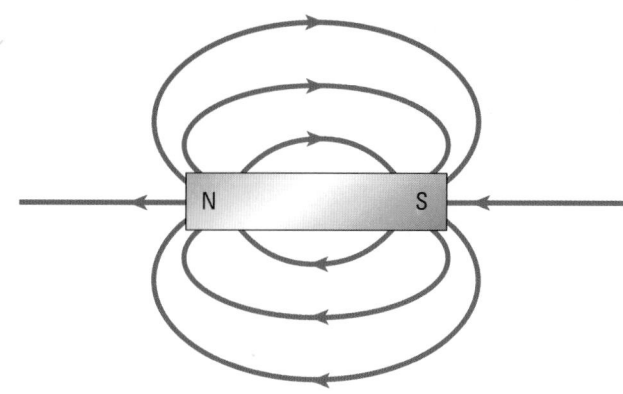

**ELECTRONIC LEARNING PARTNER**

Further information about magnetic fields is available on your Electronic Learning Partner.

## Lines of Force

You probably have noticed that as the distance from a magnet increases, the lines of force get farther apart. When an object is close to a magnet, the lines of force are more densely packed. The patterns of the lines of force in space are called **magnetic fields**. Thus, close to the magnet, many lines will act on the object causing it to experience a large magnetic force. When the object moves farther away from the magnet, the density of the lines of force is reduced, as is the magnetic force. In fact, the term **field density** is synonymous with **field strength**. When you draw magnetic fields, you must make sure that the spacing between the lines indicates the strength of the field.

Force is a vector quantity. Since the lines of force act to exert forces in particular directions, they must also be vectors. The magnetic fields from a magnet can exert forces in opposite directions, depending on the polarity of the affected object. Obviously, the field cannot point in two directions at once. To resolve this dilemma, the direction of the field lines of force has been arbitrarily assigned.

To scientists living in the Northern Hemisphere, the N-pole of the compass was a natural choice as the reference pole. By definition, the direction of a line of force is the same direction as the force that the field exerts on the N-pole of a magnet. Thus, the lines of force must point away from the N-pole of a magnet and toward the S-pole of a magnet. A magnetic S-pole, therefore, experiences a force in the opposite direction to the direction of the line of force. When a magnetic field is drawn, arrowheads must be included on the lines of force to indicate their direction and thus the direction of the magnetic field (see Figure 14.9 above). Figure 14.10 provides you with a three-dimensional view of the magnetic field.

**Figure 14.10** A suspension of iron filings in glycerine is used to demonstrate the third dimension of a magnetic field.

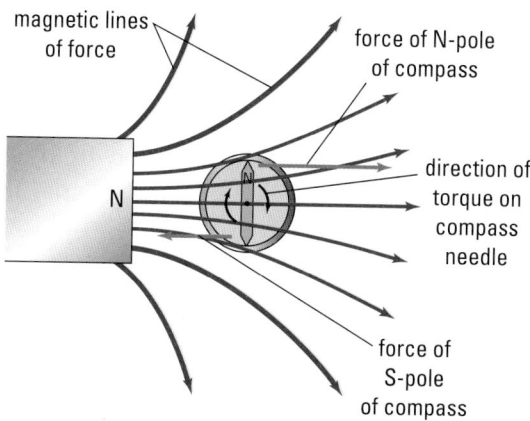

magnetic lines of force

force of N-pole of compass

N

direction of torque on compass needle

force of S-pole of compass

**Figure 14.11** A magnetic dipole, such as a compass needle, experiences a torque (twisting force) that aligns it with the surrounding magnetic field.

When a magnetic dipole (all magnets are dipoles) is placed in a magnetic field, its N-pole experiences a force in the direction of the line of force, and its S-pole experiences a force in the direction exactly opposite to the direction of the line of force. As a result, the dipole experiences a force that causes it to become aligned with the line of force, with its N-pole indicating the direction of the line of force. This explains how magnetic compasses and dip needles work. It also explains why magnets do not always point directly at the nearest magnetic N- or S-pole (see Figure 14.11).

## Earth's Magnetic Field

If you revisit Gilbert's model of Earth as a magnet, you can now imagine what the magnetic field must be like and why the magnetic compass and dip needle act as they do. The magnetic lines of force exit Earth's surface in the magnetic Southern Hemisphere and re-enter the surface in the magnetic Northern Hemisphere as shown in Figure 14.12.

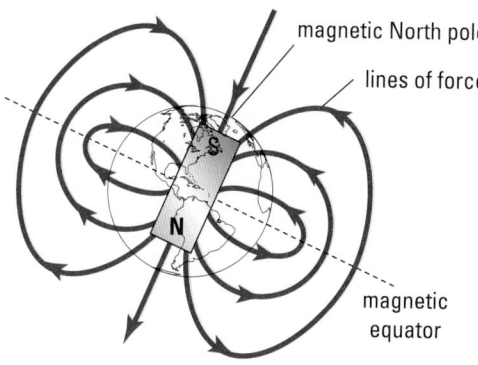

magnetic North pole

lines of force

N

magnetic equator

**Figure 14.12** Earth is enveloped by a magnetic field called the magnetosphere that extends far into space.

The location where the magnetic lines of force are parallel to Earth's surface is known as the magnet equator. At that point, a dip needle is horizontal. As you move closer to either of Earth's magnetic poles, the lines of force slope more and more steeply to the surface. At the magnetic poles, the lines of force are perpendicular to Earth's surface.

---

**PHYSICS FILE**

Michael Faraday (1791–1867) was the son of a blacksmith who had been apprenticed to a bookbinder when he was about 13 years old. Faraday began to read the books that came into the shop for binding. He was especially fascinated by the books about science and began to attend lectures in science given at the Royal Institute in London. In 1812, Faraday applied for a position at the Institute and was given a job as a research assistant.

Faraday proved to be a brilliant experimenter. However, as he had little formal education, he was not very good at mathematics. Thus, when it came to the study of magnetism, Faraday found the conceptual approach of drawing fields to explain how magnets interacted more to his liking than the mathematical approach used in the action-at-a-distance theory.

The entirety of the magnetic field around Earth is called the magnetosphere. Figure 14.13 shows an artist's rendering of the magnetosphere. The highly charged solar winds from the sun tend to "blow" the magnetosphere so that it is flattened on the side nearest the sun and elongated on the side away from the sun. It is the magnetosphere that protects Earth from the harmful effects of the solar wind. Electric charges from the solar wind can become trapped in the magnetosphere. As they spiral down inside the magnetosphere, they interact with the atoms and molecules in the atmosphere to produce an eerie light known as the aurora.

## Solar Graffiti

The year 2000 marked a banner year in our understanding of solar phenomena and Earth's weather. Until then, adequate forecasting tools remained unavailable to researchers, leaving Earth a victim of solar flares (explosions on the Sun's surface), magnetic storms, and coronal mass ejections (CMEs). CMEs are explosions in the atmosphere surrounding the Sun. For example, on March 13, 1989, during a peak in sunspot activity there was a breakdown of the Quebec hydro-electric grid causing massive power disruptions for hours throughout the province. Today, however, scientists are armed with powerful new knowledge. CMEs, believed to be the most potent force in space weather, are usually preceded by a particular S-shaped formation on the surface of the Sun (see the "S" shape on the image of the Sun, in the figure).

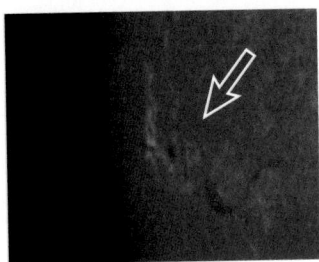

Sunspots are intense concentrations of the Sun's magnetic field on its surface. Loops of magnetism can form between pairs of sunspots of opposite magnetic polarity, gaining potential energy like a rubber band being stretched to its limit; these are observed as an S-shaped formation on the surface. When the S is fully formed, these bands of energy "snap," releasing bursts of energy that result in CMEs. A "wind" of highly energized particles ensues, travelling at such speeds that it can reach Earth in a matter of days.

Fortunately, Earth's own magnetic field (the *magnetosphere*) can protect us most of the time. It absorbs or deflects much of the Sun's released energy, but under extreme conditions, the magnetosphere can become overloaded, resulting in a *geomagnetic storm*. Here, the overloaded magnetosphere offloads excess energy in the form of a shower of highly energetic particles directed at Earth's surface. When these hit the upper atmosphere, the effects can be seen in the form of the *aurora borealis*.

Sometimes even the atmosphere can't stop the rain of energy, and parts of Earth's surface are bathed in a powerful moving magnetic field. Power lines are especially susceptible; whenever any conductor moves through a magnetic field, it induces an electric current. When a geomagnetic storm washes over power lines, it likewise induces a current spike, in exactly the same way that power is produced in an electric generator. This is what happened in Quebec. A particularly powerful geomagnetic storm produced excess current in the power lines causing a power overload.

The need for good prediction tools is clear. From now on, special attention will be paid to the "writing" on the Sun!

# Domains Revisited

A material is **ferromagnetic** if its atoms or molecules are magnets and they tend to group into domains. When a ferromagnetic material is placed in a magnetic field, the lines of the field find it easier to pass through the material than the space around the material. Ferromagnetic materials have a large magnetic permeability. When the magnetic lines of force pass through the material, they act on the domains within the material and cause them to become aligned. At this point, the ferromagnetic material is a magnet in its own right and has its own magnetic field that is aligned with and enhances the original magnetic field. Ferromagnetic materials are so permeable to magnetic lines of force that they are able to react well with even very weak external magnetic fields. Materials such as iron, nickel, cobalt and steel are ferromagnetic.

The atoms or molecules of **paramagnetic** materials are magnetic but the material is not very permeable to magnetic fields. Thus, if the external magnetic field is weak, the material shows very little or no magnetism. The magnetic effects become noticeable at room temperature, and only if the external field is very strong. With a weak field, the permeability is so small that it cannot overcome the thermal activity of the atoms in order to align the domains. As soon as the external field is gone, the small amount of magnetism that was induced in the material disappears. However, if the temperature is very low, the internal thermal activity of the material is reduced. Any external field can align with the domains more easily and the material exhibits magnetic behaviour similar to a ferromagnetic material. Aluminum is a paramagnetic material as are many gases. Will any magnetic properties be detected for aluminum?

A third type of reaction of a magnetic field is called **diamagnetism**. Diamagnetic materials are less permeable than air. They never show ferromagnetic properties but become very weakly magnetized in the *opposite* direction to the applied field. Like paramagnetic materials, they need very strong magnetic fields or very cold temperatures before they exhibit any of their characteristic behaviour. The atoms of diamagnetic materials are not magnetic and thus do not form domains. The diamagnetism results from the motion of the electrons within the atoms of the material. How this occurs will become clearer when you study Lenz's laws in the next chapter. Materials such as copper, rubber and glass are diamagnetic.

**Figure 14.13** The magnetosphere envelops Earth, protecting it from the effects of the solar wind. As electric charges from the solar wind are captured by the magnetosphere, they spiral down into the atmosphere. When they reach the atmosphere, they cause the atmospheric gases to become ionized and emit light. Depending on the hemisphere in which it occurs, that light is called the aurora borealis or the aurora australis.

## TRY THIS...

Build a chain of hanging paper clips by placing them end to end suspended from a bar magnet. Continue to add paper clips until the magnet will no longer support additional paper clips. Hold the top paper clip in the chain as still as possible, and carefully remove the magnet from above it. Have your partner record some observations. Now bring the other pole of the magnet near the top of the chain and observe. Predict what type of material must be contained within paper clips and provide justification for your predictions. Create a model to explain how a magnet might enable the paper clips to attract one another. When you removed the magnet, did all of the paper clips fall away? Does your model predict this behaviour? Finally, is your model able to explain the results observed when the opposite magnetic pole was brought near the paper clip chain?

# INVESTIGATION 14-B

## The Vector Nature of a Magnetic Field

**TARGET SKILLS**
- Performing and recording
- Analyzing and interpreting

Magnetic fields are vector quantities. Thus, if a compass is placed in a magnetic field, it aligns itself in the direction of the net magnetic field at that point in space. In this investigation, you will trace the direction of the magnetic field in the vicinity of a bar magnet, and the direction of the magnetic field at greater distances from the bar.

### Problem

What do you observe when two or more magnets create fields in the same region of space?

### Materials

- bar magnet
- compass (about 2 cm in diameter)
- large sheet of paper (approx. 60 cm × 60 cm)

### Procedure

1. If a large sheet of blank newsprint is not available, tape together several sheets of paper to form a sheet about 60 cm × 60 cm. Place the paper on the floor or on a table. Use the compass to find the direction of Earth's magnetic field and draw an arrow on the paper representing that direction.

2. Place the bar magnet horizontally at the centre of the sheet. The direction of the long axis of the bar magnet is not important. Trace an outline of the bar magnet and indicate, within the outline, the location of the magnet's N-pole.

3. Place the compass so that it touches the bar magnet near its N-pole and draw a dot at the N-pole of the compass needle.

4. Move the compass so that the S-pole of its needle is at the dot you just drew on the paper and draw another dot at the N-pole of the compass needle. Use the diagram on the upper right as a guide.

5. Continue this process until the compass returns to the magnet or goes off the paper.

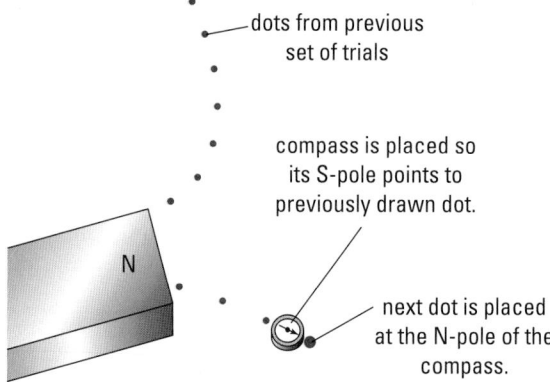

dots from previous set of trials

compass is placed so its S-pole points to previously drawn dot.

next dot is placed at the N-pole of the compass.

6. Trace a line through the series of dots you just drew on the paper. Make sure you include an arrowhead on your line to indicate the direction of the line of force.

7. Choose several more starting points at the ends of and along the sides of the bar magnet and repeat the previous steps.

### Analyze and Conclude

1. Did the compass trace out a pattern similar to the one you observed when you sprinkled iron filings on the bar magnet?

2. Using your knowledge of vectors and fields, explain why the lines you traced followed the observed paths. If you imagine how the field from the bar magnet and Earth are interacting, it will help you understand the pattern traced by the compass.

### Apply and Extend

3. If time permits, repeat the investigation using the bar magnet in a different orientation.

4. Try the investigation with two or three bar magnets placed on the sheet in various configurations.

# Field versus Action-at-a-distance

According to the Action-at-a-distance theory, Earth's Magnetic North pole was assumed to attract the N-pole of a compass. Now that Field theory has replaced the Action-at-a-distance theory, rather than being attracted by the Magnetic North pole, the magnetic field surrounding Earth causes the compass needle to become aligned with it. Earth's magnetic field, not the magnetic north-pole, acts on the compass needle. Why choose one theory over the other? Don't they both give the same result?

In science, theories come and go depending on their ability to *explain* and *predict*. As you proceed through the study of physics you will find many instances where the ability of the Field theory to explain and/or predict is far superior to the Action-at-a-distance theory. In fact, it would be very difficult to imagine modern physics without Field theory. In your Grade 12 physics course, you will explore in greater depth the concepts of Field theory as applied to gravity, electricity, and magnetism.

## 14.2    Section Review

1. **❶** The figure below shows two arrangements of equal-strength bar magnets parallel to each other. Based on your results from Investigations 14-A and B, draw what you would expect to find if you plotted the magnetic field for this arrangement. Base your diagrams on your observations of the nature of the fields when like and unlike poles were placed near each other in Investigation 14-A. You may want to verify your answers using iron filings and bar magnets.

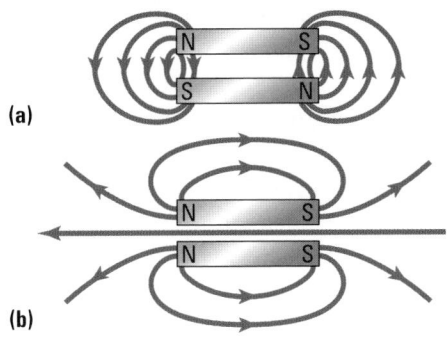

(a)

(b)

2. **❸** Magnetic lines of force are known to be passing through a circular iron plate in a direction parallel to its axis. What does this tell you about the magnetic polarity of the plate? Draw a sketch to demonstrate your answer.

3. **K/U** Using Field theory, explain why placing a ferromagnetic material between a magnet and a magnetic compass causes the compass to experience a greater magnetic force?

4. **K/U** Using Field theory, explain why placing a paramagnetic or diamagnetic material between a bar magnet and a magnetic compass causes no noticeable reaction in the compass. Under what conditions would you expect paramagnetic or dia-magnetic materials to produce a reaction in the compass?

- Describe the relationship between magnetic fields and electric current.

- Analyze and predict, by applying the right-hand rule, the direction of current produced a magnetic field.

- Interpret and illustrate, using experimental data, the magnetic field produced by a current flowing through a conductor.

## KEY
### TERMS

- electromagnetism

- right-hand rules #1 and #2

- solenoid

- magnetic monopole

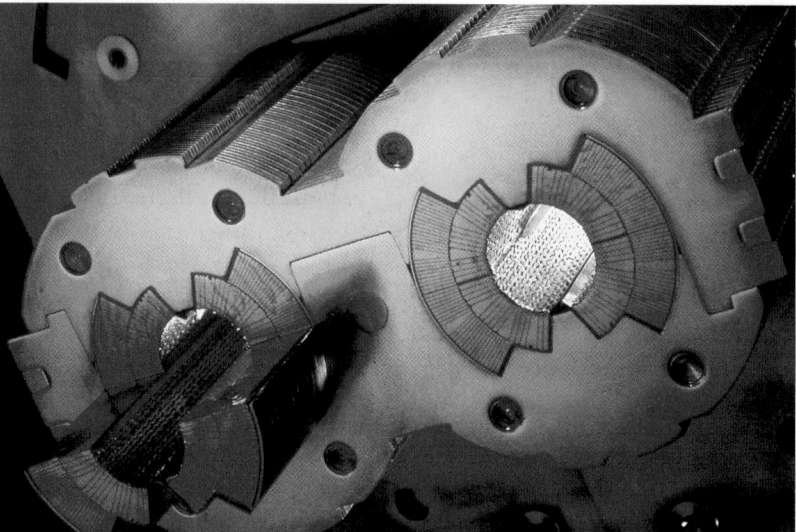

**Figure 14.14** Superconductors allow physicists to generate extremely powerful magnetic fields.

In order to complete research into atomic structure, atoms are bombarded with particles which have extremely high energy. To focus and direct particles with such great energies requires very powerful superconducting electromagnets. Superconducting electromagnets, like the one in the photograph, can produce magnetic fields more than 150 000 times stronger than Earth's magnetic field.

## Electric Charges and Magnetic Poles

In spite of the similarities between electricity and magnetism, early experimenters considered them to be two entirely separate phenomena. It is true that both electrostatic and magnetic forces of attraction and repulsion become weaker with separation. However, they display many fundamental differences. Electric charge, the source of the electric force, moves easily through conductors, while magnetic poles, the source of the magnetic force, cannot be conducted. Almost anything can be given an electric charge. However, magnetic poles are normally found in only ferromagnetic materials. Like magnetic poles, there are two kinds of electric charges. Objects displaying electric charge usually have only one type of charge on them, positive or negative. Magnetic poles seem always to come in pairs, hence the magnetic dipole. Overall, it seemed that the differences were far more significant than the similarities.

---

**PHYSICS FILE**

**Magnetic Monopoles**

Whether or not a magnetic monopole can exist is an interesting point. As you will see later in this chapter, Field theory seems to predict that a magnetic monopole is not possible. Still, it is an attractive idea and scientists are still trying to decide whether or not they are possible. In the meantime, nobody has ever seen or been able to create a magnetic monopole in a laboratory.

## Oersted's Discovery

In 1819, a Danish physicist, Hans Christian Oersted (1777–1851), was demonstrating the heating effects of an electric current in a wire to some friends and students. On his table he had some compasses ready for a demonstration he was doing later that day in magnetism. He noticed that when he closed the circuit, the needles of the compasses were deflected at right angles to the conductor. He kept this to himself until he had a chance to explore it further. It did not seem to make sense that the compass needle was neither attracted nor repelled by the current but deflected at right angles to it. Oersted had discovered that a current-carrying conductor caused the needle of a magnetic compass to deflect at right angles to the conductor. When he published his findings, it set off a flurry of research into the newly discovered phenomenon called **electromagnetism**. That is, moving electrons produce a magnetic field and a changing magnetic field will cause electrons to move.

## Right-hand Rule #1

Oersted's experiments convinced him that each point of a current-carrying conductor created a magnetic field around itself. The lines of force for that field were a set of concentric closed circles on planes perpendicular to the direction of the current. The direction of the lines of force and thus the direction of the field could be determined using a "right-hand rule" (see Figure 14.15).

**ELECTRONIC LEARNING PARTNER**

Go to your Electronic Learning Partner to learn more about electromagnetism.

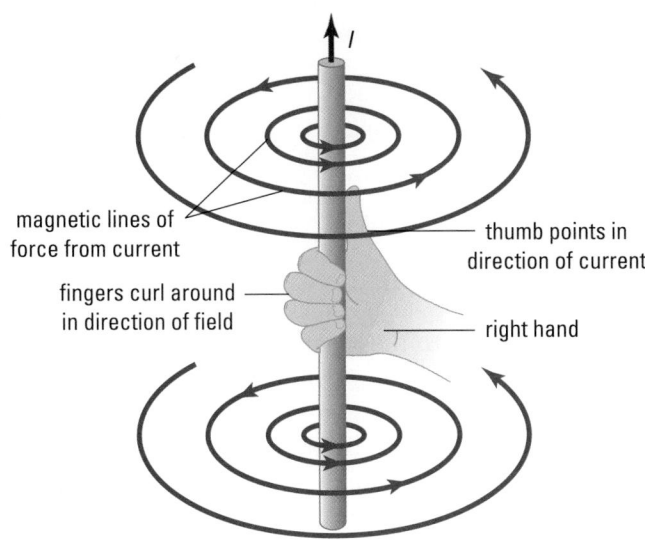

magnetic lines of force from current

thumb points in direction of current

fingers curl around in direction of field

right hand

**Figure 14.15** **Right-hand rule #1** If you grasp a current-carrying conductor with your right hand so that the thumb lies along the conductor in the direction of the current, then the fingers of your hand will be encircling the conductor in the direction of the magnetic lines of force caused by the current.

# INVESTIGATION 14-C

## Magnetic Field around a Straight Conductor

**CAUTION** In this experiment, you will be using circuits without a resistance to protect the power supply. They are *short circuits*. To protect the power supply from damage, you should only connect them for very short periods of time.

### Problem

Develop an understanding of the relationship between a current and its magnetic field.

### Prediction

In the text it is noted that Oersted observed that the needle of the compass was deflected at right angles to the current. From this information, what can you predict about the shape of the field near the conductor?

### Equipment

- ammeter (0–10 A)
- variable power supply (or fixed power supply with an external variable resistor)
- magnetic compass (2)
- wire (approximately 1 m)
- connecting leads
- cardboard square (approximately 20 cm × 20 cm)
- iron filings
- metre stick
- 1 kg mass (2)
- masking tape
- retort stand (2)
- ring clamp

### Part 1: Horizontal Conductor

### Procedure

1. Place the wire across the middle of the cardboard square and hold it in place at the edges of the cardboard with masking tape.

2. Connect the power supply and the ammeter, in series, with the wire.

3. Turn on the power supply very slightly to check that the ammeter is connected properly.

4. Once you have confirmed that the circuit is properly connected, place one compass under the wire and one compass above the wire. Rotate the cardboard until the wire lies along the direction of magnetic north and the compass needles are parallel to the conductor (see the figure below).

   (**Note:** Place the compasses far enough apart so that they do not strongly interact with each other. Tap them gently to make sure that their needles are able to move freely.)

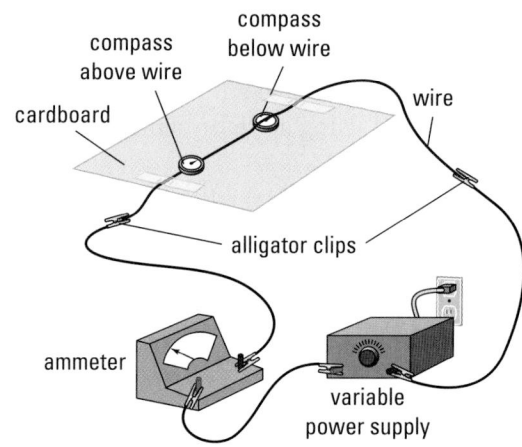

5. Turn the power on and increase the current (to about 5 A) until the needle of the compass shows a strong reaction to the current. Quickly disconnect the lead at the *anode* of the power supply. (In this way, you will not have to reset the power level each time you want to start and stop the current.) Next, momentarily touch, but do not attach, the lead to the anode of the power supply and note the reaction of the compass needles.

6. Draw a sketch of the orientation of the apparatus, including the direction of the current and the direction of the N-pole of the compasses before and after the current was turned on. (Remember that a magnetic dipole aligns itself with the net magnetic field. Also, the N-pole of a magnet is used as a reference to determine the direction of the magnetic field.) As a reference, include an arrow in your sketch to show the direction of Earth's magnetic field.

7. Draw a sketch of your observation of the reaction of the compass needles to the current.

8. Turn the cardboard 45° clockwise and repeat steps 5 through 7.

9. Turn the cardboard another 45° clockwise and repeat steps 5 through 7.

10. Turn the cardboard another 45° clockwise and repeat steps 5 through 7.

11. Without changing the power setting on the supply, reverse the connection of the leads connected to the wire on the cardboard so that the current in the wire is reversed. Re-orient the cardboard to its original position. Momentarily turn on the current and note the reaction of the compass needles to the current. Draw a sketch of your observations.

## Part 2: Vertical Conductor
### Procedure

1. Assemble the apparatus as shown below.

2. Connect the power supply and ammeter to the wire and turn the current on very slightly to check that the ammeter is connected correctly.

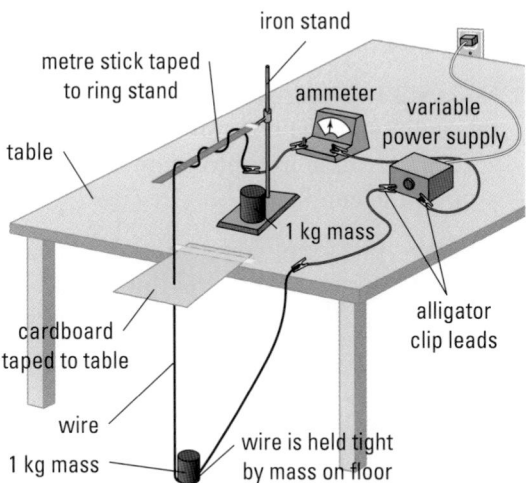

3. Place a compass on the cardboard platform to the north of the conductor so that it is very close to, but not touching, the wire.

4. As in Part 1, momentarily turn on the current and increase it until the compass reacts to the current. Once again, a current of about 5 A should be adequate. (If you want to increase the current beyond that level, consult with your teacher.) Observe and note the reaction of the compass to the current.

5. Draw a sketch of the result, showing the position of the compass needle before and after the current was turned on. As a reference, include an arrow to show the direction of Earth's magnetic field.

6. Move the magnet 45° clockwise and repeat steps 4 and 5. Record your result on the sketch you drew for the first trial.

*continued* ▶

*continued from previous page*

7. Continue to move the compass around the conductor in 45° angles until it has returned to its original position, repeating steps 4 and 5 after each move.

8. Cut a slit in a piece of paper from its edge to its centre and use it to cover the cardboard square. Sprinkle iron filings on the paper around the conductor.

9. Connect the circuit and gently tap the cardboard. Observe the effect of the current on the position of the filings. Draw a sketch of the result.

## Analyze and Conclude

1. For all trials in Part 1, determine the greatest angle that the current caused the compass needle to deflect? What is the significance of this? Remember that the compass needle points in the direction of the sum of the magnetic fields from the Earth and from the current.

2. In Part 1, what is the significance of the fact that the compass needles above and below the current deflected in opposite directions?

3. In Part 1, when you reversed the current without changing the orientation of the wire, what happened? Why is that significant?

4. In Part 1, based on the information that you have gathered, what is your prediction for the shape of the magnetic field near a current? Explain.

5. Do the results from Part 2 of the investigation confirm the hypothesis you made about the shape of the field in Part 1?

6. In the previous sections of this chapter, you saw magnetic fields pointing from the N-pole of a magnet to the S-pole. Where are the poles for this magnetic field? Where do the magnetic lines of force begin and end?

7. Does the magnetic field from the current maintain its strength as you move away from the current? What evidence is there to support your answer?

8. What is your conclusion about the shape of the magnetic field that results from a current?

## Apply and Extend

9. What would be the effect if a second wire was positioned alongside the wire in your apparatus and was carrying a current in (a) the same direction or (b) the opposite direction as the wire in your apparatus? Form a partnership with another lab group, and combine your apparatuses to run two wires side by side. Using two power supplies to produce equal currents in the wires, repeat steps 4 and 5 of the previous procedure. Does this arrangement have an effect on the strength of the field around the wires? Try it with currents running in the same and in the opposite directions.

10. What do the results tell you about the nature of magnetic fields around conductors?

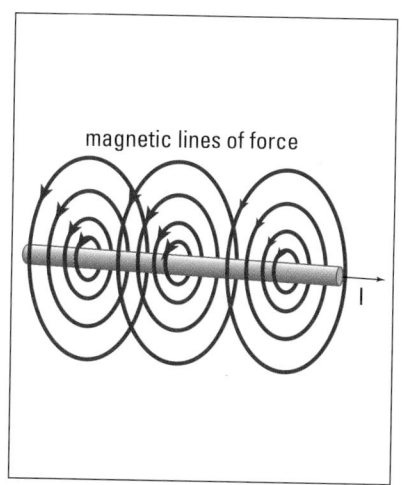

**Figure 14.16** When the current is upward, right-hand rule #1 shows the field lines as indicated.

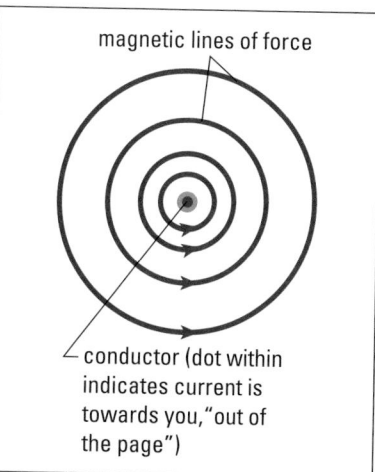

**Figure 14.17** The system in Figure 14.29 viewed directly from above, without the benefit of perspective. The dot in the cross-section of the conductor indicates that the current is flowing through the conductor directly toward you (out from the page).

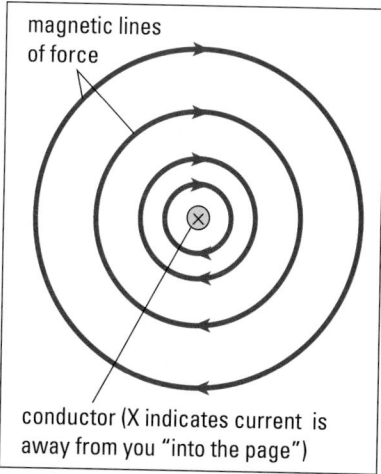

**Figure 14.18** The diagram in Figure 14.29 viewed directly from below, without the benefit of perspective. The cross in the cross section of the conductor indicates that the current is flowing through the conductor directly away from you (into the page).

## Drawing Field Diagrams

Using right-hand rule #1, diagrams of the magnetic fields around currents are easy to draw, but they do require drawing in 3-dimensions. Drawing in perspective is time consuming and often messy, so the ability to show the third dimension in a two-dimensional diagram offers the best solution.

A simple convention has been created to show a vector that is pointing directly at or directly away from you. If the arrow is pointing at you, a dot is used to represent the tip of the arrow. If the arrow is pointing directly away from you, a cross is used to represent the tail-feathers of the arrow as shown in Figure 14.16. Apply right-hand rule #1 to make sure you agree with the orientation of the current and the magnetic field in Figures 14.17 and 14.18.

## Field Theory Wins

In the previous section of this chapter, there seemed to be no compelling reason to abandon the Action-at-a-distance theory in favour of the Field theory. Now the real advantage of Field theory over the Action-at-a-distance theory becomes clear. According to the Action-at-a-distance theory, one magnet acts directly on another magnet. That works well when there are two magnets, but how can you use the Action-at-a-distance theory to explain the magnetic force on the compass needle near the current-carrying conductor when there is no magnet to exert the force on the compass?

In Field theory, a field, rather than a magnet, is the source of magnetic forces. Oersted's discovery, which preceded Field theory by a few years, clearly proved that you do not need a magnet as the source of a magnetic force. When Faraday created Field theory, it provided a framework to discuss magnetic forces on the magnets near a current-carrying conductor. According to Field theory, the current created the magnetic field around itself and the field exerted the force on the compasses. As you discovered in your investigation, Oersted proved that the direction and strength of the magnetic force on the compasses depends on the direction and strength of the current. Field theory argues that the direction and strength of the current affects the direction and strength of the magnetic field around the magnet and that, in turn, affects the direction and strength of the magnetic force.

## Magnetic Field of a Current-carrying Coil

Consider a coil of wire carrying a current as shown in Figure 14.19. What type of magnetic field should there be around the wire? Oersted's discovery predicts that there are circular magnetic lines of force perpendicular to the coil at each point on the coil. Inside the coil, all the magnetic lines of force pass through the plane of the coil in the same direction as determined using right-hand rule #1. Outside the coil, all lines of force pass through the plane of the coil in the opposite direction that they passed through inside the coil.

Examine the situation outside the coil. The lines of force from one edge of the coil point in the opposite direction to the lines of force from the edge diametrically opposite. Since they point in opposite directions, the two magnetic fields combine together destructively, making the field even weaker than it would be from a single edge.

**Figure 14.19** Using right-hand rule #1, we can show that, inside the coil, all the lines of force point in the same direction. The total strength of the magnetic field inside the coil is the vector sum of all the individual fields. Since all lines of force point in the same direction, the net magnetic field inside the coil is quite strong.

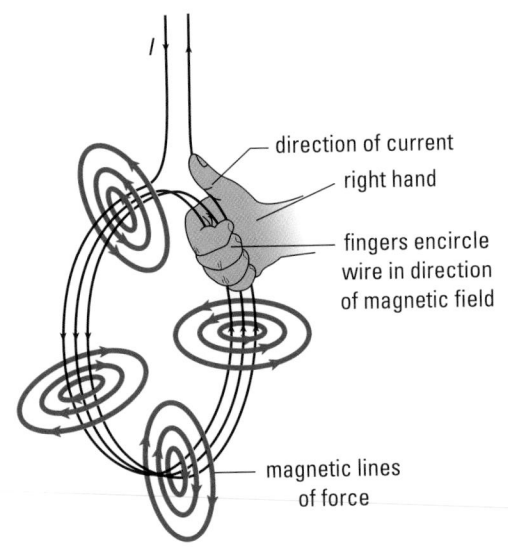

direction of current

right hand

fingers encircle wire in direction of magnetic field

magnetic lines of force

Next examine the situation inside the coil. A remarkable thing happens. The lines of force from all points on the coil pass through the plane of the coil in the same direction. Thus, the magnetic fields inside the coil combine together constructively. As you move away from one edge of the coil, along any diameter, the field strength from that edge weakens. However, as you move away from one edge, you move closer to the opposite edge, and the field from the opposite edge gets stronger at just the right rate to exactly compensate for the loss in strength from the other edge. The net effect is that the field is exactly uniform in strength.

When you draw the magnetic field for a coil, the lines of force should still be closed loops. Inside the coil, they should be uniformly spaced to represent the uniform nature of the field. Outside the coil, the magnetic lines of force should spread out to indicate the weakened field in that region. Figure 14.20 demonstrates this property if the coil was viewed in cross section. Figure 14.21 illustrates the same property of the coil when viewed in the plane of the coil from either face.

For the two current-carrying coils, dots and crosses represent the directions of the magnetic fields in the plane of the page. In the coil at right (b), the current is counter-clockwise, thus the magnetic field lines point out of the page inside the coil and into the page outside the coil. In the coil at left (a), the reverse is true. Inside the coils, the spacing of the magnetic lines of force is uniform to indicate the uniform nature of the field. Outside the coils, the spacing of the lines increases as the distance from the coil increases to indicate that the magnetic field strength is decreasing as you move away from the coil.

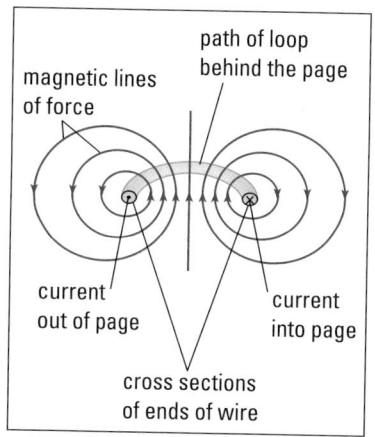

**Figure 14.20** The lines of force around the edges of a coil are closed loops. Inside the coil, the line spacing indicates that the field is uniform. Outside the coil, the line spacing indicates that the field is getting weaker as the distance from the coil increases.

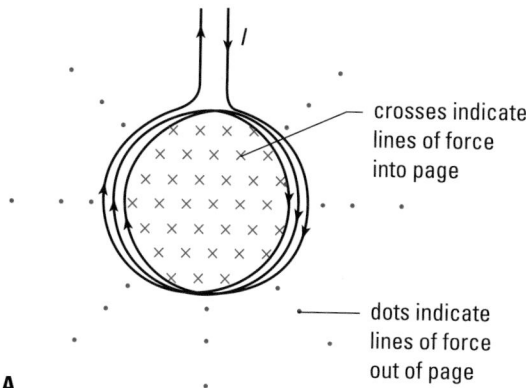

**A**

crosses indicate lines of force into page

dots indicate lines of force out of page

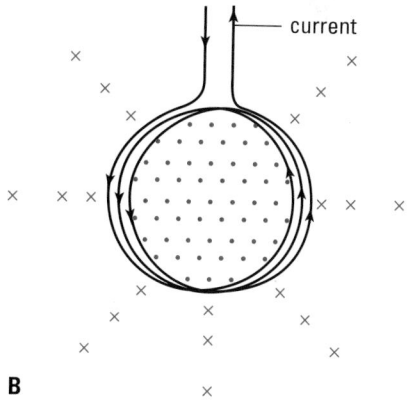

**B**

current

**Figure 14.21** (A) Magnetic fields with a clockwise current

(B) Magnetic fields with a counter-clockwise current

# INVESTIGATION 14-D

## Magnetic Field around a Helix

A helix is formed when a coil of wire is stretched so that there is space between the adjacent loops of wire. When a wire is formed into a helix and current flows through it, the magnetic field is similar to that of a coil, except that it is longer.

### Purpose

The purpose of this investigation is to plot the shape of the magnetic field around a helix.

### Hypothesis

Based on the information that you have about the magnetic field around a single coil of wire, draw a sketch to predict the shape of the magnetic field around the helix.

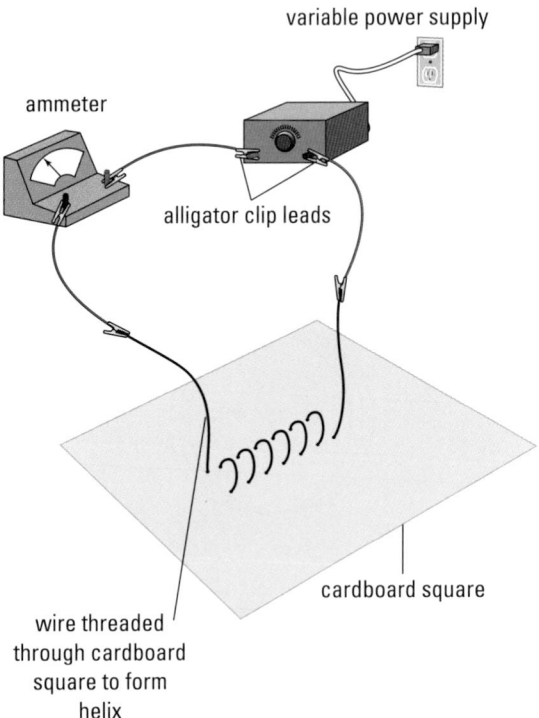

variable power supply

ammeter

alligator clip leads

cardboard square

wire threaded through cardboard square to form helix

### Equipment and Materials

- wire for helix (approx. 1 m)
- cardboard square (approx. 20 cm × 20 cm)
- power supply
- ammeter (0–10 A)
- compass
- iron filings

### Procedure

1. Thread the wire through the cardboard square as shown in Figure 14.35. A helix of about 10 turns, with loops that are two centimetres in diameter spaced about one centimetre apart, works reasonably well. Connect the power supply and the ammeter in series with the helix.

**CAUTION** Since the circuit has no resistance other than that of the coil, it is a short circuit. To protect the power supply from damage, the coil should only be connected to the circuit for very short periods of time.

2. Increase the power output of the supply very slightly to check that the ammeter has been connected correctly.

3. When the connections are correct, increase the output of the power supply until the current is about 5 A, and then disconnect the lead from the anode of the power supply.

4. Place a compass near the end of the helix and briefly connect the circuit. Notice the orientation of the compass needle.

5. Draw a sketch to record your observations. Include the coils, the direction of the current in the coils, and the orientation of the compass needle before and after the current was on. This will be used later in your analysis.

6. Repeat steps 4 and 5, placing the compass at various locations around the helix. Include all the results on the sketch made in step 5.

7. Sprinkle iron filings on the cardboard. Make sure there are filings inside the helix as well as outside it.

8. Briefly connect the circuit. Tap the cardboard to assist the filings to become aligned with the magnetic field of the helix. After a few seconds, disconnect the lead from the anode. **Do not disturb the filings on the cardboard.**

9. Give the power supply a period of time to cool down and repeat step 8 to enhance the result of the trial. (This step may be repeated again if necessary.)

10. Draw an accurate sketch of the pattern of the iron filings observed in step 9.

11. From your sketch of the pattern of iron filings, and the directions of the compasses, make a "lines of force" drawing of the magnetic field for the helix as seen in the pattern of the iron filings. Include the pattern of the lines inside the helix. Using the information from the compass sketch in step 5, place arrows on the lines of force to indicate the direction of the field.

## Analyze and Conclude

1. Does the magnetic field pattern resemble the one in your hypothesis? If not, try to explain why the actual field differs from your hypothetical field.

2. Is the magnetic field around a helix similar to any magnetic field observed previously? If yes, describe the similarities and the differences between this field and the previously observed field.

## Apply and Extend

3. What would be the effect on the magnetic field around the helix if the number of loops was increased without the helix getting longer? This would, in effect, make the turns tighter together.

## Magnetic Field around a Solenoid

A **solenoid** is a closely wound helix. The main difference between the field from a solenoid and the field from a helix is that the field from a solenoid is more uniform. Also, because there are so many coils of wire, it is much stronger for any given current. Outside the solenoid, the magnetic field closely resembles that of a bar magnet. Inside a solenoid, all the magnetic lines of force form closed loops. The lines of force leave one end of the solenoid, circle around and enter the other end of the solenoid, and then pass through the solenoid to their starting point.

With a bar magnet, Field theory predicts that the lines of force entering one end of a magnet are the same ones that exit the other end. Compare the magnetic field around the solenoid as shown in Figure 14.22 with that of the bar magnet as shown in Figure 14.9 on page 680. Compare the picture of the pattern of iron filings for a bar magnet (Figure 14.8) with the one you drew for the helix in Investigation 14-D.

If the N-pole of a compass or bar magnet were placed near the end of the solenoid at which the lines of force exit, it would experience a force pushing it away from that end, just as if that end were an N-pole of a solid. Similarly, the S-pole of a bar magnet placed near the end of the solenoid, at which the lines of force exit, would be pulled toward the solenoid. In other words, the solenoid acts just like a hollow bar magnet. The similarity between a solenoid and a bar magnet is just one of many clues that the lines of force in all magnets are closed loops that pass through the magnet.

If magnetic lines of force are always closed loops, it explains why magnets are always dipoles. As we have seen, the end of the magnet where the lines of force exit is the N-pole, and where they re-enter the magnet is the S-pole. To have a magnetic monopole, let's say an N-monopole, the magnetic lines of force would have to

**Figure 14.22** The pattern of the magnetic lines of force around a solenoid is very similar to the pattern of lines of force around a bar magnet.

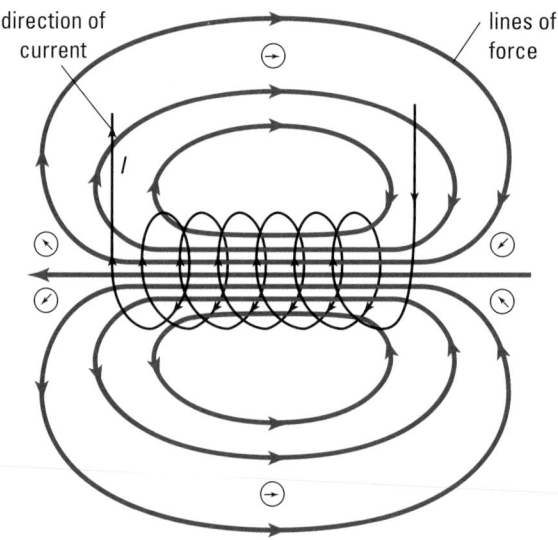

direction of current

lines of force

exit the material but never re-enter it. If, as Field theory suggests, lines of force are closed loops, they must come back and enter the material at some point. That point is the S-pole of the material. When you break a magnet into pieces, you just shorten the path of the loops for each piece. They still are loops and each piece is still a magnetic dipole.

## Right-hand Rule #2

How does the magnetic polarity of a coil or solenoid relate to the direction of the current in the coils? To find the N-pole of a coil of wire as shown in Figure 14.23, you can use right-hand rule #1. When you find the face of the coil where the lines of force exit, you have found the N-pole of the coil. Grasp the coil at some point with your right hand so that your thumb lies along the coil in the direction of the current, and your fingers encircle the wire in the direction of the magnetic lines of force. The face of the coil where your fingers exit is the N-pole of the coil.

To make this process simpler, a second right-hand rule was invented. It is actually just a variation of right-hand rule #1. **Right-hand rule #2**: Place the fingers of your right hand along the wire of the coil so that your fingers point in the direction of the current in the coil. When you extend your thumb at right angles to the plane of the coil, it will indicate the direction of the lines of force as they pass through the coil, and thus indicate the face of the coil that acts as the N-pole of the coil (Figure 14.24). The same rule will obviously apply to a solenoid or any other system where the current is moving in a circle.

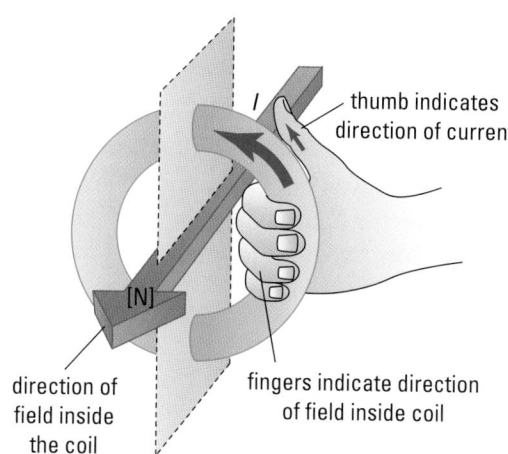

direction of field inside the coil

thumb indicates direction of current

fingers indicate direction of field inside coil

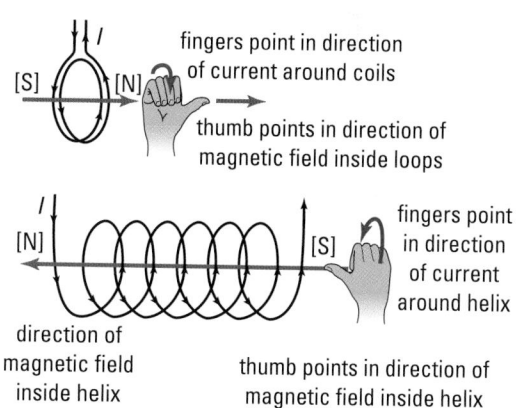

fingers point in direction of current around coils

thumb points in direction of magnetic field inside loops

fingers point in direction of current around helix

direction of magnetic field inside helix

thumb points in direction of magnetic field inside helix

**Figure 14.23** Right-hand rule #1 can be used to identify the direction of the lines of force through the coil and thus the locations of the N-pole and the S-pole of the coil.

**Figure 14.24** When the fingers of the right hand lie along a coil in the direction of the current in the coil, the thumb points in the direction of the magnetic lines of force inside the coil. The face of the coil where the magnetic lines of force exit acts like the N-pole of a magnet.

## Electromagnets

If a core of ferromagnetic material, such as iron, is placed inside a solenoid, the magnetic field strength inside the solenoid is greatly increased. Because of the permeability of the iron, the lines of force within the solenoid crowd into the iron core. This has two effects. First, the crowding concentrates the lines of force from the solenoid; the closer together the lines of force, the stronger the field. Second, the lines of force from the solenoid induce the domains of the iron core to align so that ferromagnetic material becomes a magnet whose field supplements the field of the solenoid.

The N-pole and S-pole of the electromagnet are located using right-hand rule #2. Grasp the electromagnet so that the fingers of your right hand encircle the magnet in the direction of the current around the core, and with your thumb parallel to the axis of the magnet. Your thumb points to the N-pole of the magnet (see Figure 14.25).

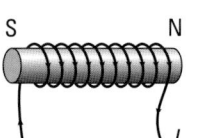

fingers circle core in direction of current

thumb points to N-pole of electromagnet

**Figure 14.25** Right-hand rule #2 is used to locate the N-pole of an electromagnet.

## Electromagnet Design

It is possible to make very strong electromagnets. Three factors affect the strength of an electromagnet: the size of the current, the number of turns, and the permeability of the core. For a core of a given material, it would seem that you just put more and more turns of wire around the core and increase the current. Unfortunately, it's not quite so simple.

Think back to your studies of electricity in the previous chapter. As the number of turns of wire around the core of a magnet increases, the resistance of the coil also increases ($R \propto l$). For a fixed potential difference, doubling the number of coils of wire around the core of an electromagnet doubles the resistance of the coils and halves the current through the coils. The result is no increase in the strength of the magnet. One solution is to use heavier wire. If the size of the magnet was not a factor, that solution might have merit. If size is a factor, then using heavier wire means that you cannot put as many turns around the magnet. Moreover, heavier wire would increase the mass of the coil and the cost of making it.

Another solution might be to increase the potential difference of the power supply to increase the current. This results in an increase in the power and thus the cost to operate the magnet. The increase in current to the coils of the electromagnet also means an increase in the amount of electrical energy that is converted to heat by the coils. Considering the importance of electromagnets to today's technology, finding the most efficient design for electromagnets is a formidable challenge.

If very strong magnetic fields are required, the magnets have to be super-cooled to the point where the coils become superconductors. At that point the coils lose their resistance and very large currents can flow through them. Many technical applications, such as high speed MAGLEV trains, magnetic resonance imaging (MRI) machines and particle accelerators, require the use of superconducting magnets.

## The Source of Earth's Magnetic Field

Not long after the discovery of electromagnetism, scientists began to wonder if a current circulating around a solenoid created a magnetic field similar to a bar magnet, perhaps Earth's rotation had something to do with its magnetic field. In 1878, H. A. Rowland, an American physicist, showed that any group of charges moving in a circle would create a magnetic field. When he placed an electrostatic charge on a rubber disk and spun it on its axis, it produced a magnetic field similar to that of a solenoid. Today, most scientific theories of Earth's magnetism include the concept of the motion of charges deep in Earth's core.

## 14.3   Section Review

1. **C** In each case, assume that the magnitudes of the currents in the conductors are the same. Indicate the relative field strengths on the diagrams by the spacing of the lines of force.

   (a) Draw a conductor in cross-section as seen end on. Indicate a current flowing directly towards you (out of the page). Draw the lines of force for the magnetic field resulting from the current in the conductor.

   (b) Draw a similar diagram to that in part (a), but showing a set of two conductors right next to each other. Indicate that the current in each conductor flows towards the viewer. Draw the lines of force diagram for the magnetic field that results from the current in the conductors.

   (c) Draw a set of two conductors, as in part (b). Indicate that the current flows toward the viewer in one conductor and away from the viewer in the other conductor. Draw the lines of force diagram for the magnetic field that results from the currents in the conductors.

2. **K/U** When electromagnets are constructed, what types of materials should be used in the core to make the strongest magnet? Explain why it is an advantage to use one of these materials for the core of the electromagnet rather than just having an air-core solenoid as the magnet.

3. **MC** A conductor is aligned with Earth's magnetic lines of force. Thus, the compass set above the conductor points in a line parallel to the conductor. A DC power supply is connected to form a closed circuit with the conductor. Explain how this set-up could be used to identify the anode and the cathode for the power supply.

- State the motor effect.

- Investigate and communicate factors that affect the force on a current carrying conductor.

- Test a device that operates using the principles of electromagnetism.

KEY
TERMS

- right-hand rule #3
- motor effect
- rotor
- armature
- commutator
- split ring commutator
- torque

**Figure 14.26** Designed to compete, the University of Waterloo's Midnight Sun solar-powered car attends races around the world.

University of Waterloo students are not alone in their effort to further solar-powered vehicle design. Undergraduate students at Queen's University set off in their solar-powered car on July 1, 2000 from Halifax and travelled 6500 km to Vancouver in less than a month. At the time this book was published, this event still held the world record for distance by a solar-powered vehicle. The electric motor was specially designed in a pancake shape to fit inside the low profile of the vehicle. Hybrid cars that use a combination of electric motors and gasoline engines are already gaining popularity.

The study of electric fields and electromagnetism in the previous section of this chapter is just the tip of the electromagnetic iceberg. Converting electrical energy to thermal energy is an easy task; you simply pass the current through a resistance and the resistance heats up. It was Oersted's discovery of electromagnetism that led to the invention of the electric motor. Currently, over half of the electricity generated in North America is used to run electric motors. It's fortunate that today's electric motors are so efficient; they typically operate at efficiencies greater than 80% in transforming electric energy into rotational motion of the motor, compared with a gasoline engine which operates at less than 30% efficiency.

**PROBEWARE**

If your school has probeware equipment, visit the **Science Resources** section of the following web site: ***www.school.mcgrawhill. ca/resources*** and follow the **Physics 11** links for several laboratory activities on electric motor efficiency.

# Electric Currents in Magnetic Fields

The first current meters, now called tangent galvanometers, consisted of a coil of wire with a compass needle at the centre. When the current flowed through the coil, the needle deflected to the east or west depending on the size and direction of the current. Because the tangent galvanometer had to be very carefully oriented, it was not particularly easy or practical to use (see Figure 14.27).

According to Newton's third law of motion, for every action there is an equal and opposite reaction. If the current in a conductor caused a magnetic field that exerted a force on a magnet, then the magnet must interact with the magnetic field from the current to exert a force on the conductor. In a tangent galvanometer, if the coil is exerting a force on the magnet at its centre to make it turn clockwise, then the magnet must be exerting a force on the coil trying to make it turn counter-clockwise.

The simplest form of the interaction of a magnet on a current-carrying conductor can be observed when a segment of wire carries a current linearly through a magnetic field. In Figure 14.28, a conductor carries a current upward past the N-pole of a bar magnet. First, let's approach this from the point of view of the conductor.

Using right-hand rule #1, as shown, the current creates a magnetic field consisting of circular lines of force around the conductor. The magnetic field of the conductor exerts a magnetic force on the N-pole of the magnet that acts in a direction tangent to the direction of the lines of force where they contact the magnet. (It is important to note that magnets do not react to their own magnetic fields.) Because of the magnetic field of the conductor, the N-pole of the bar magnet experiences a force directly into the page. If the N-pole of the magnet, in Figure 14.28, is free to move, it will move away from the observer into the page.

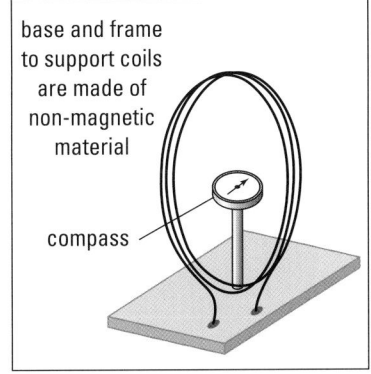

**Figure 14.27** The coils of the tangent galvanometer are set so that the plane of the coils is parallel to the direction of the compass needle. When a current is passed through the coils, the compass needle deflects to the left or the right from that plane. The tangent of the angle of deflection is proportional to the size of the current.

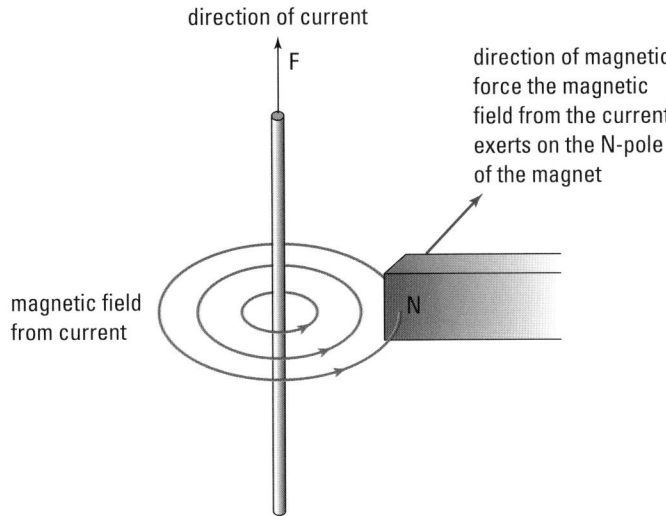

**Figure 14.28** At the position of the N-pole of the bar magnet, the lines of force from the current's magnetic field point directly into the page; therefore, the N-pole of the magnet experiences a force directly into the page.

From the perspective of the magnet, it sees the conductor carrying a current through its magnetic field (Figure 14.42). Since the magnetic field from the conductor exerts a force on the magnet, Newton's third law says that the magnet exerts an equal, but oppositely directed, force on the conductor. Right-hand rule #3, describes the direction of the force exerted on the conductor from the magnet's perspective.

## Right-hand Rule #3

Extend your right hand so that the fingers, thumb, and palm form a flat surface with the thumb at right angles to the fingers. Align the thumb along the conductor pointing in the direction of the current and the fingers pointing in the direction of the magnetic field (from the magnet) that is passing the conductor. The palm, then, is facing in the direction of the force that the field from the magnet exerts on the conductor (see Figure 14.29).

The magnetic field of the magnet points away from the N-pole of the magnet past the conductor. If the conductor were an ordinary magnet, its N-pole would experience a force directly away from the magnet (to the right). But the conductor has no N-pole or S-pole. Instead, the magnetic field from the magnet exerts a force at right angles to both the direction of the current and the direction of the magnet's field. Notice that all three directions (the current, the field, and the force) are all at right angles to each other (mutually perpendicular). It's easy to remember how to apply the rule if you think of the thumb, the fingers, and then the palm. There is one current (the thumb) passing through many lines of force (the fingers) and (the palm) "pushes" in the direction of the force on the conductor (Figure 14.29). This phenomenon is called the **motor effect** since it is the driving force that makes electric motors run.

At this point, it is important to realize that if the current crosses the field at an oblique angle, rather than at right angles, you must identify the direction of the component of the magnetic field that lies perpendicular to the current in order to apply right-hand rule #3. If the direction of the magnetic field is parallel to the current, then there is no component of the field perpendicular to the current, and as a result there is no force exerted by the field on the conductor.

**Figure 14.29** A conductor that carries a current at right angles to a magnetic field experiences a force at right angles to both the current and the direction of the field. This direction can be predicted by using right-hand rule #3.

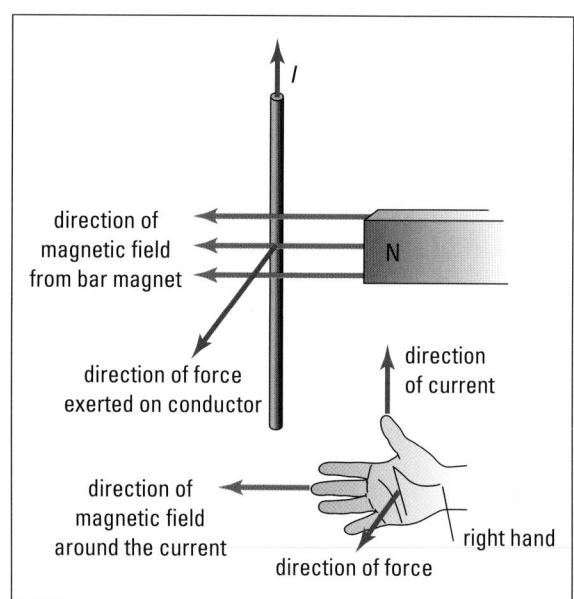

direction of magnetic field from bar magnet

N

direction of force exerted on conductor

direction of current

direction of magnetic field around the current

right hand

direction of force

When a current passes at right angles through a magnetic field, Field theory predicts that the current (and thus the conductor that carries it) experiences a force at right angles to both the direction of the current and the field through which it is passing.

## Problem

In this investigation, you will determine if that force exists and if the direction of the force is correctly predicted by right-hand rule #3.

## Equipment
- bar magnets (2) or a horseshoe magnet
- wire (about 2 metres long to make a coil)
- power supply
- ammeter (0 to 10 A)
- resistor (≈1 Ω, exact value is not important)
- retort stand (2)
- ring clamp
- string
- wax block
- elastic bands

## Part 1

### Procedure

1. Wrap the two-metre segment of wire around a block that is about 2 cm × 2 cm to make a coil with at least 20 turns. Leave the ends about 10 cm long to connect the coil to the leads from the circuit.

2. Set up the apparatus as shown on the upper right. Suspend the coil so that its bottom edge is between the poles of the two bar magnets (or the horseshoe magnet) as shown.

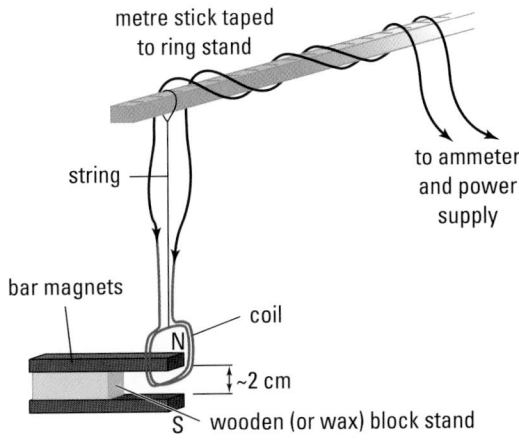

3. Turn up the potential difference of the power supply very slightly to check that the meter is connected correctly.

4. Once the circuit is connected correctly, increase the potential difference (until a current of about 5 A is reached).

5. As the current is increased, observe the coil for movement.

6. Draw a sketch showing the direction of the current, the magnetic field from the magnet, and the direction of movement.

7. Apply right-hand rule #3 to see if the movement that was observed was in the direction of the predicted force.

8. Reverse the direction of the magnetic field. Repeat steps 4 through 7.

9. Reverse the direction of the current in the coil. Repeat steps 4 through 7.

*continued* ▶

*continued from previous page*

## Part 2

### Procedure

1. Position the magnets so that both edges of the coil are between the poles of the magnet as shown below. If you attach a string to the bottom of the coil, as shown in the diagram, and hold it gently down, it will stabilize the motion of coil so that the motion caused by the field is easier to observe.

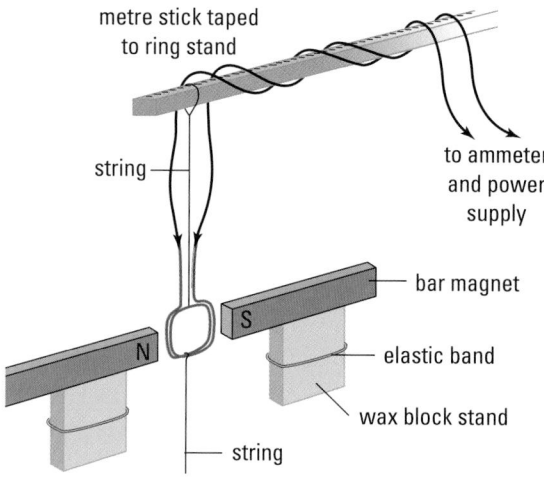

metre stick taped to ring stand

string

to ammeter and power supply

bar magnet

S

N

elastic band

wax block stand

string

2. Increase the potential difference of the power supply to about 5 A (or to a current that provided observable results during Part 1).

3. Observe the reaction of the coil to the field.

4. Draw an accurate sketch of the motion of the coil with respect to the direction of the field. Make sure you include the directions of the current and the magnetic field, as well as the forces experienced by the coil.

5. Apply right-hand rule #3 to the system to determine if the observed reaction of the coil could be predicted.

6. Reverse the direction of the magnetic field. Repeat steps 2 to 5.

7. Reverse the direction of the current in the coil. Repeat steps 2 to 5.

### Analyze and Conclude

1. In Part 1, did right-hand rule #3 correctly predict the movement of the coil when the current was (a) perpendicular, (b) oblique, and (c) parallel to the field?

2. In Part 1, did any unexpected movement of the coil occur? Try to explain these movements, if any.

3. Answer the following questions about Part 2:

   (a) Explain why the coil moves as it does? Why does it stop moving where it does?

   (b) Does right-hand rule #3 correctly predict the motion of the coil?

   (c) Apply right-hand rule #3 to the top and bottom edges of the coil. Does the rule predict that these edges experience a force? Explain why the coil does not seem to respond to the forces on the top and bottom edges of the coil.

### Apply and Extend

4. Could the coil be made to move away from the position it took up when the current was first turned on? Explain how this could happen.

---

**UNIT PROJECT PREP**

Motors rely on the interaction between electricity and magnetism.

- How important is magnetic field strength to motor design?
- How does the shape of the conductor in the magnetic field affect its operation?

# Defining the ampere

On September 4, 1820, Ampère read an account of Oersted's discovery of electromagnetism. On November 6th, he published his paper on electromagnetism which has become the basis for modern electromagnetic theory. In the paper, Ampère developed his famous mathematical law that describes the relationship between the current in a conductor and the magnetic field that results from it. In the paper Ampère also defined the unit of current, later named in his honour, and created the first "right-hand rule."

Ampère reasoned that if a current-carrying conductor created a magnetic field about itself, then two current-carrying conductors should interact by attracting or repelling each other in the same way as two magnets. To test this theory, he placed two conductors parallel and at a small distance from each other. He discovered that when the currents were in the same direction, the conductors attracted each other, and when currents were in the opposite direction, the conductors repelled each other.

Figure 14.30 shows two parallel conductors, A and B, carrying currents in the same direction. The lines of force (in blue) represent the magnetic field from conductor A. Right-hand rule #1 can be used to verify that, at the position of conductor B, these lines point vertically upward. Right-hand rule #3 can be used to verify that the direction of the magnetic force on the current in conductor B is toward conductor A.

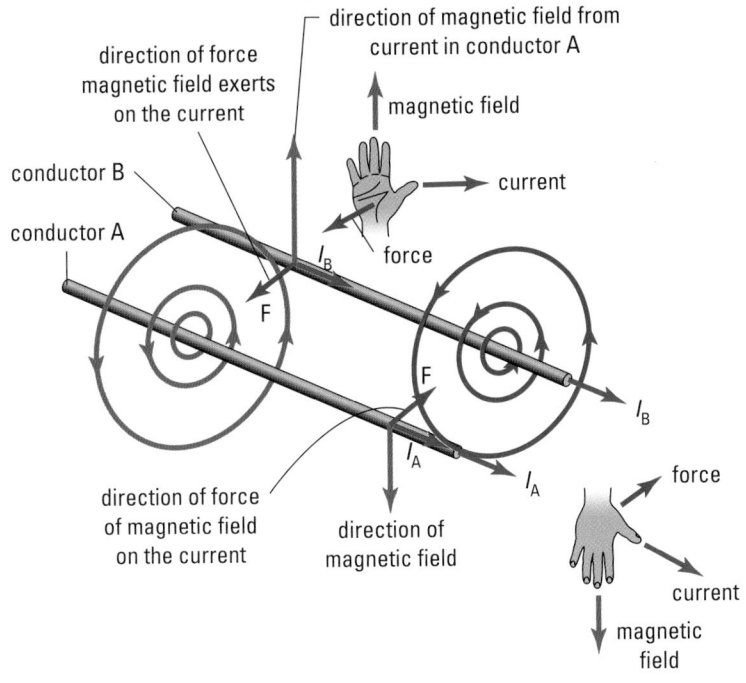

**direction of magnetic field from current in conductor A**

direction of force magnetic field exerts on the current

conductor B

conductor A

magnetic field

current

$I_B$

force

F

F

$I_B$

direction of force of magnetic field on the current

$I_A$

$I_A$

direction of magnetic field

force

current

magnetic field

**Figure 14.30** Parallel conductors carrying currents in the same direction experience a mutual force of attraction.

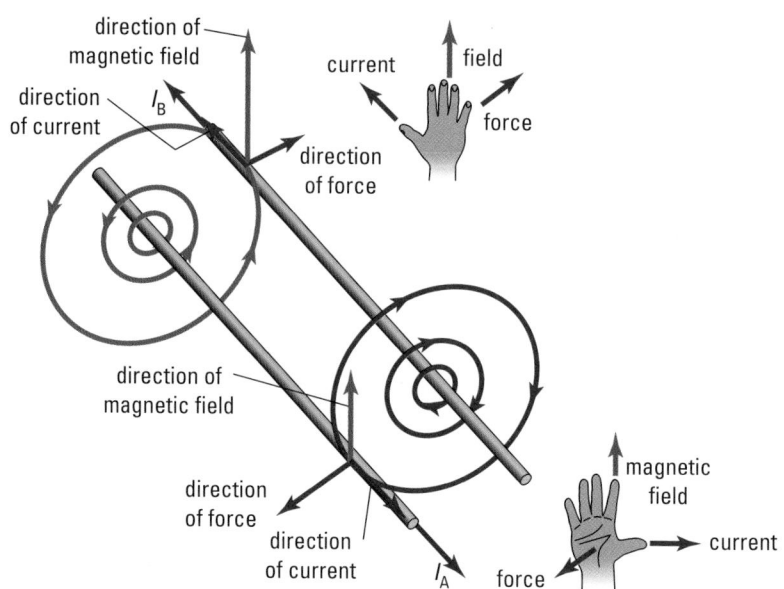

**Figure 14.31** When parallel conductors carry currents in opposite directions, the magnetic forces cause the conductors to repel each other.

Similarly, the lines of force (in red) represent the magnetic field from the current in conductor B at the position of conductor A. As they pass conductor A, these lines of force point vertically downward. Right-hand rule #3 indicates that the magnetic force on the current in conductor A is toward conductor B. Therefore, the two conductors appear to be attracted to each other.

Figure 14.31 shows two parallel conductors, A and B, carrying currents in opposite directions. Applying right-hand rule #3, in the same way as shown in Figure 14.30, will reveal why these two conductors repel each other.

Ampère discovered that the force ($F$) between the current-carrying conductors varied directly as the current in conductor A ($F \propto I_A$), directly as the current in conductor B ($F \propto I_B$), directly as the length ($F \propto L$) of the parallel conductors, and inversely as the distance ($F \propto \frac{1}{d}$) between the conductors. These relationships can be written, as shown, in one relationship.

$$F \propto \frac{I_A I_B L}{d}$$

Multiplying the right side by a proportionality constant ($K$) produces the equation:

$$F = K\frac{I_A I_B L}{d}$$

The units for force ($F$), length ($L$) and distance ($d$) had been defined prior to Ampère's investigation. The unit for current had not been defined so Ampère could now choose a unit for electric current that would produce a proportionality constant of any desired value. For example, the unit of force (1 N) was chosen to be the force that would cause one unit of mass (1 kg) to accelerate at one unit of acceleration (1 m/s²). Similarly, the unit of resistance (1Ω) was chosen so that one unit of potential difference (1 V) would cause one unit of current (1 A) to move through it.

Ampère did something that was unique in science. He chose the unit of current to be the current that, when flowing in each conductor, would cause the force of attraction or repulsion between the conductors to be exactly $2 \times 10^{-7}$ newtons per metre of conductor. Because of this choice, the value of the proportionality constant is, by defintion, exactly $2 \times 10^{-7}$ N/A$^2$. Had he chosen the current large enough to make $K = 1$, you would probably be more familiar with currents of milliamperes (mA) or microamperes (μA) rather than amperes (A).

One point that is often confusing to students of physics is the relationship between the unit of current and the unit of charge. Usually, current is defined in terms of the movement of a particular quantity per unit time. Water currents, for example, are often measured in litres per second. Therefore, it is often assumed that the unit of current (A) is defined in terms of the quantity of charge that moves per unit time, coulombs per second (C/s). But that is not the case; in fact the reverse is true. The unit of charge (one coulomb) is by definition the amount of charge that is moved by a current of one ampere in one second (1 C = 1 A·s).

Even though the coulomb is not a particularly large charge when viewed from the point of view of current electricity, it turns out that it is extremely large from the point of view of electrostatics. If you could place one coulomb of static electric charge on each of two bodies separated at a distance of one metre, they would exert an electrostatic force of about $9 \times 10^9$ N on each other.

## The Motor Force: Quantitative Analysis

When a conductor carries a current through a magnetic field, several factors affect the size of the force exerted on the conductor. First, magnitude of the force ($F$) exerted varies directly as the magnitude of the magnetic field ($B_\perp$) that acts perpendicular to the conductor. Second, the magnitude of the force varies directly as the magnitude of the current ($I$). Third, the magnitude of the force varies directly as the length of the conductor inside the field ($L$). Mathematically, the above statements can be written as

$$F \propto B_\perp$$
$$F \propto I$$
$$F \propto L$$

Combined mathematically, they become

$$F \propto B_\perp IL$$
$$\therefore F = kB_\perp IL$$

where $k$ is the proportionality constant.

When this relationship was first discovered, the units for all the quantities except magnetic field strength ($B$) had been defined. Thus, it was possible to define the **tesla**, the unit of magnetic field strength in terms of the force exerted on a current in the conductor. In this way, the value of $k$ could be made equal to one (1). Rearranging in terms of the other variables, the magnetic field strength is:

$$B_\perp = \frac{F}{IL}$$

This equation shows the relationship between the magnitudes of the variables involved. To find the directions, you must apply right-hand rule #3. Notice that the equation only works for magnetic fields that are perpendicular to the current. In the next course, it will be extended to apply to situations in which the current is not perpendicular to the field.

If a coil with n ($n$) turns of wire passes through a field, then the length of conductor inside the field is found by taking the product of number of turns in the coil ($n$) and the length of an individual turn ($\ell$). Therefore

$$L = n\ell$$

---

### MAGNETIC FIELD STRENGTH AND MOTOR FORCE

The magnetic field strength perpendicular to the conductor is the quotient of the motor force and the current and length of the conductor.

$$B_\perp = \frac{F}{IL}$$
$$L = n\ell$$

| Quantity | Symbol | SI unit |
|---|---|---|
| magnetic field strength | $B$ | tesla (T) |
| "perpendicular to" | $\perp$ | |
| motor force | $F$ | newton (N) |
| current | $I$ | amp (A) |
| length of conductor | $L$ | metre (m) |
| number of coil turns | $n$ | no unit |
| length of each turn | $\ell$ | metre (m) |

**Unit Analysis**

By definition in the first formula, $1 \text{ tesla} = \dfrac{(1 \text{ newton})}{(1 \text{ amp})(1 \text{ metre})}$

$$T = \frac{N}{A \cdot m}$$

---

## Calculating Magnetic Field Strength

A length of straight conductor carries a current of 4.8 A into the page at right angles to a magnetic field. The length of the conductor that lies inside the magnetic field is 25 cm (0.25 m). If this conductor experiences a force of 0.60 N to the right, what is the magnetic field strength acting on the current?

### Frame the Problem

- Since it is known that the current is at right angles to the field, the formula for magnetic field strength applies to this problem.

- Right-hand rule #3 can be used to find the direction of the field.

### Identify the Goal

Find the strength (size and direction) of the magnetic field acting on the current.

### Variables

| Involved in the problem | Known | Unknown |
|---|---|---|
| $F$ | $F = 0.60$ N | $B_\perp$ |
| $B_\perp$ | $I = 4.8$ A | |
| $I$ | $L = 0.25$ m | |
| $L$ | | |

> **PROBLEM TIPS**
>
> - Make sure you know that the *current* and the *magnetic field* act at right angles. This information may be given in many different ways in the problem statement so read very carefully.
> - Convert the law from the standard form into the form required to solve the problem.
> - It is always necessary to solve the direction portion of the problem separately from the calculation, using right-hand rule #3. You must always identify the directions for two of the three vectors (current, field, and force) to find the direction for the third.

### Strategy

State the equation relating magnetic field strength to force, current and conductor length.

Substitute the known values into the equation.

A $\frac{N}{A\,m}$ is equivalent to a T.

Apply right-hand rule #3 to find the direction of the magnetic field.

### Calculations

$$B_\perp = \frac{F}{IL}$$

$$B_\perp = \frac{0.60\ \text{N}}{(4.8\ \text{A})(0.25\ \text{m})}$$

$$B_\perp = 0.50\ \text{T}$$

Hold up your right hand with your thumb pointing into the page (away from you). The palm of your hand must face the right hand side of the page (the direction of the force). Then your fingers are pointing in the direction of the field.

The magnetic field strength is 0.50 T upward.

*continued* ▶

### PRACTICE PROBLEMS

1. A magnetic field has a strength of 1.2 T into the page. A current of 7.5 A flows vertically upward through a conductor that has 0.080 m inside the field. Find the force that the field exerts on the conductor.

2. A coil that consists of 250 turns of wire has an edge 12 cm long that carries a current of 1.6 A to the right. If the edge of the coil is inside a magnetic field of 0.16 T pointing out of the page, what is the force the field exerts on the coil?

3. A coil, consisting of 500 ($5.00 \times 10^2$) turns of wire has an edge that is 3.60 cm long that passes at a perpendicular angle through a magnetic field of 0.0940 T into the page. If the magnetic force on the edge of the coil is 10.8 N to the right, what was the current through the coil?

4. What magnetic field will exert a force of 22.0 N downward on a coil of 450 turns carrying a current of 3.20 A to the right through the field? The edge of the coil inside the field is 7.50 cm long.

## Torque on a Coil

In Part two of Investigation 14-E, you observed the effect on a coil when two edges were inside the magnetic field. The result was that the coil twisted in the field. The currents in the edges of the coil experience forces in the opposite direction. (right-hand rule #3) These forces create a **torque** (just like a magnetic dipole) about the axis of the coil. If the coil is free to move, the torque will reorient the coil so that its plane is perpendicular to the field. Once the plane of the coil is perpendicular to the field, the forces on the edges of the coil are linearly opposed so that the net force on the coil is zero.

Notice that the coil is now oriented so that the magnetic field due to the current through the coil (use right-hand rule #2) is in exactly the same direction as the field from the magnets. Placing an iron core inside the coil greatly increases the strength of the magnetic field from the current, and thus greatly increases the torque on the coil.

## The DC Electric Motor

You now have all the elements (in theory, at least) to build an electric motor. The motor has a coil with an iron core, called the **rotor** or **armature**, surrounded by field magnets. In many motors, the field magnets are electromagnets.

There are two obvious difficulties in electric motor design. First, when the coil turns so the magnetic forces on the edges of the coil are aligned directly opposite each other, it will no longer experience any torque. If you could now reverse the direction of the current, and thus the direction of forces, they would point inward,

**ELECTRONIC LEARNING PARTNER**

Go to your Electronic Learning Partner for a short animation about the use of electromagnets and permanent magnets in an electric motor.

rather than outward, and the edges of the coil would again experience a torque. Second, if the coil keeps turning, the leads to the coil will eventually become so twisted they will break.

The solution to both of these problems was solved by one device called a **split ring commutator**. The split ring commutator, as its name implies, is a brass or copper ring cut into two halves. The commutator is attached to the axle of the armature so that it rotates with the coil. One end of the coil is connected to each half of the commutator. Brushes (either metal or carbon) slide on the commutator to pass current from the battery to the coil. At this point, the coil can turn freely without twisting the leads to the coil. Even though a direct current comes from the battery, as the armature rotates, the brushes pass from one segment of the commutator to the other. When the brushes change contact from one half of the commutator to the other, the direction of the current in the coil is reversed. If this occurs at the instant when the coil has reached the point where its plane is perpendicular to the field, the forces on the edge of the coil will reverse and continue to cause the coil to continue its rotation through the field (see Figure 14.32).

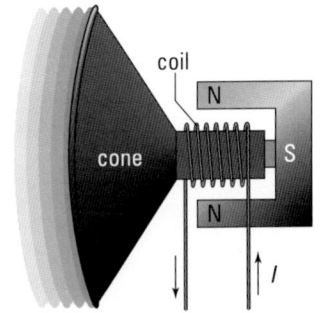
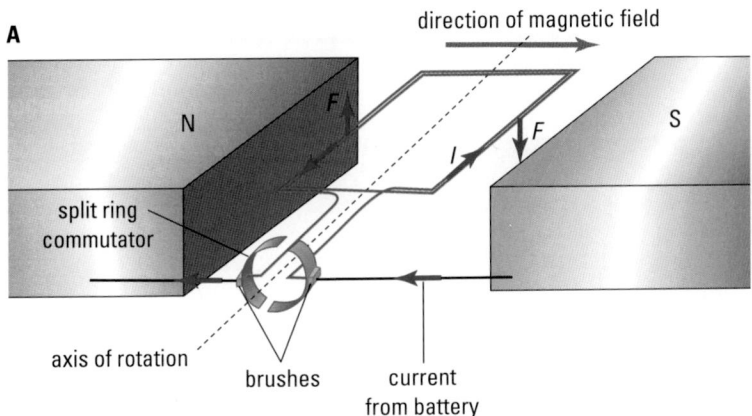

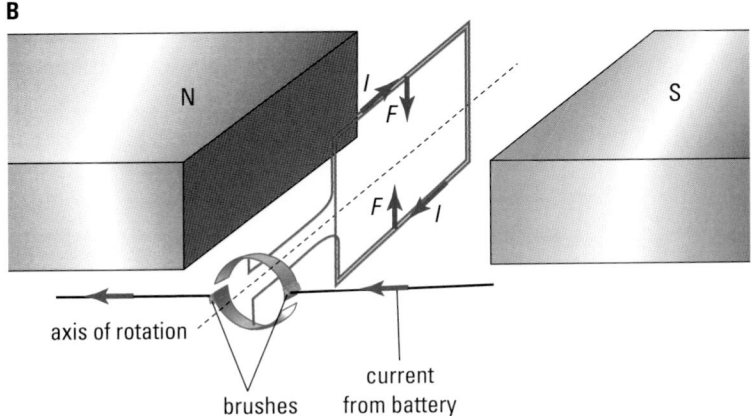

**Figure 14.32** **(A)** When the plane of the coil is parallel to the magnetic field, the torque on the coil is at a maximum. **(B)** When the plane of the coil is perpendicular to the field, the torque is at a minimum. At that point, the brushes cross over to the other segment of the commutator, reversing the current in and the forces on the edges of the coil.

### The Skeleton's Name is George.

Dr. Zahra Moussavi, biomedical engineer.

Due to revolution and war, universities in Iran were closed in 1978. As a result, it took Zahra Moussavi nine years to complete a four-year BSc in electronic engineering. In 1989, Zahra and her family moved to Canada. She obtained an MSc from University of Calgary and a PhD

from the University of Manitoba. Today, Dr. Moussavi teaches biomedical engineering and does research at the University of Manitoba. With the help of George, she studies the movement of arms and shoulders. Her research will help physicians decide how well a surgery has worked and the nature of the post-surgical rehabilitation that may be required.

Many patients, as the result of a stroke or brain tumour, have difficulty swallowing. Current techniques for detecting abnormalities are inadequate.

Dr. Moussavi is exploring the use of accelerometers as a method of detecting the sounds of swallowing and breathing. This technique has already proven valuable in the analysis of injuries to joints such as knees.

## 14.4    Section Review

1. **C** Two materials, A and B, both exhibit strong magnetic attraction when they are brought near a magnet. When the magnet is removed, material A loses its magnetism but material B retains its magnetism. Using domains, explain what happens with the two materials.

2. **K/U** A magnetic field points directly into the page. A square coil is inside the field so that the plane of the coil is parallel to the surface of the page. From your perspective, looking at the page, the current moves counter-clockwise around the coil. Find the direction of the force on each edge of the coil.

3. **K/U** A square coil lies in the plane of the page. The current flows in a clockwise direction inside the coil. If the magnetic field is to the right, identify the direction of the force on each edge of the coil.

4. **K/U** A square coil lies so that its plane is horizontal; two of its edges are parallel to the page and two of its edges are perpendicular to the page. If the magnetic field surrounding the coil points out of the page, in which direction will the coil experience a torque if the current is flowing into the page through the left-hand edge of the coil?

**UNIT PROJECT PREP**

To improve your motor design, think about the fundamental principles of magnets and motors.

- What is the motor effect?
- Why do current-carrying conductors react to magnetic fields?

## REFLECTING ON CHAPTER 14

- Magnets are always found as a dipole with both an N-pole and an S-pole. Breaking a magnet into two parts results in each part being a dipole.

- For magnetic poles, like poles repel each other while unlike pole attract. The force of attraction or repulsion varies inversely as the square of the separation of the poles.

- Materials are classified according to their magnetic behaviour. The three most common classifications are ferromagnetic, paramagnetic and diamagnetic materials.

- Domain theory is used to explain why ferromagnetic materials can be induced to display magnetic behaviour.

- Faraday invented the Field theory to explain magnetism. It postulates that it is the magnet's field rather than the magnet itself that exerts the magnetic force. The field can be represented graphically by a "lines of force" diagram.

- Lines of force are continuous loops. The point on a ferromagnetic material where the lines of force exit is an N-pole and the point where the lines of force enter the material is an S-pole.

- The relationships between the directions of electric currents and magnetic fields are defined by the right-hand rules.

- Oersted discovered the existence of magnetic fields near a current-carrying conductor. A right-hand rule is used to determine the direction of the field relative to the direction of the current.

- Ampère developed the laws describing the force exerted on a current carrying-conductor by the magnetic field in which it is located. He used the magnetic force exerted between parallel current-carrying conductors to define the ampere.

- The motor effect states that when a current passes through a magnetic field at right angles to the field, the magnetic field exerts a force on the current at right angles to both the current and the field. The direction of the motor force is found using a right-hand rule.

- Electric motors use a split-ring commutator to convert the DC current from the battery to an AC current so that the coil on the armature would always experience a force driving it in the same direction. The forces exerted on the opposite edges of the coil are in opposite directions. This results in a twisting action or torque on the coil.

### Knowledge/Understanding

1. Distinguish between the following:
   (a) the geographic North Pole and the magnetic north pole.
   (b) a compass needle and a dipping needle.
   (c) ferromagnetic, paramagnetic and diamagnetic materials.
   (d) lines of force and magnetic force.

2. Describe how magnetism is induced in a ferromagnetic material. Use references to domain theory in your explanation.

3. Describe what is meant by the magnetosphere. What is its role in creating the aurora borealis? Why are the auroras found in the polar rather than the equatorial regions of the magnetosphere?

4. Describe the two ways to use right hand rules to find the direction of the magnetic field inside a loop of wire.

5. Describe how the strength of a magnetic field is indicated in a line of force diagram.

6. A Compass needle is placed at the centre of a loop of wire. When a strong current is passed through the loop, the compass needle shows no change in position. What can you state about the orientation of the loop and the direction of the current through the loop?

7. A conductor lies in an east-to-west orientation across a table. Assume that the lines of force for Earth's magnetic field point due north across the conductor, and that two compasses are

placed so that one is above and the other is below the conductor. When the circuit is closed, a very strong electric current moves from west to east through the conductor. Use the right-hand rule to analyze what you should observe in the compass needles and why.

8. A conductor lies parallel to the lines of force from the Earth's magnetic field. A compass lies on top of the conductor so that its needle lies parallel to the conductor. As the current in the conductor is gradually increased, the compass needle gradually deflects to the west. (a) Is the current flowing north or south through the conductor? Explain. (b) At what angle would the compass needle be deflected when the magnetic field from the current is equal in magnitude to the Earth's magnetic field? Explain.

9. A solenoid is set up so that its axis is parallel to Earth's magnetic lines of force. A compass is placed at the geographic south end of the solenoid so that the N-pole of the compass points into the solenoid. When the current is turned on, the needle makes a 180° reversal in direction. Assume you are looking through the coil from its north end (due south along the axis of the coil). From your point of view, is the current moving around the coil in a clockwise or counter-clockwise direction? Explain your answer.

## Inquiry

10. Design and build a solenoid. Wrap about 200 tightly wound smooth turns of wire around a cardboard tube. The core of a roll of toilet paper or paper towel works well as a supporting structure. Keep the turns so that they are wound around a section of only 8 cm to 10 cm in length. Connect the solenoid, in series, to a variable voltage power supply, an ammeter and a resistor of about 1.0 Ω.

   (a) Increase the current to about 1.0 A. Use a compass trace a line of force from one end of the solenoid to the other. Refer to the technique used to trace lines of force in Investigation 14-B.

   (b) Place a soft iron rod so that its end is just inside the end of the solenoid. Increase the current to about 4.0 A. Why does the rod move into the solenoid? (HINT: Consider the induced magnetic polarity of the rod and the strength of the magnetic field of the solenoid at the poles of the rod.)

11. Design and build an electromagnet that, when powered by a single D-cell (flashlight battery), will support at least 2.0 kg. Connect your electromagnet to a variable voltage power supply and an ammeter. Make a graph of the weight your magnet can support versus the current. Does the strength of the electromagnet increase in direct proportion to the current? From your graph predict whether the magnet has an upper limit for the load it can support. Explain.

## Communication

12. Describe how Domain theory explains the difference between a permanent and a temporary magnet.

13. Contrast how the Action-at-a-distance theory and Field theory explain how a compass indicates directions on the Earth.

14. Describe the significance of Gilbert's spherical lodestone.

15. The table (Table 14.1) on page 673 gives the Curie point of gadolinium as 16° C. What does this mean for the magnetic properties of this material?

16. The two statements "like poles repel" and "unlike poles attract" are throwbacks to the Action-at-a-distance theory in that they imply the two poles interact with each other directly. Rewrite these two statements to reflect a Field theory perspective.

17. Explain how to apply a right-hand rule(s) to determine whether the force one long straight conductor exerts on a second conductor that runs parallel to it is an attractive or a repulsive force.

18. Draw a circular loop of wire that lies in the plane of the page. Draw an arrow to indicate that the current in the loop flows in the count-

er-clockwise direction. Draw "dots" and "crosses" to indicate the lines of force magnetic field inside and outside the loop as they pass through the plane of the page.

19. A solenoid lies in the plane of the page with its axis parallel to the edge of the page. The magnetic lines of force flow through the loop towards the bottom of the page. Draw a diagram of the solenoid, showing windings and the direction of the current through the windings.

20. A simple electric meter can be made by placing a compass at the centre of a coil of wire. How should the coil and compass be aligned so that the deflection of the compass can be used to indicate the size and direction of the current in the coil? Explain how this system could be used to indicate the size of the current.

21. Describe the role of the commutator in a DC motor. Support your description with diagrams to illustrate the function of the commutator.

22. A coil of wire, which is free to turn, is suspended so that its plane is parallel to Earth's magnetic field. When a current flows in the coil, it turns so that its plane is perpendicular to Earth's magnetic field. Explain why this happens.

## Making Connections

23. The solenoid is often described as a linear motor. If a rod is placed so that one of its ends is in a solenoid, the magnetic field of the solenoid will draw the rod into the solenoid (See Question 10). This action is employed in many industrial and home appliances. Prepare a list of the devices in the home that use solenoids. For each device, describe the purpose of the solenoid.

24. Investigate and report on the method by which information is recorded on a magnetic audio or video tape or a computer disk.

25. When you board an aircraft you are not allowed to use certain electronic devices during take off and landing. Investigate and report on which devices are not to be used and why this is so.

26. The Earth's magnetosphere at its most basic level is the reason why we can use magnetic compasses to navigate. However, its existence has many other important implications to our lives. Investigate and report on the importance of the magnetosphere to life on Earth.

## Problems for Understanding

27. Changes in magnetic field strength are shown by increasing or decreasing the number of lines. What can you deduce from a sketch showing one magnet with five times as many magnetic lines as another magnet?

28. Magnetic north continually, although slowly, changes position. Calculate the average speed with which magnetic north moves, if it moved 254 km over a 143 year period.

29. A hiker checks her compass and orients herself so that she is facing exactly north according to her compass. She then checks her map of isogonic lines and discovers that magnetic north deviates by exactly 22.0° West of the geographic North Pole.
    (a) By how many degrees and in what direction should she change her orientation so that she is facing directly at geographic North Pole?
    (b) By how many degrees and in what direction should she change her orientation so that she is facing directly east of the geographic North Pole?

30. Two parallel conductors carry current in opposite directions as shown in Figure 14.50. Describe the change in magnitude of the force for each of the following scenarios.
    (a) The current in one conductor is doubled.
    (b) The current in both conductors is tripled.
    (c) The distance between the conductors is halved.

**Answers to Numerical Problems**

**1.** 0.72 N[left] **2.** 7.7 N[down] **3.** 6.33 A[down]
**4.** 0.204 T[out of page]

H ydroelectric dams, like the one at Churchill Falls in Labrador, generate electrical power by using the potential energy of the water stored in the dam's reservoir to drive massive electric generators.

Until Faraday discovered electromagnetic induction, the only source of current electricity was from the chemical action of batteries. Generators, such as the ones in the photograph, convert mechanical energy into electrical energy at rates that would have been impossible using batteries. The Churchill Falls hydroelectric station, completed in 1971, can produce electrical energy at a rate of over 6500 megawatts (MW), with plans to add another 3200 MW by the end of the year 2006.

In this chapter, you will extend your understanding of electro-magnetism to include electromagnetic induction and the various ways it is used in generators and transformers.

# INVESTIGATION 15-A

## Faraday's Discovery

**TARGET SKILLS**
- Analyzing and interpreting
- Performing and recording

When Oersted discovered that an electric current produced magnetic effects, Faraday hypothesized that the reverse might also be true. To test his theory, he constructed a device similar to the one shown in the diagram. Faraday reasoned that when the battery caused the iron bar to become an electromagnet, the magnetic field would induce a current that would be detected by the galvanometer. In this investigation you will discover if Faraday was correct.

## Problem

Under what circumstances does a magnetic field induce an electric current?

## Hypothesis

When the switch is closed in the apparatus shown below, how do you expect the galvanometer to react, and why?

## Equipment

- iron rod (about 10 cm long)
- two 2 m lengths of insulated copper wire
- 6.0 V lantern battery
- switch
- galvanometer
- five alligator clip leads
- masking tape

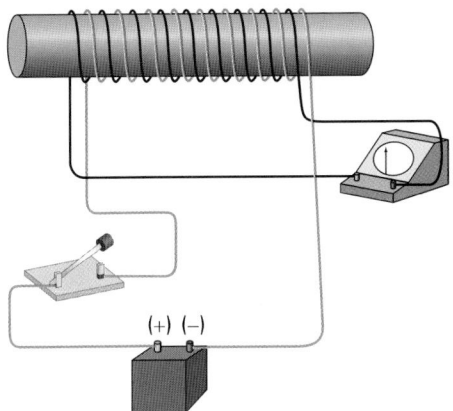

(+) (−)

## Procedure

1. Wind the iron rod with at least 50 turns of wire. Leave lengths of wire at each end of the coil to make connections. Wind a second coil of at least 50 turns of wire over the top of the first coil as shown in the diagram. A little masking tape can be used to make sure the coils do not unwind. Connect one coil to a battery with a switch, *keeping the switch open*. Connect the other coil to a galvanometer.

**CAUTION** The circuit to which the battery is connected is a short circuit. Leave the switch closed only long enough to confirm your observations in the next steps. Avoid touching connections while the switch is closed.

2. Close the switch for a second or two to complete the circuit of the coil connected to the battery. Does a current flow through the galvanometer when there is a current in the coil connected to the battery? If so, how long does the current last? Close and open the switch a few times to verify your observations.

## Analyze and Conclude

1. When a current is flowing in the coil connected to the battery, is there a current in the coil connected to the galvanometer?

2. Did the galvanometer react at any time to the current in the coil connected to the battery? If so, when and how?

3. Is the *direction* of the current in the galvanometer affected by the current in the coil connected to the battery?

4. Adjust the connection of the coil connected to the battery so that the current in the coil is reversed. Does this affect the reaction of the galvanometer? If so, describe the effect.

5. What conclusion can you make regarding a magnetic field's ability to induce an electric current in the coil connected to the galvanometer?

- Analyze and describe electromagnetic induction in quantitative terms.

- Analyze and predict the behaviour of induced currents using the right-hand rules.

- Hypothesize and test qualitative effects of electromagnetic induction.

- Explain the factors that affect the force on a current carrying conductor in a magnetic field.

## KEY
### TERMS

- generator effect
- electro-magnetic induction
- AC generator
- slip-ring commutator
- DC generator
- rectified DC current
- alternator

 **Technology Link**

Electric current is produced in microphones in three different ways. Dynamic microphones produce electricity using electromagnetic induction. Ceramic microphones use piezoelectric crystals to produce tiny electric currents. When sound waves move, the microphone diaphragm pressure is exerted on the piezoelec-tric crystals inside, causing them to produce electricity. The third type of microphone, the condenser micro-phone, uses a variable capacitor to create the fluctuations in current. Find out which types of microphone are typically used in the recording industry.

**Figure 15.1** Microphones use sound to induce an electric signal.

Microphones are just one of the thousands of devices that convert mechanical energy to electrical energy. In the most common type of microphone, sound waves striking the diaphragm of the microphone cause a tiny coil to move inside a magnetic field. This movement induces the currents that are sent to the amplifier.

## Currents from Magnetic Fields

In 1820, Ampère had shown that an electric current produces a steady magnetic field. In 1831, Faraday read of Ampère's findings and, by the principle of symmetry, predicted that a steady magnetic field should produce an electric current. The first few attempts to verify his hypothesis produced no results. On his sixth attempt, his investigations produced a rather surprising result.

On one side of an iron ring, he wound a wire solenoid that he connected to a battery and a switch. On the other side, he wound a wire solenoid that he connected to a galvanometer (Figure 15.2). Faraday thought that when he closed the switch, the current through the first solenoid would create a set of magnetic lines of force that would permeate the iron ring. These magnetic lines of force flowing through the second solenoid would cause a current that could be detected by the galvanometer.

When Faraday closed the switch, the needle on the galvanometer deflected to show a current, and then, unexpectedly, quickly dropped back to zero. As long as the switch was closed, the current remained at zero. Faraday could easily demonstrate that as long as the switch was closed, there was a magnetic field inside the ring. However, there was no accompanying current in the second solenoid. When he opened the switch, the needle on the galvanometer was again momentarily deflected, but in the opposite direction, indicating that for a brief time a current flowed in the opposite direction.

Faraday realized that a *steady* magnetic field through the ring would not generate the desired induced current. The brief pulses of current in the second solenoid, he hypothesized, must have been the result of the fluctuation in the strength of the magnetic field that occurred when the current was turned on or off.

Once again, the power of Faraday's field theory becomes evident. Faraday reasoned that when he turned the current on, the magnetic field inside the ring grew in strength. As the magnetic field grew, its lines of force expanded outward to fill space around the ring. As the lines of force expanded outward, they moved over the coils of the second solenoid. He argued that it was the motion of the lines of force across the coils that induced the current in the second solenoid.

Once the current was established in the first solenoid, the size of the magnetic field around the ring became constant. Since the lines of force were no longer changing, no current was induced in the second solenoid. However, when the current was turned off, the magnetic field collapsed. The lines of force now moved inward across the coils of the second solenoid, causing a brief current in the opposite direction.

To test his hypothesis, Faraday tried moving a magnetic field (from a bar magnet) into and out of a coil of wire. Just as he predicted, the movement of the magnetic field in the vicinity of the coil induced a current in the coil, as long as the magnet was moving. When the motion of the magnet stopped, so did the current. The direction of the induced current was dependent on the *relative motion* of the coil with respect to the magnet (Figure 15.3). Faraday had discovered the **generator effect**: the motion of magnetic lines of force past a conductor induces a current in the conductor. The process of generating an electric current in this way is now known as **electromagnetic induction**.

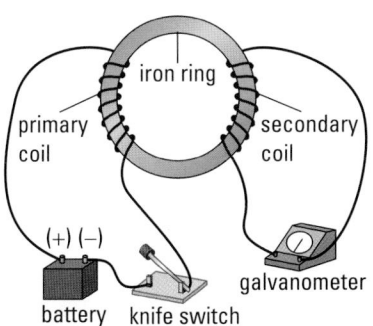

**Figure 15.2** Faraday used an iron ring wound with two solenoids to test whether the magnetic field from one solenoid could induce a current in the second solenoid.

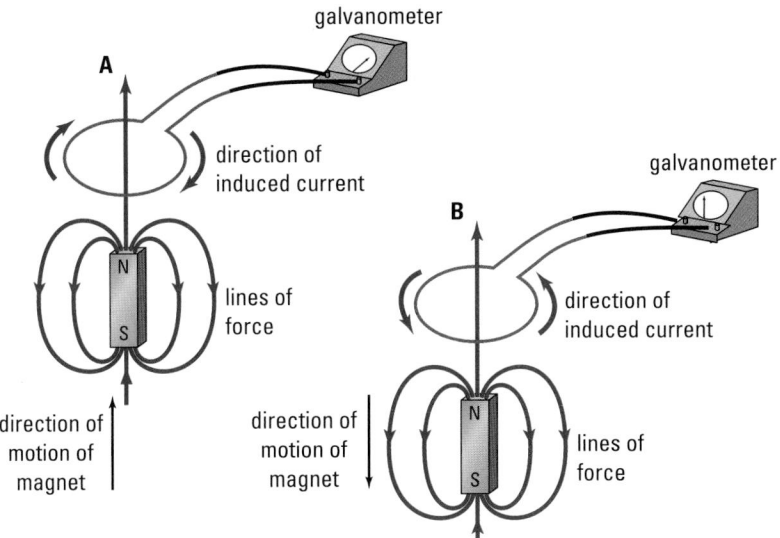

**Figure 15.3** The motion of magnetic lines of force past a conductor can induce a current in the conductor. The direction of the current depends on the relative motion of the coil with respect to the lines of force.

Electromagnetic Induction • MHR   **719**

# INVESTIGATION 15-B

## Induced Currents

When a conductor is moved inside a magnetic field, an electric current is induced in the conductor. It would seem natural to assume that the properties of the motion (speed, direction, and orientation) would affect the size and direction of the current. Other possible factors that might affect the induced current are the strength of the field and the length of the conductor inside the field.

In this investigation, you will try to determine, qualitatively, the relationship between the induced current in the coil and the factors that might affect that current.

### Problem

Determine a qualitative relationship between the induced current and (a) the motion of the coil, (b) the strength of the magnetic field, and (c) the length of the conductor.

### Prediction

For each part of the investigation, make a prediction and record it in your logbook.

**Part 1**

Predict how the speed of the coil (a) perpendicular to the field and (b) parallel to the field will affect the reading on the galvanometer.

**Part 2**

Predict the effect of the strength of the magnetic field on the reading of the galvanometer.

**Part 3**

Predict the effect of the length of the conductor that moves in the field on the reading on the galvanometer.

### Equipment

- bar magnets (6)
- wax blocks
- elastic bands
- copper coil (5 turns, 2 cm × 2 cm square)
- copper coil (10 turns, 2 cm × 2 cm square)
- copper coil (20 turns, 2 cm × 2 cm square)
- alligator clip leads
- galvanometer

### Procedure

**Note:** It is important that you make careful observations as you proceed through each part of this investigation. The observations you make in Part 1 will affect what you do in Part 2.

**Part 1**

1. Set up the magnets on the wax blocks with one pair of magnets mounted on them so that the N-pole of one magnet is about three centimetres from and facing the S-pole of the other magnet.

2. Using the coil with 20 turns, move the coil so that one edge of the coil moves across the magnetic field at right angles to the field between the magnets (see below).

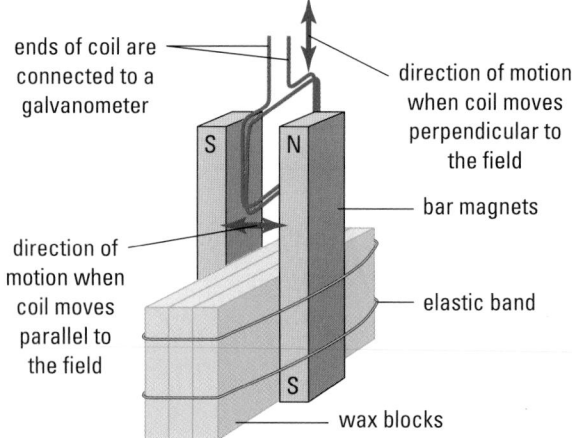

ends of coil are connected to a galvanometer

direction of motion when coil moves perpendicular to the field

bar magnets

direction of motion when coil moves parallel to the field

elastic band

wax blocks

3. Observe the motion of the galvanometer needle. In a table, like the one shown below, record your observations of the reaction (both direction and magnitude) of the galvanometer needle. Try to use consistent slow and fast speeds.

| Direction of motion | Speed of motion | Galvanometer reading (size and direction) |
|---|---|---|
| 1. Perpendicular to field, inward | slow | |
| 2. Perpendicular to field, outward | slow | |
| 3. Perpendicular to field, inward | fast | |
| 4. Perpendicular to field, outward | fast | |
| 5. Parallel to field, N- pole to S-pole | slow | |
| 6. Parallel to field, S-pole to N-pole | slow | |
| 7. Parallel to field, N-pole to S-pole | fast | |
| 8. Parallel to field, S-pole to N-pole | fast | |

**Part 2**

1. Note that the actual length of the conductor inside the field depends on the number of turns of wire that form the coil. Thus the effective length of wire for the coil with 10 turns is twice that of the coil with 5 turns.

2. Examine the response of the galvanometer to each of the tests you did in Part 1. Using your observations from Part 1, move each of the coils, in turn, through the field in directions that had significant responses. (For example, if both perpendicular and parallel

motions affected the current in the coil, then continue to test the effect of both of these motions throughout the experiment.)

3. Try to make the speed of all the coils as consistent as you can as you move them through the field. Observe the reaction of the galvanometer for each of the trials you perform, and record your observations in a table like the one below.

| Direction of movement of the coil | Number of turns | Galvanometer reading (size and direction) |
|---|---|---|
| | | |

**Part 3**

1. To test the effect the strength of the magnetic field has on the induced current, the number of magnets used to create the field will be increased. Use elastic bands to hold two bar magnets with "like" poles together. Position two sets of these magnets so that they are aligned with the N-poles of one facing the S-poles of the other. Make sure that the gap between the poles is the same as in the previous trials (see the diagram below).

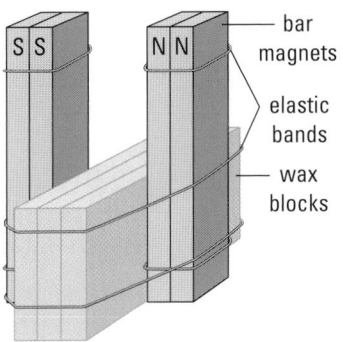

*continued* ▶

*continued from previous page*

2. Move the coil with 20 turns within the gap between the magnets, using the same orientations as in Part 2. Try to keep the speeds and orientations of the coils as constant as possible between trials. Record your observations in a table similar to the one below.

| Direction of movement of the coil | Number of magnets | Galvanometer reading (size and direction) |
|---|---|---|
|  |  |  |

3. Now, use an elastic band to hold three magnets together, just as you did in step 1 of this part of the investigation, and repeat step 2.

4. Use one magnet only on each side of the field, and repeat step 2.

## Analyze and Conclude

1. How did moving the coil (a) perpendicularly to the field, and (b) parallel to the field affect the induced current?

2. How did the speed of the coil affect the induced current?

3. How did the length of the conductor (number of turns) affect the induced current?

4. How did the strength of the magnetic field affect the induced current in the coil? (**Note:** Do not assume that the field from two pairs of magnets doubles the field strength.)

5. In summary, briefly describe the factors that affect the strength and direction of the induced current in a conductor moving through a magnetic field.

---

## TARGET SKILLS

- **Conducting research**
- **Communicating results**

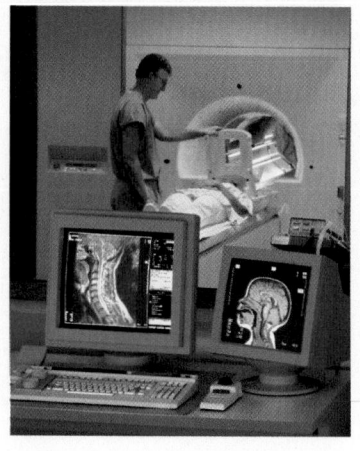

## PHYSICS & SOCIETY

Jillian was diagnosed with a brain tumour at age four. While most of the tumour could be removed by surgery, some of it was too deeply imbedded in her brain stem for removal. The doctors decided that they would need to monitor these cells for any sign of growth.

X-rays do not produce good detailed images of soft tissues. Instead, magnetic resonance imaging (MRI) is used. An MRI machine is just a large doughnut-shaped electromagnet. The nuclei of the hydrogen molecules in our tissues act like tiny magnets and become aligned with the magnetic field of the MRI machine. When these nuclei are subjected to low-energy radio waves, they are nudged out of alignment. When the radio waves are turned off, the nuclei snap back into alignment and give off a tiny electromagnetic pulse. Computer analysis transforms these pulses into detailed images of the tissue. Each image is a thin cross-section, so a series of these cross-sections creates a three-dimensional picture. Images show that, over the years, Jillian's tumour has not grown.

### Going Further

Medical imaging methods include MRI, fluoroscopy, CT scans, ultrasound, and fibre-optics. Investigate these techniques and their uses.

# Right-hand Rules for the Generator Effect

The generator effect can be explained in terms of the motor effect. Consider a conductor in the form of a straight rod, connected to a galvanometer, oriented in a magnetic field so that the rod is at right angles to the lines of force. Now, move the conductor so that it moves through the magnetic field in a direction perpendicular to both the lines of force and the orientation of the rod (Figure 15.4).

To find the direction of the induced current, Faraday devised a method using the right-hand rule in the same way as in the motor effect. He assumed that as the conducting rod moves upward through the field, each positive charge in the rod can be considered as a tiny bit of a current that is moving in the direction of the motion of the rod. Thus, by the motor effect, each charge moving within the conductor experiences a force (**F**) that acts at right angles to both the direction of the velocity (**v**) of the rod carrying the charges, and the lines of force of the magnetic field (**B**). In Figure 15.4, the magnetic field points into the page as the rod moves upward through the page. According to the right-hand rule, the direction of the force on the positive charges in the rod, and thus the current that is induced in the rod, is from right to left.

This application of the right-hand rule can be used to find the direction of the induced current in the coil as it moves through a magnetic field. In Figure 15.3(a) on page 719, the magnet is being moved upward toward the coil. As it approaches the coil, the lines of force curling outward from the magnet cut across the conductor.

To apply the right-hand rule, Faraday had to assume instead, that the conductor was moving downward toward the field. As the lines of force from the magnet loop around from the N-pole to the S-pole, the coil passes through the field so that the coil cuts across the lines of force. The detail in Figure 15.5 on the next page shows the direction of the motion of the coil, the direction of the lines of

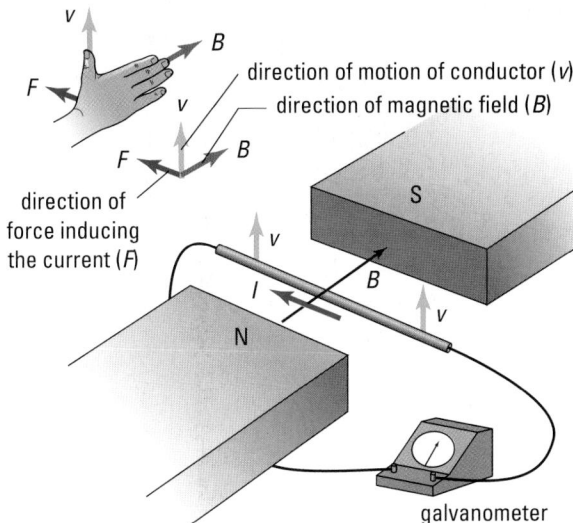

**Figure 15.4** The charges in a moving conductor experience a force that pushes them along the length of the conductor to induce a current. The right-hand rule can be used to find the direction of the force on the charges that form the induced current.

direction of motion of conductor (*v*)
direction of magnetic field (*B*)

direction of force inducing the current (*F*)

galvanometer

**Figure 15.5** To use the right-hand rule to find the direction of the current induced in a conductor, the motion of the conductor with respect to the magnetic field must be identified.

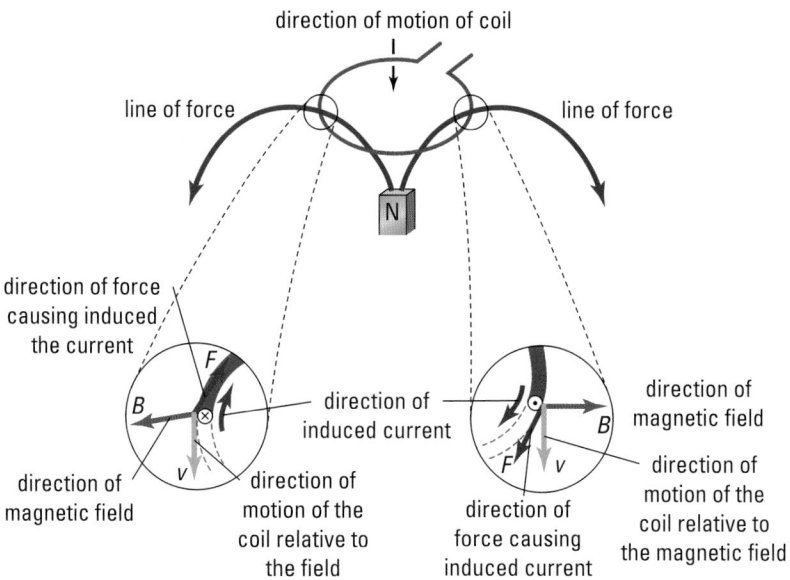

direction of motion of coil

line of force

line of force

N

direction of force causing induced the current

direction of induced current

direction of magnetic field

direction of magnetic field

direction of motion of the coil relative to the field

direction of force causing induced current

direction of motion of the coil relative to the magnetic field

induced negative pole of the conductor

(−)

conductor moving to the left through the magnetic field

magnetic field (*B*) pointing into page

induced positive pole of conductor

(+)

direction of motion

field direction

*B*

direction of force (*F*) which induces the current

*F*

*B*

*F*

**Figure 15.6** If an isolated conductor moves across a magnetic field, the generator effect induces the ends of the conductor to take on a polarity similar to a battery.

force, and the direction of the resulting force that causes the induced current. On the right edge of the coil, the direction of the force (induced current) is out of the page. If viewed from above, the current in the coil would be flowing clockwise.

## Electromotive Force

Until now, all discussion of electromagnetic induction has centred on the induced current. Since the motor effect was based on the presence of a current, it seemed natural to base the generator effect on currents as well. However, it turns out that it is more productive to discuss electromagnetic induction in terms of the electromotive force (*emf*) produced by the motion of the conductor in the circuit rather than the current.

If a rod is connected to a complete circuit, such as in Figure 15.4, then a current will be induced in the direction as shown. However, if the rod is not connected to an external circuit, the motor force on the charges within the rod still exist. The effect of the motor force is to push the charges in the rod toward the ends of the rod (Figure 15.6). Positive charges are pushed to the end of the rod in the direction of the motor force, while negative charges are pushed in the opposite direction. This action results in one end of the rod becoming positively charged and the other end becoming negatively charged. The ends of the rod, like the poles of a battery, now have a potential difference.

The action of the electromagnetic forces on the charges in a moving conductor parallels the electrochemical action inside a battery. In both cases, positive charges are moved onto the anode, leaving the cathode with a negative charge. In both cases, the potential difference between the anode and cathode can be used to move a current externally from the anode to the cathode.

In Chapter 13, you learned that if a battery is not connected to an external circuit, the potential difference is defined as the *emf*. When the battery supplied a current to a circuit, the terminal voltage was lower than the *emf* since some of the *emf* was used to move the current through the battery's internal resistance.

Like the *emf* of a battery, the induced *emf* (measured in volts) is independent of the internal resistance, since it is calculated when there is no current. Just like the motor force (see Section 14.4), the induced *emf* varies directly as the magnetic field strength (**B**); the velocity of the conductor through the field (**v**); and the length (**L**) of the conductor in the magnetic field.

As expected, the product of the units for these three quantities produces the units for *emf* (volts).

$$emf = vB_\perp L$$

$$\left(\frac{m}{s}\right)(T)(m) = \left(\frac{m}{s}\right)\left(\frac{N}{A \cdot m}\right)m$$

$$= \frac{N \cdot m}{A \cdot s}$$

$$= \frac{J}{C}$$

$$= V$$

Once the induced *emf* of a system has been found, then the induced current in a circuit of known resistance can be calculated using Ohm's law, in the same way as for a battery-driven circuit.

The polarity of the conductor moving through the field can be found if you use the right-hand rule for the generator effect. The thumb points in the direction of the motion of the conductor, the fingers point in the direction of the field, and the palm pushes charges toward the positive pole of the conductor.

## AC Generators

Until now, only the linear motion of conductors through magnetic fields has been discussed. This type of motion does allow for sustained current production; however, the solution to that problem has already been presented. Just as the electric motor produces a continuous rotation of a coil that carries a current inside a magnetic field, the electric generator produces a continuous current from a coil that rotates in a magnetic field. In fact, the two devices are essentially the same design.

In Figure 15.7, a rectangular coil is rotating counter-clockwise in the magnetic field, which points from right to left. The right edge of the coil is moving upward through the magnetic field. By Lenz's law, the *emf* in the right edge of the coil has its positive end nearest the viewer. On the left edge of the coil, which is moving downward, the positive end of its edge is farthest from the viewer. The *emf*s for the edges of the coil are like cells in series. They add together to produce an *emf* for each loop of the coil that is twice

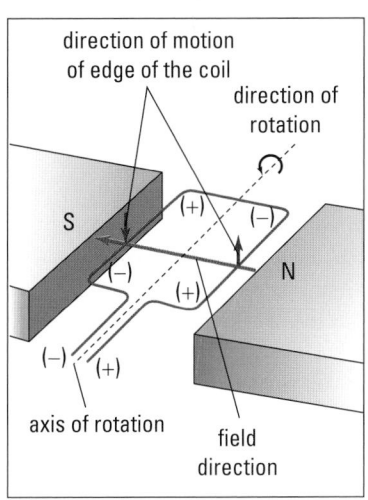

**Figure 15.7** When a coil is rotated in a magnetic field, the edges of the coil cut the magnetic lines of force to induce an *emf* in the coil.

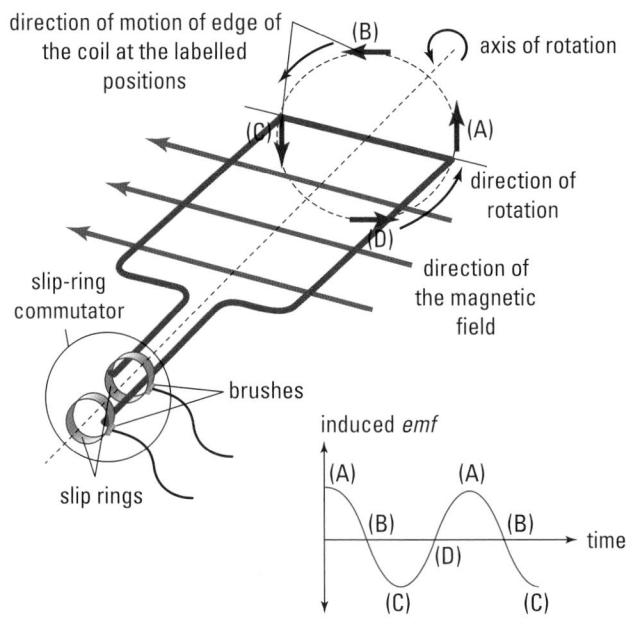

**Figure 15.8** When the armature of a generator is rotated between the poles of the field magnets, an alternating *emf* is produced.

the *emf* of a single edge. Each turn of the coil adds two more edges in series, so that the total *emf* for the generator is the *emf* of each turn multiplied by the number of turns. This makes it possible to design generators that produce any desired *emf*.

An **AC generator** has an armature, which is a copper coil wound around an iron core. As in an electric motor, the magnetically permeable iron core inside the coil greatly enhances the strength of the field inside the coil. When the armature is rotated, the edges of the coil cut across the lines of force of the magnetic field to produce an induced current in the coil. Like the motor, the current in the coil is transferred to the fixed body of the generator by brushes sliding on a **slip-ring commutator** (Figure 15.8). This consists of two unbroken brass or copper rings.

Using a slip-ring commutator, rather than a split-ring commutator, means that the brushes are always in contact with the same end of the coil. When the right edge of the coil is moving upward through the field (position A in the diagram), the slip ring nearest the viewer is the positive end of the coil. However, on the other half of the rotation, when the same edge is moving downward (through position C), the slip ring farthest from the viewer is positive.

Even though the coil is rotating at a constant speed, the induced *emf* is not constant. The edges of the coil cut across the lines of force at the greatest speed when they are moving at right angles across the lines of force. This occurs when the edges of the coil are at positions A and C in the diagram. When the edges of the coil are passing through those positions, the induced *emf* in the coil is the greatest.

As the edges of the coil move through positions B and D in Figure 15.8, they are moving parallel to the lines of force. Since, at that instant, the edges are not cutting any lines of force, the induced *emf* is zero. At positions B and D, the edges of the coil are in the process of reversing their directions of motion through the field. The edge that was moving up on the right side is now starting to move down on the left side and vice versa. As the directions of the edges of the coil reverse, so does the direction of the induced *emf* in the coil, resulting in an AC generator. The actual speed at which the edges of the coil cut the lines of force varies as the sine of the angle between the direction of the lines of force, and the direction of the motion of the edge. Thus, AC generators produce the characteristic sine wave pattern of AC electricity.

## DC Generators

As you saw in Section 14.4, in the DC motor, a split-ring commutator is used to convert the incoming DC current into an AC current in the motor coil. Similarly, in a **DC generator**, a split-ring commutator is used to replace the slip-ring commutator. As the coil rotates, the induced current reverses in the coil. At the same instant that the current changes direction in the coil, the brushes cross the gap from one half of the split ring to the other half, so that the current always leaves the generator in one direction. The current still increases and decreases depending on the angle at which the edges of the coil cut through the magnetic field. Thus, the sine wave output of the AC generator is converted into a pulsating DC output (Figure 15.9). This type of current is referred to as a **rectified** (made upright) **DC current**.

**PROBEWARE**

If your school has probeware equipment, visit the **Science Resources** section of the following web site: ***www.school.mcgrawhill.ca/resources/*** and follow the **Physics 11** links for a laboratory activity on induced electric current.

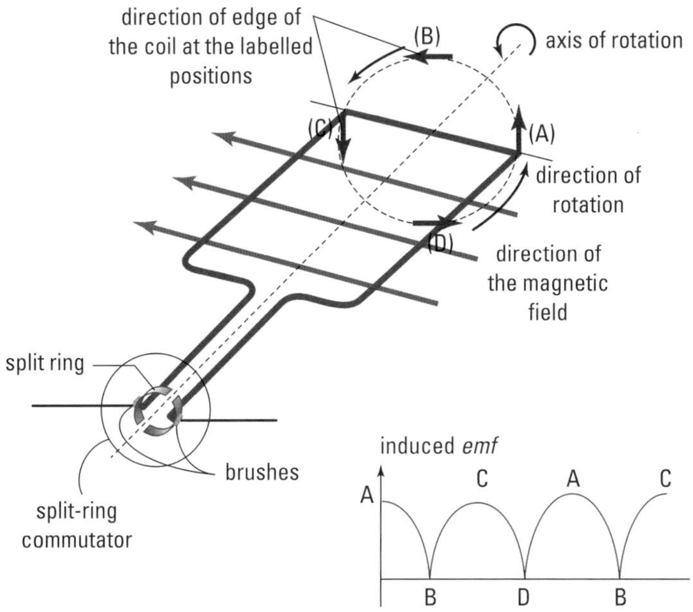


**Figure 15.9** The split-ring commutator in a DC generator rectifies the output of an AC generator to produce a pulsating DC output.

## Alternators

Every time the brush of a DC generator crosses over from one half of the split-ring commutator to the other, a tiny electric arc is formed. Eventually, this will cause the brushes to fail. Since the brushes on a slip-ring commutator slide continuously on the same ring, they last much longer than the brushes on a split-ring commutator. Today, the **alternator**, a device that uses diodes to rectify the output of an AC generator, is more common than DC generators.

## 15.1 Section Review

**1.** **K/U** A conductor is oriented horizontally parallel to the north-south direction. It is moving eastward through a magnetic field that points directly downward. In which direction does the induced current flow through the conductor?

**2.** **K/U** A loop of wire lies in the horizontal plane. A bar magnet, with its S-pole pointing downward, is lowered into the loop from above. As seen from above, in which direction will the induced current move around the loop?

**3.** **K/U** A coil is dropped into the magnetic field between the poles of a horseshoe magnet. If the current in the coil is in the direction indicated, which pole of the magnet (A or B) is its N-pole?

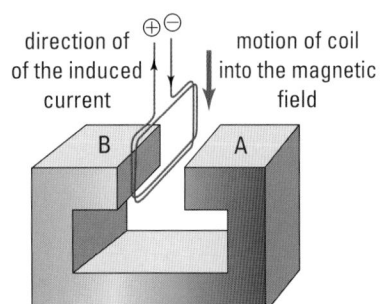

**4.** **K/U** A coil of wire lies in the same plane as the page. The pole of a bar magnet is moved toward the coil along the axis of

the coil. The induced current resulting from the motion of the magnetic field is clockwise around the coil. Which pole of the magnet approached the coil?

**5.** **K/U** The S-pole of a magnet points vertically upward below a conductor. If the magnet is moved from right to left, which pole of the conductor (A or B) will become positively charged?

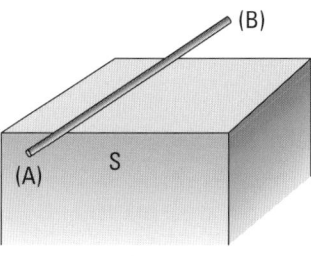

direction of the magnet's motion

### UNIT PROJECT PREP

Generators convert mechanical energy into electrical energy and motors convert electrical energy into mechanical energy.

- How can you use the results of Investigation 15-B to improve your motor design? For example, what is the effect of using a coil with a different number of windings?
- What other design ideas can you get from small, battery-operated toys?

To find the direction of the induced current using the right-hand rule and the motor effect, as in Figure 15.4 (page 723), is sometimes quite difficult. This is especially true when there is no apparent motion of the coil relative to the lines of force. An obvious example of this occurs in Faraday's original experiment with the coils wrapped around the iron ring. In 1834, a Russian physicist, Heinrich Lenz (1804–1865), devised an alternate method of finding the direction of the induced current.

## Induced Current and the Conservation of Energy

Lenz realised that when Faraday moved the bar magnet through the coil (Figure 15.3, page 719), he was generating electrical energy in the form of the induced current. By the law of conservation of energy, the gain in electrical energy had to come from the kinetic energy of the magnet moving through the coil. The transfer of energy from one object to another is, by definition, work. Work, in turn, requires a force. To remove kinetic energy from the magnet requires a force that acts in the opposite direction to the motion of the magnet. If the magnet could move unimpeded through the coil, then no work would be required to create the electrical energy. Lenz argued that, by the law of conservation of energy, whenever a conductor interacts with a magnetic field, there must be an induced current that opposes the interaction. This conclusion is known as **Lenz's law**.

> ### LENZ'S LAW
>
> When a conductor interacts with a magnetic field, there must be an induced current that opposes the interaction, because of the law of conservation of energy.

The pickup of an electric guitar, shown in Figure 15.10, is made of a permanent magnet surrounded by a tiny coil of wire. When the ferromagnetic metal string of an electric guitar is plucked, its motion near the magnetic field of the pickup causes the strength of the magnetic field inside the coil to fluctuate. By Lenz's law, a very weak current, whose magnetic field opposes the variations in magnetic field strength produced by the vibrating string, is induced in the coil. The current is then amplified and sent to a loudspeaker that converts the current back into sound.

**SECTION EXPECTATIONS**

- Define and describe Lenz's law.
- Identify the direction of induced electric current resulting from a changing magnetic field.
- Explain the relationship between induced current and the conservation of energy.

**KEY TERMS**

- Lenz's law
- back *emf*
- magnetic damping
- eddy currents

**Figure 15.10** Electric guitar pickups rely on Lenz's law.

# Lenz's Pendulum

In this investigation, a pendulum, in the form of a copper coil, is set to swing back and forth through a magnetic field. Because copper is diamagnetic, any effect the magnet has on the motion of the pendulum must be due to factors other than the interaction of the copper and the magnet.

## Problem

Observe the effect of a magnetic field on the motion of a conductor through the field, and formulate a hypothesis to account for your observations.

## Predictions

In each part of the investigation, make a prediction about the motion of the pendulum.

## Equipment

- bar magnets (2)
- copper coil  (20 turns, 2 cm × 2m)
- galvanometer
- retort stand
- masking tape
- metre stick
- 2 large paper clips
- wax blocks
- elastic bands
- alligator clip leads

## Procedure

### Part 1

1. Prepare the coil by wrapping 20 turns of copper wire around a square block to form a square coil approximately 2 cm by 2 cm. Leave 15 cm of wire on each end of the coil.

2. Use masking tape to secure the metre stick at the edge of the retort stand.

3. Suspend the coil from the metre stick using paper clips taped onto the stick as shown

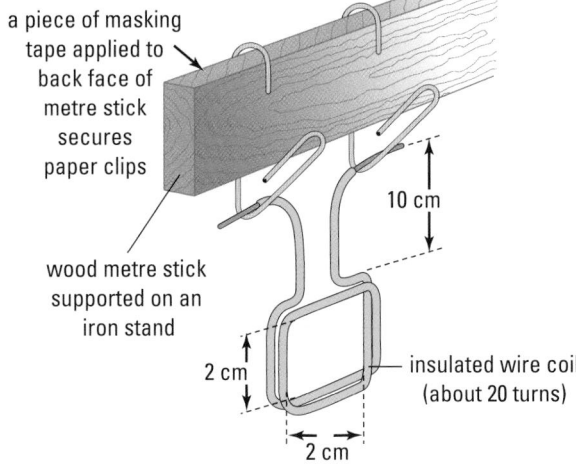

a piece of masking tape applied to back face of metre stick secures paper clips

wood metre stick supported on an iron stand

10 cm

2 cm

insulated wire coil (about 20 turns)

2 cm

in the diagram. Remove enough insulation from the ends of the coil leads so that bare copper wire is in contact with the paper clips.

4. Set the pendulum in motion. Adjust the positions of the paper clips so that the pendulum swings smoothly beneath the metre stick. Use a piece of masking tape to secure the clips on the metre stick. (**Note:** Do not place the magnets near the coil until Part 2.)

5. Move the bottom of the coil sideways about 1 cm from its vertical rest position. Allow the coil to swing back and forth until it comes to rest. Record the number of oscillations (swings) required for it to come to rest.

6. Move the coil sideways about 2 cm and release it. Count and record the number of swings required for the coil to come to rest.

### Part 2

7. Mount the bar magnets on the wax blocks. Position the magnets so that the lines of force of the magnetic field from the bar magnet cut perpendicularly across the lower edge of the coil. Make sure that the coil will swing without touching the magnets.

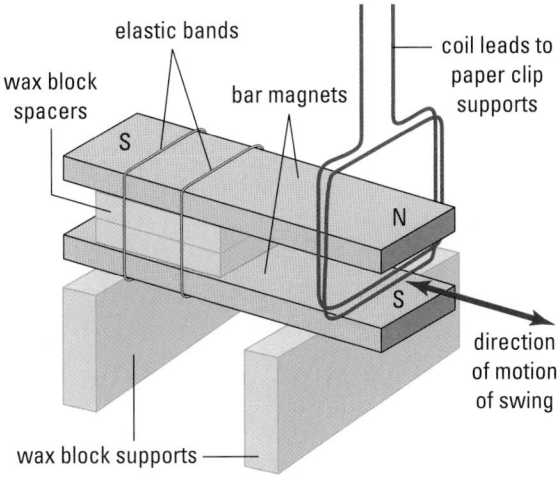

elastic bands

wax block spacers

bar magnets

coil leads to paper clip supports

S

N

S

direction of motion of swing

wax block supports

**8.** With the magnets in place, pull the coil sideways 1 cm and release it. Count and record the number of oscillations required for the coil to come to rest.

**9.** Pull the coil 2 cm to the side and release it. Count and record the number of oscillations required for the coil to come to rest.

**Part 3**

**10.** Use an alligator clip lead to connect the paper clips supporting the coil.

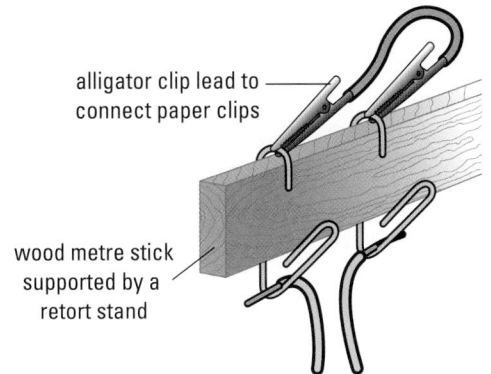

alligator clip lead to connect paper clips

wood metre stick supported by a retort stand

**11.** Pull the coil 1 cm to the side and release it. Count and record the number of swings required for the coil to come to rest.

**12.** Pull the coil 2 cm to the side and release it. Count and record the number of swings required for the coil to come to rest.

### Analyze and Conclude

**1.** Did the presence of the magnets in Part 2 affect the swing of the coil? If so, describe the difference. Was a different number of swings required for the pendulum to come to rest? Was there any noticeable qualitative difference in the swing of the pendulum? Why do you think that the pendulum behaved as it did?

**2.** In Part 3, when a wire was used to connect the paper clips that support the pendulum, did the motion of the pendulum change? If so, describe the difference.

**3.** What might have happened in the coil when the wire connected with the paper clips, that could not have happened in Parts 1 and 2?

**4.** Formulate a hypothesis that could account for your observations. Record your hypothesis in your logbook. Later, when you have completed the chapter, come back and review your hypothesis.

### Apply and Extend

**5.** Connect the paper clips supporting the coil to a galvanometer and set the coil in motion between the magnets.

**6.** Record the reaction of the galvanometer to the motion of the coil in the magnetic field.

**7.** Move the magnets away from the coil and set the coil in motion again. Is there any change in the motion of the coil? Does the motion of the coil seem to be related to the reaction of the galvanometer? Explain.

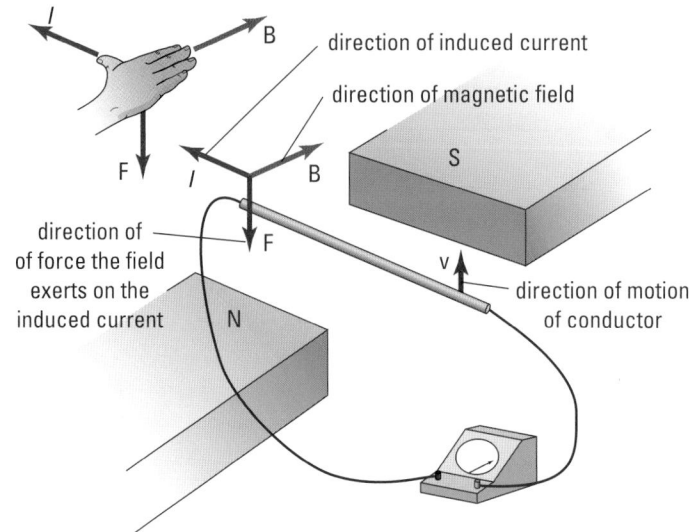

**Figure 15.11** The direction of the force exerted by the magnetic field on an induced current is opposite to the motion of the conductor through the magnetic field.

Re-examine the system in Figure 15.4 in light of Lenz's law. Firstly, recall that for the generator effect, the right-hand rule was applied in the following manner: the thumb indicated the motion of the conductor through the field; the fingers were pointed in the direction of the lines of force; and the palm pushed in the direction of the force on the positive charges in the conductor, and thus in the direction of the induced current. In Figure 15.4, this indicated that the induced current would move from right to left.

In contrast, Lenz's law states that the direction of the force the magnetic field exerts on the induced current must oppose the direction of the motion of the conductor. To apply the right-hand rule with Lenz's law, point your fingers in the direction of the lines of force and orient your palm to exert a force that opposes the motion of the conductor through the field (Figure 15.11). The thumb then must be pointing in the direction of the induced current.

In the previous case, there does not seem to be much advantage to using Lenz's law over the generator effect. But look back at how the generator effect was used in Figure 15.5 (page 724). In this case, a variation of right-hand rule #2 can be used to simplify finding the direction of the induced current in a coil or solenoid.

According to Lenz, as the bar magnet is moved upward toward the centre of the coil, the motion of the magnetic field must induce a current in the coil that interacts with the field to oppose the motion of the magnet. Since the N-pole of the magnet is approaching the coil, the induced current creates a magnetic field inside the coil with lines of force that point downward, pushing the N-pole of the magnet away from the coil.

The direction of the magnetic field for a coil is found using right-hand rule #2. Place your fingers along the edge of the coil in the direction of the current, and your thumb will point in the direction of the lines of force inside the coil. In this case, you want the magnetic field to point downward so that the N-pole of the magnet experiences a force downward. Place the fingers of your right hand along the edge of the coil so that your thumb points downward (the direction of the magnetic field through the coil). Your fingers lie along the edge of the coil in the direction of the induced current (Figure 15.12(A)).

When the N-pole of the magnet, positioned below the coil, is moved away from the coil, Lenz's law states that the magnetic field from the induced current still opposes the motion. To stop the N-pole moving away from the coil, it must try to pull the N-pole of the magnet toward the coil. The N-pole of a magnet experiences a force in the direction of the field acting on it, thus the magnetic field inside the coil would have to be directed upward. Placing your fingers along the edge of the coil, so that the thumb is pointing upward, gives the direction of the induced current in the loop (Figure 15.12(B)).

If the coil in Figure 15.12 had not been connected to the galvanometer, then the ends of the coil would have gained positive and negative charges like the anode and cathode of a battery. The end of the coil to which the current flowed would have become the anode.

## Back *emf*

When an electric motor is switched on, the magnetic field around the armature exerts a force on the current in the coils. This force causes the armature to rotate within the magnetic field. As long as a current flows through the coils, the magnetic field exerts a force on the armature. However, if forces cause accelerations, why does the armature of the motor not continue to accelerate to an increasingly faster rate of rotation? It should not be too surprising to find that the answer is found in Lenz's law.

As the armature speeds up, its coil is moving through the magnetic field that drives the motor. However, the generator effect states that the motion of the coil in the field must result in an induced *emf*. Lenz's law says that the direction of the induced *emf* (and induced current) must oppose the *emf* (and the current) supplied by the battery.

Once again, consider the simple case of a single conductor inside a magnetic field. The conductor is connected to a battery (Figure 15.13, page 734). The battery causes a current in the conductor from left to right. The right-hand rule for the motor force indicates that the direction of the force on the conductor is toward the top of the page. If this conductor is free to move, it will move in that direction.

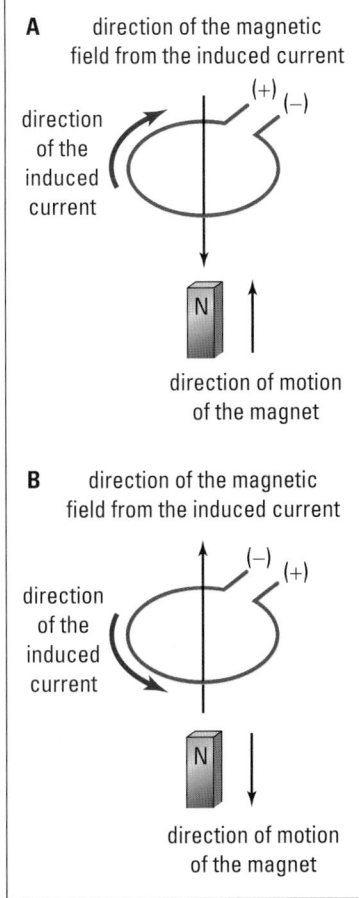

**Figure 15.12** When a conductor and a magnet move in relation to each other, by Lenz's law, the induced current creates a magnetic field that opposes the motion.

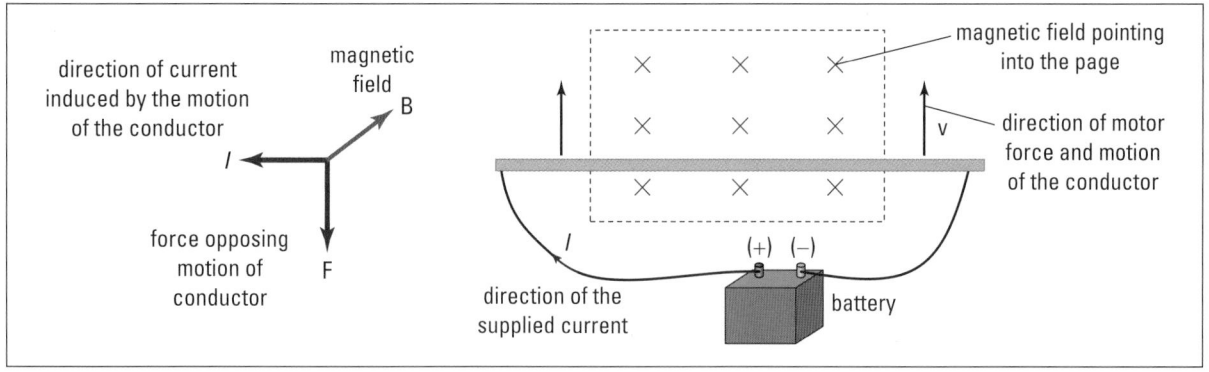

direction of current
induced by the motion
of the conductor

magnetic
field
B

I

force opposing
motion of
conductor

F

magnetic field pointing
into the page

direction of motor
force and motion
of the conductor

v

direction of the
supplied current

I

(+) (−)

battery

**Figure 15.13** As the current from the battery experiences the motor effect, the motion of the conductor in the field induces a back *emf* that opposes the supplied *emf*.

As soon as the conductor begins to move upward, the generator effect begins. By Lenz's law, the motion of the conductor, toward the top of the page through a magnetic field into the page, results in an *emf* across the conductor that pushes a current from right to left, opposing the current from the battery. This is defined as the **back *emf***. As the conductor speeds up, the back *emf* increases. In the absence of friction, the speed of the conductor would increase until the back *emf* equalled the supplied *emf* from the battery. At that point, the net *emf* (and thus the net current through the conductor) would be zero, and the conductor would be moving in equilibrium at a constant speed.

When a motor begins to run, the rate at which the armature rotates continues to increase until equilibrium is reached. Since there is always friction, the top speed of the armature is such that the back *emf* is a bit smaller than the *emf* of the battery. This leaves a net *emf* that causes just enough current through the armature so that the forward force of the motor effect equals the drag of the force of friction. The armature is now in dynamic equilibrium, and moves at a constant speed.

If the load on a motor is increased, then the rate of rotation of the armature slows down. As it slows down, the back *emf* is reduced and the net *emf*, in the forward direction, increases. The armature continues to slow down until the net current through the coils increases enough so that motor force on the resulting current is sufficient to drive the increased load.

## Eddy Currents and Magnetic Damping

**Magnetic damping** is a common application of Lenz's law. Sensitive balances, such as the equal triple-beam balances or electronic balances used in laboratories, would require considerably more time to come to rest if they were not magnetically damped (slowed down). A flat plate of diamagnetic or paramagnetic material (copper or aluminum) is attached to the arm of the scale so that it is inside a magnetic field (Figure 15.14). As the arm of the scale moves up and down, the magnetic field induces **eddy currents** (small circular currents) within the plate.

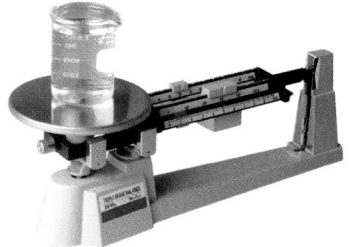

**Figure 15.14** The motion of the arm in the triple-beam balance results in eddy currents in the plate at the end of the arm. By Lenz's law, the energy to cause the eddy currents must come from the kinetic energy of the arm, damping its motion.

By Lenz's law, the direction of the eddy currents is such that the motor force on them opposes the motion of the plate, slowing down the motion of the arm. As the motion of the arm slows down, the damping effect of the eddy currents is reduced so that the final position, and thus the accuracy of the scale, is not affected.

Go back to Investigation 15-B. Can you explain your observations in terms of Lenz's law and the law of conservation of energy?

## 15.2   Section Review

1. **C** Return to the five questions at the end of section 15.1 and use Lenz's law to find the answers. Compare how you applied Lenz's law and the right-hand rule interpretation of the generator effect. For each question, compare how the right-hand rule applies using the generator effect, and how it applies using Lenz's law.

2. **K/U** A ramp is made using a U-shaped copper rod that slopes to the left. The magnetic field around the ramp points vertically downward. If a copper rod is allowed to roll down the ramp, in which direction is the induced current in the rod?

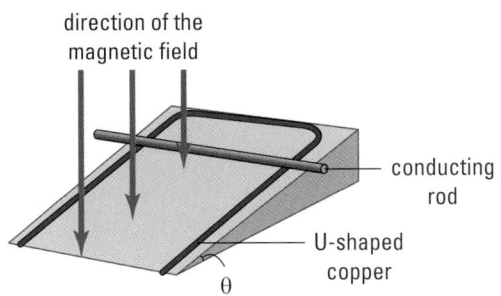

direction of the magnetic field

conducting rod

U-shaped copper

θ

3. **K/U** A coil lies in the plane of the page. When a bar magnet moves toward the coil, along the axis of the coil from behind the page, a current is induced in the coil in a counter-clockwise direction. Which pole of the magnet is approaching the coil?

4. **C** The diagram (upper right) shows a rod moving through a magnetic field pointing out of the page. Explain how to use the

right-hand rule with Lenz's law to decide which end of the rod will become positively charged.

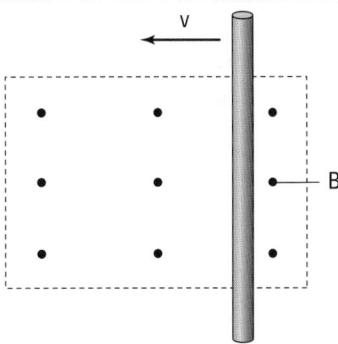

5. **C** Draw a diagram of a plate, lying in the plane of the page, and moving towards the bottom of the page, so that it will pass through a magnetic field that points directly out of the page. Apply Lenz's law to determine the direction of the eddy current that will act to damp the motion of the plate. Hint: It may help to analyze the induced currents in the plate when the eddy is just entering and leaving the field.

### UNIT PROJECT PREP

- Lenz's law predicts the effect of the back *emf* on an electric motor.
- What is the effect of the back *emf* on the motion of an electric motor?
- To what extent should you be aware of eddy currents and magnetic damping within your motor design?

## SECTION
### EXPECTATIONS

- Explain the interaction of electricity and magnetism in the operation of transformers.

- Compare direct current and alternating current in qualitative terms as they relate to the transmission of electric energy.

- Analyze the operation of industrial applications of transformers.

## KEY
### TERMS

- primary coil

- secondary coil

- step-up transformer

- step-down transformer

- transformer formula

**Figure 15.15** Electromagnetic induction in transformers delivers power efficiently and safely to your home.

A basic component of the electric power system, transformers like the one in Figure 15.15 convert power from the high voltages used in transmission to low voltages that are safe for use in your home.

## Power Transmission over Long Distances

When you do electricity experiments in the laboratory, the distances between the power supply and the load are small, so you can ignore the internal resistance of the leads used to connect the components of the circuits. However, when power companies transmit electricity over hundreds of kilometres between the generating station and the user, the internal resistance of the transmission lines becomes significant.

In all circuits, the connections between the supply and the load always have resistance, and thus always convert some of the energy of the supply to heat. In the long-distance transmissions from the generating station to your home, this becomes a major problem.

During transmission, the rate at which electrical energy is converted to heat is calculated by the equation

$$P = I^2R$$

where $I$ is the current being transmitted, and $R$ is the resistance of the transmission line. Therefore, there are only two ways to reduce this power loss. First, use conductors with the lowest possible resistance per unit of length. (For this reason, power lines are as thick as possible, and made out of materials with low resistance.) Second, reduce the amount of current. (The large distances over which transmission occurs means that the resistance of the lines is still considerable.)

If the current is reduced, why isn't the rate at which power is being transmitted also reduced? At first, it seems to be a losing battle, but there are ways to reduce the current and still maintain a high rate of power transmission. Remember that electric power can also be calculated using $P = IV$. This means that a high voltage and low current can transfer the same power as a low voltage and high current.

For example, assume that you wanted to transmit power at the rate of $1.00 \times 10^6$ W over a transmission line, with an internal resistance of only 10.0 $\Omega$. Using a voltage of $1.00 \times 10^4$ V, the power loss to heat works out to be $1.00 \times 10^5$ W. This represents a huge fraction, in fact ten percent, of the rate of transmission.

On the other hand, transmitting at $1.00 \times 10^5$ V, loss to heat would be just $1.00 \times 10^3$ W. While 1000 W may seem quite a large loss, it represents only one tenth of one percent of the transmitted power.

This argument highlights the need for a method by which voltages could be stepped up and stepped down as required. In fact, probably the strongest argument for choosing Tesla's AC electrical system over Edison's DC system, is the ease and efficiency with which transformers can change the potential difference and current of AC electricity.

Between the generator at the power station and the light bulb in your home, the potential difference of the electric supply has been stepped up (increased in voltage) and stepped down (decreased in voltage) several times. During transmission across the country, voltages in the range of 500 000 V are common, but in your home they would be lethal. The line voltage that services your home is 120 V. You now have the framework of theory to explain how transformers accomplish these voltage changes.

## Fluctuating Magnetic Fields

In the previous sections of this chapter, the relative motion of a conductor through a field was used to create the induced current. However, if you look back, the first instance of an induced current was produced without any apparent relative motion of a conductor and a field.

Faraday originally discovered the generator effect using solenoids wrapped around the edges of an iron ring. The induced current was produced only when the switch to the **primary coil** (the coil connected to the supply) was opened or closed. But Faraday was quite right when he hypothesized that the cause of the induced current was the growth and collapse of the magnetic field.

**Math Link**

In the example given, the transmission power is $1.00 \times 10^6$ W and the resistance is 10.0 $\Omega$. For a voltage of $1.00 \times 10^5$ V, calculate (a) the current in the transmission and (b) the power lost to heat, stating which formula you are using. Repeat your calculations for a voltage of $1.00 \times 10^4$ V.

**Web Link**

**www.school.mcgrawhill.ca/resources/**

Nikola Tesla and Thomas Edison were close contemporaries, perhaps the two greatest inventor-engineers of their time. They even worked together for a few years. Edison today is by far the more famous, yet Tesla's power transmission system and AC motor seem likely to last well into the new century, while Edison's best-known inventions, the incandescent light bulb and analogue recording, are steadily being replaced by newer technologies. For more information on Tesla's inventions and his roller-coaster career, go to the above address. Follow the **Science Resources** and **Physics 11** links to find out where to go next.

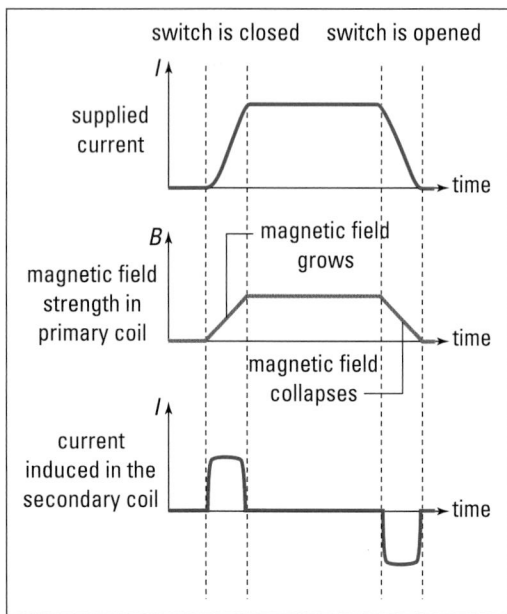

**Figure 15.16** When a current begins or stops, the increase or decrease in current causes the magnetic field to fluctuate in strength. When the magnetic field grows, it induces a current in one direction; when it collapses, it induces a current in the opposite direction.

**ELECTRONIC LEARNING PARTNER**

Review your knowledge of electricity and magnetism with a quiz available on your Electronic Learning Partner.

Faraday demonstrated that a change in the strength of the magnetic field is responsible for the induced current, as Figure 15.16 shows. Since opening and closing a switch to change the magnetic field strength is rather clumsy, a simpler way would be to change the strength of the current in the primary coil. As the current in the primary coil changes, the strength of the magnetic field in the iron ring fluctuates, and induces a current in the **secondary coil**, which is *not* directly connected to the supply. It turns out that AC current is the ideal system to do exactly that.

As an AC current alternates, it continuously increases and decreases and, in turn, causes the strength of the magnetic field produced by the primary coil to increase and decrease. The fluctuating magnetic field in the iron ring induces a current in the secondary coil that, by Lenz's law, creates a magnetic field that opposes the *changes* in the magnetic field from the primary coil.

It is important to note that the induced current in the secondary coil is not as a result of the *strength* of the magnetic field in the primary coil, but rather, as a result of the *changes* in the strength of the magnetic field. Therefore, as the magnetic field grows, it induces a current that produces a magnetic field in opposition to the growth of the magnetic field from the primary coil (Figure 15.17). Since the field is trying to increase its strength in the upward direction through the secondary coil, the magnetic field from the induced current in the secondary coil must point downward to oppose the growth.

As the current in the primary coil reduces, the magnetic field from the primary coil collapses. This collapse results in a reduction in the strength of the field in the upward direction. In essence, the direction of the field is increasing in a downward direction. The current it induces in the secondary coil produces a magnetic field that tries to sustain the strength of the weakening magnetic field from the primary coil (Figure 15.18). In order to try to stop the field from becoming weaker, the magnetic field from the induced current must also be upward.

The AC current in the primary coil, and the magnetic field generated by it, are constantly changing both strength and direction. Thus, in reaction to changes in the magnetic field, the induced current in the secondary coil is constantly changing strength and direction. Thus the secondary current is also alternating. Critically, the relative values of the voltage supplying the primary coil, and the induced voltage at the secondary coil, depend on the relative numbers of turns in their coils.

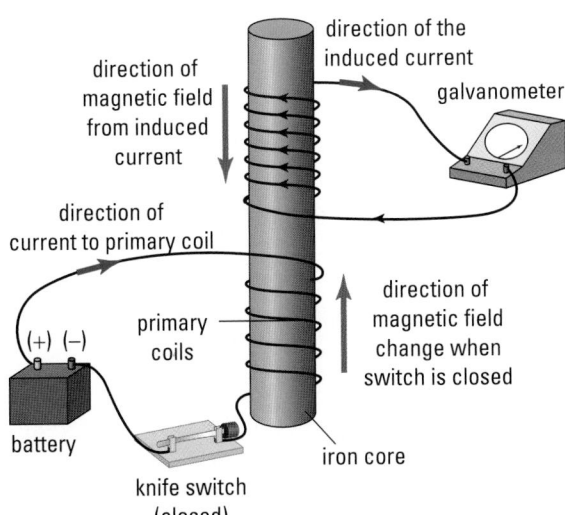

direction of the induced current

direction of magnetic field from induced current

direction of current to primary coil

galvanometer

(+) (−)

primary coils

direction of magnetic field change when switch is closed

battery

knife switch (closed)

iron core

direction of the induced current

direction of magnetic field from induced current

direction of current to primary coil

galvanometer

(+) (−)

direction of magnetic field change when switch is opened

battery

knife switch (opened)

iron core

**Figure 15.17** When the current is increasing in the primary coil, Lenz's law dictates that the direction of the magnetic field, from the current induced in the secondary coil, must oppose the growth of the field.

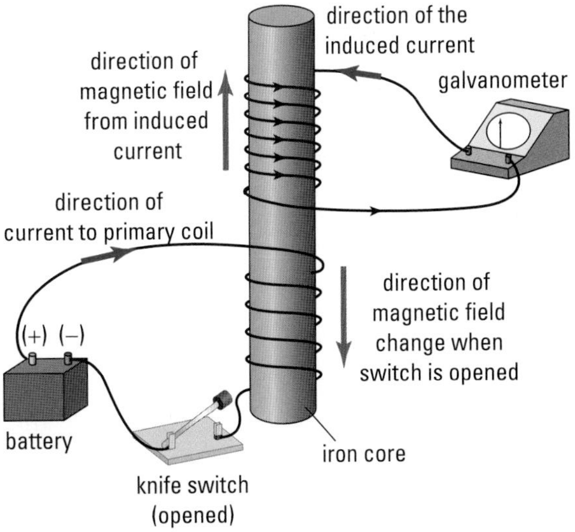

**Figure 15.18** As the current and its induced magnetic field in the primary coil diminish, the induced current in the secondary coil tries to sustain the field strength. The direction of the induced current creates a magnetic field in the same direction as the field from the primary coil.

## Step-up and Step-down Transformers

In a generator, the size of the induced *emf* depends on the strength of the magnetic field, the speed of the generator, and the length of wire in the field. In a transformer, the first two factors, field strength and rate at which the field changes, are the same for both the primary and secondary coils. However, there is no reason why the secondary coil should have the same number of turns of wire as the primary coil. If the secondary coil has ten times as many turns of wire around the core as the primary coil, then the length of wire it has in the magnetic field would be ten times as great as the primary coil.

The result is that the induced *emf* in the secondary coil would be ten times the supplied *emf* to the primary coil. This type of transformer is called a **step-up transformer** since the secondary (induced) *emf* is greater than the primary (supplied) *emf* (Figure 15.19).

A **step-down transformer** has fewer turns of wire in the secondary coil than in the primary coil. The result is that the secondary *emf* is less than the primary *emf*.

In both step-up and step-down transformers, the ratio of the secondary *emf* ($V_S$) to the primary *emf* ($V_P$) is the same as the ratio of the number of turns in the secondary coil ($N_S$) to the number of turns in the primary coil ($N_P$). Hence

$$\frac{V_S}{V_P} = \frac{N_S}{N_P}$$

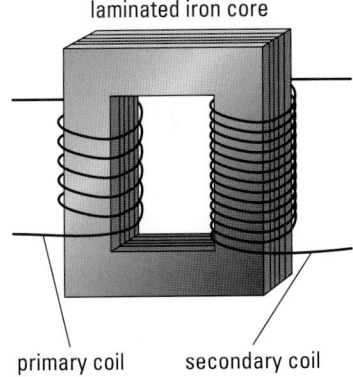

laminated iron core

primary coil          secondary coil

**Figure 15.19** A step-up transformer is created when the number of turns of wire in the secondary coil is greater than the number of turns in the primary coil. The induced voltage produced at the secondary coil is greater than the supplied voltage input at the primary coil.

At first it might seem that if voltage output from the secondary coil is greater than voltage input to the primary coil, then it should also produce a greater current. If that were true, the output energy of the transformer would be greater than the input energy. The law of conservation of energy says that is not possible. If transformers were 100% efficient, then the power out at the secondary coil ($P_S$) would exactly equal the power in at the primary coil ($P_P$). Assuming 100% efficiency,

$$P_S = P_P$$

Since $P = IV$, then $\quad I_S V_S = I_P V_P$

$$\therefore \frac{I_S}{I_P} = \frac{V_P}{V_S}$$

Note that the ratio of currents and the ratio of turns in the secondary and primary coils are both equal to the ratio of the *emf*'s. Therefore, the ratios are equal to each other. The resultant equality is known as the **transformer formula**.

---

## TRANSFORMER FORMULA

The ratio of primary and secondary voltages in a transformer is equal to the ratio of turns in the primary and secondary coils, and equal to the ratio of *secondary* and *primary* currents.

$$\frac{V_P}{V_S} = \frac{N_P}{N_S} = \frac{I_S}{I_P}$$

| Quantity | Symbol | SI unit |
|---|---|---|
| primary voltage | $V_P$ | volts (V) |
| secondary voltage | $V_S$ | volts (V) |
| primary coil turns | $N_P$ | no unit |
| secondary coil turns | $N_S$ | no unit |
| primary current | $I_S$ | amps (A) |
| secondary current | $I_P$ | amps (A) |

**Unit Analysis**

The units cancel in each ratio.

**Note:** The current ratio part of the formula only applies if the efficiency is 100%.

---

### MODEL PROBLEM

## Transforming Electricity

**The primary coils of a transformer contain 150 turns, while the secondary coil contains 1800 turns. What will be the voltage and current outputs of the transformer if the input to the primary coil is 3.60 kW of power at 250 V? Assume the efficiency is 100%.**

## Frame the Problem

- The *secondary voltage* can be found using the *primary voltage* and the *transformer formula*.

- The *secondary current* can be found in two ways. First, use the *secondary power* and the *secondary voltage* to find the secondary current. Second, use the *primary power* and the *primary voltage* to find the *primary current*. Use the *primary current* and the *transformer formula* to find the *secondary current*.

**PROBLEM TIPS**

- Make sure you have correctly identified primary coil quantities and secondary coil quantities.
- In unit analysis, remember that the numbers of turns are pure numbers with no units.
- Before you substitute numbers into it, convert the law from its given form into the form required to solve for the desired variable.
- The number of turns, *N*, is an exact number and does not affect the number of significant figures.

## Identify the Goal

Find the secondary voltage, $V_S$, and the secondary current, $I_S$.

## Variables and Constants

| Involved in the problem | | Known | Unknown |
|---|---|---|---|
| $P_P$ | $N_P$ | $P_P = 3.60$ kW | $V_S$ |
| $P_S$ | $N_S$ | $V_P = 250$ V | $P_S$ |
| $V_P$ | $I_P$ | $N_P = 150$ | $I_P$ |
| $V_S$ | $I_S$ | $N_S = 1800$ | $I_S$ |

## Strategy

Solve for the secondary voltage. State equation with voltages and numbers of turns.

Isolate the variable for secondary voltage.

Substitute values for variables and solve.

The secondary voltage is 3000 V.

Solve for the secondary current using secondary power. Since the system is 100% efficient the power input equals the power output.

Use the equation relating power, voltage and current to find the secondary current.

The secondary current is 1.20 A.

The output of the secondary coil is 3.60 kW at 3000 V and 1.20 A.

## Calculations

$$\frac{V_S}{V_P} = \frac{N_S}{N_P}$$

$$V_S = \frac{V_P N_S}{N_P}$$

$$V_S = \frac{(250 \text{ V})(1800)}{150}$$

$$V_S = 3.00 \times 10^3 \text{ V}$$

$$P_S = P_P$$
$$P_S = 3.60 \times 10^3 \text{ kW}$$

$$P_S = I_S V_S$$

$$\therefore I_S = \frac{P_S}{V_S}$$

$$I_S = \frac{3.60 \times 10^3 \text{ W}}{3.00 \times 10^3 \text{ V}}$$

$$I_S = 1.20 \text{ A}$$

*continued* ▶

*continued from previous page*

## Validate

Use the alternate method to find the secondary current. The primary current $I_P$ is given by

$$I_P = \frac{P_P}{V_P}$$

$$= \frac{3600 \text{ V}}{250 \text{ V}}$$

$$= 14.4 \text{ A}$$

The secondary current $I_S$ can be calculated directly from the primary current:

$$I_S = \frac{N_P}{N_S} I_P$$

$$= \frac{150}{1800} \times 14.4 \text{ A}$$

$$= 1.20 \text{ A}$$

as before.

### PRACTICE PROBLEMS

1. A step-up transformer increases the voltage by a factor of 25. If there are 3000 turns in the secondary coil, how many turns does the primary coil contain?

2. The transformer near your house steps the voltage down to $2.40 \times 10^2$ V. If the primary coil of the transformer has 5000 turns, and the secondary coil has 750 turns, what was the line voltage leading into the transformer?

3. The generator at a hydroelectric facility produces a voltage of $7.50 \times 10^2$ V. This is stepped up to $6.00 \times 10^5$ V for transmission to the nearest town. At the town, it is stepped down to $1.5 \times 10^4$ V at an electrical substation, and then stepped down again to $2.40 \times 10^2$ V in your neighbourhood. If the total current available from the transformer in your neighbourhood is $6.0 \times 10^3$ A, find the current that must be transmitted at each step of the journey. Assume 100% efficiency.

4. A power station has an output of $5.00 \times 10^3$ MW. This is transmitted equally over 20 lines from the generating station to the electric substation where it undergoes the first step-down in preparation for delivery into your home. The transformers at the substation have 6000 turns in their primary coils, and 400 turns in their secondary coils. Each line from the generating station supplies electricity to one transformer. The primary voltage to the transformer is $4.50 \times 10^5$ V. What is the current output of the transformer? Assume 100% efficiency.

## 15.3    Section Review

1. **G** Explain why an AC current produces a continuous current in a transformer while a DC current does not.

2. **K/U** The primary coils of two transformers, P and Q, are identical. The secondary coils have the same number of turns. However, transformer P's secondary coil uses thinner wire and hence has a greater resistance than that in Q. How will the difference in resistance affect the outputs (power, voltage, and current) of the two transformers?

3. **I** Figure 15.16 on page 738 shows the relationship between primary current, induced magnetic field, and secondary current in Faraday's iron ring experiment, which used a simple DC current and a switch. Now suppose that the primary current is AC. Sketch a similar set of graphs for this situation. Use Lenz's Law, and be careful to show where the peaks and troughs in the secondary current occur.

- The direction of the current induced by the motion of a conductor through a magnetic field can be found using the right-hand rule.

- The *emf* induced by the motion of a conductor through a magnetic field varies directly as the speed *v* of the conductor, the field strength *B*, and the length *L* of the conductor in the field.

- When a coil rotates inside a magnetic field an alternating current is induced in the coil. A slip-ring commutator takes this current off as an AC current and a split-ring commutator takes the current off as a rectified DC current.

- Lenz's law is based on the law of conservation of energy. It states that the induced current, produced when any part of an electric circuit or magnetic field around the circuit changes, must be in a direction that opposes the change.

- Lenz's law states that a motor must generate a back *emf* that opposes the applied *emf*.

- Eddy currents circulate within the body of a conducting material when it moves inside a magnetic field or if the surrounding magnetic field varies in intensity. They are often used to dampen the motion of the conductor or to dissipate energy by heating.

- Transmission of electricity over long distances uses very low currents to reduce the energy lost to heat due to the resistance of the lines.

- Transformers can be used to step up or step down the potential difference of electrical energy. By the law of conservation of energy, when a transformer increases the voltage the current must undergo a corresponding decrease.

## Knowledge and Understanding

1. Two circles, one of copper and one of glass, are placed in a fluctuating magnetic field so that the field passes through them parallel to their axes. Both glass and copper are diamagnetic materials. Compare the induced *emf* and current in each material.

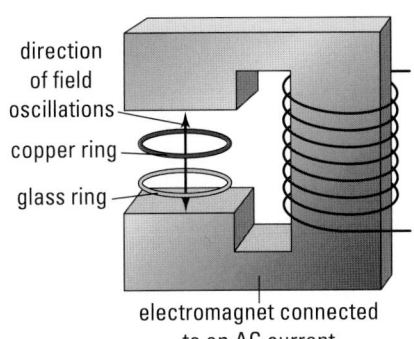

direction of field oscillations

copper ring

glass ring

electromagnet connected to an AC current

2. In a generator, what is the orientation of the coil with respect to the magnetic field when the generator output is at its peak?

3. What is the purpose of the commutator in a DC generator?

4. What is the source of energy for the alternator in your car?

5. The humidifier motor on a furnace requires a 24 V supply and a current of 3.0 A. These elements are provided by a transformer connected to a 120 V supply. Assuming 100% efficiency, calculate the current in the primary coil.

## Inquiry

6. In Investigation 15-C page 730, the swinging of the coil is damped only when a lead connects the supports of the coil. Explain why this is the case.

7. In an automobile, every time you come to a stop the brakes convert the car's kinetic energy into heat. The energy saved would be enormous if the work done to stop the motion of a car could somehow be recycled. In fact, electric trains can do just that. Investigate how the kinetic energy is transformed and recycled.

8. When a coil is connected to an AC power source, a phenomenon called self-inductance occurs. Investigate this phenomenon and how it affects the function of the coil.

## Communication

9. A bar magnet is dropped, S-pole first, into a solenoid connected to a galvanometer. Discuss the nature of the induced current in the solenoid as the magnet passes through and out the other end. As a frame of reference, use the view looking down into the solenoid.

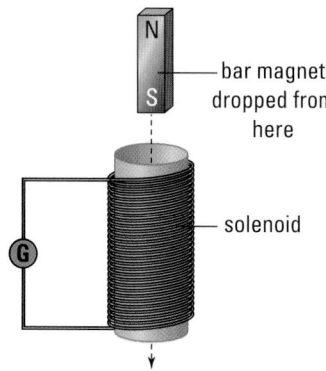

N
S
— bar magnet dropped from here

— solenoid

G

10. At the same instant as the bar magnet in Question 9 is dropped, a second bar magnet is dropped from the same height above the floor. The second magnet falls parallel to the first magnet, but does not pass through the solenoid. Explain why the second magnet would hit the floor before the first magnet.

11. Study the diagram below. The square loop of wire is moved inside the magnetic field from position P to position Q. Discuss the induced *emf* in the loop along each of it edges: **ab**, **bc**, **cd**, **da**. What is the net *emf* for the complete loop? Explain.

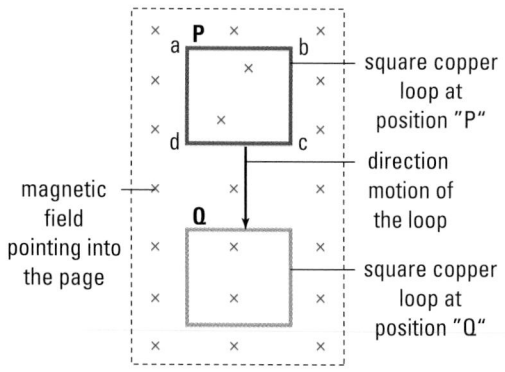

a P b
d c

square copper loop at position "P"

direction motion of the loop

magnetic field pointing into the page

Q

square copper loop at position "Q"

12. A circular copper loop lies in the plane of the page. A magnetic field, which points out of the page, moves downward in the plane of the page so that it passes over the coil from top to bottom. The loop is not inside the field at the beginning or at the end of the motion. Discuss the nature of the induced *emf* in the loop as the field moves across the coil.

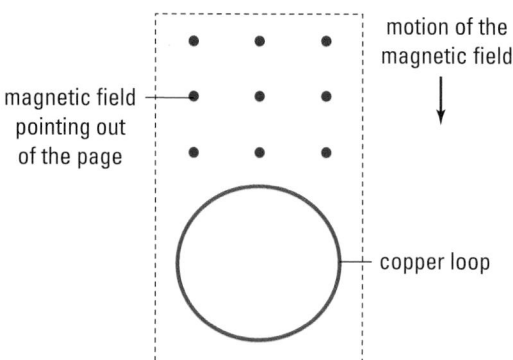

motion of the magnetic field

magnetic field pointing out of the page

copper loop

13. It has been said that "every motor is a generator". Explain why this is true. Is it also true that "every generator is a motor"? Explain.

14. A copper plate is placed between the poles of an electromagnet. When an AC current flows through the electromagnet, it creates a fluctuating magnetic field between the poles of the magnet. Discuss the effect of the changing magnetic field on the charges in the copper plate.

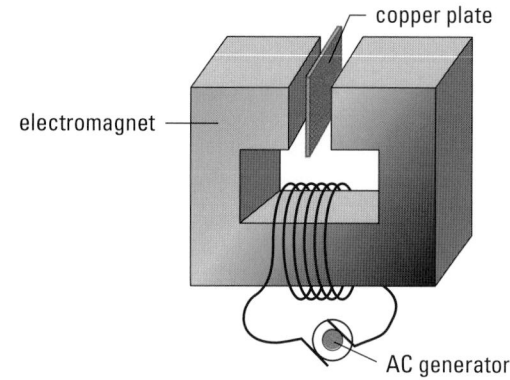

copper plate

electromagnet

AC generator

**15.** The diagram below shows two conductors, P and Q, side by side. Describe the induced current in the conductor Q, connected to the galvanometer, when the switch connecting the conductor P to the battery is closed. Does the induced current in conductor Q flow to the left or to the right? Explain how to find the direction of induced current using right-hand rule for (a) the generator effect and (b) Lenz's law.

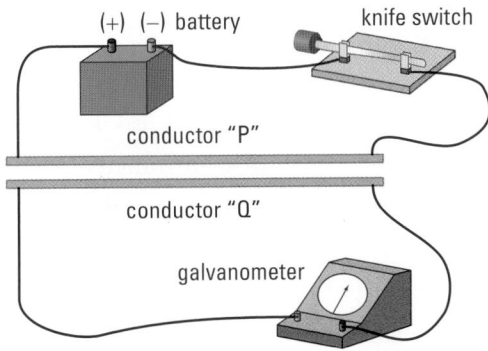

**16.** Explain the difference between the way the right-hand rule is applied for the generator effect and for Lenz's law.

## Making Connections

**17.** A 24 V DC motor runs at a speed of 3 600 rpm. At that speed, it draws 1.0 A from the battery. The resistance of the coil, measured when the motor is at rest, is found to be 6.0 Ω. Explain why the motor does not draw a current of 4.0 A when it is running. Evaluate the dangers of running at speeds much lower than its ideal operating speed.

**18.** Cable companies transmit their signals over coaxial cables. This type of conductor is very costly compared to regular conductors composed of two strands of wire running side by side, as in a household extension cord. Investigate to find out why the cable company uses the more costly coaxial cable.

## Problems for Understanding

**19.** The power rating for a transformer is $6.00 \times 10^3$ W. The current to the primary coil of the transformer is $1.20 \times 10^2$ A and the secondary voltage is $2.00 \times 10^3$ V. What is the ratio of the number of turns in the primary coil to the number of turns in the secondary coil? Assume the efficiency is 100%.

**20.** A $1.50 \times 10^3$ W step-up transformer has a 250 turns in its primary coil and 4000 turns in its secondary coil. The primary voltage is 60.0 V.
  **(a)** What is the current in the primary coil?
  **(b)** What is the voltage output at the secondary coil?
  **(c)** If the efficiency of the transformer is 100%, what is the current output at the secondary coil?
  **(d)** If the efficiency of the transformer is 97.5%, what is the power available at the secondary coil?
  **(e)** What is the current available at the secondary coil in part (d)?

**21.** A transformer runs at 95.7% efficiency. The potential difference at its primary coil is $1.60 \times 10^2$ V. The potential difference at its secondary coil is $6.40 \times 10^3$ V. The resistance of the primary coil is 0.500 Ω.
  **(a)** What is the power input at the primary coil?
  **(b)** What is the power output at the secondary coil?
  **(c)** If the secondary coil has 12 000 turns, how many turns has the primary coil?
  **(d)** At what rate is heat being produced in the transformer?
  **(e)** What is the current available at the secondary coil?

---

**Answers to Numerical Problems**

**1.** 120 turns **2.** $1.60 \times 10^3$ V **3.** $I_1 = 960$ A, $I_2 = 2.40$ A, $I_S = 1.92 \times 10^3$ A **4.** $I_S = 8.3 \times 10^3$ A

## Background

Electric energy powers most of the gadgets used by our mechanized society. An electric motor is the result of the interaction between electric current and magnetic fields. In this activity, you will design and build an electric motor. Then you will create a marketing brochure to sell your product.

## Challenge

### Part 1: A Motor That Works

In this first part, you must simply *construct a motor that works*.

### Part 2: Design a Better Motor

In the second part, you will redesign your motor. Select new materials and a design structure to construct a motor specifically designed to comply with, and compete in, one or more of the following categories.

- Fastest/slowest continually spinning motor
- Most reliable motor
- Smallest/largest functioning motor
- Motor requiring the least amount of current or voltage to operate
- Most creative overall design

## Materials

- strips of Plexiglas™ (20 cm x 3 cm)
- 1/4 inch dowelling (10 cm in length)
- 2 sewing pins
- 1.0 m thin-gauge enamelled wire (36 gauge or smaller)
- empty bathroom-tissue roll
- heat gun
- drill and 1/4 inch drill bit
- sandpaper
- 24-gauge insulated electrical wire

## Safety Precautions

- Observe caution whenever working with electric energy.
- Wear appropriate protective gloves when heating the Plexiglas™.

  **Note:** Hot Plexiglas™ looks just like cold Plexiglas™.

## Design Criteria

As a class, establish guidelines and requirements for the marketing brochure. Include

- a design blueprint of your finished product
- the category for which you are designing your motor
- a list of design improvements

Design a rubric to organize these guidelines to be used for assessment.

## Action Plan

### Part I

1. To build the base, drill a hole in each end of the Plexiglas™ strip.

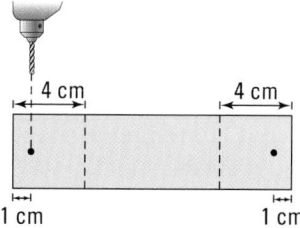

2. Carefully heat the Plexiglas™ along the dotted line drawn 4 cm from each end. Heat one end at a time, bending the Plexiglas™ to 90°.

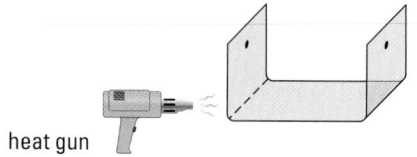

heat gun

fits the criteria for which it was designed
- **assess your technical skills during the construction phase to see if you improve**
- **use the rubric to assess your communication skills demonstrated in your brochure**

3. To build the coil, measure 1.0 m of the thin-gauge wire. Use sandpaper to sand the last 4.0 cm from both ends of the coil wire to remove the enamel.

4. Wrap the wire tightly around the tissue-paper roll. Remove the wire coil from the roll.

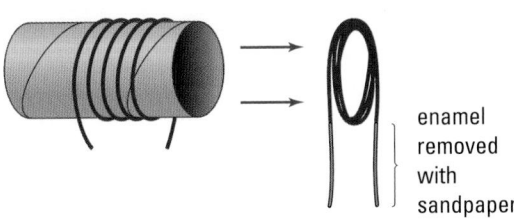

enamel removed with sandpaper

5. To build the commutators, cut two 15 cm lengths of the 24-gauge insulated wire. Carefully strip 2.0 cm of insulation off each end.

6. Individually wrap each wire strip around a pencil, leaving 4 cm unwound at each end.

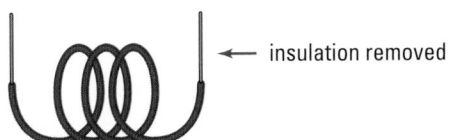

← insulation removed

7. Assembly: carefully glue the coil onto the wooden dowelling, ensuring that each sanded portion of wire runs down the length of the dowelling. Ensure that the wires are directly across from one another. Avoid getting glue on the sanded lengths of wire.

8. Carefully insert a pin into each end of the dowelling.

9. Place the pins and dowelling into the Plexiglas™ support as shown below.

10. Carefully glue the commutators into position, ensuring that they each make simultaneous contact, each with a different side of the sanded ends of the coil wire.

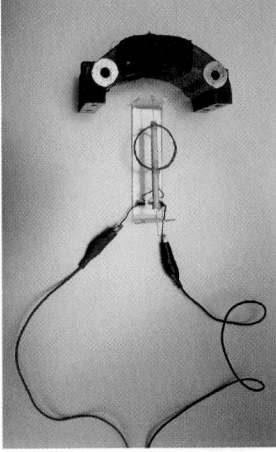

11. To make it go, place the coil in a large magnetic field.

12. Connect a direct current power supply to the free commutator ends. Turn on the power supply and watch your motor spin.

**Part 2**

Choose a category to compete in. Improve the original design, and tailor your motor to meet the specific criteria. You are encouraged to use different materials and structural layouts. Keep a log of design ideas. Include an explanation of how you solved any problems that you encountered.

**Evaluate**

1. Assess the success of your original motor. Does it operate continuously when activated?

2. Assess your technical effectiveness during the designing phase of this investigation. What skills and knowledge allowed you and your partner to accomplish the task? What aspects hindered your accomplishments?

## Knowledge and Understanding

### True/False

In your notebook, indicate if each statement is true or false. Correct each false statement.

1. All magnets have both an N-pole and an S-pole.

2. A circular coil of wire contains three turns. When a current is flowing through the coil, the adjacent turns of wire repel each other.

3. The total power output of a parallel circuit is the sum of the power outputs of the individual loads.

4. The terminal voltage of a battery is lower than its *emf*, $\mathscr{E}$, because of the internal resistance of the battery.

5. Circuit breakers and fuses protect circuits from an excessive potential difference.

6. A circuit has a single load connected to a battery. If a second load is connected in parallel to the first load, the equivalent resistance of the circuit must decrease.

7. If two equal resistors are in parallel, then the potential difference of the battery must be shared equally between them.

8. When a generator is being used to power a light bulb, increasing the power output (wattage) of the light bulb causes the generator to produce a greater *emf*, $\mathscr{E}$.

9. The magnetic North pole of the Earth is the South-seeking pole of a magnet.

### Multiple Choice

In your notebook, write the letter that gives the correct answer to each of the following questions.

10. If a 6.0 Ω and a 24 Ω load are connected in series, then the current in the 6.0 Ω load is:
    (a) four times the current in the 24 Ω load.
    (b) the same as the current in the 24 Ω load.
    (c) one quarter of the current in the 24 Ω load.
    (d) one half of the current in the 24 Ω load.

11. A circuit consists of two light bulbs connected in parallel with each other. Each branch of the circuit contains a switch connected to the bulb in that branch. If one of the switches is opened, then the other light bulb will:
    (a) glow with the same light.
    (b) glow brighter.
    (c) glow less brightly.
    (d) burn out.

12. Potential difference is a measure of:
    (a) the energy available to a current.
    (b) the energy lost when a current passes through a load.
    (c) the energy per unit of time.
    (d) the energy available to a unit of charge.

13. A copper ring lies in the plane of the page. A bar magnet is moving through the ring into the page, with its N-pole pointing away from the viewer. As the magnet moves through the ring, the current in the ring will be:
    (a) clockwise.
    (b) counter-clockwise.
    (c) first clockwise, then counter-clockwise.
    (d) first counter-clockwise, then clockwise.

14. For a transformer, the ratio of the primary voltage to the secondary voltage ($V^P/V^S$) can be increased by:
    (a) increasing the primary voltage.
    (b) increasing the number of turns in the primary coil.
    (c) decreasing the number of turns in the primary coil.
    (d) decreasing the secondary voltage.

15. A conductor carries a current eastward through a magnetic field that points directly upward. The direction of the magnetic force on the conductor is:
    (a) west            (c) north
    (b) east            (d) south

16. Which of the following statements is **not** true for the back *emf* in a motor:
    (a) The back *emf* depends on the internal resistance of the armature coil.
    (b) The back *emf* must always be less than the supplied *emf*.
    (c) The back *emf* exists whenever the motor is running.
    (d) The back *emf* increases with the speed of the motor.

**17.** Which of the following would **not** be a good material to use as the core of an electromagnet?
(a) aluminum
(b) cobalt
(c) nickel
(d) iron

**18.** Two circular coils of wire are placed on a table so that one of the coils lies on top of the other. The current in the lower coil is switched on; it is gradually increased until it is 10 A, and then the current is switched off. The upper coil is connected to a galvanometer. Which of the following statements is true for the current in the upper coil?
(a) The current in the upper coil is greatest at the instant the current in the lower coil is first turned on.
(b) The current in the upper coil continues to increase, as long as the current in the lower coil is increased.
(c) The current in the upper coil flows only when the switch to the lower coil is opened and closed.
(d) The current in the upper coil is greatest when the current in the lower coil is switched off.

**Short Answer**

**19.** Explain the similarities and differences between:
(a) current and electron flow.
(b) series and parallel circuits.
(c) Action-at-a-distance theory and Field theory.
(d) magnetic north and geographic north.
(e) Lenz's law and the generator effect.
(f) diamagnetic and paramagnetic materials.
(g) the terminal voltage and *emf* of a battery.

**20.** The diagram (top right) shows three configurations of three parallel conductors (P, Q and R) that carry currents of equal magnitude in the indicated directions. For each of the conductors, P and Q, identify the direction of the magnetic field that their current creates at the position of conductor R. Then, find the direction of the net magnetic field at the position of conductor R, and identify the direction of the force that conductor R experiences due to the net magnetic field.

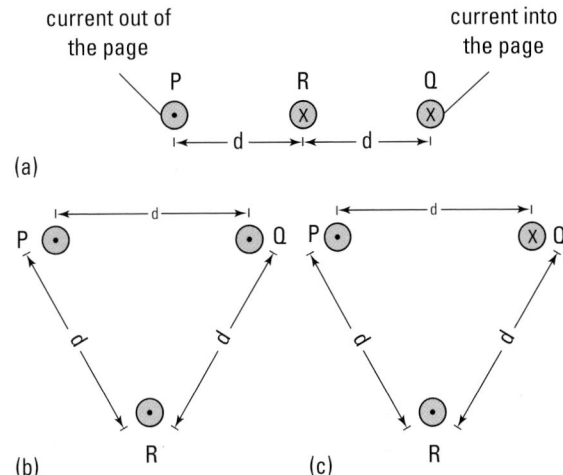

**21.** A light bulb is marked 250 W − 120 V. What is the resistance of its filament? How could you connect this type of bulb to a 240 V source so that it is not damaged by the increased potential?

**22.** An electric motor is running under a heavy load. You notice that the cover of the motor is getting quite hot. Explain why the motor runs at a relatively cool temperature at high speeds, and why, when it is trying to drive a heavy load, it seems to get hot.

**23.** When you store a horseshoe magnet, it is recommended that an iron "keeper" be placed across its poles as shown. What is the purpose of the keeper?

**Inquiry**

**24.** A voltameter and ammeter are used to measure the potential difference and current for a load. The data taken from the measurements is recorded in the table below.
(a) Make a graph of the data, so that the slope of the graph can be used to find the resistance of the load.
(b) What happens to the resistance as the current through the load rises?

| Trial | Voltage (V) | Current (A) |
|-------|-------------|-------------|
| 1 | 0 | 0 |
| 2 | 4 | 0.6 |
| 3 | 6 | 0.9 |
| 4 | 8 | 1.2 |
| 5 | 18 | 1.7 |
| 6 | 28 | 1.9 |
| 7 | 40 | 2.1 |

**25.** For a given DC motor, how does the back *emf* depend on the speed (measured in revolutions per minute/RPM) at which the motor is running? First, make a hypothesis that predicts the relationship between these two variables. Then, design an investigation to measure the back *emf* of a DC motor at various speeds. You can only use equipment that would be available in your physics laboratory. A graph of back *emf* versus motor speed should be used to test your hypothesis about the relationship between these two variables.

**26.** In the Problems for Understanding section, question 38 requests that you calculate the current in a coil from data that relates the magnetic force to the current flowing through it. Create a hypothesis that predicts a relationship between the current in a coil and the force a magnetic field exerts on the edge of the coil that lies in the field. Design an investigation that could be used to test your hypothesis. How could your experiment be used to define the strength of the magnetic field between the poles of the horseshoe magnet?

**27.** Design an investigation to show how the efficiency of a motor changes when the amount of load it carries is changed.

**28.** Design and construct a moving coil galvanometer. A moving coil galvanometer is very similar to a miniature electric motor. Investigate the structure of a galvanometer and prepare a report on the similarities and differences between motors and meters. What method is used in electric meters so that the force that the magnetic field exerts on the current in the coil, is applied perpendicularly to the edges of the coil, over the total range of the coil's motion?

## Communication

**29.** Compare the operation of a slip-ring commutator with a split-ring commutator.

**30.** Explain why the induced *emf* in a coil rotating in a magnetic field alternates in direction.

**31.** Do magnetic lines of force in a bar magnet have a beginning and an end? What are the implications of your answer with respect to the existence of magnetic monopoles?

**32.** While electric forces seem to require the existence of electric charges, magnetic forces do not require the existence of magnetic poles. Explain this apparent anomaly.

**33.** In an electric circuit, explain the difference between the "difference in potential energy" and "potential difference" across a load.

**34.** Explain why it is illogical to show lines of force in a magnetic field that cross each other?

## Making Connections

**35.** During an experiment with a bar magnet, the magnet falls on a concrete floor several times. As the experiment progresses, it is observed that the strength of the bar magnet is lower than it was at the start of the experiment. Using Domain theory, explain the probable cause of this observation.

**36.** Investigate the method used to record and read a signal from a magnetic medium, such as used in audio, video or computer disk-drive systems. Using Domain theory, explain how the information is stored. Using Lenz's law, explain how the patterns on the disk drive of a computer interact with the magnetic head of the disk drive when data is read from the disk.

**37.** Two students are challenged to design and build a transformer to operate a 6 V – 30 W light bulb at a distance of 30 m from the electrical outlet in the laboratory. They have a 30 m extension cord that is made from 18 gauge wire. First, they build a transformer to step down the voltage from the laboratory's 120 V supply. Having designed and built the transformer, they connect the bulb to the transformer in the laboratory. The bulb glows brightly, confirming that the transformer has been designed and built correctly. They then run the extension cord from the transformer to the bulb, and turn it on. The bulb does not light up. What could be the cause of this failure? Provide numerical analysis to support your argument. How could they adapt their system to make the bulb light properly at a distance of 30 m from the plug-in?

## Problems for Understanding

**38.** A coil, consisting of 600 turns of wire, has an edge that is 5.0 cm long that passes through a magnetic field. The field strength is 0.220 T east, and acts at right angles to the current in the coil. If the force on the coil was 8.25 N down, what was the magnitude and direction of the current in the coil?

**39.** An immersion heater took 18 minutes to bring one litre of water to boil, when the current through the heater was 6.0 A. How long would it take the heater to bring the same amount of water to a boil, if the current was increased to 18 A? Assume that all the heat from the coil is transmitted to the water.

**40.** The magnetic field of Earth is about $1.50 \times 10^{-4}$ T due north. A transmission line carries a current of 120 A. If the distance between two of the supporting towers is 750 m, how much force does the magnetic field from the Earth exert on the current in the line segment between the towers, when the current is due (a) east? (b) south? (**Note:** Remember that force is a vector quantity.)

**41.** Draw a circuit section that shows how to connect five resistors, each with a resistance of 2.0 Ω, such that the equivalent resistance of the section is (a) 7.0 Ω; (b) 0.40 Ω; (c) 2.5 Ω; (d) 2.4 Ω; (e) 1.0 Ω.

**42.** A 15 V battery is connected to four resistors. Two resistors, 5.00 Ω and 7.50 Ω, are connected in parallel so that they are in series with resis-

tors of 4.00 Ω and 5.00 Ω. Find the current through each resistor, and the potential drop across each resistor. What is the equivalent resistance of the circuit?

**43.** A generator has a rectangular coil, consisting of 750 turns of wire, measuring 6.0 cm by 10.0 cm. The coil is placed between the poles of a horseshoe magnet, so that the 10 cm edge is parallel to the face of the magnet. The coil is rotated so that the edges of the coil move past the face of the magnet at a speed of 8.0 m/s. The strength of the magnetic field is 0.0840 T.
  (a) What is the maximum induced *emf* in each edge of the coil as it moves past the magnet?
  (b) What is the total induced *emf* of the generator?
  (c) The coil is made of 22-gauge (measurement missing?) wire; what is its internal resistance?
  (d) If the generator is used to supply current to an external circuit with a 25.0 Ω resistance, what current will it provide?

**44.** A battery has an *emf* of 18 V. When it is connected to an external load with a resistance of $1.5 \times 10^2$ Ω, a current of 0.115 A flows.
  (a) What is the terminal voltage of the battery?
  (b) What is the internal resistance of the battery?
  (c) What will be the terminal voltage when the battery is connected to a 25 Ω load?

**COURSE CHALLENGE**

**Space-Based Power**

Consider the following as you complete the final information-gathering stage for your end-of-course project:

• Attempt to combine concepts from this unit with relevant topics from previous units.

• Verify that you have a variety of information items including conceptual organizers, useful web sites, experimental data, and unanswered questions to help you create your final assessment product.

• Scan magazines, newspapers, and the Internet for interesting information to validate previously identified content and to enhance your project.

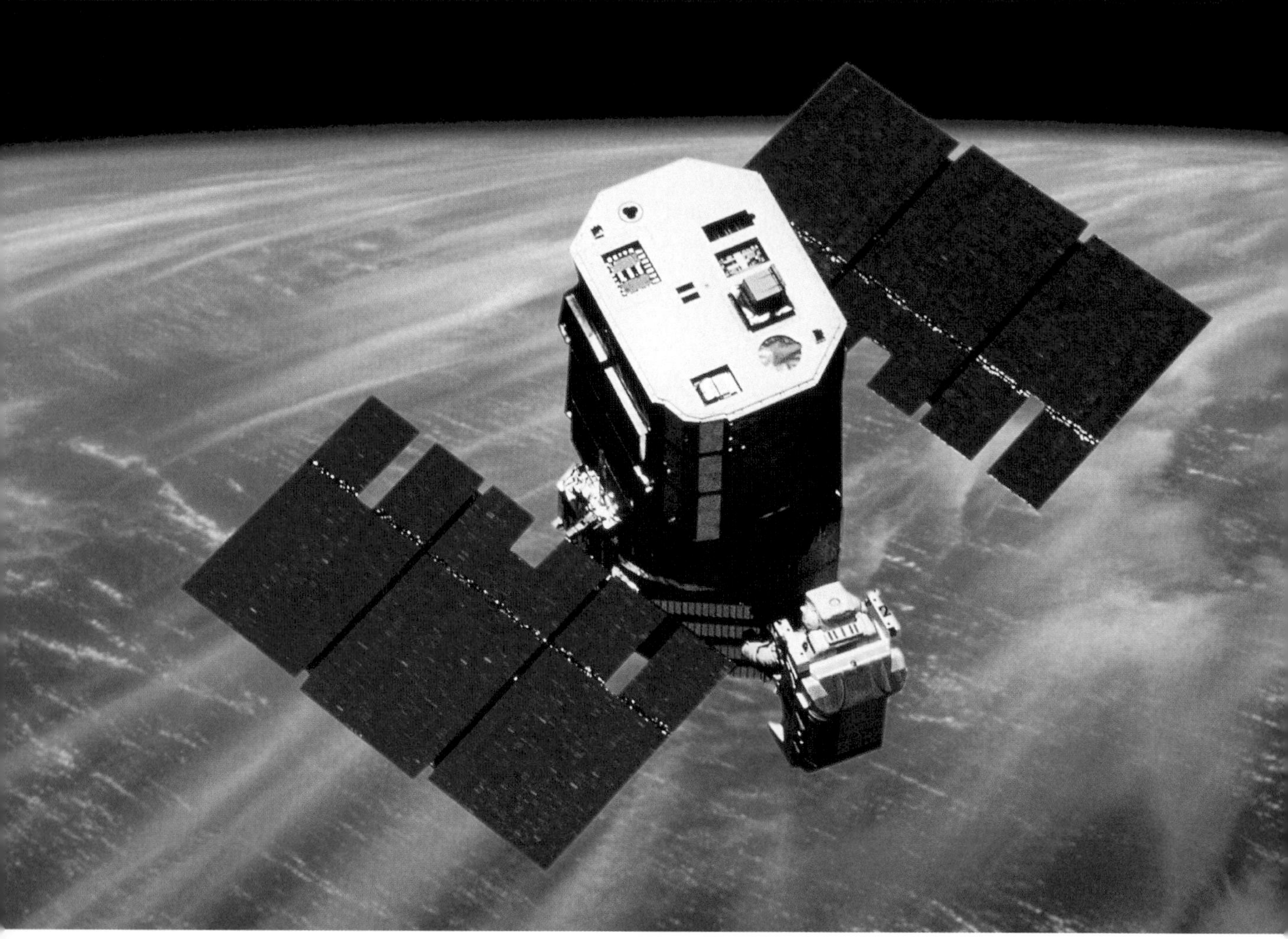

Is space-based power possible?

## Space-Based Power?

Consider the following. The world is becoming increasingly technological, and more technology means more energy use. The question about where to turn next for viable non-polluting energy sources is not a new one. It has long been well known that gases from the tailpipes of automobiles and the smokestacks of coal-fired generating stations produce noxious gases, creating smog such as that shown on the following page. Oil and natural gas reserves will not last forever, and they are not a clean source of energy. The arguments for and against use of nuclear energy have been with us for at least 30 years. What are our alternatives? Windmills? In a few places, yes, but certainly not in most. Is the use of fuel cells a partial answer? How far reaching will be the harnessing of ocean wave power to produce energy? Will fusion power be a viable alternative?

A thick brown haze of smog engulfs a city, the result of incomplete combustion from hydrocarbon fuels.

A radical new idea has been proposed to help meet the ever-increasing need for energy — one that will not negatively affect Earth's environments. The proposal is for the use of space-based power. Could an orbiting satellite capture solar energy before it is diffused by the atmosphere, convert it into microwaves to be beamed to a receiving station on Earth, and from there be transformed into useable electric power? The idea sounds as if it comes from a science-fiction novel. Yet, the concept for such a system is currently under active investigation around the world.

A space-based power delivery system requires the integration of a number of different emerging technologies. Look at just a few of the needs that must be met:

### Math Link

Only about $5 \times 10^{-8}$ % of the total energy emitted by the Sun reaches Earth's surface. Try to calculate this value by making appropriate assumptions. The necessary data and equations are provided on the inside front and back covers.

### In Space

- Satellites in geosynchronous orbit with photovoltaic panels
- Power-conversion components to convert the electricity from the panels to microwave radiation
- Transmitting antennas to direct microwave beams to Earth

### On Earth

- Receiving antennas (called rectennas) to capture the incoming microwaves and convert them into electricity that can be used
- Power-conversion components to convert the direct current (DC) output from the rectennas to alternating current (AC) which is compatible with local electrical grids

### Power Reliability

- Power-conversion efficiencies for all weather conditions

Think about these needs as you plan for the Challenge.

# Challenge

Develop and present a case either for or against the use of **space-based power** as a major source of energy to power Earth's technologies in the future. You will use knowledge of concepts investigated throughout this course, and additional research outside this resource, to develop your presentation about the feasibility of space-based power for widespread future use. Your class will together decide whether the presentations will be made:

- through a formal debate
- through role-play
- through research report presentations (either as a written report, an audiovisual presentation, or a poster presentation to an international commission (or to another group you decide upon)
- through another format of your choice

## Materials

All presentations must be supported by your portfolio of research findings, the results of supporting experiments conducted, and a complete bibliography of references used. Other materials will be considered under the heading Design Criteria.

## Design Criteria

**A.** You need to develop a system to collect and organize information that will include data, useful formulas, and even questions that you use to formulate your final recommendation near the end of the course. You can collect your own rough notes in your Physics Research Portfolio (print or electronic).

**B. Building a Physics Research Portfolio**
Your individual creativity will shape the amount, the type, and the organization of the material that will eventually fill your portfolio. Do not limit yourself to the cues scattered throughout the text; if something seems to fit, include it.

Suggested items for your Physics Research Portfolio:
- experiments you have designed yourself, and their results
- useful formulas
- specific facts
- interesting facts
- disputed facts
- conceptual explanations
- diagrams
- graphical organizers
- useful page numbers
- useful web sites
- experimental data
- unanswered questions

**C.** As a class, decide on the type(s) of assessment you will use for your portfolio and for its presentation. Working with your teacher and classmates, select which type of presentations you will use (debate, role-play, or research presentation, as outlined previously) to present your space-based power findings.

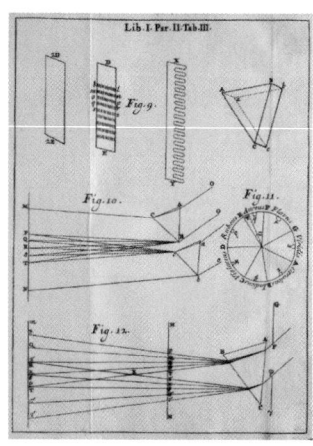

## Action Plan

1. As a class, have a brainstorming session to establish what you already know about current and possible future energy sources. For example, what are the different energy sources used around the world today, and what are the pros and cons of each of them? Why is it important to look for alternatives?

2. As a class, design a rubric or rubrics for assessing the task. (You may decide to assess facets of the challenge leading up to the presentation, as well as the presentation itself.)

3. Decide on the groupings, or assessment categories, for this task.

4. Familiarize yourself with what you need to know about the task that you chose. For example, if it's a debate, it is important to research the proper rules for debating in order to carry out the debate effectively.

5. Develop a plan to find, collect, and organize the information that is critical to your presentation and the information you record in your Physics Research Portfolio.

6. Carry out the Course Challenge recommendations that are interspersed throughout the text wherever the Course Challenge logo and heading appear, and keep an accurate record of these in your portfolio.

7. When researching concepts, designing experiments, or following a cued suggestion in the text, the McGraw-Hill Ryerson web site is a good place to begin: *www.school.mcgrawhill.ca/resources/*

8. Carry out your plan and make modifications throughout the course as necessary.

9. Present your case to your class. Review each presentation and identify Space-Based Power Supporters and their reasons, as well as Space-Based Power Opposers and their reasons.

## Evaluate Your Challenge

1. Using the rubric or rubrics you have prepared, evaluate your work and presentation. How effective were they? Were others able to follow the evidence, results, and conclusions you presented? If not, how would you revise your presentation.

2. Evaluate Course Challenges presented by your classmates.

3. After analyzing the presentations of your classmates, what changes would you make to your own project if you were to have the opportunity to do it again? Provide reasons for your proposed changes.

4. How did the organization of information for this challenge help you to think about what you have learned in this course?

**Space Link**

The Moon offers a wealth of amorphous silicon. Some supporters of space-based power foresee constructing a photovoltaic manufacturing plant on the moon. Mining the resources of the Moon is an issue of feasibility, but it is also an issue of politics and ownership. In your opinion, what is the likelihood of a Moon-mining plan proceeding if it is contrary to global political will?

## Background Information

The following information will give you some idea of the questions you should consider as you try to decide if this proposal is one that you would support or not support in your presentation. A great deal of evidence can come from your studies and investigations throughout this course. Some of the major issues are:

- How will the satellite get into orbit?
- How high should the satellite be?
- How feasible is wireless energy transmission?
- How safely can the beam be focussed?
- How much will it cost?

The following sections give you "fuel for thought." They are tied closely to Course Challenge cues in your text so that you may start to develop your plans for this culminating Course Challenge at an early stage in your course.

### Getting into Orbit

How fast does an object need to travel to get into orbit around our Earth? Does the amount of speed that a rocket attains determine both the height and shape of the orbit it will trace out? Space-based power schemes need to be able to capture sunlight and then beam the energy as microwaves to a specific spot on Earth. (See Chapter 2, page 30.)

### Staying in Orbit

A satellite, beaming energy in the form of microwaves to Earth, must have an unobstructed "line of sight" to both the sun and the receiving station on Earth. (See Chapter 3, page 104.)

### What Does "Weightless" Mean?

Weight refers to the force of gravity acting on an object near the surface of a planet or star. If astronauts were not under the influence of Earth's gravity, they would fly off in a straight line into space. Therefore, astronauts are not weightless; but their weight (the force of gravity pulling them toward Earth) is what keeps them in orbit. Is it possible for astronauts to use a spring scale to weigh themselves as they move around in Earth's orbit? (See Chapter 4, page 140.)

### Staying in Orbit

"A satellite in motion will stay in motion unless acted on by an external force." Or will it? In Chapter 2, you applied your understanding of motion to an information-gathering investigation of satellites and their orbits. A satellite has mass, perhaps a great deal of mass, and therefore has inertia. Once in orbit, what forces will act on the satellite? What will be required to keep it in a geosynchronous orbit? (See Chapter 4, page 159.)

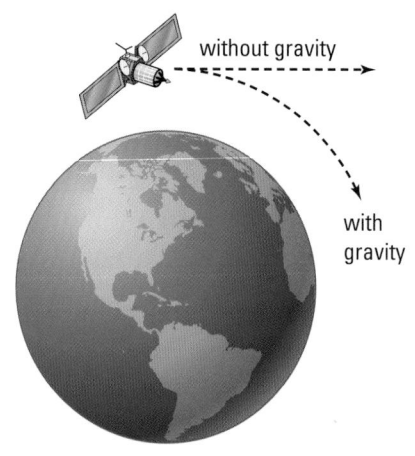

Astronauts and satellites are not "weightless" when in orbit. Their "weight" continually causes them to fall toward Earth.

## The Cost of Altitude

Space Shuttle launches are regular occurrences, as the shuttle is the main supply vehicle for the International Space Station. Newton's $2^{nd}$ law accurately predicts that the more mass that is to be sent into orbit, the greater the force that will be required. Generating this force requires rocket engines and plenty of fuel. A major drawback of any concept that involves putting objects or people into space is the cost required to generate enough acceleration to put the mass into orbit. (See Chapter 4, page 160.)

## Does It Really Work?

Wireless transmission of power has been experimentally verified. However, several challenges still exist. The first law of thermodynamics predicts that any energy transformation system developed will not be 100% efficient. Also, energy is required to fabricate the materials and equipment used to build the satellite and earthbound receiving station. (See Chapter 6, page 313.)

## Wave Frequency, Energy and Safety

How much energy can microwaves carry? Can it be dangerous to human, animal or plant life? Microwaves, part of the electromagnetic spectrum, carry energy from space to Earth. Max Plank (1858–1947) suggested that the energy contained within electromagnetic radiation was related to the radiation frequency multiplied by a constant (now called Plank's constant $h = 6.63 \times 10^{-34}$ J · s). Use Plank's energy relationship, $E = hf$, to investigate the energy differences between green light ($f = 5 \times 10^{-14}$ Hz), heat or infrared radiation ($f = 4 \times 10^{12}$ Hz), and microwaves ($f = 6 \times 10^9$). (See Chapter 7, page 337.)

## Interference: Communication Versus Energy Transmission

Microwaves make up a large part of the electromagnetic spectrum with frequencies ranging from about $10^{11}$ to $10^9$ Hz. The microwave part of the electromagnetic spectrum has been divided into small "bandwidths" for communication applications — from everyday cell phones and TV to military and space purposes. The sheer number of required bandwidth requirements has caused very tight spacing of some microwave frequencies. Signals received by an antenna from more than one source may cause effects similar to the interference patterns of superposition you saw during your study of springs. (See Chapter 7, page 335.)

## Intensity and Power Density

Sound intensity we can hear, but what about microwave intensity? Sound intensity is sometimes given in picowatts per square metre. Health Canada standards state that exposure to 5.85 GHz microwaves is safe if the power density is below 10 W/m$^2$. How

Safety concerns surrounding life near a receiving antenna must be investigated.

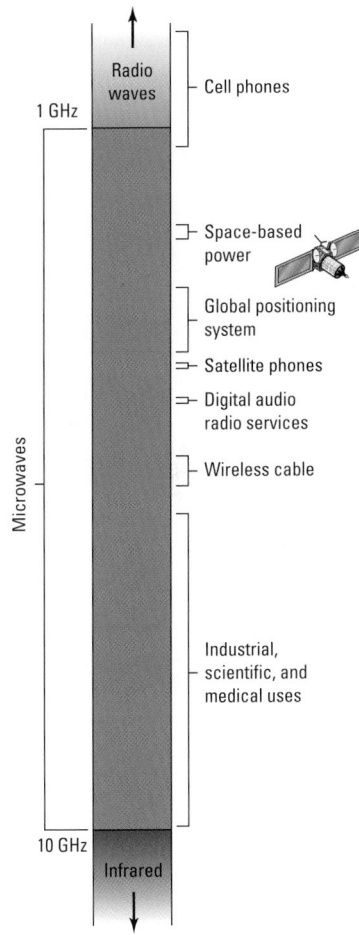

Microwaves, the portion of the electromagnetic spectrum between Radio waves and Infrared, are used in a variety of applications.

PHYSICS FILE

Astronomers at NASA once asked physicist Freeman Dyson to describe what an advanced civilization might look like. The physicist's answer points to the power of the Sun. He said:

*"I cannot tell you what an advanced civilization would look like, because they may choose to live in unpretentious circumstances. I can tell you what an advanced technology would look like. They would almost certainly have enclosed their Sun to harvest most of its energy. All that would be left for distant viewers to see would be an infrared emitter with little visible light."*

much land or water area will a receiving station require to be able to provide power to a community without endangering its nearby inhabitants? (See Chapter 8, page 377.)

## We're Losing Power

Actually, it is not power but energy that is lost. Beaming microwaves through the atmosphere is not 100% efficient. Space-based power requires an understanding of both the wave theory and the particle theory of light. The energy loss will cause the atmosphere and surrounding area to heat up. (This effect could be used to assist in aquaculture (fish farming) if a receiving station were placed on water.) (See Chapter 10, page 466.)

## Keeping the Mirror Clean

The land-based systems that focus light for solar thermal energy facilities on Earth are restricted by day-night cycles and bad weather. Some space-based power proposals suggest using parabolic mirrors to concentrate the solar radiation onto very specialized and efficient photovoltaic cells. Other systems would use larger panels of cheaper photovoltaic cells, removing the need for a parabolic reflector. The cheaper cells could be made from raw materials found on the moon! (See Chapter 10, page 488.)

## From the Equator to the Poles

The ray model of light provides a very accurate method of predicting the location of images. The same process can be used to predict how the microwave beam from the satellite will strike Earth's surface at various locations. Vacationing somewhere near the equator demands much more attention to your sun block and sunlight exposure time than a destination closer to the poles. The reason lies with Earth's shape and the linear propagation of light. (See Chapter 10, page 478.)

Photovoltaic Alternatives

| Type | Efficiency | Advantage | Application |
| --- | --- | --- | --- |
| specialized cells made from various materials | 30% | can receive concentrated light focussed by several reflectors | selected locations experiencing a great deal of direct sunlight |
| single crystal silicon | 25% | durable and long lived | satellites and industry |
| ploy-crystalline silicon | 18% | easily deposited on glass or metal surfaces; flexible cells | small scale power generation, e.g., cottages, roadway signs |
| amorphous silicon | 10% | easily manufactured inexpensive | calculators |

### Twinkling Stars Point to a Problem

Your eyes bend light, focussing an image onto your retina. If the image is blurred, corrective lenses can help. The atmosphere's continually changing density also refracts light. Stars twinkle and hot roadways shimmer due to atmospheric refraction. Unlike eyeglass lenses, the atmosphere is always changing — air temperature, pressure, and moisture content changes cause dramatic shifts in the atmospheric index of refraction. How reliably will engineers be able to focus the energy beam onto the receiving antenna? (See Chapter 11, page 517.)

### Free Energy?

Solar-powered calculators, street signs, and office buildings take advantage of the unique characteristic of semiconductor materials, such as silicon. Shining sunlight on semiconductors can generate an electric current. A space-based power system captures and converts sunlight into electric energy through the use of semiconducting photovoltaic cells. The electric energy is transformed into microwaves that are beamed to Earth where they are transformed back into useable electric energy. (See Chapter 13, page 615.)

### How Much Electric Power?

How much electric energy does a small town require in a day? You have investigated various aspects of a space-based power system. Now, it is time to investigate the feasibility of such a system in terms of power. (See Chapter 13, page 653.)

## Wrap Up

These ideas are provided to help you develop your case either for or against space-based power. You will doubtless come up with many other ideas of your own for consideration, giving your presentation its own unique characteristics.

This brief tour has included some of the factors to consider about space-based power as a possible future energy source, as well as ideas about your own presentation of a case for or against it.

We are living during a time when scientific knowledge and technology may together provide important solutions to global energy needs. The potential for sustainable power production by using a variety of energy sources has only recently become possible. Today, new choices can be considered because of advancement in technology and deeper scientific understanding of our world, from the nature of ecosystems to the operation of rockets. Your own thorough investigation of this one possible energy source — solar-based power — will give you a better understanding of how it may or may not be a viable alternative to energy sources currently being used.

**PHYSICS FILE**

Alternative energy sources, such as wind-based power, are often criticized for occupying large areas of land. Space-based power greatly reduces this problem by placing photovoltaic panels in orbit. The receiving antennas, however, still require sizable portions of land. There are cost and logistical problems associated with constructing a receiving antenna in a remote region, such as the desert shown below. However, unlike hydro-electric and wind-power installations, the location of the antenna is not dependent on geographic features. Rather, it can be placed in a location that takes into consideration environmental and other factors.

# Precision, Error, and Accuracy

A major component of the scientific inquiry process is to compare experimental results with predicted or accepted theoretical values. In conducting experiments, you must realize that all measurements have a maximum degree of certainty, beyond which there is uncertainty. The uncertainty, often referred to as error, is not a result of a mistake but rather, it is caused by the limitations of the equipment or the experimenter. The best scientist, using all possible care, could not measure the height of a doorway to a fraction of a millimetre accuracy using a metre stick. The uncertainty introduced through measurement must be communicated using specific vocabulary.

 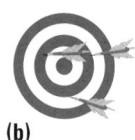 

(a)          (b)          (c)

Differentiating between accuracy and precision

Experimental results can be characterized by both their accuracy and their precision.

**Precision** describes the exactness and repeatability of a value or set of values. A set of data could be grouped very tightly, demonstrating good precision, but not necessarily be accurate. The darts in Figure **A**, above, miss the bull's-eye and yet are tightly grouped, therefore demonstrating precision without accuracy.

**Accuracy** describes the degree to which the result of an experiment or calculation approximates the true value. The darts in Figure **B**, above, miss the bull's-eye in different directions, but are all relatively the same distance away from the centre. The darts demonstrate three throws that share approximately the same accuracy, with limited precision.

The darts in Figure **C**, above, demonstrate accuracy and precision.

## Random Error

- Random error results from small variations in measurements due to randomly changing conditions (weather, humidity, equipment, level of care, etc.).
- Repeating trials will reduce but never eliminate random error.
- Random error is unbiased.

- Random error affects precision.

## Systematic Error

- Systematic error results from consistent bias in observation.
- Repeating trials will not reduce systematic error.
- Three types of systematic error are: natural errors, instrument-calibration error, and personal error.
- Systematic error affects accuracy.

## Error Analysis

Error exists in every measured or experimentally obtained value. The error could deal with extremely tiny values such as wavelengths of light or with large values such as the distances between stars. A practical way to illustrate the error is to compare it to the specific data as a percentage.

## Relative Uncertainty

Relative uncertainty calculations are used to determine the error introduced by the natural limitations of the equipment used to collect the data. For instance, measuring the width of your textbook will have a certain degree of error due to the quality of the equipment used. This error, termed **estimated uncertainty**, has been deemed by the scientific community to be half of the smallest division of the measuring device. A metre stick with only centimetres marked would have an error of $\pm 0.5$ cm. A ruler that includes millimetre divisions would have a smaller error of $\pm 0.5$ mm. The measure should be recorded showing the estimated uncertainty, such as $21.00 \pm 0.05$ cm. Use the relative uncertainty equation to convert the estimated uncertainty into a percentage of the actual measured value.

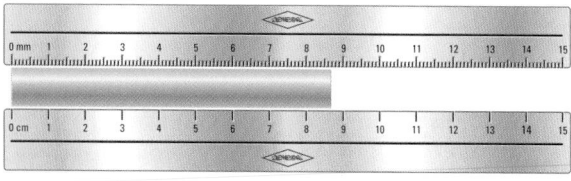

Estimated uncertainty is accepted to be half of the smallest visible division. In this case, the estimated uncertainty is $\pm 0.5$ mm for the top ruler and $\pm 0.5$ cm for the bottom ruler.

relative uncertainty $= \dfrac{\text{estimated uncertainty}}{\text{actual measurement}} \times 100\%$

**Example:**

Converting the error represented by
21.00 ±0.05 cm to a percentage.

$$\text{relative uncertainty} = \dfrac{0.05 \ \cancel{\text{cm}}}{21.00 \ \cancel{\text{cm}}} \times 100\%$$
$$= 0.2\%$$

## Percent Deviation

Frequently, in conducting experiments, it is unreasonable to expect that accepted theoretical values can be verified because of the limitations of available equipment. In such cases, percent deviation calculations are made. For instance, the standard value for acceleration due to gravity on Earth is 9.8 m/s$^2$ towards the centre of Earth in a vacuum. Conducting a crude experiment to verify this value may yield a value of 9.6 m/s$^2$. This result deviates from the accepted standard value. It is not necessarily due to error. The deviation, as with most high-school experiments, may be due to physical differences in the actual lab (for example, the experiment may not have been conducted in a vacuum). Therefore, deviation is not necessarily due to error; it may be the result of experimental conditions that should be explained as part of the error analysis. Use the percent deviation equation to determine how close the experimental results are to the accepted or theoretical value.

percent deviation =
$$\left| \dfrac{\text{experimental value} - \text{theoretical value}}{\text{theoretical value}} \right| \times 100\%$$

**Example:**

percent deviation

$$\dfrac{|9.6 \ \cancel{\text{m/s}^2} - 9.8 \ \cancel{\text{m/s}^2}|}{9.8 \ \cancel{\text{m/s}^2}} \times 100\%$$

$$\text{percent deviation} = 2\%$$

## Percent Difference

Experimental inquiry does not always involve an attempt at verifying a theoretical value. For instance, measurements made in determining the width of your textbook do not have a theoretical value based on a scientific theory. There still may exist, however, a desire to determine how precise your measurements were. Suppose you measured the width 100 times and found that the smallest width measurement was 20.6 cm, the largest was 21.4 cm, and the average measurement of all 100 trials was 21.0 cm. The error contained within your ability to measure the width of the text can be estimated using the percentage difference equation.

percentage difference =
$$\dfrac{\text{maximum difference in measurements}}{\text{average measurement}} \times 100\%$$

**Example:**

percentage difference
$$= \dfrac{(21.4 \ \cancel{\text{cm}} - 20.6 \ \cancel{\text{cm}})}{21.0 \ \cancel{\text{cm}}} \times 100\%$$
$$= 4\%$$

---

1. In Sèvres France, a piece of platinum is kept in a vacuum under lock and key. It is the standard kilogram with mass 1.0000 kg. Imagine you were granted the opportunity to experiment with this special mass, and obtained the following data: 1.32 kg, 1.33 kg, and 1.31 kg. Describe your results in terms of precision and accuracy.

2. You found that an improperly zeroed triple-beam balance affected the results obtained in question 1. If you used this balance for each measure, what type of error did it introduce?

3. Describe a fictitious experiment with obvious random error.

4. Describe a fictitious experiment with obvious systematic error.

5. (a) Using common scientific practice, find the estimated uncertainty of a stopwatch that displays up to a hundredth of a second.

   (b) If you were to use the stopwatch in part (a) to time repeated events that lasted less than 2.0 s, could you argue that the estimated uncertainty from part (a) is not sufficient? Explain.

# Rounding, Scientific Notation, and Significant Digits

When working with experimental data, follow basic rules to ensure that accuracy and precision are not either overstated or compromised. Consider the 100 m sprint race. Several people using different equipment could have timed the winner of the race. The times may not agree, but may all be accurate within the capability of the equipment used.

**Sprinter's Time with Different Devices**

| Time (s) | Estimated error of device (s) | Device |
|----------|-------------------------------|--------|
| 11.356 | ± 0.0005 | photogate timer |
| 11.36 | ± 0.005 | digital stopwatch |
| 11.4 | ± 0.05 | digital stopwatch |
| 11 | ±0.5 | second hand of a dial watch |

Using the example of the 100 m race, you will solidify ideas you need to know about exact numbers, number precision, number accuracy, and significant digits.

**Exact Numbers** If there were eight competitors in the race, then the number 8 is considered to be an exact number. Any time objects are counted, number accuracy and significant digits are not involved.

**Number Precision** If our race winner wants a very precise value of her time, she would want to see the photogate result. The electronic equipment is able to provide a time value accurate to 1/1000th of a second. The time recorded using the second hand on a dial watch is not able to provide nearly as precise a value.

**Number Accuracy and Significant Digits** The race winner goes home to share the good news. She decides to share the fastest time with her family. What timing method does she share? She would share the 11 s time recorded using the second hand of a dial watch. All of the other methods provide data that has her taking a longer time to cross the finish line. Is the 11 s value accurate?

The 11 s value is accurate to within ±0.5 s, following common scientific practice of estimating error. The 11.356 s time is accurate to within 0.0005 s. The photogate time is simply more precise. It would be inaccurate to write the photogate time as 11.35600 s. In that case, you would be adding precision that goes beyond the ability of the equipment used to collect the data, as the photogate method can measure time only to the thousandths of a second. Scientists have devised a system to help ensure that number accuracy and number precision are maintained. It is a system of significant digits. *Significant digits* require that the precision of a value does not exceed either (a) the precision of the equipment used to obtain it or (b) the least precise number used in a calculation to determine the value. The table on the left provides the number of significant digits for each measurement of the sprinter's times.

There are strict rules used to determine the number of significant digits in a given value.

### When Digits Are Significant ✔

1. All non-zero digits are significant. (159 – three significant digits)

2. Any zeros between two non-zero digits are significant. (109 – three significant digits)

3. Any zeros to the right of *both* the decimal point and a non-zero digit are significant. (1.900 – four significant digits)

4. All digits (zero or non-zero) used in scientific notation are significant.

### When Digits Are Not Significant ✘

1. Any zeros to the right of the decimal point but preceding a non-zero digit are not significant; they are placeholders. For example: 0.00019 kg = 0.19 g (two significant digits)

2. Ambiguous case: Any zeros to the right of a non-zero digit are not significant; they are placeholders. (2500 – two significant digits) If the zeros are intended to be significant, then scientific notation must be used, for example: $2.5 \times 10^3$ (two significant digits) and $2.500 \times 10^3$ (four significant digits).

**Calculations and Accuracy** As a general rule, accuracy is maintained though mathematical calculations by ensuring that the final answer has the same number of significant digits as the least precise number used during the calculations.

**Example:**

Find the product of these lengths:

12.5 m     16 m     15.88 m

Product = 12.5 m × 16 m × 15.88 m

Product = 3176 m$^3$

Considering each data point, notice that 16 has only two significant digits, therefore the answer must be shown with only two significant digits.

Total length = $3.2 \times 10^3$ m (two significant digits)

**Rounding to Maintain Accuracy** It would seem that rounding numbers would introduce error, but in fact, proper rounding is required to help maintain accuracy. This point can be illustrated by multiplying two values with differing numbers of significant digits. As you know, the right-most digit in any data point contains some uncertainty. It follows that any calculations using these uncertain digits will yield uncertain results.

---

Multiply **32** and **13.55**. The last digit, being the most uncertain, is highlighted.

13.55

× **32**

**2710** Each digit in this line is obtained using an uncertain digit.

**4065** In this line only the 5 is obtained using uncertain digits.

**433.60**

The product **433.60** should be rounded so that the last digit shown is the only one with uncertainty.

Therefore: **4.3** × 10$^2$

---

Notice that this value contains two significant digits, which follows the general rule.

Showing results of calculations with every digit obtained actually introduces inaccuracy. The number would be represented as having significantly more precision than it really has. It is necessary to round numbers to the appropriate number of significant digits.

**Rounding Rules** When extra significant digits exist in a result, rounding is required to maintain accuracy. Rounding is not simply removing the extra digits. There are three distinct rounding rules.

1. **Rounding Down**
   When the digits dropped are less than 5, 50, 500, etc., the remaining digit is left unchanged.

**Example:**

4.123 becomes

4.12       rounding based on the "3"

4.1        rounding based on the "23"

2. **Rounding Up**
   When the digits dropped are greater than 5, 50, 500, etc., the remaining digit is increased or rounded up.

**Example:**

4.756 becomes

4.76       rounding based on the "6"

4.8        rounding based on the "56"

8.649 becomes

8.65       rounding based on the "9"

8.6        rounding based on the "49"

3. **Rounding with 5, 50, 500, etc.**
   When the digits dropped are exactly equal to 5, 50, 500, etc., the remaining digit is rounded to the *closest even number*.

**Example:**

4.850 becomes

4.8        rounding based on "50"

4.750 becomes

4.8        rounding based on "50"

Always carry extra digits throughout a calculation, rounding only the final answer.

**Scientific Notation** Numbers in science are sometimes very large or very small. For example the distance from Earth to the Sun is approximated as 150 000 000 000 m and the wavelength of red light is 0.000 000 65 m. Scientific notation allows a more efficient method of writing these types of numbers.

- Scientific notation requires that a single digit between 1 and 9 be followed by the decimal and all remaining significant digits.
- The number of places the decimal must move determines the exponent.
- Numbers greater than 1 require a positive exponent.
- Numbers less than 1 require a negative exponent.
- Only significant digits are represented in scientific notation.

**Example:**

1 5 0 0 0 0 0 0 0 0 0 .    becomes $1.5 \times 10^{11}$m

0.0 0 0 0 0 0 6.5          becomes $6.5 \times 10^{-7}$m

*continued* ▶

## SET 2   Skill Review

1. There are a dozen apples in a bowl. In this case, what type of number is 12?

2. Put these numbers in order from most precise to least precise.

    (a) 3.2, 5.88, 8, 8.965, 1.000 08

    (b) 6.22, 8.5, 4.005, 1.2000 × 10 − 8

3. How many significant digits are represented by each value?

    (a) 215

    (b) 31

    (c) 3.25

    (d) 0.56

    (e) 1.06

    (f) 0.002

    (g) 0.006 04

    (h) 1.250 000

    (i) $1 \times 10^6$

    (j) $3.8 \times 10^4$

    (k) $6.807 \times 10^{58}$

    (l) $3.000 \times 10^8$

4. Round these values to two significant digits.

    (a) 1.23

    (e) 6.250

    (b) 2.348

    (c) 5.86

    (d) 6.851

    (f) 4.500

    (g) 5.500

    (h) 9.950

5. Complete the following calculations. Provide the final answer to the correct number of significant digits.

    (a) $2.358 \times 4.1$

    (b) $102 \div 0.35$

    (c) $2.1 + 5.88 + 6.0 + 8.526$

    (d) $12.1 - 4.2 - 3$

6. Write each of the following in scientific notation.

    (a) 2.5597

    (b) 1000

    (c) 0.256

    (d) 0.000 050 8

7. Write each value from question 6 in scientific notation accurate to three significant digits.

# Drawing and Interpreting Graphs

Graphical analysis of scientific data is used to determine trends. Good communication requires that graphs be produced using a standard method. Careful analysis of a graph may reveal more information than the data alone.

## Standards for Drawing a Graph

- Independent variable is plotted along the horizontal axis (include units).
- Dependent variable is plotted along the vertical axis (include units).
- Decide whether the origin (0, 0) is a valid data point.
- Select convenient scaling on the graph paper that will spread the data out as much as possible.
- A small circle is drawn around each data point to represent possible error.
- Determine a trend in the data — draw a best-fit line or best-fit smooth curve. Data points should never be connected directly when finding a trend.
- Select a title that clearly identifies what the graph represents.

### Constructing a linear graph

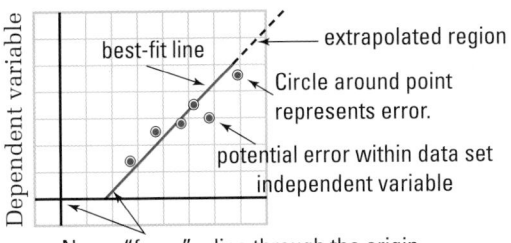

### Constructing a non-linear graph

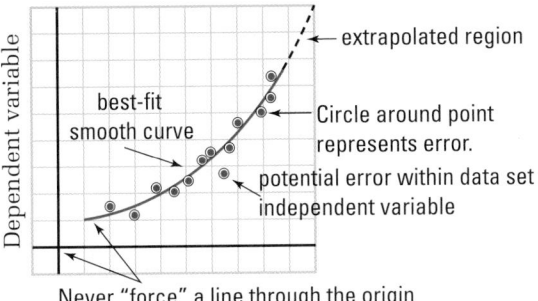

## Interpolation and Extrapolation

A best-fit line or best-fit smooth curve that is extended beyond the size of the data set should be shown as a dashed line. You are extrapolating values when you read them from the dashed line region of the graph. You are interpolating values when you read them from the solid line region of the graph.

## Find a Trend

The best-fit line or smooth curve provides insight into the type of relationship between the variables represented in a graph.

A *best-fit line* is drawn so that it matches the general trend of the data. You should try to have as many points above the line as are below it. Do not cause the line to change slope dramatically to include only one data point that does not seem to be in line with all of the others.

A *best-fit smooth curve* should be drawn so that it matches the general trend of the data. You should try to have as many points above the line as are below it, but ensure that the curve changes smoothly. Do not cause the curve to change direction dramatically to include only one data point that does not seem to be in line with all of the others.

## Definition of a Linear Relationship

A data set that is most accurately represented with a *straight line* is said to be linear. Data related by a linear relationship may be written in the form:

$$y = mx + b$$

| Quantity | Symbol | SI unit |
|---|---|---|
| y value (dependent variable) | $y$ | obtained from the vertical axis |
| x value (independent variable) | $x$ | obtained from the horizontal axis |
| slope of the line | $m$ | rise/run |
| y-intercept | $b$ | obtained from the vertical axis when $x$ is zero |

*continued* ▶

## Slope (*m*)

**Calculating the slope of a line**

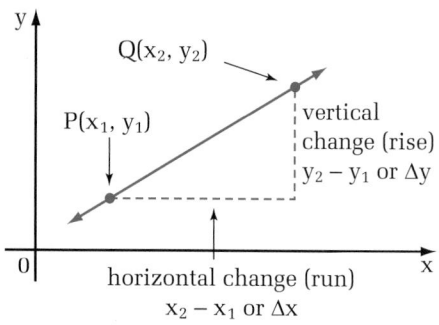

$$\text{slope}(m) = \frac{\text{vertical change (rise)}}{\text{horizontal change (run)}}$$

$$m = \Delta y / \Delta x$$

$$m = \frac{y_2 - y_1}{x_2 - x_1}, x_2 \neq x_1$$

Mathematically, slope provides a measure of the steepness of a line by dividing the vertical change (rise) by the horizontal change (run). In scientific situations, it is also very important to include units of the slope. The units will provide physical significance to the slope value.

**For example:**

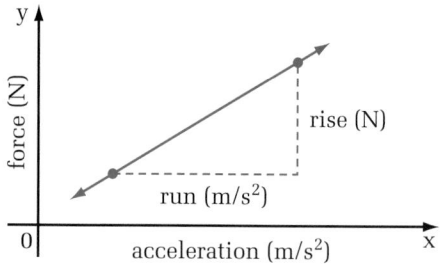

Including the units through the calculation helps verify the physical quantity that the slope represents.

$$m = \frac{\text{rise (N)}}{\text{run (m/s}^2)} \quad \text{Recall} : 1\text{N} = 1\text{kg} \cdot \text{m/s}^2$$

$$= \frac{\text{kg } \cancel{\text{m/s}^2}}{\cancel{\text{m/s}^2}}$$

$$= \text{kg}$$

In this example, the slope of the line represents the physical quantity of mass.

## Definition of a Non-linear Relationship

A data set that is most accurately represented with a smooth curve is said to be non-linear. Data related by a non-linear relationship may take several different forms. Two common non-linear relationships are:

**(a)** parabolic $\quad y = ax^2 + k$

**(b)** inverse $\quad y = 1/x$

## Area Under a Curve

Mathematically, the area under a curve can be obtained without the use of calculus by finding the area using geometric shapes.

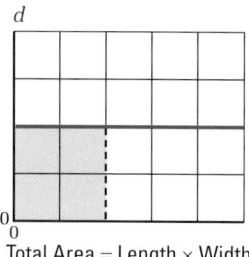

Total Area = Length × Width

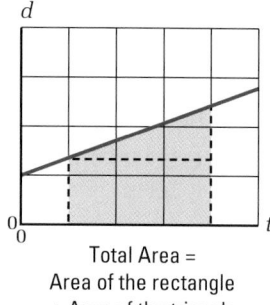

Total Area =
Area of the rectangle
+ Area of the triangle

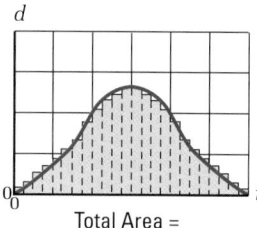

Total Area =
Area 1 + Area 2 + Area 3 ...

Always include units in area calculations. The units will provide physical significance to the area value. For example, see below:

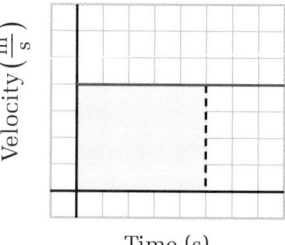

Including the units throughout the calculation helps verify the physical quantity that the area represents.

Area = $\Delta x \Delta y$
$\qquad$ = (velocity)(time)
$\qquad$ = (m/s)(s)
$\qquad$ = m (base unit for displacement)

The units verify that the area under a speed versus time curve represents displacement (m).

**1. (a)** Plot the data in Table 1 by hand, ensuring that it fills at least two thirds of the page and has clearly labelled axes including the units.

**(b)** Draw a best-fit line through the plotted data.

**(c)** Based on the data trend and the best-fit line, which data point seems to be most in error?

**(d)** Interpolate the time it would take to travel 14 m.

**(e)** Extrapolate to find how far the object would travel in 20 s.

**Table 1**

| Time (s) | Distance (m) | Time (s) | Distance (m) |
|----------|--------------|----------|--------------|
| 0 | 2 | 8 | 17 |
| 1 | 4 | 9 | 20 |
| 2 | 7 | 10 | 23 |
| 3 | 8 | 11 | 24 |
| 4 | 5 | 12 | 26 |
| 5 | 12 | 13 | 29 |
| 6 | 16 | 14 | 28 |
| 7 | 16 | 15 | 33 |

**2. (a)** Plot the data in Table 2 by hand, ensuring that it fills at least two thirds of the page and has clearly labelled axes including the units.

**(b)** Draw a best-fit smooth curve through the plotted data.

**(c)** Does this smooth curve represent a linear or non-linear relationship?

**(d)** At what distance is the force at the greatest value?

**Table 2**

| Distance (cm) | Force (N) | Distance (cm) | Force (N) |
|---------------|-----------|---------------|-----------|
| 0.0 | 0 | 2.5 | 1.1 |
| 0.5 | 0.1 | 2.5 | 1.2 |
| 0.9 | 0.2 | 2.4 | 1.3 |
| 1.3 | 0.3 | 2.2 | 1.4 |
| 1.6 | 0.4 | 2.0 | 1.5 |
| 1.9 | 0.5 | 1.7 | 1.6 |
| 2.1 | 0.6 | 1.4 | 1.7 |
| 2.3 | 0.7 | 1.1 | 1.8 |
| 2.4 | 0.8 | 0.7 | 1.9 |
| 2.5 | 0.9 | 0.2 | 2 |
| 2.6 | 1 | | |

**3.** Find the area of the shaded region under the following graphs. Use the units to determine what physical quantity the area represents.

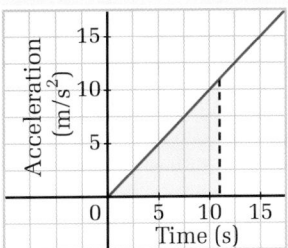

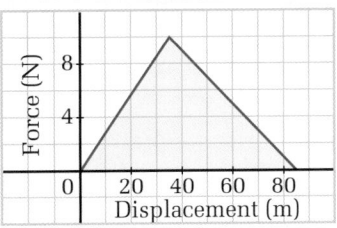

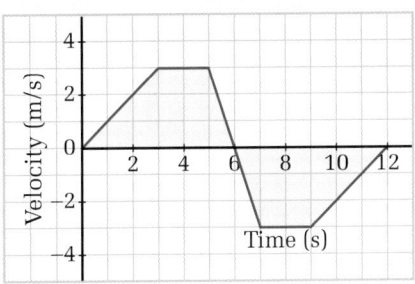

# A Math Toolbox

| | Perimeter/ circumference | Area | Surface area | Volume |
|---|---|---|---|---|
| circle (r) | $C = 2\pi r$ | $A = \pi r^2$ | | |
| square (s, s) | $P = 4s$ | $A = s^2$ | | |
| rectangle (l, w) | $P = 2l + 2w$ | $A = lw$ | | |
| triangle (h, b) | | $A = \frac{1}{2}bh$ | | |
| cylinder (r, h) | | | $SA = 2\pi rh + 2\pi r^2$ | $V = \pi r^2 h$ |
| sphere (r) | | | $SA = 4\pi r^2$ | $V = \frac{4}{3}\pi r^3$ |
| cube (s, s, s) | | | $SA = 6s^2$ | $V = s^3$ |

## Trigonometric Ratios

The ratios of side lengths from a right angle triangle can be used to define the basic trigonometric function sine (sin), cosine (cos), and tangent (tan).

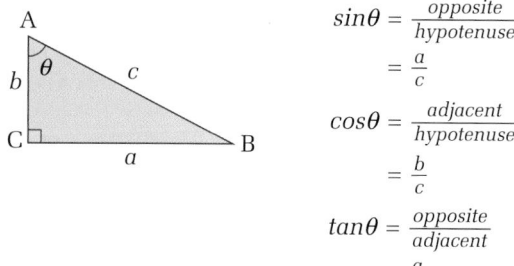

$$sin\theta = \frac{opposite}{hypotenuse}$$
$$= \frac{a}{c}$$

$$cos\theta = \frac{adjacent}{hypotenuse}$$
$$= \frac{b}{c}$$

$$tan\theta = \frac{opposite}{adjacent}$$
$$= \frac{a}{b}$$

The angle selected determines which side will be called the opposite side and the adjacent side. The hypotenuse is always the side across from the 90° angle. Picture yourself standing on top of the angle you select. The side that is directly across from your position is called the *opposite* side. The side that you could touch and is not the hypotenuse is the *adjacent* side.

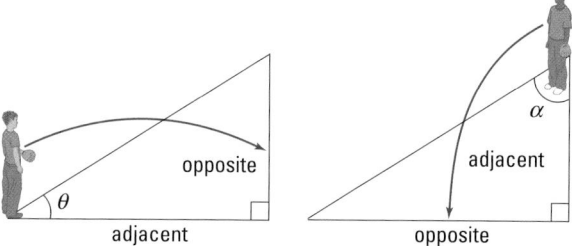

A scientific calculator or trigonometry tables can be used to obtain an angle value from the ratio result. Your calculator performs a complex calculation (Maclaurin series summation) when the $sin^{-1}$, or $cos^{-1}$, or $tan^{-1}$ operation is used to determine the angle value. $Sin^{-1}$ is not simply a 1/sin operation.

## Definition of the Pythagorean Theorem

The Pythagorean theorem is used to determine side lengths of a right angle (90°) triangle. Given a right angle triangle ABC, the Pythagorean theorem states:

$$c^2 = a^2 + b^2$$

| Quantity | Symbol | SI unit |
|---|---|---|
| hypotenuse side is opposite the 90°angle | $c$ | $m$ (metres) |
| side $a$ | $a$ | $m$ (metres) |
| side $b$ | $b$ | $m$ (metres) |

**Note:** The hypotenuse is always the side across from the right (90°) angle. The Pythagorean theorem is a special case of a more general mathematical law called the cosine law. The cosine law works for all triangles.

## Definition of the Cosine Law

The cosine law is useful when:

- determining the length of an unknown side given two side lengths and the contained angle between them;
- determining an unknown angle given all side lengths.

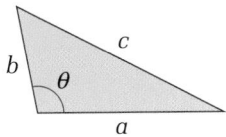

Angle θ is contained between sides a and b.

The cosine law states: $c^2 = a^2 + b^2 - 2ab\ cos\theta$

| Quantity | Symbol | SI unit |
|---|---|---|
| unknown length side $c$ opposite angle $\theta$ | $c$ | $m$ (metres) |
| length side $a$ | $a$ | $m$ (metres) |
| length side $b$ | $b$ | $m$ (metres) |
| angle $\theta$ opposite unknown side $c$ | $\theta$ | (degrees) |

**Note:** Applying the cosine law to a right angle triangle, setting $\theta = 90°$, yields the special case of the Pythagorean theorem.

## Definition of the Sine Law

The sine law is useful when:

- two angles and any one side length are known or,
- two sides lengths and any one angle are known.

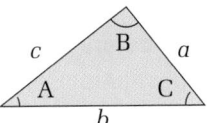

Given any triangle ABC the sine law states:

$$\frac{\sin A}{a} = \frac{\sin B}{b} = \frac{\sin C}{c}$$

| Quantity | Symbol | SI unit |
|---|---|---|
| length side a opposite angle $A$ | $a$ | $m$ (metres) |
| length side b opposite angle $B$ | $b$ | $m$ (metres) |
| length side c opposite angle $C$ | $c$ | $m$ (metres) |
| angle $A$ opposite side $a$ | $A$ | (degrees) |
| angle $B$ opposite side $b$ | $B$ | (degrees) |
| angle $C$ opposite side $c$ | $C$ | (degrees) |

**Note:** The sine law generates ambiguous results in some situations because it does not discriminate between obtuse and acute triangles. An example of the ambiguous case is shown below.

**Example:**
Use the sine law to solve for $\theta$.

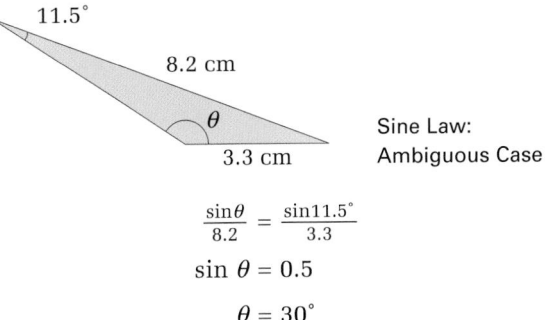

Sine Law: Ambiguous Case

$$\frac{\sin\theta}{8.2} = \frac{\sin 11.5°}{3.3}$$
$$\sin\theta = 0.5$$
$$\theta = 30°$$

Clearly, angle $\theta$ is much greater than 30°. In this case, the supplementary angle is required (180° − 30° = 150°). It is important to recognize, when dealing with obtuse angles (> 90°), that the supplementary angle may be required. Application of the cosine law in these situations will help reduce the potential for error.

## Algebra

In some situations, it may be preferable to use algebraic manipulation of equations to solve for a specific variable before substituting numbers. Algebraic manipulation of variables follows the same rules that are used to solve equations after substituting values. In both cases, to maintain equality, whatever is done to one side must be done to the other.

*continued* ▶

## Solving for "$x$" before Numerical Substitution

**(a)** $A = kx$     $x$ is multiplied by $k$,

$\therefore$ divide by $k$ to isolate $x$.

$\dfrac{A}{k} = \dfrac{\cancel{k}x}{\cancel{k}}$    Divide both sides of the equation by $k$ and

$\dfrac{A}{k} = x$     simplify.

$x = \dfrac{A}{k}$     Rewrite with $x$ on the left side.

**(b)** $B = \dfrac{x}{g}$    $x$ is divided by $g$,

$\therefore$ multiply by $g$ to isolate $x$.

$Bg = \dfrac{x\cancel{g}}{\cancel{g}}$    Multiply both sides of the equation by $g$ and

$Bg = x$     simplify.

$x = Bg$     Rewrite with $x$ on the left side.

**(c)** $W = x + f$     $x$ is added to $f$,

$W - f = x + f - f$    $\therefore$ subtract $f$ to isolate $x$.

Subtract $f$ on both sides of the equation and

$W - f = x$     simplify.

$x = W - f$     Rearrange for $x$.

**(d)** $W = \sqrt{x}$     $x$ is under a square root,

$W^2 = (\sqrt{x})^2$    $\therefore$ square both sides of the equation.

$W^2 = x$     Simplify.

$x = W^2$     Rearrange for $x$.

## Solving for "$x$" after Numerical Substitution

**(a)** $8 = 2x$     $x$ is multiplied by $k$,

$\therefore$ divide by k to isolate $x$.

$\dfrac{8}{2} = \dfrac{\cancel{2}x}{\cancel{2}}$    Divide both sides of the equation by 2 and

$4 = x$     simplify.

$x = 4$     Rewrite with $x$ on the left side.

**(b)** $8 = \dfrac{x}{4}$     $x$ is divided by 4,

$\therefore$ multiply by 4 to isolate $x$.

$(10)(4) = \dfrac{4x}{4}$    Multiply both sides of the equation by 4 and

$40 = x$     simplify.

$x = 40$     Rewrite with $x$ on the left side.

**(c)** $25 = x + 13$     $x$ is added to 13,

$25 - 13 = x + 13 - 13$    $\therefore$ subtract 13 to isolate $x$.

Subtract both sides of the equation by 13 and simplify.

$12 = x$

$x = 12$     Rewrite with $x$ on the left side

**(d)** $6 = \sqrt{x}$     $x$ is under a square root,

$6^2 = (\sqrt{x})^2$    $\therefore$ square both sides of the equation and

$36 = x$     simplify.

$x = 36$     Rewrite with $x$ on the left side.

## Definition of the Quadratic Formula

The quadratic equation is used to solve for the roots of a quadratic function. Given a quadratic equation in the form $ax^2 + bx + c = 0$, where $a$, $b$, and $c$ are real numbers and $a \neq 0$, the roots of it may be found using:

$$x = \dfrac{-b \pm \sqrt{b^2 - 4ac}}{2a}$$

## Statistical Analysis

In science, data are collected until a trend is observed. Three statistical tools that assist in determining if a trend is developing are *mean*, *median*, and *mode*.

**Mean:** The sum of the numbers divided by the number of values. It is also called the average.

**Median:** When a set of numbers is organized in order of size, the median is the middle number. When the data set contains an even number of values, the median is the average of the two middle numbers.

**Mode:** The number that occurs most often in a set of numbers. Some data sets will have more than one mode.

See examples of these on the following page.

**Example #1:**

Odd number of data points.

**Data Set #1** 12, 11, 15, 14, 11, 16, 13

$$mean = \frac{12 + 11 + 15 + 14 + 11 + 16 + 13}{7}$$

$$mean = 13$$

$$Reorganized\ data = 11, 11, 12, 13, 14, 15, 16$$

$$median = 13$$

$$mode = 11$$

**Example #2:**

Even number of data points.

**Data Set #2** 87, 95, 85, 63, 74, 76, 87, 64, 87, 64, 92, 64

$$mean = \frac{(87 + 95 + 85 + 63 + 74 + 76 + 87 + 64 + 64 + 92 + 64)}{12}$$

$$mean = 78$$

$$Reorganized\ data = 63, 64, 64, 64, 74, 76, 85, 87, 87, 87, 92, 95$$

$$median = \frac{(76 + 85)}{2}$$

$$median = 80$$

An even number of data points requires that the middle two numbers be averaged.

$$mode = 64, 87$$

In this example the data set is bi-modal (contains two modes).

---

## SET 4   Skill Review

1. Calculate the area of a circle with radius 6.5 m.

2. By how much does the surface area of the sphere increase when the radius is doubled?

3. By how much does the volume of the sphere increase when the radius is doubled?

4. Find all unknown angles and side lengths.

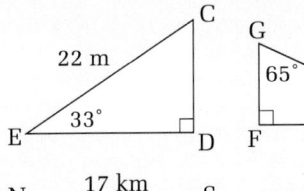

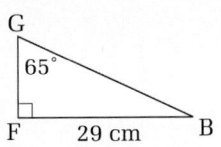

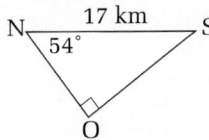

5. Use the cosine law to solve for the unknown side.

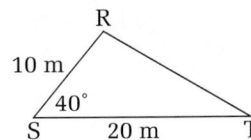

6. Use the sine law to solve for the unknown sides.

7. Solve for x in each of the following.

   (a) $42 = 7x$

   (b) $30 = x/5$

   (c) $12 = x \sin 30°$

   (d) $8 = 2x - 12^4$

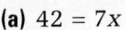

8. Solve for x in each of the following.

   (a) $F = kx$

   (b) $G = hk + x$

   (c) $a = bx \cos\theta$

   (d) $b = d \cos x$

   (e) $a = bc + x^2$

   (f) $T = 2\pi\sqrt{\frac{1}{x}}$

9. Use the quadratic equation to find the roots of the function.
   $4x^2 + 15x + 13 = 0$

10. Find the mean, median, and mode of each data set.

   (a) 25, 38, 55, 58, 60, 61, 61, 65, 70, 74, 74, 74, 78, 79, 82, 85, 90

   (b) 13, 14, 16, 17, 18, 20, 20, 22, 26, 30, 31, 32, 32, 35

# The Metric System: Fundamental and Derived Units

## Metric System Prefixes

| Prefix | Symbol | Factor |
|--------|--------|--------|
| tera | T | $1\ 000\ 000\ 000\ 000 = 10^{12}$ |
| giga | G | $1\ 000\ 000\ 000 = 10^{9}$ |
| mega | M | $1\ 000\ 000 = 10^{6}$ |
| kilo | k | $1000 = 10^{3}$ |
| hecto | h | $100 = 10^{2}$ |
| deca | da | $10 = 10^{1}$ |
| | | $1 = 10^{0}$ |
| deci | d | $0.1 = 10^{-1}$ |
| centi | c | $0.01 = 10^{-2}$ |
| milli | m | $0.001 = 10^{-3}$ |
| micro | $\mu$ | $0.000\ 001 = 10^{-6}$ |
| nano | n | $0.000\ 000\ 001 = 10^{-9}$ |
| pico | p | $0.000\ 000\ 000\ 001 = 10^{-12}$ |
| femto | f | $0.000\ 000\ 000\ 000\ 001 = 10^{-15}$ |
| atto | a | $0.000\ 000\ 000\ 000\ 000\ 001 = 10^{-18}$ |

## Fundamental Physical Quantities and Their SI Units

| Quantity | Symbol | Unit | Symbol |
|----------|--------|------|--------|
| length | $l$ | metre | m |
| mass | $m$ | kilogram | kg |
| time | $t$ | second | s |
| absolute temperature | $T$ | Kelvin | K |
| electric charge | $Q$ | coulomb | C |
| luminous intensity | $l$ | candela | cd |

## Derived Physical Quantities and Their SI Units

| Quantity | Quantity symbol | Unit | Unit symbol | Equivalent unit(s) |
|----------|-----------------|------|-------------|--------------------|
| area | $A$ | square metre | $m^2$ | |
| volume | $V$ | cubic metre | $m^3$ | |
| position | $d$ | metre | m | |
| velocity | $v$ | metre per second | m/s | |
| acceleration | $a$ | metre per second per second | $m/s^2$ | |
| force | $F$ | newton | N | $kg \cdot m/s^2$ |
| work | $W$ | joule | J | $N \cdot m,\ kg \cdot m^2/s^2$ |
| energy | $E$ | joule | J | $N \cdot m,\ kg \cdot m^2/s^2$ |
| power | $P$ | watt | W | $J/s,\ kg \cdot m^2/s^3$ |
| density | $D$ | kilogram per cubic metre | $kg/m^3$ | |
| pressure | $p$ | pascal | Pa | $N/m^2,\ kg/s^2$ |
| frequency | $f$ | hertz | Hz | $s^{-1}$ |
| period | $T$ | second | s | |
| wavelength | $\lambda$ | metre | m | |
| electric current | $I$ | ampère (amp) | A | C/s |
| electric potential | $V$ | volt | V | W/A, J/C, $kg \cdot m^2/(C \cdot s^2)$ |
| resistance | $R$ | ohm | $\Omega$ | V/A, $kg \cdot m^2/(C^2 \cdot s)$ |
| magnetic field strength | $B_L$ | tesla | T | $N \cdot s/(C \cdot m)$ |
| temperature (Celsius) | $T$ | degree Celsius | °C | $T°C = (T + 273.15)\ K$ |

# Physical Constants and Data

## Fundamental Physical Constants

| Quantity | Symbol | Accepted value |
|---|---|---|
| speed of light in a vacuum | $c$ | $2.998 \times 10^8$ m/s |
| gravitational constant | $G$ | $6.673 \times 10^{-11}$ N $\cdot$ m$^2$/kg$^2$ |
| Coulomb's constant | $k$ | $8.988 \times 10^9$ N $\cdot$ m$^2$/C$^2$ |
| charge on an electron | $e$ | $1.602 \times 10^{-19}$ C |
| rest mass of an electron | $m_e$ | $9.109 \times 10^{-31}$ kg |
| rest mass of a proton | $m_p$ | $1.673 \times 10^{-27}$ kg |
| rest mass of a neutron | $m_n$ | $1.675 \times 10^{-27}$ kg |
| atomic mass unit | u | $1.661 \times 10^{-27}$ kg |
| Planck's constant | $h$ | $6.626 \times 10^{-34}$ J $\cdot$ s |

## Electric Circuit Symbols

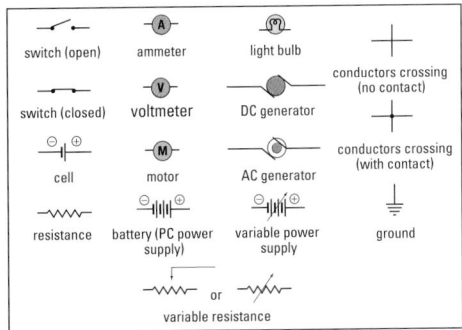

## Other Physical Data

| Quantity | Symbol | Accepted value |
|---|---|---|
| standard atmospheric pressure | $P$ | $1.013 \times 10^5$ Pa |
| speed of sound in air | | 343 m/s (at 20°C) |
| water:  density (4°C) | | $1.000 \times 10^3$ kg/m$^3$ |
| latent heat of fusion | | $3.34 \times 10^5$ J/kg |
| latent heat of vaporization | | $2.26 \times 10^6$ J/kg |
| specific heat capacity (15°C) | | 4186 J/(kg°C) |
| kilowatt hour | $E$ | $3.6 \times 10^6$ J |
| acceleration due to Earth's gravity | $g$ | 9.81 m/s$^2$ (standard value; at sea level) |
| mass of Earth | $m_E$ | $5.98 \times 10^{24}$ kg |
| mean radius of Earth | $r_E$ | $6.38 \times 10^6$ m |
| mean radius of Earth's orbit | $R_E$ | $1.49 \times 10^{11}$ m |
| period of Earth's orbit | $T_E$ | 365 days or $3.16 \times 10^7$ s |
| mass of Moon | $m_M$ | $7.36 \times 10^{22}$ kg |
| mean radius of Moon | $r_M$ | $1.74 \times 10^6$ m |
| mean radius of Moon's orbit | $R_M$ | $3.84 \times 10^8$ m |
| period of Moon's orbit | $T_M$ | 27.3 days or $2.36 \times 10^6$ s |
| mass of Sun | $m_s$ | $1.99 \times 10^{30}$ kg |
| radius of Sun | $r_s$ | $6.69 \times 10^8$ m |

## Resistor Colour Codes

| Colour | Digit represented | Multiplier | Tolerance |
|---|---|---|---|
| black | 0 | $\times 1$ | |
| brown | 1 | $\times 1.0 \times 10^1$ | |
| red | 2 | $\times 1.0 \times 10^2$ | |
| orange | 3 | $\times 1.0 \times 10^3$ | |
| yellow | 4 | $\times 1.0 \times 10^4$ | |
| green | 5 | $\times 1.0 \times 10^5$ | |
| blue | 6 | $\times 1.0 \times 10^6$ | |
| violet | 7 | $\times 1.0 \times 10^7$ | |
| gray | 8 | $\times 1.0 \times 10^8$ | |
| white | 9 | $\times 1.0 \times 10^9$ | |
| gold | | $\times 1.0 \times 10^{-1}$ | 5% |
| silver | | $\times 1.0 \times 10^{-2}$ | 10% |
| no colour | | | 20% |

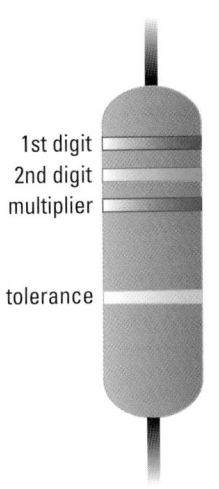

# Mathematical Formulas

## Formulas in Unit 1, Forces and Motion

| Formula | Variables | Name, if any |
|---|---|---|
| $\Delta d = v\Delta t$ <br> $v_2 = v_1 + a\Delta t$ <br> $\Delta d = v_1\Delta t + \frac{1}{2}a\Delta t^2$ <br><br> $\Delta d = v_2\Delta t - \frac{1}{2}a\Delta t^2$ <br> $v_2^2 = v_1^2 + 2\,a\Delta d$ | $\Delta d$ = displacement <br> $v$ = velocity <br> $v_1$ = initial velocity <br> $v_2$ = final velocity <br> $a$ = acceleration <br> $\Delta t$ = time interval | motion formula |
| $F_g = mg$ | $F_g$ = force of gravity <br> $m$ = mass <br> $g$ = acceleration due to gravity (on Earth) | weight |
| $F_f = \mu_s f_N$ <br> $F_f = \mu_k f_N$ | $F_f$ = force of friction <br> $\mu_s$ = coeffcient of static friction <br> $\mu_k$ = coeffcient of kinetic friction <br> $F_N$ = normal force | friction |
| $F_{Net} = ma$ <br> $W = F_{\parallel}\Delta d$ | $F_{Net}$ = net force <br> $a$ = acceleration | Newton's second law |

## Formulas in Unit 2, Energy, Work, and Power

| Formula | Variables | Name, if any |
|---|---|---|
| $W = F_{\parallel}\Delta d$ <br> $W = \lvert F\rvert\Delta d \cos\theta$ | $W$ = work done <br> $F_{\parallel}$ = force parallel to $\Delta d$ <br> $\lvert F\rvert$ = magnitude of applied force <br> $\Delta d$ = displacement <br> $\theta$ = angle of degrees | work done |
| $E_k = \frac{1}{2}mv^2$ | $E_k$ = mechanical kinetic energy <br> $m$ = mass <br> $v$ = velocity | mechanical potential energy |
| $E_g = mg\Delta h$ | $E_g$ = gravitational potential energy <br> $m$ = mass <br> $g$ = acceleration due to gravity (on Earth) <br> $\Delta h$ = change in height | gravitational potential energy |
| $E_T = E_g + E_k$ | $E_T$ = total mechanical <br> $E_g$ = mechanical gravitaional potential energy <br> $E_k$ = mechanical kinetic energy | conservation of mechanical energy |
| $W_{nc} = E_{final} - E_{initial}$ | $W_{nc}$ = work done by non-conservative forces <br> $E_{final}$ = final energy <br> $E_{initial}$ = initial energy | work done by non-conservative forces |
| $\Delta E = W + Q$ | $\Delta E$ = change in energy <br> $W$ = work done <br> $Q$ = heat | first law of thermodynamics |
| $T = T_c + 273.15$ | $T_c$ represents Celsius | Kelvin/Celsius temperature conversion |

| | | |
|---|---|---|
| $Q = mc\Delta T$ <br> $Q = mL_f$ <br> $Q = mL_v$ | $Q$ = heat <br> $m$ = mass <br> $c$ = specific heat capacity <br> $\Delta T$ = change in temperature <br> $L_f$ = latent heat of fusion <br> $L_v$ = latent hat of vaporization | heat <br> heat of fusion <br> heat of vaporization |
| $P = W/\Delta t$ <br> $P = E/\Delta t$ | $P$ = power <br> $W$ = work done <br> $E$ = energy transferred | power |
| Efficiency = $E_o/E_i \times 100\%$ <br> Efficiency = $W_o/W_i \times 100\%$ | $E_o$ = energy output <br> $E_i$ = energy input <br> $W_o$ = work output <br> $W_i$ = work input | efficiency |

## Formulas in Unit 3, Waves and Sounds

| | | |
|---|---|---|
| $T - \Delta t/N$ <br> $f = N/\Delta t$ <br> $f = 1/T$ | $N$ = number of oscillations <br> $\Delta t$ = time intervals <br> $T$ = period of oscillations <br> $f$ = frequency of oscillations | period <br> frequency <br> frequency |
| $v = f\lambda$ | $v$ = velocity of a wave <br> $f$ = frequency of a wave <br> $\lambda$ = wavelength | wave equation |
| open $resonance = \frac{n}{2}\lambda$ <br> closed <br> $resonance = \left(n - \frac{1}{2}\right)\frac{\lambda}{2}$ | $n$ = number of wavelengths <br> $\lambda$ = wavelength | open tube resonance <br> closed tube resonance |
| Beat frequency = $|f_1 - f_2|$ | $f_1$ = frequency of $1^{st}$ source <br> $f_2$ = frequency of $2^{nd}$ source | beat frequency |
| $v_s = 0.59T + 331$ | $v_s$ = speed of sound in air <br> $T$ = air temperature in °C | speed of sound in air |
| Mach number = $v_{object}/v_{sound}$ | $v_{object}$ = speed of object <br> $v_{sound}$ = speed of sound | Mach number |

## Formulas in Unit 4, Light and Geometric Optics

| | | |
|---|---|---|
| $f = R/2$ | $f$ = mirror focal length <br> $R$ = radius of curvature | focal length of a curved mirror |
| $n = c/v$ <br> $n = \sin\theta_i /\sin\theta_R$ <br> $n_1 \sin\theta_1 = n_2 \sin\theta_2$ | $n$ = index of refraction <br> $c$ = speed of light in a vacuum <br> $v$ = speed of light in medium <br> $\theta_i$ = angle of incidence <br> $\theta_R$ = angle of refraction <br> $n_1$ = index of refraction $1^{st}$ <br>     medium <br> $\theta_1$ = angle of incidence <br> $n_2$ = index of refraction $2^{nd}$ <br>     medium <br> $\theta_2$ = angle of refraction | refractive index <br><br> Snell's law (general form) |
| $\sin\theta_c = 1/n$ | $\theta_c$ = critical angle <br> $n$ = index of refraction | critical angle |

*continued* ▶

| | | |
|---|---|---|
| $d_{apparent} = d_{actual}\,(n_2/n_1)$ | $d_{apparent}$ = apparent depth<br>$d_{actual}$ = actual depth<br>$n_1$ = index of refraction 1st medium<br>$n_2$ = index of refraction 2nd medium | apparent depth |
| $\frac{1}{f} = (n-1)\left(\frac{1}{R_1} + \frac{1}{R_2}\right)$ | $f$ = lens' focal length<br>$n$ = index of refraction<br>$R_1$ = radius of curvature of one side of lens' surface<br>$R_2$ = radius of curvature of the other side of the lens' surface | lens-maker's formula |
| $M = h_i/h_o = -d_i/d_o$ | $M$ = magnification<br>$h_i$ = height of the image<br>$h_o$ = height of the object<br>$d_i$ = image distance<br>$d_o$ = object distance | magnification of a lens and a curved mirror |
| $\frac{1}{f} = \frac{1}{d_o} + \frac{1}{d_1}$ | $f$ = lens' focal length<br>$d_o$ = object distance<br>$d_i$ = image distance | thin lens and curved mirror equation |

## Formulas in Unit 5, Electricity and Magnetism

| | | |
|---|---|---|
| $Q = Ne$ | $Q$ = amount of charge<br>$N$ = number of electrons (excess or deficit)<br>$e$ = electron charge | amount of charge |
| $V = \Delta E_Q/Q$<br><br>$V = IR$ | $V$ = potential difference (voltage)<br>$\Delta E_Q$ = energy transfer<br>$Q$ = amount of charge<br>$I$ = electric current<br>$R$ = resistance | potential difference |
| $I = Q/\Delta t$ | $I$ = current<br>$Q$ = amount of charge<br>$\Delta t$ = time ihnterval | current |
| $R = r\dfrac{L}{A}$ | $R$ = resistance<br>$r$ = resistivity<br>$L$ = length of conductor<br>$A$ = cross-sectional area | resistance |
| $R_s = R_1 + R_2 + R_3 + \ldots + R_N$<br>$1/R_p = 1/R_1 + 1/R_2 + 1/R_3 + \ldots + 1/R_N$ | $R_s$ = total series resistance<br>$R_p$ = total parallel resistance<br>$R_1, R_2, R_3, R_N$ = individual resistors | series and parallel resistance |
| $Vs = \mathscr{E} - V_{int}$ | $V_s$ = terminal voltage<br>$\mathscr{E}$ = electromotive force<br>$V_{int}$ = internal potential drop | terminal voltage |
| $P = VI$<br>$P = I^2 R$<br>$P = V^2/R$ | $P$ = electric power<br>$I$ = electric current<br>$V$ = electric potential<br>$R$ = electric resistance | electric power |

| | | |
|---|---|---|
| $\frac{V_1}{V_2} = \frac{N_1}{N_2} = \frac{I_2}{I_1}$ | $V_1$ = potential 1st side<br>$V_2$ = potential 2nd side<br>$N_1$ = number of turns 1st side<br>$N_2$ = number of turns 2nd side<br>$I_1$ = current 1st side<br>$I_2$ = current 2nd side | transformer |
| $B_L = F/IL$ | $B_L$ = magnetic field strength<br>$F$ = force<br>$I$ = current<br>$L$ = length of conductor | magnetic field strength |
| $L = n\ell$ | $L$ = length of conductor<br>$n$ = number of turns<br>$\ell$ = length of one turn | conductor length |

## Appendix D

# Safety Symbols

The following safety symbols are used in this Physics 11 textbook to alert you to possible dangers. Make sure that you understand each symbol in a lab or investigation before you begin.

| | |
|---|---|
| | **Thermal Safety**<br>This symbol appears as a reminder to be careful when handling hot objects. |
| | **Sharp Object Safety**<br>This symbol appears when there is danger of cuts or punctures caused by the use of sharp objects. |
| | **Fume Safety**<br>This symbol appears when chemicals or chemical reactions could cause dangerous fumes. |
| | **Electrical Safety**<br>This symbol appears as a reminder to be careful when using electrical equipment. |
| | **Skin Protection Safety**<br>This symbol appears when the use of caustic chemicals might irritate the skin or when contact with micro-organisms might transmit infection. |
| | **Clothing Protection Safety**<br>A lab apron should be worn when this symbol appears. |
| | **Fire Safety**<br>This symbol appears as a reminder to be careful around open flames. |
| | **Eye Safety**<br>This symbol appears when there is danger to the eyes and safety glasses should be worn. |
| | **Chemical Safety**<br>This symbol appears when chemicals could cause burns or are poisonous if absorbed through the skin. |

Safety symbols used in the textbook

Look carefully at the WHMIS (Workplace Hazardous Materials Information System) safety symbols shown below. These symbols are used throughout Canada to identify dangerous materials in all work-places, including schools. Make sure that you understand what these symbols mean. When you see these symbols on containers in your classroom, at home, or in a work-place, use safety precautions.

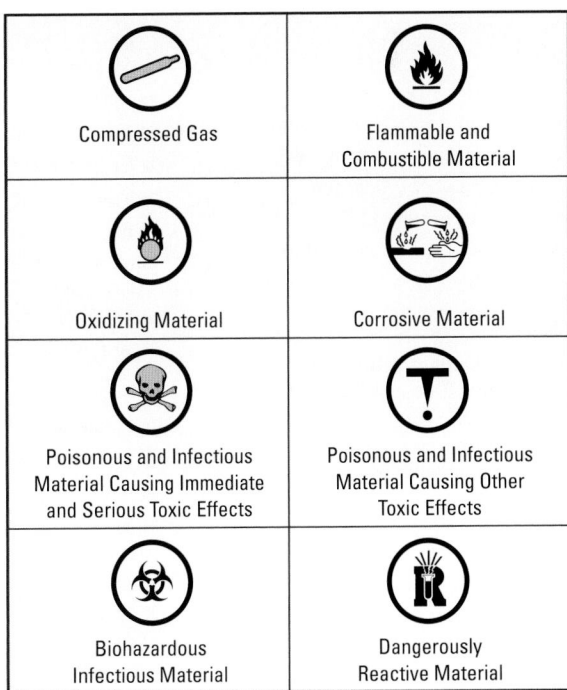

WHMIS symbols

# Chapter Review and Unit Review Numerical Problems

**Chapter 1**

**16.** 2.6%

**17. (a)** 0.3%

**(b)** Yes. The experimental measurements are very close to the theoretical value so such a small percent deviation is reasonable.

**18. (a)** 11.5 Hz

**(b)** 11 Hz

**(c)** 11 Hz

## Unit 1

**Chapter 2**

**15. (a)** with respect to the ground

**(b)** with respect to the truck

**17. (a)** 17 km    **(b)** 7.0 km[S]

**(c)** 7.0 km[N]

**18.** 26 km[W]

**19. (a)** 0.40 km downstream

**(b)** 0.53 km/h

**20.** 4.35 years

**21. (a)** 11.4 km from Vectorville

**(b)** 0.571 h or 34.2 min

**22. (a)** uniform

**(b)** non-uniform

**(c)** non-uniform

**(d)** non-uniform

**(e)** uniform

**24.** $-2.4$ m/s$^2$

**25.** 3 m/s

**26.** $-1.9$ m/s

**28. (a)** 8.3 m/s    **(b)** 1.4 m/s$^2$

**29. (a)** 27 m    **(b)** 8.0 m/s

**30. (a)** $-1.2$ m/s$^2$    **(b)** 6.9 s

**31.** $1.8 \times 10^2$ m [down]

**32. (a)** 23 s    **(b)** 550 m

**Chapter 3**

**11.** Displacement is the same; distance flown and time of travel are different. Pilots may want to fly along a component to partially avoid a headwind or to take advantage of a tailwind.

**12.** 9.40 km[N], $-9.40$ km[S], $-3.42$ km[E], 3.42 km[W]

**13. (a)** 71 km/h[SW]

**(b)** 3.9 m/s$^2$[SW] or $5.1 \times 10^4$ km/h$^2$[SW]

**14. (b)** 3.6 km[S34°W]

**15. (a)** 6.6 km[N31°W]

**(b)** 4.4 km/h[N31°W]

**16. (b)** 7.9 m/s[NW]

**17. (a)** 18 km[W24°S]

**(b)** 14 km/h[W24°S]

**18. (a)** 1.3 m/s[N]

**(b)** 3.7 m/s[S]

**19. (a)** [E26°N]

**(b)** 1.7 m/s[E]

**(c)** 47 min

**20. (a)** 8.4 m[E] and 18 m[N]

**(b)** 28 km[E] and 39 km[S]

**(c)** 12 m/s[W] and 8.6 m/s[S]

**(d)** 12 m/s$^2$[W] and 21 m/s$^2$[N]

**21.** 4.4 m/s[N5.4°E]

**22.** 12 km[W24°N]

**23. (a)** $2.0 \times 10^1$ km[N16°E]

**(b)** 9.9 km/h[N16°E]

**24.** 0.217 m/s$^2$[S19.7°W]

**25. (a)** 300 km/h[N]

**(b)** 1 h

**(c)** 0.6 h

**(d)** No

**26. (a)** He should aim upstream at an angle 41° with respect to the river bank.

**(b)** 2.3 min

**Chapter 4**

**22. (a)** $4.4 \times 10^3$ N

**(b)** $1.5 \times 10^3$ N

**(c)** 0.25

**24. (a)** 300 N    **(b)** 0.8 m/s$^2$

**25.** 4 s

**26. (a)** $-1.1$ m/s$^2$    **(b)** 0.12

**27.** $T_h = 4.1 \times 10^2$ N, $T_v = 4.6 \times 10^2$ N

**28.** 0.4 m/s$^2$

**29. (a)** $3.8 \times 10^2$ N

**(b)** 0.18 m/s$^2$

**30.** 50 N[E70°N]

**31. (a)** 0.80 m/s$^2$    **(b)** 16 N

**Unit 1 Review**

**33.** 13 km[E13°S]

**34.** 64 km/h[E51°S]

**35. (a)** 0.50 h

**(b)** 55 km[S]

**(c)** 110 km/h[S]

**36. (i)** B **(ii)** C **(iii)** A **(iv)** D

**37. (a)** 0.4 km    **(b)** 6 min

**(c)** 1 km

**38.** 2.5 m/s$^2$[N]

**39.** $5.0 \times 10^1$ m

**40.** 9.0 s

**41.** 20 s

**42. (i)** A **(ii)** C **(iii)** E

**43. (a)** 5.1 km[S28°E]

**(b)** 1.7 m/s[S28°E]

**44.** 1.8 m/s[N19°E]; $8.8 \times 10^2$ s; $5.3 \times 10^2$ m downstream

**45. (a)** 7.4 m/s[N]

**(b)** 9.5 m/s[N]

**(c)** 5.3 m/s[N]

**46.** 59 km/h[E17°S]

**47.** 45 km/h[E45°S]

**48.** Heading[N23.5°W]; 201 m/s[N30.0°W]

**49.** $1.9 \times 10^4$ m/s$^2$[N]

**50.** 6.8 m/s$^2$[NW]

**51.** 3.9 m/s[NE]

**52. (b)** It would accelerate in the horizontal direction.

**(c)** It would have constant velocity.

**(d)** It would slow down and stop.

**53. (a)** $2.7 \times 10^2$ N[W]

**(b)** 4.0 m/s$^2$[E]

**(c)** $5.0 \times 10^1$ m

**54. (a)** 0.30 m/s$^2$[N]

**(b)** 8.1 N[N]

**(c)** 12 N [at an angle of 25°]

**55.** 15 N[E19°S]

**56.** $1.2 \times 10^2$ N[up]

## Unit 2

**Chapter 5**

**14.** Air resistance slows down rain drops.

**15. (a)** Ground pushes up, gravity pulls down, engine propels car forward, ground resists backward.

**(b)** The forward force (from the car's engine) does work.

**16.** 44 N

**17.** $3.50 \times 10^2$ J

**18.** 228 N

**19.** $1.44 \times 10^4$ J

**20.** $6.2 \times 10^2$ J

**21.** 438 J

**22.** 5.0 m: $1.0 \times 10^2$ J, 13 m/s; 15.0 m: $5.8 \times 10^2$ J, 31 m/s; 25.0 m: $8.1 \times 10^2$ J, 36 m/s

**23.** 73°

**24.** the 55 kg athlete

**25. (a)** 3.2 m/s; $3.4 \times 10^2$ J
  **(b)** 1.2 m

**26.** $5.0 \times 10^1$ kg

**27.** 17 J; 4.2 m/s

**28.** 11.1 m/s

**29.** 30 m/s

## Chapter 6

**32.** $1.2 \times 10^5$ J

**33.** 7.9°C

**34.** $5.1 \times 10^3$ J

**35.** 273.15 K; 100°C; 293.15 K

**36.** $6 \times 10^4$ J, $1 \times 10^5$ J

**37.** $1.4 \times 10^6$ J

**38.** $3.5 \times 10^2$ W

**39. (a)** $2.7 \times 10^5$ J  **(c)** $4 \times 10^6$ J
  **(b)** $2 \times 10^6$ J  **(d)** $4.5 \times 10^9$ J

**40. (b)** a = 1.0 m/s$^2$
  **(c)** 4.6 m
  **(d)** 57 J
  **(e)** 19 W

**41.** 100% to 15% to 9% to 7.6%

**42.** $5 \times 10^3$ W

**43.** 2 m

## Unit 2 Review

**50. (a)** $8 \times 10^2$ N  **(b)** $1.5 \times 10^3$ J

**51.** 16.8 m/s

**52.** 49 kg

**53.** $1.8 \times 10^3$ J

**54.** 31 m/s, 22 m/s, 18 m/s

**55. (a)** $6.17 \times 10^2$ km/h
  **(b)** 99.9%

**56. (a)** $-5.8 \times 10^3$ J
  **(b)** 3.6  **(c)** yes, $\mu > 1$

**57. (a)** $6.1 \times 10^3$ N
  **(b)** $1.8 \times 10^7$ J

**58.** 0.543 kg

**59.** 1.8 h

## Unit 3

### Chapter 7

**21.** 0.25 Hz

**22.** the wavelength doubles

**23.** 0.4 m

**24.** $1.67 \times 10^{-2}$ Hz; 5.72 m

**25. (a)** 1.4 Hz  **(b)** 3.7 cm/s

**26.** 1.6 Hz

**27.** 680 km

**28. (a)** 1.2 Hz  **(b)** 0.84 s

**29. (a)** 1.02 s  **(b)** 2.56%
  **(c)** 225 h or 9.38 days
  **(d)** shorten the pendulum

## Chapter 8

**28. (a)** 307 m/s  **(c)** 343 m/s
  **(b)** $3.3 \times 10^2$ m/s  **(d)** 352 m/s

**29. (a)** 40.7°C  **(c)** 3.39°C
  **(b)** 22.0°C  **(d)** −22.0°C

**30.** 4.0°C

**31.** $7.0 \times 10^2$ m

**32. (a)** periodic increase and decrease of volume
  **(b)** 3 Hz

**33. (a)** 436.5 Hz or 443.5 Hz
  **(b)** If, as the string is tightened, the beat frequency increases, then the guitar was at 443.5 Hz, while if the beat frequency decreases, then the guitar was at 436.6 Hz.

**34. (a)** The human brain responds to harmonies, i.e. simple fraction ratios of pitch.

**35. (a)** Increases in pitch at specific, well-defined tube lengths.
  (b) $L_1$ = 0.098 m, $L_2$ = 0.29 m, $L_3$ = 0.49 m, $L_4$ = 0.68 m

**36. (a)** 0.38 m  **(b)** $9.0 \times 10^2$ Hz

## Chapter 9

**31.** The well is less then 176 m deep.

**32.** The vehicle is approaching.

**33. (a)** 0.082 m  **(b)** 0.073 m

**34.** 0.062 m

**35.** $2.8 \times 10^3$ km/h

**36.** 1.32

**37.** $1.30 \times 10^3$ m

**38.** $1.3 \times 10^2$ m

**39.** 7 m

**40.** 83 Hz, 55 Hz, 110 Hz

**41.** Yes, with 0.03 s to spare.

## Unit 3 Review

**49.** 3.0 m/s

**50.** 0.167 Hz

**51.** 0.8 m

**52.** $7.14 \times 10^9$ Hz

**53.** 0.73 m

**54.** 312 Hz

**55.** 0.69 m

**56.** 0.259 m

**57.** 382.8 Hz or 385.2 Hz

**58.** $2.16 \times 10^3$ m

**59. (a)** 1.6  **(b)** It will increase.

**60.** 0.02 m to 20 m

**61.** 2.4 s

**62.** 3.3 s

**63.** $7.55 \times 10^4$ Hz

**64.** 2.00 m

**65.** The resonance frequencies are in the range of 100 to 200 Hz, so a low voice would be more likely to appreciate resonance.

**66.** −8°C

## Unit 4

### Chapter 10

**20.** 4.6 m

**21. (a)** 55°  **(b)** 110°

**22.** 66°

**23.** 56°

**24.** 16°

**25.** 6.25 m

**26.** 7.0 m/s

**27.** 3.5 m/s

**29.** 52 cm

**30.** −30 cm

**31.** 0.850 m

**32.** $d_i = d_o$ = R; $h_o = -h_i$; M = −1

**33.** The image is real and inverted; $d_i$ = 0.75 m, $h_i$ = −1.2 cm

**34.** The image is real and inverted; $d_i$ = $1.2 \times 10^2$ cm, $h_i$ = −7.0 cm

**35.** The image is virtual and upright; $d_i$ = −2.4 m, $h_i$ = 6.0 cm

**36.** The image is virtual and upright; $d_i$ = −0.7 m, $h_i$ = 0.1 m

**37.** The image is virtual and upright; $d_i$ = −0.39 m, $h_i$ = 12 cm

**38.** 17 cm

### Chapter 11

**20.** 1.95

**21.** 22.8°

**22.** The ray exits at 30°, 5.7 cm from the bottom corner (assuming it entered 3.5 cm from the same corner).

**23.** $2.6 \times 10^{-9}$ s

**24. (a)** 1.2  **(b)** 19°  **(c)** 39°

**25.** 22°

**26.** 390 nm

**28.** 1.9 m

**29.** 68°

**30.** ± 4 cm

## Chapter 12

**27.** 1.50; crown glass

**28.** 23.1 cm

**29.** $h_i = -13$ cm; $d_i = 42$; the image is real, inverted and smaller than the object.

**30.** $h_i = 5.00$ cm; $d_i = -10.0$ cm; the image is virtual, upright and smaller than the object.

**31.** The lens is convex; the object is within the focal point.

**32.** $h_i = -4.0$ cm; $d_i = 19$ cm

**33.** $h_i = 1.1$ cm; $d_i = -5.1$ cm

**34.** $h_i = 4$ cm; $d_i = -10$ cm

**35.** 56 cm

**36.** 250

**37.** $f_{obj} = 0.441$ cm, $f_{eye} = 2.50$ cm

**38. (a)** 14.4 cm     **(b)** 10.3 cm

**39. (a)** 0.80     **(b)** 2.5

**40.** f = 0.185 m; $d_o = 0.188$ m

## Unit 4 Review

**49. (a)** $d_i = -f$, $M = 2$; the image is virtual, upright and larger.

    **(b)** $d_i = 2.33$ f, $M = -1.3$; the image is real, inverted and larger

    **(c)** $d_i = 1.5$ f, $M = -0.50$; the image is real, inverted and smaller

**50.** 48°

**51.** 12°

**52. (a)** 2.4 m/s     **(b)** 1.2 m/s

**53.** $7.0 \times 10^1$ m

**54.** −0.90 m

**55. (a)** $d_i = -36.0$ cm, $h_i = 6.00$ cm; the image is virtual, upright and larger

    **(b)** $d_i = 72.0$ cm, $h_i = -6.00$ cm; the image is real, inverted and larger

    **(c)** $d_i = -32.7$ cm, $h_i = 6.00$ cm; the image is virtual, upright and larger

**56. (a)** $d_i = -7.50$ cm, $h_i = 1.25$ cm; the image is virtual, upright and smaller

    **(b)** $d_i = -12.0$ cm, $h_i = 0.800$ cm; the image is virtual, upright and smaller

    **(c)** $d_i = -14.3$ cm, $h_i = 0.571$ cm; the image is virtual, upright

and smaller

**57.** $d_i = 24.0$ cm, $M = 0.600$

**58.** $d_i = -45.0$ cm, $M = 0.100$; the image virtual, upright and smaller

**59.** $1.60 \times 10^8$ m/s

**60.** $1.6 \times 10^{-9}$ s

**61.** 1.4

**62.** 25°

**63.** 33°

**64.** 350 nm

**65.** 15°

**66.** 4.1 m

**67.** 1.4

**68.** 60°

**69.** 38.6°

**70. (a)** $1.99 \times 10^8$ m/s

    **(b)** $1.96 \times 10^8$ m/s

**71.** -1.50; the image is real, inverted and larger

**72.** $h_o = 80.0$ cm, $d_o = 1.40$ m

# Unit 5

## Chapter 13

**24.** $3 \times 10^3$ Ω

**25. (a)** 12 A     **(b)** $2.5 \times 10^3$ C

    **(c)** $3.0 \times 10^5$ J

**26.** $5.0 \times 10^5$ J

**27.** 1.77 cents

**28.** 37.5 Ω

**29.** $I_1 = 6.0$ A, $V_1 = 150$ V, $I_2 = 1.0$ A, $V_2 = 3.0 \times 10^1$ V, $I_3 = 5.0$ A, $V_3 = 3.0 \times 10^1$ V

**30.** 9.93 s

**31. (a)** 1.9 Ω     **(b)** $1.4 \times 10^2$ Ω

    **(c)** 0.82 A     **(d)** 98 W

**32.** 24.3 V, 0.517 Ω

## Chapter 14

**27.** The magnetic field would be 5 times stronger.

**28.** $2.03 \times 10^{-4}$ km/h

**29. (a)** 22.0° E     **(b)** 112° E

**30. (a)** 2 times increase

    **(b)** 9 times increase

    **(c)** 2 times increase

## Chapter 15

**19.** 0.025

**20. (a)** 25.0 A     **(d)** $1.46 \times 10^3$ W

    **(b)** $9.60 \times 10^2$ V A **(e)** 1.52 A

    **(c)** 1.56 A

**21. (a)** $5.12 \times 10^4$ W

    **(b)** $4.90 \times 10^4$ W

    **(c)** 300 turns

    **(d)** $2.20 \times 10^3$ W

    **(e)** 7.66 A

## Unit 5 Review

**38.** 1.2 A[into the page]

**39.** 2.0 min

**40. (a)** 14 N[up]     **(b)** 0

**42.** Series: 4.00 Ω: 1.2 A, 5.0 V; 5.00 Ω: 1.2 A, 6.2 V. Parallel: 5.00 Ω, 3.8 V; 7.50 Ω : 0.50 A, 3.8 V

**43. (a)** $5.0 \times 10^1$ V     **(c)** 12 Ω

    **(b)** $1.0 \times 10^2$ V     **(d)** 2.7 A

**44. (a)** 17 V     **(b)** 6.5 Ω

    **(c)** 14 V

## A

**absorption coefficient** for sound, a property of each material that indicates the degree to which that material absorbs sound energy (9.3)

**absolute zero** the temperature at which the particles of a substance have zero kinetic energy, measured to be −273.15 °C (6.1)

**AC generator** an instrument that converts mechanical energy into electrical energy and produces alternating current (current that oscillates back and forth) (15.1)

**acceleration** the rate of change of velocity of an object (2.4)

**acceleration due to gravity** the acceleration of an object towards the centre of a celestial body when the gravitational attraction of the mass of the body is the only force acting on the object (4.2)

**accommodation** adjustment of the eye's lens by muscles in the eye in order to focus an image on the retina (12.3)

**accuracy** the degree to which the results of an experiment or calculation approximate the true value (Skill Set 1)

**acoustical design** plans for a room or building that will create desired sound characteristics (9.4)

**acoustical properties** the characteristics of a room or building that determine how sound is reflected in the room (9.4)

**acoustical shadows** regions of destructive interference of or physical barriers to sounds in an auditorium (9.4)

**action-at-a-distance** a force that two bodies exert on each other even though the bodies are not in contact (12.2)

**alternator** converts alternating current from an AC generator into direct current by the use of specially designed electric circuitry (15.1)

**ammeter** a device that measures the current in an electric circuit (13.2)

**ampere** the unit of electric current equivalent to one coulomb of charge passing a point in a circuit in one second (13.2)

**amplitude** the distance from the rest position to the maximum displacement for an object in periodic motion; or, for a wave, the distance from the rest position to the maximum point of the crest or minimum point of the trough (7.1, 7.2)

**angle of deviation** the angle between the direction of a ray incident on a prism and the direction of the emergent ray after having refracted at two surfaces of the prism (11.3)

**angle of incidence** the angle between the normal to a surface and the ray representing the incoming wave or light (7.4)

**angle of reflection** angle between the normal and the ray reflected from a surface, such as a mirror (7.4, 10.1)

**angle of refraction** angle that the refracted light ray or wave makes with the normal to the surface or boundary (11.2)

**anode** the electrode that accepts electrons; in a voltaic cell, the negative electrode; in an electrolytic cell, the positive electrode (13.1)

**antinodal line** a stationary line of points caused by constructive interference of individual waves (7.4)

**antinode** positions of maximum amplitude of a standing wave caused by the constructive interference of two individual waves travelling in opposite directions (7.3)

**aperture** part of an optical instrument through which electrons, light, or radio waves can pass (12.3)

**apparent depth** an effect observed in water in which the image of an object appears closer to the surface than the object; depends on the relative indices of refraction of air and water (11.3)

**armature** in a motor or generator, the coil with an iron core that rotates in a magnetic field (same as rotor) (14.4)

**articulators** any of the mouth and nasal cavity, the tongue and the lips, which modify specific sounds for speech and singing (9.1)

**astigmatism** the shape of the cornea in an eye is not spherical, causing vertical or horizontal lines to focus incorrectly (12.3)

**attitude of an image** formed by a mirror or lens, its orientation, upright or inverted (10.0)

**audible** sound frequencies in the range 20 to 20 000 hertz (Hz) (8.2)

**auditory canal** the part of the outer ear into which sound is funneled (9.1)

**auditory nerve** a nerve connecting the inner ear to the brain, along which signals are carried (9.1)

**average acceleration** rate of change of velocity depending only on initial and final values (2.4)

**average velocity** the quotient of the displacement and the time interval depending only on initial and final values (2.2)

**B**

**back emf** a potential difference generated by the motion of the current carrying coil in a motor, moving in the magnetic field, that opposes the potential difference that is driving the motor (15.2)

**battery** a combination of two or more voltaic cells that can convert chemical energy into electrical energy (13.1)

**beat frequency** frequency of envelope wave produced by the superposition of two waves of similar but not identical frequencies (8.3)

**beats** periodic variations in amplitude of a wave caused by superimposing two waves of nearly the same frequency (8.3)

**bel** a unit of sound intensity level; 1 bel = 10 pW/m² (8.2)

**biconvex** a lens in which both surfaces are convex, or curving outwards; also called double convex (12.1)

**Big Bang** a theoretical event considered to be the beginning of the universe (4.5)

**Big Crunch** a possible final event for the universe in which all matter and radiation recollapse into a point (4.5)

**biogas** gas created by the decay of rotting plant-matter; composed of methane, carbon dioxide and hydrogen sulphide gas (6.3)

**blend** the mixture of sounds created by the performers that is heard by the audience; in concert halls with good blends, no single instrument dominates (9.4)

**brilliance** an acoustical property of a room obtained when the reverberation time for high frequencies is longer than for low frequencies; opposite to warmth (9.4)

**C**

**cathode** the electrode that is the source of electrons; in a voltaic cell, positive; in an electrolytic cell, negative (13.1)

**centre of curvature** the centre of a sphere that would be formed if a spherical curved surface were extended into a sphere (10.3)

**chromatic aberration** an optical lens defect in which light of different wavelengths is focussed at different locations, causing colour fringes; due to the different index of refraction of different wavelengths (colours) of light (12.2)

**circuit elements** parts of a circuit, such as the loads and power supply (13.2)

**clarity** an acoustical property of a room characterized by a low intensity of reflected sound compared to the direct sound; opposite to fullness (9.4)

**classical mechanics** the study of forces and the resulting motion of macroscopic objects with velocities much less than the speed of light (4.3)

**closed air column** an air column that is closed at one end and open at the other (8.4)

**closed circuit** a complete circuit, in which current is able to flow (13.2)

**cochlea** the snail-shaped canal in the inner ear consisting of three fluid filled canals separated by membranes; one of the membranes (basilar membrane) contains thousands of tiny hair cells having their extensions (hairs) that are receptor cells for sound waves (9.1)

**coefficient of friction** for two specific materials in contact, the ratio of frictional force to the normal force between two surfaces (4.2)

**commutator** a device which passes current to or from the rotor (or armature) in an electric motor or generator (14.4)

**complex circuit** a circuit consisting of loads in a combination of parallel and series connections (13.4)

**component of a vector** part of a vector that is parallel to one of the axes of the coordinate system (3.3)

**component wave** a wave that combines with another wave to produce a resultant wave (7.3)

**compression** a region of higher pressure compared to the surrounding medium; longitudinal waves have both compressions and rarefactions (8.1)

**concave mirror** a mirror shaped similar to the inner surface of a segment of a sphere (10.3)

**concave lens** a lens that is thinner in the middle than at the ends; it causes rays to diverge when they pass through it (12.2)

**concavo-convex lens** a converging lens that has one concave and one convex surface (12.1)

**conclusion** an interpretation of the results of an experiment that relate to the hypothesis being tested (1.1)

**conductor** a material, like a metal, that allows electric charges to flow easily (13.1)

**conservation of mechanical energy** processes in which the total mechanical energy (kinetic and gravitational potential energy) is conserved (5.4)

**conservative force** a force that does work on an object in such a way that the amount of work done is independent of the path taken (5.4)

**consonance** combinations of musical notes that sound pleasant together (9.3)

**constant acceleration** acceleration that is not changing over a certain interval of time (2.4)

**constant velocity** the velocity that is unchanging in a given time interval (2.3)

**constructive interference** resultant wave has a larger amplitude than its component waves (7.3)

**contact forces** the force exerted by objects that are touching each other (4.2)

**control** a sample group to which experimental results can be compared (1.1)

**converge** to come together at a common point (10.2)

**coordinate system** a frame of reference consisting of perpendicular axes (3.1)

**converging rays** light rays that come together at the focal point after reflection or refraction from a mirror or lens (10.2)

**convex mirror** a mirror shaped similar to the outer surface of a segment of a sphere (10.3)

**convex lens** a lens which causes a parallel set of light rays that strike it to converge on a single point on the opposite side of the lens (12.1)

**convex meniscus** a lens that is thicker in the middle than at the edges and has one concave surface and one convex surface (12.1)

**convexo-concave** a lens that is thinner in the middle than at the edges and has one convex and one concave surface; also called concave meniscus (12.1)

**coulomb** amount of charge that passes a point in a circuit that is carrying a current of one ampere (13.1, 13.2)

**crest** the highest point on a wave (7.2)

**critical angle** the angle of incidence that produces a refracted light ray at an angle of 90° from the normal (11.4)

**Curie point** the temperature above which a magnet, when heated, loses its permanent magnetism and is destroyed (14.1)

**current** the net movement of electric charges (13.2)

**cycle** one complete repeat of a pattern of periodic motion, such as the crest of a wave to the next crest (7.1)

**D**

**DC generator** a device that converts mechanical energy into electrical energy and produces a direct current (15.1)

**decibel** the most common unit to describe sound intensity level; 1 decibel = 0.1 bel (8.2)

**dependent variable** the quantity that may change or respond because of changes in the independent variable (1.1)

**destructive interference** the situation when a combined or resultant wave has a smaller amplitude than at least one of its component waves (7.3)

**deviation** the change in direction of light after passing through a prism (11.3)

**diamagnetic** describes materials that become weakly magnetized in the opposite direction of an applied field (14.2)

**diffraction** the bending of waves around a barrier (7.4)

**diffuse reflection** the reflection in which the reflected light rays are not parallel to one another, as they are from a rough surface (10.1)

**dipping needle** a needle, freely suspended at its centre of gravity, used to measure the direction of the Earth's magnetic dip (14.1)

**dispersion** the separation of visible light into its range of colours (11.5)

**displacement** the change in the position of an object; the difference of the final and initial positions (2.2)

**displacement antinode** in a standing sound wave, the positions of maximum displacement of the particles (8.4)

**displacement node** in a standing sound wave, the positions of minimum displacement of the particles (8.4)

**dissonance** combinations of musical notes that sound unpleasant when played together (9.3)

**diverge** to spread out from a common point (10.2)

**diverging lens** a lens that is thinner in the middle than at the ends; it causes rays to diverge when they pass through it; also called concave lens (12.2)

**domain** small region in iron containing material in which individual "magnets" on the atomic level, are all aligned in the same direction (14.1)

**Doppler effect** change in the observed frequency (or wavelength) of a sound due to motion of the source or the observer (9.2)

**double concave** a diverging lens that has two concave surfaces (12.2)

**dynamics** the study of the motions of bodies while considering their masses and the responsible forces; simply, the study of *why* objects move the way they do (4.1)

**E**

**eardrum** a thin layer of skin stretched completely across the interior end of the auditory canal that vibrates in response to sound waves (9.1)

**echo** reflected sound (9.4)

**echolocation** determining position of obstacles and prey by emitting sound pulses and detecting time interval for sound to be reflected (9.2)

**eddy currents** electric currents induced within the body of a conductor when that conductor is subjected to a changing magnetic field (15.2)

**efficiency** the ratio of useful energy or work to the total energy or work input; describes how well a machine or device converts input energy or work into output energy or work (6.2)

**electric current** the movement of electric charge (13.2)

**electric generator** a device that converts mechanical energy into electric energy (15.3)

**electric potential difference** the difference in electric potential energy per unit charge between two points (13.0)

**electric potential energy** the energy possessed by electric charges due to their interaction with each other (13.1)

**electric resistance** inhibits the flow of electric current in a circuit (13.3)

**electrode** an electric conductor through which a current enters or leaves an electric device, such as a voltaic cell (13.1)

**electrolyte** a solution that can conduct electric current (13.1)

**electromagnet** a magnet created by placing an iron core inside a solenoid (14.3)

**electromagnetic force** the fundamental force which operates between charged particles; has an infinite range (4.5)

**electromagnetic induction** the generation of a current due to the relative motion of a conductor and a magnetic field (15.1)

**electromagnetic wave** wave consisting of changing electric and magnetic field (7.2)

**electromagnetism** phenomena associated with moving electric charges and magnetic fields (14.3)

**electromotive force (*emf*)** the potential difference produced by electromagnetic induction or by chemical reactions in a battery, that exists between the terminals when no current is flowing (15.1)

**electron flow** the net movement of negative charge (13.2)

**electrostatics** the study of electrical charges at rest (13.1)

**elementary charge** the quantity of charge on an electron or proton, equivalent to $e = 1.60 \times 10^{-19}$ C (13.2)

**energy** the ability to do work (6.3)

**ensemble** the ability of members of a performing group to hear each other play during a performance (9.3)

**equations of motion** set of mathematical equations describing the motion of an object undergoing uniform acceleration that relate velocity, displacement, acceleration, and time (2.5)

**equilibrium** the state of an object when the forces acting on it are in balance (4.4)

**equivalent resistance** the calculated total effective resistance of a group of resistors combined either in series or parallel or both (13.4)

**error analysis** the process of estimating the errors in measurements (Skill Set 1)

**estimated uncertainty** error in a measurement due to the natural limitations of the measuring device; usually described as half of the smallest division of the measuring device (Skill Set 1)

**Eustachian tube** a tube composed of bone and cartilage that drains fluid from the middle ear and allows air in or out to maintain the atmospheric pressure balance (9.1)

**exchange particle** an elementary particle thought to be responsible for the action of a fundamental forces (4.5)

**experiment** the test of a hypothesis under controlled conditions (1.1)

**eyepiece** the lens in an optical device that is used to observe the image created by the objective lens (12.0)

**F**

**fair test** an investigation in which the desired variables are adequately and objectively tested; if an investigation is a fair test, repeating it will produce similar results (1.1)

**ferromagnetic** a material whose atoms or molecules are magnets and tend to group into domains (14.2)

**field theory** a theory in which a field, rather than a magnet, electric charge, or gravitational mass, is the source of the force (14.3)

**first law of thermodynamics** energy cannot be created or destroyed; the energy of a closed system is conserved (6.1)

**fluctuating magnetic fields** magnetic fields that change in strength (15.3)

**focal length** the distance from the focal point to the vertex of the mirror or lens (10.3, 12.1)

**focal point** a point on the principal axis of a mirror or lens at which parallel light rays meet after being reflected or refracted (10.3, 12.1)

**force** an action, like a push or a pull, that causes a change in motion of an object (4.1)

**force of friction** a force that resists the motion of an object (4.4)

**force of gravity** mutual force between any two masses (4.2)

**fossil fuels** the remains of plant life (now coal) or aquatic animal life (now gasoline and natural gas) (6.3)

**frame of reference** a subset of the physical world defined by an observer in which positions or motions can be discussed or compared (2.1)

**framing** the process of setting parameters or boundaries to a problem, and organizing them in a way best suited to solve the problem (1.2)

**free body diagram** a diagram in which all the forces acting on an object are shown acting on a point representing the object (4.4)

**frequency** number of cycles of periodic motion completed in a unit of time; frequency is the inverse of the period and is measured in $s^{-1}$ or hertz (7.1)

**friction** a force that resists motion (4.2)

**fuel cell** a cell that generates electric current directly from the reaction between hydrogen and oxygen without producing thermal energy first (6.3)

**fullness** an acoustical property of a room characterized by how closely the intensity of the reflected sound compares to the direct sound; opposite to clarity (9.4)

**fundamental force** one of the four basic forces that governs the behaviour of all matter; see strong nuclear force, weak nuclear force, electromagnetic force, and gravity force (4.5)

**fundamental frequency** the lowest natural frequency able to produce resonance in a standing wave pattern (7.3)

**fundamental mode of vibration** the standing wave pattern for a medium vibrating at its fundamental frequency and displaying the fewest number of nodes and antinodes (7.3)

**G**

**Galileo telescope** uses a convex objective lens and a concave eyepiece (12.0)

**galvanometer** a device that detects and quantifies current (15.1)

**generator** a device that converts mechanical energy into electrical energy (15.0)

**generator effect** the generation of a current in a coil due to the relative motion of a magnet and a conductor (15.1)

**geometric optics** the branch of physics that uses the light ray model and rules of geometry to analyze the optical systems (Ch. 10, 11, 12)

**geothermal power** electrical power derived from the heat of Earth's core (6.3)

**gravitational force** the fundamental force which operates between masses; the gravitational force has an infinite range (4.2, 4.5)

**gravitational potential difference** the difference in gravitational potential energy per unit mass between two points, which depends only on the altitude and the acceleration due to gravity (13.1)

**gravitational potential energy** the potential energy an object has due to its location in a gravitational field; objects at higher altitudes have greater gravitational potential energy than objects at lower altitudes (5.3)

### H

**harmonic** the fundamental frequency and any overtone (8.1)

**heat** the transfer of thermal energy between two systems due to their different temperatures (6.1)

**hertz (Hz)** unit used to measure frequency, defined as $s^{-1}$ (7.1)

**hyperopia** eye condition in which images are brought to a focus beyond the retina because the eyeball is too short; also far-sightedness (12.3)

**hypothesis** a possible explanation for a question or an observation, which is subject to testing, and verification or falsification (1.1)

### I

**image** a likeness of an object as seen in a mirror or through a lens (10.2)

**independent variable** the quantity that is deliberately changed or manipulated during an experiment; compare to dependent variable (1.2)

**index of refraction** the ratio of the speed of light in a vacuum to the speed of light in a specific medium (11.1)

**induced current** a current produced in a conductor by the motion of the conductor in a magnetic field (15.1)

**induced *emf*** the potential difference induced in a circuit by a wire moving through a magnetic field or sitting in a changing magnetic field (15.1)

**inertia** the natural tendency of an object to stay at rest or uniform motion in the absence of outside forces; proportional to an object's mass (4.1)

**inertial frame of reference** a frame of reference in which the law of inertia is valid; it is is a non-accelerating frame of reference (4.3)

**infrasonic** sound frequencies lower than 20 Hz (8.2)

**inner ear** the part of the ear that transforms mechanical movement into electrical impulses that travel to the brain for interpretation (9.1)

**in phase** the periodic motion of two individual systems vibrating with the same frequency and always in the same stage of the cycle (7.1)

**instantaneous acceleration** the acceleration of an object at a particular moment in time (2.4)

**instantaneous velocity** the velocity of an object at a particular instant in time (2.3)

**insulator** a material that does not allow electric charges to move easily (13.1)

**interaction** the behaviour of objects as a result of forces (4.2)

**interference of waves** when waves meet they add algebraically (7.3, 8.3)

**internal resistance** the resistance inside a battery or power supply (13.4)

**intimacy** an acoustical property of a room characterized by short reverberation times; small performance halls are considered more intimate than large halls. (9.4)

**isogonic lines** lines on a map having constant magnetic declination (14.1)

### J

**joule (J)** the SI unit of energy or work; equivalent to applying one newton of force on an object over a distance of one metre (5.1)

### K

**kelvin (K)** the SI unit of temperature; equivalent to one degree on the Celsius scale (6.1)

**Kepler telescope** uses a convex objective lens and a convex eye piece (12.0)

**kilowatt-hour** the energy transformed by a power output of 1000 W for one hour; equivalent to $3.6 \times 10^6$ J (13.5)

**kinematics** the study of the motions of bodies without reference to mass or force; the study of *how* objects move in terms of displacement, velocity and acceleration (2.5, 4.1)

**kinetic energy** the energy of an object due to its motion (5.1, 5.2)

**kinetic frictional force** a frictional force that acts to slow the motion of an object; measured as the force required to just keep an object sliding over another object (4.2)

**kinetic molecular theory** all matter is composed of particles that are always in motion (6.1)

### L

**latent heat of fusion** the energy required to melt an amount of mass of a substance (6.1)

**latent heat of vaporization** the energy required to transform an amount of mass of a substance from the liquid state into a gaseous state (6.1)

**lateral displacement** the shifting of a light ray to one side, while maintaining the same direction, when it passes through a pane of glass with parallel sides (11.3)

**law of conservation of energy** the total energy of an isolated system remains constant; energy may be converted from one form to another, but the total energy does not change (5.4)

**law of conservation of mechanical energy** the total mechanical energy (kinetic plus gravitational potential) of a system always remains constant if work is done by conservative forces (5.4)

**law of reflection** the angle of incidence of a light ray on any surface is equal to the angle of reflection (both angles are measured with respect to the normal to the surface) (10.1)

**law of refraction** for any two media, the product of the index of refraction of the incident medium and the sine of the angle of incidence is the same as the product of the index of refraction of the refracting medium and the sine of the angle of refraction (11.2)

**law of gravitation** a gravitational force exists between all massive objects (1.1)

**law of inertia** an object remains at rest or continues in straight-line motion unless acted on by an outside force; also known as Newton's First Law (4.3)

**lens-maker's equation** relates the focal length of a lens to the index of refraction of the lens material and the radii of curvature of the two lens surfaces (12.1)

**Lenz's law** the direction of the force the magnetic field exerts on the induced current opposes the direction of the motion of the conductor (15.1, 15.2)

**light ray** an imaginary line that extends from a wave source and indicates the direction of the wave; a ray is perpendicular to a wavefront (10.1)

**light ray model** a model for finding the image of an object by using light rays to indicate the path that light travels (10.2)

**line resistance** the internal resistance in a transmission line, due to the material itself (15.3)

**lines of force** an imaginary line in a magnetic field whose direction indicates the direction of the magnetic field (14.2)

**linear propagation of light** the principle which asserts that in a uniform medium, light always travels in a straight line (10.1)

**liveness** an acoustical property of a room directly related to the reverberation time; rooms with longer reverberation times are more "live" (9.4)

**loads** devices in an electric circuit that receive power and convert electric energy to another form of energy (13.2)

**longitudinal wave** a wave in which the particles of a medium vibrate parallel to the direction of motion of the wave; for example, sound waves (7.2)

**loudness** the perceived strength of a sound (8.1)

### M

**Mach number** ratio of the speed of an object to the speed of sound (9.2)

**magnetic damping** the use of induced magnetic fields to slow down the motion of a conductor moving in a magnetic field (15.2)

**magnetic dip** the angle that a magnetized needle makes with the horizontal direction when placed in Earth's magnetic field (14.1)

**magnetic declination** the degree to which a compass needle points away from true north (14.1)

**magnetic dipole** another name for a magnet that always has two poles such as a bar magnet (14.1)

**magnetic domain** small region in iron containing material in which individual "magnets" on the atomic level, are all aligned in the same direction (14.1)

**magnetic field** region of space that will exert a force on a moving charge or magnetic field that enters that space, a vector quantity (14.2)

**magnetic field strength** the quotient of the force on a charge and charge, $Q$, and the velocity, $v$, of the moving charge at a point in the magnetic field (14.2)

**magnetic induction** the process by which an object becomes magnetized by a magnetic field (14.1)

**magnetic monopole** a theoretical magnet, never observed or created in the lab, which contains only a single pole (either north or south) (14.3)

**magnetosphere** the entirety of the magnetic field surrounding Earth (14.2)

**magnification** the ratio of the size of the image to the size of the object for a mirror or lens (12.3)

**magnification equation** an equation that relates the quotient of the object height and image height to the quotient of the object distance and image distance (10.3)

**mass** the quantity of matter an object contains (4.1, 4.2)

**mechanical energy** the sum of the kinetic and potential energy (5.1)

**mechanical wave** a wave that travels through a medium as a disturbance in that medium (7.2)

**mechanics** the branch of physics comprising kinematics and dynamics; simply, the how and the why of simple motion (4.1)

**medium** a substance, such as air or water or a solid, through which a wave disturbance travels (7.2)

**middle ear** the part of the ear that is responsible for transforming energy of sound waves into mechanical vibrations that can be transmitted to the inner ear (9.1)

**mirage** a virtual image that occurs naturally when particular atmospheric conditions cause a much greater amount of refraction of light than usual (11.3)

**mirror/lens equation** an equation that relates the focal length, object distance, and image distance (10.3)

**model** a representation of a theory (1.1)

**model problem** presents a specific physics problem and its solution, using a step-by-step approach (1.2)

**motor effect** the force exerted by a magnet on (the magnetic field of) a current-carrying conductor which drives electric motors (14.4)

**musical instrument** an object that can be used to create sounds through vibration of one or more of its parts; classified as brass, woodwind, string, or percussion (9.3)

**myopia** an eye condition in which images are brought to a focus in front of the retina because the eyeball is too long; also called near-sightedness (12.3)

### N

**nanotechnology** the technology of building mechanical devices from single atoms (1.0)

**natural frequency** the lowest resonant frequency an object undergoing periodic motion (7.1)

**net force** the vector sum of all forces acting on an object (4.4)

**newton** the unit of force required to accelerate 1 kilogram of mass at a rate of 1.0 m/s$^2$ (4.3)

**Newtonian mechanics** the study of forces and motions using Newton's laws of motion for macroscopic objects (4.3)

**Newton's laws of motion** three fundamental laws of motion which are the basis of Newtonian mechanics (4.3)

**nodal line** a stationary line in a medium caused by destructive interference of individual waves (7.4)

**node** stationary points in a medium produced by destructive interference of two waves travelling in opposite directions (7.3)

**noise** mixture of sound frequencies with no recognizable relationship to one another (8.4)

**non-conservative force** a force that does work on an object in such a way that the amount of work done is dependent on the path taken (5.4)

**non-contact forces** forces that act even though objects are separated by a distance, such as magnets (4.2)

**non-inertial frame of reference** an accelerating frame of reference (4.3)

**non-linear or non-ohmic resistance** devices or materials that do not obey Ohm's law (13.3)

**non-uniform acceleration** acceleration that is changing with time (2.4)

**non-uniform motion** the velocity is changing, either in magnitude or in direction, or both (2.3)

**normal force** a force that acts in a direction perpendicular to the common contact surface between two objects (4.2)

**normal line** a line perpendicular to a surface (10.1)

**north-seeking pole** the end of a magnet that points toward the north; commonly known as the north pole (14.1)

**nuclear power** power produced by the process of splitting extremely large atoms such as uranium into two or more pieces (6.3)

### O

**object** in geometric optics, anything that is a source of light (10.2)

**objective lens** the lens in an optical instrument that lies nearest the object viewed (12.0)

**observation** information gathered by the senses (1.1)

**Oersted effect** an electric current can affect the orientation of a compass needle, just as a permanent magnet can (14.3)

**ohm** the unit of electric resistance that will allow one ampere of current to move through the resistor when a potential difference of one volt is applied across the resistor, $1\,\Omega = 1V/1A$ (13.3)

**Ohm's law** the law the relates electric resistance to potential different and current (13.3)

**open air column** an air column that is open at both ends (8.4)

**open circuit** an incomplete circuit, in which current is unable to flow (13.2)

**optical fibre** a very fine strand of glass; when light shines into one end of an optical fibre, total internal reflection causes the energy to be confined within the fibre (11.4)

**optically dense** a refractive medium in which the speed of light is low in comparison to the speed of light in other media (11.1)

**ossicles** tiny bones, malleus, incus, and stapes, (commonly the hammer, anvil, and stirrup) in the middle ear that work together to amplify soft sounds entering the ear and to protect against excessively loud sounds (9.1)

**outer ear** the part of the ear that captures and focuses sound waves; composed of the external ear and auditory canal (9.1)

**out of phase** the periodic motion of two individual systems vibrating with the same frequency is said to be out of phase if they both don't reach the same amplitude at the same instant (7.1)

**oval window** the membrane-covered opening into the inner ear that transmits mechanical vibrations from the ossicles into longitudinal waves in the fluid of the cochlea (9.1)

**overtone** all natural frequencies higher than the fundamental frequency in a standing wave pattern (7.3)

**P**

**parabola** a geometrical figure formed by slicing a cone with a plane that is parallel to the axis of the cone (10.3)

**parabolic reflector** a mirror that focuses all the parallel rays of incident light at the focal point (10.3)

**parallel** a connection in a circuit in which there is more than one path for the current to follow (13.2, 13.4)

**paramagnetic** a material, whose magnets are initially randomly oriented, which becomes weakly magnetized in the same direction as an applied field (14.2)

**paraxial ray** a light ray that is close to the principal axis (12.1)

**partial reflection and refraction** phenomenon in which some of the energy light rays travelling from one medium into another is reflected and some of the energy is refracted at the interface between the media (11.3)

**percent deviation** a description of the accuracy of a measured value as compared to a theoretical value; calculated as (experimental value – theoretical value)/theoretical value times 100% (Skill Set 1)

**percent difference** a description of the precision of a set of observations; calculated as (maximum value – minimum value)/(average value of data) times 100% (Skill Set 1)

**period** the time required for an object to complete one cycle of its repeated pattern of motion (7.1)

**periodic motion** the motion of an object in a repeated pattern over regular time intervals (7.1)

**permanent magnet** a magnet that maintains its magnetic properties when removed from an external magnetic field (14.1)

**permeability** a number which characterizes a material's tendency to become magnetized (is permeable to magnetic lined of force) (14.2)

**phase change** the change in an object's state of matter, for example, from solid to liquid or liquid to gas (6.1)

**phase difference** the angular difference between two systems in periodic motion that are not in phase (7.1)

**photovoltaic cells** solar cells composed of semi-conductors such as silicon, that convert light energy directly into electric energy (6.3)

**physics** the study of the relationships between matter and energy (1.1)

**pitch** an attribute of a sound that determines its position in a musical scale; pitch is measured in frequency (8.1)

**place theory of hearing** the theory that suggests that the location of the hair cells on the basilar membrane detect a specific range of frequencies of sound (9.1)

**plane mirror** a flat, polished surface that reflects light (10.2)

**plano-concave** a lens in which one surface is flat and the other curves inwards (12.1)

**plano-convex** a lens in which one surface is flat and the other surface curves outwards (12.1)

**point of divergence** point from which light rays diverge (10.3, 12.2)

**position vector** a vector which points from the origin of a coordinate system to the location of an object at a particular instant in time (2.2)

**potential difference** the difference in potential the potential energy per unit of mass, charge, etc, of an object due to its position or condition

**potential energy** energy stored by an object (5.1, 5.3)

**power** the rate at which work is done, measured in watts (W), or joules per second; also defined as the rate at which energy is transferred or transformed (6.2)

**power output** the rate at which an appliance can transform electric energy into a desired form such as light, heat, or sound (13.5)

**power supply** a device such as a cell, battery, or generator which supplies electrical energy to a circuit (13.2)

**precision** describes the exactness and repeatability of a value or set of values (Skill Set 1)

**prediction** a statement of what you would expect to observe in an experiment, based on the hypothesis (1.1)

**presbyopia** an eye condition equivalent to far-sightedness due to the lens losing flexibility which occurs with advanced age (12.3)

**primary coil** the coil in a transformer that is connected to the initial fluctuating or alternating current (i.e., the power supply) (15.3)

**principal axis** a straight line that passes through the vertex, the centre of curvature and the mirror or lens (10.3, 12.1)

**principal focus** the point on the principal axis which light rays parallel to the principle axis converge on or appear to diverge from; also called the principal focal point (10.3, 12.1)

**principle of reversibility of light** when a light ray is reversed, it travels back along its original path (11.2)

**principle of superposition** a combined or resultant wave is the sum of its component waves (7.3)

**prism** a transparent object used for refracting, dispersing, or reflecting light rays (11.3)

**pure** a description of the quality of a sound, such as that of a flute or whistle; pure sounds typically have few overtones; compare to rich (8.1)

## Q

**qualitative observation** a verbal description of an object or phenomena; for example, "the book is heavy" (1.1)

**quality** an attribute of a sound used to distinguish between sounds having the same fundamental frequency but a different set of overtones (8.1)

**quantitative observation** a numerical description of an object or phenomena; for example, "the book has a mass of 5 kg"; quantitative observations typically involve measurements of a particular quantity (1.1)

**quantum mechanics** a branch of modern physics that deals with matter and energy on atomic scales (4.5)

**quintessence** a fifth substance that celestial objects were made of, according to the Greek scholar, Aristotle (1.1)

## R

**radius of curvature** any straight line drawn from the centre of curvature of a mirror or lens to the curved surface (10.3)

**rainbow** an arc of colours of the visible spectrum appearing opposite the sun, caused by reflection, refraction, and dispersion of the sun's rays as they pass through drops of rain (11.5)

**random error** results from small variations in measurements due only to chance (Skill Set 1)

**rarefaction** a region of lower air pressure compared to the surrounding medium; longitudinal waves have both compressions and rarefactions (8.1)

**ray** a line drawn perpendicular to the wavefronts of a wave (7.4)

**real image** an optical image that would appear on a screen if a screen were placed at the image location (10.2)

**recombination** the process whereby a dispersed spectrum of light can be put back together into white light (11.5)

**rectified current** an alternating current that is transformed into a pulsating direct current (15.1)

**refraction** the change in the speed of a wave due when it travels from one medium to another (7.4, 11.1)

**refracting telescope** a telescope which uses only lenses and no mirrors to create an image (12.3)

**regular reflection** reflection in which the reflected light rays are parallel to one another, as from the surface of a mirror (10.1)

**relative velocity** a velocity relative to a frame of reference, such as an air current, that itself is moving with a velocity relative to another frame of reference, such as the ground (3.2)

**relative uncertainty** the ratio of the estimated uncertainty to the actual measured value, written as a percentage (Skill Set 1)

**resistance** in electricity, the resistance to the flow of electric current in a circuit; similar to frictional resistance in motion (13.3)

**resistivity** a property of a material that describes the ease with which it permits the flow of electric current (13.3)

**resolved vectors** components of a vector that are at right angles to each other; the components lie parallel to the axes of the coordinate system (3.2)

**resonance** phenomena that occurs when energy is added to a vibrating system at the same frequency as its natural frequency (7.1)

**resonance length** the specific lengths of a column at which resonance occurs, typically measured in fractions of wavelength (8.4)

**resonant cavity** an open or closed air column in which a standing wave pattern and varied sounds can be created (9.1)

**rest position** the position of an object, such as a simple pendulum or a mass on a spring, when it is allowed to hang freely and is not moving (7.1)

**resultant vector** a vector obtained by adding or subtracting two or more vectors (3.1)

**resultant wave** a wave produced by combining or superimposing two or more individual waves (7.3)

**retroreflector** an optical device used to reflect light directly back parallel to its original path (11.4)

**reverberation time** the time it takes for all echoes in a room to fade away and become inaudible (9.4)

**rich** a description of the quality of a sound, such as that of a cello or organ; rich sounds typically have many overtones; compare to pure (8.1)

**right-hand rules** the rules which help to visualize the directions of vectors describing properties involved in electromagnetism by using the fingers and thumb of your right hand. (14.3)

**S**

**scalar** a physical quantity that has only a magnitude or size (2.2)

**scale diagram** a diagram in which the relative sizes and directions of objects or vectors are preserved with respect to a particular coordinate system (4.3)

**scientific method** a procedure used to understand the natural and physical world; it consists of several steps: 1) observations of phenomena; 2) formulating a hypothesis that describes these phenomena and is consistent with present knowledge; 3) testing the hypothesis by making new observations, analyzing experiments, and using it to predict new phenomena; 4) accepting, modifying, or rejecting the hypothesis (1.1)

**science** the process of creating a system of principles and laws that describe phenomena in the natural and physical world (1.1)

**scientific law** a principle that has been thoroughly tested and observed that scientists are convinced that it will always be true (1.1)

**second law of thermodynamics** a process of transferring heat and doing mechanical work in which some thermal energy will always leave the system without doing work; thermal energy is always transferred from an object at a higher temperature to an object at a lower temperature (6.2)

**secondary axis** any line other than the vertical axis that passes through the intersection of the vertical axis and the principal axis (12.1)

**secondary coil** the coil in a transformer in which currents are induced (15.3)

**secondary focus** a point on the secondary axis which parallel light rays converge on or appear to diverge from; also called the secondary focal point (12.1)

**sensorineural** hearing loss due to damage to the hair cells within the cochlea (sensori) or to the auditory nerve and auditory receptors within the brain (neural) (9.1)

**series** a connection in a circuit in which there is only one path for the current to follow (13.2, 13.4)

**slip-ring commutator** a circular conductor attached to the coil of a motor or AC generator that is in contact with brushes that allows the continuous electric connection to the rest of the circuit (15.1)

**Snell's law** for any two media, the product of the index of refraction of the incident medium and the sine of the angle of incidence is the same as the product of the index of refraction of the refracting medium and the sine of the angle of refraction (11.2)

**solenoid** a closely wound helix of wire that acts as a magnet when current runs through the wire (14.3)

**sonic boom** an acoustic pressure wave caused by an object moving faster than the speed of sound (9.2)

**sound intensity level** the rate of energy flow across a unit area; measured in an exponential scale in units of bels (B) (8.2)

**sound spectrum** a plot of intensity versus frequency for the various frequencies that make up a sound (8.4)

**source of light** anything that has light coming from it, either from reflection or radiation (10.2)

**south-seeking pole** the end of a magnet that points toward the magnetic pole near the geographic South Pole; commonly known as the south pole of a magnet (14.1)

**specific heat capacity** the amount of energy that must be added to a substance to raise 1.0 kg of its material by a temperature of 1.0 degree kelvin (6.1)

**spectrometer** an optical instrument used to measure the precise frequencies or wavelengths produced by a light source (11.5)

**speed** the distance an object travels divided by the time the object was travelling; speed is a scalar quantity (2.2)

**speed of light** the speed at which light travels; the speed of light in a vacuum is a fundamental physical constant (11.1)

**spherical aberration** an optical problem in spherical mirrors and lenses in which parallel rays far from the principal axis are not brought to the same focus as parallel rays close to the principal axis (10.3)

**spherical mirror** a mirror with the shape of a section sliced from the surface of a sphere; a mirror, either convex or concave, whose surface forms part of a sphere (10.3)

**split-ring commutator** a device which allows continuous connection of the rotating rotor (or armature) to the rest of the circuit in a motor or generator; used in DC motors and generators to reverse the current direction (14.4, 15.1)

**standing wave** a stationary wave consisting of nodes and antinodes, formed when two equal travelling waves pass through one another in opposite directions (7.3)

**static frictional force** a frictional force that acts to keep an object at rest; measured as the force required to start to move an object from rest (4.2)

**step-down transformer** a transformer that has fewer windings on its secondary coil, and acts to decrease voltages (15.3)

**step-up transformer** a transformer that has more windings on its secondary coil, and acts to increase voltages (15.3)

**strong nuclear force** the fundamental force that holds the parts of the nuclear together; the strong nuclear force has a short range (4.5)

**super force** a theoretical force, thought to exist early in the history of the universe, in which all four of the fundamental forces are unified (4.5)

**superconductors** materials that at extremely low temperatures conduct electricity with absolutely no resistance (13.3)

**systematic error** results from bias in observations that won't be reduced by repeating the measurement (Skill Set 1)

## T

**tangent** a line that intersects a curve at only one particular point (2.3)

**tectorial membrane** in the cochlea, the organ that touches the projections of hair cells and causes them to bend when the basilar membrane vibrates (9.1)

**temperature** a measure of the average kinetic energy of the atoms or molecules of a substance (6.1)

**temporal theory of hearing** a theory that suggests that groups of hair cells in the cochlea (in the inner ear) discharge signals in time with the frequency of the incident sound (9.1)

**temporary magnet** a magnet that loses its magnetic properties when removed from an external magnetic field (14.1)

**terminal voltage** the potential difference across the poles of a battery (13.4)

**tesla (T)** unit of magnetic field strength, equivalent to the magnetic field that exerts a force of one newton (1 N) on a one-metre-long (1 m) conductor carrying a current of one ampere (1 A); equivalent to a newton per ampere metre (14.1)

**texture** an acoustical property of a room which refers to how rapidly reflected sounds from different directions reach a listener (9.4)

**theory** a collection of ideas and principles, validated by many scientists, that have been demonstrated to describe and predict a natural phenomenon (1.1)

**thermal energy** the kinetic energy of the particles of a substance due to their constant, random motion (6.1)

**thermal equilibrium** the state in which the energy transfer between bodies in a system is equal; bodies in thermal equilibrium have the same temperature (6.1)

**thermosphere** the highest layer of the atmosphere, beginning at approximately 100 km above Earth's surface, where the temperature rises continuously with altitude (6.1)

**thin-lens equations** the mirror/lens and magnification equations, relating focal length, image distance, and object distance; accurate, in the case of lenses, only if the thickness of the lens is small compared to its diameter (12.1)

**tidal power** power derived by capturing high tide waters and releasing them through turbines during low tide (6.3)

**timbre** the difference in quality of sound between two instruments playing the same note; due to the different harmonic structure of the sounds (9.3)

**time interval** the amount of time that passes between two instants of time (2.1)

**torque** similar to force but causes a change in the rotation of an object (4.4)

**total internal reflection** phenomenon in which light incident on the boundary of an optically less-dense medium is not refracted at all but is entirely reflected back from the boundary into the optically more-dense medium; occurs when the angle of incidence is greater than the critical angle (11.4)

**transformer** a device used to convert power from the high voltages used in transmission to low voltages safe for use in homes; increases or decreases AC voltages with little loss of energy (15.3)

**transverse wave** a wave in which the particles of a medium vibrate at right angles to the direction of motion; for example, water waves (7.2)

**trough** the lowest point on a wave (7.2)

## U

**Ultrasonic** sound frequencies higher than 20 000 Hz (8.2)

**uniform acceleration** acceleration that is constant throughout a particular time interval (2.4)

**uniform motion** moving at constant velocity (2.3)

**universal wave equation** the fundamental equation governing the motion of waves that relates the velocity of the wave to its frequency and wavelength (7.2)

## V

**variable** a quantity that may change in an experiment (1.2)

**vector** a quantity that has a magnitude and a direction; vectors must be defined in terms of a frame of reference (2.2)

**vector components** parts of a vector that are parallel to the axes of a coordinate system, into which a vector can be resolved; they are scalar quantities (3.3)

**vector diagram** a diagram, with a coordinate system, in which all quantities are represented by vectors (3.1)

**velocity** the rate of change of position of an object; a vector quantity (2.2)

**vertex** the geometric centre of a curved mirror or lens surface (10.3)

**vertical axis** the axis of the mirror or lens which passes through the vertex and is perpendicular to the principal axis (12.1)

**virtual focus** a point which light rays appear to converge to or diverge from (10.3)

**virtual image** an image that can only be seen by looking *into* the mirror or lens that is creating it; virtual images will not appear on a screen when a screen is placed at the apparent image location as light rays do not actually pass through a virtual image (10.2, 12.1)

**virtual object** an apparent image, not yet formed, used as an object for a second lens; this happens by placing a second lens or mirror in the path of the converging rays of the first lens or mirror, before a real image is formed (12.4)

**visible spectrum** the range of colours of light that human eyes can see; from long wavelength to short wavelength, the visible spectrum comprises the colours red, orange, yellow, green, blue, indigo, and violet; includes wavelengths from 400 to 700 nm (11.5)

**vocal cords** two thin folds of muscle and elastic tissue that can be opened and closed to restrict air flow entering and leaving the lungs; oscillations of the vocal chords are responsible for speech (9.1)

**volt (V)** the SI unit of potential difference and emf (13.1)

**voltage** the potential difference between two points in a circuit (13.0)

**voltaic cell** a cell consisting of two different metals, called electrodes, placed in an electrolytic solution in which chemical reactions produce an electric charge on the electrodes (13.1)

**voltmeter** a device that measures the potential difference across a circuit element (13.2)

## W

**warmth** an acoustical property of a room obtained when the reverberation time for low frequencies is longer than for high frequencies; opposite to brilliance (9.4)

**watt (W)** a unit of power, equivalent to 1 joule per second (13.5)

**wave** a disturbance that transfers energy through a medium (7.2)

**wave power** electrical power derived by harnessing the energy of water waves (6.3)

**wave theory of light** a theory that proposes that light travels as a wave and has all of the properties of waves (10.1)

**wavefront** a group of adjacent points in a wave that all have the same phase, usually indicated by a line drawn along the crests of a wave (7.4, 10.1)

**wave equation** the fundamental equation governing the motion of waves that relates the velocity of the wave to its frequency and wavelength (7.2)

**wavelength** the shortest distance between any two points in a medium that are in phase; commonly measured from one trough to the next trough, or one crest to the next crest (7.2)

**weak nuclear force** the fundamental force that causes radioactive decay; the weak force has an extremely short range (4.5)

**weight** the force that gravity exerts on an object due to its mass (4.2)

**work** the transfer of mechanical energy; equivalent to a force acting through a distance (5.1)

**work-energy theorem** the relationship between the work done on an object and the resulting change in any of the object's forms of energy, $W = \Delta E$ (5.2)

**work-kinetic energy theorem** the relationship between the work done on an object and the resulting change in kinetic energy, $W = \Delta E_k$ (5.2)

The page numbers in **boldface** type indicate the pages where terms are defined. Terms that occur in investigations (*inv*), Model Problems (*MP*), MultiLabs (*ML*), and QuickLabs (*QL*), are also indicated.

# Credits

**474** (top right), From *Glencoe Physics Principles and Problems* © 1999 The McGraw-Hill Companies Inc.; **477** (center left), © Bob Daemmerich/Boston Stock; **479** (bottom left), © Joyce Photographics/Photo Researchers, Inc.; **479** (bottom center), © Mark E. Gibson/Visuals Unlimited, Inc.; **480** (top left), NASA; **480** (center left), NASA/John Krist, Karl Stapelfeldt, Jeff hester, Chris Burrows; **480** (bottom left), NASA/John Krist, Karl Stapelfeldt, Jeff hester, Chris Burrows; **483** (top right), © Massimo Mastroillo/Firstlight.ca; **484** (bottom left), © George Hall/Check Six; **485** (center left), Dr. Neil Tyson/Andrew Brusso; **486** (bottom left), © Jay Lurie/Bruce Coleman Inc.; **486** (bottom right), © Yoav Levy/Phototake; **488** (bottom right), © Tom Pantages; **502** (top), © William Whitehurst/The Stock Market/Firstlight.ca; **502** (top left), © Deep Light Productions/Science Photo Library/Photo Researchers, Inc.; **503** (bottom right), © Peter Menzel/Stock Boston; **503** (center left), © Bill Beatty/Visuals Unlimited, Inc.; **504** (center right), © Warren Stone/Visuals Unlimited, Inc.; **505** (top center), © Jerome Wexler/Photo Researchers, Inc.; **509** (center left), © Rich Treptow/Photo Researchers, Inc.; **522** (bottom left), © Ron Stroud/Masterfile; **527** (bottom right), © Jerome Wexler/Photo Researchers, Inc.; **529** (center left), © Ellis Herwig/Boston Stock; **531** (top left), © Alfred Pasieka/Science Photo Library/Photo Researchers, Inc.; **531** (center right), © Richard Megna/FUNDAMENTAL PHOTOGRAPHS, NEW YORK; **533** (top left), © Damien Lovegrove/Science Photo Library/Photo Researchers, Inc.; **533** (top right), Jun Park; **540** (center left), George B. Diebold/The Stock Market/Firstlight.ca; **543** (top left), © Russell D. Curtis/Photo Researchers, Inc.; **543** (center left), ©1998 Seth Resnick/Stock Boston ; **544** (center right), © Irwin Barrett/First Light; **545** (bottom left), © David Parker/Science Photo Library/Photo Researchers, Inc.; **545** (bottom right), From *Glencoe Physics Principles and Problems* © 1999 The McGraw-Hill Companies Inc.; **546** (center left), From *Glencoe Physics Principles and Problems* © 1999 The McGraw-Hill Companies Inc.; **546** (center right), © Glenn M. Oliver/Visuals Unlimited, Inc.; **550** (top), © Bill Ross/Firstlight.ca; **550** (top left), Steve Grand/Science Photo Library/Photo Researchers Inc.; **552** (top center), © Tom Pantages; **552** (top center), © Tom Pantages; **569** (center right), © Jerome Wexler/Photo Researchers, Inc.; **576** (top right), Royal Observtory, Edinburgh/Science Photo Library/Photo Researchers, Inc.; **580** (top left), © Stephen Frisch/Stock Boston; **582** (center left), Kaz Mori/The Image Bank; **582** (bottom left), © Norman Owen Tomalin/Bruce Coleman Inc.; **583** (top right), © M. Carr/Firstlight.ca; **584** (center left), © Charles D. Winters/Photo Researchers, Inc.; **587** (center right), © Charles Gupton/Stock Boston; **601** (bottom right), Artbase Inc.; **602** (bottom left), © Bettmann/CORBIS/MAGMA; **602–603** (background), © E.R. Degginger/Bruce Coleman, Inc.; **604** (top right), © Garry Black/ Masterfile; **604** (top), © Stephen Studd/Stone; **606** (top center), From *Sciencepower 9* © 1999 McGraw-Hill Ryerson Ltd.; **607** (top right), © J-L Charmet/Science Photo Library/Photo Researchers, Inc.; **612** (bottom right), © Miles Ertman/Masterfile; **615** (top right), © Daryl Benson/Masterfile; **615** (center left), © James L. Amos/CORBIS/MAGMA; **616** (center left), Science Photo Library/Photo Researchers, Inc.; **623** (center left), © Inga Spence/Visuals Unlimited, Inc.; **633** (top left), © Gunter Marx Photography/CORBIS/MAGMA; **643** (top left), © Jeff J. Daly/Visuals Unlimited, Inc.; **643** (bottom left), © Terry Gleason/Visuals Unlimited, Inc.; **647** (top right), Artbase Inc.; **652** (center), © Henry Grossman/TimePix; **656** (top left), © Sheila Terry/Science Photo Library/Photo Researchers, Inc.; **662** (bottom right), © Steve Callahan/Visuals Unlimited, Inc.; **668** (top), © Grantpix/Photo Researchers, Inc.; **668** (top right), Artbase Inc.; **670** (top right), © Brent P. Kent/Animals Animals; **673** (center right), © Tom Pantages; **676** (top right), From *Glencoe Physics Principles and Problems* © 1999 The McGraw-Hill Companies Inc.; **680** (center left), From *Glencoe Physics Principles and Problems* © 1999 The McGraw-Hill Companies Inc.; **682** (center right), NOAA Space Environment Center (SEC); **683** (top right), © Kevin Kelly/Corbis/Magma; **686** (top right), © CERN/Science Photo Library/Photo Researchers, Inc.; **700** (top right), Photo Courtsey Midnight Sun/University Of Waterloo; **716** (top), Artbase Inc.; **716** (top right), © Lester Lefkowitz/Firstlight.ca; **718** (top right), © Nick White/Masterfile; **720** (bottom right), From *Glencoe Physics Principles and Problems* © 1999 The McGraw-Hill Companies Inc.; **722** (top left), © Reuters New Media/CORBIS/MAGMA; **722** (bottom left), © Lester Lefkowitz/FPG; **726** (top left), From p. 45 *Sciencepower 10* © 1999 McGraw-Hill Ryerson Ltd.; **729** (bottom right), Artbase Inc.; **731** (center left), From *Sciencepower10* © 1999 McGraw-Hill Ryerson Ltd.; **731** (bottom center), From *Sciencepower 10* © 1999 McGraw-Hill Ryerson Ltd.; **731** (center right), From *Sciencepower 10* ©1999 McGraw-Hill Ryerson Ltd.; **734** (center left), From *Glencoe Physics Principles and Problems* © 1999 The McGraw-Hill Companies Inc.; **736** (top center), © Tom Myers/Photo Researchers, Inc.; **751** (bottom right), Artbase Inc.; **752** (top), Artbase Inc.; **752** (background), Artbase Inc.; **753** (top left), © Zig Leszczynski/Earth Scenes; **754** (bottom left), © Archivo Iconografico, S.A./CORBIS/MAGMA; **755** (center right), Artbase Inc.; **759** (center right), Artbase Inc.; **773** (top right), Stephen Studd/Stone; **N/A** All corner banners (top), Artbase Inc.

## Fundamental Physical Constants

| Quantity | Symbol | Accepted value |
|---|---|---|
| speed of light in a vacuum | $c$ | $2.998 \times 10^8$ m/s |
| gravitational constant | $G$ | $6.673 \times 10^{-11}$ N · m$^2$/kg$^2$ |
| Coulomb's constant | $k$ | $8.988 \times 10^9$ N · m$^2$/C$^2$ |
| charge on an electron | $e$ | $1.602 \times 10^{-19}$ C |
| rest mass of an electron | $m_e$ | $9.109 \times 10^{-31}$ kg |
| rest mass of a proton | $m_p$ | $1.673 \times 10^{-27}$ kg |
| rest mass of a neutron | $m_n$ | $1.675 \times 10^{-27}$ kg |
| atomic mass unit | u | $1.661 \times 10^{-27}$ kg |
| Planck's constant | $h$ | $6.626 \times 10^{-34}$ J · s |

## Metric System Prefixes

| Prefix | Symbol | Factor |
|---|---|---|
| tera | T | $1\ 000\ 000\ 000\ 000 = 10^{12}$ |
| giga | G | $1\ 000\ 000\ 000 = 10^9$ |
| mega | M | $1\ 000\ 000 = 10^6$ |
| kilo | k | $1000 = 10^3$ |
| hecto | h | $100 = 10^2$ |
| deca | da | $10 = 10^1$ |
| | | $1 = 10^0$ |
| deci | d | $0.1 = 10^{-1}$ |
| centi | c | $0.01 = 10^{-2}$ |
| milli | m | $0.001 = 10^{-3}$ |
| micro | $\mu$ | $0.000\ 001 = 10^{-6}$ |
| nano | n | $0.000\ 000\ 001 = 10^{-9}$ |
| pico | p | $0.000\ 000\ 000\ 001 = 10^{-12}$ |
| femto | f | $0.000\ 000\ 000\ 000\ 001 = 10^{-15}$ |
| atto | a | $0.000\ 000\ 000\ 000\ 000\ 001 = 10^{-18}$ |

## Other Physical Data

| Quantity | Symbol | Accepted value |
|---|---|---|
| standard atmospheric pressure | $P$ | $1.013 \times 10^5$ Pa |
| speed of sound in air | | 343 m/s (at 20°C) |
| water:  density (4°C) | | $1.000 \times 10^3$ kg/m$^3$ |
| latent heat of fusion | | $3.34 \times 10^5$ J/kg |
| latent heat of vaporization | | $2.26 \times 10^6$ J/kg |
| specific heat capacity (15°C) | | 4186 J/(kg°C) |
| kilowatt hour | $E$ | $3.6 \times 10^6$ J |
| acceleration due to Earth's gravity | $g$ | 9.81 m/s$^2$ (standard value; at sea level) |
| mass of Earth | $m_E$ | $5.98 \times 10^{24}$ kg |
| mean radius of Earth | $r_E$ | $6.38 \times 10^6$ m |
| mean radius of Earth's orbit | $R_E$ | $1.49 \times 10^{11}$ m |
| period of Earth's orbit | $T_E$ | 365 days or $3.16 \times 10^7$ s |
| mass of Moon | $m_M$ | $7.36 \times 10^{22}$ kg |
| mean radius of Moon | $r_M$ | $1.74 \times 10^6$ m |
| mean radius of Moon's orbit | $R_M$ | $3.84 \times 10^8$ m |
| period of Moon's orbit | $T_M$ | 27.3 days or $2.36 \times 10^6$ s |
| mass of Sun | $m_s$ | $1.99 \times 10^{30}$ kg |
| radius of Sun | $r_s$ | $6.69 \times 10^8$ m |